GENETIC MARKERS IN HUMAN BLOOD

GENETIC MARKERS IN HUMAN BLOOD

ELOISE R. GIBLETT M.D.

Research Professsor of Medicine at
the University of Washington
School of Medicine

Head of Immunogenetics,
King County Central Blood Bank
Seattle, Washington

BLACKWELL SCIENTIFIC PUBLICATIONS

OXFORD AND EDINBURGH

Printed in Great Britain by

SPOTTISWOODE, BALLANTYNE AND CO. LTD.
LONDON AND COLCHESTER

and bound by

WILLIAM CLOWES AND SONS LTD.
BECCLES, SUFFOLK

CONTENTS

LIST OF FIGURES xix

LIST OF TABLES xxiii

PREFACE xxv

PART I

GENETIC MARKERS IN PLASMA

I THE IMMUNOGLOBULINS AND THEIR ALLOTYPES: Gm AND Inv

Characteristics of the Immunoglobulins 4
  General features 4
  Immunoglobulin metabolism 7
    Concentration in plasma 7
    Synthesis 7
    Turnover 9
  Immunoglobulin structure 11
    General features of IgG molecules 11
    Polypeptide chain structure 13
      The light chains 13
        Electrophoretic patterns of light chains 14
        Light chains of antibodies 14
        Amino acid sequence of light chains 15
      The heavy chain (γ) of IgG 17
        Heterogeneity of Fd fragment 17
        The 'hinge' region 18
        IgG subclasses 18
      Heavy chains of the other immunoglobulins 20
      Evolution of immunoglobulin chains 20
      Genetic basis of immunoglobulin variability 22

Inherited Antigenic Determinants (Allotypes) 23
  The Gm system 23
    Gm antigens and their detection 23
    Gm genetics 26
      Family studies 26
      Comparison of Gm and Rh inheritance 27
      Gm heterogeneity in single serum specimens 27

CONTENTS

CHAPTER I—*Continued*

Assignment of Gm antigens to IgG subclasses 28
Two genetic theories 31
Studies of Gm structure 33
Geographic distribution of Gm antigens 36
Anti-Gm antibodies 37
The Inv System 38
Inv antigens 38
Studies of Inv structure 39
Inv gene frequencies 40

Methods 40
The IgG antibody coat 41
The agglutinator 42
The agglutination-inhibition test 42
The tube technique 42
The tile technique 43

References 44

References to Gm and Inv Geographic Distribution 60

2 HAPTOGLOBIN

Haptoglobin Physiology 64
Synthesis 64
Haptoglobin levels 65
Catabolism 66
Physiological role of haptoglobin 68
Retention of iron by HpHb complex formation 68
Prevention of methemalbumin formation 69
Suggested role in heme catabolism 70

The HpHb Complex: Hemoglobin Binding 70
Relation of hemoglobin conformation to haptoglobin binding 71
The 'intermediate' complex 71

The Structure of Haptoglobin 72
General features 72
Haptoglobin glycopeptides 73
Subunit structure: haptoglobin α and β chains 74
Structure of Hp 1-1 and Hp 2-2 76
Hemoglobin binding by haptoglobin polymers 77
The binding site of haptoglobin: immunological and chemical studies 78

Haptoglobin Genetics 79
Hp gene duplication and its genetic consequences 79
Haptoglobin variants 84
Quantitative variants 84
The Hp 2-1(mod) phenotypes 84

CHAPTER 2—*Continued*

Other quantitative variation of Hp 2-1 85
Hypohaptoglobinemia and anhaptoglobinemia 86
Family studies 86
Genetic basis of hypohaptoglobinemia 87
An Hp° allele 88
Hypohaptoglobinemia and the possible localization of the Hp structural gene locus 88
Qualitative variants 89
Alpha chain variants 89
Beta chain variants 91
Other rare phenotypes, possibly β chain variants 92

Geographic Distribution of Hp Genes 93
Difficulties in data interpretation 93
General summary of gene frequencies shown in Table 1 99
Gene frequencies determined by subtyping 100

Methods 100
Starch gel electrophoresis 101
Apparatus: horizontal and vertical 101
Source of starch 102
Buffer systems 102
Gel preparations 103
Sample insertion 104
Electrophoresis conditions 105
Gel slicing 105
Staining 105
Measurements of haptoglobin concentration 106

References 107

References to Haptoglobin Geographic Distribution 118

3 TRANSFERRIN

Physico-chemical Properties of Transferrin 127
Preparation 127
Molecular weight 127
Carbohydrate moiety 127
Possible subunit structure 128
Iron binding by transferrin 129

Transferrin Metabolism and Function 131
Transferrin turnover 131
Iron turnover; role of transferrin 132
Possible role of transferrin in infection 134
Physiological effect of atransferrinemia 134

Inherited Variants of Transferrin 135

CHAPTER 3—*Continued*

Detection of Tf polymorphism 135
Family studies and linkage data 139
Chemical structure of Tf variants 140
Function of Tf variants 142
Difficulties in identifying variant transferrins 143
Frequency of variants 144
Geographic distribution of variants 144

Methods 147
Phenotype differentiation by electrophoresis 147
Starch gel electrophoresis 147
Fe^{59} radioautography 148
Iron staining 148
Rivanol precipitation 148

References 149

References to Tf Geographic Distribution 156

4 THE Gc SYSTEM

General Properties 160

Inheritance of Gc Phenotypes 161

Electrophoretic Patterns 161

Serum Concentration of Gc Protein 162

Phenotypic Variants 162

Chemical Studies 164

Geographic Distribution of Gc Genes 165

Methods 167
Immunoelectrophoresis 167
Starch gel electrophoresis 168

References 169

References to Gc Geographic Distribution 172

5 BETA LIPOPROTEIN ALLOTYPES: THE Ag AND Lp SYSTEMS

Structure of Beta Lipoprotein 177

Inherited Variations in β Lipoprotein Structure 178

The Ag System 178
Serum C de B 178
Other Ag antisera 178
Ag phenotypes 179
Genetic interpretation 180

CHAPTER 5—*Continued*
Frequencies of Ag phenotypes in various populations 181

The Lp System 182
Association of β lipoprotein concentration and Lp phenotype 183
Lp phenotype frequencies in various populations 184

Factors Influencing the Formation of Antibodies Against Lp and Ag Antigens 185

Methods 186

References 187

6 PSEUDOCHOLINESTERASE

The Cholinesterases in Blood 192

Pseudocholinesterase Physiology 193
   Metabolism 193
   Function 194

Structure of Pseudocholinesterase 195
   Heterogeneity of serum cholinesterase 195
   Anionic and esteratic sites 197

Genetics 197
   The E_1 locus 197
      Early studies 197
      Evidence for inherited structural variation 198
      Relative amount of atypical enzyme in serum 200
      The 'dibucaine number' 201
      Inheritance of the 'atypical', dibucaine-resistant enzyme 203
      The fluoride-resistant enzyme variant 203
      The silent allele, E_1^s 204
      Evidence for heterogeneity of the E_1^u allele 206
      Nomenclature and characteristics of the E_1 locus phenotypes 206
      Frequencies of the E_1 locus genes 207
      Linkage of E_1 and Tf loci 210
   The E_2 locus 210
      The C_5 component 210
      Inheritance 211
      Effect of E_1 locus genes on C_5 characteristics 211
      Frequency of the C_5^+ phenotype 212
      Other possible E_2 locus variants 212

Methods 213
   Screening tests for E_1 variants 213
      Agar diffusion test for I and A phenotypes 213
      Rapid tube test for I and A phenotypes 214

CHAPTER 6—*Continued*
    The NaCl screening test 215
Determination of percent inhibition 216
    Measurement of dibucaine number (DN) 216
    Measurement of fluoride number (FN) 218
Detection of E_2 locus variants 218

References 219

7   PLASMA ALKALINE PHOSPHATASE

Intestinal Alkaline Phosphatase in Plasma 227
Electrophoretic phenotypes 227
Association of phosphatase phenotypes, ABO blood group and
secretor status 228
Inheritance of alkaline phosphatase phenotypes 230
Possible biological significance 231

Placental Alkaline Phosphatase 232

Methods 233
Starch gel electrophoresis 233
Inhibition of the intestinal enzyme with L-phenylalanine 235
Effect of neuraminidase on electrophoretic mobility 235

References 235

8   OTHER GENETIC MARKERS IN PLASMA

Albumin 240
Bisalbuminemia (alloalbuminemia) 240
An unusual albumin variant 243
Genetic linkage 245
Methods 246

The Protease-Inhibitor (Pi) System: α_1- Antitrypsin 247
Physical properties of α_1- antitrypsin 247
Phenotypic variants of α_1- antitrypsin 248
Frequency of the Pi genes 250
Physiological role of α_1- antitrypsin 252
Methods 252

Ceruloplasmin 252
Physical properties of ceruloplasmin 253
Genetic variants of ceruloplasmin 253
Methods 255

The Xm System (α_2 Macroglobulin) 256
Inheritance of Xm(a) 256
Xm gene frequencies 257

Alpha$_1$-Acid Glycoprotein 257
Types I, II and III 258

References 259

PART 2

GENETIC MARKERS IN BLOOD CELLS

9 THE RED CELL ANTIGENS: BLOOD GROUPS

General Considerations 268
  Blood group antigens 268
  Transglycosylase enzymes as blood group gene products........ 270
  The problem of multiple antigenic determinants 271
  Nomenclature 272
  Inheritance of blood group antigens 276
  'Minus-minus' phenotypes 278

The ABO System and its Genetic Interactions 278
  The ABO, Hh, Secretor and Lewis systems 278
    A, B and H antigens 278
      Secretion of water-soluble A, B and H substances 280
      Cellular localization of A, B and H substances 280
    The Lewis system 281
    Structural relationships of the A, B, H and Lewis antigens........... 284
    Chemical characterization of secreted blood group substances 284
    Sequential gene activity in blood group antigen biosynthesis 288
      Biosynthesis of secreted substances 288
      Biosynthesis of red cell H, A and B antigens 290
      Blockage of biosynthesis: 'Bombay' and other rare pheno-
      types 290
    Changes in phenotype associated with disease 292
  Interactions of ABO and Secretor genes with gene determining
  plasma alkaline phosphatase phenotype 292
    Summary of observations 292
    Suggested mechanisms 294
  The Ii system and its relationship to ABH antigens............... 296
    The antigens I and i 296
    Serological association of I and ABH 297
    Possible association of I antigen and leukemia 298

Possible Biosynthetic Pathways in Other Blood Group Systems 298
  The P system 298
    Relationship of P, Ii and ABH antigens 300
  The MNSs system.. 301
    Chemical characteristics of M and N substances 302
    The antigen U 302
  The Rh system 303
    The LW antigen 304
    The Rh_{null} phenotypes 305

CONTENTS

CHAPTER 9—*Continued*

Studies on Rh antigen structure 309
Quantitative studies on Rh reactive sites................................ 310

Blood Groups and Genetic Linkage 311
Autosomal linkage 311
X chromosome linkage: Xg^a 312

Blood Groups and Selection 314
Disease association 314
Maternal-fetal incompatibility 314
Prezygotic selection 316

Some Blood Group Gene Frequencies 317

Blood Grouping Methods 320
General principles.. 321
The antiglobulin technique 322
Neutralization tests 323
Pitfalls 323
Automated blood grouping.. 324

References 325

10 HEMOGLOBIN

The Normal Hemoglobins.. 347
General structure 347
The allosteric effect: Hemoglobin as an enzyme 353
Hb A$_2$.. 355
Hb F 356
Embryonic hemoglobins 358
Heterogeneity of Hb A and Hb F 359

The Abnormal Hemoglobins 360
Hemoglobin homotetramers 360
The Lepore hemoglobins.. 362
Hemoglobins with amino acid substitution or deletion 363
Hemoglobins causing red cell structural distortion.. 371
Sickle hemoglobin: Hb S.. 371
Hb C$_{Georgetown}$ 374
Hb C$_{Harlem}$ 374
Hb I 374
Hb C 375
Hb D$_{Punjab}$ and Hb E 375
Hemoglobins associated with methemoglobinemia 375
Hemoglobins with increased oxygen affinity 377
Hb Chesapeake.. 377
Hb J$_{Capetown}$ 378
Hb Yakima 378

CONTENTS

xiii

CHAPTER 10—*Continued*

Hb Rainier 379
Hb Ypsi 379
Hemoglobin with decreased oxygen affinity: Hb Kansas 380
The unstable hemoglobins 380
   Unstable beta chain variants 380
      Hb Zürich 380
      Hb Köln 381
      Hb Genova 381
      Hb Sydney 382
      Hb Hammersmith 382
      Hb Freiburg 383
      Hb Gun Hill 383
      Other beta chain variants associated with molecular
      instability 384
   Unstable alpha chain variants 384
      Hb Sinai (Sealy) 384
      Hb Bibba 384
      Hb Torino 385
   Hb Porto Alegre: *in vitro* polymerization 385

Hemoglobin Genetics 386
   Inheritance of structural genes 386
   Synthesis of hemoglobin 388
   Normal synthesis 388
      Synthesis of Hb A 388
   Hemoglobin synthesis during development 389
   Synthesis of structurally abnormal variants 390
   Defects in globin chain synthesis: Thalassemia 390
   Defective synthesis of β or δ chains 391
      β-thalassemia (A_2 thalassemia) 391
      $\beta\delta$-thalassemia (F thalassemia) 393
      Hb Lepore syndrome 393
      Hereditary persistence of Hb F 394
      δ-thalassemia 395
   Defective synthesis of α chains 395
      α-thalassemia 395
      Hb H disease 397
      α-thalassemia in subjects with β chain abnormalities 398
   Genetic mechanisms in thalassemia 399
   Evolution of hemoglobin 401
   Hemoglobin and balanced polymorphism: Geographic distribution 404

Methods 405

References 406

11 RED CELL ACID PHOSPHATASE

Early Studies 424

CHAPTER 11—*Continued*
Red Cell Acid Phosphatase Phenotypes 425
    Common types 426
    Rare types 427

Quantitative Studies 428
    Relationship of phenotype to enzyme activity 428
    Increased phosphatase activity in megaloblastic anemia 431
    Phosphatase activity and G6PD deficiency 431

Physico-Chemical Properties of Acid Phosphatase Isozymes 432

Acid Phosphatase Gene Frequencies 434

Methods 435
    Electrophoresis using a tris-succinic-citrate buffer system 438
    Electrophoresis using phosphate buffer 438
    Electrophoresis using formate buffer 439
    Staining 439

References 440

12 GLUCOSE-6-PHOSPHATE DEHYDROGENASE

Physiology of G6PD 444
    Glycolysis in the red cell 444
    Tissue distribution of G6PD 447

Molecular Structure of G6PD 448

Structural Variants of G6PD 449
    G6PD in Negroes 450
        G6PD deficiency 450
        The B, A and A- enzyme types 452
    G6PD deficiency in non-Negroes 453
    G6PD variants with altered electrophoretic mobility 459
    G6PD deficiency with congenital nonspherocytic hemolytic disease   459

G6PD Genetics 460
    Lyon hypothesis 460
        Demonstration of G6PD cellular mosaicism 460
        G6PD as an indicator of the embryonic X inactivation event 463
    Linkage studies 463
    Heterogeneity of G6PD gene expression 464
    G6PD and natural selection 465

Methods 465
    Brilliant cresyl blue dye screening test 466
    Measurement of G6PD enzymatic activity 466
    Starch gel electrophoresis 467

References 469

CONTENTS

XV

13 6 PHOSPHOGLUCONATE DEHYDROGENASE

Inherited Variation of 6PGD in Man	485
Electrophoretic variants	485
Phenotypes I, II and III	485
Phenotypes IV, V and VI	487
The Richmond variant (IV)	487
The Hackney variant (V)	487
The Friendship variant (VI)	488
Inherited 6PGD deficiency	489
Ilford and Newham variants	489
Whitechapel and Dalston variants	489
Other phenotypes	490
Geographic Distribution of 6PGD Phenotypes	490
Methods	491
Measurement of 6PGD activity	491
Differentiation of 6PGD phenotypes by starch gel electrophoresis	494
References	495

14 PHOSPHOGLUCOMUTASE

Properties of PGM	497
The PGM Phenotypes	499
Isozymes of the PGM_1 locus genes	499
Common types	499
Rare PGM_1 locus variants	500
Linkage studies	501
Isozymes of the PGM_2 locus genes	501
The PGM_3 locus	504
Distribution of PGM isozymes in blood cells	505
Geographic Distribution of PGM Genes in Various Populations	505
Methods	508
Determination of PGM types by starch gel electrophoresis	508
References	510

15 ADENYLATE KINASE

Electrophoretic Variants of AK	512
Inheritance of AK Phenotypes	514
Geographic Distribution of AK Genes	515
Methods	517
Demonstration of AK phenotypes by starch gel electrophoresis	517
References	518

16 OTHER ENZYME MARKERS IN BLOOD CELLS

Lactate Dehydrogenase ... 521
   Subunit structure ... 521
   Genetic variants ... 522
   Methods ... 523

Malate Dehydrogenase ... 524
   Cytoplasmic MDH ... 524
   Mitochondrial MDH ... 525
   Methods ... 527

Phosphohexose Isomerase ... 527
   Variants found at random ... 527
   Variants associated with enzyme deficiency ... 528
   Methods ... 529

Carbonic Anhydrase ... 529
   Genetic variation ... 530
   Quantitative variation ... 531
   Methods ... 533

Red Cell Esterases ... 533
   Isozyme classification ... 533
   Variants ... 535
   Methods ... 535

Peptidases ... 535
   Isozyme characteristics ... 535
   Variants ... 537
   Methods ... 539

Glutathione Reductase ... 540
   Heterogeneity ... 540
   Methods ... 541

Methemoglobin Reductase: NADH and NADPH Diaphorases ... 541
   Evidence for heterogeneity ... 542
   Methods ... 543

Catalase ... 543
   Molecular variation ... 544
   Methods ... 545

Galactose-1-Phosphate Uridyl Transferase ... 546
   Enzyme variability in galactosemia ... 546
   Methods ... 547

'Indophenol Oxidase' ... 547
   The achromatic regions in tetrazolium-stained gels ... 547

CONTENTS   xvii

CHAPTER 16—*Continued*
    A variant phenotype... 548
    Methods.. 549

References 549

PART 3

MISCELLANY

17 CONTRIBUTIONS OF BLOOD GENETIC MARKERS TO STUDIES OF HUMAN BIOLOGY

Mutation Mechanisms 557
    Point mutation and the genetic code 557
    Small deletions from intragenic crossing-over 563
    Unequal crossing-over with hybrid molecule formation 563
    Duplications 563

Molecular Organization.. 565
    Association of polypeptide subunits 565
    Association of oligosaccharide subunits and a peptide or lipid 'backbone'... 566

X Chromosome Inactivation.. 566
    Origin of tumours... 567

Origin of Developmental Abnormalities.. 567
    X chromosome anomalies 567
    Cellular mosaicism 568
        Blood group chimerism 568
        Generalized tissue mosaicism 568
        Artificial mosaicism from surgical grafting or in cell cultures 569

Genetic Linkage 570

Non-identity and Intra-family Relationships 571

Genetic Polymorphism and Selection 572

References 573

GLOSSARY 576

ADDENDA... 587

INDEX 602

LIST OF FIGURES

1.1 Diagram of immunoglobulin synthesis 8

1.2 Diagram of the IgG molecule 12

1.3 Diagram of the kappa light chain of the immunoglobulins 16

1.4 Diagram showing evolutionary development of the immunoglobulin chains 21

1.5 Diagram of the serological system used for detecting Gm and Inv allotypes 26

2.1 Starch gel electrophoretic patterns of hemoglobin A, haptoglobin (Hp 1-1) and their complexes, intermediate and saturated 72

2.2 Diagram of starch gel electrophoretic patterns of the three common haptoglobin types: Hp 1-1, 2-1 and 2-2 73

2.3 Diagram of the common Hp subtype patterns 74

2.4 Diagram of Hp^2 gene formation by random breakage and reunion of Hp^{1F} and Hp^{1S} 75

2.5 Diagram of Hp^2 gene formation by non-homologous crossover between Hp^{1F} and Hp^{1S} 80

2.6 Diagram of amino acids and corresponding codons in the region of crossover between Hp^{1F} and Hp^{1S} 81

2.7 Diagram of Hp^{2FF} and Hp^{2SS} formation by unequal crossover in two Hp 2-1 heterozygotes 83

2.8 Diagram of Hp^1 formation, representing partial triplication of the Hp^1 gene 83

2.9 Starch gel electrophoretic subtype patterns of Hp 2-1, Hp 2-1(mod) and Hp 1-J 84

2.10 Spectrum of Hp 2-1 and Hp 2-1(mod) electrophoretic patterns 85

2.11 Electrophoretic patterns of Hp 2-1 'quantitative' variants 86

2.12 Electrophoretic patterns of haptoglobin 'qualitative' variants 90

2.13 Photograph of a simple starch gel electrophoresis apparatus 102

2.14 Components of vertical electrophoresis apparatus 102

2.15 Diagram of slot-forming cover for starch gels 103

3.1 Electrophoretic patterns of transferrin with and without added iron 129

3.2 Diagram of two-dimensional electrophoresis of serum containing Tf CD_1 and Tf B_2C 136

3.3 Autoradiograph of Fe^{59}-labelled transferrin in five serum specimens with different Tf phenotypes 137

3.4 Diagram of the 28 reported human transferrin phenotypes 138

3.5 Immunodiffusion patterns of several human transferrins 143

4.1 Immunoelectrophoretic patterns of Gc 1-1, 2-1 and 2-2 162

4.2 Starch gel electrophoretic patterns of Gc 1-1, 2-1 and 2-2 163

xix

5.1   Diagram of rosette cutter for detecting Ag and Lp types by immuno-diffusion 186
5.2   Photograph of antigen-antibody precipitates in agar immunodiffusion test for Ag or Lp 186

6.1   Photograph of serum esterase electrophoretic pattern and diagram of two-dimensional electrophoretic pattern 196
6.2   Inhibition of usual, intermediate and atypical pseudocholinesterase by RO2-0683 201
6.3   Inhibition of usual and atypical pseudocholinesterase by dibucaine..... 202
6.4   Starch gel electrophoretic patterns of four pseudocholinesterase phenotypes 205
6.5   Starch gel electrophoretic patterns of C_5- and C_5^+ phenotypes 211

7.1   Starch gel electrophoretic patterns of the alkaline phosphatase phenotypes: p^{++}, p^+ and p° 227

8.1   Diagram of electrophoretic patterns of six albumin variants 243
8.2   Diagram comparing the usual two-dimensional electrophoretic patterns with an unusual serum containing an X component.................... 244
8.3   Agarose electrophoretic patterns of an unusual albumin variant compared with the usual and bisalbuminemia patterns 244
8.4   Diagram of crossed-electrophoresis patterns of three protease-inhibitor (Pi) phenotypes 249
8.5   Diagram of twelve Pi phenotypes 250
8.6   Diagram of seven ceruloplasmin (Cp) phenotypes 254

9.1   Comparison of the Rh multiple-allele and linked-gene theories 273
9.2   Molecular structure of the four sugars which determine A, B, H, Le^a and Le^b blood group specificity 285
9.3   Terminal and subterminal sugars associated with blood group specificity.. 287
9.4   Sequential steps in the biosynthesis of the blood group specific substances: H, A, B, Le^a and Le^b 289
9.5   Association of blood group and secretor status with serum alkaline phosphatase phenotypes 293
9.6   Sequential steps in biosynthesis of antigens in the P system 299
9.7   Rh types in the family of a woman with Rh_{null} phenotype 306
9.8   Rh types in the family of a boy with Rh_{null} phenotype apparently due to an amorph at the Rh locus 307
9.9   Alternative pathways in biosynthesis of Rh and LW antigens 308

10.1   Diagram of the hemoglobin beta chain 349
10.2   Amino acid sequences of hemoglobin alpha and beta chains arranged according to their three-dimensional positions 350
10.3   The prosthetic group of hemoglobin, iron protoporphyrin IX 351
10.4   Orientation of the two beta chains in the hemoglobin molecule.......... 352
10.5   Successive oxygenation of the $\alpha\beta$ dimers of hemoglobin 354

10.6   Diagram of the two Lepore hemoglobin genes and their corresponding peptide chains 362

10.7   Sites of amino acid substitution associated with altered function of the hemoglobin molecule 371

10.8   Diagram of the internal bonding at the N-terminus of the Hb S beta chain 372

10.9   Diagram of the five pairs of genes controlling the synthesis of normal human hemoglobins 386

10.10   Diagram of changes in non-alpha globin chain production during early human development 389

10.11   Diagram showing evolution of the globin chains of hemoglobin 402

11.1   Starch gel electrophoretic patterns of six acid phosphatase phenotypes 426

11.2   Comparison of electrophoretic patterns of seven acid phosphatase phenotypes in three different buffer systems 427

11.3   Red cell acid phosphatase activity associated with five different phenotypes 429

11.4   Unimodal distribution curve of acid phosphatase activity 430

12.1   Glycolytic pathways in the red cell 445

12.2   Starch gel electrophoretic patterns of six G6PD phenotypes 452

12.3   Principle of G6PD stain 468

13.1   Starch gel electrophoretic patterns of 6PGD phenotypes from red cell and white cell lysates 486

13.2   Diagram of electrophoretic patterns of six 6PGD phenotypes 488

14.1   Starch gel electrophoretic patterns of the three common PGM types 498

14.2   Diagram of the common PGM types compared with nine rare variants 499

14.3   Starch gel electrophoretic patterns of six different PGM types representing variation at the PGM_1 and PGM_2 loci 500

14.4   Pedigree of family with the PGM 'Atkinson' type 502

14.5   Diagram of PGM phenotypes reflecting two mutant genes at the PGM_2 locus 503

14.6   Diagram of PGM components determined by genes at PGM_1, PGM_2 and PGM_3 loci 504

14.7   Comparison of PGM 1 and PGM 2 patterns in lysates of red cells and white cells 504

14.8   Principle of PGM stain 509

15.1   Diagram of four adenylate kinase (AK) phenotypes 513

15.2   Starch gel electrophoretic patterns of AK 1, 2-1 and 4-1 514

15.3   Principle of AK stain 518

16.1   Starch gel electrophoresis of lactate dehydrogenase (LDH) isozymes in red cells and white cells 522

16.2   Diagram of cytoplasmic malate dehydrogenase (MDH): usual and variant patterns ... 524

16.3   Diagram of cytoplasmic and mitochondrial MDH ... 526

16.4   Starch gel electrophoretic patterns of nine phosphohexose isomerase (PHI) phenotypes ... 528

16.5   Starch gel electrophoretic patterns of PHI in the family of a child with PHI deficiency ... 529

16.6   Diagram of red cell carbonic anhydrase patterns of CA I and CA II ... 531

16.7   Diagram of red cell esterase patterns ... 534

16.8   Diagram of the common phenotypes of five red cell peptidases ... 537

16.9   Diagram of five peptidase A and four peptidase B phenotypes ... 538

17.1   The genetic code ... 558

17.2   Diagram of an example of point mutation ... 559

17.3   Diagram showing the amino acid changes of the abnormal hemoglobins, representing single-step point mutation ... 561

17.4   Diagram of an example of partial gene deletion due to unequal intragenic crossover ... 562

17.5   Diagram of an example of gene hybridization due to unequal crossover between Hb_β and H_δ genes ... 564

LIST OF TABLES

1.1 General properties of the human immunoglobulin classes 5
1.2 Metabolism of human immunoglobulins 10
1.3 Subclasses of IgG 19
1.4 Antigens of the Gm and Inv systems 24
1.5 Gm type and IgG subclass of 160 myeloma proteins 29
1.6 Assignment of Gm antigens to IgG subclass 30
1.7 Racial distribution of inherited Gm antigens arranged according to IgG subclass 31
1.8 C-terminal sequences of the γ chains of seven human myeloma proteins 35
1.9 Slide test (Steinberg) for Gm and Inv antigens 43

2.1 Geographic distribution of the Hp^1 gene 94

3.1 The human transferrins listed in order of electrophoretic mobility 137

4.1 Electrophoretic mobility of the Gc variants 163

5.1 Distribution of Ag^x and Ag^y genes in a small number of populations 181

6.1 Behavior of the usual and atypical cholinesterases with two substrates and five classes of inhibitors 199
6.2 Characteristics of the pseudocholinesterase types 204
6.3 Frequency of the E_1^a allele in various populations 208

7.1 Lewis, ABO and alkaline phosphatase types in 800 people 229

8.1 Pi gene frequencies in three populations 251
8.2 Distribution of Xm phenotypes in four populations 257

9.1 Blood group antigens of human red cells 269
9.2 Comparison of Rh nomenclature according to the multiple-allele and linked-gene theories 274
9.3 Numerical designation of Rh antigens 274
9.4 Numerical designation of Kell antigens 275
9.5 Relationships of phenotypes and genotypes in the H, ABO, Lewis and Secretor systems 283
9.6 Reactions of guinea pig anti-LW serum with red cells of various Rh and LW phenotypes 305
9.7 Frequency of genes which appear to determine a single antigenic determinant in ten blood group systems 318
9.8 Frequency of 'genes' and 'gene complexes' of three blood group systems in which multiple antigenic determinants are inherited as a unit 319

10.1 Amino acid differences of β, γ and δ chains of hemoglobin 356

10.2 Abnormal hemoglobins with amino acid substitution or deletion at one or more sites 364

10.3 Amino acid substitutions in the M hemoglobins 376

10.4 Beta thalassemia and related syndromes 391

10.5 Hereditary persistence of fetal hemoglobin 394

10.6 Alpha thalassemia syndromes 398

11.1 Geographic distribution of genes in the acid phosphatase system 436

12.1 Association of drugs with hemolysis in G6PD deficiency 455

12.2 G6PD variants 456

13.1 6PGD phenotypes 484

13.2 Population distribution of 6PGD phenotypes 492

14.1 Distribution of the common PGM_1 genes and the unusual PGM phenotypes in various populations 506

15.1 Distribution of adenylate kinase types in various populations 516

16.1 Relative activity of five red cell peptidases with dipeptides and tripeptides 536

16.2 Incidence of peptidase A and B phenotypes in two populations 539

17.1 Genetic systems useful for distinguishing between two random samples of blood from western Europeans 572

PREFACE

The purpose of this book is to bring together into one volume descriptions of those components of human blood which exhibit heritable variation sufficient to be classified as genetic polymorphism. As defined by E.B. Ford, the term polymorphism applies only to those phenotypes which occur in a population more frequently than would be expected on the basis of recurrent mutation. Although the rarer molecular variants are of considerable importance from the standpoint of biochemistry and physiology, their usefulness as genetic markers is very limited. Thus, those so-called inborn errors of metabolism which are associated with clinical abnormalities are usually not included here unless their phenotype frequencies exceed one or two per cent in large population samples.

Interest in human genetic markers began with the blood group antigens and received a major stimulus when Oliver Smithies introduced the technique of starch gel electrophoresis. This sensitive method permits the separation of molecules on the basis of their size as well as electrical charge. Initially, it was used for differentiating serum protein variants, such as haptoglobin and transferrin. However, largely due to the work of Harry Harris and his colleagues, the technique was extended to the study of cellular enzymes, which proved a fruitful source of genetic markers.

It is very likely that many molecular variants escape detection because they have the same size and electrical charge as the common types. Nevertheless, when tests for the electrophoretic variants are combined with tests for antigenic types, the probability that the blood of any two random human beings of European origin will have identical phenotypes is now about one in 350,000 (See p. 571). The certain discovery of more polymorphic systems will increase these odds even further.

Since the topics discussed in this book are the objects of extensive investigation, it is inevitable that much important work must be omitted. In some cases, the omissions are deliberate. For example, since other authors have written well-known texts about the blood group antigens and the hemoglobins, Chapters 9 and 10 are limited mainly to considerations of molecular structure and biosynthesis. Furthermore, no attempt has been made to include in this book descriptions of the white cell and platelet antigenic systems, because of the author's limited experience in this important field.

Undoubtedly there are many unintentional omissions which can only be defended by personal limitations in covering the rapidly expanding literature. To reduce the number of errors and exclusions, I requested and received a great deal of help. Especial thanks are due to Marie and John Crookston, who provided me with enormous support. They not only encouraged me to write this book and maintained that enthusiasm, but also read much of the manuscript at several stages of its development, suggesting numerous changes and additions. Marie Crookston then read the final typescript and saved me from a multitude of errors.

To my former and constant teachers, C.A. Finch, P.L. Mollison, A.G. Motulsky and A.G. Steinberg, I am very grateful for widening my scope of interests and giving me advice about the preparation of this book: in particular, the chapters on transferrin, blood groups, pseudocholinesterase, glucose-6-phosphate dehydrogenase and immunoglobulins. I also wish to express my indebtedness for very constructive comments and suggestions to M. Mannik, who read the chapter on immunoglobins; to Winifred Watkins and T.J. Greenwalt (blood group chapter); to E.R. Huehns and G. Stamatoyannopoulos (hemoglobin) and N.D. Carter (6-phosphogluconate dehydrogenase). However, for errors and omissions which remain in the book, I alone am responsible.

My thanks are also due to many others who provided me with data from manuscripts in preparation or in press. They include Jill Luffman, Elizabeth Robson, F.H. Allen, E. Beutler, J.V. Dacie, J.C. Detter, G.H. Dixon, P. Fialkow, H. Harris, D.A. Hopkinson, K.G. Jensen, C.B. Laurell, G. Modiano, D.C. Shreffler, H.E. Sutton and A. Yoshida.

The officers and directors of the King County Central Blood Bank have been most generous in permitting me the time to do research and write this book as a part of my responsibilities as immunogeneticist. I also wish to acknowledge my indebtedness to the United States Public Health Service, from which I received grants HE 5780 and AM 09745 for carrying out much of my work during the past several years.

The laboratory management and assistance of Miss Lucy Brooks and technical contributions of Miss Jeanne Anderson are acknowledged with many thanks. To Miss Nancy Scott, who worked both in the laboratory and in the darkroom, I am particularly grateful for the photographs of starch gels and other figures. Miss Dorothy Wolf made the line drawings and typed the entire manuscript from my often illegible handwriting without discontinuing her other duties at the blood bank. Her resourcefulness and tireless efforts were of tremendous help. Thanks are due also to Mrs. Margaret Harrison.

Per Saugman, of Blackwell Scientific Publications, was a most helpful and patient adviser during the book's preparation.

For permission to reproduce, in whole or in part, figures and tables from their published papers, I wish to thank the numerous authors acknowledged in the text, and the publishers of the *American Journal of Human Genetics*, the *Annals of Human Genetics*, *Biochemical Pharmacology*, *Biochimica et Biophysica Acta*, the *Canadian Journal of Biochemistry and Physiology*, *Clinica Chimica Acta*, the *Journal of Clinical Investigation*, the *Journal of Experimental Medicine*, the *Journal of Human Genetics*, the *Lancet*, the *Proceedings of the National Academy of Science* (U.S.), the *Proceedings of the Royal Society B*, *Progress in Medical Genetics*, *Science*, *Series Haematologica*, *Transfusion* and *Vox Sanguinis*.

Finally, a note of deepest thanks to my mother, who had the strength, in spite of serious illness, to bear the thousand inconveniences which my preoccupation created. I hope this book is worthy of her fortitude.

PART 1

GENETIC MARKERS
IN PLASMA

CHAPTER 1

THE IMMUNOGLOBULINS AND THEIR ALLOTYPES: Gm AND Inv

Characteristics of the Immunoglobulins 4
General features.......... 4
Immunoglobulin Metabolism 7
Concentration in plasma..... 7
Synthesis 7
Turnover 9
Immunoglobulin Structure 11
General features of IgG molecules 11
Polypeptide chain structure 13
The light chains 13
Electrophoretic patterns of light chains 14
Light chains of antibodies.......... 14
Aminoacidsequence of light chains 15
The heavy chain (γ) of IgG 17
Heterogeneity of Fd fragment 17
The 'hinge' region..... 18
IgG subclasses 18
Heavy chains of the other immunoglobulins 20
Evolution of immunoglobulin chains 20
Genetic basis of immunoglobulin variability..... 22
Inherited Antigenic Determinants (Allotypes).......... 23

The Gm system 23
Gm antigens and their detection 23
Gm genetics 26
Family studies.......... 26
Comparison of Gm and Rh inheritance.......... 27
Gm heterogeneity in single serum specimens 27
Assignment of Gm antigens to IgG subclasses 28
Two genetic theories 31
Studies of Gm structure 33
Geographic distribution of Gm antigens.......... 36
Anti-Gm antibodies.......... 37
The Inv system 38
Inv antigens 38
Studies of Inv structure 39
Inv gene frequencies.......... 40

Methods 40
The IgG antibody coat.......... 41
The agglutinator.......... 42
The agglutination-inhibition test 42
The tube technique 42
The tile technique.......... 43
References.......... 44
References to Gm and Inv Geographic distribution.......... 60

The inherited antigenic determinants of the Gm and Inv systems are located on molecules of the immunoglobulins, a family of serum proteins with antibody activity. Among the many recent papers reviewing immunoglobulin structure, function and inheritance are those of Fahey (1962), Fudenberg & Franklin (1963), Cohen (1963a), Nisonoff & Thorbecke (1964), Cohen & Porter (1964), Bernier & Putnam (1964a), Tomasi (1965), Ropartz, Rousseau & Rivat (1965a), Steinberg (1962a, 1964, 1967), Martensson (1966), Fleischman (1966), Schwartz (1966), Poulik (1967), Muir & Steinberg (1967), Litwin & Kunkel (1967a), Cohen & Milstein (1967a, b), Lennox & Cohn (1967) and Natvig & Kunkel (1968). Several excellent symposia have appeared; especially recommended are those in the *Proceedings of the Royal Society*, Series B, vol. 166, 1966; the *New England Journal of Medicine*, vol. 275, 1966; the *Third Nobel Symposium*, 1967; and the *Cold Spring Harbor Symposium on Quantitative Biology*, vol. 32, 1967.

CHARACTERISTICS OF THE IMMUNOGLOBULINS

GENERAL FEATURES

Five different molecular classes of immunoglobulins are currently recognized: IgG, IgM, IgA, IgD and IgE (also known as γG, γM, γA, γD and γE globulins, respectively). In order to function as antibodies with widely varying specificity, the immunoglobulin molecules must possess remarkable heterogeneity, even within the individual classes. However certain general features have been recognized, and these are listed in Table 1.1.

Since IgD and IgE were discovered only recently, and since they normally have low concentrations in the plasma, much less is known about them than the other immunoglobulins. Thus, the function of IgD has not been determined (Rowe & Fahey, 1965a, b), and, while IgE is known to carry reaginic activity (Ishizaka *et al.*, 1966, 1967a, b; Ishizaka & Ishizaka, 1967), its structure is not well characterized.

All of the immunoglobulins are assumed to have a basic structural unit which consists of two heavy chains (M.W. 50,000–70,000) unique to the molecular class and two light chains (M.W. 20,000) common to all classes (Edelman & Benacerraf, 1962; Fleischman *et al.*, 1962; Small *et al.*, 1963; Carbonara & Heremans, 1963; Cohen & Porter, 1964; Olins & Edelman, 1964). IgM has a pentamer structure containing five of these units held together by disulfide bonds, thus accounting for its 900,000 molecular weight (Miller & Metzger, 1965; Chaplin *et al.*,

TABLE 1.1. General properties of the human immunoglobulin classes.

	IgG, γG	IgM, γM	IgA, γA	IgD, γD	IgE, γE
Synonyms	7Sγ, γ₂, γss	19Sγ, β₂M, γ₁M	β₂A, γ₁A		
Molecular weight	150,000	900,000	180,000, 400,000*	150,000	150,000
Sedimentation constant	7S	19S	7S, 11S	7S	8S
Carbohydrate content	3%	12%	8%		
Plasma level, mg/100 ml	900–1500	70–160	150–400	0·3–40	0·01–0·06
Heavy chain, mol. wt.	γ 50,000	μ 70,000	α 64,000	δ	ε
Light chains	κ, λ	κ, λ	κ, λ	κ, λ	κ, λ
Molecular formula	$\gamma_2\kappa_2$	$(\mu_2\kappa_2)_5$	$\alpha_2\kappa_2.\ \alpha_2\lambda_2$	$\delta_2\kappa_2$	
	$\gamma_2\lambda_2$	$(\mu_2\lambda_2)_5$	$(\alpha_2\kappa_2)_2T,\ (\alpha_2\lambda_2)_2T^*$	$\delta_2\lambda_2$	
Gm specificity (γ chain)	Yes	No	No	No	
Inv specificity (κ chain)	Yes	Yes	Yes	?	
Number of subclasses known	4	2	2		
Antibody activity	Yes	Yes	Yes	?	Yes
Passes placenta	Yes	No	No	No	
Present in CSF	Yes	No	Yes		
Secreted by exocrine glands	No	No	Yes		
Fixes complement	Yes	Yes	No		

* In exocrine gland secretions, IgA has a dimeric structure and an attached T chain.

1965; Lamm & Small, 1966) and its susceptibility to sulfhydryl reagents (Deutsch & Morton, 1957). Two molecular weight classes of IgA are detectable in plasma, the major component (90 per cent) weighing about 180,000, and the other 400,000. This larger-sized component is apparently produced by exocrine glands, and consists of two structural units with an attached 'T' chain which may or may not be involved in the transport of IgA across cell membranes (Tomasi & Zigelbaum, 1963; Tomasi *et al.*, 1965; South *et al.*, 1966; Stiehm *et al.*, 1966b). IgD is thought to have a molecular weight similar to that of IgG (Rowe & Fahey, 1965b), while IgE may be somewhat larger (Ishizaka *et al.*, 1966). (See Addenda.)

Presumably all of the immunoglobulins contain a carbohydrate moiety attached to their heavy chains, although quantitative data are available only for the three best known classes, showing considerably more carbohydrate in IgM and IgA than in IgG (Müller-Eberhard *et al.*, 1956; Heremans *et al.*, 1959; Nolan & Smith, 1962; Heimburger *et al.*, 1964). Attempts to assign a functional role to the sugars have been largely unsuccessful (Smiley & Horton, 1965). However, in a recent study, Melchers & Knopf (1967) found a difference in the tryptic glycopeptide of mouse light chains inside plasma cells as compared with the secreted protein. They proposed that the attachment of carbohydrate is required for secretion of immunoglobulin from the cell.

Certain biological properties are characteristic of each immunoglobulin class. For example, only IgG is known to cross the placental barrier from mother to infant, so that hemolytic disease of the newborn is presumably always due to IgG antibodies (Hemmings, 1961). The property of complement fixation is characteristic of IgG and IgM antibodies, but not of IgA (Taranta & Franklin, 1961; Adinolfi *et al.*, 1966). The secretion of IgA into seromucinous fluids such as saliva, colostrum and tears as well as into the intestinal, nasal and bronchial secretions has been well documented (Tomasi & Zigelbaum, 1963; Chodirker & Tomasi, 1963; Tomasi *et al.*, 1965). The participation of IgE antibodies in hypersensitivity phenomena was mentioned previously. All of these findings as well as the predilection of antibodies specific for certain antigens to belong to a given class of immunoglobulins strongly support the assumption that although their functions may overlap, the members of the individual classes are specialized for certain functional capacities as antibodies.

The antigenic properties of the immunoglobulin molecules within a given species have been used to differentiate three degrees of specificity related to structural variation (Oudin, 1966a, b). The *isotypic* specificities are common to all the members of the species, but differentiate the total population of immunoglobulin molecules into classes and subclasses. The *allotypic* specificities differ within the species, and since they reflect inherited variation of molecular structure, they are useful as genetic markers. The *idiotypic* specificities characterize the product of a single cell or clone of cells, exemplified by the myeloma proteins.

IMMUNOGLOBULIN METABOLISM

CONCENTRATION IN PLASMA

The synthesis of antibodies does not normally occur in the absence of an antigenic stimulus. Thus, animals raised under germ-free conditions have very low immunoglobulin levels (Wostman, 1959), and newborn infants emerge from their uterine environment with only trace amounts of IgM and IgA in their serum (Franklin & Kunkel, 1958; van Furth *et al.*, 1965; Stiehm & Fudenberg, 1966). However, in infants born with certain viral infections, the amount of both of these globulins is increased over that in normal infants, consistent with an *in utero* response to antigenic stimulus (Stiehm *et al.*, 1966a). IgG is actively secreted across the placenta from mother to infant, particularly in the third trimester. Thus, the IgG level of premature infants is often lower than that of their mothers (Hobbs & Davis, 1967), but in normal infants, the concentration of IgG may actually exceed that of the mother, even though it is mainly of maternal origin. Nevertheless, babies born to hypogammaglobulinemic mothers have low IgG levels, while babies of mothers with hypergammaglobulinemia have elevated levels (Bridges *et al.*, 1959; Rosen & Janeway, 1966). That some fetal synthesis of IgG can occur was indicated by the work of Fudenberg & Fudenberg (1964), who showed that mothers can form antibodies directed against the Gm antigenic groups on the IgG molecules synthesized by their unborn infants.

SYNTHESIS

The major antibody-producing cell is the plasmacyte (Coons *et al.*, 1955; Solomon *et al.*, 1963a), although cells having the appearance of lymphocytes have also been shown to contain, and presumably to make, immunoglobulin (Hummeler *et al.*, 1966). Mechanisms underlying the induction of antibody synthesis are incompletely understood. However, there is evidence for the interaction of antigen and phagocyte, resulting in the production of either RNA or an RNA-antigen complex which can initiate lymphoid cell proliferation, differentiation into plasma cells, and antibody production (Franzl, 1962; Fishman & Adler, 1963; Schoenberg *et al.*, 1964; Askonas & Rhodes, 1965; Adler *et al.*, 1966; Gottlieb *et al.*, 1967).

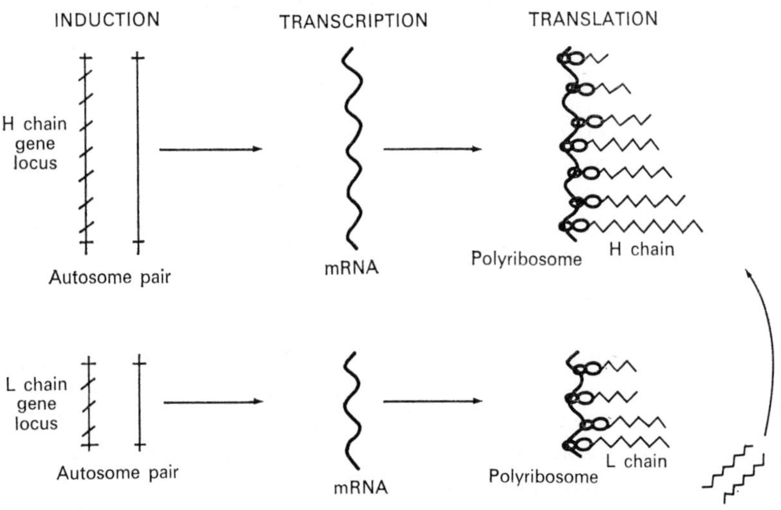

FIGURE I.I. Schematic representation of the synthesis of immunoglobulin molecules. The DNA in only one member of each pair of genes coding for heavy and light chains, respectively, is transcribed into complementary messenger RNA (mRNA), which is then translated on the polyribosomes into the finished polypeptide chains. Released light chains then combine with heavy chains to form the complete molecule.

As part of this process, structural genes at two different loci which control the synthesis of heavy and light chains respectively, are stimulated into activity, as shown in Fig. I.I. However in any given cell only one of each pair of alleles is active, and there is further limitation to production of a single kind of heavy chain (γ, μ, α, δ, or ϵ) and light chain (κ or λ) (Mellors & Korngold, 1963; Bernier & Cebra, 1964, 1965; Colberg & Dray, 1964; Gell & Sell, 1965; Pernis *et al.*, 1965, 1966; Burtin & Buffe, 1967, Green *et al.*, 1967). Although a similar exclusion of alleles is known to occur in the cells of females because of random inactivation of one or the other of their X chromosomes (see chapter on G 6 PD), the immunoglobulins provide the only known example of autosomal allele exclusion. According to Gell & Sell (1965), the lymphocytes are 'primed' prior to the antigenic stimulus so that only one of a pair of alleles is activated.

Transcription into individual messenger RNA molecules is followed by translation on polyribosomes of the appropriate sizes for the two chains (Scharff & Uhr, 1966; Shapiro *et al.*, 1966; Becker & Rich, 1966). Synthesis proceeds from N- to C-terminal ends, as in the construction of other proteins (Lennox *et al.*, 1967). Conceivably, the completed heavy chains remain attached to their respective polyribosomes until they combine with light chains which have been released after synthesis (Shapiro *et al.*, 1966; Askonas & Williamson, 1966a, b; 1967a, b). The final assembly of H_2L_2 molecules from HL half-molecules then occurs. In the case of IgM, pentamers are formed from these basic units. If the suggestion is correct that the light chains of IgM molecules are not homogeneous (Costea *et al.*, 1966), then pentamer formation would have to occur after the release of $\mu_2\kappa_2$ or $\mu_2\lambda_2$ units from their originating cells.

There appears to be a feedback effect of IgG antibodies on the synthesis of both IgG and IgM of the same specificity (Uhr & Baumann, 1961; Neiders *et al.*, 1962; Schwartz, 1966; Dixon *et al.*, 1967). Thus, while passive administration of IgM antibodies has little or no effect on the production of either IgM or IgG, similar injections of specific IgG profoundly depress IgG production and partially curtail IgM antibody production (Finkelstein & Uhr, 1964). Conversely, when the inhibitory effect of IgG is removed by immunosuppressive agents, IgM antibody production is prolonged (Sahiar & Schwartz, 1965).

TURNOVER

Studies of immunoglobulin catabolism based on measurements of radioactivity after the injection of labelled purified protein, indicate that each of the molecular classes is independently metabolized. Table 1.2 presents a summary of average turnover data based on such catabolic studies. The accuracy of the calculations is dependent on a steady state of immunoglobulin production and destruction, as well as a state of equilibrium between the intra- and extravascular components of each system. The latter condition may not be applicable in the case of IgA, where the 11S molecules found in the seromucinous secretions may comprise a separate functional compartment. Nevertheless, certain prominent differentiating features are indicated by the data.

TABLE 1.2. Metabolism of human immunoglobulins

	IgG	IgM	IgA	IgD
Average plasma level, mg/100 ml	1100	100	220	3
% of total in plasma	45	75	? 40	75
Catabolic rate, $T\frac{1}{2}$ in days	23	5	6	3
Amount metabolized, mg/kg/day	40	4	? 40	0·40
Proportion of total pool metabolized daily	0·03	0·13	0·13	0·25
Fractional catabolic rate*	0·07	0·18	0·33	0·33

* Fractional catabolic rate = proportion of intravascular component catabolized daily

As indicated in Table 1.2, the half-life ($T\frac{1}{2}$) of IgG is considerably longer than that of the other immunoglobulins, so that even though the amount of IgG synthesized daily does not greatly exceed that of IgA, normal levels of IgG in the plasma are about four or five times those of IgA. IgM, and particularly IgD, are produced in much smaller amounts than IgG or IgA, and their turnover times are even more rapid than that of IgA. However, instead of having the approximate 50:50 distribution of IgG (? and IgA) in intra- and extravascular compartments, IgM and IgD are largely localized in the intravascular space (Cohen & Freeman, 1960; Solomon *et al.*, 1963b; Cohen, 1963b; Barth *et al.*, 1964; Rogentine *et al.*, 1966; Stiehm *et al.*, 1966b).

The plasma level of IgG exerts a curious effect on the IgG catabolic rate. Thus, the $T\frac{1}{2}$ of injected IgG is found to be appreciably longer than normal in hypogammaglobulinemic patients, while in subjects with very high IgG levels, the $T\frac{1}{2}$ is considerably shorter than normal (Andersen, 1963; Solomon *et al.*, 1963b; Fahey & Robinson, 1963; Freeman, 1965). In other words, the higher the IgG level, the shorter the IgG half-life. A possible explanation for this phenomenon was suggested by Brambell *et al.*, (1964) and Brambell (1966) who proposed that in the catabolism of IgG, the molecules are taken up by pinocytosis into the phagosomes of intestinal mucosal cells. Some of the IgG molecules are protected from subsequent lysosomal degradation because of their attachment, by means of the Fc fragment, to specific sites on the phagosome wall. Such sites, being limited in number, can protect only a corresponding number of molecules. Thus, regardless of the total amount of IgG taken up by the cell, a fixed quantity, rather than a fixed proportion, can be preserved for re-entry into circulation.

IMMUNOGLOBULIN STRUCTURE

GENERAL FEATURES OF IgG MOLECULES

In each molecule of IgG antibody, there are two identical heavy (γ) chains and two identical light (κ or λ) chains covalently linked by interchain disulfide bonds (Edelman & Poulik, 1961; Edelman & Benacerraf, 1962; Cohen & Porter, 1964). Only one covalent bond links light and heavy chains. The number of bonds between the two heavy chains is still a matter of controversy (Fleischman *et al.*, 1962; Fleischman, 1966; Nisonoff *et al.*, 1961). However, the recent studies of Frangione & Milstein (1967) and Pink & Milstein (1967b) indicate that there are probably two inter-heavy chain bridges, at least in some molecules of IgG. (See Addenda.) When the heavy-light chain disulfide bonds are cleaved, the molecule is still held together by non-covalent forces, the existence of which permits recombination of separated heavy and light chains (Olins & Edelman, 1964; Grey & Mannik, 1964; Inman & Nisonoff, 1966).

Early studies of rabbit gamma globulin by Porter (1959, 1960) showed that treatment of IgG with papain and cysteine produced two fragments (now call Fab) capable of combining with specific antigen, and a third crystallizable fragment (now called Fc) with no antigen-binding ability but necessary for other functions of the antibody molecule such as complement fixation (Taranta & Franklin, 1961), microsomal binding during antibody synthesis (Kern *et al.*, 1961), skin sensitization (Franklin & Ovary, 1963; Terry, 1965), trans-placental passage (Hemmings, 1961), and binding to phagosome sites during passage through intestinal epithelial cells (Brambell,

1966). The molecular weight of Fc is about 48,000 (Noelken *et al.*, 1965).

Similar separation of human IgG into two kinds of fragments was achieved by Franklin (1960) and Edelman *et al.* (1960). It was subsequently shown by Cohen & Porter (1964) that papain cleaves the heavy chains at about their midpoint, so that each of the two Fab fragments contains the amino terminal half of the heavy chain (known

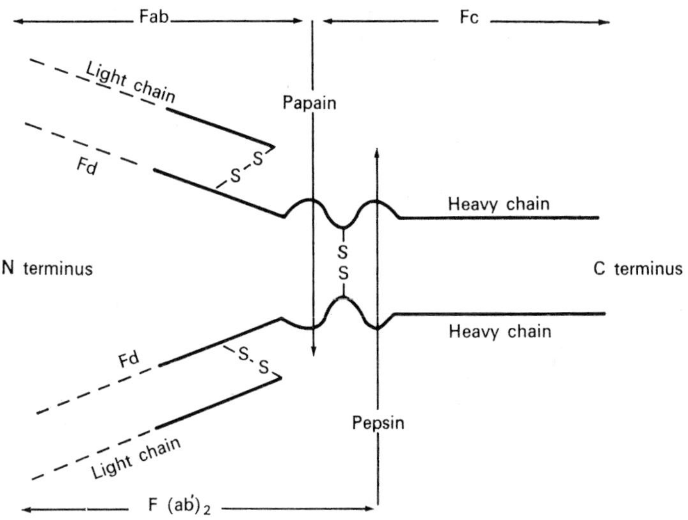

FIGURE I.2. Diagram of the IgG molecule, showing the inter-chain disulfide bonds and the products of cleavage by the proteolytic enzymes, papain and pepsin. The broken lines in the N-terminal portions of the light chains and Fd fragments of the heavy chains represent heterogeneity in amino acid sequence. (Adapted from Cohen & Milstein, 1967a. Reprinted with permission from *Nature*, **214**, 449.)

as the Fd fragment) and a complete light chain (Fougereau & Edelman, 1965). The Fc fragment thus consists of the carboxy-terminal half of the two heavy chains, and can be split into two identical halves by disulfide bond cleavage (Merler *et al.*, 1964).

Treatment of IgG with pepsin at pH 4 or 5 breaks the Fc fragment into a number of smaller peptides but leaves a large 5S fragment with a molecular weight of 105,000 called F(ab′)$_2$, since it can be split by sulfhydryl reagents into two equal parts which resemble Fab but contain some additional peptides (Franklin, 1960; Nisonoff *et al.*, 1960, 1961; Utsumi & Karush, 1964; Fougereau & Edelman, 1964, 1965). Figure 1.2 presents a diagram of the molecule and its components as modified from Porter (1962) by Cohen & Milstein (1967a) to show the interchain disulfide bonds and their relation to the sites of enzymatic cleavage. Several other molecular models have been proposed to emphasize various structural features (Edelman & Gally, 1964; Noelken *et al.*, 1965; Ovary, 1966).

Since fragments similar to Fc are recovered in the urine of patients with 'heavy chain disease' (Osserman & Takatsuki, 1964; Franklin *et al.*, 1964), and since Fd is more variable in structure than Fc, it has been suggested that the two fragments may represent structurally distinct units (Frangione & Franklin, 1965; Feinstein *et al.*, 1963). A similar possibility exists in the case of the variant and invariant portions of the light chains to be described below. However, there is currently no firm evidence for more than four polypeptide chains in the IgG molecule.

POLYPEPTIDE CHAIN STRUCTURE
Wide structural diversity is required in order that the immunoglobulin molecules of normal serum can function as antibodies with varying specificity. Thus, in order to obtain homogeneous samples of intact molecules, it is necessary to isolate the 'monoclonal' product from the serum of patients with myeloma (IgG, IgA or IgD) or macroglobulinemia (IgM). Similarly, homogeneous light chains are obtained by extracting Bence Jones protein from the urine of such patients (Edelman & Gally, 1962; Migita & Putnam, 1963; Bernier & Putnam, 1964b).

The light chains
The two kinds of light chains, K and L, designated by Greek letters κ and λ, respectively, are distinguishable by their reactions with anti-κ and anti-λ antibodies (Korngold & Lipari, 1956; Mannik & Kunkel, 1962, 1963a; Fahey, 1963a, b). The currently accepted terminology for these two chains, 'type K and type L', refers to their isotypy and not their allotypy, since all members of the species have light chains of both kinds (WHO, 1964). Among the immunoglobulins in normal serum the ratio of molecules with κ chains to those with λ chains is about 2:1. A similar distribution is observed among individual

myeloma and macroglobulinemia proteins (Fahey & Solomon, 1963; Laurell & Snigurowicz, 1967). The Bence Jones proteins, which are light chain monomers or dimers (Edelman & Gally, 1962; Bernier & Putnam, 1964b) nearly always correspond in K or L type to the myeloma protein in any given case (Migita & Putnam, 1963; Schwartz & Edelman, 1963; Gross & Epstein, 1964).

Electrophoretic patterns of light chains. When the light chains isolated from normal serum are subjected to electrophoresis in 8 M urea starch gel in an alkaline glycine buffer, ten nearly equidistant bands are observed which differ by a single charge. The bands of this pattern do not vary from one person to another and are independent of K or L antigenic type (Cohen & Porter, 1964, Feinstein, 1966). The fact that these light chains exhibit a random quantitative distribution has been offered as evidence that they represent a very large number of variants each with one of the ten different charges, originating from a single parent sequence (Cohen & Gordon, 1965). Under the same electrophoretic conditions, Bence Jones proteins and the light chains of myeloma proteins consist of a single major component and one or two minor components. A similar pattern is observed at pH 3·5 (Edelman & Poulik, 1961). However, if the light chains of myeloma proteins are reduced and alkylated by 7 M guanidine hydrochloride prior to urea gel electrophoresis at acid pH, a variable but larger number of bands is seen (Terry *et al.*, 1966; Sjöquist & Vaughan, 1966). The latter observation remains to be explained, but there is a strong likelihood that the heterogeneity does not reflect a difference in amino-acid sequence coded by the DNA in individual cells.

Light chains of antibodies. IgG antibodies obtained from individual donors and eluted after combination with their specific antigens tend to show a predominance of molecules with either K or L type of light chains (Mannik & Kunkel, 1963b; Franklin & Fudenberg, 1964; Leddy & Bakemeier, 1965). Similarly, a tendency to electrophoretic homogeneity is found in some preparations of 'single specificity' antibodies (Edelman & Kabat, 1964; Koshland, 1966; Miller *et al.*, 1967). On the other hand considerable heterogeneity has been reported in other preparations (Doolittle & Singer, 1965; Hughes-Jones, 1965; Hong & Nisonoff, 1966; Reisfield & Small, 1966).

Several investigators have made the observation that almost invariably light chains of K type but not of L type are found in the IgM with anti-I specificity which is produced by patients with autoimmune cold-agglutinin disease (Harboe et al., 1965b; Costea *et al.*, 1966; Feizi, 1967). On the other hand, IgM anti-I produced by patients with pneumonia due to *M. pneumoniae,* and the anti-i found in the serum of patients with infectious mononucleosis have light chains of both types (Costea *et al.*, 1966; Harboe & Lind, 1966).

Antibodies in the sera of many subjects detect 'hidden' idiotypic antigenic determinants on isolated normal light chains and on Bence Jones proteins, but not on the parent IgG molecules (Nachman & Engle, 1964; Williams, 1964; Epstein *et al.*, 1964). In addition, there is evidence for antigenic uniqueness shared by the light chains of individual myeloma proteins and their corresponding Bence Jones proteins but not demonstrable in pooled normal light chains (Nachman & Engle, 1965). This finding could be ascribed to an intrinsic abnormality of Bence Jones protein, but it is more likely to be due to the dilution of molecules having one antigenic structure by molecules having many other antigenic structures.

Amino acid sequence of light chains. Many detailed studies on the amino acid sequence of light chains have been carried out, using Bence Jones proteins of mouse as well as of human origin (Hilschmann & Craig, 1965; Milstein, 1965, 1966a, b; Milstein *et al.*, 1967; Easley & Putnam, 1966; Putnam *et al.*, 1966; Titani *et al.*, 1965, 1966; Baglioni & Cioli, 1966; Baglioni *et al.*, 1966; Hill *et al.*, 1966b; Hood *et al.*, 1966; Gray *et al.*, 1967). In both κ and λ chains there are about 214 amino acids, of which the 107 residues of the carboxy-terminal half are essentially invariant, while those of the amino-terminal half are highly variable, particularly in certain positions. This (variable) half, which also varies slightly in number of amino acids, contains an intrachain disulfide bond linking the two cysteine residues at positions 23 and 88, while in the invariant half, another intrachain bond links the cysteines at positions 134 and 194 (See Fig. 1.3, adapted from Cohen & Milstein, 1967a). In the κ chains, a fifth cysteine occupies the C-terminus at position 214; in λ chains the C-terminal amino acid is serine, and the fifth cysteine is at the subterminal position, 213. In both instances, this cysteine forms half of the interchain disulfide bond linking light and heavy chains.

The N-terminal amino acid of κ chains is either aspartic or glutamic acid; in λ chains the N terminus is blocked. About 40 per cent of the amino acids in the C-terminal half of κ and λ chains occupy homologous positions, suggesting a common ancestral chain (Milstein *et al.*,

1967; Titani *et al.*, 1967). Certain portions of the N-terminal (variable) half of κ and λ appear to differ almost as much within a given type as between the two types; and in comparisons of mouse and man, this half of the chain is found to contain very few species-specific residues (Gray *et al.*, 1967; Titani *et al.*, 1967; Kabat, 1967a, b). On the contrary, there is considerable inter-species similarity which may

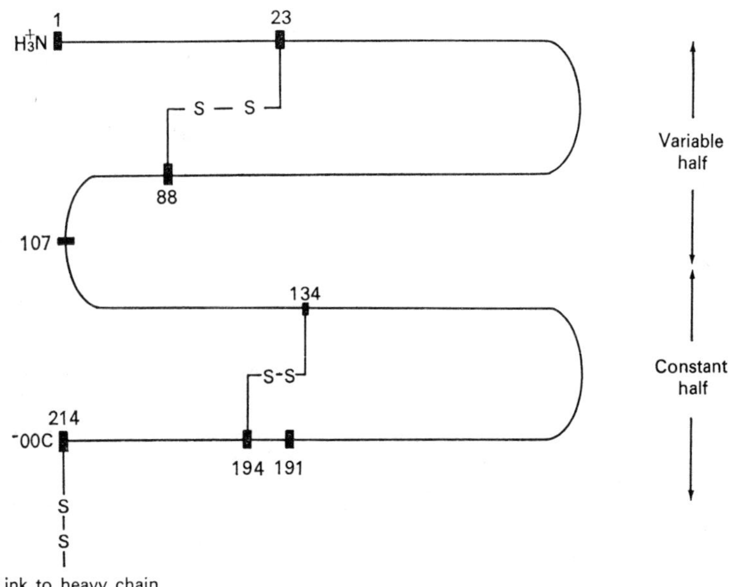

FIGURE 1.3. Diagram of the kappa light chain, indicating the location of the two intra-chain disulfide bridges in the variable and constant halves, the C-terminal link to the heavy chain, and the site (at position 191) of amino acid substitution associated with Inv specificity. (Adapted from Cohen & Milstein, 1967a. Reprinted with permission from *Nature* **214**, 449.)

indicate that certain structural features of the antibody combining site, unrelated to antigen specificity, have been preserved throughout mammalian evolution (Titani *et al.*, 1967; Kabat, 1967a, b). Some direct evidence for this assumption was provided by Singer & Doolittle (1966) who found tyrosine residues at fixed positions in the active sites of rabbit antibodies which they studied by affinity labelling. They suggested the existence of a 'conservative' region necessary for the conformational regularity of all antibody sites, regardless of their specificity.

A large proportion of the amino acid differences found in the variable half of the κ chains (and presumably also of the λ chains) could have arisen by a one-step mutation involving either transitions (i.e. purine to purine or pyrimidine to pyrimidine) or transversions (i.e. purine to pyrimidine or pyrimidine to purine) (Putnam *et al.*, 1966; Cohen & Milstein, 1967b). In the invariant half of the κ chain, point mutations affecting the amino acid at position 191 resulted in the inherited Inv determinants (Hilschmann & Craig, 1965; Milstein, 1966b; Titani *et al.*, 1966; Baglioni *et al.*, 1966). There are also two forms of λ chain, called Oz (+) and Oz (−), which represent amino acid substitution at position 190 (Ein & Fahey, 1967; Appella & Ein, 1967; Quattrocchi *et al.*, 1967).

The heavy chain (γ) of IgG

Heterogeneity of Fd fragment. Cohen (1963c) observed that the heavy (γ) chains from individual IgG myeloma proteins have variable electrophoretic mobility, and Fahey (1964) localized the variability on the Fd fragment of the γ chain. Similarly, Frangione & Franklin (1965) observed much more variation in the peptides contributed by Fd to Fab fingerprints than in the peptide maps of Fc fragments from the same molecules. Subsequently, Frangione *et al.*, (1967) and Pink & Milstein (1967a) presented evidence for an invariant region which, within a given subclass (see below) may involve as much as a half of the Fd fragment. Thus the Fd fragment may resemble its light chain partner in the Fab fragment in possessing both an invariant and a variant region, the latter (as shown in Fig. 1.2) presumably occupying the N terminus of the fragment, where it takes part in antigen binding. However, according to Koshland (1967), the variability occurs in non-consecutive segments throughout the length of the Fd fragment in rabbit IgG heavy chains. Some of the variability in the N terminus of both chains is undoubtedly involved in the specificity of the antibody, but there is additional idiotypic variability which is probably related to overall molecular configuration (Press *et al.*, 1966a). Structural interdependence between the heavy and light chains of a given antibody is shown by the fact that after chemical separation, homologous pairs of chains (i.e. chains from the same parent molecules) recombine by non-covalent bonding more efficiently than heterologous pairs (Grey & Mannik, 1965; Dorrington *et al.*, 1967). Indeed, Burnet (1966) has postulated that the portions of the two chains involved in antigen binding are identical. However, the studies of Singer & Doolittle (1966) indicate similarity but not identity.

The 'hinge' region. The 'hinge' region which couples the Fc and Fd fragments of the γ chain is present in the $F(ab')_2$ fragment, but not in Fab. Its amino acid sequence has been studied in rabbits by Smyth & Utsumi (1967) and in human subjects by Frangione & Milstein (1967) and Pink & Milstein (1967b). In the rabbit, a single, highly reactive disulphide bridge appears to link the two γ chains in the intact molecule. The associated cysteine residue (in the hinge region) is flanked on one side by threonine, and on the other by a sequence of three proline residues which probably confer an essential structural element on this important region. The site of cleavage by papain is variable, depending upon the presence or absence of a galactosamine residue bound through glycosidic linkage to the hydroxyl group of the threonine.

In man, the same region of the heavy chain (investigated in only two myeloma proteins of differing subclass) contains two cysteine residues separated by two proline residues and followed by a third. In this case, there appear to be two inter-heavy chain disulfide bonds, and papain splits the γ chain on the N-terminal side of the first cysteine in the hinge region sequence. Additional studies are required to determine whether all human IgG subclasses have the same hinge region structure and whether the Pro-Pro-Pro tripeptide is confined to the rabbit hinge region. (See Addenda.)

IgG sublcasses. The existence of isotypic subdivisions within the IgG fraction based on differences in the γ chain was noted by Grey & Kunkel (1964), who injected into rabbits the IgG myeloma proteins of several patients and absorbed the antisera with other myeloma proteins. By immunodiffusion tests, they were able to distinguish four different IgG molecular subclasses, which they named We, Ne, Vi and Ge. These subclasses are now known as γG1, γG2, γG3 and γG4, respectively, on the basis of their relative concentration in normal serum (WHO, 1966). Similar observations were made independently by other investigators (Terry & Fahey, 1964; Ballieux *et al.*, 1964) who applied different nomenclature systems, as shown in Table 1.3.

Several authors have used the term 'subgroup' or 'subtype' to refer to the subdivisions of IgG. However, such terms tend to imply allotypy (i.e. that the molecular variants are allelic products) as the basis for the observed differences, while the term 'subclass' does not. Since molecules of all four IgG subclasses are present in the serum of normal subjects, the particular antigenic determinants by which they are characterized are isotypic, in that they are not subject to inherited variation. In the case of γG1 and γG2 this determinant appears to be located on the Fc fragment (Terry & Fahey, 1964) but there is evidence that Fd is involved in the case of γG3, where the antigenic configuration is maintained by an interchain disulfide bond (Grey & Kunkel, 1964).

TABLE 1.3. Subclasses of IgG

| Present name | Previous nomenclatures* | | | Proportion of total IgG in serum | Location on γ chain of specific determinant |
	(1)	(2)	(3)		
γG1	We	γ2b	C	0·70	Fc
γG2	Ne	γ2a		0·18	Fc
γG3	Vi	γ2c	Z	0·08	Fd
γG4	Ge	γ2d		0·03	

* (1) Grey & Kunkel, 1964; (2) Terry & Fahey, 1964; (3) Ballieux *et al.*, 1964.

The Fc fragments derived from each of the subclasses share from 17 to 22 of their 24 peptides, the order of their degree of relationship being γG1, γG2, γG3 and γG4 (Grey & Kunkel, 1967). Molecules of subclass γG3 are much more susceptible to breakdown by papain in the presence of cysteine than are γG1 molecules (Takatsuki & Osserman, 1964; Poulik & Shuster, 1964, 1965). Also, according to Williams & Lawrence (1966), the Fd fragments of γG2 and γG4 are deficient in some of the antigenic determinants of γG1 and γG3 which are revealed by pepsin digestion.

A difference in function among the subclasses has been described by Terry (1965) who showed that molecules of the γG2 subclass do not fix to skin. Recently, Müller-Eberhard has found a deficiency in the interaction of γG2 and γG4 molecules with the C'_{1q} component of complement (cited in Pickering *et al.*, 1967). There is also evidence that antibody specificity is related to IgG subclass. For example, anti-Rh antibodies apparently lack γG2 molecules, while many examples of anti-dextran are composed entirely of γG2 molecules (Natvig & Kunkel, 1968). One example of an antibody against clotting factor VIII (the antihemophilic globulin) has been found in the γG4 subclass (Andersen & Terry, 1968). Since this subclass comprises a very small proportion of the total γG molecules, its participation in the immune response is difficult to assess. According to some workers (H. Fudenberg, personal communication), γG4 may represent another class of immunoglobulin molecules.

Individual cells produce IgG of only one subclass (Bernier *et al.*, 1967), so that allele exclusion applies not only to the homologous chromosome, but also to stretches of DNA on the same chromosome as the active gene.

Heavy chains of the other immunoglobulins

The α and μ chains differ from each other and from the γ chain in molecular weight, carbohydrate content and electrophoretic mobility. None of the particular antigenic determinants which characterize the IgG subclasses have been found on the IgA or IgM molecules, but subclasses are known to exist within these two immunoglobulin categories. By preparing antisera in animals against selected IgA myeloma proteins, Feinstein & Franklin (1966) obtained evidence for at least two kinds of IgA, one being about six times as common as the other. Presumably these two IgA subclasses correspond with those which were independently described by Kunkel & Prendergast (1966), Vaerman & Heremans (1966) and Terry & Roberts (1966). Data indicating the existence of IgM subclasses have been similarly provided by Harboe *et al.* (1965a), Wollheim & Williams (1966) and Franklin & Frangione (1967). (See Addenda.)

There is also immunological evidence for the presence of common regions in the heavy chains of the three major immunoglobulins. For example, Kunkel, Grey & Solomon (1966) reported some cross reactivity between IgA and IgM molecules which appeared to involve their Fd fragments. A similar cross reactivity between the Fd fragments of IgG and IgM was described by Harboe & Deverill (1966) and by Seligmann *et al.* (1966).

Evolution of immunoglobulin chains

Antibodies obtained from primitive vertebrates such as the dogfish and the bullfrog appear to have a single molecular class with heavy chains similar to those of human μ chains, suggesting the subsequent evolution of γ and α chains from a precursor μ chain (Marchalonis & Edelman, 1966). Furthermore, there is considerable evidence, based on amino acid sequence homology, to suggest that all of the immuno-globulin heavy and light chains are evolutionary products of a common precursor chain with a molecular weight of about 10,000, which

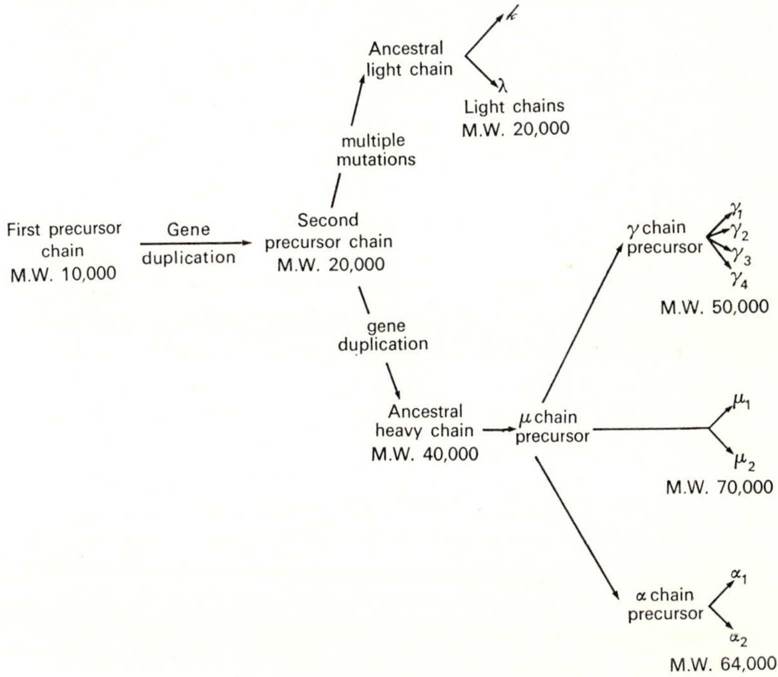

FIGURE I.4. Diagram of the gene duplications and point mutations which may have occurred during the evolutionary development of the immuno-globulin chains, based on hypotheses summarized by Cohen & Milstein (1967a) and Lennox & Cohn (1967).

existed at some early stage of mammalian phylogeny (Hill *et al.*, 1966a, b; Singer & Doolittle, 1966; Hood *et al.*, 1966; Titani *et al.*, 1967). That the gene determining the μ chain evolved from a light chain precursor gene was supported by the work of Doolittle *et al.* (1966). They found a carboxy-terminal or subterminal cysteine on the μ chain, analogous to that of the light chains, suggesting an evolutionary survivor of a gene duplication which led to the development of μ

chains on one hand and to light chains on the other. Abel & Grey (1967) confirmed this finding and obtained data suggesting the C-terminal sequence of the μ chain to be Ala-Gly-Thr-Cys-Tyr COOH. The existence of a sub-terminal cysteine residue was also determined for the α chain. On the other hand, the γ chain does not have an analogous cysteine residue, its C-terminal amino-acid sequence being Leu-Ser-Leu-Ser-Pro-Gly COOH (Press *et al.*, 1966b). Thus, the more recent evolution of its gene is a reasonable assumption.

Figure 1.4 shows a tentative scheme for the evolutionary development of the immunoglobulin chains adapted from the diagrams of Hill *et al.* (1966a) and Cohen & Milstein (1967a).

Genetic basis of immunoglobulin variability
While evolutionary models provide a reasonable explanation for inter- and intra-chain similarities, they shed no light on mechanisms for the synthesis of peptide chains in which the amino acid sequence is individually unique in the first half and essentially invariant in the other half. It is apparent from their C-terminal structure that each of the heavy chain classes and sub-classes, as well as the light chain types, is controlled by its own structural allele, accounting for a minimum of 12 genes. However, the remarkable diversity of the N-terminal portion would require the existence of hundreds or even thousands of additional genes. Thus, even if one were to assume that the C- and N-terminal sections are controlled by separate genes whose peptide products are synthesized on the same polyribosome, it would still be necessary to account for the presence in the germ line of a very large number of genes coding for similar, but not identical peptides.

According to one theory (Dreyer & Bennett, 1965), the members of this independent set of 'N-terminal' alleles are evolutionary products in the form of DNA rings, which are inserted during cell maturation in front of the chromosomal DNA segment coding for the invariant region. However, most theories propose that the N-terminal variability is not inherited, but represents a somatic event, occurring during cell differentiation by one of several mechanisms. For example, Potter *et al.* (1965) proposed that the translation of certain DNA triplets might vary because of small alterations in transfer RNA or amino acid activating enzyme. Alternatively, Brenner & Milstein (1966) postulated that a DNA-degrading enzyme breaks the DNA coding for the N-terminal half of the chain, and errors created by a defective repair enzyme account for the hypermutability. In the more elaborate theory of Lennox & Cohn (1967), there are two light chain genes and a gene for the variable region of the heavy chain which undergo somatic variation, following which the latter gene is trans-located into the heavy chain constant region gene.

Several theories embody the concept of crossing-over. For example, Edelman & Gally (1967) proposed the existence of several adjacent genes which arose, one from another, by a process of tandem duplication followed by multiple point mutations. Crossing over between these genes during lymphocyte maturation could bring about a wide diversity of sequences in the variable region. Whitehouse (1967) suggested a somewhat similar mechanism.

Smithies (1967a) supported a crossing-over mechanism from a consideration of the N-terminal amino acid sequence of thirty different light chains. He presented evidence that the K and L types each have two different 'linkage' groups with similar sequences and that apparent mixtures of the two sequences in some light chains is a manifestation of recombination. He suggested the existence of antibody gene pairs, consisting of a 'master' and a 'scrambler' gene, the latter being an inverted partial duplication of the former with differences in a relatively small number of positions. For example, with only twenty differences in base sequence, there would be 2^{20} possible recombinants.

In a subsequent analysis including additional light chains, Milstein (1967) implicated at least three basic sequences analogous to Smithies' two linkage groups. He concluded from this study that somatic crossing over may be involved in creating variability, but some additional mechanism of somatic hypermutation is still required to fit the data. Alternatively, Smithies (1967b) has proposed that the additional variability may be a consequence of germ-line polymorphism in the antibody gene pairs.

INHERITED ANTIGENIC
DETERMINANTS (ALLOTYPES)

THE GM SYSTEM

Gm ANTIGENS AND THEIR DETECTION

In 1956, Grubb reported that Rh positive red cells coated with some examples of incomplete (IgG) anti-D antibodies were agglutinated by the serum of some patients with rheumatoid arthritis, and that the agglutination could be inhibited by the serum of about 60 per cent of random normal donors. Somewhat similar observations were also made by Waller & Vaughan (1956) as well as by Milgrom *et al.* (1956). Subsequent studies by Grubb & Laurell (1956) showed that this blocking property in normal serum is inherited as a simple autosomal dominant character. Since the factor responsible for inhibiting agglutination was found in the GaMma globulin fraction now known as γG or IgG, it was given the name Gm(a).

TABLE I.4. Antigens of the Gm and Inv systems

Number *Gm*	Original name	Reference
I	a	Grubb & Laurell (1956)
2	x	Harboe & Lundevall (1959)
3	$b^w = b^2$	Steinberg & Wilson (1963a)
4	f	Gold *et al.* (1965b)
5	$b = b^1$	Harboe (1959)
6	$c = c^5$	Steinberg *et al.* (1960)
7	r	Brandtzaeg *et al.* (1961)
8	e	Ropartz *et al.* (1964)
9	p	Waller *et al.* (1963)
10	α	Ropartz *et al.* (1963)
11	$\beta = b^0$	Ropartz *et al.* (1963)
12	γ	Ropartz *et al.* (1963)
13	b^3	Steinberg & Goldblum (1965)
14	b^4	Steinberg & Goldblum (1965)
15	s	Martensson *et al.* (1966)
16	t	Martensson *et al.* (1966)
17	z	Litwin & Kunkel (1966b)
18	Rouen 2	Ropartz *et al.* (1966)
19	Rouen 3	Ropartz *et al.* (1966)
20	20	Klemperer *et al.* (1966)
21	g	Natvig (1966)
22	y	Litwin & Kunkel (1967a)
23	n	Kunkel, Yount & Litwin (1966)
	b^5	van Loghem & Martensson (1967)
	c^3	van Loghem & Martensson (1967)
Inv		
I	l	Ropartz *et al.* (1964)
2	a	Ropartz *et al.* (1961)
3	b	Steinberg *et al.* (1962)

During the intervening years, a large number of additional antigens in the Gm genetic system have been detected by the use of various combinations of agglutinating and inhibiting sera. Table 1.4 presents their original names and their new names under the currently accepted terminology, which was adopted by a WHO committee (WHO, 1965). Not all of these factors have been proven to be separate entities. Thus, Gm(3) and Gm(4) may be the same; so may Gm(10) and Gm(13), Gm(5) and Gm(12), and Gm(9) and Gm(22). No numbers have yet been assigned to Gm(b^5) and Gm(c^5), described by van Loghem & Martensson (1967). Those authors also introduced the designation b^0 and c^5, but their investigations indicated that b^0 is the same as Gm(11), and c^5 is probably the same as Gm(6).

While most of the Gm antigenic determinants are located on the Fc fragment of the IgG heavy chains, three of them: Gm(3), Gm(4) and Gm(17) are found on the Fd fragment (Harboe, *et al.*, 1962; Franklin *et al.*, 1962; Polmar & Steinberg, 1964; Gold *et al.*, 1965a; Grubb *et al.*, 1965; Litwin & Kunkel, 1966b, 1967a). There is conflicting evidence about the separate identities of Gm(3) and Gm(4) (Steinberg, 1965; Gold *et al.*, 1965a, b). Since Gm(3) was described first, further mention of it in this chapter can be taken to refer also to Gm(4).

Although there was initially a controversy about the nature of the reactions observed in Gm grouping, it now appears certain that the agglutinating activity of rheumatoid sera (often called Ragg) as well as that of certain normal sera (called SNagg) is due to the presence of anti-Gm antibodies which are usually in the IgM fraction, but also have been detected in IgG and IgA (Fudenberg & Kunkel, 1961; Steinberg, 1962a; Martensson, 1962a; Allen, 1967). Thus, specific Gm determinants on the IgG anti-D molecules coating the red cells react (as antigens) with their corresponding anti-Gm antibodies in the Ragg or SNagg agglutinator serum, causing the red cells to agglutinate (see Fig. 1.5). However, if the IgG molecules in the serum specimen being tested contain that specific Gm factor, addition of that serum to the agglutinator neutralizes the anti-Gm antibodies and thus prevents agglutination of the coated cells. Conversely, serum lacking the specific Gm determinant has no neutralizing capability and thus does not interfere with agglutination.

Unfortunately, this cumbersome method of Gm grouping is still widely used in one of several modifications, since most of the human

Ig allotypes, unlike those of the rabbit (Dray & Young, 1959) are not easily demonstrated by simpler techniques, such as immunodiffusion. Even when the anti-Gm antibodies are prepared in another species such as the monkey (Hess & Bütler, 1962; Alepa & Steinberg, 1964) or the rabbit (Litwin & Kunkel, 1966a) Gm antigens are often not detectable by precipitation techniques, and require the use of antibody-coated red cells in an agglutination inhibition system.

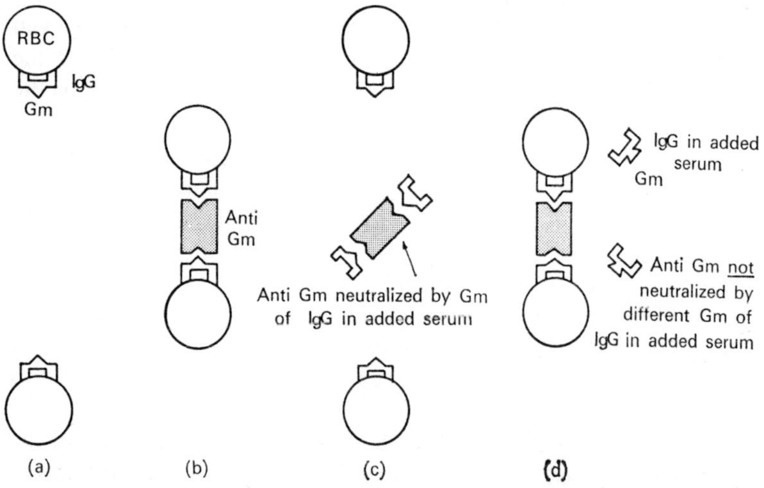

FIGURE I.5. Diagram of the serological system used for detecting the Gm (and Inv) allotypes: a. Red cells are coated with IgG molecules containing a specific Gm antigenic determinant. b. Specific anti-Gm antibodies in the 'agglutinator' react with the Gm antigens of the red cell 'coat', causing agglutination. c. The IgG in serum being tested contains the specific Gm determinant which combines with and neutralizes the anti-Gm antibodies; no agglutination of the coated cells occurs. d. The IgG in serum being tested contains another Gm determinant which does not combine with the anti-Gm antibodies in the agglutinator; thus the coated cells are agglutinated.

Gm GENETICS

Family studies

Since antisera for determining most of the Gm factors are not readily available, there are no family studies in which all of the Gm factors have been determined. The most extensive study is that of Steinberg & Goldblum (1965), in which 214 Brazilian families were tested for Gm(1), Gm(2), Gm(3), Gm(5), Gm(6), Gm(13) and Gm(14). According to Steinberg (1967), over 800 additional families from the same Brazilian population were tested for all but Gm(13) and Gm(14). In Germany, Ritter *et al.* (1964a, b) tested a large number of families for Gm(1), Gm(2), Gm(5), Gm(7) and Gm(8). These and other less extensive investigations show that the Gm factors are inherited as co-dominant characters representing either a single complex locus or loci so closely linked that examples of crossing-over are very rare. A large scale study has yielded no evidence for linkage between the Gm locus and the loci of several other autosomally determined characters (Corcoran *et al.*, 1966).

Comparison of Gm and Rh inheritance
In some respects, the genetic complexity of the Gm locus resembles that of other complex systems, such as the Rh and MNSs blood groups. For example, the Rh antigens, c, D and e are very commonly found in coupling phase in Negro subjects, while among Caucasians, the combination is fairly unusual. Similarly, the Gm antigens 1, 3 and 5 are virtually never inherited from a single parent by Caucasians, while in Orientals, these three determinants are not infrequently inherited together. Furthermore, in most subjects, C and c appear to be antithetical Rh antigens, but there is evidence that they can be inherited together. In a similar fashion, Gm(1) and Gm(5) have an antithetical relationship in European subjects (i.e. their respective genes behave as alleles), but in African Negroes, they are virtually always inherited as a unit.

The similarity stops at this point, however, because the expression by individual cells of the Gm genes, unlike that of any other known autosomal locus, is limited to one or the other of the homologous chromosomes, and this expression is in turn related to the IgG molecular subclass. Furthermore, the distribution of the Gm types in various racial groups is much more sharply defined than in any other known polymorphism.

Gm heterogeneity in single serum specimens
The heterogeneity in expression of Gm by individual IgG molecules within a given serum has been noted by several observers (Grubb, 1959; Harboe, 1960; Martensson, 1961, 1962b; Martensson & Kunkel, 1965; Fahey & Lawler, 1961; Nilsson, 1964; Kunkel *et al.*, 1964a, b; Ropartz, Rousseau & Rivat, 1965b). For example, in

individuals with the phenotype Gm(1,5) there is a much larger proportion of IgG molecules carrying Gm(1) than of those carrying Gm(5). Furthermore, an IgG myeloma protein does not possess the same phenotype as the patient's whole serum. Thus, IgG myeloma proteins obtained from Caucasian patients whose whole serum is typed as Gm(1,3,5) contain only one or none of these determinants. However, if the serum phenotype is Gm(1,2,3,5) and the myeloma protein contains Gm(1), it also contains Gm(2). In such a case, the patient can be shown to have inherited Gm(1) and Gm(2) from one parent and Gm(3) and Gm(5) from the other parent. In Negro myeloma patients with the serum phenotype Gm(1,5), it can be shown that although the patient inherited Gm(1) and Gm(5) from both parents, his myeloma protein contains either Gm(1) or Gm(5) or neither determinant, but never both.

These observations indicate that (1) In any given plasma cell, only one member of the paired genes on homologous chromosomes is active, and (2) When genes determining two or more Gm factors are inherited on a single chromosome, they may be expressed together, separately or not at all on a given IgG molecule. This expression is not a matter of chance, but is dependent on the molecular subclass.

Assignment of Gm antigens to IgG subclasses

A large number of IgG myeloma proteins have been tested for their subclass and Gm specificity. The results obtained in three laboratories are listed in Table 1.5 (Terry *et al.*, 1965; Martensson & Kunkel, 1965; Frangione *et al.*, 1966). The numerical distribution is somewhat disturbed by the tendency in some of the work to select cases belonging to the γG3 subclass. It appears that a more correct frequency ratio of the four subclasses would be about 70:18:8:3, which also applies to the proportions of these molecules in normal serum (Natvig & Kunkel, 1968), as was shown in Table 1.3.

From studies of this kind, most of the known Gm factors have been assigned to the γ chains in one of three IgG subclasses. These are listed in Table 1.6 along with the appropriate references compiled by Muir & Steinberg (1967) and Natvig & Kunkel (1968). Table 1.7, adapted from Litwin & Kunkel (1967a), van Loghem & Martensson (1967) and Steinberg (1967), indicates the more common combinations of the Gm determinants inherited as a unit by the members of three racial groups. For example, the γ1 chains (in γG1 molecules) of

TABLE 1.5. Gm type and IgG subclass of 160 myeloma proteins

Reference	γG1		γG2		γG3		γG4	
	No.	Gm	No.	Gm*	No.	Gm	No.	Gm
Terry *et al.*, 1965	33	(1)	17		10	(5)	3	(–)
	24	(3)	5					
	2	?	4					
Martensson & Kunkel, 1965	4	(1)	5		11	(5)	3	(–)
	8	(3)			2	(–)†		
Frangione *et al.*, 1966	8	(1)	4		10	(5)	3	(–)
	12	(3)						
	1	(–)						
Total	92		26		33		9	
Percentage	58%		16%		21%‡		5%	

* Gm(23) not known at time of testing. On the basis of a subsequent study (Kunkel *et al.*, 1966) about two-thirds of the γG2 proteins would be Gm(23+).

† Gm(21) not known at time of testing. These Gm(–) proteins were probably Gm(21+).

‡ This higher than expected percentage of γG3 myeloma proteins is an artefact due to selection for Gm(5).

Caucasian subjects contain either Gm(1) and Gm(17) on the Fc and Fd fragment, respectively, or Gm(22) and Gm(3). In some instances, Gm(2) accompanies Gm(1) on the Fc fragment. However, in Orientals, both Gm(1) and Gm(22) factors can occupy the same Fc fragment when Gm(3) is on the Fd fragment of the molecule. Thus, while Gm(1) and Gm(3) are commonly inherited in coupling phase in Orientals, they are nearly always in repulsion in Caucasians. An exception was found among the inbred Hutterites, several of whom carry the genetic determinants for Gm(1), (3) and (17) on the same chromosome (Steinberg *et al.*, 1968).

The Gm factors characteristic of γG1 molecules are not found on the γG3 molecules. In this subclass, there are several antigenic determinants, all on the Fc fragment, belonging to the so-called 'b group' (Gm 5, 11, 13, 14, b^5, 15, 16, c^3 and 6). Combinations of these antigens are commonly inherited as a unit, as shown in Table 1.7. Another Gm factor on γG3 molecules is Gm(21), which has an antithetical relationship to the Gm(b) complex. It occurs mainly in Caucasians and Mongoloids, and is rarely found in Negroes (Natvig, 1966).

TABLE I.6. Assignment of Gm antigens to IgG subclass

Gm antigen*		Reference
WHO	Original	
$\gamma G1$		
1	a	Kunkel *et al.* (1964a)
2	x	Terry *et al.* (1965)
3 (Fd)	b^w	Terry *et al.* (1965)
4 (Fd)	f	Kunkel *et al.* (1964a)
? 7	r	Muir & Steinberg (1967)
8	e	Ropartz *et al.* (in press)
9	p	Kronvall (1965)
17 (Fd)	z	Litwin & Kunkel (1966b)
18	Rouen 2	Ropartz *et al.* (in press)
20	20	Klemperer *et al.* (1966)
22	y	Litwin & Kunkel (1967a)
$\gamma G2$		
23	n	Kunkel, Yount & Litwin (1966)
$\gamma G3$		
5	b^1	Kunkel *et al.* (1964a)
6	c	Terry *et al.* (1965)
10	α	Ropartz *et al.* (1963)
11	β	Ropartz *et al.* (1963)
12	γ	Ropartz *et al.* (1963)
13	b^3	Terry *et al.* (1965)
14	b^4	Terry *et al.* (1965)
15	s	Martensson *et al.* (1966)
16	t	Martensson *et al.* (1966)
19	Rouen 3	Ropartz *et al.* (1966)
21	g	Natvig (1966)
	b^5	van Loghem & Martensson (1967)
	c^3	van Loghem & Martensson (1967)

* All antigens are on Fc fragment except Gm(3), Gm(4) and Gm(17).

Only one Gm factor has been assigned with certainty to molecules of the $\gamma G2$ subclass. It was given the designation Gm(n) by Kunkel, Yount & Litwin (1966) and subsequently, Gm(23). Unlike most of the other Gm factors, Gm(23) was detected by precipitin analysis using animal antisera. If γG4 molecules possess Gm determinants, they have yet to be found.

TABLE 1.7. Racial distribution of inherited Gm antigens, arranged according to IgG (heavy chain) subclass. (Adapted from Litwin & Kunkel, 1967a; van Loghem & Martensson, 1967; Steinberg, 1967)

	Caucasoid				Negroid				Mongoloid			
γG1 (Fd)			3	3								3
	17	17			17	17	17	17	17	17	17	
(Fc)	1	1			1	1	1	1	1	1	1	1
		2									2	
			22	22								22
γG3 (Fc)	21	21								21	21	
			11	11	11	11	11	11	11			11
			5	5	5	5	5					5
			13	13	13			13	13			13
			14	14	14		14					14
			b^5	b^5	b^5		b^5	b^5	b^5			b^5
								15	15			
									16			
						c^3	c^3					
						6						
γG2 (Fc)			23									23

Two genetic theories
Observations of Gm distribution and inheritance led Kunkel *et al.* (1964a, b), Martensson (1966) and Litwin & Kunkel (1967a) to postulate a set of closely linked loci containing the structural genes coding for γ chains possessing the various Gm complexes characteristic of the individual IgG subclasses. According to this hypothesis, the antigens arranged in the vertical columns of Table 1.7 represent the expression of the commonest gene combinations occupying the three linked loci in different racial groups. Thus, in Caucasians, γ chains of γG1 molecules are determined by the allelic genes $Gm^{17\cdot1}$ (i.e. Gm^{za}) and $Gm^{3\cdot22}$ (i.e. Gm^{fy}), so that their respective gene products contain as genetic markers either Gm(1) or Gm(22) on the Fc fragment and either Gm(17) or Gm(3) on the Fd fragment. However, in

Mongoloid populations, there is a $Gm^{3,22,1}$ (Gm^{fya}) gene, the product of which contains both Gm(22) and Gm(1) on the same Fc fragment. The latter gene is postulated to have arisen either by intragenic crossing over or by point mutation. In a similar fashion, another series of alleles governs the synthesis of γG3 molecules, but in this case, only the Fc fragment appears to be involved. A third set of alleles, only one of which is currently detectable, occupies the locus governing the γ chain of γG2.

In the horizontal alignment of Table 1.7, unlike numbers are on different lines to prevent the implication that their genetic determinants are necessarily homoalleles (i.e. that they are alternative nucleotide sequences at a given position on the chromosomal DNA). There is no definitive information about the amino acid sequence of Gm determinants except for the Gm(1) and Gm(−1) peptides discussed on p. 34.

Steinberg (Steinberg, 1965, 1967; Muir & Steinberg, 1967) has reserved his decision as to whether or not three separate but linked loci are involved. He pointed out that if there were separate loci, then their products should have a random distribution with respect to one another in various populations unless there were very potent selection pressures. Furthermore, the limitation of Gm expression to specific IgG subclass might simply reflect variations in molecular structure related to the subclass, permitting only certain conformations. Muir & Steinberg (1967) also mentioned some preliminary observations made with Ropartz indicating that the factors Gm(8) and Gm(9), which sometimes occur on γG1 molecules carrying Gm(3), may also occur on γG2 molecules. If this association is borne out, it may indicate a closer relation between γ chains of γG1 and γG2 than between either of these subclasses and γG3. The studies of Grey & Kunkel (1967) on Fc fragment structure suggest that γG2 is more closely related than γG1 to γG3.

Steinberg currently proposes that the transmission of Gm factors be considered as due to a genetic system, defined by Mi & Morton (1966) as a DNA sequence determining a series of phenotypes which are not randomly distributed among various populations and which are inherited in a manner consistent with a series of alleles. In accordance with this definition, the Gm alleles as listed by Steinberg are not separated on the basis of molecular subclass. Thus for example, in considering the inherited combination of Gm antigens listed in the last vertical column of Table 1.7, Litwin & Kunkel (who, in fact, prefer to use letter, rather than numerical notation) designate three closely linked but separate genes. $Gm^{3,22,1}$, $Gm^{11,5,13,14,b5}$ and Gm^{23}, while Steinberg would designate a single gene: $Gm^{1,3,22,5,11,13,14,b5,23}$. As pointed out by Steinberg *et al.* (1968) neither the linked-gene nor the multiple allele theory can provide an adequate explanation for the Gm^- gene, which they found in a Hutterite population. The IgG of the single Gm^- homozygote did not contain any of the known Gm factors, indicating some kind of molecular alteration affecting both Fc and Fd regions of γ1 and Fc of γ3.

Yount *et al.* (1967) observed that the concentration of γG3 molecules in a given serum specimen is correlated with the presence of the Gm(5) antigen so that individuals lacking a gene coding for Gm(5) have only about half as many γG3 molecules as those who are homozygous for the gene. A similar effect of Gm genotype was noted by Steinberg *et al.* (1968) who found that in individuals carrying the rare Gm^- gene there is an increase in γG3 concentration.

STUDIES OF GM STRUCTURE

Papain digestion brings about some loss of Gm antigenic activity on the Fc fragment of γG3 molecules, but not of γG1 molecules (Steinberg, 1967). A susceptibility of the Fc fragment of γG3 molecules to papain had been determined in other ways by Fougereau & Edelman (1965) and by Frangione & Franklin (1965). Papain digestion decreases the activity of Gm(3) on the Fd fragment of γG1 molecules, but pepsin does not, so Fab' has more Gm activity than Fab (Steinberg, 1967). The fact that Fab is antigenically deficient (Franklin, 1960; Williams & Lawrence, 1966) and chemically deficient (Fougereau & Edelman, 1965) when compared to Fab' is consistent with this finding, and suggests that the Gm(3) antigenic determinant is close to the enzymatic cleavage point.

Polmar & Steinberg (1964, 1967) found that although cleavage of γG1 molecules into their constituent heavy and light chains moderately decreased activity of Fc fragment Gm antigens, the Fd fragment antigen, Gm(3), was completely lost. However, it could be restored by recombining the specific heavy chain with nine out of twelve Bence Jones proteins (light chains). This restoration capacity of the light chains was independent of their K and L subtype or the Gm type of their individual myeloma proteins, so that the predominant function of the light chain in restoring Gm(3) was apparently to provide the necessary conformation of the molecule. Some similar findings with light chain antigens restored by heavy chains are described on p. 40.

Peptide mapping of the heavy chains and Fc fragments of IgG obtained from normal serum and from patients with myeloma has been reported by Meltzer *et al.* (1964), Fudenberg *et al.* (1964, 1966), Frangione & Franklin (1965), Frangione *et al.* (1966) and Grey & Kunkel (1967). These studies show that the Fc fragments of γG1 myeloma proteins are very similar except for the presence of one peptide characteristic of molecules containing Gm(1) and another peptide found in molecules lacking Gm(1). The latter peptide was also detected in the Fc fragments of γG2 and γG3 proteins, but neither peptide was present in γG4 proteins. Both Gm(1) and Gm(− 1) peptides were found to be pentapeptides differing by two amino acids. Thorpe & Deutsch (1966) showed the sequence of Gm(1) peptide to be Asp-Glu-Leu-Thr-Lys, while that of Gm(− 1) was *Met*-Glu-*Glu*-Thr-Lys. The two italicized differences could not have arisen by single point mutations, since both would require at least two changes in codon nucleotides (Muir & Steinberg, 1967).

Prahl (1967) determined the sequence of the 19 C-terminal amino acids in the heavy chains of seven myeloma proteins and found differences at positions C2, C11 and C12. Table 1.8 presents the portion of his data which shows these differences. The single examples of γ1 and γ2 subclass had the same C-terminal sequence, which differed from that of the two γ4 proteins by the subterminal residue only. The γ3 protein of Gm(21) phenotype had tyrosine at position C11, in common with the other subclasses. However, the two γ3 proteins of Gm(5,13,14) type had phenylalanine at this position. All three γ3 proteins had arginine at position C12, which was occupied by histidine in γ chains of the other three subclasses. Prahl (1967) pointed out that this substitution, which introduces a new point of tryptic cleavage and the formation of two peptides, probably distinguishes the γ3 subclass. He suggested that the phenylalanine for tyrosine replacement at position C11 in γ3 might be related to the Gm type, but a direct relationship could not be demonstrated by inhibition of anti-Gm antibodies by the Vil and Zuc C-terminal octadecapeptides.

TABLE 1.8. C-terminal sequences of the γ chains of seven human myeloma proteins. The sites of amino acid replacement are indicated in italics. C1 indicates the carboxyterminal residue. (Adapted from Prahl, 1967.)

Protein	γ chain subclass	Gm type	C-terminal residues		
			C10	C5	C1
Daw Wan	$\gamma1$ $\gamma2$	Gm(1) ?	*His-Tyr*-Thr-Gln-Lys-Ser-Leu-Ser-Leu-Ser-*Pro*-Gly		
Zuc Mar	$\gamma3$ $\gamma3$	Gm(5, 13, 14)	*Arg-Phe*-Thr-Gln-Lys-Ser-Leu-Ser-Leu-Ser-*Pro*-Gly		
Vil	$\gamma3$	Gm(21)	*Arg-Tyr*-Thr-Gln-Lys-Ser-Leu-Ser-Leu-Ser-*Pro*-Gly		
She Ger	$\gamma4$ $\gamma4$	—	*His-Tyr*-Thr-Gln-Lys-Ser-Leu-Ser-Leu-Ser-*Leu*-Gly		

GEOGRAPHIC DISTRIBUTION OF GM ANTIGENS

In a previous paragraph, it was pointed out that the Gm polymorphism shows a most remarkable tendency toward characteristic patterns within various racial groups. The discussion of Gm inheritance and Table 1.7 presented some examples of this phenomenon.

Detailed reviews on the geographic distribution of Gm have been published by Ropartz, Rousseau & Rivat (1965a), and by Steinberg (1967); a list of references is included at the end of this chapter. Most of these studies have been limited to Gm(1), and only a few have presented data on two or more Gm factors. However, sufficient data are available to enable some general statements to be made. The reader is referred to the papers by Steinberg (1967), Muir & Steinberg (1967), van Loghem & Martensson (1967) and Natvig & Kunkel (1968) from which the following summary was abstracted.

The Gm(1) antigen is found in nearly 100 per cent of all except Caucasian populations, in which genes determining Gm(1) or Gm(1,2) have a frequency of less than 0·50. In Europe, Gm(1) is most prevalent in the north, decreasing with latitude from Sweden and Finland to Italy. There is a similar cline for Gm(2) which, when present, is nearly always accompanied on the same Fc fragment by Gm(1). An opposite cline is observed for antigens of the 'b complex', Gm(5), (11), (13), (14), (b⁵), which, in Caucasians, is inherited independently of Gm(1) (i.e. in trans-position).

Among Negro populations, the outstanding features are the co-existence of genetic determinants for Gm(1), Gm(17) and Gm(5) on the same chromosome, and the virtual absence of Gm(2), (3), (21), (22) and (23). The 'b complex' is frequently the same as that of Caucasians, but the antigens Gm(6) and (c⁵) are sometimes substituted for Gm(13) and Gm(b⁵), respectively.

Mongoloid subjects, like Negroes, are nearly always Gm(1 +), but Gm(2), Gm(3), Gm(21), Gm(22) and Gm(23) have a variable frequency. Gm(17) has a lower frequency than in Negroes, because it does not invariably accompany Gm(1), unlike other racial groups tested. Gm(21) is commonly inherited with Gm(1) and Gm(17), but the latter two antigens may alternatively be inherited with Gm(11), (13), (b⁵), (15), (16).

Very unusual alleles have been detected among certain population isolates. For example, there is a gene among the Ainus of Hokkaido that determines Gm(2) but not Gm(1) or Gm(5). No unique alleles have been found so far among inhabitants of the Pacific Islands; they possess a mixture of certain alleles found commonly in Caucasians, in Negroes, and in Mongoloids.

The frequency of Gm(7), Gm(8) and Gm(9) has not been widely determined in population studies. Gm(7), like Gm(2), appears to accompany Gm(1), at least in Europeans. Gm(8), first thought to be a part of the Gm^b complex in Caucasians, was shown to be inherited along with Gm(3) in certain European families. Gm(9) has not been clearly related to the other Gm factors by inheritance studies. Its frequency in Caucasians appears to be about twice that in Negroes.

A few Caucasian families have been described in which the phenotypes could be explained only on the basis of an allele having no representative Gm activity (Steinberg, 1962b; Ropartz *et al.*, 1962); the frequency of this gene is probably not very high in any of the major populations of the world, although it has been detected in over 70 members of the Hutterite population isolate, including one apparent Gm^- homozygote (Steinberg *et al.*, 1968).

As pointed out by Steinberg (1967), a striking feature of Gm gene distribution is not only that four of the races tested (i.e. Caucasians, Negroes, Bushman and Ainus) have alleles not present in other races, but also that no two racial groups have the same array of alleles. There is no other known human genetic system in which the distribution of alleles is so distinctive.

ANTI-Gm ANTIBODIES

Since $\gamma G1$ molecules in normal serum outnumber those of the $\gamma G3$ subclass by about 6:1, Gm(1) and Gm(3) antigens are much more prevalent than Gm(5). Gm(1) is a potent antigen, so that anti-Gm(1) is the most common Gm antibody.

Antibodies reacting with the Gm antigens have been found under several circumstances: (1) In patients with rheumatoid arthritis and certain other diseases. Such Ragg antibodies usually have a high titer, but they commonly have autoantibody activity, as well as troublesome prozones and multiple specificity (Grubb, 1957, 1958; Harboe, 1960; Steinberg, 1962a). (2) In patients who have received transfusions or have been pregnant. These SNagg antibodies have lower titers than most Raggs, but they do not have prozones and are usually monospecific (Allen & Kunkel, 1963, 1966; Vierucci, 1965; Fudenberg & Fudenberg, 1964; Martensson & Fudenberg, 1965). (3) In

3

nontransfused subjects, especially children, who in many instances can be shown to have received their Gm antigenic stimulus, while *in utero*, from their mothers' IgG globulin (Steinberg & Wilson, 1963b; Speiser, 1963; Wilson & Steinberg, 1965). (4) In animals injected with human serum (Hess & Bütler, 1962; Alepa & Steinberg, 1964; Litwin & Kunkel, 1966a).

There is some question about the likelihood of inducing untoward reactions by the transfusion of blood, plasma or gamma globulin preparations into subjects who have anti-Gm antibodies in their serum. One febrile episode described by Fudenberg *et al* (1964) appeared to have been due to such an antigen-antibody reaction, but previous studies described by Grubb (1961) and by Harris & Vaughan (1961) as well as a subsequent one by Steinberg (1967) suggest that anti-Gm antibodies probably have little clinical significance.

THE Inv SYSTEM

Inv ANTIGENS

Ropartz, Lenoir & Rivat (1961) and Ropartz, Rousseau *et al.* (1961) found that the serum of a healthy donor, Virm., possessed agglutinating activity for Rh positive red cells coated with certain examples of anti-D antibodies, and this agglutination could be inhibited by about 19 per cent of serum specimens obtained from French donors. The inhibiting factor was first localized in the IgG fraction, but family and population studies indicated that it was inherited independently of Gm, so it was named InV (later changed to Inv). Subsequent studies by other investigators indicated that Inv was present not only on IgG molecules, but also on IgM and IgA. This activity, localized on the Fab fragment, was found to be due to a specific determinant on the light chain (Gross *et al.*, 1962; Harboe *et al.*, 1962; Franklin *et al.*, 1962; Lawler & Cohen, 1965) which Terry *et al.* (1965) further localized to the κ chain. Thus, while Gm is a genetic marker for the heavy (γ) chain of IgG, Inv is an inherited characteristic of the κ light chain, found in all immunoglobulin classes.

Only three Inv factors have been detected thus far. They are named Inv(1), Inv(2) and Inv(3). Their inheritance is consistent with the presence in all populations studied of three alleles, Inv^1, $Inv^{1,2}$ and Inv^3 (Ropartz *et al.*, 1961, 1964; Steinberg *et al.*, 1962; Ritter & Wendt, 1964; Ritter *et al.*, 1964c). If an Inv^2 allele exists, it must be extremely rare. There is, however, evidence for a 'silent' Inv gene of unknown frequency.

STUDIES OF Inv STRUCTURE

A previous section of this chapter described the structure of the κ and λ light chains and pointed out that the C terminal half of the κ chain contains only one variable amino acid residue, at position 191. Kappa chains with leucine in this position have Inv(1,2) specificity, while those with valine have Inv(3) specificity (Milstein, 1966 a, b; Titani et al., 1966; Easley & Putnam, 1966; Baglioni et al., 1966). It is not known whether both Inv(1) and Inv(2) antigens are dependent on the presence of leucine at position 191. Conceivably, one or the other of these two Inv specificities may reflect an alteration in the variable portion of the κ chain. It is of particular importance to determine the precise location of the altered amino acid sequence, just as in the case of the heavy chain Fd fragment antigens, Gm(3) and Gm(17). If they prove to be in the variable portion of their respective chains, the likelihood of separate genes controlling different segments of the same chain (γ1 or κ) will become even smaller than the evidence now indicates (Litwin & Kunkel, 1967b).

In the λ chains, an amino acid substitution of lysine for arginine at position 190 is reflected in the Oz (+) and Oz (−) types, respectively. (Appella & Ein, 1967; Quattrocchi et al., 1967). However there is some evidence that Oz(+) and Oz(−), unlike Inv, may be isotypes than than allotypes, Appella & Ein (1967) reported that of 12 type L Bence Jones proteins analyzed chemically and serologically, the ratio of Oz(+) to Oz(−) was nearly 50:50. Thus, if the genes determining the two types were alleles, they would have a similar frequency, and about 25 per cent of people would be expected to be Oz(−). But when Ein & Fahey (1967) examined normal human serum specimens, they found that all of 40 tested apparently possessed the Oz(+) antigenic determinant.

On the other hand, Quattrocchi et al. (1967) analysed 40 type L proteins and found that 35 had arginine at position 190 (i.e. were Oz(−)) and only five had lysine in that position (i.e. were Oz(+)). Thus, it is possible that the allele determining Oz(+) has a frequency as low as 0·125. In that case, the expected incidence of Oz(+) individuals (homozygous and heterozygous for this gene) would be only 23 per cent. Family studies have not yet been reported. They should resolve the problem of whether Oz(+) and Oz(−) represent λ chain subclasses analogous to γ1, γ2, etc. or phenotypes reflecting allelic variation analogous to Inv. (See Addenda.)

An interesting relationship between the Inv determinant on the κ chain and the IgG subclass on the γ chain was noted by Terry *et al.* (1965), who were unable to find Inv activity of any known specificity on γG2 or γG4 molecules. Subsequently, Williams & Lawrence (1966) showed that these two molecular subclasses are also deficient in the 'pepsin sites' found on the Fd fragments of γG1 and γG3 molecules.

The dependence of heavy-light chain interaction for full expression of Inv activity has been documented by the studies of Litwin & Kunkel (1967b) and Polmar & Steinberg (1967). Thus, κ chains isolated from myeloma proteins lose appreciable Inv activity, and Bence Jones proteins have less Inv activity than the parent myeloma molecule. Furthermore, the activity of both isolated κ chains and Bence Jones proteins can be enhanced by combining them with either homologous or non-homologous γ chains. According to Litwin & Kunkel (1967b), enhancement even occurs in combinations with γ2 chains, whose natural light chain partners have no Inv activity. The considerable variability found in extent of heavy chain dependence appeared to be a function of the Inv typing system used for titration, including both the antigen 'coat' and the anti-Inv serum.

Inv GENE FREQUENCIES

The Inv polymorphism is not very useful for the differentiation of populations. The *Inv*[3] gene has a higher frequency than the sum of the other two alleles in nearly all races tested, ranging from about 0·70 in some Negro tribes to about 0·95 in several Caucasian samples (Salzano & Steinberg, 1965). However, there is a paucity of data because of the scarcity of useful anti-Inv reagents, which do not occur as Raggs in subjects with rheumatoid disease. The examples of anti-Inv so far reported were obtained from normal sera, and like most SNaggs, were usually of low titer. Some papers giving gene frequencies are listed at the end of this chapter.

METHODS

Antibody neutralization is the serological basis for the tests most commonly used for detecting Gm and Inv antigens. As shown in Fig. 1.5, the presence of a specific antigen in the tested serum is reflected by its ability to prevent the agglutination of IgG-coated red cells by specific antibodies. Methods using passive hemagglutination or immuno-diffusion are also employed when the proper reagents are available and the results are unequivocal. Descriptions of these methods can be found in papers by Epstein & Fudenberg (1962), Kunkel *et al.*, (1966b) and Natvig & Kunkel (1967). They are not included in the following paragraphs.

THE IgG ANTIBODY COAT

The red cells are coated with an IgG 'incomplete' antibody—i.e. an antibody which reacts with an antigen on the red cells, but does not bring about agglutination. For this purpose, anti-Rh (and particularly anti-D) antibodies are customarily used, because they are both readily available and sufficiently potent. An anti-D titer (by the antiglobulin technique) of $1:64$ or more is usually adequate, and the donor of this antibody-containing serum should if possible be homozygous for the gene determining the Gm antigen, to ensure a high concentration of IgG molecules carrying the antigen. The latter requirement is particularly desirable in tests for Gm(5) and the rest of the 'Gm^b complex' which are found only on γG3 molecules. In normal serum, this molecular subclass is greatly outnumbered by γG1 and γG2 molecules, which carry Gm antigens of other specificities (see p. 29). Since Rh antibodies do not contain γG2 molecules, it is not possible to use the agglutination inhibition method to detect Gm(23) (Natvig & Kunkel, 1968).

For reasons which are still unclear, even if a serum contains high-titered anti-D of the appropriate Gm type, it is frequently found to be unacceptable as a coating reagent, or it reacts successfully with some, but not other, agglutinators within the same apparent specificity. Furthermore, the specificity of the 'coat' can vary with time (Natvig, 1965). Consequently, it is desirable to test a given anti-D serum with the indirect antiglobulin technique against as many agglutinators as possible. For practical purposes, several dilutions of the anti-D should be used to coat the cells. Thus, if the serum is still effective when diluted, its utility is increased.

In choosing the Rh positive red cells for the coating process, it is well to avoid phenotypes CcDe and CDe, since the number of D reacting sites is usually decreased on such cells. After the cells are coated and thoroughly washed, they should be shown to react strongly with anti-human globulin (Coombs) serum.

THE AGGLUTINATOR

Ragg sera (i.e. those obtained from patients with rheumatoid arthritis) are not desirable because of their multispecificity and prozones. Thus, even though they have lower titers, the monospecific SNagg sera from normal donors are preferable. Such sera are screened for agglutinating activity by diluting them 1:2 and 1:8 and then testing them against appropriately coated red cells. When agglutination is noted, the serum is then tested for its specificity by observing the inhibitory effect of serially diluted normal sera of known Gm or Inv specificity. Finally, the titer is determined, and the optimal dilution to be used for testing purposes is ascertained.

THE AGGLUTINATION-INHIBITION TEST

The technique for this test is not standardized, because of individual preferences of the relatively few investigators in this field. However, regardless of the technique used, the employment of both positive and negative controls is a strict requirement.

THE TUBE TECHNIQUE

The tube technique (Grubb & Laurell, 1956) is performed in a series of Kahn tubes (80 mm × 8 mm) containing 0·12 ml of serially diluted serum to be tested and 0·12 ml of the agglutinator (at its pre-determined dilution). For each determination, there are three additional (control) tubes containing (1) 0·12 ml diluted agglutinator plus 0·12 ml saline, (2) 0·12 ml of a 1:4 dilution of the serum being tested plus 0·12 ml saline, and (3) 0·25 ml saline. (The agglutinator should also be tested with uncoated red cells to be certain that no agglutination occurs.)

After 15 minutes at room temperature, each tube receives 0·25 ml of a 0·4 per cent suspension of red cells coated (i.e. sensitized) with anti-D. Incubation at room temperature for 4 hours is recommended, after which the tubes are examined for agglutination.

An improvement of this technique was described by Martensson (1964), utilizing a mixture of 0·06 ml of the serum being tested, 0·03 ml of the diluted agglutinator and 0·02 ml of a 2 per cent suspension of sensitized cells, along with appropriate controls. After 1 hour of incubation at room temperature, the tubes are centrifuged and then observed for agglutination.

THE TILE TECHNIQUE

The tile technique (Steinberg, 1962a, and personal communication) is preferred by many workers, being performed on a glass slide with twelve round concavities of the type used for micro-flocculation tests. As shown in Table 1.9, one drop of a 1 : 16 dilution of the serum being

TABLE 1.9. Steinberg slide test for Gm and Inv: Initial contents of concavities

	1	2	3
Row 1	Dil. unknown serum Agglutinator	Dil. unknown serum Agglutinator	Dil. unknown serum Saline
Row 2	Dil. pos. control Agglutinator	Dil. pos. control Agglutinator	Dil. pos. control Saline
Row 3	Dil. neg. control Agglutinator	Dil. neg. control Agglutinator	Dil. neg. control Saline
Row 4	Saline Agglutinator	Saline Agglutinator	Saline Saline

As described in the text, IgG-coated red cells are added to all 12 holes. The agglutination of these cells by agglutinator is inhibited (in holes 1 and 2 of row 1) if the unknown serum contains a Gm or Inv factor recognized by the antigen-antibody detector system.

tested is placed in the three holes of the top row; the holes of rows 2 and 3 receive a drop of diluted positive and negative control serum, respectively (i.e. serum known to contain and not to contain the antigen). A drop of saline is added to the holes of the bottom row and to the third hole in each row. A drop of diluted agglutinator is added to the first two holes in each row and the slide is agitated in a shaker for about 5 minutes. Then one drop of a 0·2 per cent suspension of well-washed, freshly sensitized red cells is added to all 12 holes. After another 5 minutes of shaking, the slide is incubated at room temperature in a moist chamber for 40 minutes. Prior to reading the slide, it is again shaken for 10 minutes, and then examined under a stereoscopic dissecting microscope. The degree of agglutination is recorded up to a maximum of 4.

If the serum being tested agglutinates the cells in the absence of agglutinator, it contains an IgM anti-Gm or anti-Inv antibody (i.e. it is itself a SNagg). The antibody must be removed before the Gm or Inv group can be determined. For this purpose, the serum is diluted 1:8 and then dialysed against water overnight in a cold room or refrigerator.

Lawler (1960) devised a similar slide test which, however, is based on reversibility of the agglutination by the inhibitor. In this technique, the sensitized cells are first mixed with the agglutinator so that agglutination occurs during shaking. Then the serum being tested is added in dilutions, and the slide is again agitated. Dispersal of the cells indicates that the unknown serum contains the specific Gm factor. Appropriate controls, similar to those used in the previous test, are included. In general, this test requires more potent reagents, and it may not be sufficiently sensitive for use with certain combinations of sera.

For further details on the rationale and performance of the tests briefly described here, the reader is urged to consult the papers mentioned above as well as those by Harboe *et al.* (1962), Natvig & Kunkel (1968) and (for a micro-titer technique), Borel *et al.* (1967).

REFERENCES

ABEL C.A. & GREY H.M. (1967) Carboxy-terminal amino acids of γA and γM heavy chains. *Science* **156**, 1609.

ADINOLFI M., MOLLISON P.L., POLLEY M.J. & ROSE J.M. (1966) γA-blood group antibodies. *J. exp. Med.* **123**, 951.

ADLER F.L., FISHMAN M. & DRAY S. (1966) Antibody formation initiated *in vitro*. III. Antibody formation and allotypic specificity directed by ribonucleic acid from peritoneal exudate cells. *J. Immunol.* **97**, 554.

ALEPA F.P. & STEINBERG A.G. (1964) The production of anti-Gm reagents by rhesus monkeys immunized with pooled human gamma globulin. *Vox Sang.* **9**, 333.

ALLEN J.C. (1967) Evidence for anti-Gm(a) antibody in γA immunoglobulins. *Proc. Soc. exp. Biol.* **124**, 138.

ALLEN J.C. & KUNKEL H.G. (1963) Antibodies to genetic types of gamma globulin after multiple transfusions. *Science* **139**, 418.

ALLEN J.C. & KUNKEL H.G. (1966) Antibodies against γ-globulin after repeated blood transfusions in man. *J. clin. Invest.* **45**, 29.

ANDERSEN B.R. & TERRY W.D. (1968) Gamma G4-globulin antibody causing inhibition of clotting factor VIII. *Nature* **217**, 174.

ANDERSEN S.B. (1963) Metabolism of gamma$_{ss}$-globulin in secondary hypogamma-globulinemia. *Am. J. Med.* **35**, 708.

APPELLA E. & EIN D. (1967) Two types of lambda polypeptide chains in human immunoglobulins based on an amino acid substitution at position 190. *Proc. natn. Acad. Sci.* **57**, 1449.

ASKONAS B.A. & RHODES J.M. (1965) Immunogenicity of antigen-containing ribonucleic acid preparations from macrophages. *Nature* **205**, 470.

ASKONAS B.A. & WILLIAMSON A.R. (1966a) Biosynthesis of immunoglobulins on polyribosomes and assembly of the IgG molecule. *Proc. roy. Soc.* B **166**, 232.

ASKONAS B.A. & WILLIAMS A.R. (1966b) Biosynthesis of immunoglobulins. Free light chain as an intermediate in the assembly of γ-G molecules. *Nature* **211**, 369.

ASKONAS B.A. & WILLIAMSON A.R. (1967a) Balanced synthesis of light and heavy chains of immunoglobulin G. *Nature* **216**, 264.

ASKONAS B.A. & WILLIAMSON A.R. (1967b) Biosynthesis and assembly of immuno-globulin G. *Cold Spring Harbor Symposia on Quantitative Biology.* **32**, 223.

BAGLIONI C. & CIOLI D. (1966) A study of immunoglobulin structure. II The comparison of Bence Jones proteins by peptide mapping. *J. exp. Med.* **124**, 307.

BAGLIONI C., ZONTA L.A., CIOLI D. & CARBONARA A. (1966) Allelic antigenic factor Inv(a) of the light chains of human immunoglobulins: chemical basis. *Science* **152**, 1517.

BALLIEUX R.E., BERNIER G.M., TOMINAGA K. & PUTNAM F.W. (1964) Gamma globulin antigenic types defined by heavy chain determinants. *Science* **145**, 168.

BARTH W.F., WOCHNER D., WALDMANN T.A. & FAHEY J.L. (1964) Metabolism of human gamma macroglobulins. *J. clin. Invest.* **43**, 1036.

BECKER M.J. & RICH A. (1966) Polyribosomes of tissues producing antibodies. *Nature* **212**, 142.

BERNIER G.M., BALLIEUX R.E., TOMINAGA K.T. & PUTNAM F.W. (1967) Heavy chain subclasses of human γG globulin. Serum distribution and cellular local-ization. *J. exp. Med.* **125**, 303.

BERNIER G.M. & CEBRA J.J. (1964) Polypeptide chains of human gamma-globulin: cellular localization by fluorescent antibody. *Science* **144**, 1590.

BERNIER G.M. & CEBRA J.J. (1965) Frequency distribution of α, γ, κ and λ poly-peptide chains in human lymphoid tissues. *J. Immun.* **95**, 246.

BERNIER G.M. & PUTNAM F.W. (1964a) Myeloma proteins and macroglobulins: hallmarks of disease and models of antibodies. *Prog. Hemat.* **4**, 160.

BERNIER G.M. & PUTNAM F.W. (1964b) Polymerism, polymorphism and impurities in Bence-Jones proteins. *Biochim. biophys. Acta* **86**, 295.

BOREL H., PRYCE S. & ALLEN F.H. (1967) Gm typing with microtiter plates. *Vox Sang.* **12**, 319.

BRAMBELL F.W.R. (1966) The transmission of immunity from mother to young and the catabolism of immunoglobulin. *Lancet* **ii**, 1087.

BRAMBELL F.W.R., HEMMINGS W.A. & MORRIS I.G. (1964) A theoretical model of γ-globulin catabolism. *Nature* **203**, 1352.

Brandtzaeg B., Fudenberg H. & Mohr J. (1961) The Gm(r) serum group. *Acta genet.* **11**, 170.

Brenner S. & Milstein C. (1966) Origin of antibody variation. *Nature* **211**, 242.

Bridges R.A., Condie R.M., Zak S.J. & Good R.A. (1959) Morphologic basis of antibody formation development during the neonatal period. *J. Lab. clin. Med.* **53**, 331.

Burnet M. (1966) A possible genetic basis for specific pattern in antibody. *Nature* **210**, 1308.

Burtin P. & Buffe D. (1967) Synthesis of human immunoglobulins in germinal centers of lymphoid organs. *J. Immun.* **98**, 536.

Carbonara A.O. & Heremans J.F. (1963) Subunits of normal and pathological γ_{1A}-globulins (β_{2A}-globulins). *Arch. Biochem. Biophys.* **102**, 137.

Chaplin H., Cohen S. & Press E.M. (1965) Preparation and properties of the peptide chains of normal 19S γ-globulin (IgM). *Biochem. J.* **95**, 256.

Chodirker W.B. & Tomasi T.B. (1963) Gamma globulins: quantitative relationships in human serum and nonvascular fluids. *Science* **142**, 1080.

Cohen S. (1963a) Antibodies, *in* Cruickshank R. (ed.) *Modern Trends in Immunology*, p. 25. Butterworth, Washington.

Cohen S. (1963b) γ-globulin metabolism. *Br. med. Bull.* **19**, 202.

Cohen S. (1963c) Properties of the peptide chains of normal and pathological γ globulins. *Biochem. J.* **89**, 334.

Cohen S. (1966) General structure and heterogeneity of immunoglobulins. *Proc. roy. Soc. B.* **166**, 114.

Cohen S. & Freeman T. (1960) Metabolic heterogeneity of human γ-globulin. *Biochem. J.* **76**, 475.

Cohen S. & Gordon S. (1965) Dissociation of κ and λ chains from reduced human immunoglobulins. *Biochem. J.* **97**, 460.

Cohen S. & Milstein C. (1967a) Structure of antibody molecules. *Nature* **214**, 449.

Cohen S. & Milstein C. (1967b) Structure and biological function of immunoglobulins, *in* Dixon F.J. & Humphrey J.H. (eds.) *Advances in Immunology*, vol. 7, p. 1. Academic Press, New York & London.

Cohen S. & Porter R.R. (1964) Structure and biological activity of immunoglobulins, *in* Dixon F.J. & Humphrey J.H. (eds.) *Advances in Immunology*, vol. 4, p. 287. Academic Press, New York & London.

Colberg, J.E. & Dray S. (1964) Localization by immunofluorescence of gamma-globin allotypes in lymph node cells of homozygous and heterozygous rabbits. *Immunology* **7**, 273.

Coons A.H., Leduc E.H. & Connolly J.M. (1955) Studies on antibody production I. Method for histochemical demonstration of specific antibody and its application to the study of the hyperimmune rabbit. *J. exp. Med.* **102**, 49.

Corcoran P.A., Weiner J., Vitagliano E., Hoagland E.M., Holbrook E.R., Keller J.R., Moulton H.E. & Diamond L.K. (1966) Linkage study of gamma globulin groups. *Vox Sang.* **11**, 620.

Costea N., Yakulis V. & Heller P. (1966) Light-chain heterogeneity of cold agglutinins. *Science* **152**, 1520.

Deutsch H.F. & Morton J.I. (1957) Dissociation of human serum macroglobulins. *Science* **125**, 600.

DIXON F.J., JACOT-GUILLARMOD H. & MCCONAHEY P.J. (1967) The effect of passively administered antibody on antibody synthesis. *J. exp. Med.* **125**, 1119.

DOOLITTLE R.F. & SINGER S.J. (1965) Tryptic peptides from the active sites of antibody molecules. *Proc. natn. Acad. Sci.* **54**, 1773.

DOOLITTLE R.F., SINGER S.J. & METZGER H. (1966) Evolution of immunoglobulin polypeptide chains: carboxy-terminal of an IgM heavy chain. *Science* **154**, 1561.

DORRINGTON K.J., ZARLENGO M.H. & TANFORD C. (1967) Conformational change and complementarity in the combination of H and L chains of immunoglobulin-G. *Proc. natn. Acad. Sci.* **58**, 996.

DRAY S. & YOUNG G.O. (1959) Two antigenically different γ globulins in domestic rabbits revealed by isoprecipitins. *Science* **129**, 1023.

DREYER W.J. & BENNETT J.C. (1965) The molecular basis of antibody formation: a paradox. *Proc. natn. Acad. Sci.* **54**, 864.

EASLEY C.W. & PUTNAM F.W. (1966) Structural studies of the immunoglobulins. III. Aminoethylated type κ Bence-Jones proteins. *J. biol. Chem.* **241**, 3671.

EDELMAN G.M. & BENACERRAF B. (1962) On structural and functional relations between antibodies and proteins of the gamma-system. *Proc. natn. Acad. Sci.* **48**, 1035.

EDELMAN G.M. & GALLY J.A. (1962) The nature of Bence-Jones proteins. Chemical similarities to polypeptide chains of myeloma globulins and normal γ globulins. *J. exp. Med.* **116**, 207.

EDELMAN G.M. & GALLY J.A. (1964) A model for the 7S antibody molecule. *Proc. natn. Acad. Sci.* **51**, 846.

EDELMAN G.M. & GALLY J.A. (1967) Somatic recombination of duplicated genes: an hypothesis on the origin of antibody diversity. *Proc. natn. Acad. Sci.* **57**, 353.

EDELMAN G.M., HEREMANS J.F., HEREMANS M.T. & KUNKEL H.G. (1960) Immunological studies of human γ-globulins. *J. exp. Med.* **112**, 203.

EDELMAN G.M. & KABAT E.A. (1964) Studies on human antibodies. I. Starch gel electrophoresis of the dissociated polypeptide chains. *J. exp. Med.* **119**, 443.

EDELMAN G.M. & POULIK M.D. (1961) Studies on structural units of the γ-globulins. *J. exp. Med.* **113**, 861.

EIN D. & FAHEY J.L. (1967) Two types of lambda polypeptide chains in human immunoglobulins. *Science* **156**, 947.

EPSTEIN W.V. & FUDENBERG H.H. (1962) Demonstration of Gm 1(a) and anti-Gm 1(a) specificities by tanned cells coated with individual γ-globulins. *J. Immun.* **89**, 293.

EPSTEIN W.V., TAN M. & GROSS D. (1964) Blocked antigenic sites on the L-chain of human gamma globulin. *Nature* **202**, 1175.

FAHEY J.L. (1962) Heterogeneity of γ-globulins, *in* TALIAFERRO W.H. & HUMPHREY J.H. (eds.) *Advances in Immunology*, vol. 2, p. 41. Academic Press, New York & London.

FAHEY J.L. (1963a) Structural basis for the differences between type I and type II human γ-globulin molecules. *J. Immunol.* **91**, 448.

FAHEY J.L. (1963b) Heterogeneity of myeloma proteins. *J. clin. Invest.* **42**, 111.

FAHEY J.L. (1964) Contribution of γ globulin subunits to electrophoretic heterogeneity; identification of a distinctive group of 6.6S γ-myeloma proteins. *Immunochemistry* **1**, 121.

FAHEY J.L. & LAWLER S.D. (1961) Gm factors in normal γ-globulin fractions, myeloma proteins and macroglobulins. *J. natn. Cancer Inst.* **27**, 973.

FAHEY J.L. & ROBINSON A.G. (1963) Factors controlling serum γ-globulin concentration. *J. exp. Med.* **118**, 845.

FAHEY J.L. & SOLOMON A. (1963) Two types of γ-myeloma proteins, β_{2A} proteins, γ_1-macroglobulins, and Bence-Jones proteins identified by two groups of common antigenic determinants. *J. clin. Invest.* **42**, 811.

FEINSTEIN A. (1966) Use of charged thiol reagents in interpreting the electrophoretic patterns of immune globulin chains and fragments. *Nature* **210**, 135.

FEINSTEIN D. & FRANKLIN E.C. (1966) Two antigenically distinguishable subclasses of human A myeloma proteins differing in their heavy chains. *Nature* **212**, 1496.

FEINSTEIN A., GELL P.G. & KELUS A.S. (1963) Immunochemical analysis of rabbit γ globulin allotypes. *Nature* **200**, 653.

FEIZI T. (1967) Lambda chains in cold agglutinins. *Science* **156**, 1111.

FINKELSTEIN N.S. & UHR J.W. (1964) Specific inhibition of antibody formation by passively administered 19S and 7S antibody. *Science* **146**, 67.

FISHMAN M. & ADLER F.L. (1963) Antibody formation initiated *in vitro*. II. Antibody synthesis in X-irradiated recipients of diffusion chambers containing nucleic acids derived from macrophages incubated with antigen. *J. exp. Med.* **117**, 595.

FLEISCHMAN J.B. (1966) Immunoglobulins. *A. Rev. Biochem.* **35**, 835.

FLEISCHMAN J.B., PAIN R.H. & PORTER R.R. (1962) Reduction of γ globulins. *Arch. Biochem. Biophys.* suppl. 1, 174.

FOUGEREAU M. & EDELMAN G.M. (1964) Resemblance of the gross arrangement of polypeptide chains in reconstituted and native γ globulins. *Biochemistry* **3**, 1120.

FOUGEREAU M. & EDELMAN G.M. (1965) Corroboration of recent models of the γG immunoglobulin molecule. *J. exp. Med.* **121**, 373.

FRANGIONE B. & FRANKLIN E.C. (1965) Structural studies of human immunoglobulins. Differences in the Fd fragments of the heavy chains of G myeloma proteins. *J. exp. Med.* **122**, 1.

FRANGIONE B., FRANKLIN E.C., FUDENBERG H.H. & KOSHLAND M.E. (1966) Structural studies of human γG myeloma proteins of different antigenic subgroups and genetic specificities. *J. exp. Med.* **124**, 715.

FRANGIONE B. & MILSTEIN C. (1967) Disulphide bridges of immunoglobulin G1 heavy chains. *Nature* **216**, 939.

FRANGIONE B., PRELLI F. & FRANKLIN E.C. (1967) The structure of Fd fragments of G myeloma proteins. *Immunochemistry* **4**, 95.

FRANKLIN E.C. (1960) Structural units of human 7S gamma globulin. *J. clin. Invest.* **39**, 1933.

FRANKLIN E. C. & FRANGIONE B. (1967) Structural difference between two subclasses of μ chains of human macroglobulins. *J. clin. Invest.* (abstr.) **46**, 1057.

FRANKLIN E.C. & FUDENBERG H.H. (1964) Antigenic heterogeneity of human Rh antibodies, rheumatoid factors and cold agglutinins. *Arch. Biochem. Biophys.* **104**, 433.

FRANKLIN E.C., FUDENBERG H.H., MELTZER M. & STANWORTH D.S. (1962) The structural basis for genetic variations of normal γ-globulins. *Proc. natn. Acad. Sci.* **48**, 914.

FRANKLIN E.C. & KUNKEL H.G. (1958) Comparative levels of high molecular weight (19S) gamma-globulin in maternal and umbilical cord sera. *J. Lab. clin. Med.* **52**, 724.

FRANKLIN E.C., LOWENSTEIN J., BIGELOW B. & MELTZER M. (1964) Heavy chain disease—a new disorder of serum γ globulins. *Am. J. Med.* **37**, 332.

FRANKLIN E.C. & OVARY Z. (1963) On the sensitizing properties of some normal and pathologic human immune globulins and fragments obtained by papain or pepsin digestion. *Immunology* **6**, 434.

FRANZL R.E. (1962) Immunogenic sub-cellular particles obtained from spleens of antigen-injected mice. *Nature* **195**, 457.

FREEMAN T. (1965) Gamma globulin metabolism in normal humans and in patients. *Series Haematologica* **4**, 76.

FUDENBERG H.H., FEINSTEIN D., McGEHEE W., & FRANKLIN E.C. (1966) Molecular localization of Gm(a) and Gm(b) factors in Negroes. *Vox Sang.* **11**, 45.

FUDENBERG H.H. & FRANKLIN E.C. (1963) Human gamma globulin. Genetic control and its relation to disease. *Ann. intern. Med.* **58**, 171.

FUDENBERG H.H. & FUDENBERG B.R. (1964) Antibody to hereditary human gamma-globulin (Gm) factor resulting from maternal-fetal incompatibility. *Science* **145**, 170.

FUDENBERG H.H. & KUNKEL H.G. (1961) Specificity of the reactions between rheumatoid factor and gamma-globulins. *J. exp. Med.* **114**, 257.

FUDENBERG H.H., STIEHM E.R., FRANKLIN E.C., MELTZER M. & FRANGIONE B. (1964). Antigenicity of hereditary human gamma globulin (Gm) factors—biological and biochemical aspects. *Cold Spring Harbor Symposia on Quantitative Biology* **29**, 463.

GELL P.G.H. & SELL S. (1965) Studies on rabbit lymphocytes *in vitro*. II. Induction of blast transformation with antisera to six IgG allotypes and summation with mixtures of antisera to different allotypes. *J. exp. Med.* **122**, 813.

GOLD E.R., MANDY W.J. & FUDENBERG H.H. (1965a) Relation between Gm(f) and the structure of the γ-globulin molecule. *Nature* **207**, 1099.

GOLD E.R., MARTENSSON L., ROPARTZ C., RIVAT L. & ROUSSEAU P.Y. (1965b) Gm(f)—a determinant of human γ globulin; preliminary communication. *Vox Sang.* **10**, 299.

GOTTLIEB A.A., GLISIN V.R. & DOTY P. (1967) Studies on macrophage RNA involved in antibody production. *Proc. natn. Acad. Sci.* **57**, 1849.

GRAY W.R., DREYER W.J. & HOOD L. (1967) Mechanism of antibody synthesis: size differences between mouse kappa chains. *Science* **155**, 465.

GREEN I., VASSALLI P., NUSSENZWEIG V. & BENACERRAF B. (1967) Specificity of the antibodies produced by single cells following immunization with antigens bearing two types of antigenic determinants. *J. exp. Med.* **125**, 511.

GREY H.M. & KUNKEL H.G. (1964) H chain subgroups of myeloma proteins and normal 7S γ globulin. *J. exp. Med.* **120**, 253.

GREY H.M. & KUNKEL H.G. (1967) Heavy chain subclasses of human γG-globulin. Peptide and immunochemical relationships. *Biochemistry* **6**, 2326.

GREY H.M. & MANNIK M. (1965) Specificity of recombination of H and L chains from human γG-myeloma proteins. *J. exp. Med.* **122**, 619.

GROSS D. & EPSTEIN W.V. (1964) Macroglobulinemia with Bence Jones proteinuria: comparison of urinary protein and L chain of serum proteins. *J. clin. Invest.* **43**, 83.

GROSS D., TERRY W. & EPSTEIN W.V. (1962) The association of hereditary gamma globulin (a) activity with a fragment of human gamma globulin produced by papain digestion. *Biochem. biophys. Res. Commun.* **7**, 259.

GRUBB R. (1956) Agglutination of erythrocytes coated with 'incomplete' anti-Rh by certain rheumatoid arthritic sera and some other sera. The existence of human serum groups. *Acta path. microbiol. scand.* **39**, 195.

GRUBB R. (1957) A relationship between blood group serology and rheumatoid arthritis serology. Serum protein groups. *Vox Sang.* **2**, 305.

GRUBB R. (1958) Interaction between rheumatoid arthritis sera and human gamma globulin. *Acta haemat.* **20**, 246.

GRUBB R. (1959) Hereditary gamma globulin groups in man, *in* WOLSTENHOLME G.E.W. & O'CONNOR C.M. (eds.) *Ciba Foundation Symposium on Biochemistry of Human Genetics*, p. 264. Little, Brown, Boston.

GRUBB R. (1961) The Gm groups and their relation to rheumatoid arthritis serology. *Arthritis Rheum.* **4**, 195.

GRUBB R., KRONVALL G. & MARTENSSON L. (1965) Some aspects of the relations between rheumatoid arthritis, anti-gamma-globulin factors and the polymorphism of human gamma-globulin. *Ann. N.Y. Acad. Sci.* **124**, 865.

GRUBB R. & LAURELL A.B. (1956) Hereditary serological human serum groups. *Acta path. microbiol. scand.* **39**, 390.

HARBOE M. (1959) A new hemagglutinating substance in the Gm system, anti-Gm^b. *Acta path. microbiol. scand.* **47**, 191.

HARBOE M. (1960) Relation between Gm types and hemagglutinating substances in rheumatoid sera. *Acta path. microbiol. scand.* **50**, 89.

HARBOE M. & DEVERILL J. (1966) Structural relationship between γG and γM globulin in man. *Acta med. scand.* suppl. 445, 74.

HARBOE M., DEVERILL J. & GODAL H.C. (1965a) Antigenic heterogeneity of Waldenström type γM globulins. *Scand. J. Haemat.* **2**, 137.

HARBOE M. & LIND K. (1966) Light chain types of transiently occurring cold haemagglutinins. *Scand. J. Haemat.* **3**, 269.

HARBOE M. & LUNDEVALL J. (1959) A new type in the Gm system. *Acta path. microbiol. scand.* **45**, 357.

HARBOE M., OSTERLAND C.K. & KUNKEL H.G. (1962) Genetic γ globulin (Gm and Inv) characteristics of myeloma proteins and their localization in the split products produced by papain. *Ann. N.Y. Acad. Sci.* **101**, 235.

HARBOE M., VAN FURTH R., SCHUBOTHE H., LIND K. & EVANS R.S. (1965b) Exclusive occurrence of κ chains in isolated cold haemaglgutinins. *Scand. J. Haemat.* **2**, 259.

HARRIS J. & VAUGHAN J.H. (1961) Transfusion studies in rheumatoid arthritis. *Arthritis Rheum.* **4**, 47.

HEIMBURGER N., HEIDE H., HAUPT H. & SCHULTZE H.E. (1964) Bausteinanalysen von Humanserum Proteinen. *Clin. chim. Acta* **10**, 293.

HEMMINGS W.A. (1961) Protein transfer across the foetal membranes. *Br. med. Bull.* **17**, 96.

HEREMANS J.F., HEREMANS M.T. & SCHULTZE H.E. (1959) Isolation and description of a few properties of β_2A-globulins of human serum. *Clin. chim. Acta* **4**, 96.

HESS M. & BÜTLER R. (1962) Anti-Gm specificities in sera of rhesus monkeys immunized with human gamma-globulin. *Vox Sang.* **7**, 93.

HILL R.L., DELANEY R., FELLOWS R.E. & LEBOVITZ H.E. (1966a) The evolutionary origins of the immunoglobulins. *Proc. natn. Acad. Sci.* **56**, 1762.

HILL R.L., DELANEY R., LEBOVITZ H.E. & FELLOWS R.E. (1966b) Studies on the amino acid sequence of heavy chains from rabbit immunoglobulin G. *Proc. roy. Soc.* B **166**, 159.

HILSCHMANN N. & CRAIG L.C. (1965) Amino acid sequence studies with Bence-Jones proteins. *Proc. natn. Acad. Sci.* **53**, 1403.

HOBBS J.R. & DAVIS J.A. (1967) Serum γG-globulin levels and gestational age in premature babies. *Lancet* i, 757.

HONG R. & NISONOFF A. (1966) Heterogeneity in the complementation of polypeptide subunits of a purified antibody isolated from an individual rabbit. *J. Immun.* **96**, 622.

HOOD L.E., GRAY W.R. & DREYER W.J. (1966) On the mechanism of antibody synthesis: a species comparison of L-chains. *Proc. natn. Acad. Sci.* **55**, 826.

HUGHES-JONES N.C. (1965) Iodine-125-labelled L chains of human blood group antibodies. *Nature* **207**, 989.

HUMMELER K., HARRIS S. & HARRIS T.N. (1966) Fine structure of some antibody-producing cells. *Fedn. Proc.* **25**, 1734.

INMAN F.P. & NISONOFF A. (1966) Localization of noncovalent interactions between the heavy chains of rabbit γG-globulin. *Proc. natn. Acad. Sci.* **56**, 542.

ISHIZAKA K. & ISHIZAKA T. (1967) Identification of γE antibodies as a carrier of reagenic activity. *J. Immun.* **99**, 1187.

ISHIZAKA K., ISHIZAKA T. & HORNBROOK M.M. (1966) Physico-chemical properties of human reaginic antibody. IV. Presence of a unique immunoglobulin as a carrier of reaginic activity. *J. Immun.* **97**, 75.

ISHIZAKA K., ISHIZAKA T. & HORNBROOK M.M. (1967a) Allergen-binding activity of γE, γG and γA antibodies in sera from atopic patients: *in vitro* measurements of reaginic antibody. *J. Immun.* **98**, 490.

ISHIZAKA K., ISHIZAKA T. & TERRY W.D. (1967b) Antigenic structure of γE-globulin and reaginic antibody. *J. Immun.* **99**, 849.

KABAT E.A. (1967a) The paucity of species-specific amino acid residues in the variable regions of human and mouse Bence-Jones proteins and its evolutionary and genetic implications. *Proc. natn. Acad. Sci.* **57**, 1345.

KABAT E.A. (1967b) A comparison of invariant residues in the variable and constant regions of human K, human L and mouse K Bence-Jones proteins. *Proc. natn. Acad. Sci.* **58**, 229.

KERN P.M., HELMREICH E. & EISEN H.N. (1961) The solubilization of microsomal antibody by the specific interaction between crystallizable fraction of γ-globulin and lymph node microsomes. *Proc. natn. Acad. Sci.* **47**, 767.

KLEMPERER M.R., HOLBROOK E.R. & FUDENBERG H.H. (1966) Gm(20), a new hereditary gamma globulin factor. *Am. J. hum. Genet.* **18**, 433.

KORNGOLD L. & LIPARI R. (1956) Multiple myeloma proteins. III. The antigenic relationship of Bence-Jones proteins to normal γ globulins and multiple myeloma serum proteins. *Cancer* **9**, 262.

KOSHLAND M.E. (1966) Primary structure of immunoglobulins and its relationship to antibody specificity. *J. cell Physiol.* **67**, suppl. 1, 33.

KOSHLAND M.E. (1967) Location in antibody Fd fragments of amino acid residues associated with immunological and allotypic specificities. *Cold Spring Harbor Symposia on Quantitative Biology*, **32**, 119.

KRONVALL G. (1965) Gm(f) activity of human gamma globulin fragments. *Vox Sang.* **10**, 303.

KUNKEL H.G., ALLEN J.C. & GREY H.M. (1964a) Genetic characters and the polypeptide chains of various types of gamma globulin. *Cold Spring Harbor Symposia on Quantitative Biology*, **29**, 443.

KUNKEL H.G., ALLEN J.C., GREY H.M., MARTENSSON L. & GRUBB R. (1964b) A relationship between the H chain groups of 7S γ-globulin and the Gm system. *Nature* **203**, 413.

KUNKEL H.G., GREY H.M. & SOLOMON A. (1966) Antigenic variations among myeloma proteins and antibodies, *in* GRABAR P. & MIESCHER P. (eds.) *Fourth Intern. Symp. on Immunopathology*, p. 220. Schwabe, Basel.

KUNKEL H.G. & PRENDERGAST R.A. (1966) Subgroups of γA immune globulins. *Proc. Soc. exp. Biol.* **122**, 910.

KUNKEL H.G., YOUNT W.J. & LITWIN S.D. (1966) Genetically determined antigen of the Ne subgroup of gamma-globulin: detection by precipitin analysis. *Science*, **154**, 1041.

LAMM M.E. & SMALL P.A. (1966) Polypeptide chain structure of rabbit immunoglobulins. II. γM immunoglobulin. *Biochemistry* **5**, 267.

LAURELL C-B. & SNIGUROWICZ J. (1967) The frequency of kappa and lambda chains in pathologic serum γG, γA, γD and $\gamma\mu$ immunoglobulins. *Scand. J. Haemat.* **4**, 46.

LAWLER S.D. (1960) A genetical study of the Gm groups in human serum. *Immunology* **3**, 90.

LAWLER S.D. & COHEN S. (1965) Distribution of allotypic specificities on the peptide chains of human gamma-globulin. *Immunology* **8**, 206.

LEDDY J.P. & BAKEMEIER R.F. (1965) Structural aspects of human erythrocyte autoantibodies. I. L chain types and electrophoretic dispersion. *J. exp. Med.* **121**, 1.

LENNOX E.S. & COHN M. (1967) Immunoglobulins. *A. Rev. Biochem.* **36**, 365.

LENNOX E.S., KNOPF P.M., MUNRO A.J. & PARKHOUSE R.M.E. (1967) A search for biosynthetic subunits of light and heavy chains of immunoglobulins. *Cold Spring Harbor Symposia on Quantitative Biology*, **32**, 249.

LITWIN S.D. & KUNKEL H.G. (1966a) Genetic factors of human gamma globulin detected by rabbit antisera. *Transfusion* **6**, 140.

LITWIN S.D. & KUNKEL H.G. (1966b) A γ globulin genetic factor related to Gm(a) but localized to a different portion of the same heavy chains. *Nature* **210**, 866.

LITWIN S.D. & KUNKEL H.G. (1967a) The genetic control of γ-globulin heavy chains. Studies of the major heavy chain subgroup utilizing multiple genetic markers. *J. exp. Med.* **125**, 847.

LITWIN S.D. & KUNKEL H.G. (1967b) The relationship between the Inv(1) and (2) genetic antigens of kappa human light chains. *J. Immun.* **99**, 603.

MANNIK M. & KUNKEL H.G. (1962) Classification of myeloma proteins, Bence Jones proteins and macroglobulins into two groups on the basis of antigenic characters. *J. exp. Med.* **116**, 119.

MANNIK M. & KUNKEL H.G. (1963a) Two major types of 7S γ-globulin. *J. exp. Med.* **117**, 213.

MANNIK M. & KUNKEL H.G. (1963b) Localization of antibodies in group I and group II γ-globulins. *J. exp. Med.* **118**, 817.

MARCHALONIS J. & EDELMAN G.M. (1966) Phylogenetic origins of antibody structure. II. Immunoglobulins in the primary immune response of the bullfrog, *Rana catesbiana, J. Exp. Med.* **124**, 901.

MARTENSSON L. (1961) Gm characters of M-components. *Acta med. scand.* **170**, suppl. 367, 87.

MARTENSSON L. (1962a) Anti-Gm molecules with distinctly different physico-chemical properties. *Acta path. microbiol. scand.* **56**, 352.

MARTENSSON L. (1962b) Distribution of Gm specificities among the gamma globulin molecules. *Acta path. microbiol. scand.* **54**, 343.

MARTENSSON L. (1964) On the relationships between the γ-globulin genes of the Gm system. *J. exp. Med.* **120**, 1169.

MARTENSSON L. (1966) Genes and immunoglobulins. *Vox Sang.* **11**, 521.

MARTENSSON L. & FUDENBERG H.H. (1965) Gm genes and γG synthesis in the human fetus. *J. Immun.* **94**, 514.

MARTENSSON L. & KUNKEL H.G. (1965) Distribution among the γ-globulin molecules of different genetically determined antigenic specificities in the Gm system. *J. exp. Med.* **122**, 799.

MARTENSSON L., VAN LOGHEM E., MATSUMOTO H. & NIELSEN J. (1966) Gm(s) and Gm(t): genetic determinants of human γ-globulin. *Vox Sang.* **11**, 393.

MELCHERS F. & KNOPE P.M. (1967) The biosynthesis of the carbohydrate portion of immunoglobulins: possible relation to secretion. *Cold Spring Harbor Symposia on Quantitative Biology*, **32**, 255.

MELLORS R.C. & KORNGOLD L. (1963) The cellular origin of human immuno-globulins. *J. exp. Med.* **118**, 387.

MELTZER M., FRANKLIN E.C., FUDENBERG H.H. & FRANGIONE B. (1964) Single peptide differences between γ-globulins of different genetic (Gm) types. *Proc. natn. Acad Sci.* **51**, 1007.

MERLER E., NELSON C.A. & TANFORD C. (1964) The polypeptide chains of rabbit gamma-globulin and its papain-cleaved fragments. *Biochemistry* **3**, 279.

MI M.P. & MORTON N.E. (1966) Blood factor association. *Vox Sang.* **11**, 434.

MIGITA S. & PUTNAM F.W. (1963) Antigenic relationships of Bence Jones proteins, myeloma globulins and normal γ-globulin. *J. exp. Med.* **117**, 81.

MILGROM F., DUBISKI S. & WOZNICZKO G. (1956) Human sera with 'anti-antibody'. *Vox Sang.* **1**, 172.

MILLER F. & METZGER H. (1965) Characterization of human macroglobulin. I. Molecular weight of its subunit. *J. biol. Chem.* **240**, 3325.

MILLER E.J., OSTERLAND C.K., DAVIE J.M. & KRAUSE R.M. (1967) Electrophoretic analysis of polypeptide chains isolated from antibodies in the serum of immunized rabbits. *J. Immun.* **98**, 710.

MILSTEIN C. (1965) Interchain disulfide bridge in Bence-Jones proteins and in γ globulin B chains. *Nature.* **205**, 1171.

MILSTEIN C. (1966a) Variation in amino-acid sequence near the disulfide bridges of Bence-Jones proteins. *Nature* **209**, 370.

MILSTEIN C. (1966b) Chemical structure of light chains. *Proc. roy. Soc.* B. **166**, 138.

MILSTEIN C. (1967) Linked groups of residues in immunoglobulin κ chains. *Nature* **216**, 330.

MILSTEIN C., CLEGG J.B. & JARVIS J.M. (1967) C terminal half of immunoglobulin λ chains. *Nature* **214**, 270.

MUIR W.A. & STEINBERG A.G. (1967) On the genetics of the human allotypes, Gm and Inv. *Sem. Hematol.* **4**, 156.

MÜLLER-EBERHARD H.J., KUNKEL H.G. & FRANKLIN E.C. (1956) Two types of γ-globulin differing in carbohydrate content. *Proc. Soc. exp. Biol.* **93**, 146.

NACHMAN R.L. & ENGLE R.L. (1964) Gamma globulin: unmasking of hidden antigenic sites in light chains. *Science* **145**, 167.

NACHMAN R.L. & ENGLE R.L. (1965) Antigenic uniqueness of Bence Jones proteins. *Nature* **205**, 290.

NATVIG J.B. (1965) Incomplete anti-D antibody with changed Gm specificity. *Acta path. microbiol. scand.* **65**, 467.

NATVIG J.B. (1966) Gm(g)—a 'new' gamma-globulin factor. *Nature* **211**, 318.

NATVIG J.B. & KUNKEL H.G. (1967) Detection of genetic antigens utilizing gamma globulins coupled to red blood cells. *Nature* **215**, 68.

NATVIG J.B. & KUNKEL H.G. (1968) Genetic markers of human immunoglobulins. The Gm and Inv Systems. *Series Haematologica* 1, **1**, 66.

NATVIG J.B., KUNKEL H.G. & GEDDE-DAHL T. (1967) Genetic studies of the heavy chain subgroups of γG globulin. Recombination between the closely linked cistrons, *in* KILLANDER J. (ed.) *IIIrd Nobel Symposium, Gamma Globulins*, p. 313. Almquist-Wiksell, Stockholm.

NEIDERS M.E., ROWLEY D.A. & FITCH F.W. (1962) The sustained suppression of hemolysin in passively immunized rats. *J. Immun.* **88**, 718.

NILSSON U. (1964) Gm characters of 7S myeloma proteins and corresponding individual normal γ-globulins. *Acta path. microbiol. scand.* **61**, 181.

NISONOFF A., MARKUS G. & WISSLER F.C. (1961) Separation of univalent fragments of rabbit antibody by reduction of a single, labile disulphide bond. *Nature* **189**, 293.

NISONOFF A. & THORBECKE C.J. (1964) Immunochemistry. *A. Rev. Biochem.* **33**, 355.

NISONOFF A., WISSLER F.C. & LIPMAN L.N. (1960) Properties of the major component of a peptic digest of rabbit antibody. *Science* **132**, 1770.

NOELKEN M.E., NELSON C.A., BUCKLEY C.E. & TANFORD C. (1965) Gross confirmation of rabbit 7S γ-immunoglobulin and its papain-cleaved fragments. *J. biol. Chem.* **240**, 218.

NOLAN C. & SMITH E.L. (1962) Glycopeptides. III. Isolation and properties of glycopeptides from a bovine globulin of colostrum and from fraction II-3 of human globulin. *J. biol. Chem.* **237**, 453.

OLINS D.E. & EDELMAN G.M. (1964) Reconstitution of 7S molecules from L and H polypeptide chains of antibodies and γ-globulins. *J. exp. Med.* **119**, 789.

OSSERMAN E.F. & TAKATSUKI K. (1964) Clinical and immunochemical studies of four cases of heavy ($H^{\gamma 2}$) chain disease. *Am. J. Med.* **37**, 351.

OUDIN J. (1966a) Genetic regulation of immunoglobulin synthesis. *J. cell. Physiol.* **67**, suppl. 1, 77.

OUDIN J. (1966b) The genetic control of immunoglobulin synthesis. *Proc. roy. Soc.* B. **166**, 207.

OVARY Z. (1966) The structure of various immunoglobulins and their biologic activity. *Ann. N.Y. Acad. Sci.* **129**, 776.

PERNIS B., CHIAPPINO G., KELUS A.S. & GELL P.G.H. (1965) Cellular localization of immunoglobulins with different allotypic specificities in rabbit lymphoid tissues. *J. exp. Med.* **122**, 853.

PERNIS B., CHIAPPINO G. & ROWE D.S. (1966) Cells producing IgD immunoglobulins in human spleen. *Nature* **211**, 424.

PICKERING R.J., HONG R. & GOOD R.A. (1967) Deficient complement fixation by aggregated gamma globulin from hypogammaglobulinemic patients. *Science* **157**, 454.

PINK J.R.L. & MILSTEIN C. (1967a) Inter heavy-light chain disulphide bridge in immune globulins. *Nature* **214**, 92.

PINK J.R.L. & MILSTEIN C. (1967b) Disulphide bridges of a human immunoglobulin G protein. *Nature* **216**, 941.

POLMAR S.H. & STEINBERG A.G. (1964) Dependence of a Gm(b) antigen on the quaternary structure of human gamma globulin. *Science* **145**, 928.

POLMAR S.H. & STEINBERG A.G. (1967) The effect of the interaction of heavy and light chains of IgG on the Gm and Inv antigens. *Biochem. Genet.* **1**, 117.

PORTER R.R. (1959) Hydrolysis of rabbit γ-globulins and antibodies with crystalline papain. *Biochem. J.* **73**, 119.

PORTER R.R. (1960) γ-globulin and antibodies, *in* PUTNAM F.W. (ed.) *The Plasma Proteins*, vol. 1, p. 241. Academic Press, New York.

PORTER R.R. (1962) The structure of gamma globulins and antibodies, *in* GELLHORN A. & HIRSCHBERG E. (eds.) *Symposium on Basic Problems in Neoplastic Disease*, p. 177. Columbia University Press, New York.

POTTER M., APPELLA E. & GEISSER S. (1965) Variations in the heavy polypeptide chain structure of gamma myeloma immunoglobulins from an inbred strain of mice, and a hypothesis as to their origin. *J. mol. Biol.* **14**, 361.

POULIK M.D. (1967) Structure and pathologic variations of immunoglobulins, *in* GREENWALT T.J. (ed.) *Advances in Immunogenetics*, p. 31. Lippincott, Philadelphia.

POULIK M.D. & SHUSTER J. (1964) Heterogeneity of H chains of myeloma proteins: susceptibility to papain and trypsin. *Nature* **204**, 577.

POULIK M.D. & SHUSTER J. (1965) Role of cysteine in the production of Fc and F'c sub-components of γG myeloma globulins. *Nature* **207**, 1092.

PRAHL J.W. (1967) The C-terminal sequences of the heavy chains of human IgG myeloma proteins of differing isotypes and allotypes. *Biochem. J.* **105** 1019.

PRESS E.M., GIVOL D., PIGGOT P.J., PORTER R.R. & WILKINSON J.M. (1966a) The chemical structure of the heavy chains of rabbit and human immunoglobulin G (IgG). *Proc. roy. Soc.* B. **166**, 150.

PRESS E.M., PIGGOT P.J. & PORTER R.R. (1966b) The N and C terminal amino acid sequence of heavy chain from a pathological human immunoglobulin (γG). *Biochem. J.* **99**, 356.

PUTNAM F.W., TITANI K. & WHITLEY E. (1966) Chemical structure of light chains: amino acid sequence of type K Bence-Jones proteins. *Proc. roy. Soc.* B. **166**, 124.

QUATTROCCHI R., CIOLI D. & BAGLIONI C. (1967) Distribution of an argenine-lysine interchange in the invariable half of human L-type Bence-Jones proteins. *Nature* **216**, 56.

REISFELD R.A. & SMALL P.A. (1966) Electrophoretic heterogeneity of polypeptide chains of specific antibodies. *Science* **152**, 1253.

RITTER H., ROPARTZ C., BAITSCH H., ROUSSEAU P.Y., RIVAT L. & REMY K. (1964a) Zur Formalgenetik des Gammaglobulin-Polymorphismus Gm (Merkmale GM(a), GM(b), Gm(e)); Untersuchungen an 387 Familien. *Acta genet.* **14**, 4.

RITTER H., ROPARTZ C., ROUSSEAU P.Y., RIVAT L. & SATI A. (1964b) Studies on the formal genetics of the gamma globulin polymorphism Gm (Characters Gm(a), Gm(b), Gm(x)); A study of 386 families. *Vox Sang.* **9**, 340.

RITTER H., ROPARTZ C., ROUSSEAU P.Y., RIVAT L. & BÄHR M.L. (1964c) Zur Formalgenetik und Populationsgenetik des Gammaglobulin-Polymorphismus InV (Merkmale InV(1) und InV(a)). *Acta genet.* **14**, 15.

RITTER H. & WENDT G.G. (1964) Untersuchung von 223 Familien zur Formalen Genetik des INV-Polymorphismus. *Humangenetik* **1**, 123.

ROGENTINE G.N., ROWE D.S., BRADLEY J., WALDMANN T.A. & FAHEY J.L. (1966) Metabolism of human immunoglobulin D (IgD). *J. clin. Invest.* **45**, 1467.

ROPARTZ C., LENOIR J. & RIVAT L. (1961) A new inheritable property of human sera: the InV factor. *Nature* **189**, 586.

ROPARTZ C., RIVAT L. & ROUSSEAU P.Y. (1963) Le Gm(b) et ses problèmes. *Vox Sang.* **8**, 717.

ROPARTZ C., RIVAT L. & ROUSSEAU P.Y. (1964) Deux nouveaux facteurs dans les systèmes héréditaires de gamma-globuline le Gm(e) et l'InV(1). *Proc. 9th Cong. int. Soc. Blood Transf.* p. 455.

ROPARTZ C., RIVAT L., ROUSSEAU P.Y., FUDENBERG H.H., MOLTER R. & SALMON C. (1966) Seven new human serum factors presumably supported by the gamma globulins. *Vox Sang.* **11**, 99.

ROPARTZ C., ROUSSEAU P.Y., RIVAT L. (1962) Mise en évidence d'une allèle silencieux au locus Gm. *Vox Sang.* **7**, 233.

ROPARTZ C., ROUSSEAU P.Y. & RIVAT L. (1965a) Hypothèses sur la génétique for-malle due système Gm chez les Caucasians. *Humangenetik* **1**, 483.

ROPARTZ C., ROUSSEAU P.Y. & RIVAT L. (1965b) The influence of the anti-Rh coat and red cells on the manifestation of the Gm(b) phenotype. *Vox Sang.* **10**, 583.

ROPARTZ C., ROUSSEAU P.Y., RIVAT L. & LENOIR J. (1961) Etude génétique du facteur serique Inv: fréquence dans certaines populations. *Rev. fr. Etud. clin. biol.* **6**, 374.

ROSEN F.S. & JANEWAY C.A. (1966) The gamma globulins. III. The antibody deficiency syndromes. *New Engl. J. Med.* **275**, 709, 769.

ROWE D.S. & FAHEY J.L. (1965a) A new class of human immunoglobulins. I. A unique myeloma protein. *J. exp. Med.* **121**, 171.

ROWE D.S. & FAHEY J.L. (1965b) A new class of human immunoglobulins. II. Normal serum IgD. *J. exp. Med.* **121**. 185.

SAHIAR K. & SCHWARTZ R.W. (1965) The immunoglobulin sequence. I. Arrest by 6-mercaptopurine and restitution by antibody, antigen or splenectomy. *J. Immun.* **95**, 345.

SCHARFF M.D. & UHR J.W. (1966) Functional ribosomal unit of gamma-globulin synthesis. *Science* **148**, 646.

SCHOENBERG M.D., MUMAW V.R., MOORE R.D. & WEISBERGER A.S. (1964) Cytoplasmic interaction between macrophages and lymphocytic cells in antibody synthesis. *Science* **143**, 964.

SCHWARTZ J.H. & EDELMAN G.M. (1963) Comparisons of Bence-Jones proteins and L polypeptide chains of myeloma globulins after hydrolysis with trypsin. *J. exp. Med.* **118**, 41.

SCHWARTZ R.S. (1966) Immunoglobulin metabolism. *Med. Clins N. Am.* **50**, 1487.

SELIGMANN M., MIHAESCO C. & MESHAKA G. (1966) Antigenic determinants common to human immunoglobulins G and M: importance of conformational antigens. *Science* **154**, 790.

SHAPIRO A.L., SCHARFF M.D., MAIZEL J.V. & UHR J.W. (1966) Polyribosomal synthesis and assembly of the H and L chains of gamma globulin. *Proc. natn. Acad. Sci.* **56**, 216.

SINGER S.J. & DOOLITTLE R.F. (1966) Antibody active sites and immunoglobulin molecules. *Science* **153**, 13.

SJÖQUIST J. & VAUGHAN M.H. (1966) Heterogeneity of H and L chains of normal and myeloma γG-globulin. *J. molec. Biol.* **20**, 527.

SMALL P.A., KEHN J.E. & LAMM M.E. (1963) Polypeptide chains of rabbit γ-globulin. *Science* **142**, 393.

SMILEY J.D. & HORTON H. (1965) Isolation and study of the function of human γ_2-globulin glycopeptide. *Immunochemistry* **2**, 61.

SMITHIES O. (1967a) Antibody variability. *Science* **157**, 267.

SMITHIES O. (1967b) The genetic basis of antibody variability. *Cold Spring Harbor Symposia on Quantitative Biology.* **32**, 161.

SMYTH D.G. & UTSUMI S. (1967) Structure at the hinge region in rabbit immunoglobulin-G. *Nature* **216**, 332.

SOLOMON A., FAHEY J.L. & MALMGREN R.A. (1963a) Immunohistologic localization of gamma-1-macroglobulins, beta-2A-myeloma proteins, 6.6S gamma-myeloma proteins and Bence Jones proteins. *Blood* **21**, 403.

SOLOMON A., WALDMANN T.A. & FAHEY J.L. (1963b) Metabolism of normal 6.6S γ-globulin in normal subjects and in patients with macroglobulinemia and multiple myeloma. *J. Lab. clin. Med.* **62**, 1.

SOUTH M.A., COOPER M.D., WOLLHEIM F.A., HONG R. & GOOD R.A. (1966) The IgA system. I. Studies of the transport and immunochemistry of IgA in the saliva. *J. exp. Med.* **123**, 615.

SPEISER P. (1963) Über Antikörperbildung von Säuglingen und Kleinkindern gegen mütterliches $\gamma 2$ globulin. *Wien. med. Wschr.* **113**, 966.

STEINBERG A.G. (1962a) Progress in the study of genetically determined human gamma globulin types (the Gm and Inv groups), *in* STEINBERG A.G. & BEARN A.G. (eds.) *Progress in Medical Genetics*, vol. 2, p. 1. Grune & Stratton, New York & London.

STEINBERG A.G. (1962b) Evidence for a Gm allele negative for both Gm(a) and Gm(b). *Vox Sang.* **7**, 89.

STEINBERG A.G. (1964) Population, immunogenetic, and biochemical studies on the Gm(b) factors of human gamma globulin. *Cold Spring Harbor Symposia on Quantitative Biology* **29**, 449.

STEINBERG A.G. (1965) Comparison of Gm(f) with Gm(b^2) [Gm(b^w)] and a discussion of their genetics. *Am. J. hum. Genet.* **17**, 311.

STEINBERG A.G. (1967) Genetic variations in human immunoglobulins. The Gm and Inv types, *in* GREENWALT T.J. (ed.) *Advances in Immunogenetics*, p. 75. Lippincott, Philadelphia.

STEINBERG A.G., GILES B.D. & STAUFFER R. (1960) A Gm-like factor present in Negroes and rare or absent in whites; its relation to Gm^a and Gm^x. *Am. J. hum. Genet.* **12**, 46.

STEINBERG A.G. & GOLDBLUM R. (1965) A genetic study of the antigens associated with the Gm(b) factor of human gamma globulin. *Am. J. hum. Genet.* **17**, 133.

STEINBERG A.G., MUIR W.A. & MCINTIRE S.A. (1968) Two unusual Gm alleles: their implications for the genetics of the Gm antigens. *Am. J. hum. Genet.* **20**, 258.

STEINBERG A.G. & WILSON J.A. (1963a) Studies on hereditary gamma globulin factors: evidence that Gm(b) in whites and Negroes is not the same and that Gm-like is determined by an allele at the Gm locus. *Amer. J. hum. Genet.* **15**, 96.

STEINBERG A.G. & WILSON J.A. (1963b) Hereditary globulin factors and immune tolerance in man. *Science* **140**, 303.

STEINBERG A.G., WILSON J.A. & LANSET S. (1962) A new human gamma globulin factor determined by an allele at the Inv locus. *Vox Sang.* **7**, 151.

STIEHM E.R., AMMANN A.J. & CHERRY J.D. (1966a) Elevated cord macroglobulins in the diagnosis of intrauterine infections. *New Engl. J. Med.* **275**, 971.

STIEHM E.R. & FUDENBERG H.H. (1966) Serum levels of immune globulins in health and disease: a survey. *Pediatrics* **37**, 715.

STIEHM E.R., VAERMAN J.P. & FUDENBERG H.H. (1966b) Plasma infusions in immunologic deficiency states: metabolic and therapeutic studies. *Blood* **28**, 918.

TAKATSUKI K. & OSSERMAN E.F. (1964) Demonstration of two types of low molecular weight γ globulins in normal human urine. *J. Immunol.* **92**, 100.

TARANTA A. & FRANKLIN E.C. (1961) Complement fixation by antibody fragments. *Science* **134**, 1981.

TERRY W.D. (1965) Skin-sensitizing activity related to γ-polypeptide chain characteristics of human IgG. *J. Immunol.* **95**, 1041.

TERRY W.D. & FAHEY J.L. (1964) Subclasses of human γ_2-globulin based on differences in the heavy polypeptide chains. *Science* **146**, 400.

TERRY W.D., FAHEY J.L. & STEINBERG A.G. (1965) Gm and Inv factors in subclasses of human IgG. *J. exp. Med.* **122**, 1087.

TERRY W.D. & ROBERTS M.S. (1966) Antigenic heterogeneity of human immunoglobulin A proteins. *Science* **153**, 1007.

TERRY W.D., SMALL P.A. & REISFELD R.A. (1966) Electrophoretic heterogeneity of the polypeptide chains of human G-myeloma proteins. *Science* **152**, 1628.

THORPE N.O. & DEUTSCH H.F. (1966) Studies on papain produced subunits of human γG globulins. II. Structures of peptides related to the genetic Gm activity of γG-globulin Fc fragments. *Immunochemistry* **3**, 329.

TITANI K., WHITLEY E., AVOGARDO L. & PUTNAM F.W. (1965) Immunoglobulin structure: partial amino acid sequence of a Bence Jones protein. *Science* **149**, 1090.

TITANI K., WHITLEY E. & PUTNAM F.W. (1966) Immunoglobulin structure: variation in the sequence of Bence Jones proteins. *Science* **152**, 1513.

TITANI K., WIKLER M. & PUTNAM F.W. (1967) Evolution of immunoglobulins: structural homology of kappa and lambda Bence Jones proteins. *Science* **155**, 828.

TOMASI T.B. (1965) Human gamma globulin. *Blood* **25**, 382.

TOMASI T.B., TAN E.M., SOLOMON A. & PRENDERGAST R.A. (1965) Characteristics of an immune system common to certain external secretions. *J. exp. Med.* **121**, 101.

TOMASI T.B. & ZIGELBAUM S. (1963) The selective occurrence of γ1A globulins in certain body fluids. *J. clin. Invest.* **42**, 1552.

UHR W. & BAUMANN J. (1961) Antibody formation. I. The suppression of antibody formation by passively administered antibody. *J. exp. Med.* **113**, 935.

UTSUMI S. & KARUSH F. (1964) The subunits of purified rabbit antibody. *Biochemistry* **3**, 1329.

VAERMAN J.P. & HEREMANS J.F. (1966) Antigenic heterogeneity of human immunoglobulin A proteins. *Science* **153**, 647.

VAN FURTH R., SCHUIT H.R.E. & HIJMANS W. (1965) The immunological development of the human fetus. *J. exp. Med.* **122**, 1173.

VAN LOGHEM E. & MARTENSSON L. (1967) Genetic (Gm) determinants of the γ2c (Vi) sublcass of human IgG immunoglobulins. *Vox Sang.* **13**, 369.

VIERUCCI A. (1965) Gm groups and anti-Gm antibodies in children with Cooley's anemia. *Vox Sang.* **10**, 82.

WALLER M., HUGHES R.D., TOWNSEND J.I., FRANKLIN E.C. & FUDENBERG H.H. (1963) New serum group, Gm(p). *Science* **142**, 1321.

WALLER M.V. & VAUGHAN J.H. (1956) Use of anti-Rh sera for demonstrating agglutination activating factors in rheumatoid arthritis. *Proc. Soc. exp. Biol.* **92**, 198.

WHITEHOUSE H.L.K. (1967) Crossover model of antibody variability. *Nature* **215**, 371.

WILLIAMS R.C. (1964) Reaction of human and rabbit anti-gamma-globulin factors with heavy and light chains of gamma globulin. *Arthritis Rheum.* **7**, 368.

WILLIAMS R.C. & LAWRENCE T.G. (1966) Variations among γ-globulins at the antigenic site revealed by pepsin digestion. *J. clin. Invest.* **45**, 714.

WILSON J.A. & STEINBERG A.G. (1965) Antibodies to gamma globulin in the serum of children and adults. *Transfusion* **5**, 516.

WOLLHEIM F.A. & WILLIAMS R.C. (1966) Studies on the macroglobulins of human serum. II. Heterogeneity of antigenic determinants among M-components in Waldenström's macroglobulinemia. *Acta med. scand.* suppl. 445, 115.

WORLD HEALTH ORGANIZATION (1964) Nomenclature for human immunoglobulins. *Bull. Wld. Hlth. Org.* **30**, 447.

WORLD HEALTH ORGANIZATION (1965) Notation for genetic factors of human immunoglobulins. *Bull. Wld. Hlth. Org.* **33**, 721.

WORLD HEALTH ORGANIZATION (1966) Notation for human immunoglobulin subclasses. *Bull. Wld. Hlth. Org.* **35**, 953.

WOSTMAN B.S. (1959) Serum proteins in germ-free vertebrates. *Ann. N.Y. Acad. Sci.* **78**, 254.

YOUNT W.J., KUNKEL H.G. & LITWIN S.D. (1967) Studies of the Vi (γ2C) subgroup of γ-globulin. A relationship between concentration and genetic type among normal individuals. *J. exp. Med.* **125**, 177.

REFERENCES TO Gm AND Inv GEOGRAPHIC DISTRIBUTION

BOYER S.H., ISEKI S. & MAYEDA A. (1961) Serum gamma globulin in Japanese. *Ann. hum. Genet.* **25**, 69.

BOYER S.H. & WATSON-WILLIAMS E.J. (1961) The γ-globulin, Gm^{ab}, in Nigerians. *Nature* **190**, 456.

BROCTEUR J. (1961) La fréquence du facteur Gm^a en Belgique. *C. r. Soc. Biol.* **1**, 192.

BROMAN B., HEIKEN A. & HIRSCHFELD J. (1963) Gm(a) frequencies in Sweden. *Acta genet.* **13**, 132.

DEICHER H. (1962) Frequenz von Gm-Gruppen in Deutschland. *Klin. Wschr.* **40**, 655.

EYQUEM A., PODLIACHOUK L. & PRESLES J. (1961) Le facteur sérique Gm(a) chez les Africains de Port-Novo (Dahomey). *Vox Sang.* **6**, 120.

GALLANGO M.L. & ARENDS T. (1963) Incidence of *Gm^a* and *Gm^x* genes among Venezuelan Indians and Mestizos. *Hum. Biol.* **35**, 361.

GALLANGO M.L. & ARENDS T. (1964) Incidence of Gm factors among various South American populations. *Proc. int. Congr. Blood Transf.* **10**, 472.

GALLANGO M.L. & ARENDS T. (1965) Frecuencia de los factores Gm(a) y Gm(x) en españoles residentes en Venezuela. *Acta cient. venez.* **16**, 18.

GALLANGO M.L. & ARENDS T. (1965) Inv(a) serum factor in Venezuelan Indians. *Transfusion* **5**, 457.

GILES E., OGAN E. & STEINBERG A.G. (1965) Gamma-globulin factors (Gm and Inv) in New Guinea: anthropological significance. *Science* **150**, 1158.

HERZOG P. & DRODVÁ A. (1961) Inv factor in CSSR. *Vox Sang.* **6**, 636.

HESS M., BÜTLER R. & ROSIN S. (1961) Gm-Gruppen bei 500 Berner Blutspendern. *Vox Sang*, **6**, 366.

IZATT M.M. (1966) Frequency of the Gm(a) factor in Scotland. *Vox Sang.* **11**, 103.

JENKINS T. & STEINBERG A.G. (1966) Some serum protein polymorphisms in Kalahari Bushmen and Bantu: gamma globulins, haptoglobins and transferrins. *Am. J. hum. Genet.* **18**, 399.

KLUGE A. & KRAH E. (1962) Serumeiweissgruppe Gm(a). Frequenzuntersuchungen an 1100 Seren aus dem Raum Heidelberg. *Klin. Wschr.* **40**, 57.

KOBIELA J., TUROWSKA B. & MOSTOWSKI J. (1961) The Hp and Gm groups in Poland. *Vox Sang.* **7**, 638.

López V. & Bütler R. (1965) The Inv groups in Switzerland with remarks on the methods of detection. *Vox Sang.* **10**, 314.

Moullec J., Henry C. & Silverie C. (1958) Le facteur de groupe sérique humain Gm^a chez les Africains. *C. r. Acad. Sci.* **246**, 3701.

Natvig J.B. & Kunkel H.G. (1968) Genetic markers of human immunoglobins. The Gm and Inv systems. *Series Haematologica* 1 **1**, 66.

Neel J.V., Salzano F.M., Junqueira, P.C., Keiter F. ; Maybury-Lewis D. (1964) Studies on the Xavante Indians of the Brazilian Mato Grosso. *Am. J. hum. Genet.* **16**, 52.

Podliachouk L., Angelopoulos G. & Eyquem A. (1961) Le facteur sérique Gm(a) chez les Greco. *Vox Sang.* **6**, 123.

Podliachouk L., Eyquem A., Choaripour R. & Eftekhari M. (1961) Serum factor Gm^a among the Iranians. *Nature* **191**, 717.

Popwassilew J. & Prokop O. (1962) Die Frequenz der Faktoren Gm^a und Gm^x in Bulgarien. *Zschr. ärztl Fortibild.* **56**, 775.

Ritter H. (1963) Formalgenetik und Populationsgenetik der Gamma-globulin-Serumgruppen. *Artz. Forsch.* **17**, 417.

Ritter H., Ropartz C., Rousseau P.Y., Rivat L. & Bähr M.L. (1964 Zur Formalgenetik und Populations-genetik des Gammaglobulin-Polymorphismus InV (Merkmale InV(1) und InV(a)). *Acta genet.* **14**, 15.

Ropartz C. & Lenoir J. (1960) Les antiglobulines humaines présentes dans les sérums de sujets normaux. II. Fréquence et caractéristiques sérologiques. *Rev. hémat.* **15**, 40.

Ropartz C., Rivat L., Rousseau P.Y., Baitsch H. & van Loghem J. (1963) Les systèmes Gm et InV en Europe. *Acta genet.* **13**, 109.

Ropartz C., Rivat L., Rousseau P.Y., Choaripour R. & Eftekhari M. (1962) Répartition des groupes de gammaglobulines Gm et InV chez les Iranians. *Acta genet.* **12**, 45.

Ropartz C., Rivat L., Rousseau P.Y., & Lenoir J. (1961) Les facteurs Gma, Gmb, Gmx, 'Gm-like' et InV chez les Japonais. *Rev. fr. Étud. clin. biol.* **6**, 813.

Ropartz C., Rousseau P.Y. & Rivat L. (1963) Intéret des groupes de γ globuline Gm et Inv dans l'appréciation du metissage des populations: étude de ces groupes sériques dans l'ouest africain et l'extrême-orient. *Rev. fr. Étud. clin. biol.* **8**, 465.

Ropartz C., Rousseau P.Y., Rivat L., Baitsch H., Ritter H., Pinkerton E.J. & Mermod L.E. (1964) Les groupes de γ-globulines Gm et Inv parmi la population d'Honolulu (Hawaii). *Acta genet.* **14**, 25.

Ropartz C., Rousseau P.Y., Rivat L. & Kirk R. (1962) Fréquence du facteur Inv(a) chez les aborigenes de l'Ouest australien. *Rev. hémat.* **2**, 86.

Ropartz C., Rousseau P.Y., Rivat L. & Lenoir J. (1961) Étude génétique du facteur sérique Inv: fréquence dans certaines populationes. *Rev. fr. Étud. clin. biol.* **6**, 374.

Ropartz C., Rousseau P.Y. & Rivat L. (1965) Hypothèses sur la génétique formalle du système Gm chez les Caucasians. *Humangenetik* **1**, 483.

Ropartz C., Walter H., Arndt-Hanser A., Rivat L., Rousseau P.Y. & Bernhard W. (1964) On the frequency of the Gm and Inv serum groups in South Western Germany. *Acta genet.* **14**, 298.

Salzano F.M. & Steinberg A.G. (1965) The Gm and Inv groups of Indians from Santa Catarina, Brazil. *Am. J. hum. Genet.* **17**, 273.

Steinberg A.G. (1962) Progress in the study of genetically determined human gamma globulin types (the Gm and Inv groups), *in* Steinberg A.G. & Bearn A.G. (eds.) *Progress in Medical Genetics*, vol. 2, p. 1. Grune & Stratton, New York and London.

STEINBERG A.G. (1966) Gm and Inv studies of a Hokkaido population: evidence for a *Gm²* allele in the Ainu. *Am. J. hum. Genet.* **18**, 459.

STEINBERG A.G. (1967) Genetic variation in human immunoglobulins. The Gm and Inv types, *in* GREENWALT T.J. (ed.) *Advances in Immunogenetics*, p. 75. Lippincott, Philadelphia.

STEINBERG A.G., LAI L.Y.C., VOS G.H., SINGH R.B. & LIM T.W. (1961) Genetic and population studies of the blood types and serum factors among Indians and Chinese from Malaya. *Am. J. hum. Genet.* **13**, 355.

STEINBERG A.G. & MATSUMOTO H. (1964) Studies on the Gm, Inv, Hp and Tf serum factors of Japanese populations and families. *Hum. Biol.* **36**, 77.

STEINBERG A.G., STAUFFER R., BLUMBERG B.S. & FUDENBERG H.H. (1961) Gm phenotypes and genotypes in U.S. whites and Negroes; in American Indians and Eskimos; in Africans; and in Micronesians. *Am. J. hum. Genet.* **13**, 205.

STEINBERG A.G., STAUFFER R. & BOYER S.H. (1960) Evidence for a *Gm*ᵃᵇ allele in the Gm system of American Negroes. *Nature* **188**, 169.

STEINBERG A.G., STAUFFER R. & FUDENBERG H.H. (1960) Distribution of Gmᵃ and Gm-like among Javanese, Djuka Negroes, and Oyana and Carib Indians. *Nature* **185**, 324.

UENO N. & YOKOYAMA M. (1964) Population study of gamma globulin types in Japanese. *Z. Immun-Forsch.* **127**, 58.

VAN LOGHEM E. & MARTENSSON L. (1967) Genetic (Gm) determinants of the γ2c (Vi) subclass of human IgG immunoglobulins. *Vox Sang.* **13**, 369.

VOS G.H., KIRK R.L. & STEINBERG A.G. (1963) The distribution of the gamma globulin types Gm(a), Gm(b), Gm(x) and Gm-like in South and Southeast Asia and Australia. *Am. J. hum. Genet.* **15**, 44.

YOKOYAMA M. & UENO N. (1961) The Gmᵃ factor in Japanese. *Nature* **190**, 455.

CHAPTER 2

HAPTOGLOBIN

Haptoglobin Physiology 64
  Synthesis 64
  Haptoglobin levels 65
  Catabolism 66
  Physiological role of hapto-
  globin 68
    Retention of iron by HpHb
    complex formation 68
    Prevention of methemalbu-
    min formation 69
    Suggested role in heme cata-
    bolism 70

The HpHb Complex: Hemoglobin
Binding 70
  Relation of hemoglobin con-
  formation to haptoglobin bind-
  ing 71
  The 'intermediate' complex 71

The Structure of Haptoglobin 72
  General features 72
  Haptoglobin glycopeptides 73
  Subunit structure: haptoglobin
  α and β chains 74
  Structure of Hp 1-1 and Hp 2-2   76
  Hemoglobin binding by hapto-
  globin polymers 77
  The binding site of haptoglo-
  bin: immunological and chem-
  ical studies 78

Haptoglobin Genetics 79
  Hp gene duplication and its
  genetic consequences 79
  Haptoglobin variants 84
    Quantitative variants 84
      The Hp 2-1 (mod)
      phenotypes 84
      Other quantitative vari-
      ation of Hp 2-1 85
    Hypohaptoglobinemia
    and anhaptoglobinemia   86

    Family studies 86
    Genetic basis of hy-
    pohaptoglobinemia   87
    An Hp⁰ allele 88
    Hypohaptoglobin-
    emia and the pos-
    sible localization of
    the Hp structural
    gene locus 88
  Qualitative variants 89
    Alpha chain variants 89
    Beta chain variants 91
    Other rare phenotypes,
    possibly β chain vari-
    ants 92

Geographic Distribution of Hp
Genes 93
  Difficulties in data interpre-
  tation 93
  General summary of gene fre-
  quencies shown in Table 1 99
  Gene frequencies determined
  by subtyping 100

Methods 100
  Starch gel electrophoresis 101
    Apparatus: horizontal and
    vertical 101
    Source of starch 102
    Buffer systems 102
    Gel preparations 103
    Sample insertion 104
    Electrophoresis conditions 105
    Gel slicing 105
    Staining 105
  Measurements of haptoglobin
  concentration 106

References 107

References to Haptoglobin Geographic Distribution 118

The presence in serum of a protein having the property of binding hemoglobin was first detected by Polonovski & Jayle (1938), when they observed that the addition of serum to hemoglobin (Hb) had an enhancing effect on its peroxidase activity. Subsequently, these authors (Polonovski & Jayle, 1940) characterized the binding substance as an α-2 glycoprotein and gave it the name of haptoglobin (Hp). The extensive studies of this group of investigators, summarized by Jayle & Moretti (1962), showed that measurement of the serum Hp level, expressed as Hb-binding capacity, had clinical usefulness. Furthermore, they found that Hp is not a single molecular species, but a group of closely related molecules. An explanation for this phenomenon was supplied by Smithies (1955) who described the separation of three distinct Hp patterns by starch gel electrophoresis, and, with Walker (Smithies & Walker, 1956) demonstrated that these patterns represent phenotypic variants in a genetic system. In the following year, Laurell & Nyman (1957) and Allison & ap Rees (1957) reported that since the HpHb complex cannot pass the normal glomerular membrane, haptoglobin is the major determinant of hemoglobin excretion, thus accounting for what was previously considered to be a renal threshold function.

The stimulus of these combined observations is evident in the very large number of subsequent papers dealing with the physiology, biochemistry, inheritance and geographic distribution of the Hp types. One or more of these subjects have been reviewed by Smithies (1959a, 1964), Smithies & Connell (1959), Nyman (1959), Harris *et al.* (1959), Barnicot (1961), Baitsch & Liebrich (1961), Baitsch *et al.* (1964), Galatius-Jensen (1960, 1962), Jayle & Moretti (1962), Laurell & Grönvall (1962), Prokop & Bundschuh (1963), Parker & Bearn (1963), Shultze & Heremans (1966), Javid (1967a) and Giblett (1961, 1968).

HAPTOGLOBIN PHYSIOLOGY

SYNTHESIS

Haptoglobin is probably synthesized in the liver. Although there is no direct evidence in man, several studies of mice, rats, rabbits and dogs support that conclusion (Williams *et al.*, 1963; Krauss & Sarcione, 1964; Mouray *et al.*, 1964; Alper *et al.*, 1965; Peters & Alper, 1966).

The amount of haptoglobin produced is subject to a number of influences. For example, the haptoglobin level rises in diseases with associated inflammation or tissue necrosis (Jayle & Boussier, 1955; Nyman, 1959; Robert *et al.*, 1961; Owen *et al.*, 1964). This rise was at one time considered to be due to a local release of haptoglobin from the diseased tissue itself or from a preformed pool. However, more recent evidence indicates a direct stimulation of haptoglobin synthesis, probably in the liver (Miale & Kent, 1962; Maung *et al.*, 1964; Krauss *et al.*, 1966; Peters & Alper, 1966). The nature of this stimulus is unknown; it involves the increased production of several other plasma proteins (Jayle *et al.*, 1962). Hormonal substances may play an intermediary role; for example, androgens tend to increase, and estrogens to decrease the haptoglobin level (Nyman, 1959; Borglin & Nyman, 1961). Furthermore, Krauss (1963) has shown that adrenalectomy in rats interferes with the increased production of haptoglobin which normally accompanies inflammation. Since acute depletion of haptoglobin by intravenous hemoglobin injection does not stimulate haptoglobin synthesis, there is little evidence for a feedback controlling mechanism (Laurell & Nyman, 1959). However, the data of Noyes & Garby (1967) indicate that with repeated injections of hemoglobin, the rate of haptoglobin synthesis tends to increase.

In liver disease, haptoglobin is either increased or decreased (Nyman 1959; Owen *et al.*, 1959). Low levels may reflect either severe hepatocellular damage or increased red cell destruction, so measurement of haptoglobin in such cases is not in itself adequate to differentiate between intra- and extrahepatic disease. Nevertheless, it can be usefully incorporated as one of a battery of tests in following the course of hepatitis (Hever & Vadesz, 1965; Badr-El-Din *et al.*, 1966).

HAPTOGLOBIN LEVELS

Only 10 to 20 per cent of newborn infants have detectable haptoglobin in their plasma, and comparison of fetal with maternal Hp phenotype indicates that this haptoglobin is fetal, not maternal, in origin (Galatius-Jensen, 1960; Giblett, 1961; Rausen, *et al.*, 1961; Hirschfeld & Lunell, 1962; Siniscalco *et al.*, 1963; Beckman & Grivea, 1964). It was suggested (Siniscalco *et al.*, 1963) that the lowest levels are found in babies whose Hp phenotype is 'incompatible' with that of their mothers. However, the low levels usually observed probably indicate either that there is general impairment of haptoglobin synthesis (possibly due to liver immaturity) or that the decreased lifespan of fetal red cells (Garby *et al.*, 1964) is associated with sufficient intravascular release of hemoglobin to remove haptoglobin. By the age of one year, most infants have haptoglobin levels similar to those of adults (Bergstrand *et al.*, 1961).

The normal level of plasma haptoglobin varies considerably, ranging from about 40 to 180 mg per 100 ml, expressed as Hb-binding capacity (Nyman, 1959; Bayani-Sioson *et al.*, 1962; Shinton *et al.*, 1965; Ferris *et al.*, 1966; Murray *et al.*, 1966). The quantity of hemoglobin bound is related to the Hp genetic type (See p. 78), and a second genetic influence is indicated by the greater correlation in the haptoglobin levels of identical twins than in fraternal twins of the same phenotype (Bayani-Sioson *et al.*, 1962).

In any healthy individual, the haptoglobin level remains fairly constant (Jayle & Boussier, 1955; Nyman, 1959). Thus observation of a marked rise or fall usually has some clinical significance. For example, Venezialc *et al.* (1966), Anderson *et al.* (1966) and Cullhed (1967) have used the haptoglobin level before and after insertion of cardiac prosthesis as an indication of the degree of red cell destruction. Conversely, sickle cell anemia patients (nearly always anhaptoglobinemic) were shown to have a marked rise in haptoglobin level after hypertransfusion with normal red cells, due to suppression of the patients' erythropoiesis (Mehta & Jensen, 1960; Whitten, 1962).

Plasma is not the only body fluid containing haptoglobin. Variable amounts have been detected in lymph, in pleural, synovial, ascitic and spinal fluids, and in the urine (Laurell, 1959; Neuhaus & Sogoian, 1961; Marnay, 1961; Berggard & Bearn, 1962; Ng *et al.*, 1963; Blau *et al.*, 1963; Berggard *et al.*, 1964; Alper *et al.*, 1965; Laurent & Panelius, 1965). The level in urine is partly dependent on Hp phenotype, since only the smaller sized Hp molecules are filtered through the glomerular membrane.

CATABOLISM

Whenever hemoglobin is released into the circulation, it combines immediately with haptoglobin, and the complex is removed by the reticuloendothelial system (Laurell & Nyman, 1957; Franklin *et al.*, 1960; Garby & Obara, 1960; Murray *et al.*, 1961; Keene & Jandl,

1965; Engler *et al.*, 1966 & 1967). The half-life of the complex varies with its concentration (Garby & Noyes, 1959; Noyes & Garby, 1967): the higher the concentration, the longer the half-life. Thus, when normal subjects are given sufficient intravenous hemoglobin to nearly saturate the plasma haptoglobin (i.e. the amount of Hb in about 8 ml of red cells), the resulting HpHb complex is removed at the rate of about 15 mg (of Hb) per 100 ml per hour (Laurell & Nyman, 1957; Faulstick *et al.*, 1962). However, this linear removal does not occur at low, physiological levels of about 0·3 mg per cent (Hanks *et al.*, 1960), where the clearance curve is exponential (Garby & Noyes, 1959). Freeman (1964) found that tracer amounts of isotope-labelled complex had a $T\frac{1}{2}$ of only 9 minutes, consistent with a single passage through the liver.

A normal man catabolizes about 6 g of hemoglobin daily, but only about 10 to 20 per cent passes through the intravascular compartment (Garby & Noyes, 1959; Freeman, 1964). Hemoglobin released from non-viable red cells into the extravascular compartment does not deplete plasma haptoglobin (Gydell, 1960; Langley *et al.*, 1962). Thus, haptoglobin plays only a minor role in normal hemoglobin catabolism, and about half of the haptoglobin produced is catabolized independently, with a $T\frac{1}{2}$ of about 4 to 5 days (Nyman, 1959; Laurell & Grönvall 1962; Moretti *et al.*, 1963; Freeman, 1964; Noyes & Garby, 1967). In patients with elevated Hp, the $T\frac{1}{2}$ may be unchanged (Krauss *et al.*, 1966).

In hemolytic states, there is usually an increase in the amount of hemoglobin presented to the plasma, and thus much or all of the haptoglobin is removed at an accelerated rate. For this reason, anhaptoglobinemia is usually observed in patients with increased red cell turnover, including those with ineffective erythropoiesis where turnover in the marrow is excessive (Allison & ap Rees, 1957; Nyman, 1957; Nosslin & Nyman, 1958; Nyman *et al.*, 1959; Brus & Lewis, 1959; Owen *et al.*, 1960b, 1960c; Herman, 1961b; Reerink-Brongers *et al.*, 1962). However, this is not an invariable finding, so that in patients with hemolytic states whose haptoglobin level is normal or elevated, it is necessary to postulate either that the intravascular release of hemoglobin is not increased or that the rate of haptoglobin production is considerably higher than normal.

Haptoglobin measurement can be helpful in confirming that an acute hemolytic episode has taken place, even when there is no longer any demonstrable free hemoglobin or methemalbumin in the plasma, because the haptoglobin level does not return to its usual level for several days after complete depletion (unless the patient has a disease associated with increased Hp synthesis). However, interpretation of a low haptoglobin level in terms of increased hemolysis is not reliable unless a specimen obtained before the hemolytic episode is available for comparison. Isolated observation of a low haptoglobin level in any patient is virtually meaningless, because the normal range is so wide. The observation of anhaptoglobinemia in an adult Caucasian is usually indicative of increased red cell destruction or liver impairment. However, about 4 per cent of normal adult Negroes and a higher proportion of Negro children have little or no demonstrable serum haptoglobin (Giblett, 1959; Giblett & Steinberg, 1960; Sutton & Karp, 1964).

The liver is the principal site of HpHb removal, although the marrow, and to a lesser extent, the spleen, are capable of some uptake (Jandl *et al.*, 1957; Keene & Jandl, 1965; Garby & Obara, 1960; Murray *et al.*, 1961; Engler *et al.*, 1966). The complex is too large to pass the glomerular membrane, so that renal localization is not observed when there is no free hemoglobin.

PHYSIOLOGICAL ROLE OF HAPTOGLOBIN

RETENTION OF IRON BY HpHb COMPLEX FORMATION
Murray *et al.* (1961) and Keene & Jandl (1965) showed that in animals depleted of haptoglobin, injected hemoglobin is deposited in the same sites as the HpHb complex in normal animals; but in addition, there is rapid uptake of hemoglobin by the kidneys. Thus, as Keene & Jandl pointed out, haptoglobin is not truly a transport protein, first because it does not deliver hemoglobin to specific receptor sites and second, because neither component of the HpHb complex re-enters the blood stream after clearance (Noyes & Garby, 1967). Therefore, it appears that the role of haptoglobin is to prevent hemoglobin loss (and thus, iron loss) by renal excretion, as originally proposed by Allison (1958).

The effectiveness of haptoglobin in conserving iron has been questioned on the basis that only a small amount of hemoglobin is catabolized as the HpHb complex (Nyman, 1959). Another objection lies in the fact that there are additional mechanisms for retaining heme

(Whitten, 1962). For example, when hemoglobin is injected intravenously, the amount must be appreciably in excess of the plasma Hb-binding capacity before hemoglobinuria occurs (Sears *et al.*, 1966). A proportion of the unbound hemoglobin is broken down, and the released heme forms a complex with the β globulin, hemopexin, and also with albumin, to form methemalbumin. Neither of these complexes is excreted, so their heme iron is conserved. For passage through the glomerular membrane, it may be necessary for free hemoglobin to be in a dissociated form (i.e. as $\alpha\beta$ half-molecules), according to Bunn *et al.* (1967). Of this hemoglobin, a variable proportion is retained in the kidneys by tubular reabsorption (Lathem, 1959; Lathem & Jensen, 1962; Lowenstein *et al.*, 1961; Sears *et al.*, 1966). However, it is not clear whether this retained hemoglobin or its breakdown products are transferred back to plasma. In short-term experiments on rabbits, Neustein (1966) found no evidence for the return of hemoglobin products metabolized by tubular epithelial cells. Also, Sears *et al.* (1966) reported the development of iron deficiency due to renal excretion of hemosiderin and ferritin in a few patients with chronic intravascular hemolysis. Thus it appears likely that much of the hemoglobin iron retained by the kidney is subsequently lost in the urine.

PREVENTION OF METHEMALBUMIN FORMATION

Bunn & Jandl (1966) suggested another possible role of haptoglobin in iron conservation. They observed that the free exchange of heme between molecules of methemoglobin (ferrihemoglobin) and oxyhemoglobin did not occur when the methemoglobin was first bound to haptoglobin. The transfer of heme from methemoglobin to albumin (to form methemalbumin) was similarly prevented by haptoglobin binding. The latter observation is in accord with the previous work of Nyman (1960) who showed that methemalbumin formation does not occur unless the amount of hemoglobin in the plasma exceeds the Hb-binding capacity of haptoglobin. Bunn & Jandl (1966) suggested that haptoglobin may help to prevent 'indiscriminate loss of heme and its iron' by blocking the dissociation of heme from methemoglobin. However, the reported exchanges of heme were studied in a non-cellular system, and it seems very unlikely that heme is freely exchanged between the intact red cell and its plasma environment. Thus, it is not clear what advantage is gained by retaining heme in the plasma as the

4

Hp-methemoglobin complex rather than as methemalbumin, when neither substance is excreted by the normal kidney, and both must be metabolized before the heme iron can be re-used.

SUGGESTED ROLE IN HEME CATABOLISM

Another haptoglobin function, the facilitation of heme catabolism, was suggested by Nakajima *et al.* (1963), who observed that the liver enzyme, heme-α-methenyl oxygenase, could catalyse oxidative cleavage of the heme ring only when the Hb molecule was combined with Hp. However, as noted in the previous section, only a fraction of the total catabolized hemoglobin is normally removed as the HpHb complex. Furthermore, even when Hb breakdown is accelerated, binding with haptoglobin has no observable *in vivo* effect on either the conversion of Hb to bile pigments or the rate of re-utilization of hemoglobin iron (Murray *et al.*, 1961; Ostrow *et al.*, 1962; Keene & Jandl, 1965).

THE HpHb COMPLEX:
HEMOGLOBIN BINDING

Haptoglobin readily combines with globin (Laurell, 1960; Laurell & Grönvall, 1962; Rowe & Soothill, 1960; Giblett, 1966) but not with heme (Nyman, 1960) or myoglobin (Javid *et al.*, 1959; Wheby *et al.*, 1960). The precise nature of the linkage is not known, although it is thought to involve electrostatic and van der Waals forces (Nyman, 1959). Covalent linkage is unlikely since blockage of the Hb reactive sulfhydryl groups does not interfere with HpHb binding (Robert *et al.*, 1957; Bunn, 1967). In addition, Rafelson *et al.* (1961) and Blumberg & Warren (1961) reported that neuraminic acid, which comprises about 5 per cent of the haptoglobin molecule, can be removed without loss of Hb-binding ability. Once formed, the complex is very stable, no exchange of bound with unbound Hb being demonstrable (Noyes & Laurell, 1961).

The visible spectra of oxyHb A and oxyHbHp are identical, but some other properties of hemoglobin are considerably altered by combination with haptoglobin. For example, the complex has a much higher oxygen affinity than Hb A, and both the sigmoid shape of the oxygen dissociation curve and the Bohr effect are lost (Nagel *et al.*, 1965b). Nagel & Gibson (1966) showed that the kinetic basis for the high ligand (i.e. oxygen) affinity of the complex is a greatly increased combination rate, and that the velocity constant varies with the Hp genetic type.

RELATION OF HEMOGLOBIN CONFORMATION TO HAPTOGLOBIN BINDING

Deoxyhemoglobin does not combine with haptoglobin (Nagel *et al.*, 1965a). When oxyhemoglobin is bound by haptoglobin, subsequent reduction does not disrupt the HpHb bond, and the conformational change characteristic of deoxyhemoglobin is not observed (Nagel *et al.*, 1965b). However, when hemoglobin is treated with carboxypeptidase A, which removes the last two amino acids from the β chain, both its oxy and deoxy forms are bound by haptoglobin (Chiancone *et al.*, 1966).

The dependence of binding on molecular configuration was also strongly suggested by the studies of Nagel & Ranney (1964). They found that although carboxyhemoglobin, methemoglobin and cyanmethemoglobin, as well as a variety of animal hemoglobins and abnormal hemoglobins of human origin, form a stable complex with haptoglobin, there is no demonstrable binding of the β and γ chain tetramers, Hb H and Hb Barts. Both of these hemoglobins have a molecular configuration and resistance to dissociation resembling that of deoxyhemoglobin (see also p. 360). Similarly, hemoglobin treated with bis-(*N*-maleimidomethyl)ether becomes more resistant to dissociation (Simon & Konigsberg, 1966), and has an impaired ability to form a complex with haptoglobin (Bunn, 1967).

The ability of α and β chain monomers to combine with haptoglobin has been more difficult to determine (Nagel & Ranney, 1964; Chiancone *et al.*, 1966; Smith & Beck, 1966). However, recent studies by Nagel & Gibson (1967) suggest that the haptoglobin site (discussed on p. 79) binds α globin chains specifically, and that the normal binding of hemoglobin involves either the consecutive binding of α and β monomers or attachment of $\alpha\beta$ dimers through the α chain. (See Addenda.)

THE 'INTERMEDIATE' COMPLEX

When Hp of type 1-1 is only partly saturated with hemoglobin, two complexes are formed, one of which was called 'intermediate' by Laurell (1959) because of its electrophoretic migration rate and its disappearance upon the addition of further hemoglobin (see Fig. 2.1). Laurell suggested that the intermediate compound might consist of one molecule of haptoglobin and one half-molecule of hemoglobin. This interpretation is apparently correct, since Shim *et al.* (1965) have shown that its smaller molecular size is in the range expected of such a complex, and Giblett (1966) demonstrated that haptoglobin can combine with single $\alpha\beta$ dimers prepared from hemoglobin. A number of investigators have reported that in the saturated complex of hemoglobin with haptoglobin of type 1-1, the molecular ratio is 1:1 (Laurell, 1959; Jayle & Moretti, 1962; Shim & Bearn, 1964a; Korngold, 1965). The binding ratio of hemoglobin with the other genetic types of haptoglobin is more difficult to determine, as will be seen in the next section. Not only do these haptoglobins exist as multiple components, but each component forms more than one intermediate complex when only partially saturated (Laurell & Grönvall, 1962; Robson *et al.*, 1964).

THE STRUCTURE OF HAPTOGLOBIN

GENERAL FEATURES

From the early studies of Jayle and his associates, summarized by Jayle & Moretti (1962), haptoglobin was characterized as an α-2 glycoprotein with a carbohydrate content of about 20 per cent, consisting of hexoses, glucosamine, fucose and sialic acid. Two different molecular species, I and II, separable on the basis of ammonium sulfate solubility, were isolated and studied by these investigators. This heterogeneity was confirmed when Smithies (1955) described three different Hp phenotypic patterns demonstrable by starch gel electrophoresis of serum or plasma. Smithies & Walker (1956) then showed that the inheritance of these Hp types is determined by two allelic autosomal genes, called Hp^1 and Hp^2; the resultant phenotypes were named Hp 1-1, Hp 2-2 and Hp 2-1. They are shown in Fig. 2.2. Several subsequent studies on large numbers of families have confirmed these original observations of inheritance (Galatius-Jensen, 1958a,b; Harris *et al.*, 1959; Mäkelä *et al.*, 1959; Prokop *et al.*, 1961; Fleischer & Mohr, 1962; Matsunaga *et al.*, 1962; Baitsch *et al.*, 1964; Goedde *et al.*, 1965; Ritter & Hinkelmann, 1966).

Connell & Smithies (1959) isolated and purified the three kinds of

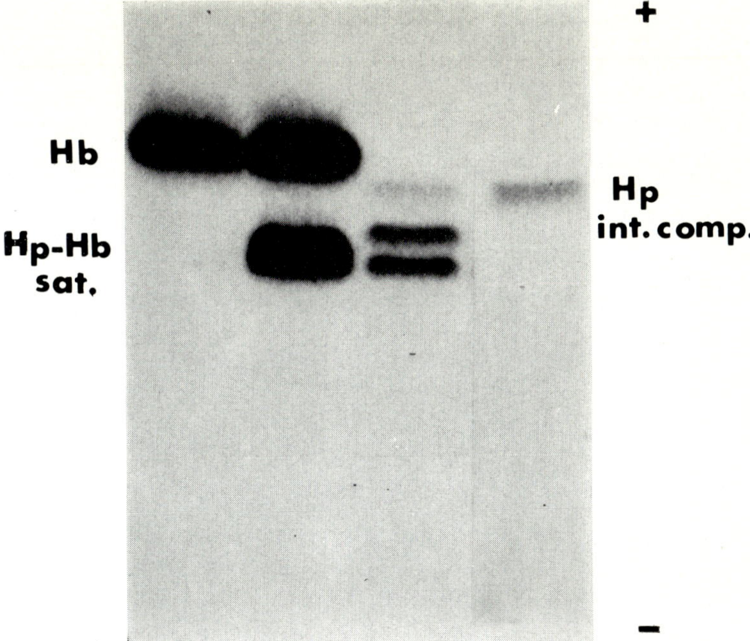

FIGURE 2.1. Starch gel electrophoretic patterns of Hb A (left), Hp 1-1 (right), and their complexes (middle). When hemoglobin is in excess (second from left), a single, saturated complex is formed; it migrates more slowly than hemoglobin. When haptoglobin is in excess (second from right), two complexes are formed: the saturated complex and a faster-moving, intermediate complex (int. comp.).

Hp, and showed that the multiple bands characteristic of Hp 2-1 and Hp 2-2 were not artifacts, even though their presence appeared to contradict the genetic doctrine of one gene, one polypeptide. Bearn & Franklin (1958, 1959) also demonstrated the molecular heterogeneity by ultracentrifugation, and pointed out that in mixtures of serum containing the two homozyogous types ,1-1 and 2-2, the resultant pattern was not the same as that of the Hp 2-1 heterozygote. Smithies & Connell (1959), Allison (1959) and Parker & Bearn (1963) suggested that the Hp 2-2 phenotype represents a series of stable polymers containing the Hp^2 gene product, while the Hp 2-1 pheno-

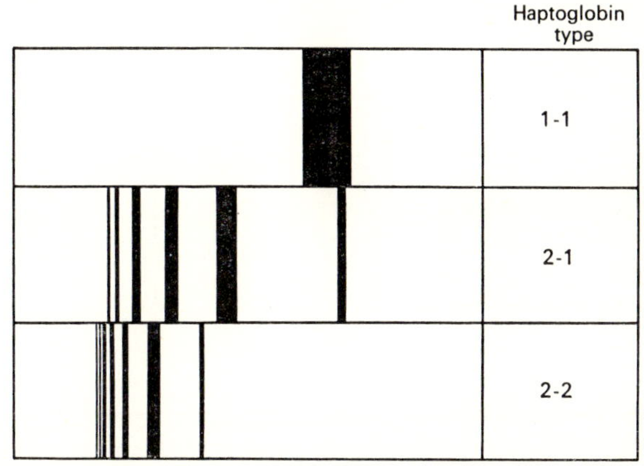

FIGURE 2.2. Diagram of starch gel electrophoretic patterns of the three common haptoglobin types. The origin is on the left, and migration is from left to right.

type represents a similar polymeric series, each member of which contains products of both Hp^2 and Hp^1 genes, except for the fastest-migrating band, which does not have any Hp^2 gene product. Javid (1964) provided the confirmatory evidence from immunological and biochemical cleavage studies of sephadex-separated fractions.

HAPTOGLOBIN GLYCOPEPTIDES

In studies on the haptoglobin carbohydrate moiety, pronase digestion was shown to cleave the molecule into glycopeptides of 200–300 molecular weight, which were separable into two major classes, on the basis of amino acid content. In each type of glycopeptide, the carbohydrate composition was very similar: 5–6 moles of hexose (galactose and mannose), 3–4 moles of N-acetylglucosamine and 0–3 moles of N-acetylneuraminic acid, with fucose present in some units but not in others (Cheftel *et al.*, 1965; Gerbeck *et al.*, 1967). No qualitative differences associated with Hp genetic type were observed.

SUBUNIT STRUCTURE: HAPTOGLOBIN α AND β CHAINS

Smithies, Connell & Dixon (1962a, b, 1966) and Connell *et al.* (1962, 1966) succeeded in breaking down the haptoglobin molecules into

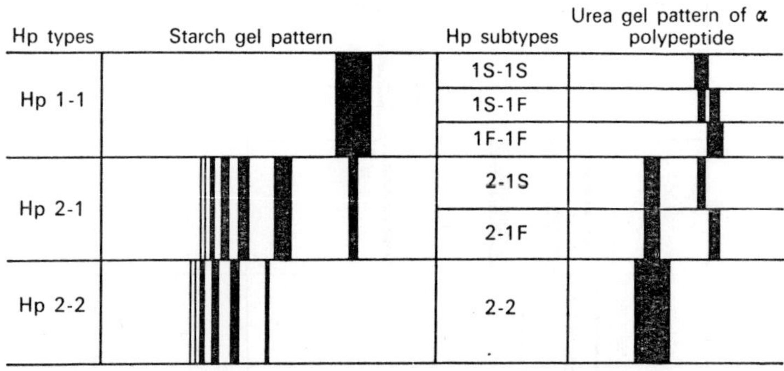

Hp types	Starch gel pattern	Hp subtypes	Urea gel pattern of α polypeptide
Hp 1-1		1S-1S	
		1S-1F	
		1F-1F	
Hp 2-1		2-1S	
		2-1F	
Hp 2-2		2-2	

FIGURE 2.3. Diagram of the common Hp subtype patterns as revealed by reductive cleavage and electrophoresis in acid-urea gel. Only the alpha polypeptide chains are shown: α^{1F}, α^{1S} and α^{2}.

their component polypeptides by reductive cleavage with 8 M urea and mercaptoethanol, followed by electrophoresis in acid-urea starch gel. This procedure revealed that haptoglobin consists of two kinds of polypeptide chains, called α and β, and that variations in the structure of the α chain are responsible for the three different electrophoretic patterns of the intact molecules. Thus, Hp 2-2 contains hp α^{2} chains, Hp 1-1 contains hp α^{1} chains, and Hp 2-1 contains both hp α^{2} and hp α^{1} chains, while all three of these common Hp types appear to contain the same hp β chains. (The nomenclature used here was recently suggested by Black & Dixon, 1968). There are two alternative hp α^{1} chains; the faster-migrating is called hp α^{1F} and the slower, hp α^{1S}. Their corresponding genes, Hp^{1F} and Hp^{1S}, are alleles of Hp^{2}

(Smithies *et al.*, 1962a; Goedde *et al.*, 1965), but their products cannot be differentiated unless the protein is reduced and cleaved. As shown in Fig. 2.3, this procedure permits the subdivision of Hp 2-1 into two subtypes, Hp 2-1F and Hp 2-1S, and Hp 1-1 into three subtypes, Hp 1F-1F, 1S-1S and 1F-1S.

Smithies *et al.* (1962b) and Connell *et al.* (1966) found that the molecular weight of hp α^{1F} and hp α^{1S} is 8900 ± 400; Black & Dixon (1968) reported a chain length of 83 amino acids. The two polypeptides

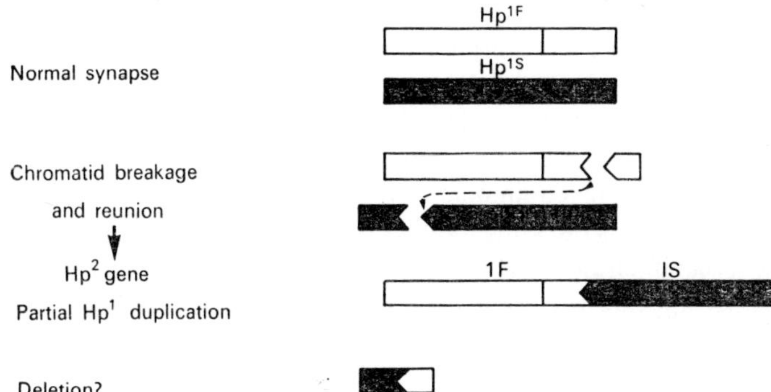

FIGURE 2.4. Diagram of a possible mechanism for Hp^2 gene formation by breakage of Hp^{1S} and Hp^{1F} at two widely separated regions, followed by reunion. This event would be purely random, because it would not be dependent upon homology of the base sequences in the two regions. A second, 'deletion' gene might also be formed by this mechanism as well as by the unequal crossing-over mechanism shown in Figure 2.5. (The vertical line in Hp^{1F} represents the site of base substitution which differentiates it from Hp^{1S}.)

differ by a single amino acid substitution: lysine in hp α^{1F} for glutamic acid in hp α^{1S} (Smithies *et al.*, 1962b), at position 54 (Black & Dixon, 1968). This substitution, representing a one-step mutation, is responsible for an alteration in the peptide fingerprints of the two hp α^1 chains, resulting in peptide spots called F and S, respectively. In the peptide fingerprint of hp α^2, both of these peptides are present, as well as a characteristic peptide given the name J. Furthermore, the α^2 chain has a molecular weight of 16,000, nearly twice that of α^{1F} or α^{1S}. Sequence analysis of its 142 amino acids by Dr. Gordon Dixon, reported in part by Smithies (1964) and in full by Black & Dixon (1968) showed that hp α^2 contains all but a short segment of both hp α^1 chains, and that the J peptide represents the junction of the C-terminal portion of one α^1 chain at position 70 or 71 with the N-terminal portion of the other α^1 chain at position 12 or 13 (see Fig. 2.4). Thus, the Hp^2 gene product, hp α^2, can also be called hp α^{1F1S} (or α^{1S1F}) since it contains the sequence of amino acids expected from a fusion of the Hp^{1F} and Hp^{1S} genes, except for a short deletion at the site of their junction. The genetic event responsible for this partial duplication of Hp^1 is discussed on page 79.

The β chain of haptoglobin has not been subjected to fine structural analysis, although the peptide maps and total amino acid content of isolated β chains from the three common types were shown by Cleve et al. (1967) to be very similar, if not identical. According to Smith et al. (1962), the N-terminal amino acids of haptoglobin are valine and isoleucine, in equimolar amounts. Thus, in the haptoglobin molecule, there is the same number of α and β chains. Since Smithies et al. (1962b) showed that valine is the N-terminal residue of the α chain, it appears that isoleucine occupies a similar position in the β chain.

STRUCTURE OF Hp 1-1 AND Hp 2-2

Shim and Bearn (1964a) estimated that Hp 1-1 consists (by weight) of about 20 per cent α and 80 per cent β chains, the latter containing the carbohydrate moiety (as does the heavy chain of the γ globulins). Since the molecular weight of Hp 1-1 is about 85,000 (Jayle & Boussier, 1955), and the α chain molecular weight is 9100, Shim & Bearn (1964a) deduced that the Hp 1-1 molecular formula is probably $\alpha_2^1\beta_2$, and that the molecular weight of the β chain is about 36,000 (Shim et al., 1965). A similar estimate of 40,000 was reported by Cheftel & Moretti (1966), and Waks et al. (1967) gave a figure of $42,000 \pm 3000$ for the combined weight of haptoglobin α and β chains in $\alpha\beta$ half-molecules they obtained from Hp 1-1 and 2-2 by reversible dissociation at pH 9·5 to 11·5.

In a study of Hp 2-2, Smithies (1965) demonstrated that reversible depolymerization could be induced by a disulfide interchange reaction. The final product, still capable of binding hemoglobin, was shown to be the basic unit of α and β chains, i.e. the 'monomer' from which the Hp 2-2 polymers are constructed. Its electrophoretic migration rate in starch gel and its sedimentation coefficient were found to resemble those of the intact Hp 1-1 molecule, which does not form a polymeric series. If the Hp 2-2 'monomer' has the molecular formula $\alpha_2^2 \beta_2$, then one would expect that its molecular weight would exceed that of Hp 1-1 by about 14,000 (i.e. twice the difference in molecular weight between hp α^1 and hp α^2). The fastest-moving member of the Hp 2-2 polymeric series may represent a doubling of the basic 'monomer', the next, a tripling, and so forth (Marinis & Ott, 1965). Since there are at least ten or twelve members in the series, the molecular weight of the largest, slowest-moving components is probably well over a million.

Black & Dixon (1968) pointed out that there are structural and functional similarities between haptoglobin and immunoglobulin molecules: both consist of two heavy and two light chains joined by disulfide bonds, and both have the property of bivalent combination with specific sites on other proteins. An extensive computer analysis by these authors compared the codons of sequential amino acids in hp α^1 and several Bence Jones proteins of both K and L types. Their data showed that in certain regions, there is as much similarity between the haptoglobin sequence and any given Bence Jones protein as there is between different Bence Jones proteins. Their conclusion that the two kinds of peptide chains probably have a common evolutionary origin seems very likely to be correct.

An interesting incidental finding in their analysis of the hp α^1 amino acid sequence was the presence, near a cysteine residue at position 21, of three prolines in a row, at positions 17, 18 and 19. This tripeptide was previously reported adjacent to a cysteine in the 'hinge' region of rabbit IgG heavy chain by Smyth & Utsumi (1967), who stated that the Pro-Pro-Pro sequence had not been previously found among proteins of known structure. However, since this tripeptide does not appear to exist in the human IgG 'hinge region' (see p. 18), there is no reason to assume from current evidence that the hp α^1 chain might have an even greater resemblance to the IgG heavy chain than it has to the light chains.

HEMOGLOBIN BINDING BY HAPTOGLOBIN POLYMERS

As mentioned previously, the Hp 1-1 protein binds hemoglobin with a one-to-one molecular ratio, when fully saturated. There is conflicting evidence about the molecular ratio of hemoglobin with the Hp 2-2 (and Hp 2-1) components. The studies reported by Jayle & Moretti (1962) and by Cloarec & Moretti (1966) indicated stoichiometric binding, regardless of phenotype. However, Javid (1965) has shown that the larger polymers bind less hemoglobin, weight for weight, than do the smaller polymers. Since it is customary to express the haptoglobin level on the basis of Hb-binding capacity rather than by direct measurement, falsely low estimates of protein concentration would thereby be obtained for Hp 2-2 and Hp 2-1 protein, as compared with Hp 1-1. Numerous studies have shown that the haptoglobin level, expressed as Hb-binding capacity, is generally higher in individuals with Hp 1-1 than in those with Hp 2-2 and Hp 2-1 (Allison & ap Rees, 1957; Nyman, 1958, 1959; Smith & Owen, 1961; Kahlich-Koenner & Weippl, 1961; Planas *et al.*, 1965; Ferris *et al.*, 1966). However, as noted by Javid (1965) these levels may not reflect a difference in the amount (by weight) of Hp protein present. Some observations made in the author's laboratory tend to support Javid's assumption. However, contradictory evidence was presented by Kluthe *et al.* (1965), who found that there was very good correlation between Hb-binding (expressed as peroxidase activity) and haptoglobin protein weight (measured immunologically), independent of the Hp phenotype.

THE BINDING SITE OF HAPTOGLOBIN: IMMUNOLOGICAL AND CHEMICAL STUDIES

Some informative data about the three-dimensional structure of Hp and the HpHb complex have been derived from immunological studies. For example, Korngold (1963) reported that antisera prepared in rabbits against Hp 2-2 could be used to distinguish Hp 2-2 from Hp 1-1 by immunodiffusion in agar. He also described some unusual immunological variants, but in a later paper, Korngold (1965) showed that the presence of hemoglobin blocks one or more of the binding sites of native haptoglobin, and suggested that the apparent variants may have been due simply to contamination with hemoglobin. Subsequently, Beuing *et al.* (1965) and Eichmann *et al.* (1966) reported the demonstration of at least three haptoglobin antigenic determinants. The first, called A, can be shown in all three genetic types, regardless of the presence or absence of hemoglobin. The second, called B, is also found in all three types. However, the addition of hemoglobin to Hp 1-1 completely blocks this determinant, and partially blocks it on the Hp 2-1 and Hp 2-2 molecules. The third, called C, is characteristic of Hp 2-1 and 2-2, but not of 1-1, so it is a property of polymerized haptoglobin or, more specifically, of hp 2α polypeptide. Partial blockage of this determinant is also achieved by prior addition of hemoglobin.

According to Shim *et al.* (1965), an antiserum prepared in rabbits by injecting purified β chains of human haptoglobin reacted only with free haptoglobin, and not with the HpHb complex, while anti-α chain antiserum reacted with both. They therefore concluded that hemoglobin is bound by the haptoglobin β polypeptide chain, and that the bound hemoglobin covers the antigenic determinant characteristic of the hp β, but not the hp α chain. Confirmatory evidence that the hp β chain probably contains the Hb-binding site was presented by Gordon & Bearn (1966), who observed an appreciable difference in the protein profile of the hp β chain preparation eluted from a G-100 Sephadex column before and after mixing it with hemoglobin. In addition, Dobryszcha & Lisowska (1966) showed that a tryptic fragment containing the carbohydrate moiety of haptoglobin (and thus from the β chain) could bind hemoglobin. Finally, the unusual hemoglobin-binding behavior of the rare hp β chain variants (see p. 91) further supports the location of the binding site on the haptoglobin β chain. Thus it appears that in the HpHb complex, the α chains of hemoglobin are bound to the β chains of haptoglobin. (See Addenda.)

HAPTOGLOBIN GENETICS

Hp GENE DUPLICATION AND ITS GENETIC CONSEQUENCES

It was pointed out in a previous paragraph that there are three common alleles at the Hp locus: Hp^{1F}, Hp^{1S} and Hp^2. Smithies *et al.* (1962b) showed that Hp^{1F} and Hp^{1S} differ from each other at only one site. But from the structure of the hp α^2 polypeptide, they deduced that its corresponding allele, Hp^2, represents a partial duplication of Hp^1 genes arising during meiosis in an individual heterozygous for Hp^{1F} and Hp^{1S}. Because the junction of the two α^1 chains occurs at a region of apparently very limited homology, Smithies (1964) suggested that the underlying genetic mechanism was a unique and purely random event, involving chromosomal breakage at non-homologous points, followed by repair (see Fig. 2.4, p. 75).

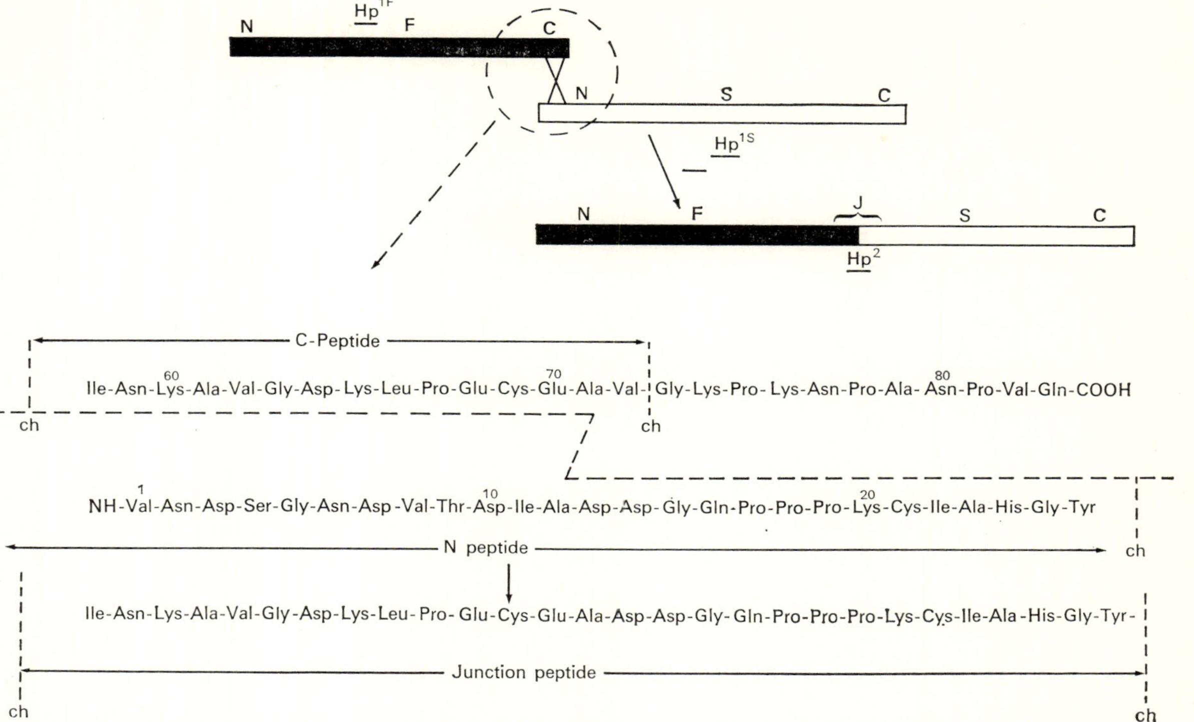

FIGURE 2.5. Diagram of non-homologous crossing-over between two regions of Hp^1 with very little apparent homology. The letters N, C, F, S and J refer to regions of the Hp genes corresponding to the peptides in the α^{1F}, α^{1S} and α^2 chains. The amino acid sequences of the C-terminal and N-terminal peptides and of the J peptide resulting from their junction are given at the bottom. The interrupted line denotes the site of crossover and those amino acids thereby included in the J peptide of hp α^2. Sites of chymotryptic action are indicated by Ch. (Adapted from Dixon, 1966 and Black & Dixon,

Smithies *et al.* (1962b) had previously suggested that the gene duplication was the result of unequal crossing-over between Hp^{1F} and Hp^{1S}. The possibility of such a mechanism, shown in Fig. 2.5, was reconsidered by Black & Dixon (1968). They examined the base sequences corresponding to the amino acids in positions 9–17 of the N-terminal portion, and 67–75 of the C-terminal portion of the α^1 chains. (Since the single site of amino acid substitution occurs at position 54, the N- and C-terminal regions are the same in both

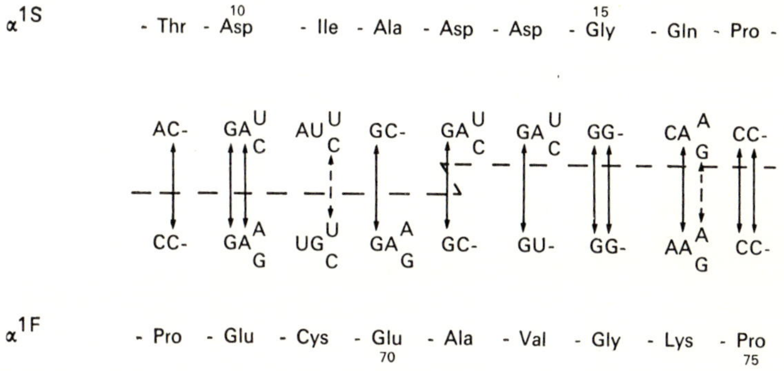

FIGURE 2.6. Diagram of the amino acids and RNA codons corresponding to positions 9–17 of the α^{1S} polypeptide and 67–75 of the α^{1F} polypeptide (or vice versa; the two chains are identical except for position 54). Complete identity of two bases is indicated by a full arrow, possible identity by a broken arrow. (For a brief description of the genetic code, see Chapter 17; for definitions, see Glossary.) The interrupted line shows the position of crossing-over. (Adapted from Black & Dixon, 1968. Reprinted with permission from *Nature*, **218**, 736).

α^{1F} and α^{1S}.) Considering only the 18 unambiguous positions, that is, the first two bases in each of the nine corresponding codons, Black & Dixon found that there were 11 identities, as shown in Fig. 2.6. They pointed out that from the presence of six guanines and three cytosines, one would expect that strong hydrogen bonding could occur between the base sequence coding for positions 9–17 and the DNA strand complementary to that coding for positions 67 to 75 (or vice versa). Thus, there could be misplaced synapsis between these two regions, and unequal crossing-over could account for the origin of the Hp^2 gene.

Both of the proposed mechanisms might produce, in addition to the duplicated gene, a small chromosomal segment or 'deletion gene' represented by the amino acids in positions 1-12 and 73-83. If such a gene were able to produce a peptide, it would almost certainly not combine with hemoglobin, and therefore it would be thought to represent a 'silent' allele at the Hp locus.

The fact that the two halves of hp α^2 are identical, except for the amino acid differences characteristic of α^{1F} and α^{1S}, suggests that the partial gene duplication responsible for its production is a fairly recent evolutionary event (Black & Dixon, 1968). Moreover, the present prevalence of the Hp^2 gene in many human populations is presumably indicative of a selective advantage.

Nakajima *et al.* (1963) found that Hp 2-2 bound to hemoglobin is more effective than Hp 1-1 or Hp 2-1 in bringing about the oxidative cleavage of heme by the enzyme, heme α-methenyl oxygenase. However, since haptoglobin does not appear to facilitate hemoglobin degradation *in vivo* (Keene & Jandl, 1965) a selective advantage on this basis is questionable. A possible association of Hp^1 and the development of leukemia has been reported by Peacock (1966), but this observation requires further documentation.

Another possible disadvantage of Hp^1 is the fact that Hp 1-1 molecules can pass the glomerular membrane more readily than the Hp 2-1 and 2-2 polymers (Berggaard & Bearn, 1962). Thus, individuals of type 1-1 might be less capable of retaining, as HpHb, the iron from red cells subject to rapid destruction. While it is true that there is a high frequency of Hp^2 in most populations of Asia, where hemolytic diseases are prevalent, the highest Hp^1 gene frequencies occur in Africa, where hemolytic diseases have been at least as common as in any other area of the world.

Smithies *et al.* (1962b) pointed out that once the Hp^1 gene duplication occurred, there would then be a good opportunity for unequal crossing-over to occur at the Hp locus. For example, since the Hp^2 gene is nearly twice as long as the Hp^1 gene, synapsis cannot be perfect in an individual heterozygous for Hp^2 (i.e. Hp^{2FS}) and Hp^1 (i.e. Hp^{1F} or Hp^{1S}). Thus, the shorter gene is probably aligned with one or the other of the two nearly homologous segments of the Hp^2 gene. Consequently, as shown in Fig. 2.7, crossing-over, homologous but always unequal, would lead to the production of new kinds of Hp^2 genes: Hp^{2FF}, Hp^{2SS} and Hp^{2SF} (the latter is not shown in the figure). The frequencies of such genes would be determined by the frequency of crossing-over as well as by environmental selective factors. By subtyping Hp2-1 from numerous serum donors, Nance & Smithies (1963) found hp α^2 chains with migration rates indicating their

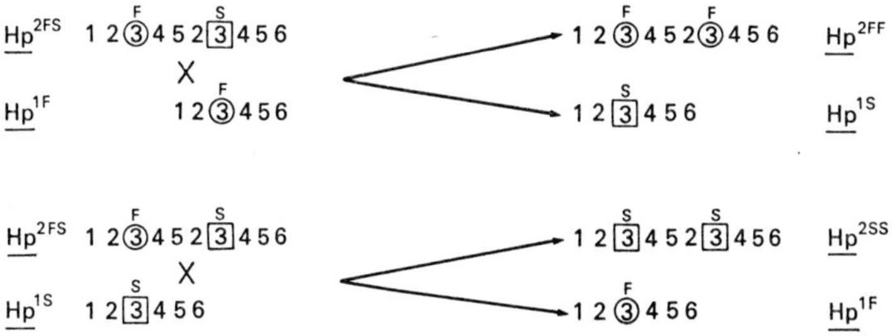

$$Hp^{2FS} \quad 1\ 2\ \text{③}\ 4\ 5\ 2\ \boxed{3}\ 4\ 5\ 6 \quad\longrightarrow\quad 1\ 2\ \text{③}\ 4\ 5\ 2\ \text{③}\ 4\ 5\ 6 \quad Hp^{2FF}$$
$$\times$$
$$Hp^{1F} \qquad 1\ 2\ \text{③}\ 4\ 5\ 6 \qquad\qquad 1\ 2\ \boxed{3}\ 4\ 5\ 6 \quad Hp^{1S}$$

$$Hp^{2FS} \quad 1\ 2\ \text{③}\ 4\ 5\ 2\ \boxed{3}\ 4\ 5\ 6 \quad\longrightarrow\quad 1\ 2\ \boxed{3}\ 4\ 5\ 2\ \boxed{3}\ 4\ 5\ 6 \quad Hp^{2SS}$$
$$\times$$
$$Hp^{1S} \quad 1\ 2\ \boxed{3}\ 4\ 5\ 6 \qquad\qquad 1\ 2\ \text{③}\ 4\ 5\ 6 \quad Hp^{1F}$$

FIGURE 2.7. Diagram of unequal crossing-over in two Hp 2-1 heterozygotes, producing two new Hp^2 alleles, Hp^{2FF} and Hp^{2SS}. Each number represents a segment of DNA in a haptoglobin gene. A single base-pair substitution in segment 3 is shown as ③ in Hp^{1F} and $\boxed{3}$ in Hp^{1S}. (From Giblett, 1968.)

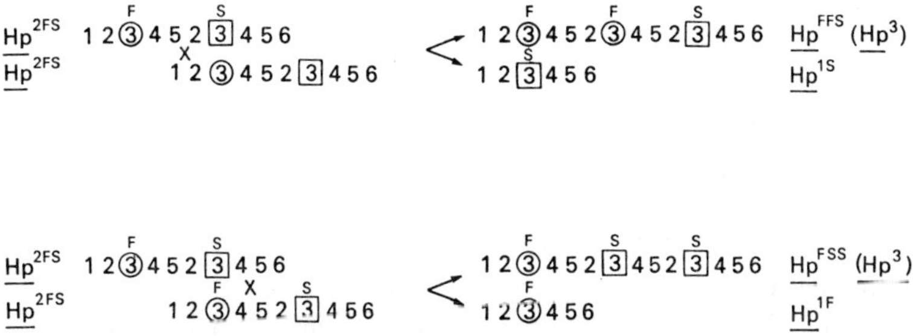

$$Hp^{2FS} \quad 1\ 2\ \text{③}\ 4\ 5\ 2\ \boxed{3}\ 4\ 5\ 6 \qquad 1\ 2\ \text{③}\ 4\ 5\ 2\ \text{③}\ 4\ 5\ 2\ \boxed{3}\ 4\ 5\ 6 \quad Hp^{FFS}\ (Hp^3)$$
$$Hp^{2FS} \qquad 1\ 2\ \text{③}\ 4\ 5\ 2\ \boxed{3}\ 4\ 5\ 6 \qquad\qquad 1\ 2\ \boxed{3}\ 4\ 5\ 6 \quad Hp^{1S}$$

$$Hp^{2FS} \quad 1\ 2\ \text{③}\ 4\ 5\ 2\ \boxed{3}\ 4\ 5\ 6 \qquad 1\ 2\ \text{③}\ 4\ 5\ 2\ \boxed{3}\ 4\ 5\ 2\ \boxed{3}\ 4\ 5\ 6 \quad Hp^{FSS}\ (Hp^3)$$
$$Hp^{2FS} \qquad 1\ 2\ \text{③}\ 4\ 5\ 2\ \boxed{3}\ 4\ 5\ 6 \qquad\qquad 1\ 2\ \text{③}\ 4\ 5\ 6 \quad Hp^{1F}$$

FIGURE 2.8. Diagram of partial triplication, probably exemplified by the Hp^J gene, here indicated as Hp^3. See legend of Fig. 2.7 for explanation of symbols. The two examples of Hp^3 shown here vary slightly because of a difference in the site of crossing-over.

probable identities as hp 2FF and 2SS. However, the limited data available show that their respective genes have low fequencies (Shim & Bearn, 1964b).

Another opportunity for homologous but unequal crossing-over would arise from displaced synapse in the Hp^2 homozygote (Smithies $et\ al.$, 1962b). As shown in Fig. 2.8, crossing-over occurring after displaced synapse could produce a partial triplication, such as FFS. A probable example of such triplication is the so-called Johnson Hp. On subtyping, it contains, in addition to the usual hp α^{1S} chain, a second, highly aggregated chain with electrophoretic mobility much slower than that of hp α^2 (see Fig. 2.9). Smithies *et al.* (1962b) noted that several different triplet combinations of α^{1S} and α^{1F} could be produced, depending upon the crossover site; two examples are given in Fig. 2.8. Hp Johnson variants have, indeed, been found (see p. 90).

HAPTOGLOBIN VARIANTS

A large number of variant Hp phenotypes have been described, but since their polypeptide chains have not yet been subjected to fine structural analysis, the actual basis of their individual alterations is still a matter of conjecture. Nevertheless, it is convenient to divide them into two categories depending on whether their phenotypic patterns show a change in relative quantity of the usual Hp 2-1 and Hp 2-2 components (i.e. quantitative variants) or an actual alteration in electrophoretic migration indicative of structural change (i.e. qualitative variants).

QUANTITATIVE VARIANTS

The Hp 2-1(mod) *phenotypes*
The quantitative variants consist largely of modifications in the Hp 2-1 phenotype. As shown in Fig. 2.10, there is a spectrum of patterns, ranging from a heavy concentration of the slower-moving polymers with no visible representation of the fastest-moving (Hp 1) band, to a heavy concentration of the fastest-moving band, with only one accompanying member of the polymeric series. The types shown on the right are collectively known as Hp 2-1(mod) or Hp 2-1M (Connell & Smithies, 1959). They are found in about 10 per cent of American Negroes (Giblett, 1959) and only occasionally in individuals with no African ancestry.

Harris *et al.* (1960) described a Caucasian family having this phenotype and, on the basis of its mode of inheritance, suggested that the modification was due either to (1) an allele, Hp^{2M} at the Hp locus, which produced Hp 2-1(mod) when partnered with Hp^1 but, when partnered with Hp^2, produced a phenotypic pattern indistinguishable

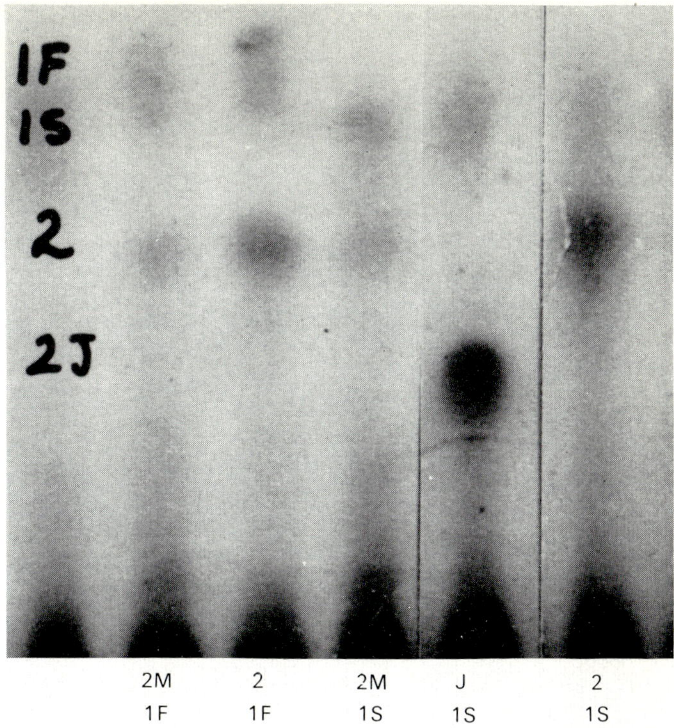

FIGURE 2.9. Subtype patterns of Hp 2-1, Hp 2-1 (mod) and Hp 1-J on acid-urea starch gel. The alpha chains are labelled IF, IS, 2 and 2J. The appearance of the hp β chain, partially shown at the bottom of the picture, does not vary in these phenotypes.

[*Facing page* 84]

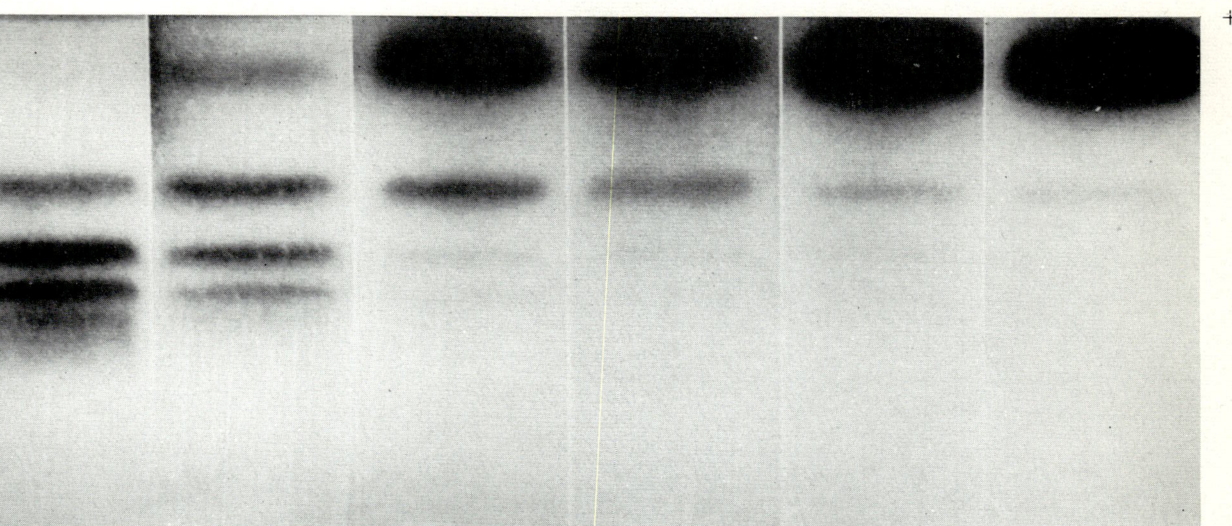

FIGURE 2.10. A spectrum of Hp 2-1 and Hp 2-1 (mod) patterns on starch gel. The first two patterns represent Hp 2-1, and the last four, Hp 2-1 (mod). There is a progressive loss of slow-moving components and increase in the fastest-moving component from left to right.

[*Facing page* 85]

from the usual Hp 2-2; or (2) a modifying gene at another locus altering only the heterozygous phenotype.

Giblett & Steinberg (1960) reported their findings in 491 members of 92 Negro families, indicating that the first hypothesis of Harris *et al.*—that is, an allele Hp^{2M} at the Hp locus—appeared to be the correct explanation for the Hp 2-1(mod) variant. Smithies & Connell (1960) originally thought that their subtyping procedure showed a slight difference in electrophoretic mobility of the α polypeptide products of Hp^2 and Hp^{2M}. However, they subsequently reported that the hp α² preparation was made from the plasma of two people, one of whom had the unusual hp 2SS α chain produced by the Hp^{2SS} gene discussed on p. 83 (Connell *et al.*, 1966). Thus, the characteristic subtype pattern of the Hp 2-1(mod) variants contains either hp $α^{1F}$ or hp $α^{1S}$ and an α polypeptide indistinguishable from hp α² except for its fainter stain (see Fig. 2.9).

Sutton & Karp (1964) divided Hp 2-1(mod) into four classes, b, c, d and e, on the basis of the extent of the shift in concentration toward the fastest-moving bands. From studies of the 579 members of 92 Negro families, they concluded that the designation Hp^{2M} includes at least two different alleles, Hp^{2cd} and Hp^{2e}, producing different amounts of hp α² polypeptide.

Thus there may be a series of Hp structural isoalleles whose peptide products differ from each other only in quantity. An alternative explanation based on controller genes was offered by Parker & Bearn (1963). While the model they originally proposed could not be reconciled with the family data of either Giblett & Steinberg (1960), Sutton & Karp (1964) or Higashi & Lubs (1966), it seems likely that some kind of quantitative regulation exists, either as a part of the Hp^2 gene itself, or closely linked with it.

Other quantitative variation of Hp 2-1

Evidence for similar variability in amount of the Hp^1 gene product lies in the Carlberg phenotype shown in Fig. 2.11, which was originally described by Galatius-Jensen (1958b), and subsequently observed in very low frequency in various populations (Harris *et al.*, 1959; Giblett, 1964; Nance & Smithies, 1964). The pattern of Hp Carlberg (Hp Ca) resembles a mixture of Hp 2-2 and Hp 2-1 in variable proportions. On subtyping, the Hp^1 gene product is reduced in quantity, relative to the Hp^2 gene product, although this ratio varies, even within the same family (Nance & Smithies, 1964). Sutton (1965) proposed that Hp Ca may occur as the result of genetic mosaicism, involving the production of Hp 2-2 by one cell population, and of Hp 2-1 by the other population. Indeed, deGrouchy *et al.* (1964) found the Hp Ca type in an individual known to have cellular mosaicism on the basis of double fertilization. However, since Hp Ca is also an inherited phenomenon, one would have to postulate an inherited tendency toward cellular somatic mosaicism in families with Hp Ca.

Gilbett (1964) described two other quantitative variants, Hp 2-1 (Trans) and Hp 2-1(Haw) (see Fig. 2.11). The first of these variants appears to be transitional between the phenotypes Hp 2-1 and Hp 2-1(mod), while Hp 2-1 (Haw) has the appearance of a mixture of Hp 2-1 and Hp 1-1. It was first observed in a serum specimen from random blood donors in Hawaii, and its subtype resembled that of Hp 2-1(mod). Subsequently, the same pattern was found in the serum of a boy with generalized cellular mosaicism (again, due to double fertilization) described by Corey *et al.* (1967).

Hypohaptoglobinemia and
anhaptoglobinemia
Most perplexing of the quantitative variants is Hp 0, representing hypo- or anhaptoglobinemia. Since it is difficult to detect haptoglobin by electrophoresis if the hemoglobin binding capacity is less than 15–20 mg per 100 ml, the term 'anhaptoglobinemia' merely designates that the level of any haptoglobin present is too low to be detected by the method used. The absence or near absence of haptoglobin is a common feature in patients with increased red cell destruction, so that the genetically determined phenotype is often masked in such cases.

Allison *et al.* (1958) found no demonstrable haptoglobin in about a third of the Nigerians they tested, and similar high frequencies of anhaptoglobinemia have been observed in certain other African tribal groups (Allison & Barnicot, 1960; Blumberg & Gentile, 1961; Giblett *et al.*, 1966a). However, malaria is endemic in these areas, as well as other diseases known to be associated with hemolysis and/or hepatocellular damage.

Family studies. Among American Negroes, Hp 0 is observed in about 3–5 per cent of apparently normal adults and in about 12–15 per cent of the children (excluding infants, who commonly lack Hp; see p. 65). Furthermore, children of matings with one or both parents

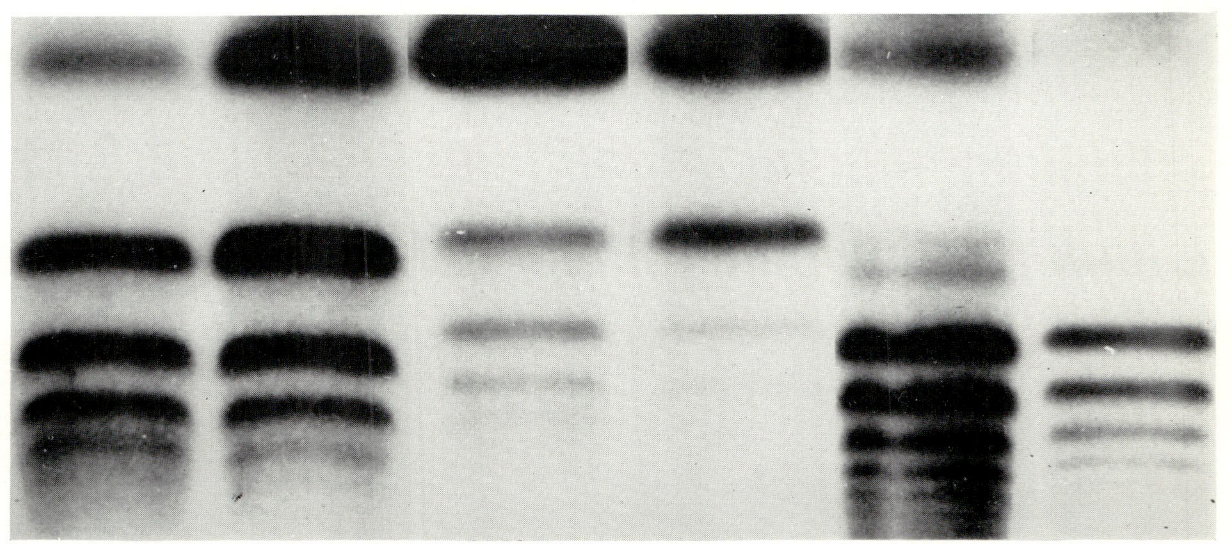

FIGURE 2.11. Starch gel electrophoretic patterns of Hp 2-1 'quantitative' variants compared with Hp 2-2 and Hp Ca, which contains components of both Hp 2-1 Hp 2-2.

Hp 2-1(mod) or Hp o are much more likely to have Hp o than children of other matings, particularly Hp 1-1 × Hp 1-1 (Giblett & Steinberg, 1960; Sutton & Karp, 1964). The genotype Hp^{2M}/Hp^2, when expressed, cannot be differentiated from that of Hp^2/Hp^2. In matings of Hp 2-1(mod) × Hp 2-1(mod), the offspring have either Hp1-1, Hp 2-1 (mod) or Hp o phenotypes. The phenotypic expression of Hp^{2M}/Hp^{2M} is therefore usually, and perhaps always, Hp o. The data appear to indicate that Negroes with no demonstrable Hp usually have the genotypes Hp^{2M}/Hp^1 or Hp^{2M}/Hp^2 (Giblett & Steinberg, 1960) and, less commonly, Hp^2/Hp^2 or Hp^2/Hp^1 (Gottlieb *et al.*, 1963; Sutton & Karp, 1964). The effect of age on the expression of Hp genotype may have a humoral origin, but there is no direct proof for this assumption.

The data of Mehta & Jensen (1960) indicated that sickle trait (i.e. heterozygosity for Hb_β^S) might be a factor responsible for increased anhaptoglobinemia among American Negroes. In the family data collected by Giblett & Steinberg (1960), there was no increase in the incidence of anhaptoglobinemia among adults with sickle trait. However, there was a moderate increase among their children with sickle trait (unpublished observation), indicating the possibility of a combined effect of age and Hb SA. Such a combination may have been responsible for the observations of Blumberg *et al.* (1964), who found a greater tendency to anhaptoglobinemia in Greek children with Hb SA than those with Hb A. Other factors which in combination may depress haptoglobin levels include female sex (Barnicot *et al.*, 1960), pregnancy (Neel *et al.*, 1961; Giblett *et al.*, 1966a) and G6PD deficiency (Giblett *et al.*, 1966a).

Genetic basis of hypohaptoglobinemia. Parker & Bearn (1963) proposed that Negro populations carry some mutant controller genes which suppress haptoglobin formation, and the limited data of Murray *et al.* (1966) suggest a similar low frequency mutant among Caucasians, which is expressed mainly as a recessive trait. Although there is as yet no answer to the problem of genetically influenced anhaptoglobinemia, it seems likely that the same factors discussed above in considering Hp 2-1(mod) and Hp Ca apply to Hp o. In other words, the quantitative expression of haptoglobin may be dependent on controller elements which are either intrinsic to the structural gene locus itself or closely linked with it. Shim & Lee (1966) suggested that too little attention has been paid to defects in β chain production in considering the genetic basis for anhaptoglobinemia. However, it is difficult to accept this explanation for the Hp 0 observed in Negroes. The subtype of Hp 2-1(mod) clearly implicates a defect at some stage of hp α^2 chain production. In order to explain the strong familial associations of Hp 2-1(mod) with Hp 0, Shim & Lee had to assume that the Hp 2-1(mod) phenotype represents heterozygosity for operator genes closely linked to both α and β structural gene loci. If such were the case, one would expect Negroes to have a fairly high frequency of the Hp 2-1(Haw) phenotype, which Shim & Lee ascribe to heterozygosity at the β operator locus. However, no instance of Hp 2-1(Haw) has been observed among the thousands of Negroes tested.

An Hp^0 allele. Evidence for a very rare amorph at the Hp α chain locus is available in the pedigrees studied by Harris *et al.* (1958), Matsunaga (1962), Cann & van West (1966) and Schwerd & Sander (1967). In these families (Italian, Japanese, Danish and German in origin), a parent of Hp 1-1 phenotype has one or more children of type Hp 2-2, or vice versa. Subtyping (unpublished) of the families reported by Harris *et al.* and Cann & van West indicates the strong likelihood of a true Hp^0 allele with no demonstrable gene product. When partnered with Hp^0, the Hp^1 or Hp^2 allele on the homologous chromosome is sometimes not expressed (i.e. the phenotype is Hp 0). However, in other cases, the phenotype has the same appearance as that of the respective homozygote patterns, Hp 1-1 and Hp 2-2, in spite of the single parental contribution. Thus, it would be desirable to revise Hp nomenclature so that the phenotypes are named Hp 1, Hp 2 and Hp 2-1, in order to avoid the ambiguity of implied double parental contribution. Thanks largely to the influence of Dr Harry Harris and his colleagues, the terminology used for phenotypes in several other serum and enzyme groups expresses only the observed component (e.g. Tf C, PGM 1, acid phosphatase B) rather than the assumed contribution of both parental genes. The 'silent' genes in the various red cell antigen (blood group) systems present a similar complication in phenotypic designation (see Chapter 9).

Hypohaptoglobinemia and the possible localization of the Hp structural gene locus. Gerald *et al.* (1964, 1967) also described a family with the mother Hp 1-1, father Hp 2-2 and child Hp 1-1. However, in this case, the child had multiple congenital malformations and a karyotype containing a ring chromosome of the 'D' group (13–15). These findings suggested that if non-paternity was not a factor (the blood groups showed no evidence for this), then either the child had inherited the very rare Hp^0 gene from his father or else the deleted portion of the ring 'D' chromosome, possibly the end of the short arm (Bloom *et al.*, 1966) contained the Hp locus. Recently, Bloom *et al.* (1967) have supplied additional evidence for assigning the Hp locus to a D chromosome, identified as chromosome 13. A second unrelated child with a ring D chromosome was found to have no detectable Hp, while his father had the Carlberg phenotype with very little hp α^1 chain production. The authors postulated that the maternal Hp gene was lost during formation of the ring chromosome and that the child possessed only the paternal variant Hp α^1 gene. deGrouchy *et al.* (1966) described a family in which a father and three of his children had a short arm D-group chromosome deletion. The father had no demonstrable haptoglobin, but one of the affected children was typed as Hp 2-1, so that particular D chromosome does not carry the Hp locus on its short arm. Bias & Migeon (1967) described a pedigree in which there were 14 persons in three generations with deletion of the short arm of a late-replicating D chromosome, presumably number 13, since the replication pattern was similar to that in the case reported by Gerald *et al.* (1964). However, in four of their pedigree members with the deleted short arm, the Hp type was 2-1. Thus, although there is very suggestive evidence that the Hp α chain locus is on chromosome 13, further family studies are needed for confirmation and for localization to the long or short arm.

QUALITATIVE VARIANTS

Alpha chain variants
The qualitative variants (i.e. those which contain components not present in the common phenotypes) are very rare. They are shown diagrammatically in Fig. 2.12. Best known is Hp Johnson (or Hp 1-J) in which the subtype reveals one or the other of the hp 1F or 1S α polypeptides and a much slower-moving, heavily stained α chain thought to be the result of partial gene triplication (see Fig. 2.8). This peptide chain was originally named hp 2J α (Smithies *et al.*, 1962a), but a better name would be either hp α^J or hp α^3, particularly if its triplicate nature is eventually confirmed. When its gene is partnered with Hp^2 the haptoglobin level is usually so low that a clearcut phenotype is very difficult to discern, even when the serum is concentrated (Ramot

et al., 1962). However, in one specimen from a family studied in the author's laboratory, the pattern was sufficiently clear that it could be differentiated from either Hp 2-2 or Hp 1-J, appearing as shown diagrammatically in Fig. 2.12. Dr. Elizabeth Robson (personal communication) has identified a similar phenotype.

Isolated examples of 1-J phenotypes have been found in the families of an American Negro, a Kurdish Jew, an Australian aborigine, a Hawaiian Chinese and several individuals of northern and southern

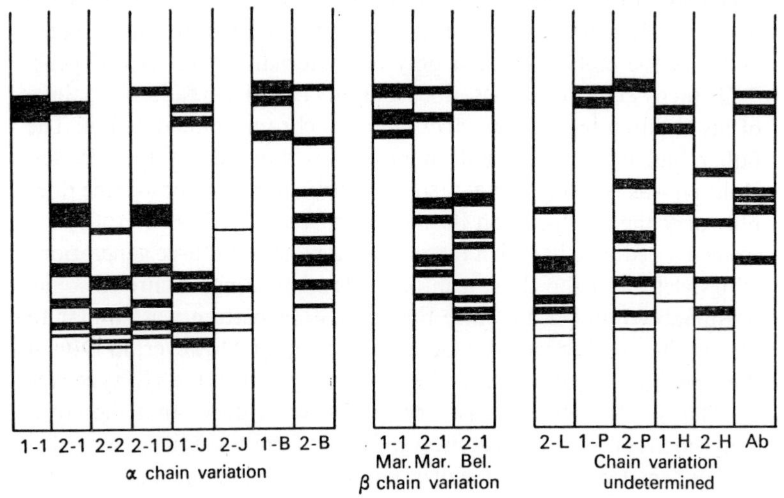

FIGURE 2.12. Diagram of the starch gel electrophoretic patterns of hapto-globins with structural variations in the hp α and hp β chains. In the six phenotypes shown on the right, the source of the variation has not been determined, although some are thought to be β chain variants.

European origin (summarized by Smithies *et al.*, 1962a; also Cooper *et al.*, 1965). The fact that their electrophoretic patterns have not been identical is consistent with the postulated origin of the Johnson-type genes from unequal but homologous crossing-over at different sites in Hp^2/Hp^2 homozygotes, as outlined by Smithies *et al.* (1962a).

The other rare Hp variants shown in Fig. 2.12 are more difficult to explain, although they obviously are due to structural alterations. The Ba types, 1-B and 2-B, appear to represent the combination of Hp^1 and Hp^2 genes, respectively, with Hp^B, another allele at the Hp locus. Its α polypeptide, hp α^B, migrates between hp α^2 and hp α^{1S}

(Giblett *et al.*, 1966). However the nature of the rare mutation responsible for this Hp^B gene in a Caucasian family is completely unknown, and further studies are required to determine whether it is the β chain, rather than the α chain, which is altered. Hp 2-1D (Renwick & Marshall, 1966) is indistinguishable from Hp 2-1 when combined with Hb, but in its absence, the fastest-moving band has increased mobility. Subtyping reveals the common hp α^2 band accompanied by a second band, called hp α^{1D}, with slightly faster mobility than hp α^{1F}. These findings could be the result of single amino acid substitution.

Beta chain variants

Two phenotypes which resemble the Ba types are Hp 2-1 Mb and 1-1 Mb (for Marburg) originally found in a German family by Aly *et al.* (1962) and studied extensively by Cleve & Deicher (1965) and Weerts *et al.* (1966). These authors reported that all of the electrophoretic components of Hp 2-1 Mb demonstrated an atypical immunological reaction, in that antigenic determinant B (see p. 78), normally blocked by hemoglobin, still reacted with its specific antibody when the Hp was fully complexed with Hb. Furthermore, although on subtyping, the purified Hp contained no unusual α polypeptide components, there was a slight difference in the appearance of the β chain. Cleve & Deicher (1965) concluded that these properties of the Marburg types were consistent with mutation at the locus determining the β, rather than the α chain of Hp. Recent studies by Bowman & Cleve (1967) have demonstrated that the peptide fingerprint of the Marburg β chain is, indeed, altered.

Javid (1967b) studied another phenotype, Hp 2-1 Bellevue, which showed the same kind of immunological behavior as Hp 2-1 Mb, but its electrophoretic pattern was distinctly different (see Fig. 2.12). On subtyping in urea gel, the α chains had the mobility of hp α^2 and hp α^{1F}, but in the hp β region, there was, in addition to the usual diffuse band, a faster-moving component. Treatment of the Hp prior to reductive cleavage with a mixture of glycosidases prepared from *Clostridium perfringens* resulted in faster migration of both β components toward the cathode, due to removal of negatively charged sugar residues. It thus appeared that the observed phenotype reflected a change in the protein moiety which was due to heterozygosity at the β chain locus. The Negro male propositus had three sons, all of whom had greatly reduced levels of haptoglobin. The (faint) phenotypic pattern of one could not be distinguished from Hp 1-1, while the Hp bands of the other two children resembled those of Hp 2-2 but were faster-moving. The haptoglobin in all three sons had the same immunological behavior as Hp 2-1 Bellevue so it was inferred that the children had inherited their father's aberrant β chain gene. (The mother's phenotype was Hp 2-1.) Since the mobility of the anodal (Hp 1) band was not altered in either the father's Hp 2-1 Bellevue or one son's Hp 1-1 Bellevue phenotype, it was suggested that the altered mobility of the polymer bands (of Hp 2-1 Bellevue and Hp 2-2 Bellevue) reflected interference with their regular polymerization rather than a difference in electrical charge. Another possibility, that the aberrant β chain might be unable to take part in the formation of the unpolymerized (Hp 1) protein, seems quite unlikely from the normal amount of that protein present in Hp 2-1 Bellevue.

Javid (1967b) showed that the loci determining haptoglobin α and β chain structure are not closely linked, and suggested that the locus involving hp β chain synthesis be named Bp for the (globin) peptide-binding characteristic of hp β. The common allele would then be Bp^A, and the mutant allele Bp^B. This nomenclature may be adopted, but a preferable alternative would be to change the first Hp locus name to Hp_α, analogous to the Hb_α locus, thus permitting the second locus to have the more descriptive name of Hp_β.

Other rare phenotypes, possibly β chain variants
Robson *et al.* (1964) described five rare phenotypes found in three different families, which they named Hp 1-P, 2-P, 2-H and 2-L (see Fig. 2.12). Phenotypes 1-P and 2-P resembled Hp 1-1 and 2-1, respectively, except that their individual components had a faster migration rate, and some of them appeared to be doubled. Hp 1-H and 2-H were very similar to Hp 2-1 and 2-2, with the exception of an additional band characteristic of each type, migrating in positions between the two fastest bands of Hp 2-1. Hp 2-L superficially resembled Hp 2-2 but showed broader bands with increased mobility.

All five variants were detected in their Hb-complexed forms. When no Hb was added, 1-H (and probably, 2-H) could be differentiated from the common phenotypes, but 1-P, 2-P and 2-L could not be distinguished from Hp 1-1, 2-1 and 2-2, respectively. These findings suggested some kind of anomaly in the manner of their complex formation with Hb. A further indication of this anomaly was provided by the fact that no intermediate zones of Hp 2-P were observed when the protein was only partially saturated with Hb.

Subtyping of these variants proved unsuccessful, in that the only α polypeptides observed had the same migration rates as the common hp 1F, 1S and 2 α chains. Nevertheless, it seemed likely at the time that the variant phenotypes were due to mutation of the Hp^1 and Hp^2 genes which could not be demonstrated in their peptide products by electrophoresis. Similar inability to find new components was reported by Giblett (1964), on examining another rare variant, Hp Ab, sent from Boston by Dr. Umansky (Fig. 2.12). A possible explanation for these failures was offered by Smithies & Nance (1964), namely, that the very low pH in the acid-urea gel used for subtyping would suppress ionization of carboxylic side chains, thus preventing a change in migration rate. Alternatively, a substitution involving equally charged or neutral amino acids might be reflected only in the tertiary folding of the molecule. Shim *et al.* (1965) and Shim & Lee (1966) proposed as another explanation, that the P and L variants represent additional mutations at the β chain structural locus. Further biochemical analysis is required to resolve this problem.

GEOGRAPHIC DISTRIBUTION OF
Hp GENES

A very large number of papers have been written about the Hp gene frequencies in various populations throughout the world. Planas (1963) provided a nearly complete summary of the data available at that time, and the purpose of Table 2.1 is to indicate trends and to provide newer data, rather than repeat the entire list in Planas' paper. The references are listed separately at the end of the chapter.

DIFFICULTIES IN DATA INTERPRETATION

There are two major problems in interpreting the data. First, in areas consisting of fairly small isolated populations, there is usually considerable heterogeneity, which may reflect either genetic drift or local selective factors. This problem is common to all population genetics studies, and, in the absence of clear evidence favoring selective influences, one is inclined to ascribe such differences to drift. The second problem, which is peculiar to the Hp system, is failure of gene

TABLE 2.1. Geographic distribution of Hp^1 gene

Population	No.	Hp^1	Reference
Europe			
Norway	1000	0·36	Fleischer & Lundevall, 1957
Sweden	1003	0·38	Tarukoski, 1959
Sweden (Lapps)	329	0·32	Beckman & Mellbin, 1959
Iceland	188	0·39	Walter & Pálsson, 1962
Iceland	400	0·42	Beckman & Johannsson, 1967
Iceland	193	0·52	Pálsson & Walter, 1967
Denmark	2050	0·40	Galatius-Jensen, 1958
Finland	891	0·36	Mäkelä *et al.*, 1959
Holland (in Surinam)	476	0·42	Joe *et al.*, 1965
France (N.E.)	1900	0·43	Michon *et al.*, 1962
France	406	0·40	Moullec & Fine, 1959
Belgium	416	0·43	van Sande *et al.*, 1963
Germany	1467	0·40	Wichman & Schleyer, 1961
Germany	33000	0·38	Baitsch *et al.*, 1962
Austria	621	0·40	Baitsch *et al.*, 1960
Switzerland	920	0·40	Bütler *et al.*, 1959
England	218	0·41	Allison *et al.*, 1958
Scotland	100	0·36	Kamel *et al.*, 1963
Italy (No.)	119	0·41	Harris *et al.*, 1959
Italy (No.)	595	0·38	Benerecetti *et al.*, 1964
Italy (So.)	788	0·37	LaTorretta *et al.*, 1962
Italy (So.)	877	0·36	Bernini *et al.*, 1963
Italy (So.)	752	0·32	Modiano *et al.*, 1965
Sicily	107	0·40	Harris *et al.*, 1959
Sicily	384	0·35	Bernini *et al.*, 1963
Sardinia	147	0·37	Harris *et al.*, 1959
Sardinia	440	0·40	Bernini *et al.*, 1963
Spain	2763	0·40	Planas *et al.*, 1966
Spain (Basques)	107	0·37	Allison *et al.*, 1958
Spain (Basques)	314	0·43	Planas *et al.*, 1966
Portugal	1000	0·39	Torrinha, 1967
Portugal	838	0·39	Barros, 1963
Greece	798	0·36	Baitsch *et al.*, 1962
Greece	1311	0·35	Glowatski & Paidoussis, 1964
Greece	2026	0·34	Angelopoulos *et al.*, 1966
Cyprus	197	0·22	Plato *et al.*, 1964
Poland	151	0·36	Murawski & Miszczak, 1961
Poland	3809	0·38	Kobiela *et al.*, 1961

TABLE 2.1. Geographic distribution of Hp^1 gene (continued)

Population	No.	Hp^1	Reference
Europe (continued)			
Czechoslovakia	1720	0·37	Baitsch *et al.*, 1962
Jugoslavia	459	0·37	Grünwald & Herman, 1964
Australia (European)	322	0·38	Kirk *et al.*, 1960a
USA (European)	409	0·38	Giblett & Brooks, 1963
Tristan da Cunha	215	0·45	Harris & Robson, 1963
Gypsies (Sweden)	115	0·12	Beckman *et al.*, 1965
Africa (North)			
Liberia*	614	0·72	Sutton *et al.*, 1959
Liberia	356	0·70	Neel *et al.*, 1961
Nigeria: Yoruba*	99	0·87	Allison *et al.*, 1958
Fulani*	111	0·76	Barnicot *et al.*, 1960
Fulani	84	0·73	Blumberg & Gentile, 1961
Habe*	120	0·60	Barnicot *et al.*, 1960
Ibo*	70	0·49	Harris *et al.*, 1959
Ibadan	63	0·73	Shim & Bearn, 1964
Senegal	398	0·63	Moullec *et al.*, 1960
Gambia*	157	0·70	Harris *et al.*, 1959
Ethiopia*	312	0·40	Barnicot *et al.*, 1962
Africa (East)			
Uganda*	165	0·63	Allison & Barnicot, 1960
Tanganyika	60	0·46	Allison & Barnicot, 1960
Kenya*	50	0·48	Allison & Barnicot, 1960
Africa (Central)			
Congo: Metropolitan	151	0·77	Sonnet & Michaux, 1960
Tutsi	86	0·52	van Sande *et al.*, 1963
Hutu	96	0·52	van Sande *et al.*, 1963
Barundi	182	0·52	van Sande *et al.*, 1963
Pygmies*	125	0·40	Giblett *et al.*, 1966
Mixed tribes*	660	0·59	Giblett *et al.*, 1966
Africa (South)			
Zulu	113	0·53	Barnicot *et al.*, 1959
Zulu	116	0·53	Jenkins & Steinberg, 1966
Hottentot	59	0·51	Barnicot *et al.*, 1959
Ngalagadi	55	0·53	Barnicot *et al.*, 1959
Xhosa	265	0·53	Giblett *et al.*, 1966
Msutu	218	0·54	Giblett *et al.*, 1966
Cape Colored	88	0·47	Barnicot *et al.*, 1959
Bushmen	113	0·29	Barnicot *et al.*, 1959
Bushmen	125	0·31	Jenkins & Steinberg, 1966

TABLE 2.1. Geographic distribution of Hp^1 gene (continued)

Population	No.	Hp^1	Reference
North American Negroes	1657	0·55	Giblett & Brooks, 1963
Non-Negro African			
Canary Islanders	343	0·42	Fusté *et al.*, 1965
Egyptians	219	0·21	Hashem *et al.*, 1966
Jewish: North African	104	0·29	Ramot *et al.*, 1961; 1962
Iraqi	118	0·29	Ramot *et al.*, 1961; 1962
Askenazi	170	0·34	Ramot *et al.*, 1961; 1962
Askenazi	669	0·30	Fried *et al.*, 1963
Oriental	706	0·28	Fried *et al.*, 1963
Asia			
China: Hong Kong	122	0·39	Sanford *et al.*, 1966
Taiwan	299	0·40	Fraser *et al.*, 1965
Taiwan	172	0·28	Blackwell *et al.*, 1962
China: Malaya	103	0·28	Kirk *et al.*, 1960b
Malaya	90	0·29	Steinberg *et al.*, 1961
Hawaii	124	0·37	Baitsch *et al.*, 1962
New York	112	0·34	Parker & Bearn, 1961a
Thailand	682	0·24	Blackwell & Thephusdin, 1963
Thailand	472	0·23	Kirk & Lai, 1961
Malaya	236	0·23	Kirk & Lai, 1961
Malaya	266	0·24	Lie-Injo *et al.*, 1967
Ceylon	462	0·14	Kirk *et al.*, 1962c
Korea	120	0·32	Shim & Bearn, 1964
Japan	349	0·24	Matsunaga & Morai, 1960
Japan	822	0·28	Steinberg & Matsumoto, 1964
Japan	498	0·27	Baitsch *et al.*, 1962
Japan	170	0·23	Shim & Bearn, 1964
South India: Tamils	291	0·09	Kirk *et al.*, 1962c
Todas	89	0·35	Kirk *et al.*, 1962c
Korumbas	49	0·18	Kirk *et al.*, 1962c
Irulas	74	0·07	Kirk *et al.*, 1962c
Marathas	145	0·14	Baxi & Hakim, 1966
North India: Oraons	125	0·15	Kirk *et al.*, 1962b
Pakistan	392	0·21	Kirk & Lai, 1961
Iran: Moslem	429	0·28	Bowman, 1964
Zoroastrian	145	0·19	Bowman, 1964
Ghashghai	117	0·33	Bowman, 1964

TABLE 2.1. Geographic distribution of Hp^1 gene (continued)

Population	No.	Hp^1	Reference
Greenland			
Eskimos (west)	444	0·35	Persson, 1962
Eskimos (west)	74	0·30	Galatius-Jensen, 1960
Eskimos (east)	737	0·50	Persson & Tingsgaard, 1966
North America			
Alaska: Eskimos	418	0·30	Blumberg et al., 1959
Eskimos	220	0·32	Scott, 1966
Aleuts	64	0·54	Scott, 1966
Athabascan	104	0·37	Scott, 1966
Athabascan	202	0·42	Blumberg et al., 1959
Indian	284	0·43	Blumberg et al., 1959
Arizona Navajo	263	0·45	Parker & Bearn, 1961b
New Mexico Apache	98	0·59	Sutton et al., 1959
Idaho & Wyoming			
Nez Perce	180	0·48	Flory, 1963
Alabama Coushatta	143	0·37	Shim & Bearn, 1964
Mexico (several tribes)	711	0·33 to 0·70	Matson et al., 1963
Mexico Lacandon	89	0·92	Sutton et al., 1960
Central America			
Guatemala (several tribes)	555	0·61	Matson et al., 1963
British Honduras Mayan	212	0·66	Matson et al., 1965
Honduras (several tribes)	428	0·49	Matson et al., 1963
Nicaragua (several tribes)	423	0·50 to 0·71	Matson et al., 1963
Costa Rica (4 tribes)	139	0·20 to 0·52	Matson et al., 1965
Panama: Cuna	174	0·38	Matson et al., 1965
Choco	74	0·45	Matson et al., 1965
South America			
Columbia: Ica	114	0·56	Gallango & Arends, 1966
Paez	103	0·73	Gallango & Arends, 1966
Venezuela (10 tribes)	1276	0·21 to 0·90	Arends & Gallango, 1962; 1964
Venezuela (hybrids)	208	0·55	Galango & Arends, 1959
British Guiana (3 tribes)	325	0·47 to 0·69	Arends & Gallango, 1965
Surinam (hybrids)	253	0·67	Joe et al., 1965
(hybrids)	181	0·70	Joe et al., 1965
Ecuador Quechua	192	0·78	Matson et al., 1965
Peru (7 tribes)	661	0·44 to 0·69	Matson et al., 1966a
(2 tribes)	173	0·73	Giblett & Best, 1961

TABLE 2.1. Geographic distribution of Hp^1 gene (continued)

Population	No.	Hp^1	Reference
South America (continued)			
Bolivia Aymara	71	0·70	Matson *et al.*, 1966b
Brazil: Xavante	78	0·46	Neel *et al.*, 1964
Xavante	440	0·48	Shreffler & Steinberg, 1967
Caingang	326	0·74	Salzano & Sutton, 1963
Others	87	0·52	Salzano & Sutton, 1965
Chile: Pehuenches	113	0·72	Nagel & Etcheverry, 1963
Mapuche	116	0·78	Nagel & Etcheverry, 1963
Alacaluf	43	0·48	Matson *et al.*, 1967
Atacameños	79	0·67	Matson *et al.*, 1967
Pacific Islands			
Australia aborigines			
North Queensland	100	0·18	Flory, 1964
North Queensland	493	0·17	Kirk *et al.*, 1962a
Carpenteria Gulf	136	0·24	Curtain *et al.*, 1966
Central area	331	0·20	Nicholls *et al.*, 1965
New Guinea	82	0·66	Bennett *et al.*, 1961
New Guinea (5 tribes)	1866	0·66 to 0·75	Curtain *et al.*, 1965
New Britain Is.	821	0·73	Curtain *et al.*, 1965
New Hebrides Is.*	199	0·73	Douglas *et al.*, 1964
Solomon Is.*	183	0·64	Douglas *et al.*, 1962
Philippines	403	0·39	Fraser *et al.*, 1964
Philippines	293	0·38	Blackwell *et al.*, 1964
Guam	193	0·37	Plato & Cruz, 1967
Saipan	142	0·37	Plato *et al.*, 1966
Marshall Is.	176	0·58	Blumberg & Gentile, 1961
Gilbert Is.	236	0·45	Douglas *et al.*, 1961
Truk	283	0·41	Plato & Cruz, 1966
Yap	73	0·29	Plato & Cruz, 1966
Ellice Is.	108	0·50	Douglas *et al.*, 1961
Samoa	80	0·59	Douglas & Staveley, 1960
Tonga	200	0·60	Douglas & Staveley, 1960
Hawaii	131	0·69	Baitsch *et al.*, 1962
Hawaii	75	0·68	Beckman *et al.*, 1964
Easter Is. (mixed)	123	0·80	Nagel *et al.*, 1964
('pure' poly-nesian)	36	0·86	Nagel *et al.*, 1964

* Hp o phenotype greater than 10 per cent

expression (in this case, so-called anhaptoglobinemia) due to a number of causes. Although there is evidence supporting the existence of some genetic mechanisms for quantitative variation, the resulting effect on Hp^1 and Hp^2 gene frequencies within a given population is difficult to determine. In addition, there are physiological factors such as hemolytic or hepatic disease which are associated with depressed or absent Hp levels. It is customary to exclude those individuals with Hp 0 from the total sample before calculating the Hp gene frequencies. However, if there is differential suppression among the various phenotypes, the resulting figures cannot be correct, particularly when anhaptoglobinemia is frequent. Thus, in Table 2.1, those populations with an incidence of Hp 0 greater than 10 per cent are marked with an asterisk, and the reported Hp^1 frequencies are accordingly subject to error of unknown magnitude.

GENERAL SUMMARY OF GENE FREQUENCIES SHOWN IN TABLE 2.1

The Hp^1 (and therefore, Hp^2) gene frequency is remarkably similar throughout the European continent, ranging from 0·35–0·43 in most areas, the exceptions being the Swedish Lapps (0·32) in the North, and the southernmost Italians in the South (also 0·32).

On the African continent, the Hp^1 frequency is generally much higher than it is in Europe, but there are clearly two, and possibly three, exceptions, namely, the natives of Ethiopia (0·40), the Bushmen of South Africa (0·30) and the pygmies of the Ituri forest (0·40). The latter figure is open to question because of the very high incidence of Hp 0 among the pygmies (over 30 %) which could conceivably mask a higher Hp^1 frequency. However, a similar proportion of anhaptoglobinemics was found in some of the Nigerian tribes, and their Hp^1 frequencies were generally higher than those observed in most other regions of Africa.

In most of the countries of Asia, Hp^1 frequencies are appreciably lower than those in Europe. Some heterogeneity is seen in the Chinese, most of whom have been tested in countries other than China. It appears that the inhabitants of India are generally more deficient in Hp^1 than any other large population group, but at least one exception has been found in the Todas, whose Hp^1 frequency is 0.35. No doubt, other exceptions exist in that vast country.

In the Americas, the Eskimos of Alaska (and Western Greenland) resemble the Asiatics in their fairly low Hp^1 frequencies, while the Alaskan Indian tribes vary considerably, as do the other tribes of North, Central and South America. Nevertheless, the tendency is toward a generally high level of Hp^1. A north to south cline in Hp gene frequencies has been suggested for the aboriginal peoples of the Americas, but, as pointed out by Salzano & Sutton (1965), the range of variability within given regions is so great that the originally conceived cline is becoming blurred by increasing data.

Racial heterogeneity in the Pacific area is reflected in the Hp gene distribution. However, with the exception of the Australian aborigines, whose Hp^1 frequency is in the range of 0·17–0·24, most of the tested natives resemble the South Americans in having high levels of Hp^1, rather than the low levels found in the Asiatic mainlanders.

GENE FREQUENCIES DETERMINED BY SUBTYPING

Haptoglobin subtyping has been performed on only a small proportion of the serum specimens typed by conventional starch gel electrophoresis. However, from the studies of Giblett & Brooks (1963), Flory (1963), Shim & Bearn (1964b) and Angelopoulos *et al.* (1966) it appears that the Hp^{1F} gene is virtually non-existent in Mongoloid populations (including American Indians), while its frequency in certain African Negroes may be higher than that of Hp^{1S}. Among the Caucasian groups tested, Hp^{1S} outnumbers Hp^{1F} by about 2:1. In all of the populations studied, the Hp^2 gene is nearly always Hp^{2FS} (or Hp^{2SF}); only rare examples of Hp^{2SS} and Hp^{2FF} have been observed. In view of the suggestion of Smithies *et al.* (1962b) that unequal crossing-over in Hp^{2FS}/Hp^1 heterozygotes is responsible for the production of Hp^{2SS} (and Hp^{2FF}), it is rather surprising that in populations lacking the Hp^{1F} gene, Hp^2 is virtually always Hp^{2FS}. However, many unknown factors, particularly relating to crossing-over frequencies and selective advantage, may be responsible for maintaining Hp^{2FS} as the predominant Hp^2 gene throughout the world.

METHODS

Demonstration of the haptoglobin phenotypes depends upon separating the individual components on the basis of their molecular size as well as their electrical charge. This 'molecular sieving' effect, first described by Smithies (1955, 1959a) from experiments with starch gel electrophoresis, was later employed by Davis (1964), Ornstein (1964) and Raymond (1964) in acrylamide gel electrophoresis systems. Both media can be used for Hp phenotype differentiation, and each has certain advantages. However, since the author's experience with acrylamide is limited, this description of technique will be confined to starch gel electrophoresis, and the reader is urged to read also the more detailed directions of Smithies (1955, 1959a,b). Subtyping is not described, since the original paper of Smithies *et al.* (1962a) should be studied carefully before this somewhat difficult procedure is undertaken. (The only important omission from that paper is a note of caution about the heat-lability of urea). For those who wish to use acrylamide gel electrophoresis for Hp typing, methods are described by Peacock *et al.* (1965), Ritchie *et al.* (1966), Ferris *et al.* (1966), Margolis & Kenrick (1967), Woodworth & Clark (1967) and Holmes (1967).

STARCH GEL ELECTROPHORESIS

APPARATUS: HORIZONTAL AND VERTICAL

Horizontal electrophoresis requires a simple basic system which is quite adequate for distinguishing many of the plasma proteins, including haptoglobin (Smithies, 1955). The equipment consists of (1) electrophoresis tanks with a baffle and either platinum or silver-silver chloride electrodes (2) plastic trays with an inside depth of 6 to 7 mm and plastic lids (3) a power supply capable of providing 400 to 500 volts (4) a plastic cutting tray with either a blade or wire cutting device, (5) glass or plastic staining tanks and (6) a voltmeter with thin metal probes for inserting into gel.

Several commercial models are available, some of which are more elaborate than is necessary for determining Hp types. A very simple, inexpensive horizontal system constructed in the author's laboratory is shown in Fig. 2.13. The four tanks, containing bridge buffer, are plastic food-storage boxes, and the staining trays are glass baking dishes. Platinum wire is used for the electrodes, which run the length of the outer tanks and are attached by cables to a Heathkit Voltage Supplier. The electrode tanks are connected by flannel (or filter paper) wicks to adjacent tanks, which are in turn connected by wicks to two spanning plastic trays containing the starch gel.

5

Figure 2.14 shows the components of the vertical electrophoresis apparatus described by Smithies (1959b), which can be purchased from Otto Hiller, Post Office Box 1294, Madison, Wisconsin. A modification of this apparatus, available from Buchler Corp., employs less starch and differs in the method of forming the slots for sample insertion into the gel. A third, water-cooled system, made by EC Apparatus Corp. permits the use of higher voltage, and is designed for both starch and acrylamide gel electrophoresis.

SOURCE OF STARCH

Although it is possible to prepare hydrolysed potato starch in the laboratory (Smithies, 1955), the method is laborious and individual batches vary in both pH and viscosity. Thus it is preferable to purchase several kilograms of starch with the same lot number from a commercial source. The major supplier has been the Connaught Laboratories of Toronto; their product is marketed by Fischer Corp. in the United States. Recently, Otto Hiller has begun to market 'Electrostarch' which is less expensive and forms gels of firmer consistency.

BUFFER SYSTEMS

Several different buffer systems can be used for Hp typing. The original, low ionic strength borate buffer described by Smithies (1955) gives reliable results. In the bridge (tank or electrode) buffer, the boric acid is 0·3 M and sodium hydroxide 0·06 M. The gel buffer contains 0·03 M boric acid and 0·012 M sodium hydroxide. The higher pH of the gel buffer is decreased by admixture with the starch, which usually has enough residual acid (from its hydrolysis) to make a final pH of about 8·5. With this buffer, there is marked spreading of the albumin zone during electrophoresis. Also, free hemoglobin is often not well separated from the Hp 1-1:Hb complex, particularly in horizontal electrophoresis.

Poulik (1957) introduced a discontinuous buffer system in which the bridge buffer is 0·3 M boric acid and 0·05 M sodium hydroxide, but the gel buffer is 0·076 M tris (tris-(hydroxymethyl)-aminomethane) and 0·005 M citric acid, pH 8·6. During electrophoresis, a visible yellow-brown boundary moves anodally through the gel. It is the site of a marked change in voltage, which apparently accounts for the improved resolution of some of the proteins. The migration of haptoglobin is not greatly altered, but the free hemoglobin is better

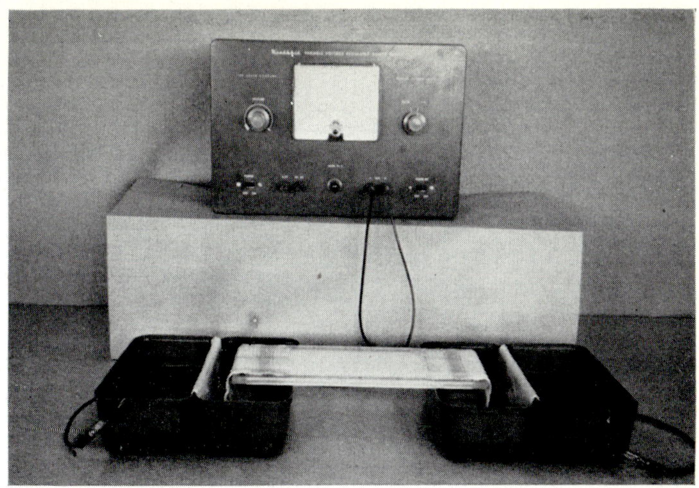

FIGURE 2.13. Photograph of simple horizontal starch gel electrophoresis employing a Heathkit voltage source, platinum wire electrodes, flannel wicks (pattern reveals source as old printed pajamas) and plastic storage boxes for buffer tanks. The tray used for staining is a glass baking dish.

FIGURE 2.14. Components of apparatus for vertical electrophoresis, including a wooden (or plastic) stand, buffer compartments, starch gel tray with cover, and slicing frame with blade. Silver-silver chloride electrodes are shown, but platinum wire can also be used. A spirit-level is necessary for placing the gel in an absolutely upright position.

[Facing page 102]

separated from the Hp 1-1 : Hb complex. Thus, serum samples which lack demonstrable haptoglobin are more readily detected.

GEL PREPARATION

The concentration of starch in the gel varies somewhat with the individual lots, although for haptoglobin phenotype identification, the concentration is not critical. Usually 12 to 14 g of starch per 100 ml of buffer provides a gel of suitable strength. The higher the gel

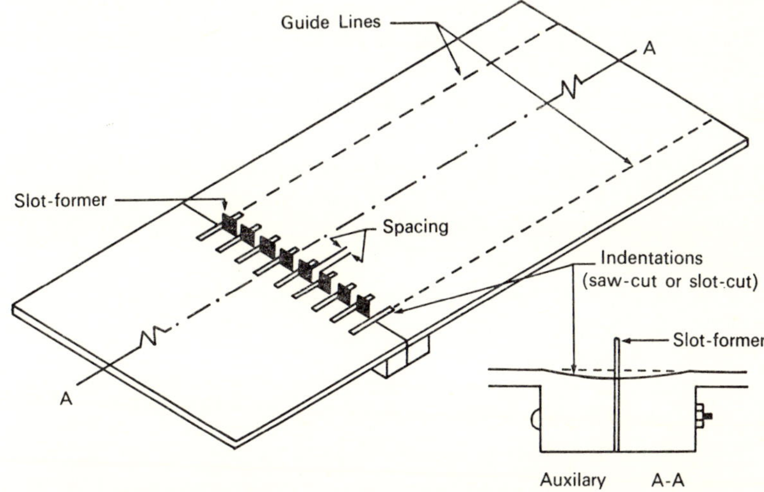

FIGURE 2.15. Diagram of slot-forming plastic cover used for making evenly-spaced insertion holes in starch gel. The indentations between slots are useful for keeping the specimens separated, but they can be omitted. (From catalog of Otto Hiller, Madison, Wisconsin.)

concentration, the slower the migration of serum components, depending upon their relative sizes. (Smithies (1962) has used this principle as the basis for determining molecular size of proteins.)

It is customary to mix the starch and buffer together in a flask and to heat the contents by swirling over a gas flame. However, large flasks are necessary for preparing 500–1000 ml of gel, and an alternative procedure is to mix and heat the contents in a large beaker immersed in a stationary bath of hot water or preferably, a liquid with a high boiling point, such as glycerol, maintained at about 100° C. An electrical stirrer with a large paddle for viscous solutions can then be substituted for manual rotation.

Smithies (1959a) described the N-shaped viscosity curve which starch gel displays during heating and cooling, the first peak of the 'N' being reached at 65° C and the sharp drop occurring between 65 and 80° C. The gel must be heated until well past the initial maximum viscosity; it is then readily poured into the plastic trays. Inadequately cooked gels do not provide the sieving effect necessary for differentiating the proteins.

Permitting the flask or beaker to stand upright for a few minutes after heating is usually sufficient to rid the gel of large air bubbles. However, for more careful work, it is possible to 'de-gas' the solution by applying vacuum pressure and releasing it suddenly on two or three occasions. After the gel is poured into the tray, it may be covered with a smooth piece of lightly oiled thin plastic or with a plastic lid containing a mold of several 'teeth' to form the slots for sample insertion. An example of such a slot-former is shown in Fig. 2.15. To avoid air bubbles, an excess of gel is necessary, and weights on each corner are required to hold the cover on tightly during cooling.

SAMPLE INSERTION

For horizontal electrophoresis, the simplest method of insertion employs small pieces of filter paper. With a razor blade or surgical knife, a cut is first made across the gel about two inches from the end, and this piece of gel is pushed back, preferably by temporarily removing the terminal inch. Distortion or breakage of the gel is avoided. A piece of filter paper (Whatman no. 3) about 10 × 15 mm is then dipped into a tube containing the serum specimen which contains sufficient added hemoglobin to oversaturate the haptoglobin (i.e. a concentration of about 200–300 mg Hb per 100 ml serum). This piece of saturated paper is blotted slightly and carefully applied to the cut edge of the starch. Eight to twelve pieces of paper (i.e. 8–12 serum specimens) can be applied with a tray width of 10–15 cm; only a one mm space is necessary between adjacent insertion papers. The end piece of gel is replaced (if removed), and carefully pressed back against the pieces of paper to remove air bubbles. The gel is then covered with cellophane or Saran wrap.

Vertical electrophoresis methods usually employ a slot-forming mold, so that the serum is placed in its individual slot with a thin

pipet. The narrow region of starch containing the slots (from 8–20)
is covered with a thin film of nearly-cooled melted petroleum jelly,
which can also be used to cover the rest of the gel; however, a cellophane or Saran wrap layer is equally effective for the non-insertion
region.

The final arrangement of the vertical apparatus must be in a strictly
upright, level position. For this purpose, a plastic or wooden platform
is used, and the gel is positioned with a spirit level and plumbob. The
cathode is at the top of the gel, so that migration of most components
is in a downward direction.

ELECTROPHORESIS CONDITIONS

With either of the buffer systems described above, an average voltage
gradient of four to six volts per cm is maintained. (This measurement
is made by determining the distance between two points at either end
of the gel, inserting the voltmeter probes at those points and dividing
the voltage read by the distance.) The current should not exceed 2
milliamperes per cm of gel width. Refrigeration is not required for
haptoglobin phenotyping, and with horizontal gels, a period of four
hours is usually adequate for electrophoresis at room temperature.
With vertical gels, overnight runs are required unless a water-cooled
system at higher voltage (15–20 volts per cm) is employed.

GEL SLICING

The gel is placed in a plastic tray with sides 2–3 mm high, and sliced
with either a microtome knife or a fine piano wire stretched across a
metal frame in the same fashion as a cheese cutter. Alternatively, a
regular starch gel tray is partially filled by a 2–3 mm piece of plastic
cut to the same dimensions. Placed in this tray, the gel is sufficiently
elevated above the sides to permit its being cut into two slices.

STAINING

The zones containing hemoglobin can be seen as faint pink bands, but
much better identification is made with one of several stains containing
either benzidine, guaiacol, o-dianisidine or leucomalachite green. In
the simplest procedure, about a half g of benzidine base is placed in a
test tube about one-third filled with 95 per cent ethyl alcohol, and a
nearly saturated solution is prepared by stirring with a glass rod. An

equal volume of 3 per cent hydrogen peroxide is added, and the resulting precipitate is just dissolved by dropwise addition of glacial acetic acid, while stirring. This solution is dropped by pipet over the cut gel surface. The Hb and Hp-Hb zones stain dark blue for about 10 minutes, after which they begin to fade, especially when too much peroxide has been used. These zones, in addition to Hb A and the Hp-Hb complex(es) may also include Hb A_2 and/or methemoglobin, both of which migrate appreciably more slowly than Hb A, and can be confused with zones of haptoglobin. Other heme-containing components which react with benzidine are methemalbumin (trailing edge of albumin) and the heme-hemopexin complex, with a mobility similar to that of transferrin, near free hemoglobin at this alkaline pH.

Preservation of the heme-containing zones as black bands against a light blue background can be obtained by first washing the gel, then immersing it in a very dilute solution of amido black stain (see below), which stains other proteins blue. A similar method has been described by Singh (1967).

The benzidine stain used by Smithies (1959a) contains 100 ml water, 0·5 ml glacial acetic acid, 0·2 g benzidine (warm to dissolve) and 0·2 ml 30 per cent hydrogen peroxide. Its sensitivity is excellent and reproducible, but eventual fading requires the preservation treatment outlined above. (*Note:* Since benzidine is a carcinogenic agent, care must be taken to avoid contact with skin.)

It is usually desirable to stain the other half of the sliced gel with a protein stain. Amido black is particularly useful for this purpose, the dye being dissolved in a methanol-water-acetic acid solution, proportions 50-50-10, respectively. Staining requires only a few minutes, after which the gel is decolorized in the same solution without the dye. Automatic destaining devises are commercially available. For photography, a red filter is useful to accentuate either the amido black-stained proteins or prior to protein staining, the benzidine-stained, heme-containing zones.

MEASUREMENTS OF HAPTOGLOBIN CONCENTRATION

A variety of methods have been developed for determining the haptoglobin content of serum, based on the formation of the HpHb complex. For this purpose, an excess of hemoglobin is added to the serum (or plasma) samples, and the amount of HpHb is estimated on the basis of its peroxidase activity (Jayle, 1951; Nyman, 1959; Connell & Smithies, 1959; Owen *et al.*, 1960a; Smith & Owen, 1961; Tarukoski, 1966; Burrows & Hosten, 1966). Alternatively, the complex is separated from free Hb, and the proportion of the total Hb bound to Hp is taken as the Hb-binding capacity of the Hp. The separation is achieved either by electrophoresis (Laurell & Nyman, 1957; Hommes, 1959; Brus & Lewis, 1959; Owen *et al.*, 1959; Lathem & Worley, 1959; Javid & Horowitz, 1960; Rowe, 1961; Herman, 1961a; Sass & Spear, 1962; Valeri *et al.*, 1965) or by Sephadex gel filtration (Lionetti *et al.*, 1964; Killander, 1964; Ratcliff & Hardwicke, 1964). An immuno-chemical method for the direct measurement of haptoglobin concentration with anti-haptoglobin was used by Kluthe *et al.* (1965), and radial diffusion methods are also feasible, although the larger size of the Hp polymers alters their diffusion rate. (Unpublished observation).

REFERENCES

ALLISON A.C. (1958) The genetical and clinical significance of the haptoglobins. *Proc. roy. Soc. Med.* **51**, 641.

ALLISON A.C. (1959) Genetic control of human haptoglobin synthesis. *Nature* 183, 1312.

ALLISON A.C. & AP REES W. (1957) The binding of haemoglobin by plasma proteins. *Brit. med. J.* **ii**, 1137.

ALLISON A.C. & BARNICOT N.A. (1960) Haptoglobins and transferrins in some East Africa peoples. *Acta genet.* **10**, 17.

ALLISON A.C., BLUMBERG B.S. & AP REES W. (1958) Haptoglobin types in British, Spanish Basque and Nigerian African populations. *Nature* **181**, 824.

ALPER C.A., PETERS J.H., BIRTCH A.G. & GARDNER F.H. (1965) Haptoglobin synthesis. I. *In vivo* studies of the production of haptoglobin, fibrinogen and γ globulin by the canine liver. *J. clin. Invest.* **44**, 574.

ALY F.W., BRINKER G., DEICHER H., HARTMANN F. & NIX W. (1962) Inherited immunologically atypical haptoglobin. *Nature* **194**, 1091.

ANDERSON M.N., MOURITZEN C.V. & GABRIELI E. (1966) Mechanisms of plasma hemoglobin clearance after acute hemolysis: studies in open-heart surgical patients. *Ann. Surg.* **163**, 529.

ANGELOPOULOS B., TSOUKANTAS A. & DANOPOULOS E. (1966) Distribution of haptoglobin subtypes in Greeks. *J. med. Genet.* **3**, 276.

BADR-EL-DIN M.K., KHALIL M. & KASSEM A.S. (1966) Haptoglobin level in some diseases of infancy and childhood. *Acta paed. scand.* **55**, 584.

BAITSCH H. & LIEBRICH K.G. (1961) Die Haptoglobintypen. *Blut* 7, 27.

BAITSCH H., RITTER H., GOEDDE H.W. & RIEDEL V. (1964) Formalgenetische Untersuchungen über den Haptoglobin-Polymorphismus. *Z. Morph. Anthrop.* **55**, 175.

BARNICOT N.A. (1961) Haptoglobins and transferrins, *in* HARRISON G.A. (ed.) *Genetical Variation in Human Population*, p. 41. Pergamon Press, Oxford.

BARNICOT N.A., GARLICK J.P. & ROBERTS D.F. (1960) Haptoglobin and transferrin inheritance in northern Nigerians. *Ann. hum. Genet.* **24**, 171.

BAYANI-SIOSON P.S., LOUCH J., SUTTON H.E., NEEL J.V., HORNE S.L. & GERSHOWITZ H. (1962) Quantitative studies on the haptoglobin of apparently healthy adult male twins. *Am. J. hum. Genet.* **14**, 210.

BEARN A.G. & FRANKLIN E.C. (1958) Some genetical implications of physical studies of human haptoglobins. *Science* **128**, 596.

BEARN A.G. & FRANKLIN E.C. (1959) Comparative studies on the physical characteristics of the heritable haptoglobin groups of human serum. *J. exp. Med.* **109**, 55.

BECKMAN L. & GRIVEA M. (1964) Haptoglobin variations in newborn children. *Acta genet.* **14**, 159.

BERGGARD I. & BEARN A.G. (1962) Excretion of haptoglobin in normal urine. *Nature* **195**, 1311.

BERGGARD I., CLEVE H. & BEARN A.G. (1964) The excretion of five plasma proteins previously unidentified in normal human urine. *Clin. chim. Acta* **10**, 1.

BERGSTRAND C.G., CZAR B. & TARUKOSKI P.H. (1961) Serum haptoglobin in infancy. *Scand. J. clin. Lab. Invest* **13**, 576.

BEUING H., CLEVE H. & DEICHER H. (1965) Immunologische Untersuchungen menschlicher Haptoglobine. *Z. Infek. Kr.* **151**, 291.

BIAS W.B. & MIGEON B.R. (1967) Haptoglobin: a locus on the D_1 chromosome? *Am. J. hum. Genet.* **19**, 393.

BLACK J.A. & DIXON G.H. (1968) Amino acid sequence of the alpha chains of human haptoglobins and their possible relation to the immunoglobulin light chains. *Nature* **218**, 736.

BLAU J.N., HARRIS H. & ROBSON E.R. (1963) Haptoglobins in cerebrospinal fluid. *Clin. chim. Acta* **8**, 202.

BLOOM G.E., GERALD P.S. & DIAMOND L.K. (1966) Autoradiographic analysis of the specific D chromosome bearing the haptoglobin locus. Presentation at the 36th annual meeting of the *Soc. Pediat. Res.*

BLOOM G.E., GERALD P.S. & REISMAN L.E. (1967) Ring D chromosome: a second case associated with anomalous haptoglobin inheritance. *Science* **156**, 1746.

BLUMBERG B.S. & GENTILE Z. (1961) Haptoglobins and transferrins of two tropical populations. *Nature* **189**, 897.

BLUMBERG B.S., MURRAY R.F., ALLISON A.C., BARNICOT N.A., HIRSCHFELD J. & KRIMBAS C. (1964) Serum protein polymorphisms in Greek populations. *Ann. hum. Genet.* **28**, 189.

BLUMBERG B.S. & WARREN L. (1961) The effect of sialidase on transferrins and other serum proteins. *Biochim. biophys. Acta* **50**, 90.

BORGLIN N.E. & NYMAN M. (1961) Effect of estrogens on the haptoglobin level in the blood. *Scand. J. clin. Lab. Invest.* **13**, 107.

Bowman B.H. & Cleve H. (1967) Haptoglobin Marburg, an inherited variant of the beta chain of human haptoglobin (Abstract 43). *Meeting of Amer. Soc. hum. Genet.*, Toronto.

Brus I. & Lewis S.M. (1959) The haptoglobin content of serum in haemolytic anemia. *Br. J. Haemat.* **5**, 348.

Bunn H.F. (1967) Effect of sulfhydryl reagents on the binding of human hemoglobin to haptoglobin. *J. Lab. clin. Med.* **70**, 606.

Bunn H.F., Bull R.W. & Jandl J.H. (1967) A proposed mechanism for the glomerular filtration of hemoglobin. *Clin. Res.* **15**, 272 (abstract).

Bunn H.F. & Jandl J.H. (1966) Exchange of heme among hemoglobin molecules. *Proc. natn. Acad. Sci.* **56**, 974.

Burrows S. & Hosten E.B. (1966) Serum haptoglobin. Automated method applied to patient screening. *Am. J. clin. Path.* **36**, 634.

Cann H.M. & van West B. (1966) Atypical segregation of haptoglobin types. *Int. Congr. hum. Genet.*, Chicago (abstract No. 47).

Cheftel R., Cloarec L., Moretti J. & Jayle M.F. (1965) Structure des glycopeptides obtenus par protéolyse de l'haptoglobine humaine. *Bull. Soc. chim. Biol.* **47**, 385.

Cheftel R.I. & Moretti J. (1966) Sur la structure des haptoglobines humaines. *C. r. Acad. Sci.* **262**, 1982.

Chiancone E., Wittenberg J.B., Wittenberg B.A., Antonini E. & Wyman J. (1966) The combination of haptoglobin 2-2 with the oxy and deoxy forms of human hemoglobin before and after digestion by carboxypeptidase A, and with isolated α chains. *Biochim. biophys. Acta* **117**, 379.

Cleve H. & Deicher H. (1965) Haptoglobin 'Marburg'; Untersuchungen über eine seltene erbliche Haptoglobin-Variante mit zwei verschiedenen Phänotypen innerhalb einer Familie. *Humangenetik* **1**, 537.

Cleve H., Gordon S., Bowman B.H. & Bearn A.G. (1967) Comparison of the tryptic peptides and amino acid composition of the β-polypeptide chains of the three common haptoglobin phenotypes. *Am. J. hum. Genet.* **19**, 713.

Cloarec L. & Moretti J. (1966) Combinaison de l'hémoglobine avec les divers polymèrs constitutifs de l'haptoglobine. *C. r. Acad. Sci.* **262**, 2081.

Connell G.E., Dixon G.H. & Smithies O. (1962) Subdivision of the three common haptoglobin types based on 'hidden' differences. *Nature* **193**, 505.

Connell G.E. & Smithies O. (1959) Human haptoglobins: estimation and purification. *Biochem. J.* **72**, 115.

Connell G.E., Smithies O. & Dixon G.H. (1966) Gene action in the human haptoglobins. II. Isolation and physical characterization of alpha polypeptide chains. *J. molec. Biol.* **21**, 225.

Cooper D.W., Lewis H.B.M. & Nicholls E.M. (1965) Haptoglobin Johnson in Australian aborigines. *Nature* **208**, 694.

Corey M.J., Miller J.R., MacLean J.R. & Chown B. (1967) A case of XX/XY mosaicism. *Am. J. hum. Genet.* **19**, 378.

Cullhed I. (1967) Serum haptoglobin in cases with Starr-Edwards ball-valve prosthesis. *Acta med. scand.* **181**, 321.

Davis B.J. (1964) Disc electrophoresis. II. Method and application to human serum proteins. *Ann. N. Y. Acad. Sci.* **121**, 404.

DIXON G.H. (1966) Mechanisms of protein evolution. *Essays in Biochem. II.* 148.

DOBRYSZYCHA W. & LISOWSKA E. (1966) Effect of degradation on the chemical and biological properties of haptoglobin. I. Products of tryptic digestion. *Biochim. biophys. Acta* **121**, 42.

EICHMANN K., DEICHER H. & CLEVE H. (1966) Immunologische Analyse der Beziehungen zwischen den drei verschiedenen Typen von Antigendeterminanten normaler menschlicher Haptoglobine. *Humangenetik* **2**, 271.

ENGLER R., BESCOL-LIVERSAC J. & MORETTI J. (1966) Catabolisme du complexe haptoglobine-hémoglobine dans le système reticuloendothélial. *C. r. Acad. Sci.* **263**, 1636.

ENGLER R., MORETTI J. & JAYLE M.F. (1967) Catabolisme du complexe haptoglobine-hémoglobine. *Bull. Soc. chim. Biol.* **49**, 263.

FAULSTICK D.A., LOWENSTEIN J. & YIENGST M.J. (1962) Clearance kinetics of haptoglobin-hemoglobin complex in the human. *Blood* **20**, 65.

FERRIS T.G., EASTERLING R.E., NELSON K.J. & BUDD R.E. (1966) Determination of serum hemoglobin binding capacity and haptoglobin type by acrylamide gel electrophoresis. *Am. J. clin. Path.* **46**, 385.

FLEISCHER E.A. & MOHR J. (1962) Concerning the genetics of the human haptoglobins. 126 Norwegian families with 428 children. *Acta genet.* **12**, 281.

FLORY L.L. (1963) Haptoglobin types of the Nez Perce Indians. *Nature* **197**, 578.

FRANKLIN E.C., ORATZ M., ROTHSCHILD M.A. & ZUCKER-FRANKLIN D. (1960) Haptoglobin-hemoglobin metabolism in rabbits studied with I^{131} and Fe^{59} labelled complexes. *Proc. Soc. exp. Biol.* **105**, 167.

FREEMAN T. (1964) Haptoglobin metabolism in relation to red cell destruction. *Proc. XII Colloq. Prot. Biol. Fluids*, Bruges, p. 344.

GALATIUS-JENSEN F. (1958a) On the genetics of the haptoglobins. *Acta genet.* **8**, 232.

GALATIUS-JENSEN F. (1958b) Rare phenotypes in the Hp system. *Acta genet.* **8**, 248.

GALATIUS-JENSEN F. (1960) The haptoglobins. A genetic study. *Dansk Videnskabs Forlag*, Copenhagen.

GALATIUS-JENSEN F. (1962) The use of serum haptoglobin patterns in cases of disputed paternity, *in* LUNDQUIST F. (ed.) *Methods of Forensic Science*, vol. 1, p. 497. J. Wiley & Sons, New York.

GARBY L. & NOYES W.D. (1959) Studies on hemoglobin metabolism. I. The kinetic properties of the plasma hemoglobin pool in normal man. *J. clin. Invest.* **38**, 1479.

GARBY L. & OBARA J. (1960) Organ uptake and plasma transportation kinetics of hemoglobin in rats. *Blut* **6**, 143.

GARBY L., SJÖLIN S. & VUILLE J.C. (1964) Studies on erythro-kinetics in infancy. V. Estimations of the life span of red cells in the newborn. *Acta paediat.* **53**, 165.

GERALD P.S., WARNER S., SINGER J.D., CORCORAN P.A. & UMANSKY I. (1964) Possible identification of the chromosome bearing the haptoglobin locus. *J. clin. Invest.* **43**, 1297.

GERALD P.S., WARNER S., SINGER J.D., CORCORAN P.A. & UMANSKY I. (1967) A ring D chromosome and anomalous inheritance of haptoglobin type. *J. Pediat.* **70**, 172.

GERBECK C.M., BEZKOROVAINY A. & RAFELSON M.E. (1967) Glycopeptides obtained from human haptoglobin 2-1 and 2-2. *Biochemistry* **6**, 403.

GIBLETT E.R. (1959) Haptoglobin types in American Negroes. *Nature* **183**, 192.

GIBLETT E.R. (1961) Haptoglobin: a review. *Vox Sang.* **6**, 513.

GIBLETT E.R. (1964) Variant haptoglobin phenotypes. *Cold Spring Harbor Symposia on Quantitative Biology* **29**, 321.

GIBLETT E.R. (1966) Recent advances in haptoglobin and transferrin genetics. *Plenary Session, XIth int. Cong. Blood Transf.*, Sydney, p. 8.

GIBLETT E.R. (1968) The haptoglobin system. *Series Haematologica 1*, **1**, 3.

GIBLETT E.R. & BROOKS L.E. (1963) Haptoglobin subtypes in three racial groups. *Nature* **197**, 576.

GIBLETT E.R., MOTULSKY A.G. & FRASER G.R. (1966a) Population genetics in the Congo. IV. Haptoglobin and transferrin serum groups in the Congo and in other African populations. *Am. J. hum. Genet.* **18**, 553.

GIBLETT E.R. & STEINBERG A.G. (1960) The inheritance of serum haptoglobin types in American Negroes: evidence for a third allele, Hp^{2M}. *Am. J. hum. Genet.* **12**, 160.

GIBLETT E.R., UCHIDA I. & BROOKS L.E. (1966b) Two rare haptoglobin phenotypes 1-B and 2-B, containing a previously undescribed α polypeptide chain. *Am. J. hum. Genet.* **18**, 448.

GOEDDE H.W., RITTER H. & WEYRAUCH U. (1965) Zum Polymorphismus der Haptoglobine: Methodik der Untergruppen bestimmung (subtyping) und Formalgenetik. *Humangenetik* **1**, 414.

GORDON S. & BEARN A.G. (1966) Hemoglobin binding capacity of isolated haptoglobin polypeptide chains. *Proc. Soc. exp. Biol.* **121**, 846.

GOTTLIEB A., WISCH N. & ROSS J. (1963) Familial hypohaptoglobinemia. A genetically determined trait segregating from G6PD deficiency. *Blood* **21**, 129.

GROUCHY J. DE, MOULLEC J., SALMON C., JOSSO N., FREZAL J. & LAMY M. (1964) Hermaphrodisme avec caryotype XX/XY étude génétique d'un cas. *Ann. Génét.* **7**, G25.

GROUCHY J. DE, SALMON CH., SALMON D. & MAROTEAUX P. (1966) Délétion du bras court d'un chromosome 13–15, hypertélorisme et phenotype haptoglobine Hp o dans une même famille. *Ann. Génét.* **9**, 80.

GYDELL K. (1960) Hyperbilirubinemia and hypersideremia following intravenous injection of stored blood, nicotinic acid and hemoglobin solution. *Acta med. scand.* **166**, 433.

HANKS G.E., CASSELL M., RAY R.N. & CHAPLIN H. (1960) Further modification of the benzidine method for measurement of hemoglobin in plasma. *J. Lab. clin. Med.* **56**, 486.

HARRIS H., LAWLER S.D., ROBSON E.B. & SMITHIES O. (1960) The occurrence of two unusual serum protein phenotypes in a single pedigree. *Ann. hum. Genet.* **24**, 63.

HARRIS H., ROBSON E.B. & SINISCALCO M. (1958) Atypical segregation of haptoglobin types in man. *Nature* **182**, 1324.

HARRIS H., ROBSON E.B. & SINISCALCO M. (1959) Genetics of the plasma protein variants, *in* WOLSTENHOLME G.E.W. & O'CONNOR C.M. (eds.) *Biochemistry of Human Genetics*, p. 151. Little, Brown, Boston.

HERMAN E.C. (1961a) Serum haptoglobins: their semiquantitative estimation by a paper electrophoretic technique. *J. Lab. clin. Med.* **57**, 825.

HERMAN E.C. (1961b) Serum haptoglobins in hemolytic disorders. *J. Lab. clin. Med.* **57**, 834.

HEVER O. & VADESZ G. (1965) Haptoglobin in children with infectious hepatitis. *J. Pediat.* **67**, 1156.

HIGASHI G.I. & LUBS H.A. (1966) Quantitative variations of haptoglobins in a Caucasian family. *J. med. Genet.* **3**, 281.

HIRSCHFELD J. & LUNELL N.-O. (1962) Serum protein synthesis in foetus: haptoglobins and group-specific components. *Nature* **196**, 1220.

HOLMES R. (1967) Discontinuous acrylamide-gel plate electrophoresis. *Biochim. biophys. Acta,* **133**, 174.

HOMMES F.A. (1959) A new method for the quantitative determination of haptoglobin. *Clin. chim. Acta* **4**, 707.

JANDL J.H., JONES A.R. & CASTLE W.B. (1957) The destruction of red cells by antibodies in man. I. Observations on the sequestration and lysis of red cells altered by immune mechanisms. *J. clin. Invest.* **36**, 1428.

JAVID J. (1964) The nature of the difference between haptoglobin polymers in the phenotypes Hp 2-1 and Hp 2-2. *Proc. natn. Acad. Sci.* **52**, 663.

JAVID J. (1965) The effect of haptoglobin polymer size on hemoglobin binding capacity. *Vox Sang.* **10**, 320.

JAVID J. (1967a) Human serum haptoglobins: a brief review. *Sem. Hemat.* **4**, 35.

JAVID J. (1967b) Haptoglobin 2-1 Bellevue, a haptoglobin β chain mutant. *Proc. natn. Acad. Sci.* **57**, 920.

JAVID J., FISCHER D.S. & SPAET T.H. (1959) Inability of haptoglobin to bind myoglobin. *Blood* **14**, 683.

JAVID J. & HOROWITZ H.I. (1960) An improved technic for the quantitation of serum haptoglobin. *Am. J. clin. Path.* **34**, 35.

JAYLE M.F. (1951) Méthode de dosage de l'haptoglobine sérique. *Bull. Soc. Chim. biol.* **33**, 876.

JAYLE M.F. & BOUSSIER G. (1955) Les séromucoides du sang; leur relations avec les mucoprotéines de la substance fondamentale du tissu conjonctiv. *Exp. Ann. biochim. Med.* **17**, 172.

JAYLE M.F., MARNAY A. & POINTIS J. (1962) Relation entre l'haptoglobine, l'orosomucoide et le fibrinogène. *Nouv. Revue fr. Hémat.* **2**, 483.

JAYLE M.F. & MORETTI J. (1962) Haptoglobin: biochemical, genetic and physiopathologic aspects, *in* MOORE C.V. & BROWN E.B. (eds.) *Progress in Hematology* Vol. 3, p. 342. Grune & Stratton, New York.

KAHLICH-KOENNER D.M. & WEIPPL G. (1961) Abhängigkeit der Haptoglobin-Typenbestimmung von der Haptoglobin-Konzentration. *Klin. Wschr.* **39**, 1025.

KEENE W.R. & JANDL J.H. (1965) The sites of hemoglobin catabolism. *Blood* **26**, 705.

KILLANDER J. (1964) Separation of human heme and hemoglobin-binding plasma proteins, ceruloplasmin and albumin by gel filtration. *Biochim. biophys. Acta* **93**, 1.

KLUTHE R., FAUL J. & HEIMPEL H. (1965) Quantitative estimation of human serum haptoglobins by an immunological method. *Nature* **205**, 93.

KORNGOLD L. (1963) Antigenic differences among human haptoglobins. *Int. Arch. Allergy appl. Immun.* **23**, 268.

KORNGOLD L. (1965) The effect of hemoglobin on the haptoglobin–anti-haptoglobin reaction. *Immunochemistry* **2**, 103.

KRAUSS S. (1963) Response of serum haptoglobin to inflammation in adrenalectomized rat. *Proc. Soc. exp. Biol.* **112**, 552.

KRAUSS S. & SARCIONE E.J. (1964) Synthesis of haptoglobin by the isolated perfused rat liver. *Biochim. biophys. Acta* **90**, 301.

KRAUSS S., SCHROTT M. & SARCIONE E.J. (1966) Haptoglobin metabolism in Hodgkin's disease. *Am. J. med. Sci.* **252**, 184.

LANGLEY G.R., OWEN J.A. & PADANYI R. (1962) Effect of blood transfusions on serum haptoglobins. *Br. J. Haemat.* **8**, 392.

LATHEM W. (1959) The renal excretion of hemoglobin: regulatory mechanisms and the differential excretion of free and protein-bound hemoglobin. *J. clin. Invest.* **38**, 652.

LATHEM W. & JENSEN W.N. (1962) The renal excretion of hemoglobin in sickle cell anemia, with observations on spontaneously occurring hemoglobinemia and methemalbuminemia. *J. Lab. clin. Med.* **59**, 137.

LATHEM W. & WORLEY W.E. (1959) The distribution of extracorpuscular hemoglobin in circulating plasma. *J. clin. Invest.* **38**, 474.

LAURELL C.B. (1959) Purification and properties of different haptoglobins. *Clin. chim. Acta* **4**, 79.

LAURELL C.B. (1960) Metal-binding plasma proteins and cation transport, *in* PUTNAM F.W. (ed.) *The Plasma Proteins*, vol. 1, p. 459. Academic Press, New York.

LAURELL C.B. & GRÖNVALL C. (1962) Haptoglobins, *in* SOBOTKA H. & STEWART C.P. (eds.) *Advances in Clinical Chemistry*, vol. 5, p. 135. Academic Press, New York.

LAURELL C.B. & NYMAN M. (1957) Studies on the serum haptoglobin level in hemoglobinemia and its influence on renal excretion of hemoglobin. *Blood* **12**, 493.

LAURENT B. & PANELIUS M. (1965) Occurrence of haptoglobin and hemopexin in human cerebrospinal fluid. *Scand. J. clin. lab. Invest.* **17**, suppl. 86, 160.

LIONETTI F.J., VALERI C.R., BOND J.C. & FORTIER N.L. (1964) Measurement of hemoglobin binding capacity of plasma by means of dextran gels. *J. Lab. clin. Med.* **64**, 519.

LOWENSTEIN J., FAULSTICK D.A., YIENGST M.J. & SHOCK N.W. (1961) The glomerular clearance and renal transport of hemoglobin in adult males. *J. clin. Invest.* **40**, 1172.

MÄKELÄ O., ERIKSSON A.W. & LEHTOVAARA R. (1959) On the inheritance of the haptoglobin serum groups. *Acta genet.* **9**, 149.

MARGOLIS J. & KENRICK K.G. (1967) Electrophoresis in polyacrylamide concentration gradient. *Biochem. biophys. Res. Commun.* **27**, 68.

MARINIS S. & OTT H. (1965) Natürliche und Künstliche polymeren von Plasmaproteinen: Untersuchungen in acrylamid-gel. *Proc. XII Colloq. Prot. Biol. Fluids*, Bruges, p. 420.

MARNAY A. (1961) Haptoglobinuria in nephrotic syndromes. *Nature* **191**, 74.

MATSUNAGA E. (1962) An inert allele Hp^0 at the Hp locus. *Jap. J. hum. Genet.* **7**, 133.

MATSUNAGA E., MURAI K. & MATSUDA E. (1962) Inheritance of haptoglobin types in 51 Japanese families. *Acta genet.* **12**, 262.

MAUNG M., BAKER D.G. & MURRAY R.K. (1964) Effect of puromycin on the plasma haptoglobin level of rats during experimental inflammation. *Life Sci.* **3**, 1349.

MEHTA S.R. & JENSEN W.N. (1960) Haptoglobins in haemoglobinopathy: a genetic and clinical study. *Br. J. Haemat.* **6**, 250.

MIALE J.B. & KENT J.W. (1962) Serum haptoglobin in rabbits after subcutaneous injection of Freund's adjuvant or turpentine. *Proc. Soc. exp. Biol.* **III**, 589.

MORETTI J., BOREL J., DOBRYSZYCKA W. & JAYLE M.F. (1963) Détermination de la demivie de l'haptoglobine plasmatique humaine. *Biochim. biophys. Acta* **69**, 205.

MOURAY H., MORETTI J. & JAYLE M. (1964) Biosynthèse de l'haptoglobine par perfusion du foie isolé de Lapin. *C. r. acad. Sci.* **259**, 2721.

MURRAY R.F., ROBINSON J.C. & VISNICH S. (1966) Observations on the inheritance of hypohaptoglobinemia. *Acta genet.* **16**, 113.

MURRAY R.K., CONNELL G.E. & PERT J.H. (1961) The role of haptoglobin in the clearance and distribution of extracorpuscular hemoglobin. *Blood* **17**, 45.

NAGEL R.L. & GIBSON Q.H. (1966) Kinetics of the reaction of carbon monoxide with the hemoglobin-haptoglobin complex. *J. molec. Biol.* **22**, 249.

NAGEL R.L. & GIBSON Q.H. (1967) Kinetics and mechanism of complex formation between hemoglobin and haptoglobin. *J. biol. Chem.* **242**, 3428.

NAGEL R.L. & RANNEY H.M. (1964) Haptoglobin binding of certain abnormal hemoglobins. *Science* **144**, 1014.

NAGEL R.L., ROTHMAN M.C., BRADLEY T.B. & RANNEY H.M. (1965a) Comparative haptoglobin binding properties of oxyhemoglobin and deoxyhemoglobin. *J. biol. Chem.* **240**, PC4543.

NAGEL R.L., WITTENBERG J.B. & RANNEY H.M. (1965b) Oxygen equilibria of the hemoglobin-haptoglobin complex. *Biochim. biophys. Acta* **100**, 286.

NAKAJIMA H., TAKEMURA T., NAKAJIMA O. & YAMAOKA K. (1963) Studies on heme α-methenyl oxygenase. I. The enzymatic conversion of pyridine hemochromogen and hemoglobin-haptoglobin into a possible precursor of biliverdin. *J. biol. Chem.* **238**, 3784.

NANCE W.E. & SMITHIES O. (1963) New haptoglobin alleles: a prediction confirmed. *Nature* **198**, 869.

NANCE W.E. & SMITHIES O. (1964) Discussion of Giblett E.R. *Cold Spring Harbor Symposia on Quantitative Biology* **29**, 326.

NEEL J.V., ROBINSON A.R., ZUELZER W.W., LIVINGSTONE F.B. & SUTTON H.E. (1961) The frequency in the A_2 and fetal hemoglobin fractions in the natives of Liberia and adjacent regions, with data on haptoglobin and transferrin types. *Am. J. hum. Genet.* **13**, 262.

NEUHAUS O.W. & SOGOIAN V.P. (1961) Presence of haptoglobin in synovial fluid. *Nature* **192**, 558.

NEUSTEIN H.B. (1966) The kinetics of clearance of plasma hemoglobin in the rabbit. *Lab. Invest.* **14**, 2133.

NG A., OWEN J.A. & PADANYI R. (1963) Haptoglobins in pleural and ascitic fluids. *Clin. chim. Acta* **8**, 145.

NOSSLIN B.F. & NYMAN M. (1958) Haptoglobin determination in diagnosis of haemolytic diseases. *Lancet* **i**, 1000.

NOYES W.D. & GARBY L. (1967) Rate of haptoglobin synthesis in normal man. *Scand. J. clin. lab. Invest.* **20**, 33.

NOYES W.D. & LAURELL C.B. (1961) The *in vitro* stability of the haptoglobin-hemoglobin complex. *Scand. J. clin. lab. Invest.* **13**, 625.

NYMAN M. (1957) Haptoglobin in pernicious anemia. *Scand. J. clin. lab. Invest.* **9**, 168.

NYMAN M. (1958) Über Haptoglobinbestimmung im Serum, Normalkonzentration und Verhältnis zu Smithies Serumgruppen. *Clin. chim. Acta* **3**, 111.

NYMAN M. (1959) Serum haptoglobin: methodological and clinical studies. *Scand. J. clin. lab. Invest.* suppl. 39.

NYMAN M. (1960) On plasma proteins with heme or hemoglobin binding capacity. *Scand. J. clin. lab. Invest.* **12**, 121.

NYMAN M., GYDELL K. & NOSSLIN B. (1959) Haptoglobin und Erythrokinetik. *Clin. chim. Acta* **4**, 82.

ORNSTEIN L. (1964) Disc electrophoresis. I. Background and theory. *Ann. N.Y. acad. Sci.* **121**, 321.

OSTROW J.D., JANDL J.H. & SCHMID R. (1962) The formation of bilirubin from hemoglobin *in vivo*. *J. clin. Invest.* **41**, 1628.

OWEN J.A., BETTER F.C. & HOBAN J. (1960a) A simple method for the determination of serum haptoglobins. *J. clin. Path.* **13**, 163.

OWEN J.A., CAREW J.P., COWLING D.C., HOBAN J.P. & SMITH H. (1960b) Serum haptoglobins in megaloblastic anaemia. *Br. J. Haemat.* **6**, 242.

OWEN J.A., DE GRUCHY G.C. & SMITH H. (1960c) Serum haptoglobins in haemolytic states. *J. clin. Path.* **13**, 478.

OWEN J.A., MACKAY I.R. & GOT C. (1959) Serum haptoglobins in hepatobiliary disease. *Br. med. J.* **ii**, 1454.

OWEN J.A., SMITH H., PADANYI R. & MARTIN J. (1964) Serum haptoglobin in disease. *Clin. Sci.* **26**, 1.

PARKER W.C. & BEARN A.G. (1963) Control gene mutation as a possible explanation of certain haptoglobin phenotypes. *Am. J. hum. Genet.* **15**, 159.

PEACOCK A.C., BUNTING L.S. & QUEEN K.G. (1965) Serum protein electrophoresis in acrylamide gel: patterns from normal human subjects. *Science* **147**, 1451.

PEACOCK A.C. (1966) Serum haptoglobin type and leukemia: an association with possible etiological significance. *J. natn. Cancer Inst.* **36**, 631.

PETERS J.H. & ALPER C.A. (1966) Haptoglobin synthesis. II. Cellular localization studies. *J. clin. Invest.* **45**, 314.

PLANAS J. (1963) Los tipos de haptoglobinas en el hombre. Datos sobre una poblacion Castellana. *Genet. Iber.* **15**, 103.

PLANAS J., VIÑAS J., DE CASTRO S., ARRIBAS J.M. & MARTIN-MATEO M.C. (1965) The values of haptoglobins and their relation to the genetic type in a group of donors. *Revta esp. Fisiol.* **21**, 15.

POLONOVSKI M. & JAYLE M.F. (1938) Existence dans le plasma sanguin d'une substance activant l'action peroxydasique de l'hémoglobine. *C. r. Soc. Biol.* **129**, 457.

POLONOVSKI M. & JAYLE M.F. (1940) Sur la préparation d'une nouvelle fraction des proteines plasmatiques, l'haptoglobine. *C. r. acad. Sci.* **211**, 517.

POULIK M.D. (1957) Starch gel electrophoresis in a discontinuous system of buffers. *Nature* **180**, 1477.

PROKOP O. & BUNDSCHUH G. (1963) *Die Technik und die Bedeutung der Haptoglobine und Gm-gruppen in Klinik und Gerischtsmedizin.* Walter de Gruyter & Co. Berlin.

PROKOP O., BUNDSCHUH G. & FALK H. (1961) Neue Ergebnisse auf dem Gebiete der Haptoglobine. *Dtsch. Z. ges. gericht. Med.* **51**, 480.

RAFELSON M.E., CLOAREC L., MORETTI J. & JAYLE M.F. (1961) Action of neuraminidase on haptoglobin. *Nature* **191**, 279.

RAMOT B., KENDE G. & ARNON A. (1962) Johnson type haptoglobin. *Nature* **196**, 176.

RATCLIFF A.P. & HARDWICKE J. (1964) Estimation of serum haemoglobin-binding capacity (haptoglobin) on Sephadex G100. *J. clin. Path.* **17**, 676.

RAUSEN A.R., GERALD P.S. & DIAMOND L.K. (1961) Haptoglobin patterns in cord blood serums. *Nature* **191**, 717.

RAYMOND S. (1964) Acrylamide gel electrophoresis. *Ann. N.Y. Acad. Sci.* **121**, 350.

REERINK-BRONGERS E.E., PRINS H.K. & KRIJNEN H.W. (1962) Haptoglobin and increased haemolysis. *Vox Sang.* **7**, 619.

RENWICK J.H. & MARSHALL H. (1966) A new type of human haptoglobin, Hp 2-1D. *Ann. hum. Genet.* **29**, 389.

RITCHIE R.F., HARTER J.G. & BAYLES T.B. (1966) Refinements of acrylamide electrophoresis. *J. Lab. clin. Med.* **68**, 842.

RITTER H. & HINKELMANN K. (1966) Zur Balance des Polymorphismus der Haptoglobine. *Humangenetik.* **2**, 21.

ROBERT L., BOUSSIER G. & JAYLE M.F. (1957) Groupements sulfhydrique de l'haptoglobine et de sa combinaison hémoglobinique. *Experientia* **13**, 111.

ROBERT L., MOMBELLONI P. & CROSTI P. (1961) Studies on serum haptoglobin in experimental connective tissue disorders. *Proc. Soc. exp. Biol.* **107**, 499.

ROBSON E.B., GLEN-BOTT A.M., CLEGHORN T.E. & HARRIS H. (1964) Some rare haptoglobin types. *Ann. hum. Genet.* **28**, 77.

ROWE D.S. (1961) A rapid method for the estimation of serum haptoglobin. *J. clin. Path.* **14**, 205.

ROWE D.S. & SOOTHILL J.F. (1960) Observations on globin, haemoglobin and haptoglobin using rabbit antiglobulin serum. *Nature* **186**, 975.

SASS M.D. & SPEAR P.W. (1962) Estimation of haptoglobin, using haemoglobin labelled with radioactive iron-59. *Nature* **193**, 285.

SCHULTZE H.E. & HEREMANS J.F. (1966) *Molecular Biology of Human Proteins*, vol. 1, p. 384. Elsevier, Amsterdam.

SCHWERD W. & SANDER I. (1967) Gen-Defekte im Haptoglobin-System. *Blut* **15**, 99.

SEARS D.A., ANDERSON P.R., FOY A.L., WILLIAMS H.L. & CROSBY W.H. (1966) Urinary iron excretion and renal metabolism of hemoglobin in hemolytic diseases. *Blood* **28**, 708.

SHIM B.S. & BEARN A.G. (1964a) Immunological and biochemical studies on serum haptoglobin. *J. exp. Med.* **120**, 611.

SHIM B.S. & BEARN A.G. (1964b) The distribution of haptoglobin subtypes in various populations, including subtype patterns in some non-human primates. *Am. J. hum. Genet.* **16**, 477.

SHIM B.S. & LEE T.H. (1966) Genetic control mechanism of haptoglobin synthesis: an operator gene mutation hypothesis for the explanation of certain phenotypes. *New Med. J.* **9**, 77.

SHIM B.S., LEE T.H. & KANG Y.S. (1965) Immunological and biochemical investigations of human serum haptoglobin: composition of haptoglobin-haemoglobin intermediate, haemoglobin-binding sites and presence of additional alleles for β chain. *Nature* **207**, 1264.

SHINTON N.K., RICHARDSON R.W. & WILLIAMS J.D.F. (1965) Diagnostic value of serum haptoglobins. *J. clin. Path.* **18**, 114.

SIMON S.R. & KONIGSBERG W.H. (1966) Chemical modification of hemoglobins: a study of conformation restraint by internal bridging. *Proc. natn. Acad. Sci.* **56**, 749.

SINGH P.J. (1967) A revised staining procedure for haptoglobin. *Vox Sang.* **12**, 78.

SINISCALCO M., BERNINI L., LA TORRETTA G., DEL BIANCO C. & MARSICO S. (1963) Preliminary data suggesting a possible influence of the mother's genotype on foetal haptoglobin synthesis. *Acta genet.* **13**, 235.

SMITH H., EDMAN P. & OWEN J.A. (1962) N-terminal amino-acids of human haptoglobins. *Nature* **193**, 286.

SMITH H. & OWEN J.A. (1961) The determination of haptoglobin in normal human serum. *Biochem. J.* **78**, 723.

SMITH M.J. & BECK W.S. (1966) Interaction between haptoglobin and isolated α and β subunits of human hemoglobin. *J. clin. Invest.* **45**, 1074 (abstract).

SMITHIES O. (1955) Zone electrophoresis in starch gels: group variations in the serum proteins of normal human adults. *Biochem. J.* **61**, 629.

SMITHIES O. (1959a) Zone electrophoresis in starch gels and its application to studies of serum proteins, *in* ANFINSEN C.B., ANSON M.L., BAILEY K. & EDSALL J.T. (eds.) *Advances in Protein Chemistry*, vol. 14, p. 65. Academic Press, New York.

SMITHIES O. (1959b) An improved procedure for starch-gel electrophoresis: further variations in the serum proteins of normal individuals. *Biochem. J.* **71**, 585.

SMITHIES O. (1962) Molecular size and starch electrophoresis. *Arch. Biochem. Biophys.* suppl. 1, 125.

SMITHIES O. (1964) Chromosomal rearrangements and protein structure. *Cold Spring Harbor Symposia on Quantitative Biology* **29**, 309.

SMITHIES O. (1965) Disulfide-bond cleavage and formation in proteins. *Science* **150**, 1595.

SMITHIES O. & CONNELL G.E. (1959) Biochemical aspects of the inherited variations in human serum haptoglobins and transferrins, *in* WOLSTENHOLME G.E.W. & O'CONNOR C.M. (eds.) *Biochemistry of Human Genetics*, p. 178. Little, Brown, Boston.

SMITHIES O. & CONNELL G.E. (1960) *Transactions of First Conference on Genetics*, SUTTON H.E. (ed.) p. 129. Josiah Macy Foundation, New York.

SMITHIES O., CONNELL G.E. & DIXON G.H. (1962a) Inheritance of haptoglobin subtypes *Am. J. hum. Genet.* **14**, 14.

SMITHIES O., CONNELL G.E. & DIXON G.H. (1962b) Chromosomal rearrangements and evolution of haptoglobin genes. *Nature* **196**, 232.

SMITHIES O., CONNELL G.E. & DIXON G.H. (1966) Gene action in the human haptoglobins. I. Dissociation into constituent polypeptide chains. *J. mol. Biol.* **21**, 213.

SMITHIES O. & WALKER N.F. (1956) Notation for serum protein groups and the genes controlling their inheritance. *Nature* **178**, 694.

SMYTH D.S. & UTSUMI S. (1967) Structure at the hinge region in rabbit immunoglobulin-G. *Nature* **216**, 332.

SUTTON H.E. (1965) Biochemical genetics and man: accomplishments and problems. *Science* **150**, 858.

SUTTON H.E. & KARP G.W. (1964) Variations in heterozygous expression at the haptoglobin locus. *Am. J. hum. Genet.* **16**, 419.

TARUKOSKI P.H. (1966) Quantitative spectrophotometric determination of haptoglobin. *Scand. J. clin. lab. Invest.* **18**, 80.

VALERI C.R., BOND J.C., FOWLER K. & SOBUCKI J. (1965) Quantitation of serum haptoglobin-binding capacity using cellulose acetate membrane electrophoresis *Clin. Chem.* **11**, 581.

VENEZIALE C.M., McGUCKIN W.F., HERMANS P.E. & MANKIN H.T. (1966) Hypohaptoglobinemia and valvular heart disease: association with hemolysis after insertion of valvular prostheses and in cases in which operation had not been performed. *Mayo Clinic Proc.* **41**, 657.

WAKS M., ALFSEN A. & CITTANOVA N. (1967) The subunit structure of haptoglobulins. *Biochem. biophys. Res. Comm.* **27**, 693.

WEERTS G., NIX W. & DEICHER H. (1966) Isolierung und nähere charakterisierung eines neuen Haptoglobins: Hp-Marburg. *Blut* **12**, 65.

WHEBY M.S., BARRETT O. & CROSBY W.H. (1960) Serum protein binding of myoglobin, hemoglobin and hematin. *Blood* **16**, 1579.

WHITTEN C.F. (1962) Studies on serum haptoglobin: a functional inquiry. *New Engl. J. Med.* **266**, 529.

WILLIAMS C.A., ASOFSKY R. & THORBECKE G.J. (1963) Plasma protein formation *in vitro* by tissues from mice infected with staphylococci. *J. exp. Med.* **118**, 315.

WOODWORTH R.C. & CLARK L.G. (1967) An improved vertical polyacrylamide gel electrophoresis apparatus. Application to typing and subtyping of haptoglobins. *Analyt. Biochem.* **18**, 295.

REFERENCES TO HAPTOGLOBIN
GEOGRAPHIC DISTRIBUTION

ADAM A., BAT-MIRIAM M., BARNICOT N.A., LEHMANN H., MOURANT A.E., RAMOT B., SHEBA C. & SZEINBERG A. (1961) A survey of some genetical characters in Ethiopian tribes. *2nd Int. Conf. Hum. Genet.*, p. E54. Excerpta Medica Found., New York.

ALLISON A.C. & BARNICOT N.A. (1960) Haptoglobins and transferrins in some East Africa peoples. *Acta genet.* **10**, 17.

ALLISON A.C., BLUMBERG B.S. & AP REES W. (1958) Haptoglobin types in British, Spanish Basque and Nigerian African populations. *Nature* **181**, 824.

ALLISON A.C., BLUMBERG B.S. & GARRY B. (1960) Haptoglobins and haemoglobins of Alaska Eskimos and Indians. *Ann. hum. Genet.* **23**, 349.

ANGELOPOULOS B., TSOUKANTOS A. & DANOPOULOS E. (1966) Distribution of haptoglobin subtypes in Greeks. *J. med. Genet.* **3**, 276.

ARENDS T. & GALLANGO M.L. (1960) Haptoglobin types in a Paraujano Indian population. *Vox Sang.* **5**, 452.

ARENDS T. & GALLANGO M.L. (1962) Frecuencia de haptoglobinas en varias poblaciones Suramericanas. *Acta cient. venez.* **13**, 116.

ARENDS T. & GALLANGO M.L. (1964) Frequencies of haptoglobin types in various South American populations. *Proc. 9th Cong. int. Soc. Blood Transf.*, Mexico City, 1962, p. 463.

ARENDS T. & GALLANGO M.L. (1965) Haemoglobin types and blood serum factors in British Guiana Indians. *Br. J. Haemat.* **11**, 350.

BAITSCH H., LIEBRICH K.G., PINKERTON F.J. & MERMOD L.E. (1962) Zur Populationsgenetik der Haptoglobinserum-Gruppen Allelenhaufigkeit in Europa und Ozeanien. *Acta Genet. med. Gemell.* **11**, 308.

BAITSCH H. & MEIER G. (1959) Zur Verteilung der Haptoglobintypen in Bayern. *Blut* **5**, 302.

BAITSCH H., MEIER G., SCHOELLER L. & KAHLICH-KOENNER D.M. (1960) Frequencies of the haptoglobin serum groups among blood donors from Austria and Germany. *Nature* **186**, 976.

BAITSCH H., RITTER H., GOEDDE H.W. & ALTLAND K. (1963) Zur Genetik der Serumproteine: Hp-serumgruppen, Gc-faktor, Gm-serumgruppen und Pseudocholinesterase-varienten in Europäischen Populationen. *Vox Sang.* **8**, 594.

BARNICOT N.A., GARLICK J.P., ADAM A. & BAT-MIRIAM M. (1962) A survey of some genetical characters in Ethiopian tribes. III. Haptoglobins and transferrins. *Am. J. phys. Anthrop.* **20**, 175.

BARNICOT N.A., GARLICK J.P. & ROBERTS D.F. (1960) Haptoglobin and transferrin inheritance in northern Nigerians. *Ann. hum. Genet.* **24**, 171.

BARNICOT N.A., GARLICK J.P., SINGER R. & WEINER J.S. (1959) Haptoglobin and transferrin variants in Bushmen and some other South African peoples. *Nature* **184**, 2042.

BARNICOT N.A. and KARIKS J. (1960) Haptoglobin and transferrin variants in peoples of the New Guinea highlands. *Med. J. Aust.* **ii**, 859.

BARROS F. (1965) Frèquence des haptoglobines dans la population du Portugal metropolitain. *9th Cong. Europ. Soc. Hemat.* 1963, p. 26. Karger, Basel.

BAXI A.J. & HAKIM S.M.A. (1966) Haptoglobin types in Marathas of Bombay. *Indian J. med. Res.* **54**, 1150.

BECKMAN L. & JOHANNSSON E.O. (1967) Haptoglobins and transferrins in the Icelandic population. *Acta genet.* **17**, 341.

BECKMAN L., JOHNSON F.M., SAKAI R.K. & WOODS J.L. (1964) Serum protein variations in Hawaiian population groups. *Acta genet.* **14**, 309.

BECKMAN L. & MELLBIN T. (1959) Haptoglobin types in the Swedish Lapps. *Acta genet.* **9**, 306.

BECKMAN L., TAKMAN J. & ARFORS K.E. (1965) Distribution of blood and serum groups in a Swedish

Gypsy population. *Acta genet.* **15**, 134.

BENERECETTI-SANTACHIARA A.S. & MODIANO G. (1964) The frequencies of haptoglobin and transferrin types in some villages of the Milan province. *Acta genet.* **14**, 36.

BENNETT J.H. (1961) Haptoglobin types in natives from the Kuru region and other parts of Melanesia. *Br. med. J.* **ii**, 428.

BENNETT J.H., AURICHT C.O., GRAY A.J., KIRK R.L. & LAI L.Y. (1961) Haptoglobin and transferrin types in the Kuru region of Australian New Guinea. *Nature* **189**, 68.

BERNINI L., LATTE B., MODIANO G., POLOSA P. & SINISCALCO M. (1963) Ulteriori dati sulla distribuzione dei tipi aptoglobinici in Italia. *Rc. Accad. naz. Lincei.* **34**, 308.

BLACKWELL R.Q., CHEN H.H. & CHEN H.C. (1964) Haptoglobin distribution in a Filipino population. *Nature* **202**, 814.

BLACKWELL R.Q. & THEPHUSDIN C. (1963) Distribution of haptoglobin among Thais. *Nature* **197**, 503.

BLACKWELL R.Q., TSAU-YEN L., SHIAO D.D.F. (1962) Distribution of haptoglobins among Chinese in Taiwan. *Nature* **193**, 284.

BLUMBERG B.S., ALLISON A.C. & GARRY B. (1959) The haptoglobins and haemoglobins of Alaskan Ekimos and Indians. *Ann. hum. Genet.* **23**, 349.

BLUMBERG B.S. & GENTILE Z. (1961) Haptoglobins and transferrins of two tropical populations. *Nature* **189**, 897.

BOWMAN J.E. (1964) Haptoglobin and transferrin differences in some Iranian populations. *Nature* **201**, 88.

BUDTZ-OLSEN O.E. (1958) Haptoglobins and haemoglobins in Australian aborigines, with a simple method for the estimation of haptoglobins. *Med. J. Aust.* **ii**, 689.

BÜTLER R., METAXAS-BÜHLER M., ROSIN S. & WANDREY R. (1959) Untersuchungen über die Haptoglobingruppen von Smithies. *Schweiz. med. Wschr.* **89**, 1041.

CLEVE, H. (1966) Die Verteilung der Haptoglobin-Untergruppen in einer Stichprobe gesunder Blutspender aus Hessen. *Humangenetik* **21**, 115.

CURTAIN C.C., GAJDUSEK D.C., KIDSON C., GORMAN J.G., CHAMPNESS L. & RODRIQUE R. (1965) Haptoglobins and transferrins in Melanesia: relation to hemoglobin, serum haptoglobin and serum iron levels in population groups in Papua-New Guinea. *Am. J. phys. Anthrop.* **23**, 363.

CURTAIN C.C., TINDALE N.B. & SIMMONS R.T. (1966) Genetically determined blood protein factors in Australian aborigines of Bentinck, Mornington and Forsyth Islands and the mainland, Gulf of Carpentaria. *Arch. Phys. Anthrop. Oceania* **1**, 74.

DOUGLAS R., JACOBS J., GREENHOUGH R. & STAVELEY J.M. (1964) Blood groups, serum genetic factors and hemoglobins in New Hebrides Islanders. *Transfusion* **4**, 177.

DOUGLAS R., JACOBS J., HOULT G.E. & STAVELEY J.M. (1962) Blood groups, serum genetic factors and hemoglobins in western Solomon Islanders. *Transfusion* **2**, 413.

DOUGLAS R., JACOBS J., SHERLIKER J. & STAVELEY J.M. (1961) Blood groups, serum genetic factors and haemoglobins in Ellice Islanders. *N.Z. med. J.* **60**, 259.

DOUGLAS R., JACOBS J., SHERLIKER J. & STAVELEY J.M. (1961) Blood groups, serum genetic factors and haemoglobins in Gilbert Islanders. *N.Z. med. J.* **60**, 146.

DOUGLAS R. & STAVELEY J.M. (1960) Haptoglobins in Tongans (Polynesia) *N.Z. med. J.* **ii**, 391.

FLEISCHER E.A. & LUNDEVALL J. (1957) Inheritance of serum groups. *Proc. 6th Cong. europ. Soc. Haemat.* Copenhagen, vol. 2, p. 906. Karger, Basel.

FLORY L.L. (1963) Haptoglobin types of the Nez Perce Indians. *Nature* **197**, 578.

FLORY L.L. (1964) Serum factors of Australian aborigenes from North Queensland. *Nature* **201**, 508.

FRASER G.R., GIBLETT E.R., LEE T.C. & MOTULSKY A.G. (1965) Blood and serum groups in Taiwan. *J. med. Genet.* **2**, 21.

FRASER G.R., GIBLETT E.R., STRANSKY E. & MOTULSKY A.G. (1964) Blood groups in the Philippines. *J. med. Genet.* **1**, 107.

FRIED K., BLOCH N., SUTTON E., NEEL J.V., BAYANI-SIOSON P., RAMOT B. & DUVDEVANI P. (1963) Haptoglobins and transferrins, *in* GOLDSCHMIDT E. (ed.) *The Genetics of Migrant and Isolate Populations*, p. 266. Williams & Wilkins, Baltimore.

FUSTÉ M., PLANAS J. & DIAZ J.M. (1965) Avance de un estudio sobre la distribución de los tipos de haptoglobinas en la población de las Islas Canarias. *El Museo Canario, Las Palmas de Gran Canaria*, p. 1.

GALATIUS-JENSEN F. (1958) On the genetics of haptoglobins. *Acta genet.* **8**, 232.

GALATIUS-JENSEN F. (1960) The haptoglobins. A genetic study. *Dansk Videnskabs, Forlag*, Copenhagen.

GALLANGO M.L. & ARENDS T. (1959) Distribution of haptoglobins in native Venezuelans. *Nature* **183**, 1465.

GALLANGO M.L. & ARENDS T. (1966) Haemoglobin types and blood serum factors in Columbian Indians *Acta genet.* **16**, 162.

GIBLETT E.R. (1962) Haptoglobins and transferrins in Pacific populations. *Eugen. Q.* **9**, 45.

GIBLETT E.R. & BEST W.R. (1961) Haptoglobin and transferrin types in Peruvian Indians. *Nature* **192**, 1300.

GIBLETT E.R. & BROOKS L.E. (1963) Haptoglobin subtypes in three racial groups. *Nature* **197**, 576.

GIBLETT E.R., MOTULSKY A.G. & FRASER G.R. (1966) Population genetics in the Congo. IV. Haptoglobin and transferrin serum groups in the Congo and in other African populations. *Am. J. hum. Genet.* **18**, 553.

GLOWATZKI G. & PAIDOUSSIS J. (1964) Die Verteilung der Haptoglobintypen in Greichenland, Untersuchungen von 1311 Seren gesunder Blutspender. *Z. biol.* **114**, 285.

GOLDSCHMIDT E., BAYANI-SIOSON P., SUTTON H.E., FRIED K., SANDOR A. & BLOCK N. (1962) Haptoglobin frequencies in Jewish communities. *Ann. hum. Genet.* **26**, 39.

GRÜNWALD P. & HERMAN C. (1964) Study of several gene frequencies in the Yugoslav population. *Nature* **199**, 830.

HARRIS H. & ROBSON E.B. (1963) Haptoglobins in Tristan da Cunha. *Vox Sang.* **8**, 226.

HARRIS H., ROBSON E.B. & SINISCALCO M. (1959) Distribution of serum haptoglobin types in some Italian populations. *Ciba Foundation Symposium: Medical Biology & Etruscan Origins*, p. 220. Little, Brown, Boston.

HASHEM N., KAMEL K. & HAMMOND E.I. (1966) Haptoglobin phenotypes among Egyptians. *J. med. Genet.* **3**, 279.

JENKINS T. & STEINBERG A.G. (1966) Some serum protein polymorphisms in Kalahari Bushmen and Bantu: gamma globulins, haptoglobins and transferrins. *Am. J. hum. Genet.* **18**, 399.

JOE J.T.T., PRINS H.K. & NIJENHUIS L.E. (1965) Hereditary and acquired blood factors in the Negroid population of Surinam. I. Origin, collection and transport of the blood samples; characteristic blood groups. *Trop. geogr. Med.* **5**, 56.

KAMEL K., DAVIS S.H. & CUMMING R.A. (1963) A comparison of haptoglobin phenotypes in haemophilics and normals in Scotland. *Vox Sang.* **8**, 219.

KIRK R.L. (1965) Population genetic studies of the indigenous peoples of Australia and New Zealand, *in* STEINBERG A.G. & BEARN A.G. (eds.) *Progress in Medical Genetics*, vol. 4, p. 202. Grune & Stratton, New York.

KIRK R.L. & LAI L.Y. (1961) The distribution of haptoglobin and transferrin groups in South and Southeast Asia. *Acta genet.* **11**, 97.

KIRK R.L., LAI L. & HOGBEN D.L. (1960a) Haptoglobin groups of white Australians. *Med. J. Aust.* **i**, 45.

KIRK R.L., LAI L.Y.C. & HORSFALL W.R. (1962a) The haptoglobin and transferrin groups among Australian aborigines from North Queensland. *Aust. J. Sci.* **24**, 486.

KIRK R.L., LAI L.Y.C., MAHMOOD S. & SINGH R.B. (1960b) Haptoglobin types in south-east Asia. *Nature* **185**, 185.

KIRK R.L., LAI L.Y.C., VOS G.H. & VIDYARTHI L.P. (1962b) A genetical study of the Oraons of the Nagpur Plateau (Bihar, India). *Am. J. phys. Anthrop.* **20**, 375.

KIRK R.L., LAI L.Y.C., VOS G.H., WICKREMASINGHE R.L. & PERERA D.J.B. (1962c) The blood and serum groups of selected populations in South India and Ceylon. *Am. J. phys. Anthrop.* **20**, 485.

KOBIELA J. (1961) Eigene Untersuchingen über Haptoglobine: Die Frequenz der Hp-Gruppen in Polen. *Z. hyg. Grenz.* **7**, 399.

KOBIELA J., TUROWSKA B. & MOSTOWSKI J. (1961) The Hp and Gm groups in Poland. *Vox Sang.* **7**, 638.

LATORRETTA F., DEL BIANCO C. & MARSICO S. (1962) Distribuzione dei fenotipi aptoglobinici nella populazione napoletana. *Archo. Ostet. Ginec.* **67**, 574.

LIE-INJO L.E., BOLTON J.M. & FUDENBERG H.H. (1967) Haptoglobins, transferrins and serum gamma globulin types in Malayan aborigines. *Nature* **215**, 777.

MÄKELÄ O., ERIKSSON A.W. & LEHTOVARRA R. (1959) On the inheritance of the haptoglobin serum groups. *Acta genet.* **9**, 149.

MATSON G.A., SUTTON H.E., ETCHEVERRY R.B., SWANSON J. & ROBINSON A. (1967) Distribution of hereditary blood groups among Indians in South America. IV. In Chile. *Am. J. phys. Anthrop.* **27**, 157.

MATSON G.A., SUTTON H.E., SWANSON J. & ROBINSON A.R. (1963) Distribution of haptoglobin, transferrin, and hemoglobin types among Indians of Middle America: Southern Mexico, Guatemala, Honduras and Nicaragua. *Hum. Biol.* **35**, 474.

MATSON G.A., SUTTON H.E., SWANSON J. & ROBINSON A.R. (1965) Distribution of haptoglobin, transferrin and hemoglobin types among Indians of Middle America: in British Honduras, Costa Rica and Panama. *Am. J. phys. Anthrop.* **23**, 123.

MATSON G.A., SUTTON H.E., SWANSON J. & ROBINSON A. (1966a) Distribution of hereditary blood groups among Indians in South America. II. In Peru. *Am. J. phys. Anthrop.* **24**, 325.

MATSON G.A., SWANSON J. & ROBINSON A. (1966b) Distribution of hereditary blood groups among Indians in South America. III. In Boliva. *Am. J. phys. Anthrop.* **25**, 13.

MATSUNAGA E. & MURAI K. (1960) Haptoglobin types in a Japanese population. *Nature* **186**, 320.

MICHON P., STREIFF F., KLING C. & NOEL B. (1962) Répartition des groupes d'haptoglobine dans le Nord-Est de la France. *Nouv. Revue fr. Hémat.* **2**, 506.

MODIANO G., BENERECETTI-SANTACHIARA A.S., GONANO F., ZEI G., CAPALDO A. & CAVALLI-SFORZA

L.L. (1965) An analysis of ABO, MN, Rh, Hp, Tf and G-6-PD types in a sample from the human population of the Lecce province. *Ann. hum. Genet.* **29**, 19.

MOULLEC J. & FINE J.M. (1959) Frequencies of the haptoglobin groups in 406 French blood donors. *Nature* **184**, 196.

MOULLEC J., FINE J.M. & LINHARD J. (1960a) Les groupes d'haptoglobine dans un échantillon de population Africaine de Dakar. *Rev. Hémat.* **15**, 74.

MOULLEC J., FINE J.M. & LINHARD J. (1960b) Haptoglobin types in 398 blood donors from Dakar (Senegal). *Nature* **187**, 517.

MURAWSKI K. & MISZCZAK T. (1961) Haptoglobin types in Poland. *Science* **133**, 1427.

NAGEL R. & ETCHEVERRY R. (1963) Types of haptoglobin in Araucanian Indians of Chile. *Nature* **197**, 187.

NAGEL R., ETCHEVERRY R. & GUZMAN G. (1964) Haptoglobin types in inhabitants of Easter Island. *Nature* **201**, 216.

NEEL J.V., ROBINSON A.R. ZUELZER W.W., LIVINGSTONE F.B. & SUTTON H.E. (1961) The frequency of the A_2 and fetal hemoglobin fractions in the natives of Liberia and adjacent regions, with data on haptoglobin and transferrin types. *Am. J. hum. Genet.* **13**, 262.

NEEL J.V., SALZANO F.M., JUNQUEIRA R.C., KEITER F. & MAYBURY-LEWIS D. (1964) Studies on the Xavante Indians of the Brazilian Mato Grosso. *Am. J. hum. Genet.* **16**, 52.

NICHOLLS E.M., LEWIS H.B.M., COOPER D.W. & BENNETT J.H. (1965) Blood and serum protein differences in some central Australian aborigines. *Am. J. hum. Genet.* **17**, 293.

PÁLSSON J. & WALTER H. (1967) Untersuchungen zur Population-genetik von Island, insbesondere der Region Dalasýsla. *Human-genetik* **4**, 352.

PARKER W.C. & BEARN A.G. (1961a) Haptoglobin and transferrin variation in humans and primates: two new transferrins in Chinese and Japanese populations. *Ann. hum. Genet.* **25**, 227.

PARKER W.C. & BEARN A.G. (1961b) Haptoglobin and transferrin gene frequencies in a Navajo population: a new transferrin variant. *Science* **134**, 106.

PERSSON I. (1962) The main haptoglobin types in Greenland Eskimos. *Acta genet.* **12**, 292.

PERSSON I. & TINGSGAARD P. (1966) Serum types in East Greenland Eskimos. *Acta genet.* **16**, 84.

PLANAS J. (1963) Los tipos de hapto-globinas en el hombre. Datos sobre una poblacion Castellana. *Genet. Iber.* **15**, 103.

PLANAS J., DE CASTRO S. & ARRIBAS J.M. (1965) Haptoglobin types in the population of Castilla. *Acta genet.* **15**, 140.

PLANAS J., FUSTÉ M., VIÑAS J. & IRIZAR J.L. (1966) Haptoglobin types in the Iberian peninsula. *Acta genet.* **16**, 371.

PLATO C.C. & CRUZ M. (1966) Blood group and haptoglobin frequencies of the Trukese of Micronesia. *Acta genet.* **16**, 74.

PLATO C.C. & CRUZ M. (1967) Blood group and haptoglobin frequencies of the Chamorros of Guam. *Am. J. hum. Genet.* **19**, 722.

PLATO C.C., RUCKNAGEL D.L. & GERSHOWITZ H. (1964) Studies on the distribution of glucose-6-phosphate dehydrogenase deficiency, thalassemia and other genetic traits in the coastal and mountain villages of Cyprus. *Am. J. hum. Genet.* **16**, 267.

PLATO C.C., RUCKNAGEL D.L. & KURKLAND L.T. (1966) Blood group investigations on the Carolinians and Chamorros of Saipan. *Am. J. phys. Anthrop.* **24**, 147.

RAMOT B., ZIKERT-DUVDEVANI P. & TAUMAN G. (1961) Distribution of haptoglobin types in Israel. *Nature* **192**, 765.

RAMOT B., ZIKERT-DUVDEVANI P. & KENDE G. (1962) Haptoglobin and transferrin types in Israel. *Ann. hum. Genet.* **25**, 267.

SALZANO F.M. & SUTTON H.E. (1963) Haptoglobin and transferrin types in Southern Brazilian Indians. *Acta genet.* **13**, 1.

SALZANO F.M. & SUTTON H.E. (1965) Haptoglobin and transferrin types of Indians from Santa Catarina, Brazil. *Am. J. hum. Genet.* **17**, 280.

SANFORD R., GRIMMO A.E.P. & LEE S.K. (1966) The haptoglobin and transferrin types of some Cantonese in Hong Kong. *Vox Sang.* **11**, 106.

SCOTT E.M., DUNCAN I.W., EKSTRAND V. & WRIGHT R.C. (1966) Frequency of polymorphic types of red cell enzymes and serum factors in Alaskan Eskimos and Indians. *Am. J. hum. Genet.* **18**, 408.

SHIM B.S. & BEARN A.G. (1964) The distribution of haptoglobin subtypes in various populations, including subtype patterns in some non-human primates. *Am. J. hum. Genet.* **16**, 477.

SHREFFLER D.C. & STEINBERG A.G. (1967) Further studies on the Xavante Indians. IV. Serum protein groups and the SC_1 trait of saliva in the Simões Lopes and São Marcos Xavantes. *Am. J. hum. Genet.* **19**, 514.

SINGER R. (1961) Serum haptoglobins in Africa. *S.A. med. J.* **35**, 520.

SONNET J. & MICHAUX J.L. (1960) Glucose-6-phosphate-dehydrogenase deficiency, haptoglobin groups and sickle cell trait in the Bantus of West Belgian Congo. *Nature* **188**, 504.

STEINBERG A.G., LAI L.Y.C., VOS G.H., SINGH R.B. & LIM T.W. (1961) Genetic and population studies of the blood types and serum factors among Indians and Chinese from Malaya. *Am. J. hum. Genet.* **13**, 355.

STEINBERG A.G. & MATSUMOTO H. (1964) Studies on the Gm, Inv, Hp and Tf serum factors of Japanese populations and families. *Hum. Biol.* **36**, 77.

SUTTON H.E., MATSON G.A., ROBINSON A.R. & KOUCKY R.W. (1960) Distribution of haptoglobin, transferrin, and hemoglobin types among Indians of Southern Mexico and Guatemala. *Am. J. hum. Genet.* **12**, 338.

SUTTON H.E., NEEL J.V., BINSON G. & ZUELZER W.W. (1956) Serum protein differences between Africans and Caucasians. *Nature* **178**, 1287.

SUTTON H.E., NEEL J.V., LIVINGSTONE F.G., BENSON G., KUNSTADTER P. & TROMBLEY L. (1959) The frequencies of haptoglobin types in five populations. *Ann. hum. Genet.* **23**, 175.

TARUKOSKI P.H. (1959) Haptoglobingruppernas Frekvens i en Mellansvensk Befolkning. *Nord. Med.* **62**, 1425.

TIWARI S.C. (1960) Haptoglobin and transferrin variants in some upper castes of Bengal. *Anthropologist* **7**, 17.

TONDO C.V., MUNDT C. & SALZANO F.M. (1963) Haptoglobin types in Brazilian Negroes. *Ann. hum. Genet.* **26**, 325.

TORRINHA J.A.F. (1967) Haptoglobin frequencies in the North of Portugal *Acta genet.* **17**, 74.

VAN ROS G., VAN SANDE M. & DRUET R. (1963) Groupes d'haptoglobine et de transferrine dans des populations africaines et européenes. *Ann. Soc. belge. Méd. trop.* **4**, 511.

VAN SANDE M., VAN ROS G. & DRUET R. (1963) Determination of haptoglobin group frequencies by starch gel and agar gel electrophoresis: application to Belgian and Barundi populations. *Nature* **197**, 603.

WALTER H. & PÁLSSON J. (1962) Zur

Häufigkeit der serumgruppen in Island. *Vox Sang.* **7**, 732.

WICHMANN D. (1961) Populations-genetische Untersuchungen über das Haptoglobinsystem. *Z. Konst. forsch. menschl. Vererb.* **36**, 93

WICHMANN D. & SCHLEYER F. (1961) Untersuchungen über die Häufigkeit der Haptoglobintypen in Nordrhein-Westfalen und ihre Forensische Verwertbarkeit. *Z. Immun-Forsch. exp. Ther.* **121**, 196.

CHAPTER 3

TRANSFERRIN

Physico-chemical Properties of Transferrin 127
   Preparation 127
   Molecular weight 127
   Carbohydrate moiety 127
   Possible subunit structure 128
   Iron binding by transferrin 129

Transferrin Metabolism and Function 131
   Transferrin turnover 131
   Iron turnover: role of transferrin 132
   Possible role of transferrin in infection 134
   Physiological effect of atransferrinemia 134

Inherited Variants of Transferrin 135
   Detection of Tf polymorphism 135
   Family studies and linkage data 139

Chemical structure of Tf variants 140
Function of Tf variants 142
Difficulties in identifying variant transferrins 143
Frequency of variants 144
Geographic distribution of variants 144

Methods 147
   Phenotype differentiation by electrophoresis 147
   Starch gel electrophoresis 147
   Fe^{59} radioautography 148
   Iron staining 148
   Rivanol precipitation 148

References 149

References to Tf geographic distribution 156

Transferrin (or siderophilin) is a β globulin which has the vital function of transporting iron from the plasma to receptor cells of the bone marrow and tissue storage compartment. Over 30 years ago, Starkenstein & Harvalik (1933) localized the iron binding protein in the serum globulin fraction, and a few years later, Vahlquist (1941) identified it as a β globulin by electrophoresis. However, it was the independent work of Holmberg & Laurell (1945) and Schade & Caroline (1946) which established the physiological role of the protein as well as the usefulness of its quantitation in clinical medicine. Thus the importance of transferrin (Tf) was well known when Smithies (1957) demonstrated the existence of heritable variation in its molecular structure. This discovery stimulated numerous genetic and anthropological studies which were summarized by Barnicot (1961), Giblett (1962), Parker (1963) and Bowman (1968). Reviews of the physico-chemical properties of Tf have been presented by Schultze *et al* (1957), Ramsay (1958), Laurell (1960) and Putnam (1965), while Tf physiology, summarized by Laurell (1947, 1960), has been more recently reviewed in the book by Bothwell & Finch (1962) and in the papers of Pollycove (1966) and Brown (1966).

PHYSICO-CHEMICAL PROPERTIES OF TRANSFERRIN

PREPARATION

Surgenor *et al.* (1949) prepared the first crystalline transferrin from Cohn's fraction IV-7, and a similar method of ethanol fractionation was used by Koechlin (1952). Inman (1956) and Inman *et al.* (1961) started with fraction IV-7 and IV-4, respectively, utlilizing ethanol and zinc precipitation followed by chromatographic separation. Kistler *et al.* (1960) precipitated out most of the extraneous plasma proteins with Rivanol (2-ethoxy-6,9-diaminoacridine lactate), and removed the gamma globulin from the transferrin in the supernatant solution by alcohol precipitation and DEAE cellulose column fractionation.

MOLECULAR WEIGHT

The purified protein has a molecular weight of between 86,000 and 92,000 according to a number of workers, including Koechlin (1952), Schultze *et al.* (1957), and Bezkorovainy & Rafelson (1964). Thus, the figure of 90,000 has been used in most calculations. However, the work of Charlwood (1963) suggested that the molecular weight is closer to 70,000, and the recent studies of Roberts *et al.* (1966) were in essential agreement with those of Charlwood. They obtained values within a range of 73,000–76,000 on four different human Tf preparations tested by sedimentation and diffusion, sedimentation equilibrium, osmotic pressure and iron-binding capacity measurements. The authors were unable to explain the large discrepancy between the two reported molecular weight ranges, and the problem is currently unresolved.

CARBOHYDRATE MOIETY

Schultze *et al.* (1958) characterized Tf as a glycoprotein with a 5·5 per cent carbohydrate moiety composed of eight galactose, eight glucosamine, four mannose, and four sialic acid residues as well as approximately one residue of fucose. Schultze & Schwick (1957) showed that sialic acid is an important determinant of Tf electrophoretic mobility, a stepwise pattern of slow-moving components being observed with progressive loss of the four residues (Poulik, 1961; Parker & Bearn, 1961a; Schultze, 1962). Jamieson (1964) suggested that the Tf molecule contains two branched carbohydrate chains with a molecular weight of 2350, each consisting of four galactose, four *N*-acetylglucosamine, and two mannose residues, with sialic acid in the terminal positions. According to Robinson & Pierce (1964), sialic acid is the terminal, and galactose the subterminal residue in each of these chains.

POSSIBLE SUBUNIT STRUCTURE

The amino-acid composition of Tf has been determined in a number of laboratories (Parker & Bearn, 1962a; Bezkorovainy *et al.*, 1963; Heimburger *et al.*, 1964; Roop & Putnam, 1967). Eriksson & Sjoquist (1960) and Parker (1963) reported that there is only a single N-terminal amino-acid residue (valine) per Tf molecule, indicating that it probably consists of a single long polypeptide chain. Nevertheless, the fact that Tf binds two iron atoms per molecule and has two apparently identical carbohydrate prosthetic groups is more consistent with subunit structure. The molecule contains about 17 (Bowman, 1968) to 22 (Schultze & Schwick, 1957) disulfide bonds and no free sulfhydryl groups. Roop and Putnam (1967) found that their peptide maps of reduced and alkylated transferrin contained about 70 of the 85–100 peptides expected from the amino-acid data, but they ascribed the peptide deficit to overlapping of spots, insolubility, or poor staining characteristics of some peptides. Bowman (1968) reported that there were 85 tryptic peptides in her preparations, making subunit structure unlikely unless the two component polypeptide chains were so different that their individual peptides would have little or no homology. Some attempts to separate the molecule into two or more components have been unsuccessful (Charlwood, 1963; Parker, 1963). However, Jeppsson (1967a, b) obtained two chromatographic fractions of similar molecular weight (about 40,000) by gradient elution of reduced and alkylated transferrin with urea-phosphate buffer. Since the peptide fingerprint of his reduced and alkylated transferrin contained only 34–38 peptides, he postulated that the molecule consists of two similar subunits held together by noncovalent bonds, and that one of these subunits contains a blocked N-terminal amino

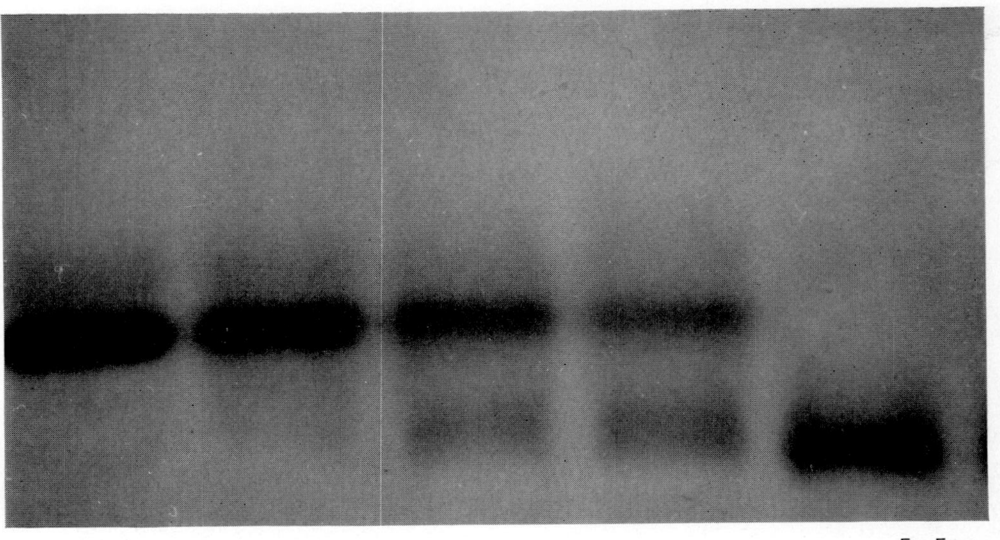

FIGURE 3.1. Electrophoretic patterns on iron-free starch gel of purified transferrin (phenotype Tf C), iron-free on the right and iron-saturated on the left. The three intervening specimens contain increments of iron from right to left. Failure to observe a third component migrating between the iron-free and iron-saturated protein suggests that the binding of the second atom of iron is facilitated by the binding of the first; however, see text. (From Giblett, 1966.)

acid. However, according to Bezkorovainy & Grohlich (1967), the apparent subunits formed when transferrin is treated with urea are actually denaturation products which do not represent two separate polypeptide chains. In view of the discordant findings, further data are required to resolve the questions about both the molecular weight and the number of polypeptides in the transferrin molecule. (See Addenda.)

IRON BINDING BY TRANSFERRIN

The two atoms of ferric iron are linked to the Tf molecule by ionic bonding which is highly dependent on pH (Schade *et al.*, 1949; Ehrenberg & Laurell, 1955). The complex is stable between pH 7·2 and 10 (Laurell, 1947; Schade *et al.*, 1954). Measurable dissociation occurs at pH 6·5 and is complete at about pH 5, but the reaction is reversible, the complex being restored by raising the pH to 7·0 (Surgenor *et al.*, 1949). When fully saturated, Tf is more soluble and heat resistant than when it is free of iron (Azari & Feeney, 1958).

The nature of the ionic bond between Tf and iron is still not entirely clear. For each atom of iron bound, one bicarbonate ion is taken up (Schade *et al.*, 1949) and three protons are liberated (Aasa *et al.*, 1963). The latter authors suggested that the chelating site of Tf involves two imidazole groups in addition to three tyrosyl residues. Komatsu & Feeney (1967) were able to demonstrate that the loss of iron-binding capability following treatment of human transferrin with *N*-acetylimidazole could be reversed by deacylation with hydroxylamine. They considered this finding as strong support for the essential role of tyrosine in the metal-binding property of Tf.

There is, however, a difference of opinion concerning the possibility of interaction between the two binding sites. Warner & Weber (1953) reported that in studies with conalbumin, a protein very similar to Tf, the equilibrium constant for the reaction binding the second iron atom was much greater than that for the first atom, thus creating a strong tendency for the metal ions to attach in pairs. Electrophoretic patterns of free and iron-bound Tf favoring this concept were obtained by Sass-Kortsak & Vamosi (1964) and Giblett (1966) on starch gel, and Yoshioka *et al.*, (1966) on acrylamide gel. In these instances, it was possible to demonstrate only one Fe-Tf complex (i.e. the saturated complex Fe_2Tf) when increments of iron were added to the iron-free protein (see Fig. 3.1). However, the free electrophoretic studies of

Aisen *et al.* (1966) provided evidence for two different Fe-Tf complexes (Fe$_1$Tf and Fe$_2$Tf), and the work of Aasa *et al.* (1963) indicated that the molecular binding sites are equivalent and independent. The experimental conditions varied widely in these investigations, and further studies will be required to resolve their differences.

Fletcher & Huehns (1967) studied the availability of iron from plasma Tf at various levels of iron saturation, before and after the plasma was incubated with reticulocytes. They concluded that the availability of iron differed from one Tf molecule to another, and that this difference was dependent in part upon whether the Tf molecule carried one or two atoms of iron. The (indirect) evidence obtained from their experiments tends to support the view of Aasa *et al.* (1963) that the binding of iron occurs at random.

In vitro, the Tf molecule is capable of binding other metallic ions, such as those of cobalt, copper, manganese and zinc; the binding site(s) for these ions is not the same as the iron-binding site (Jones & Perkins, 1965). It appears unlikely that Tf is the physiological transport protein for these ions (Laurell, 1960), although one study showed that trivalent chromium fed to rats could subsequently be detected in the Tf fraction of their plasma (Hopkins & Schwarz, 1964). *In vitro*, certain salts with a high affinity for iron, such as phosphate and citrate, show some ability to compete with Tf (Schade *et al.*, 1954). However, the very high stability constants of the Tf binding sites for iron reported by Saltman and Charley (1960) and Davis *et al.* (1962) make it very unlikely that such a competition exists *in vivo*.

Iron-free Tf is colorless, and a progressive change from light to dark salmon-pink occurs as increments of iron are added. In the presence of a complexing agent such as α,α'dipyridyl, iron added in excess of the Tf saturation point forms a colored complex. Thus, if the iron content of a given serum or plasma specimen is known, its total iron-binding capacity can be calculated from the amount of additional iron required to reach saturation (Holmberg & Laurell, 1945). Other color reagents, such as α,α',α'' tripyridyl and 4,7 diphenyl-phenanthroline (bathophenanthroline) are also used for this purpose (Schade *et al.*, 1954; Peters *et al.*, 1956; Ramsay, 1957). A more direct measurement of Tf content can be made by immuno-chemical techniques, using anti-human transferrin (Jager & Gubler, 1952; Goodman *et al.*, 1958; Bambach, 1966). A simple method employing radial immunodiffusion in agar is commercially available.

TRANSFERRIN METABOLISM AND FUNCTION

TRANSFERRIN TURNOVER

The liver is probably the major site of Tf synthesis (Dancis *et al.*, 1957; Clausen *et al.*, 1960; Asofsky & Thorbecke, 1961; McCarter *et al.*, 1966; Lane, 1967), although some uptake of labelled amino acids into Tf has also been demonstrated in cultures of mouse spleen as well as monkey lymph nodes and ileum (Asofsky & Thorbecke, 1961; Hochwald *et al.*, 1961). Production of Tf begins at an early stage in human fetal development (Scheidegger *et al.*, 1956) and most, if not all, of the Tf in the blood of newborn infants is of fetal, rather than maternal origin (Rausen *et al.*, 1961).

The average concentration of Tf in human plasma is about 250 mg/ 100 ml, indicating that a 70 kg man has a total plasma Tf of around 7 g. An approximately equal amount is distributed in the extravascular compartment. Since the half-life of Tf is about 7–10 days, the daily Tf turnover amounts to about 12–16 mg/kg (Gitlin *et al.*, 1956; Katz, 1961; Awai & Brown, 1963).

The two iron atoms bound per molecule of Tf correspond to 1·25 mg of iron per g of protein (if the molecular weight of Tf is actually 90,000). Thus, the total iron-binding capacity (TIBC) of plasma is about 300 μg/100 ml. However, in normal subjects, the Tf is only a third saturated (Laurell, 1947; Rath & Finch, 1949; Cartwright & Wintrobe, 1949), so that the mean (morning) plasma iron level is about 125 μg/100 ml in men and 110 μg in women (Bothwell & Finch, 1962), leaving a latent or unsaturated binding (UIBC) of about 200 μg.

Although the plasma iron level is subject to considerable diurnal variation, similar fluctuations in plasma Tf are not normally observed. Administration of albumin to normal subjects is accompanied by a marked decrease in Tf concentration (Schedl & Bartter, 1959). However, this phenomenon is due partly to a dilution effect from expansion of plasma volume and partly to a shift of Tf from the intravascular to the extravascular space (Awai & Brown, 1963). In newborn infants, the Tf level is generally lower than that of the mother and undergoes a gradual rise, until the level exceeds that of the normal adult (Laurell, 1947; Hagberg, 1953; Sturgeon, 1954; Smith, 1954).

Plasma Tf is characteristically raised during the last few months of pregnancy as well as in patients with iron-deficiency anemia (Laurell, 1947; Rath & Finch, 1949; Fay *et al.*, 1949). Low levels are associated with many diseases, in most of which there is concurrent albumin depletion, due either to impaired synthesis (e.g. cirrhosis, starvation and chronic infection) or to increased excretion (e.g. nephrosis).

In most instances, alterations in the plasma Tf concentration are due to changes in synthesis, rather than to changes in catabolism or shifts of Tf to or from the extracellular space. Thus, the half-life of the individual molecules tends to remain fairly constant (Morgan, 1966). However, Awai & Brown (1963) reported that in patients with high Tf levels, as in iron deficiency anemia, the average breakdown of Tf was increased, in terms of absolute amounts, but slightly decreased in terms of the expanded Tf pool (i.e. the proportion of molecules catabolized was decreased). As a result, the average survival of a Tf molecule appeared to be somewhat longer than normal in iron deficient patients. The treatment of such patients with oral iron has been reported to cause a decrease in plasma Tf (Mitchell *et al.*, 1960; Lane, 1966). However, Awai & Brown (1963) were unable to demonstrate this change in Tf concentration. The studies of Hillman *et al.*, (1967) were in agreement with those of Awai & Brown, and they indicated that the reported decrease in Tf in this situation may have been due to the method used for measuring total iron-binding capacity.

Unlike haptoglobin, Tf is a true carrier protein in that a Tf molecule picks up and delivers iron atoms to specific receptor sites many times before being itself catabolized. Thus, while the half-time clearance rate of radioiron bound to Tf is normally less than two hours, the $T\frac{1}{2}$ of I^{131} labelled Tf is about 7–10 days (Huff *et al.*, 1950; Katz, 1961; Awai & Brown, 1963). At any given time, the total amount of iron present in the plasma of a normal adult is only about 4 mg, representing a thousandth of the total body iron. However, because of the rapid and efficient delivery of iron by Tf, the marrow is able to utilize about 25 mg daily for hemoglobin synthesis, and this amount can be greatly increased under the stress of hemorrhage or hemolysis.

IRON TURNOVER; ROLE OF TRANSFERRIN

Since only a negligible quantity of iron (i.e, about 1 mg/day) is normally absorbed and excreted, iron from catabolized hemoglobin is repeatedly taken up from the reticuloendothelial cells by Tf and transported to the marrow for reincorporation into hemoglobin. Dietary iron, after intestinal absorption, is also transported by Tf molecules. Although the actual process of iron absorption is not directly dependent upon the presence of Tf in the plasma, Tf is required for the selective transport of iron to red cell precursors in the marrow (Heilmeyer *et al.*, 1961; Wheby & Jones, 1963). Thus, in pathological conditions associated with iron saturation of the circulating Tf, additional iron, in excess of the Tf binding capacity, is rapidly cleared by the liver (Wheby & Jones, 1963; Fawwaz *et al.*, 1967). Under normal circumstances, a blocking mechanism in the intestinal mucosa prevents over-absorption of iron (Hahn *et al.*, 1943, Crosby, 1966). If this barrier is bypassed by injection of iron, the amount of unsaturated Tf available for combining with iron is sufficient to accommodate only a few mg of iron.

The liver is a secondary site for the deposition of Tf-bound iron, so that when erythropoiesis is depressed (as in hypoplastic anemia) or when excessive iron is presented to the marrow, Tf deposits its load preferentially in liver parenchymal cells. However, under normal conditions, the liver is only a minor target, the principal iron recipient being the marrow red cell precursors (Hosain *et al.*, 1962).

Since it is possible to demonstrate the presence of Tf on the surface of human immature red cells, such as reticulocytes, by their agglutination with anti-human Tf antibodies, Tf is probably briefly bound to the receptor cell during the transfer of iron (Jandl, 1960; Jandl & Katz, 1963). According to some investigators, the cells bind iron-carrying Tf molecules in preference to iron-free Tf molecules, so that exchange is facilitated (Pollycove & Maqsood, 1962; Jandl & Katz, 1963). However, other workers have found no evidence for a difference in the cellular uptake of the two forms of Tf (Morgan & Laurell, 1963; Morgan *et al.*, 1966). While the precise mechanism of iron uptake is not clear, it is known to be an active, energy-requiring process (Jandl *et al.*, 1959; Mazur *et al.*, 1960).

Attempts have been made, *in vitro*, to relate the degree of Tf saturation with iron to the rate of iron uptake by receptor sites of immature red cells and other tissues. Initial studies with reticulocytes indicated a rise in iron uptake at higher levels of saturation (Jandl *et al.*, 1959). However, with improvements in methods, the apparent effect of saturation was found to be an artifact of incubation time (Katz & Jandl, 1964). Nevertheless, it is possible that the level of plasma Tf

6

and/or the degree of its saturation with iron has physiological importance. For example, in iron deficiency anemia, where the Tf level is
high and the per cent saturation is low, nearly all of the plasma iron
is utilized by the bone marrow for hemoglobin synthesis (Bainton &
Finch, 1964). Similarly, in pregnancy, the increased Tf and decreased
saturation percentage may facilitate iron transfer to the placenta.
Conversely, in conditions with high serum iron and per cent saturation, such as chronic uremia, much of the iron is deposited in the
parenchymal tissues. However, in thalassemia, the Tf level is low, and
the per cent saturation is high, so that the amount of iron available to
support the greatly hyperplastic marrow is limited by the decrease in
total iron-binding capacity (C. A. Finch, personal communication).

POSSIBLE ROLE OF TRANSFERRIN IN INFECTION

Schade & Caroline (1946) and Schade (1961) suggested another
possible physiological role for Tf when they showed that Tf has
bacteriostatic activity *in vitro* due to its successful competition for
iron with various micro-organisms. Martin *et al.* (1963) found that
intraperitoneal injections of ferric iron in rats and mice enhanced the
virulence of some bacteria but not others. Rogers (1967) and Bullen
et al. (1967a, b) recently reported that the protective effect of *Clostridium welchii* antitoxin in guinea pigs injected with that organism was
abolished by ferric iron injection. From the results of these and other
studies reviewed in their papers, these authors concluded that the
binding of iron by transferrin may indeed be important in resistance
to infection, and that the normal low saturation of plasma Tf may thus
be regarded as a protective factor. Of possible relevance was the
report by Rossen *et al.* (1966) that human nasal washings contain a
small amount of Tf which, the authors suggested, could inhibit the
development of pathogens with restrictive nutritional iron requirements.

PHYSIOLOGICAL EFFECT OF ATRANSFERRINEMIA

As would be expected, permanent absence of Tf from the plasma is not
compatible with life. Very severe depletion secondary to disease has
been reported in a few instances (Riegel & Thomas, 1956; Hitzig *et al.*,
1960), but there is only one well-documented case of congenital

atransferrinemia (Heilmeyer *et al.*, 1961). The patient was found to have severe hypochromic, microcytic anemia when she was 3 months old. There was no response to iron, steroid or vitamin B_{12} therapy, so that she required frequent blood transfusions. Her clinical course was characterized by recurrent infections and growth impairment. Ferrokinetic studies, performed when she was 7 years old, revealed a very low serum iron level (1–14 μg/100 ml) and UIBC of 19 μg/100 ml. Injected radioiron was cleared from the plasma at a remarkably rapid rate ($T\frac{1}{2}$ less than 5 minutes), with localization mainly in the liver; only a minute amount was utilized by the marrow for hemoglobin synthesis. Orally administered radioiron was absorbed at what appeared to be an increased rate. The bone marrow had few developing red cells and no iron stores visible as hemosiderin granules. Immunological studies showed virtual absence of plasma transferrin; the minute quantity detected probably was derived from transfused blood. Death from heart failure was attributed to severely damaged cardiac muscle, which, like the liver, kidneys and pancreas, was found at necropsy to be massively infiltrated with iron.

The Tf content of the parents' serum, measured by immunoprecipitation, was about half that of normal controls; furthermore, the maternal history included three pregnancies with inviable fetuses. These findings, while hardly conclusive, support the contention of the authors that the patient's atransferrinemia was due to homozygosity for a very rare gene, recessive in the clinical, but not in the biochemical sense.

INHERITED VARIANTS OF TRANSFERRIN

DETECTION OF Tf POLYMORPHISM

Poulik & Smithies (1958) performed two-dimensional electrophoresis on specimens of human serum, separating the proteins first on filter paper and then, at right angles, on starch gel. They noted several components in the β-globulin region, and since the most heavily stained component was third in its migration rate, it was called β-globulin C. During subsequent studies (Smithies, 1957, 1958; Horsfall & Smithies, 1958), a small proportion of sera obtained from Negro donors and a larger proportion from Australian aborigines were found to have an additional β-globulin with a somewhat slower migration rate, which they called β-globulin D. Furthermore, about 1 per cent of sera taken from Canadian Caucasians had a slightly faster protein given the name β-globulin B. The identity of these three proteins as transferrin was subsequently established independently by Poulik, Allison and Sutton (cited in Smithies & Hiller, 1959). The appearance of Tf C, B and D on two-dimensional electrophoresis is shown in Fig. 3.2.

Smithies & Hiller (1959) studied a number of families, and proposed that the three Tf molecules, B, C and D, are the products of three allelic autosomal structural genes having equal dominance: Tf^B, Tf^C and Tf^D. There appeared to be little, if any, consistent difference in the amount of the two proteins observed in heterozygotes, the total quantity being about the same as that observed in homozygotes. Furthermore, in mixtures of serum obtained from Tf^C and Tf^D homozygotes, the electrophoretic pattern could not be differentiated from that of a heterozygote.

It soon became apparent that the list of Tf variants was by no means complete. For example, Harris *et al.*, (1958) examined the sera of European and African subjects and found not only the two variants described by Smithies but also one migrating more slowly than Tf D and one more rapidly than Tf B. Then Giblett *et al.* (1959) reported still another fast-moving variant and two with slow mobility. These authors also described a simple technique for locating Tf on starch gel by Fe^{59} radioautography, as shown in Fig. 3.3 and described on p. 148.

Until that time, it had been possible to number the Tf variants moving faster than Tf C as B_2, B_1 and B_0, in order of increasing mobility, and those moving more slowly as D_3, D_2, D_1 and D_0. However, Harris *et al.*, (1960) then found a variant having a migration rate between that of D_1 and D_0. Subsequently, several other aberrant transferrins with intermediate mobilities were found, such as those described by Parker & Bearn (1961b, c). Table 3.1 contains a list of the 18 currently known molecular forms of Tf in the order of their electrophoretic mobility. In some instances, the phenotype name indicates the relative electrophoretic rate (e.g. B_{1-2} migrates between B_1 and B_2). In others, the B (fast) or D (slow) designation is followed by the geographic origin of the sample (e.g. $D_{Montreal}$ and B_{Lae}). Several efforts have been made to establish some uniformity in nomenclature, but these efforts are largely defeated by the fact that

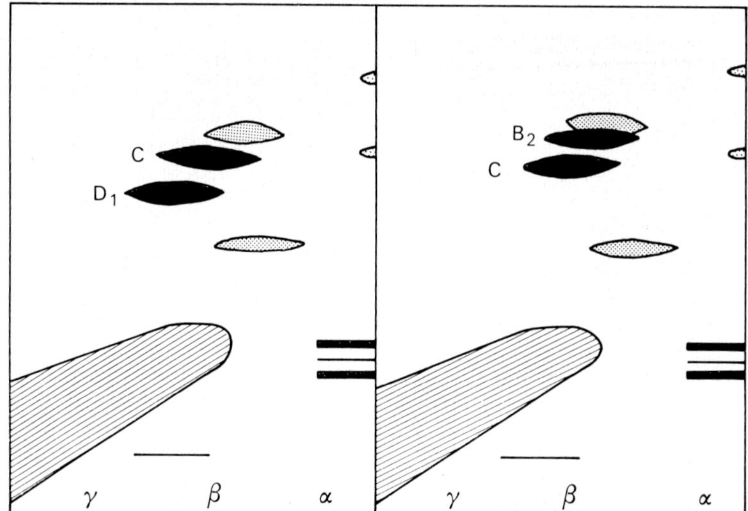

FIGURE 3.2. Diagram of two-dimensional electrophoresis (first on filter paper, then on starch gel) of two serum specimens with the transferrin phenotypes CD$_1$ (left) and B$_2$C (right). (Adapted from Smithies & Hiller, 1959.)

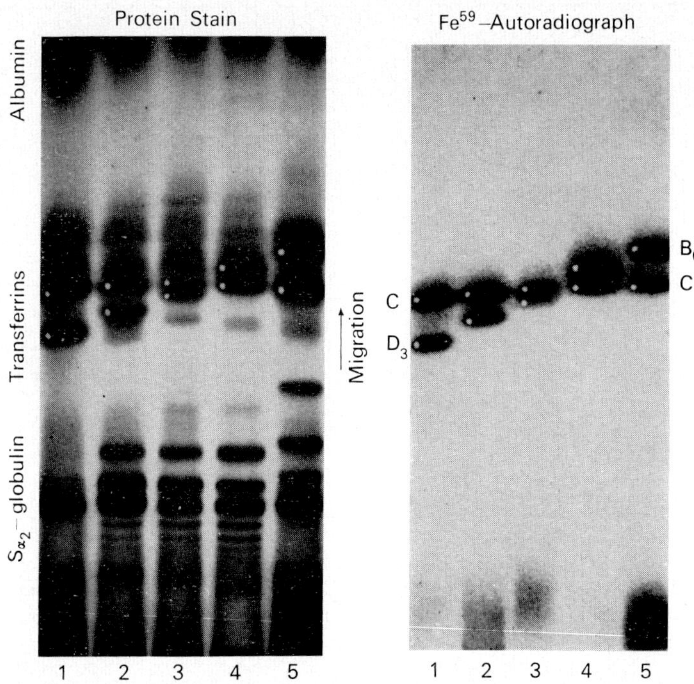

FIGURE 3.3. Starch gel patterns of five serum specimens to which Fe^{59} was added before electrophoresis. One half of the gel was stained with amido black (left) and the other half placed on X-ray film for the autoradiograph on the right. The white dots indicate the individual transferrins in serum of types CD_3, CD_1, CD_0, B_2C and B_0C. (From Giblett, Hickman & Smithies, 1959).

allowance must be left for new variants as they are found and also, that variants having the same electrophoretic migration rate do not necessarily have the same amino-acid sequence. A reasonable solution might be to eliminate the numerical subscripts, to assign D and B to slow and fast moving variants, respectively, and to give each variant a geographic subscript, as is done with the hemoglobin variants.

TABLE 3.1. The human transferrins listed in order of electrophoretic mobility

Transferrins	References	Population source, first sample
D_3	Giblett *et al.* (1959)	American Negro
D_2	Harris *et al.* (1958)	African Negro
D_{Fin}*	Seppälä (1965)	Finnish Caucasian
D_1	Smithies (1958)	American Negro
D_{Chi}	Parker & Bearn (1961c)	Chinese
$D_{Montreal}$	Parker & Bearn (1962b)	Canadian Caucasian
D_{0-1}	Harris *et al.* (1960)	English Caucasian
D_{Wigan}	Glen-Bott *et al.* (1964)	English Caucasian
D_0	Giblett *et al.* (1959)	American Negro
$D_{Adelaide}$	Cooper *et al.* (1964)	Australian Caucasian
C	Poulik & Smithies (1958)	Canadian Caucasian
B_3	Parker & Bearn (1961c)	Japanese
B_2	Smithies (1958)	Canadian Caucasian
B_{1-2}	Arends *et al.* (1962)	Italo-African
B_1	Harris *et al.* (1958)	English Caucasian
$B_{Atalanti}$	Murray *et al.* (1964)	Greek
B_{0-1}	Parker & Bearn (1961b)	Navajo Indian
B_0	Giblett *et al.* (1959)	American Caucasian
B_{Lae}	Lai (1963)	New Guinea native

* Not yet differentiated from D_2.

Eight of the variants listed in Table 3.1 have a faster mobility than Tf C and nine are slower. A possible tenth, D_{Fin} described by Seppälä (1965) has not yet been distinguished from D_2. In most instances, the frequencies of the variants are so low that the listed population source provides the only available information about their overall geographic location. Exceptions are noted in a later section.

Because the aberrant Tf genes have generally low frequencies, and because they have shown no tendency to co-exist within a given geographic area, only the 28 phenotypes shown in Fig. 3.4 have been

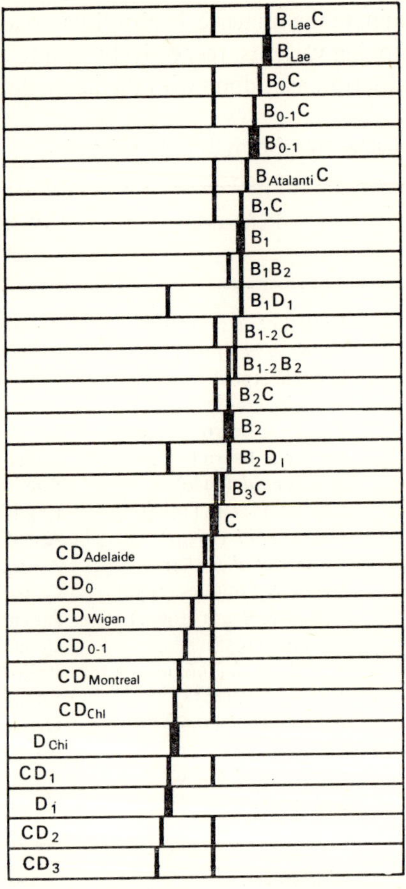

FIGURE 3.4. Diagram of the 28 reported transferrin phenotypes in human serum, representing Tf C and 17 variants with slower or faster mobility. (From Giblett, 1967.)

identified. If the Tf polymorphism is due to a series of (at least) 18 allelic genes at a single autosomal locus, there are 171 theoretically possible phenotypes. The assumption of multiple allelism is not proven but it can be supported by several lines of evidence. First of all, no individuals with more than two forms of Tf have been reported. Second, in family studies, the variant transferrins appear to be inherited as simple Mendelian co-dominant traits. Third, the various Tf molecules differ from each other in electrical charge, rather than molecular weight. Finally, the three variants which have been subjected to chemical analysis appear to differ from Tf C by a single amino acid substitution. These points are discussed in more detail in the following sections.

FAMILY STUDIES AND LINKAGE DATA

Nearly all of the variants detected by routine screening of populations have subsequently been found in other members of the propositus' family. When parents have been available, one or the other parent has invariably possessed the variant. Only two families have been described in which two Tf variants were segregating. In the first, reported by Beckman (1962), the proband, of Tf type CD_1, had a sister of type B_2C. While the father had the common Tf C phenotype, the mother was found to have the very rare type, Tf B_2D_1. Thus the mother gave one or the other of her rare genes to each of her offspring. In the second family, reported by Robinson et al. (1963), a B_2C father had a daughter of type $B_{1-2}B_2$. In this instance, differentiation of the two fast-moving variants was difficult to achieve. However, segregation of the two alleles was further clarified by the fact that the daughter's children were of phenotype $B_{1-2}C$ and B_2C (their father having the common Tf C type).

Giblett (1962) summarized the data of Smithies & Hiller (1959), Barnicot et al. (1960) and Giblett & Steinberg (unpublished) which provided information about the distribution of transferrin types in 13 African Negro, 15 American Negro, seven Australian Aborigine and 60 American Caucasian familes having either B_2 or D_1 variants. No statistically significant departure from expectation was observed. In a later review of 86 Finnish families having the molecular variants B_2, B_{0-1}, D_{Chi} or D_{Fin}, Seppälä (1965) also found no well-defined disturbance in the segregation ratios observed.

Family studies have failed to reveal close linkage of the Tf locus with the loci of most of the blood-group antigens, as well as Hp, Gm and Gc. However, a recent report by Robson et al. (1966) presented good evidence for linkage with the serum cholinesterase E_1 locus.

Their data indicated only a 1 in 20 chance that the recombination fraction for this linkage lies outside the range of 0·07–0·29.

CHEMICAL STRUCTURE OF THE Tf VARIANTS

Smithies & Connell (1959) noted that Tf C, D_1 and B_2 had varying mobilities on filter paper, as well as on starch gel electrophoresis, suggesting a difference in electrical charge rather than molecular weight. Furthermore, Parker & Bearn (1962a) found no difference in the ultracentrifugation patterns of purified Tf C and Tf B_2, and more significantly, were unable to detect any antigenic differences when six purified variants were compared by agar immunodiffusion with anti-human Tf serum.

Sialic acid content of the aberrant Tfs was also measured by Parker & Bearn (1961a, 1962a) to determine if their electrophoretic differences might reflect alterations in the number of sialic acid residues. However, the stepwise removal of sialic acid by neuraminidase occurred in the same manner in all of the variants tested, and subsequent analysis showed that they, like Tf C, contained four residues per molecule. Since the loss of sialic acid did not appear to interfere with iron-binding capacity of the molecules, it was possible to demonstrate the neuraminidase-treated Tf bands by Fe^{59} radioautography. An interesting by-product of this study was the observation (Parker *et al.*, 1963) that the starch gel electrophoretic patterns of serum obtained from newborn infants contained, in addition to the major band in the Tf C position, four faint slower-moving bands with mobilities similar to, but not exactly the same as, the components observed as the result of treating purified Tf (from adults) with neuraminidase. The authors concluded that newborn infants may be deficient in an enzyme responsible for adding the carbohydrate prosthetic groups to the Tf molecule. Thus, not only sialic acid, but other carbohydrates contained in the prosthetic groups would be missing from a small proportion of the protein molecules.

Wang & Sutton (1965) prepared tryptic digests of purified C and D_1 transferrin and found on fingerprinting, a difference in one peptide. Amino-acid analysis indicated that an aspartic acid residue in Tf C was replaced by another residue, probably glycine, in Tf D_1. Such an amino-acid substitution would coincide with the substitution of a single base in their respective codons. The authors pointed out that the difference in charge resulting in an aspartic acid to glycine change would be sufficient to account for the electrophoretic behavior of the two Tfs, but they did not rule out the possibility of additional minor differences in the amino acid sequence of the two molecules.

Jeppsson & Sjöquist (1966) and Jeppsson (1967c) prepared fragments of transferrin by degradation with cyanogen bromide and confirmed the glycine for aspartic acid substitution in Tf D_1. However, according to his theory that the transferrin molecule consists of two similar polypeptide chains controlled by separate gene loci, Jeppsson (1967b) postulated that the Tf D_1 molecule contains one chain in common with Tf C and another chain which has the glycine for aspartic acid substitution. He also suggested that the genes determining Tf B and Tf D may not be alleles. Thus, if the structure of Tf C were controlled by two similar but non-allelic genes, $Tf^{C'}$ and $Tf^{C''}$, a mutation in one gene might result in a fast-moving variant (BC″), while a mutation in the other gene could result in a slow-moving variant (C′D). However, if this were the case, molecular hybrids should occur in 'double heterozygotes'. Thus, in the rare type Tf B_2D_1 described by Beckman (1962) three transferrin bands would be expected (B_2C'', $C'D_1$ and B_2D_1), but only two were observed. It therefore appears that if transferrin is the product of two non-allelic genes, either molecular hybridization of two variant products is suppressed, or else no individual has yet been found who actually is heterozygous at both loci.

Wang *et al.* (1966) compared the peptide maps of Tf B_2 and Tf C and detected a single peptide difference which was found to be due to the substitution of a glycine residue in Tf C by a glutamic acid residue in Tf B_2. Glycine to glutamic acid can also be accounted for by a one-step mutation, and would be consistent with the characteristic difference in electrophoretic mobility of their respective protein molecules.

Recently, Wang *et al.* (1967a) have found that D_{Chi}, which has an electrophoretic migration rate very similar to that of D_1 is probably also the product of a single point mutation. In this case the substitution of histidine in a Tf C peptide by lysine or, more likely, by arginine, in Tf D_{Chi} produced a new point of attack for tryptic digestion. Thus, the Tf C peptide was shown to have the sequence Asp-Ser-Ala-His-Gly-Phe-Leu-Lys, while its corresponding peptide in Tf D_{Chi} contained only Gly, Phe, Leu and Lys. The other half of this peptide was not localized on the fingerprint of the trypsin digest, possibly because of peptide overlap.

FUNCTION OF Tf VARIANTS

Since the aberrant Tfs are distinguished from each other on the basis of difference in electrical charge, Turnbull & Giblett (1961) reasoned that they might differ in their ability to bind and transport iron. They added increments of Fe^{59} to the serum of heterozygotes and then measured the radioactivity in the two Tfs, separated by electrophoresis. They also determined the rate of iron uptake by measuring the radio-activity of the two Tfs sampled at intervals after the addition of a constant amount of Fe^{59}. These *in vitro* experiments failed to reveal any measureable differences in the amount and rate of iron binding by the two proteins. *In vivo* tests were then performed by intravenous injection of mixtures of different Tfs tagged with Fe^{55} and Fe^{59} respectively. The disappearance rates of the two isotopes were deter-mined, as well as their utilization by the marrow for hemoglobin synthesis. Again, no differences were observed, even though the (then) fastest-moving variant, B_0 and the slowest-moving D_3 were compared.

It could be argued that the results obtained from both kinds of experiments were invalid because of rapid exchange of the tagged iron atoms between the Tf molecular species. However, Turnbull & Giblett (1961) found that when one Tf, labelled with Fe^{59}, was incubated for several hours with another, unlabelled, Tf, very little exchange of radioiron occurred. Furthermore, Morgan *et al.* (1967) confirmed the very slow *in vitro* iron exchange between Tf C and its slow-moving, neuraminidase product. They also found that when rabbits were injected with a mixture of Fe^{55}-tagged rabbit Tf and Fe^{59}-tagged human Tf, a definite difference in their clearance rates was observed, and thus a rapid iron exchange between the two forms of Tf seemed unlikely.

Recently, Kornfeld & Brown (1967) have shown that the ability of transferrin to combine with iron is relatively insensitive to blockage of free amino groups and to the net charge of the protein. Furthermore, although binding of Tf to reticulocyte iron transfer sites is affected by both of these changes, the release of iron to reticulocytes (and pre-sumably, to other acceptors) is not decreased. These findings are

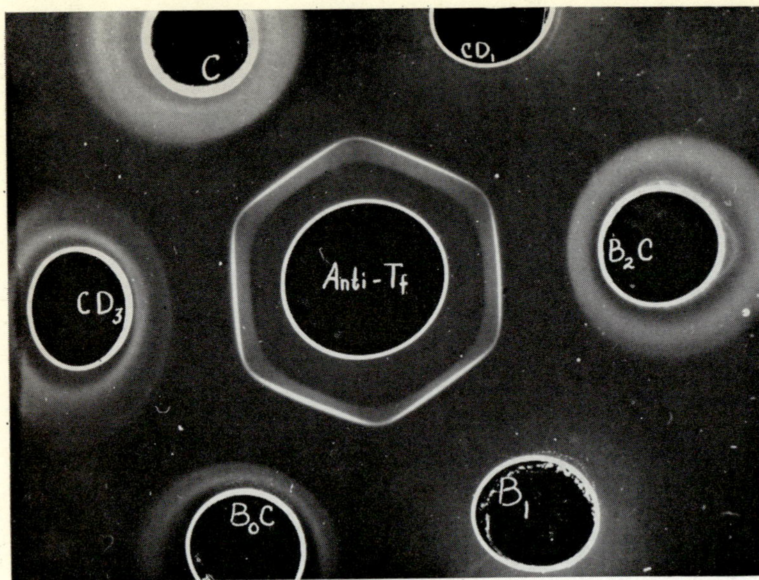

FIGURE 3.5. Immunodiffusion in agar with rabbit anti-human transferrin in the central well and human sera of several Tf phenotypes in the six peripheral wells. Absence of spur formation indicates that the antiserum cannot detect the small differences in amino acid sequence of the various transferrin molecules. (From Giblett, 1966.)

consistent with the observations described in the previous two paragraphs, and suggest that only amino acid substitutions occurring at the molecular site of iron attachment to Tf may alter the iron-combining capacity of the protein.

Another functional difference between Tf molecules could conceivably be related to the bacteriostatic activity known to be a characteristic of Tf because of its successful competition for iron with microorganisms requiring the metal for growth (See p. 134). However, there have been no systematic studies to show whether the Tfs vary in their ability to compete in this fashion.

Many of the difficulties associated with studying the Tf polymorphism in population samples are due to the fact that differentiation of the phenotypes is based on small, sometimes minute, differences in the electrophoretic mobility of a single protein band. Thus, it is necessary to make careful comparisons of serum suspected to contain a variant with other serum specimens of known Tf phenotypes. However, even this procedure does not ensure the variant's identity, because amino acid substitution is not invariably accompanied by a change in electrical charge. Also, as demonstrated by certain of the abnormal hemoglobins, molecules of variants with different amino acid substitutions but similar charges cannot be differentiated by electrophoresis.

There are other problems in this type of identification. For example, as previously noted, bacterial contamination may bring about a loss of sialic acid residues with a consequent decrease in electrophoretic mobility. Furthermore, since free Tf and the FeTf complex have different isoelectric points (Keller & Pennell, 1959), the pattern shown in Figure 3.1 may be misinterpreted as that of a heterozygote when the Tf is not sufficiently saturated with iron. In most instances, there is enough iron present in the starch to bring this about. However, addition of small amounts of an iron salt to the serum or plasma prevents this source of error.

It would be highly desirable if specific antisera could be developed as a more precise means of identifying the Tf variants. Unfortunately, as exemplified by the immunodiffusion plate shown in Figure 3.5, the antigenic determinant(s) of human Tf which react with rabbit anti-human Tf antibodies either do not contain the substituted amino acids or else there is insufficient difference in the precipitin bands to be detected as spur formation. The best theoretical source of such antibodies would be human subjects of the common Tf C phenotype injected with molecular variants of human Tf. However, such 'experiments' have unwittingly been carried out on many occasions by transfusion, and diligent search in numerous laboratories has failed to detect precipitating antibodies with Tf specificity in human serum.

FREQUENCY OF VARIANTS

There is no lack of variability in the Tf polymorphism, as attested by the 18 electrophoretic species. However, most variants are very rare, so that of the 171 possible phenotypes, only 21 heterozygous and seven (presumably) homozygous patterns have been reported. The latter include Tf C, D_1, D_{Chi}, B_2, B_1, B_{Lae} and B_{0-1}. In most of the heterozygous types, the variant Tf is accompanied by the common Tf C. However, rare 'double heterozygotes' are also documented, such as Tf B_2D_1 (Beckman, 1962), B_1B_2, B_1D_1 (Beckman & Holmgren, 1963) and $B_{1-2}B_2$ (Robinson *et al.*, 1963). Some other reported instances of random serum specimens whose Tf pattern showed two slow-moving variants probably represented artifacts caused by neuraminidase. The author has seen two specimens with the apparent Tf D_1D_3 phenotype which proved to be Tf C when blood was again obtained from the donors. It is not sufficient in such cases to show that the observed slow-moving bands are capable of binding iron, because treatment of Tf with neuraminidase does not alter the iron-binding capacity to any appreciable extent, even when nearly all of the sialic acid residues have been removed (Blumberg & Warren, 1961; Parker, 1963).

GEOGRAPHIC DISTRIBUTION OF VARIANTS

Giblett (1962) reviewed in a lengthy table the reports available at that time on the frequencies of the various Tf phenotypes in several parts of the world. The data showed that only four of the aberrant types, Tf CD_1, CD_{Chi}, B_2C and $B_{0-1}C$, have a frequency of 1 per cent or more in the populations tested. Subsequent papers have confirmed the earlier reports and have clarified some, but by no means all, of the confusion arising from the similarities in electrophoretic mobility.

There is a separate list of references on Tf gene frequencies at the end of this chapter.

By far the most common Tf molecular variants are D_1 and D_{Chi}. Tf D_1 appears to have two different population reservoirs, in both of which there is considerable variability in frequency from tribe to tribe. The first is among the aborigines of Australia and New Guinea (Barnicot & Kariks, 1960; Bennett *et al.*, 1961; Kirk & Lai, 1961; Kirk *et al.*, 1962; Flory, 1964; Kirk, 1965; Nicholls *et al.*, 1965; Curtain *et al.*, 1965, 1966). The second is in a large part of Africa south of the Sahara Desert (Barnicot *et al.*, 1959, 1960, 1962; Allison & Barnicot, 1960; Neel *et al.*, 1961; Blumberg & Gentile, 1961; van Ros *et al.*, 1963; Jenkins & Steinberg, 1966; Giblett *et al.*, 1966). Among American Negroes, the phenotype Tf CD_1 has a frequency of about 8–10 per cent (Giblett, 1962).

Recent studies by Wang *et al.* (1967b) indicate that the D_1 variant found in Negroes has the same amino-acid substitution as the D_1 of Australian aborigines. Thus, it appears almost certain that the same mutation in a very long DNA segment occurred independently in these two widely separated regions. If, on the other hand, the gene was introduced from one population to the other at some very early stage, or the gene existed in a very early common ancestor of both populations, one would still have to postulate that some selective advantage was operating in both geographic regions to maintain the elevated frequency of the gene (Wang *et al.*, 1967b).

The variant D_{Chi}, described by Parker & Bearn (1961c) migrates only slightly faster than D_1 on starch gels. Failure to differentiate the two variants originally created the impression that D_1 had a remarkably widespread distribution, for example in natives of Central America (Sutton *et al.*, 1960); South America (Arends & Gallango, 1964); Sweden (Beckman & Holmgren, 1961); The Philippines and Formosa (Giblett, 1962); Ceylon (Kirk & Lai, 1961; Kirk *et al.*, 1962c); India (Kirk *et al.*, 1962b); Fiji and New Hebrides (Douglas *et al.*, 1964). However, Kirk *et al.* (1964) obtained a number of serum specimens containing slow-moving Tfs from the various authors for comparative purposes. They found that Tf variants indistinguishable (by electrophoresis) from D_{Chi} were the only slow variants encountered in specimens from Malaya, North Thailand, Ceylon, Formosa, China and an area of India bordering on China. However, both D_{Chi} and D_1 were found in Central and South American sera, probably reflecting admixture of Mongoloid and Negroid populations, respectively. Subsequent reports by Salzano & Sutton (1965) and by Gallango & Arends (1966) confirmed that both variants occur in variable proportion among some tribes of Central and South America.

Kirk *et al.* (1964) could not differentiate the variant found in Swedish Lapps by Beckman & Holmgren (1961) from D_{Chi}, thus suggesting a Mongoloid origin. However, Seppälä (1965) subsequently reported the presence of both D_{Chi} and D_{Fin} in a large Finnish population sample. The latter variant has a mobility slightly slower than that of D_1, and its geographic origin is currently unclear. If D_{Fin} is actually the same as D_2, which has only been reported in an African native, independent mutation in two widely separated populations, such as that which must have occurred in the case of D_1, is the probable explanation.

The fast-moving variants are virtually absent from the populations of Africa and the Pacific islands, with the possible exception of small isolates such as the one described by Kirk *et al.* (1962) in North Queensland, Australia. No population has been reported to have a Tf B_2C phenotype frequency of more than 4 per cent, but it does occur in about 1 per cent of most western European groups sampled and in their counterparts on the North American continent (Giblett, 1962). Presumably the fast-moving variant described in 3 per cent of Poles by Prochnicka (1966) and in 1–2 per cent of southern Europeans by Benerecetti-Santachiara & Modiano (1964) and by Modiano *et al.* (1965) may also be B_2, but comparative tests were not performed.

Another fast-moving variant, B_{0-1}, was described by Parker & Bearn (1961b), who found the Tf B_{0-1}C phenotype in 16 out of 230 members of a Navajo Indian tribe in Arizona. Subsequently, Matson *et al.* (1963, 1965) reported the same variant in some Indian tribes in Mexico, Guatemala and British Honduras.

Selective mechanisms which might be responsible for maintaining these four phenotypes (Tf CD_1, CD_{Chi}, B_2C and B_{0-1}C) in certain populations are completely unknown. Because of the previously mentioned bacteriostatic activity of Tf, it is tempting to associate the Tf polymorphism with resistance to infectious disease. However, as in the case of nearly all of the polymorphic systems described in this book, such speculation is not currently supported by any actual data.

METHODS

PHENOTYPE DIFFERENTIATION BY ELECTROPHORESIS

Although the various transferrins appear to differ in their electrical charge rather than molecular weight, the use of a molecular sieve such as starch gel or acrylamide gel is apparently necessary for clearly distinguishing most of the types. On filter paper and cellulose acetate electrophoresis, the fastest and slowest variants are visibly separated from the usual transferrin, Tf C (Turnbull & Giblett, 1961). However, the most common variants, Tf D_1, Tf D_{Chi} and Tf B_2, cannot be differentiated from Tf C in these media, except as a slight widening of the Tf zone in the electrophoretic pattern of heterozygotes. Most of the genetic and population studies have been performed with starch gel electrophoresis. However acrylamide gels, prepared either in discs or on plates provide excellent separation, and the reader is referred to the papers of Raymond (1964) and Davis (1964) for descriptions of these procedures.

STARCH GEL ELECTROPHORESIS

The general methods described in the previous chapter for differentiating the haptoglobin types are also applicable to the transferrin types. However, with horizontal electrophoresis, the discontinuous buffer system of Poulik (1957) is much more satisfactory than the original borate buffer system. In Poulik's system, the gels are made from a buffer containing 0·076 M tris and 0·005 M citric acid (pH 8·6), while the bridge buffer consists of 0·3 M boric acid and 0·05 M sodium hydroxide. With vertical electrophoresis, either buffer system gives good results. The electrophoretic conditions are the same as those used for haptoglobin. Thus, it is possible to determine both the haptoglobin and transferrin phenotypes from a single gel. However, since transferrin migrates with a mobility similar to that of several other proteins (especially ceruloplasmin and hemopexin) it is not possible to be certain about the identity of transferrin from a gel stained for protein. Thus, an additional procedure is required so that Tf can be distinguished by either its iron-binding or physico-chemical characteristics. Three methods are briefly described.

Fe[59] RADIOAUTOGRAPHY

In the method of Giblett *et al.* (1959), Fe[59] is added to the serum specimen to contain about 5 μc/ml, using high specific activity Fe[59]Cl$_3$ or Fe[59] citrate. (If the transferrin is already iron-saturated this method is of no value, since free exchange of iron atoms between transferrin molecules does not occur.) After electrophoresis, the gel is sliced in the usual fashion. One gel slice is wrapped in a transparent material such as Saran Wrap and, in a dark room, placed in contact with Kodak No-Screen X-ray film for about 12 hours. The film is then developed, and, if desired, a print is made by placing the film in a photographic enlarger. For longer incubation periods, it is necessary to prevent diffusion by freezing the gel rapidly in a dry ice-acetone bath, and exposing the X-ray film in the deep freeze.

IRON STAINING

The iron bound to transferrin can also be detected by staining it with a sensitive iron indicator such as Nitroso R salt. In the method of Mueller *et al.* (1962), ferric ammonium sulfate is first added to the serum in an amount equivalent to about 5 μg of Fe^{+++} per ml. After electrophoresis, the bottom half of the sliced gel is immersed for 15 minutes in a solution which contains 1 g hydroxylamine hydrochloride, 2·7 g sodium acetate . 3H$_2$O, 1·5 ml glacial acetic acid and 0·5 g Nitroso R salt (Fisher Corp.) in 100 ml of distilled water.

The gel is then washed with several changes of a solution containing distilled water, methanol and glacial acetic acid in proportions 50:50:10. The transferrin zones appear in about an hour as faint, but distinct, green bands.

RIVANOL PRECIPITATION

Sutton (in Matson *et al.*, 1966) described the use of rivanol (2-ethoxy-6,9-diaminoacridine lactate) for precipitating most of the serum proteins, leaving in the supernatant mainly transferrin and γ globulin which are readily separated by electrophoresis. In this method, one drop of serum is diluted with three drops of ferric ammonium citrate solution (0·15 per cent in the gel buffer of 0·03 M boric acid and 0·006 M sodium hydroxide, pH 8·4). Five drops of a 0·6 per cent solution of rivanol in gel buffer is added, and a precipitate forms. After centrifugation, the supernatant solution is subjected to starch gel electrophoresis and subsequently stained with amido black as described in the section on haptoglobin staining.

REFERENCES

AASA R., MALSTRÖM B.G., SALTMAN P. & VÄNNGARD T. (1963) The specific binding of iron and copper to transferrin and conalbumin. *Biochim. biophys. Acta* **75**, 203.

AISEN P., LEIBMAN A. & REICH H.A. (1966) Studies on the binding of iron to transferrin and conalbumin. *J. biol. Chem.* **241**, 1666.

ARENDS T., GALLANGO M.L., PARKER W.C. & BEARN A.G. (1962) A new variant of human transferrin in a Venezuelan family. *Nature* **196**, 477.

ASOFSKY R. & THORBECKE G.J. (1961) Sites of formation of immune globulins and of a component of C'_3. II. Production of immunoelectrophoretically identified serum proteins by human and monkey tissues *in vitro. J. exp. Med.* **114**, 471.

AWAI M. & BROWN E.B. (1963) Studies of the metabolism of I^{131} labelled human transferrin. *J. Lab. clin. Med.* **61**, 363.

AZARI P.R. & FEENEY R.E. (1958) Resistance of the metal complexes of conalbumin and transferrin to proteolysis and thermal denaturation. *J. biol. Chem.* **232**, 293.

BAINTON D.F. & FINCH C.A. (1964) The diagnosis of iron deficiency anemia. *Am. J. Med.* **37**, 62.

BAMBACH M.N. (1966) Untersuchungen zur Transferrin-Bestimmung. *Klin. Wschr.* **44**, 1276.

BARNICOT N.A. (1961) Haptoglobins and transferrins, *in* HARRISON G.A. (ed.) *Genetical Variation in Human Populations*, p. 41. Pergamon Press, Oxford.

BARNICOT N.A., GARLICK J.P. & ROBERTS D.F. (1960) Haptoglobin and transferrin inheritance in Northern Nigerians. *Ann. hum. Genet.* **24**, 171.

BECKMAN L. (1962) Slow and fast transferrin variants in the same pedigree. *Nature*, **194**, 796.

BECKMAN L. & HOLMGREN G. (1963) On the genetics of the human transferrin variants B_1, B_2, and D_1. *Acta genet.* **13**, 361.

BEZKOROVAINY A. & GROHLICH D. (1967) The behavior of native and reduced alkylated human transferrin in urea and guanidine-HCl solutions. *Biochim. biophys. Acta* **147**, 497.

BEZKOROVAINY A. & RAFELSON M.E. (1964) Some hydrodynamic properties of human transferrin. *Arch. Biochem. Biophys.* **107**, 302.

BEZKOROVAINY A., RAFELSON M.E., Jr. & LIKHITE V. (1963) Isolation and partial characterization of transferrin from normal human plasma. *Arch. Biochem. Biophys.* **103**, 371.

BLUMBERG B.S. & WARREN L. (1961) The effect of sialidase on transferrins and other serum proteins. *Biochim. biophys. Acta* **50**, 90.

BOTHWELL T.H. & FINCH C.A. (1962) *Iron Metabolism*. Little, Brown, Boston.

BOWMAN B.H. (1968) Serum transferrin. *Series Haematologica* 1 1, 97.

BROWN E.B. (1966) Clinical aspects of iron metabolism. *Semin. Hematology* 3, 314.

BULLEN J.J., CUSHNIE G.H. & ROGERS H.J. (1967a) The abolition of the protective effect of *Clostridium welchii* type A antiserum by ferric iron. *Immunology* 12, 303.

BULLEN J.J., ROGERS H.J. & CUSHNIE G.H. (1967b) Abolition of passive immunity to bacterial infections by iron. *Nature* 214, 515.

CARTWRIGHT G.E. & WINTROBE M.M. (1949) Chemical, clinical and immunological studies on the products of human plasma fractionation 39. The anemia of infection; studies on the iron-binding capacity of serum. *J. clin. Invest.* 28, 86.

CHARLWOOD P.A. (1963) Ultracentrifugal characteristics of human, monkey and rat transferrins. *Biochem. J.* 88, 394.

CLAUSEN J., RASK-NIELSEN R., CHRISTENSEN H.E. & MUNKNER T. (1960) Two transplantable mouse hepatomas associated with an increase of metal-combining β-globulin (transferrin) in serum. *Cancer Res.* 20, 178.

COOPER D.W., LANDER H. & KIRK R.L. (1964) $D_{Adelaide}$—a new transferrin variant in man. *Nature* 204, 102.

CROSBY W.H. (1966) Mucosal block: an evaluation of concepts relating to control of iron absorption. *Semin. Hematology* 3, 299.

DANCIS J., BRAVERMAN N. & LIND J. (1957) Plasma protein synthesis in the human fetus and placenta, *J. clin. Invest.* 36, 398.

DAVIS B.J. (1964) Disc electrophoresis. II. Method and application to human serum proteins. *Ann. N.Y. acad. Sci.* 121, 404.

DAVIS B., SALTMAN P. & BENSON S. (1962) The stability constants of the iron-transferrin complex. *Biochem. biophys. Res. Commun.* 8, 56.

EHRENBERG A. & LAURELL C.B. (1955) Magnetic measurements on crystallized Fe-transferrin isolated from the blood plasma of swine. *Acta chem. scand.* 9, 68.

ERIKSSON S. & SJOQUIST J. (1960) Quantitative determination of N terminal amino acid in some serum proteins. *Biochim. biophys. Acta* 45, 290.

FAWWAZ R.A., WICHELL H.S., POLLYCOVE M. & SARGENT T. (1967) Hepatic iron deposition in humans. I. First-pass hepatic deposition of intestinally absorbed iron in patients with low plasma latent iron-binding capacity. *Blood* 30, 417.

FAY J., CARTWRIGHT G.E. & WINTROBE M.M. (1949) Studies on free erythrocyte protoporphyrin, serum iron, serum iron-binding capacity and plasma copper during normal pregnancy. *J. clin. Invest.* 28, 487.

FLETCHER J. & HUEHNS E.R. (1967) Significance of the binding of iron by transferrin. *Nature* 215, 584.

GIBLETT E.R. (1962) The plasma transferrins, *in* BEARN A.G. & STEINBERG A.G. (eds.) *Progress in Medical Genetics*, vol. 2, p. 34. Grune & Stratton, New York.

GIBLETT E.R. (1966) Recent advances in haptoglobin and transferrin genetics. *Plenary Sessions.* 11th int. *Cong. Blood Transf.*, Sydney.

GIBLETT E.R., HICKMAN C.G. & SMITHIES O. (1959) Serum transferrins. *Nature* 183, 1589.

GITLIN D., JANEWAY C.A. & FARR L.E. (1956) Studies on the metabolism of plasma proteins in the nephrotic syndrome. I. Albumin, γ-globulin and iron-binding globulin. *J. clin. Invest.* 35, 44.

GLEN-BOTT A.M., HARRIS H., ROBSON E.B., BEARN A.G. & PARKER W.C. (1964) Transferrin D_{Wigan}. *Acta genet.* **14**, 52.

GOODMAN M., NEWMAN H.S. & RAMSAY D.S. (1958) The use of chicken antiserum for the rapid determination of plasma protein components. III. The assay of human serum transferrin. *J. Lab. clin. Med.* **51**, 814.

HAGBERG B. (1953) The iron-binding capacity of serum in infants and children. *Acta paediat.* **42**, 589.

HAHN P.F., BALE W.F., ROSS J.F., BALFOUR W.M. & WHIPPLE G.H. (1943) Radioactive iron absorption by gastrointestinal tract. *J. exp. Med.* **78**, 169.

HARRIS H., PENINGTON D.G., ROBSON E.B. & SCRIVER C.R. (1960) A further genetically determined transferrin variant in man. *Ann. hum. Genet.* **24**, 327.

HARRIS H., ROBSON E.B. & SINISCALCO M. (1958) β-globulin variants in man. *Nature,* **182**, 452.

HEILMEYER L., KELLER W., VIVELL O., KEIDERLING W., BETKE K., WÖHLER F. & SCHULTZE H.E. (1961) Kongenitale Atransferrinämie bei einem sieben Jahre alten Kind. *Dt. med. Wschr.* **86**, 1745.

HEIMBURGER N., HEIDE K., HAUPT H. & SCHULTZE H.E. (1964) Bausteinanalysen von Humanserum proteinen. *Clin. chim. Acta,* **10**, 293.

HILLMAN R.S., MORGAN E.H. & FINCH C.A. (1967) A comparison of total iron binding capacity methods in the iron deficiency state. *J. Lab. clin. Med.* **60**, 874.

HITZIG W.H., SCHMID M., BETKE K. & ROTHSCHILD M. (1960) Erythroleukämie mit Hämoglobinopathie und Eisenstoffwechselstörung. *Helv. paediat. Acta* **15**, 203.

HOCHWALD G.M., THORBECKE G.J. & ASOFSKY R. (1961) Sites of formation of immune globulins and of a component of C'_3. I. A new technique for the demonstration of the synthesis of individual serum proteins by tissues *in vitro.* *J. exp. Med.* **114**, 459.

HOLMBERG C.G. & LAURELL C.B. (1945) Studies on the capacity of serum to bind iron. A contribution to our knowledge of the regulation mechanism of serum iron. *Acta physiol. scand.* **10**, 307.

HOPKINS L.L. & SCHWARZ K. (1964) Chromium binding to serum proteins, specifically siderophilin. *Biochim. biophys. Acta* **90**, 484.

HORSFALL W.R. & SMITHIES O. (1958) Genetic control of some human serum β-globulins. *Science* **128**, 35.

HOSAIN F., MARSAGLIA G., NOYES W. & FINCH C.A. (1962) The nature of internal iron exchange in man. *Trans. Ass. Am. Phys.* **75**, 59.

HUFF R.L., HENNESSEY T.G., AUSTIN R.E., GARCIA J.F., ROBERTS B.M. & LAWRENCE J.H. (1950) Plasma and red cell iron turnover in normal subjects and in patients having various hematopoietic disorders. *J. clin. Invest.* **29**, 1041.

INMAN J.K. (1956) Preparation of crystalline β_1 metal-combining protein. *Proc. 10th Conf. on Plasma Protein and Cellular Elements of the Blood.* Protein Foundation, Cambridge, p. 46.

INMAN J.K., CORYELL F.C., McCALL K.B., SGOURIS J.T. & ANDERSON H.D. (1961) A large scale method for the purification of human transferrin. *Vox Sang.* **6**, 34.

JAGER B.V. & GUBLER C.J. (1952) An immunologic study of the iron-binding protein. *J. Immunol.* **69**, 311.

JAMIESON G.A. (1964) The isolation of sialoglycopeptides from human transferrin. *Biochem. biophys. Res. Commun.* **17**, 775.

JANDL J.H. (1960) The agglutination and sequestration of immature red cells. *J. Lab. clin. Med.* **55**, 663.

JANDL J.H., INMAN J.K., SIMMONS R.L. & ALLEN D.W. (1959) Transfer of iron from serum iron-binding protein to human reticulocytes. *J. clin. Invest.* **38**, 161.

JANDL J.H. & KATZ J.H. (1963) The plasma-to-cell cycle of transferrin. *J. clin. Invest.* **42**, 314.

JEPPSSON J.O. (1967a) Subunits of human transferrin. *Acta chem. scand.* **21**, 1686.

JEPPSSON J.O. (1967b) Structural studies on human transferrin. *Dissertation*. Dept. of Microbiology, Univ. of Umea, Sweden.

JEPPSSON J.O. (1967c) Structural studies of fragments resulting from cyanogen bromide degradation of human transferrin. *Biochim. biophys. Acta* **140**, 477.

JEPPSSON J.O. & SJÖQUIST J. (1966) Structural studies on genetic variants of human transferrin. *Proc. XIV Colloq. Prot. Biol. Fluids*, p. 87. Elsevier, Amsterdam.

JONES H.D.C. & PERKINS D.J. (1965) Metal-ion binding of human transferrin. *Biochim. biophys. Acta* **100**, 122.

KATZ J.H. (1961) Iron and protein kinetics studied by means of doubly-labeled human crystalline transferrin. *J. clin. Invest.* **40**, 2143.

KATZ J.H. & JANDL J.H. (1964) The role of transferrin in the transport of iron into the developing cell, *in* GROSS F. (ed.) *Ciba Symposium on Iron Metabolism*. Springer-Verlag, Berlin.

KELLER W. & PENNELL R.B. (1959) Sterilization of plasma components by heat. I. β-1 metal combining protein. *J. Lab. clin. Med.* **53**, 638.

KIRK R.L., PARKER W.C. & BEARN A.G. (1964) The distribution of the transferrin variants D_1 and D_{Chi} in various populations. *Acta genet.* **14**, 41.

KISTLER P., NITSCHMANN H.S., WYTTENBACH A., STUDER M., NIEDERÖST C. & MAUERHOFER M. (1960) Humanes Siderophilin: Isolierung mittels Rivanol aus Blutplasma und Plasmafractionen, analytische Bestimmung und Kristallisation. *Vox Sang.* **5**, 403.

KOECHLIN B.A. (1952) Preparation and properties of serum and plasma proteins. XXVIII. The β-1 metal combining protein of human plasma. *J. Am. chem. Soc.* **74**, 2649.

KOMATSU S.K. & FEENEY R.E. (1967) Role of tyrosyl groups in metal binding properties of transferrins. *Biochemistry* **6**, 1136.

KORNFELD S. & BROWN E.B. (1967) The effects of chemical modifications on the biologic activity of human transferrin. *Blood* **30**, 873.

LAI L.Y.C. (1963) A new transferrin in New Guinea. *Nature* **198**, 589.

LANE R.S. (1966) Changes in plasma transferrin levels following the administration of iron. *Br. J. Haemat.* **12**, 249.

LANE R.S. (1967) Localization of transferrin in human and rat liver by fluorescent antibody technique. *Nature* **215**, 161.

LAURELL C.B. (1947) Studies on the transportation and metabolism of iron in the body. *Acta physiol. scand.* suppl. 46.

LAURELL C.B. (1960) Metal-binding plasma protein and cation transport, *in* PUTNAM F.W. (ed.) *The Plasma Proteins*, vol. 1, p. 349. Academic Press, New York.

MARTIN C.M., JANDL J.H. & FINLAND M. (1963) Enhancement of acute bacterial infections in rats and mice by iron and their inhibition by human transferrin *J. infect. Dis.* **112**, 158.

MATSON G.A., SUTTON H.E., SWANSON J., ROBINSON A.R. & SANTIANA A. (1966) Distribution of hereditary blood groups among Indians in South America. *Am. J. phys. Anthrop.* **24**, 51.

MAZUR A., GREEN S. & CARLETON A. (1960) Mechanism of plasma iron incorporation into hepatic ferritin. *J. biol. Chem.* **235**, 595.

MCCARTER H.R., GOLDSWORTHY P.D., McGUIGAN J.E., MacMARTIN M.P., WOOD P.A. & VOLWILER W. (1966) The liver as a principal site of transferrin synthesis. *Clin. Res.* **14**, 136.

MITCHELL J., HALDEN E.R., JONES F., BRYAN B.A., STIRMAN J.A. & MUIRHEAD E.E. (1960) Lowering of transferrin during iron absorption in iron deficiency states. *J. Lab. clin. Med.* **56**, 555.

MORGAN E.H. (1966) Transferrin and albumin distribution and turnover in the rat. *Am. J. Physiol.* **211**, 1486.

MORGAN E.H., HUEHNS E.R. & FINCH C.A. (1966) Iron reflux from reticulocytes and bone marrow cells *in vitro. Am. J. Physiol.* **210**, 579.

MORGAN E.H. & LAURELL C.B. (1963) Studies on the exchange of iron between transferrin and reticulocytes. *Br. J. Haemat.* **9**, 471.

MORGAN E.H., MARSAGLIA G., GIBLETT E.R. & FINCH C.A. (1967) A method of investigating internal iron exchange utilizing two types of transferrin. *J. Lab. clin. Med.* **69**, 370.

MUELLER J.O., SMITHIES O. & IRWIN M.R. (1962) Transferrin variation in Columbidae. *Genetics* **47**, 1385.

MURRAY R.F., ROBINSON J.C. & BLUMBERG B.S. (1964) A new variant of transferrin from Greece. *Nature* **204**, 382.

PARKER W.C. (1963) Genetic and biochemical studies on the haptoglobins, transferrins and group-specific components of human serum. Regulatory mechanisms in human genetics. *Doctoral Dissertation.* Rockefeller Institute, New York.

PARKER W.C. & BEARN A.G. (1961a) Alterations in sialic acid content of human transferrin. *Science* **133**, 1014.

PARKER W.C. & BEARN A.G. (1961b) Haptoglobin and transferrin gene frequencies in a Navajo population: a new transferrin variant. *Science* **134**, 106.

PARKER W.C. & BEARN A.G. (1961c) Haptoglobin and transferrin variation in humans and primates: two new transferrins in Chinese and Japanese populations. *Ann. hum. Genet.* **25**, 227.

PARKER W.C. & BEARN A.G. (1962a) Studies on the transferrins of adult serum, cord serum and cerebrospinal fluid: the effect of neuraminidase. *J. exp. Med* **115**, 83.

PARKER W.C. & BEARN A.G. (1962b) Additional genetic variation of human serum transferrin. *Science* **137**, 854.

PARKER W.C., HAGSTROM W.C. & BEARN A.G. (1963) Additional studies on the transferrins of cord serum and cerebrospinal fluid. Variation in carbohydrate prosthetic groups. *J. exp. Med.* **118**, 975.

PETERS T., GIOVANNIELLO T.J., APT L. & ROSS J.F. (1956) A simple improved method for the determination of serum iron. *J. Lab. clin. Med.* **48**, 280.

POLLYCOVE M. (1966) Iron metabolism and kinetics. *Semin. Hematology*, 3, 235.

POLLYCOVE M. & MAQSOOD M. (1962) Existence of an erythropoietic labile iron pool in animals. *Nature* 194, 152.

POULIK M.D. (1957) Starch gel electrophoresis in a discontinuous system of buffers. *Nature* 180, 1477.

POULIK M.D. (1961) Interaction of transferrin, haptoglobin and other serum proteins with neuraminidase of diphtheria toxin. *Clin. chim. Acta.* 6, 493.

POULIK M.D. & SMITHIES O. (1958) Comparison and combination of the starch gel and filter paper electrophoretic methods applied to human sera: two dimensional electrophoresis. *Biochem. J.* 68, 636.

PUTNAM F.W. (1965) Structure and function of the plasma proteins, *in* NEURATH E.H. (ed.) *The Proteins*, vol. 3, p. 154. Academic Press, New York.

RAMSAY W.N.M. (1957) The determination of the total iron-binding capacity of serum. *Clin. chim. Acta.* 2, 221.

RAMSAY W. (1958) Plasma iron, *in* SOBOTKA H. & STEWART C.P. (eds.) *Advances in Clinical Chemistry*, vol. 1, p. 1. Academic Press, New York.

RATH C.E. & FINCH C.A. (1949) Chemical, clinical and immunological studies on the products of human plasma fractionation. XXXVIII. Serum iron transport. Measurement of iron-binding capacity of serum in man. *J. clin. Invest.* 28, 79.

RAUSEN A.R., GERALD P.S. & DIAMOND L.K. (1961) Genetical evidence for transferrin synthesis in the foetus. *Nature* 192, 182.

RAYMOND S. (1964) Acrylamide gel electrophoresis. *Ann. N.Y. Acad. Sci.* 121, 350.

RIEGEL C. & THOMAS D. (1956) Absence of β-globulin fraction in the serum protein of a patient with unexplained anemia. *New Engl. J. Med.* 255, 434.

ROBERTS R.C., MAKEY D.G. & SEAL U.S. (1966) Human transferrin; molecular weight and sedimentation properties. *J. biol. Chem.* 241, 4907.

ROBINSON J.C., BLUMBERG B.S., PIERCE J.E., COOPER A.J. & HAMES C.G. (1963) Studies on inherited variants of blood proteins. II. Familial segregation of transferrin $B_{1-2}B_2$. *J. Lab. clin. Med.* 62, 762.

ROBINSON J.C. & PIERCE J.E. (1964) Studies on inherited variants of blood proteins. III. Sequential action of neuraminidase and galactose oxidase on transferrin $B_{1-2}B_2$. *Arch. Biochem. Biophys.* 106, 348.

ROBSON E.R., SUTHERLAND I. & HARRIS H. (1966) Evidence for linkage between the transferrin locus (Tf) and the serum cholinesterase locus (E_1) in man. *Ann. hum. Genet.* 29, 325.

ROGERS H.J. (1967) Bacteriostatic effects of horse sera and serum fractions on *Clostridium welchii* type A, and the abolition of bacteriostasis by iron salts. *Immunology*, 12, 285.

ROOP W.E. & PUTNAM F.W. (1967) Purification and properties of human transferrin C and a slow moving genetic variant. *J. biol. Chem.* 242, 2507.

ROSSEN R.D., SCHADE A.L., BUTLER W.T. & KASEL J.A. (1966) The proteins in nasal secretion: a longitudinal study of the γA-globulin, γG-globulin, albumin, siderophilin, and total protein concentrations in nasal washings from adult male volunteers. *J. clin. Invest.* 45, 768.

SALTMAN P. & CHARLEY P.J. (1960) The regulation of iron metabolism by equilibrium binding and chelation, *in* SEVEN M.J. & JOHNSON L.A. (eds.) *Metal Binding in Medicine*, p. 241. Lippincott, Philadelphia.

Sass-Kortsak A. & Vamosi J. (1964) Separation of Fe-saturated and Fe-free forms of transferrin by starch gel electrophoresis. *Fed. Proc.* **23**, 171.

Schade A.L. (1961) The microbiological activity of siderophilin. *Proc. 8th Colloq. Prot. Biol. Fluids. Elsevier, Amsterdam*, p. 261.

Schade A.L. & Caroline L. (1946) An iron-binding component in human blood plasma. *Science* **104**, 340.

Schade A.L., Oyama J., Reinhart R.W. & Miller J.R. (1954) Bound iron and unsaturated iron-binding capacity of serum; rapid and reliable quantitative determination. *Proc. Soc. exp. Biol.* **87**, 443.

Schade A., Reinhart R. & Levy H. (1949) Carbon dioxide and oxygen in complex formation with iron and siderophilin, the iron-binding component of human plasma. *Arch. Biochem. Biophys.* **20**, 170.

Schedl H.P. & Bartter F.C. (1959) Serum iron-binding protein levels after infusion of human serum albumin. *Lancet* **i**, 1163.

Scheidegger J.J., Martin E. & Riotton G. (1956) L'apparition des diverses composantes antigéniques du sérum au cours du développement foetal. *Schweiz med. Wschr.* **86**, 224.

Schultze H.E. (1962) Influence of bound sialic acid on electrophoretic mobility of human serum proteins. *Arch. Biochem. Biophys.* suppl. 1, 290.

Schultze H.E., Heide K. & Müller H. (1957) Über Transferrin/Siderophilin. *Behringwerk-Mitt.* **32**, 25.

Schultze H.E., Schmidtberger R. & Haupt H. (1958) Untersuchungen über die gebundenen Kohlenhydrate in isolierten Plasmaproteiden. *Biochem. Z.* **329**, 490.

Schultze H.E. & Schwick G. (1957) Immunchemischer nachweis von Profeinveränderungen unter besonderer Berücksichtigung fermentativer Einwirkungen auf Glyko und Lipoproteine. Immunoelektrophoretische Studien. *Behringwerk. Mitt.* **33**, 11.

Seppälä M. (1965) Distribution of serum transferrin groups in Finland and their inheritance. *Annl. Med. exp. Biol. Fenn.* **43**, Suppl. 4.

Smith C.H. (1954) Anemias in infancy and childhood: diagnostic and therapeutic considerations. *Bull. N.Y. acad. Med.* **30**, 155.

Smithies O. (1957) Variations in human serum β-globulins. *Nature* **180**, 1482.

Smithies O. (1958) Third allele at the serum β-globulin locus in humans. *Nature* **181**, 1203.

Smithies O. & Connell G.E. (1959) Biochemical aspects of the inherited variations in human serum haptoglobins and transferrins, *in* Wolstenholme G.E.W. & O'Connor C.M. (eds.) *Ciba Foundation Symposium on Biochemistry of Human Genetics*, p. 178. Little, Brown, Boston.

Smithies O. & Hiller O. (1959) The genetic control of transferrin in humans. *Biochem. J.* **72**, 121.

Starkenstein S. & Harvalik Z. (1933) Über einem intermediären Eisenstoffweschel enstehende Ferriglobulinverbindung. *Arch. exp. Path. Pharmak.* **12**, 75.

Sturgeon P. (1954) Studies on iron requirements in infants and children; normal values for serum iron, copper, and free erythrocyte protoporphyrin. *Pediatrics* **13**, 107.

SURGENOR D.M., KOECHLIN B.A. & STRONG L.E. (1949) Chemical, clinical and immunological studies on the products of human plasma fractionation. XXXVII. The metal-combining globulin of human plasma. *J. clin. Invest.* **28**, 73.

TURNBULL A. & GIBLETT E.R. (1961) The binding and transport of iron by transferrin variants. *J. Lab. clin. Med.* **57**, 450.

VAHLQUIST B. (1941) Das Serumeisen, eine padiatrischklinische und experimentelle Studie. *Acta paediat.* **28**, Suppl. 5.

WANG A.C. & SUTTON H.E. (1965) Human transferrins C and D_1: chemical difference in a peptide. *Science* **149**, 435.

WANG A.C., SUTTON H.E. & HOWARD P.N. (1967a) Human transferrins C and D_{Chi}: an amino acid difference. *Biochem. Genet.* **1**, 55.

WANG A.C., SUTTON H.E. & RIGGS A. (1966) A chemical difference between transferrins B_2 and C. *Am. J. hum. Genet.* **18**, 454.

WANG A.C., SUTTON H.E. & SCOTT I.D. (1967b) Transferrin D_1: identity in Australian aborigines and American Negroes. *Science* **156**, 936.

WARNER R.C. & WEBER I. (1953) The metal combining properties of conalbumin. *J. Am. chem. Soc.* **75**, 5094.

WHEBY M.S. & JONES L.G. (1963) Role of transferrin in iron absorption. *J. clin. Invest.* **42**, 1007.

YOSHIOKA R., FUJII T. & ITO Y. (1966) An electrophoretic resolution and densitometric determination of apo-transferrin and iron-bound transferrin. *Biochem. biophys. Res. Commun.* **24**, 203.

REFERENCES TO TRANSFERRIN
GEOGRAPHIC DISTRIBUTION

ALLISON A.C. & BARNICOT N.A. (1960) Haptoglobins and transferrins in some East African peoples. *Acta genet.* **10**, 17.

ANGELOPOULOS B., KALOS A. & DANOPOULOS E. (1967) Transferrin variants in Greece. *J. med. Genet.* **4**, 31.

ARENDS T. & GALLANGO M.L. (1962) Haptoglobin and transferrin groups in Venezuela. *Proc. 8th int. Cong. Blood Transf.*, p. 379. Tokyo.

ARENDS T. & GALLANGO M.L. (1964) Transferrins in Venezuelan Indians: high frequency of a slow-moving variant. *Science* **143**, 367.

ARENDS T. & GALLANGO M.L. (1965) Haemoglobin types and blood serum factors in British Guiana Indians. *Br. J. Haemat.* **11**, 350.

ARENDS T. & GALLANGO M.L. (1965) Transferrin groups in South American Indians. *Proc. 10th Cong. int. Soc. Blood Transf.*, p. 405. Stockholm.

BARNICOT N.A. (1961) Haptoglobins and transferrins, *in* HARRISON G.A. (ed.) *Genetical Variation in Human Populations*, p. 41. Pergamon Press, Oxford.

BARNICOT N.A., GARLICK J.P., ADAM A. & BAT-MIRIAM M. (1962) A survey of some genetic characters in Ethiopian tribes. III. Haptoglobins and transferrins. *Am. J. phys. Anthrop.* **20**, 175.

BARNICOT N.A., GARLICK J.P. & ROBERTS D.F. (1960) Haptoglobin and transferrin inheritance in Northern Nigerians. *Ann. hum. Genet.* **24**, 171.

BARNICOT N.A., GARLICK J.P., SINGER R. & WEINER J.S. (1959) Haptoglobin and transferrin variants in Bushmen and some other South African peoples. *Nature*, **184**, 2042.

BARNICOT N.A. & KARIKS J. (1960) Haptoglobin and transferrin variants in peoples of the New Guinea Highlands. *Med. J. Aust.* ii, 859.

BECKMAN L. & HOLMGREN G. (1961) Transferrin variants in Lapps and Swedes. *Acta genet.* **11**, 106.

BECKMAN L., HOLMGREN G. & MARTENSSON E.H. (1962) Transferrin types in the Swedish population. *Nature* **193**, 185.

BECKMAN L. & JOHANNSSON E.O. (1967) Haptoglobins and transferrins in the Icelandic population. *Acta genet.* **17**, 341.

BECKMAN L., JOHNSON F.M., SAKAI R.K. & WOODS J.L. (1964) Serum protein variations in Hawaiian population groups. *Acta genet.* **14**, 309.

BECKMAN L., TAKMAN J. & ARFORS K.E. (1965) Distribution of blood and serum groups in a Swedish gypsy population. *Acta genet.* **15**, 134.

BENERECETTI-SANTACHIARA A.S. & MODIANO G. (1964) The frequencies of haptoglobin and transferrin types in some villages of the Milan province. *Acta genet.* **14**, 36.

BENNETT J.H., AURICHT C.O., GRAY A.J., KIRK R.L. & LAI L.Y.C. (1961) Haptoglobin and transferrin types in the Kuru region of Australian New Guinea. *Nature* **189**, 68.

BLUMBERG B.S. & GENTILE Z. (1961) Haptoglobins and transferrins of two tropical populations. *Nature* **189**, 897.

BRAEND M., EFREMOV G., FAGERHOL M.K. & HARTMANN O. (1965) Albumin and transferrin variants in Norwegians. *Hereditas* **53**, 137.

CURTAIN C.C., GAJDUSEK D.C., KIDSON C., GORMAN J.G., CHAMPNESS L. & RODRIGUE R. (1965) Hapto-globins and transferrins in Melanesia: relation to hemoglobin, serum haptoglobin and serum iron levels in population groups in Papua, New Guinea. *Am. J. phys. Anthrop.* **23**, 363.

CURTAIN C.C., TINDALE N.B. & SIMMONS R.T. (1966) Genetically determined blood protein factors in Australian aborigines of Bentinck, Mornington and Forsyth Islands and the mainland, Gulf of Carpentaria. *Arch. phys. Anthrop. Oceania.* **1**, 74.

DOUGLAS R., JACOBS J., GREENHOUGH R. & STAVELEY J.M. (1964) Blood groups, serum genetic factors and hemoglobins in New Hebrides islanders. *Transfusion* **4**, 177.

DOUGLAS R., JACOBS J., HOULT G.E. & STAVELEY J.M. (1962) Blood groups, serum genetic factors and hemoglobins in Western Solomon islanders. *Transfusion* **2**, 413.

DOUGLAS R., JACOBS J., SHERLIKER J. & STAVELEY J.M. (1961) Blood groups, serum genetic factors and haemoglobins in Gilbert islanders. *N. Zeal. med. J.* **60**, 146.

DOUGLAS R., JACOBS J., SHERLIKER J. & STAVELEY J.M. (1961) Blood groups, serum genetic factors and haemoglobins in Ellice islanders. *N. Zeal. med. J.* **60**, 259.

FLORY L.L. (1964) Serum factors of Australian aborigines from North Queensland. *Nature* **201**, 508.

FRASER G.R., GIBLETT E.R., STRANSKY E. & MOTULSKY A.G. (1964) Blood groups in the Philippines. *J. med. Genet.* **1**, 107.

FRASER G.R., GIBLETT E.R., TING-CHIEN L. & MOTULSKY A.G. (1965) Blood and serum groups in Taiwan. *J. med. Genet.* **2**, 21.

GALLANGO M.L. & ARENDS T. (1966) Haemoglobin types and blood serum factors in Columbian Indians *Acta Genet.* **16**, 162.

GIBLETT E.R. (1962) The plasma transferrins, *in* STEINBERG A.G. & BEARN A.G. (eds.) *Progress in*

Medical Genetics, vol. 2, p. 34. Grune & Stratton, New York.

GIBLETT E.R. (1962) Haptoglobins and transferrins in Pacific populations. *Eugen. Quart.* **9**, 45.

GIBLETT E.R., MOTULSKY A.G. & FRASER G.R. (1966) Population genetics in the Congo. IV. Haptoglobin and transferrin serum groups in the Congo and in other African populations. *Am. J. hum. Genet.* **18**, 553.

JENKINS T. & STEINBERG A.G. (1966) Some serum protein polymorphisms in Kalahari Bushmen and Bantu: gamma globulins, haptoglobins and transferrins. *Am. J. hum. Genet.* **18**, 399.

KIRK R.L. (1965) Population genetic studies of the indigenous peoples of Australia and New Zealand, *in* STEINBERG A.G. & BEARN A.G. (eds.) *Progress in Medical Genetics*, vol. 4, p. 202. Grune & Stratton, New York.

KIRK R.L. & LAI L.Y.C. (1961) The distribution of haptoglobin and transferrin groups in South and Southeast Asia. *Acta genet.* **11**, 97.

KIRK R.L., LAI L.Y.C. & HORSFALL W.R. (1962) The haptoglobin and transferrin groups among Australian aborigines from North Queensland. *Aust. J. Sci.* **24**, 486.

KIRK R.L., LAI L.Y.C., VOS G.H. & VIDYARTHI L.P. (1962) A genetical study of the Oraons of the Chota Nagpur Plateau (Bihar, India). *Am. J. phys. Anthrop.* **20**, 375.

KIRK R.L., LAI L.Y.C., VOS G.H., WICKREMASINGHE R.L. & PERERA D.J.B. (1962) The blood and serum groups of selected populations in South India and Ceylon. *Am. J. phys. Anthrop.* **20**, 485.

KIRK R.L., PARKER W.C. & BEARN A.G. (1964) The distribution of the transferrin variants D_1 and D_{Chi} in the various populations. *Acta genet.* **14**, 41.

LIE-INJO L.E., BOLTON J.M. & FUDENBERG H.H. (1967) Haptoglobins, transferrins and serum gamma-globulin types in Malayan aborigines. *Nature* **215**, 777.

LISKER R., LORIA A. & ZARATE G. (1967) Studies on several genetic hematological traits of the Mexican population. XIII. Red cell and serum polymorphisms in Spanish immigrants. *Acta genet.* **17**, 524.

MATSON G.A., SUTTON H.E., ETCHEVERRY R., SWANSON J. & ROBINSON A. (1967) Distribution of hereditary blood groups among Indians in South America. IV. In Chile. *Am. J. phys. Anthrop.* **27**, 157.

MATSON G.A., SUTTON H.E., SWANSON J. & ROBINSON A.R. (1963) Distribution of haptoglobin types among Indians of Middle America: Southern Mexico, Guatemala, Honduras and Nicaragua. *Hum. Biol.* **35**, 474.

MATSON G.A., SUTTON H.E., SWANSON J. & ROBINSON A.R. (1965) Distribution of haptoglobin, transferrin and hemoglobin types among Indians of Middle America: in British Honduras, Costa Rica and Panama. *Am. J. phys. Anthrop.* **23**, 123.

MATSON G.A., SUTTON H.E., SWANSON J., ROBINSON A.R. & SANTIANA A. (1966) Distribution of hereditary blood groups among Indians in South America. I. In Ecuador. *Am. J. phys. Anthrop.* **24**, 51.

MATSON G.A., SWANSON J. & ROBINSON A. (1966) Distribution of hereditary blood groups among Indians of South America. III. In Bolivia. *Am. J. phys. Anthrop.* **25**, 13.

MELARTIN L. & KAARSALO E. (1965) The distribution of transferrin variants in Southwestern Finland and in Finnish Lappland. *Acta genet.* **15**, 63.

MODIANO G., BENERECETTI-SANTACHIARA A.S., GONANO F., ZEI G., CAPALDO A. & CAVALLI-SFORZA L.L. (1965) An analysis of ABO, MN, Rh, Hp, Tf, and G-6-PD types in a sample from the human population of the Lecce province. *Ann. hum. Genet.* **29**, 19.

NEEL J.V., ROBINSON A.R., ZUELZER W.W., LIVINGSTONE F.B. & SUTTON H.E. (1961) The frequency of elevations in the A_2 and fetal hemoglobin fractions in the natives of Liberia and adjacent regions, with data on haptoglobin and transferrin types. *Am. J. hum. Genet.* **13**, 262.

NICHOLLS E.M., LEWIS H.B.M., COOPER D.W. & BENNETT J.H. (1965) Blood group and serum protein differences in some Central Australian aborigines. *Am. J. hum. Genet.* **17**, 293.

PARKER W.C. & BEARN A.G. (1961) Haptoglobin and transferrin variation in humans and primates; two new transferrins in Chinese and Japanese populations. *Ann. hum. Genet.* **25**, 227.

PARKER W.C. & BEARN A.G. (1961) Haptoglobin and transferrin gene frequencies in a Navajo population: a new transferrin variant. *Science* **134**, 106.

PERSSON I. (1962) Transferrins in Greenland Eskimos. *Acta genet.* **11**, 41.

PLATO C.C., RUCKNAGEL D.L. & GERSHOWITZ H. (1964) Studies on the distribution of glucose-6-phosphate dehydrogenase deficiency, thalassemia, and other genetic traits in the coastal and mountain villages of Cyprus. *Am. J. hum. Genet.* **16**, 267.

PROCHNICKA B. (1966) The frequencies of transferrin types in the population of Cracow (Poland). *Acta genet.* **16**, 248.

RAMOT B., DUVDEVANI-ZIRKERT P. & Kende G. (1962) Haptoglobin and transferrin types in Israel. *Ann. hum. Genet.* **25**, 267.

SALZANO F.M. & SUTTON H.E. (1963) Haptoglobin and transferrin types in southern Brazilian Indians. *Acta genet.* **13**, 1.

SALZANO F.M. & SUTTON H.E. (1965) Haptoglobin and transferrin types of Indians from Santa Catarina, Brazil. *Am. J. hum. Genet.* **17**, 280.

SANFORD R., GRIMMO A.E.P. & SHUN-KEUNG L. (1966) The haptoglobin and transferrin types of some Cantonese in Hong Kong. *Vox Sang.* **11**, 106.

SEPPÄLÄ M. (1965) Distribution of serum transferrin groups in Finland and their inheritance. *Ann. med. exp. Biol. Fenn.* **43**, suppl. 4.

SEPPÄLÄ M. & MÄKELÄ O. (1963) Transferrin variants in Finland. *Nature*, **199**, 831.

SHIM B.S. (1964) Occurrence of transferrin D_1 in Korea. *Nature*, **203**, 432.

SHREFFLER D.C. & STEINBERG A.G. (1967) Further studies on the Xavante Indians. IV. Serum protein groups and the SC_1 trait of saliva in the Simões Lopes and São Marcos Xavantes. *Am.J.hum.Genet.* **19**, 514.

STAVELEY J.M. & DOUGLAS R. (1960) Transferrins in Tongans. *N. Zeal. med. J.* **59**, 546.

STEINBERG A.G., LAI L.Y.C., VOS G.H., SINGH R.B. & LIM T.W. (1961) Genetic and population studies of the blood types and serum factors among Indians and Chinese from Malaya. *Am.J.hum.Genet.* **13**, 355.

STEINBERG A.G. & MATSUMOTO H. (1964) Studies on the Gm, Inv, Hp and Tf serum factors of Japanese populations and families. *Hum. Biol.* **36**, 77.

SUTTON H.E., MATSON G.A., ROBINSON A.R. & KOUCKY R.W. (1960) Distribution of haptoglobin, transferrin, and hemoglobin types among Indians of Southern Mexico and Guatemala. *Am. J. hum. Genet.* **12**, 338.

TIWARI S.C. (1960) Haptoglobin and transferrin variants in some upper castes of Bengal. *Anthropologist* **7**, 17.

VAN ROS G., VAN SANDE M. & DRUET R. (1963) Groupes d'haptoglobine et de transferrine dans des populations africaines et européenes. *Ann. Soc. belge. Med. Trop.* **4**, 511.

CHAPTER 4

THE Gc SYSTEM

General Properties 160

Inheritance of Gc Phenotypes 161

Electrophoretic Patterns.................. 161

Serum Concentration of Gc Protein 162

Phenotypic Variants........................ 162

Chemical Studies 164

Geographic Distribution of Gc Genes.. 165

Methods 167

Immunoelectrophoresis.............. 167

Starch gel electrophoresis.......... 168

References.. 169

References to Gc Geographic distribution 172

Hirschfeld (1959) reported that in human serum specimens subjected to immunoelectrophoresis, the precipitation pattern of one of the α-2 globulins besides haptoglobin showed systematic variation in appearance and migration rate. The three phenotypes so defined consisted of a fast, a slow and an intermediate (bimodal) arc; mixtures of sera containing the slow and fast types resulted in a pattern indistinguishable from the intermediate type. This variable globulin was given the name 'group-specific component' or Gc. It has been the subject of reviews by Hirschfeld (1962), Cleve & Bearn (1962), Bearn *et al.* (1964a), Nerstrøm (1965) and Reinskou (1968).

GENERAL PROPERTIES

The function of Gc protein has yet to be determined. It is synthesized by the liver (Prunier *et al.*, 1964), and synthesis begins in the human fetus as early as the tenth to thirteenth week of gestation (Melartin *et al.*, 1966). A decrease in serum Gc concentration is observed in liver disease (Kitchin & Bearn, 1965), and an increase occurs during pregnancy (Reinskou, 1968), resembling the behavior of another liver-synthesized protein, transferrin. Several fluids of the body contain Gc, including urine, ascitic fluid and spinal fluid (Hirschfeld, 1962; Nielsen *et al.*, 1963; Berggaard *et al.*, 1964).

Cleve & Bearn (1961a,b, 1962) found that the Gc globulin is soluble in 40 per cent saturated ammonium sulfate but is precipitated at 60 per cent saturation as well as by rivanol at concentrations of 0·3 and 0·4 per cent. The protein is not soluble in 0·6 M perchloric acid or in 5 per cent trichloracetic acid, and it is not precipitated at pH 5·4 or 5·8 at 0·02 or 0·005 M concentrations, so it is not a euglobulin. Its electrophoretic migration rate is not affected by heating the purified protein to 56°C for 24 hours, nor by incubation with neuraminidase.

INHERITANCE OF Gc PHENOTYPES

Family studies by Hirschfeld *et al.* (1960) showed that the Gc types are determined by a pair of autosomal codominant alleles called Gc^1 and Gc^2. Thus, individuals homozygous for Gc^1 have the fast-moving component; those homozygous for Gc^2 have the slow-moving component; and heterozygotes have the intermediate arc containing both components. The three phenotypes, Gc 1-1, Gc 2-2 and Gc 2-1, are shown in Fig. 4.1.

Subsequent studies by a number of investigators have confirmed the mode of inheritance reported in Hirschfeld's original studies, and have shown the usefulness of the Gc system as a genetic marker (Cleve & Bearn, 1961a, 1962; Hirschfeld, 1962; Reinskou & Mohr, 1962; Hirschfeld & Heiken, 1963; Marek *et al.*, 1963; Mansa *et al.*, 1963; Nerstrøm, 1963a; Mohr & Reinskou, 1963; Baitsch *et al.*, 1963, 1964; Reinskou, 1965a, 1966a; and Seppälä *et al.*, 1967). Furthermore, genetic linkage of the Gc and albumin loci has been established, as discussed on p. 245 (Weitkamp *et al.*, 1966; Kaarsalo *et al.*, 1967).

ELECTROPHORETIC PATTERNS

The variability in migration rate demonstrated by immunoelectrophoresis can also be shown by electrophoresis in starch gel and acrylamide gel. Several years ago, Smithies (1959) described differences in migration of a serum protein in the post-albumin region and speculated about a possible genetic polymorphism. However, the fact that this post-albumin protein was Gc was not recognized until 1962 when Schultze *et al.* (1962a) described the starch gel pattern of the Gc phenotypes. These findings were extended and clarified by Arfors &

Beckman (1963), Parker *et al.* (1963) and Bearn *et al.* (1964b), using different electrophoretic conditions. These authors also noted heterogeneity in the homozygote phenotypes, as shown in Fig. 4.2 and discussed on p. 164.

SERUM CONCENTRATION OF Gc PROTEIN

Kitchin & Bearn (1965) found that in eight sera of type 1-1, ten of type 2-1 and three of type 2-2, the level of Gc protein did not vary with the phenotype. On the other hand, Reinskou (1967) tested 20 serum specimens of each phenotype and observed a significantly higher level of Gc 1-1 over Gc 2-2. The difference in results may be due to the relatively small number of samples tested in the first investigation or to the different kinds of immunological measurements used in the two studies. The average level of Gc protein reported by Kitchin & Bearn (1965) was $74\cdot8 \pm 6\cdot5$ mg per 100 ml.

PHENOTYPIC VARIANTS

As indicated in Table 4.1, several phenotypes other than Gc 1-1, 2-2 and 2-1 have been described. Hirschfeld (1962) found a bimodal precipitin arc in the serum of an African Negro which appeared to represent Gc 1 and a faster-moving component, Gc Y. A second unusual component found in a Caucasian serum, had a slightly faster mobility than Gc 2; it was called Gc X. Parker *et al.* (1963) also reported a faster-moving variant in a Negro and a variant with intermediate mobility in a white subject. Family members were not available to prove that any of the four variants represented alleles at the Gc locus.

During a study of 150 serum specimens obtained from Chippewa Indians, Cleve *et al.* (1963a) provisionally classified 59 as Gc 2-1. However, in five of these 59 sera, there was greater than usual separation of the fast and slow Gc components because the fast component had a more rapid migration rate than Gc 1. Some of the sera with an apparent Gc 1-1 phenotype had slightly extended arcs of precipitation, but the distinction between Gc 1 and a faster component was difficult to demonstrate by immunoelectrophoresis. However, on starch gel electrophoresis, a band migrating faster than Gc 1 was clearly

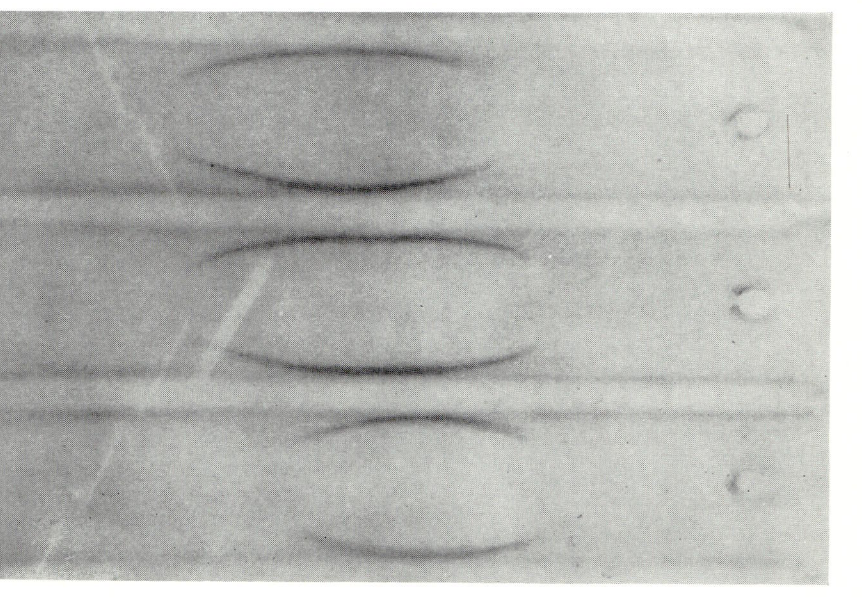

FIGURE 4.1. Immunoelectrophoretic patterns of Gc 1-1, 2-1 and 2-2, made from serum specimens of the three Gc types and developed with anti-Gc serum in the troughs. The bright diagonal lines on the left are due to imperfections in the photographic film.

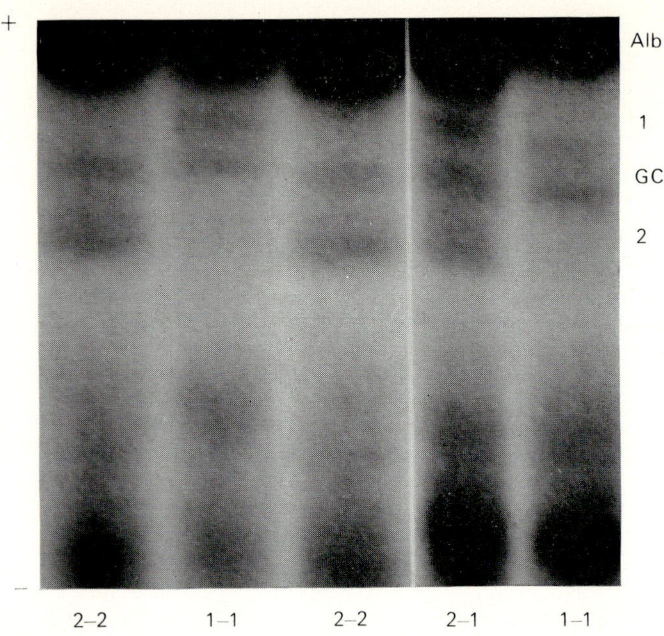

FIGURE 4.2. Starch gel electrophoretic patterns of five serum specimens with the three common Gc phenotypes. Albumin is at the top of the picture. Both Gc 1-1 and Gc 2-2 consist of two bands, of which the fast band of Gc 1-1 and the slow band of Gc 2-2 are indicated by the numbers 1 and 2. Three bands are seen in the heterozygote pattern. (From Bearn *et al.*, 1964b. Reprinted with permission from the *J. exp. Med.*, **120**, 83).

[Facing page 163]

distinguishable, and it was designated Gc Chippewa. By this method, ten of 38 sera originally thought to be Gc 1-1 were re-classified as Gc Chip-1. Tests on the sera of 41 members of a large kindred showed that inheritance of the Chippewa variant was consistent with a co-dominant allele, Gc^{Chip} at the autosomal Gc locus.

TABLE 4.1. Electrophoretic mobility of the Gc variants. The three variants marked with an asterisk have the same mobility and might represent the same structural variation. (Kitchin & Bearn, 1966)

Gc Variant	Electrophoretic Migration	Reference
X	Faster than Gc 2; slower than Gc 1	Hirschfeld (1962)
Y*	Faster than Gc 1 and Gc Chip	Hirschfeld (1962)
Ab*	Same as Gc Y	Cleve *et al.* (1963a)
Unnamed*	Same as Gc Y	Persson & Tingsgaard (1965)
Chip	Faster than Gc 1; slower than Gc Ab	Cleve *et al.* (1963a)
Negro	Faster than Gc 1 by iep; slower in st. gel	Parker *et al.* (1963)
Caucasian	Faster than Gc 2; slower than Gc 1	Parker *et al.* (1963)
Z	Slower than Gc 2	Hennig & Hoppe (1965)
Norwegian	Faster than Gc 2; slower than Gc 1	Reinskou (1965b)

In immunoelectrophoretic tests of 74 serum specimens obtained from Australian aborigines, Cleve *et al.* (1963a) found three previously undescribed phenotypes. These were due to a fourth allele, Gc^{Ab} in combination with itself and the two common alleles, Gc^1 and Gc^2. The migration rate of Gc Ab was faster than either Gc 1 or Gc Chip, extending well into the α-1 region. On starch gel electrophoresis, this zone was covered by the trailing edge of albumin. However, Kitchin & Bearn (1966) subsequently showed that if they used the method of acrylamide gel electrophoresis described by Kitchin (1965), the fast moving Gc Ab component could be visualized. Kitchin & Bearn found no differences by either electrophoresis or immunoelectrophoresis among Gc Ab, Gc Y (found in Negroes) and a fast-moving

Eskimo variant described by Persson & Tingsgaard (1965). Chemical analysis of the individual proteins is required to determine whether there are differences in amino acid sequence not reflected by electrophoretic rate in these three variants.

Gc Z, with slower mobility than Gc 2, was found in a North German family by Hennig & Hoppe (1965). Studies of this family and another family reported by Cleve *et al.* (1966) indicated the existence of an allele, Gc^z at the Gc locus. Another variant, Gc Norw., with electrophoretic mobility between Gc 1 and Gc 2, was described by Reinskou (1965b). The existence of a rare Gc^0 'silent' allele was suggested by Henningsen (1966) from studies of a family with anomalous Gc inheritance. Most of the Gc variants have a very low frequency. In fact, of the aberrant Gc genes, only Gc^{Chip} and Gc^{Ab} occur with a frequency of 1 per cent or more in any known population. (See Addenda.)

CHEMICAL STUDIES

The sedimentation constant of both Gc 1 and Gc 2 is $S_{20} = 4 \cdot 1$ S, and the average molecular weight of both proteins is 50,800 ± 2900 (Bearn *et al.*, 1964a). Cleve *et al.* (1963b) and Heimburger *et al.* (1964) found no obvious difference in amino-acid composition between Gc 1 and Gc 2. Furthermore, none of the many examples of anti-Gc antibodies prepared in various animals has been capable of detecting an immunological difference, although one antiserum appeared to differentiate two kinds of Gc 1 protein (Ruoslahti, 1965).

Bearn *et al.* (1964a,b) reported that better electrophoretic resolution of the Gc proteins could be achieved by incorporating lithium salts into the buffer, in accordance with the method of Arfors & Beckman (1963). As shown in Fig. 4.2, heterogeneity of both Gc 1 and Gc 2 was revealed in this manner, in agreement with an earlier report of Reinskou (1963). Specimens of purified Gc 1-1 and Gc 2-2 as well as serum samples with these phenotypes were found to consist of two bands each, of which the faster was more faintly stained than the slower. Each type had one characteristic band; fast in Gc 1 and slow in Gc 2. However, the slower-moving of the two Gc 1 bands and the faster-moving of the two Gc 2 bands had the same electrophoretic mobility. In each homozygous type, the two components could not be separated by Sephadex gel filtration, suggesting that they differed in charge rather than size. In the serum of heterozygotes, three bands were observed.

Since these results suggested the possibility of two or more poly-peptide subunits in Gc, attempts were made by Bearn *et al.* (1964a) and Bowman & Bearn (1965) to split the molecule. First, tryptic digests of Gc 2-1 and 1-1 were subjected to peptide fingerprinting, which revealed 23–27 peptides rather than the 54 expected from the number of lysine and arginine residues. Also, one-half the number of residues predicted by amino-acid analysis was found for tyrosine, arginine and half-cystines. Furthermore, although the molecular weight of unaltered protein was calculated to be about 50,000, treatment with 6 M guanidium chloride and 8 M urea in the presence of 0·1 M mercaptoethanol brought about a decrease to about 25,000, as calculated from ultracentrifugation data. The authors concluded that they had confirmed the presence of two subunits, each with a molecular weight of 25,000, held together by disulfide bonds.

Bowman (1967) performed enzymatic digestion on fractions of the three phenotypes isolated by polyacrylamide column electrophoresis (Simons & Bearn, 1967). She found a striking similarity in the peptide patterns of the fast and slow components of Gc 1-1, but an amino-acid substitution in at least one peptide could be inferred. The peptide patterns of Gc 1-1 and 2-2 were also nearly identical, but there was evidence for at least one amino-acid substitution.

From these studies, it appears that Gc protein consists of two very similar subunits, the structure of which could be controlled by very similar genes at two different loci. However, the precise relationship between the two separable components of Gc 1 and of Gc 2 has not yet been determined; they may represent differences which do not have a genetic origin.

GEOGRAPHIC DISTRIBUTION
OF Gc GENES

Considering the fact that Gc is a relative newcomer to the growing collection of polymorphic serum proteins, a remarkably large number of studies on its geographic distribution have been performed. These have been summarized in several reports (Wendt & Theile, 1963; Gallango & Arends, 1965; Schultze & Heremans, 1966). The references listed separately at the end of this chapter provided the basis for the following brief summary.

In nearly all of the populations studied so far, the Gc^2 gene has a lower frequency than Gc^1. The single exception is a figure of 0·69 for the Gc^2 gene frequency among 524 members of three Xavante Indian tribes in Brazil and of 0·56 in another Brazilian tribe, the Caingang. Among Europeans, the Gc^2 gene frequency is fairly constant, averaging about 0·26; it has the high frequency of 0·39 in the Finnish island of Kokär, and among the Swedish and Finnish Lapps, it drops to about 0·12 to 0·14. African Negroes generally have a low Gc^2 gene frequency, rarely exceeding 0·10; while the Gc^2 frequencies in the tested populations of Asia are similar to or lower than those of Europe, with the exception of the Israeli Ashkenazi (0·34) and the Kurumbas of India (0·35). The Indian tribes of South America (except the Xavantes of Brazil) tend to have Gc^2 frequencies between 0·30 and 0·35, while North American Indians vary considerably from tribe to tribe, the Navajos having the lowest recorded level of 0·02. Among the Chippewas, Gc^2 has a 0·21 frequency, and the variant allele Gc^{Chip}, 0·10. Evidence for this allele has not been found in other population samples, including various Indian tribes.

In Australian aborigines, Gc^2 has a relatively low frequency in most tribes, although there is an apparent increase proceeding from the central desert area toward the north and east coast. The Gc^{Ab} allele is present in some central and coastal tribes of Australia as well as New Guinea and at least one island of the New Hebrides group, but not in adjacent areas of southeast Asia.

In a report on the Gc types in East Greenland Eskimos, Persson & Tingsgaard (1966) showed that a group of serum specimens sent under 'unfavorable conditions' had a quite different distribution of Gc phenotypes than another group of specimens which were not delayed in transit. They cited the previous studies of Nerstrøm (1963b), Nerstrøm & Skafte-Jensen (1963) and Nerstrøm et al. (1964), which showed that under a number of circumstances, the Gc components were shifted toward the anode on electrophoresis, or so altered that the precipitation arc was elongated, extending into the α-1 globulin zone. Such alterations were observed in grossly hemolysed samples, in eluates of dried blood, in sera incubated with bacteria or with disintegrated white cells and platelets, and in the sera of infants with hyperbilirubinemia. This susceptibility of the Gc protein to structural degradation, also described by Reinskou (1966b) should be recalled whenever there is an unexpected distribution of phenotypes, particularly a decrease of Gc 2-2 and 2-1, in a population.

Although Gc is a glycoprotein with about 4 per cent carbohydrate content, it does not contain any sialic acid (Schultze *et al.*, 1962b; Cleve & Bearn, 1962). Thus, it would not be expected to undergo the cathodal shift characteristic of such glycoproteins as haptoglobin, transferrin and α-1 acid glycoprotein upon exposure to the enzyme, neuraminidase.

METHODS

IMMUNOELECTROPHORESIS

The performance of immunoelectrophoresis involves an initial electrophoretic separation of the serum protein components followed by immunodiffusion against anti-human serum antibodies. The insoluble antigen–antibody precipitates appear as one or more of a series of arcs, depending on the number of antibody specificities (Grabar & Williams, 1953). A gel medium is usually used for this procedure, the most common being high grade agar or agarose. The general principles and details of immuno-electrophoretic procedures have been described in several books and papers, including those of Grabar & Burtin (French ed., 1960; English ed., 1964), Wunderly (1961), Crowle (1961), Ouchterlony (1962), Peetom (1963) and Schultze & Heremans (1966).

For the differentiation of Gc precipitation arcs, a very satisfactory procedure was described by Hirschfeld (1960, 1961) employing a micro-method of Scheidegger (1955) and the discontinuous buffer system of Laurell *et al.* (1956). Two per cent Difco Noble Agar or Ion Agar No. 2 in distilled water is heated to dissolve the agar and stored in airtight flasks at 4°C. One volume of this melted agar is mixed with an equal volume of buffer containing 1·11 g diethylbarbituric acid, 6·99 g sodium veronal and 1·02 g calcium lactate per liter, pH 8·6. The tank buffer contains the same chemicals, but in amounts per liter of 1·38, 8·76 and 3·8 g, respectively. The (melted) agar is poured onto thoroughly cleaned microscope slides (about 2 ml per slide) and permitted to harden. Two 1·2 mm antigen wells, 14 mm apart, are made in the agar 28 mm from one end, using a plastic template or commercially available punch. A 2 mm trough, equidistant between the holes, is made to run nearly the length of the agar, except for 8 mm at each end.

With the Immunophor apparatus produced by the LKB Company, 15 slides (five rows of three) can be set up at a time. These are arranged so that the cathodal end is nearest to the antigen wells. Five μl of serum is placed in each well, and electrophoresis is performed at room temperature at 7–8 volts/cm for about 2 hours. In a modification of the method (Reinskou, 1963), longer glass plates are used, permitting prolonged electrophoresis for several hours, or until the albumin, marked with bromphenol blue, has travelled 10 cm from the origin (Kitchin & Bearn, 1966).

Anti-human Gc serum (commercially available) is placed in the trough of each slide, and the slides are then incubated in a humid chamber for 24 to 36 hours. The precipitation arcs are visible in the α-2 region. They may be stained after soaking (to remove unprecipitated protein) and drying at 37°C with a piece of filter paper laid over the surface. Several protein stains have been used. Hirschfeld (1960) recommended amido black, one g dissolved in a mixture of 100 ml glacial acetic acid, 700 ml methanol and 200 ml distilled water. After 5 minutes, the slides are destained with the acetic acid–methanol–water solution.

For photography, it is convenient to place the slide in the negative holder of a photographic enlarger and project the image directly on a sheet of light-sensitive paper. In this way, the size of the image can be adjusted over a wide range.

STARCH GEL ELECTROPHORESIS

Separation of the Gc components by starch gel electrophoresis is performed by the method of Parker *et al.* (1963) modified by Bearn *et al.* (1964b); it employs a buffer system described by Ashton & Braden (1961), first suggested for this purpose by Arfors & Beckman (1963). The bridge buffer consists of 11·9 g boric acid and 1·2 g lithium hydroxide dissolved in 1 liter of water. For the gel, 50 ml of this solution is mixed with 450 ml of a solution prepared by dissolving 6·3 g Tris (Sigma) and 1·6 g citric acid in 1 liter of water. Vertical electrophoresis is performed for 3–4 hours at high voltage (20 volts/cm), using a power source capable of supplying up to 1000 volts. The apparatus is preferably water-cooled, although a cold room or refrigerator at 4°C is adequate if the gel is also kept cool with an electric fan on each side. A cellophane cover instead of petroleum jelly promotes heat exchange.

The gels are stained with amido black, as described in the chapter on haptoglobin. In each of the homozygous types, the Gc protein is visualized as two bands, the slow band of Gc 1 migrating at the same rate as the fast band of Gc 2. Thus, in heterozygotes, three bands are seen. Similar results are obtained using acrylamide gel electrophoresis, the details of which are described by Kitchin (1965), Kitchin & Bearn (1966) and Raunio *et al.* (1966).

REFERENCES

ARFORS K.E. & BECKMAN L. (1963) Genetic variations of the human serum postalbumins (the Gc groups). *Acta genet.* **13**, 231.

ASHTON G.C. & BRADEN A.W.H. (1961) Serum β-globulin polymorphism in mice. *Aust. J. exp. Biol. med. Sci.* **14**, 248.

BAITSCH H., RITTER H., GOEDDE H.W. & ALTLAND K. (1963) Zur Genetik der Serum-proteine: Hp Serumgruppen, Gc-faktor, Gm-Serumgruppen und Pseudocholinesterase-varianten in Europaischen Populationen. *Vox Sang.* **8**, 594.

BAITSCH H., RITTER H. & SOMMER R. (1964) Zur formalen Genetik des Gc-polymorphismus; Untersuchungen an 339 Familien. *Anthrop. Anz.* **27**, 63.

BEARN A.F., BOWMAN B.H. & KITCHIN F.D. (1964a) Genetic and biochemical considerations of the serum group-specific component. *Cold Spring Harb. Symp. quant. Biol.* **29**, 435.

BEARN A.G., KITCHIN F.D. & BOWMAN B.H. (1964b) Heterogeneity of the inherited group-specific component of human serum. *J. exp. Med.* **120**, 83.

BERGGAARD I., CLEVE H. & BEARN A.G. (1964) The excretion of five plasma proteins previously unidentified in normal human urine. *Clin. chim. Acta* **10**, 1.

BOWMAN B.H. (1967) Biochemical characterization of the inherited group-specific protein. *Fed. Proc.* **26**, 724.

BOWMAN B.H. & BEARN A.G. (1965) The presence of subunits in the inherited group-specific protein of human serum. *Proc. natn. Acad. Sci.* **53**, 722.

CLEVE H. & BEARN A.G. (1961a) Inherited variations in human serum proteins: studies on the group-specific component. *Ann. N.Y. Acad. Sci.* **94**, 218.

CLEVE H. & BEARN A.G. (1961b) Genetic and chemical aspects of the group specific component. *Proc. 2nd Cong. int. hum. Genet.* Rome. p. 703.

CLEVE H. & BEARN A.G. (1962) The group specific component of serum; genetic and chemical consideration, *in* STEINBERG A.G. & BEARN A.G. (eds.) *Progress in Medical Genetics*, vol. 2, p. 64. Grune & Stratton, New York.

CLEVE H., KIRK R.L., PARKER W.C., BEARN A.G., SCHACHT L.E., KLEINMAN H. & HORSFALL W.R. (1963a) Two genetic variants of the group-specific component of human serum: Gc Chippewa and Gc Aborigine. *Am. J. hum. Genet.* **15**, 368.

CLEVE H., KRÜPE M. & ENSGRABER A. (1966) Zur Vererbung der Gc-variante Gc Z. Bericht über eine weitere Familie. *Humangenetik.* **3**, 46.

CLEVE H., PRUNIER J.H. & BEARN A.G. (1963b) Isolation and partial characterization of the two principal inherited group-specific components of human serum. *J. exp. Med.* **118**, 711.

CROWLE A.J. (1961) *Immunodiffusion.* Academic Press, New York.

GALLANGO M.L. & ARENDS T. (1965) El sistema de grupos séricos Gc: sus variantes fenotípicas y su distribución mundial. *Acta cient. venez.* **16**, 120.

GRABAR P. & BURTIN P. (eds.) (1964) *Immunoelectrophoretic Analysis.* Elsevier, Amsterdam.

GRABAR P. & WILLIAMS C.A. (1953) Méthode permettant l'étude conjugée des propriétés électrophorétiques et immunochimiques d'un mélange de protéines. Application au sérum sanguin. *Biochim. biophys. Acta,* **10**, 193.

HEIMBURGER N., HEIDE K., HAUPT H. & SCHULTZE H.E. (1964) Bausteinanalysen von Humanserumproteinen. *Clin. chim. Acta.* **10**, 293.

HENNIG W. & HOPPE H.H. (1965) A new allele in the Gc system, Gc^2. *Vox Sang.* **10**, 214.

HENNINGSEN K. (1966) A silent allele within the Gc system. *Proc. 4th Cong. int. Forensic Med.* Copenhagen.

HIRSCHFELD J. (1959) Immunoelectrophoretic demonstration of qualitative differences in normal human sera and their relation to the haptoglobins. *Acta path. microbiol. scand.* **47**, 160.

HIRSCHFELD J. (1960) Immunoelectrophoresis-procedure and application to the study of group specific variations in sera. *Sci. Tools,* **7**, 18.

HIRSCHFELD J. (1961) The use of immunoelectrophoresis in the analysis of normal sera and in studies of the inheritance of certain serum proteins. *Sci. Tools,* **8**, 17.

HIRSCHFELD J. (1962) The Gc system. Immunoelectrophoretic studies of normal human sera with special reference to a new genetically determined serum system (Gc). *Prog. Allergy* **6**, 155.

HIRSCHFELD J. & HEIKEN A. (1963) Application of the Gc system in paternity cases. *Am. J. hum. Genet.* **15**, 19.

HIRSCHFELD J., JONSSON B. & RASMUSON M. (1960) Inheritance of a new group-specific system demonstrated in normal human sera by means of an immuno-electrophoretic technique. *Nature* **185**, 931.

KAARSALO E., MELARTIN L. & BLUMBERG B.S. (1967) Autosomal linkage between the albumin and Gc loci in humans. *Science* **158**, 123.

KITCHIN F.D. (1965) Demonstration of the inherited serum group specific protein by acrylamide electrophoresis. *Proc. Soc. exp. Biol. Med.* **119**, 1153.

KITCHIN F.D. & BEARN A.G. (1965) Quantitative determination of the group specific protein in normal human serum. *Proc. Soc. exp. Biol. Med.* **118**, 304.

KITCHIN F.D. & BEARN A.G. (1966) The electrophoretic patterns of normal and variant phenotypes of the group specific (Gc) components in human serum. *Am. J. hum. Genet.* **18**, 201.

LAURELL C.B., LAURELL S. & SKOOG N. (1956) Buffer composition in paper electrophoresis. Considerations on its influence, with special reference to the interaction between small ions and proteins. *Clin. chem.* **2**, 99.

MANSA B., NERSTRØM B. & HAUGE M. (1963) The Gc types in a twin material. *Acta genet.* **13**, 247.

MAREK Z., BUNDSCHUH G., KERDE C. & GESERIK G. (1963) Untersuchungen über die Anwendbarkeit der menschlichen Gc-Komponenten in der forenischen Serologie. *Arztl. Lab.* **9**, 228.

MELARTIN L., HIRVONEN T., KAARSALO E. & TOIVANEN P. (1965) Group-specific components and transferrins in human fetal sera. *Scand. J. Haemat.* **3**, 117.

MOHR J. & REINSKOU T. (1963) Genetics of the Gc serum types: associations and linkage relations. *Acta genet.* **13**, 328.

NERSTRØM B. (1963a) Further investigation on inheritance of the Gc system. A Danish mother-child material. *Acta genet.* **13**, 150.

NERSTRØM B. (1963b) Experimental transformation of the group-specific components (Gc) of serum into a single alpha-1 globulin immunologically identical with the Gc. *Acta path. microbiol. scand.* **57**, 495.

NERSTRØM B. (1965) *Gc-serumtypsesystemet og dets anvendelse i Retsmedicinen.* Eget forlag, Copenhagen.

NERSTRØM B., MANSA B. & FREDERIKSEN W. (1964) Alteration of the Gc patterns in human sera incubated with bacteria. *Acta path. microbiol. scand.* **61**, 474.

NERSTRØM B. & SKAFTE-JENSEN J. (1963) Immunoelectrophoretic analysis of blood stains with special reference to Gc grouping. *Acta path. microbiol. scand.* **58**, 257.

NIELSEN J.C., NERSTRØM B. & FELDBO M. (1963) On the presence of group specific Gm and Gc substances in urine. *Acta path. microbiol. scand.* **58**, 264.

OUCHTERLONY O. (1962) Diffusion-in-gel methods for immunological analysis. *Prog. Allergy* **6**, 30.

PARKER W.C., CLEVE H. & BEARN A.G. (1963) Determination of phenotypes in the human group-specific component (Gc) system by starch gel electrophoresis. *Am. J. hum. Genet.* **15**, 353.

PEETOOM F. (1963) *The Agar Precipitation Technique and Its Application As a Diagnostic and Analytical Method.* Oliver & Boyd, London.

PERSSON I. & TINGSGAARD P. (1965) A deviating Gc type. *Acta genet.* **15**, 51.

PERSSON I. & TINGSGAARD P. (1966) Serum types in East Greenland Eskimos. *Acta genet.* **16**, 84.

PRUNIER J.H., BEARN A.G. & CLEVE H. (1964) Site of formation of the group-specific component and certain other serum proteins. *Proc. Soc. exp. Biol. Med.* **115**, 1005.

RAUNIO V., RUOSLAHTI E. & KRAUSE U. (1966) Determination of the Gc groups by disc electrophoresis. *Acta path. microbiol. scand.* **67**, 424.

REINSKOU T. (1963) A heterogeneity of the fast-moving component of the Gc system. *Acta path. microbiol. scand.* **59**, 526.

REINSKOU T. (1965a) Genetics of the Gc serum types: family and mother-child studies. *Acta genet.* **15**, 234.

REINSKOU T. (1965b) A new variant in the Gc system. *Acta genet.* **15**, 248.

REINSKOU T. (1966a) Application of the Gc system in 1338 paternity cases. *Vox Sang.* **11**, 59.

REINSKOU T. (1966b) On the confidence of Gc type determination. *Vox Sang.* **11**, 70.

REINSKOU T. (1967) Quantitative studies of the group specific component (Gc) of human serum. *Proc. 10th Cong. europ. Soc. Hematol.*, Karger, Strasbourg, p. 941.

REINSKOU T. (1968) The Gc system. *Series Haematologica* **1**, 1, 21.

REINSKOU T. & MOHR J. (1962) Inheritance of the Gc types: 95 Norwegian families with 343 children. *Acta genet.* **12**, 51.

RUOSLAHTI E. (1965) Further polymorphism in Gc system. Evidence for the occurrence of subgroups. *Annl. Med. exp. Biol. Fenn.* **43**, 213.

SCHEIDEGGER J.J. (1955) Une micro-méthode de l'immunoélectrophorèse. *Int. Arch. Allergy* **7**, 103.

SCHULTZE H.E., BIEL H., HAUPT H. & HEIDE K. (1962a) Über die Gc-Komponenten von Hirschfeld. I. Lage im Starkegel-Elektrophoresebild. *Naturwissenschaften* **49**, 16.

SCHULTZE H.E., BIEL H., HAUPT H. & HEIDE K. (1962b) Über die Gc-Komponenten von Hirschfeld. II. Darstellung und Eigenschaften. *Naturwissenschaften* **48** 108.

SCHULTZE H.E. & HEREMANS J. (1966) *Molecular Biology of Human Proteins*, vol. 1, p. 424. Elsevier, New York.

SEPPÄLÄ M., RUOSLAHTI E. & MÄKELÄ O. (1967) Inheritance and genetic linkage of Gc and Tf groups. *Acta genet.* **17**, 47.

SIMONS K. & BEARN A.G. (1967) The use of preparative polyacrylamide column electrophoresis in isolation of electrophoretically distinguishable components of the serum group-specific protein. *Biochim. biophys. Acta*, **133**, 499.

SMITHIES O. (1959) An improved procedure for starch-gel electrophoresis: further variations in the serum proteins of normal individuals. *Biochem. J.* **71**, 585.

WEITKAMP L.R., RUCKNAGEL D.L. & GERSHOWITZ H. (1966) Genetic linkage between structural loci for albumin and group specific component (Gc). *Am. J. hum. Genet.* **18**, 559.

WENDT G.G. & THEILE U. (1963) Untersuchungen über den Gc-Faktor. *Dt. med. Wschr.* **88**, 696.

WUNDERLY C. (1961) Immunoelectrophoresis: methods, interpretation, results, *in* SOBOTKA H. & STEWART C.P. (eds.) *Advances in Clinical Chemistry*, vol. 4, p. 208. Academic Press, New York.

REFERENCES TO PAPERS ON GEOGRAPHIC DISTRIBUTION OF Gc TYPES

ARENDS T., BREWER G., CHAGNON N., GERSHOWITZ H., LAYRISSE M., NEEL J., SHREFFLER D., TASHIAN R. & WEITKAMP L. (1967) Intratribal genetic differentiation among the Yanomama Indians of Southern Venezuela. *Proc. natn. Acad. Sci.* **57**, 1252.

Baitsch H., Ritter H., Goedde H.W. & Altland K. (1963) Zur genetic der Serum-proteine: Hp Serum-gruppen, Gc-factor, Gm-Serum-gruppen und Pseudocholinesterase-Varianten in europaischen Populationen. *Vox. Sang.* **8**, 594.

Baitsch H. & Jenssen W. (1962) Zur Population-genetik des Gc systems: Allelenhäufigkeit in einen Stichprobe bayrischer Blutspender. *Anthrop. Anz.* **25**, 185.

Baitsch H. & Ritter H. (1963) Untersuchungen zur Genetic der Serum-proteine: Der Gc-Faktor nach Hirschfeld und seine Allelenhäufigkeit in Südwestdeutschland. *Blut* **9**, 278.

Baumgarten A., Giles E. & Curtain C.C. (1967) Distribution of the group specific (Gc) serum component in the populations of the Markam Valley, New Guinea. *Am. J. phys. Anthrop.* **26**, 79.

Blumberg B.S., Workman P.L. & Hirschfeld J. (1964) Gamma globulin, group specific and lipoprotein groups in a U.S. white and Negro population. *Nature.* **202**, 561.

Cleve H. & Bearn A. (1961) Studies on the 'group specific component' of human serum. Gene frequencies in several populations. *Am. J. hum. Genet.* **13**, 372.

Cleve H. & Bearn A.G. (1962) The group specific component of serum: genetic and chemical considerations, *in* Steinberg A.G. & Bearn A.G. (eds.) *Progress in Medical Genetics*, vol. 2, p. 64. Grune & Stratton, New York.

Cleve H., Kirk R.L., Gajdusek D.C. & Guiart J. (1967) On the distribution of the Gc variant Gc Aborigine in Melanesian populations: determination of Gc-types in sera from Tongariki Island, New Hebrides. *Acta genet.* **17**, 511.

Cleve H., Kirk R.L., Parker W.C., Bearn A.G., Schacht L.E., Kleinman H. & Horsfall W.R. (1963) Two genetic variants of the group specific component of human serum: Gc Chippewa and Gc Aborigine. *Am. J. hum. Genet.* **15**, 368.

Cleve H., Ramot B. & Bearn A.G. (1962) Distribution of the serum group specific components (Gc) in Israel. *Nature* **195**, 86.

Gallango M.L. & Arends T. (1965) El sistemo de grupes séricos Gc: sus variantes fenotípicas y su distribución mundial. *Acta cient. venez.* **16**, 120.

Gilberg A. & Persson I. (1967) Serum protein types in Polar Eskimos. *Acta genet.* **17**, 422.

Hallerman W. & Stürner K.H. (1963) Die Verteilung der Gc-(postalbumin)-Typen in Schleswig-Holstein. *Blut* **9**, 185.

Hennig W. & Hoppe H.H. (1964) Häufigkeitsverteilung und Mutter/Kind-Kombinationen bei den Hp-, Gm- und Gc-Serumgruppen am Hamburger Material und ihre Brauchbarkeit im Blutgruppengutachen. *Blut* **10**, 361.

Herbich J. (1963) Häufigkeit der Gc-Gruppen in der Bevölkerung von Wien und Umgebung: Brauchbarkeit dieses Systems in der forensischen Serologie. *Wien. klin. Wschr.* **75**, 803.

Hess M. & Bütler R. (1962) Untersuchunger über die Gc-gruppen von Hirschfeld. *Schweiz. med. Wschr.* **92**, 1351.

Hirschfeld J. (1962) The Gc System, *in* Kallod P. & Waksman B.H. (eds.) *Progress in Allergy*, vol. 6, p. 155. Karger, Basel.

Hirschfeld J. & Beckman L. (1961) Distribution of the Gc-serum groups in Northern and Central Sweden. *Acta genet.* **11**, 185.

Hirschfeld J. & Sonnet J. (1961) Distribution of group specific components (Gc) in the sera of native Africans. *Nature* **192**, 766.

Kaarsalo E. & Melartin L. (1967) Distribution of the Gc serum types in Finland. *Acta genet.* **17**, 120.

Kenrick K.G. (1967) Gc-aborigine in

a New Guinea population. *Acta genet.* **17**, 222.

KENRICK K.G. & DOUGLAS R. (1967) The distribution of Gc-types in selected populations from South-East Asia, Polynesia and Australia. *Acta genet.* **17**, 518.

KIRK R.L., CLEVE H. & BEARN A.G. (1963) The distribution of the group-specific component (Gc) in selected populations from South and Southeast Asia and Oceania. *Acta genet.* **13**, 140.

KIRK R.L., CLEVE H. & BEARN A.G. (1963) The distribution of Gc-types in sera from Australian aborigines. *Am. J. phys. Anthrop.* **21**, 215.

KITCHIN F.D. & BEARN A.G. (1964) Distribution of serum group-specific component (Gc) in Afghanistan, Korean, Nigerian and Israeli populations. *Nature* **202**, 827.

KOBIELA J., MAREK Z. & TUROWSKA B. (1964) The Gc serum groups in populations of Cracow (Poland). *Vox Sang.* **9**, 634.

KORINEK J. & KOUT M. (1964) The Gc system in human serum. *Vnitrni lek.* **10**, 900.

LENDERINK-VAN ITALLIE M.I., PEETOM F. & NIJENHUIS L.E. (1965) The distribution of Gc-types in the Netherlands. *Vox Sang.* **10**, 349.

LOVETT C.A. (1967) Haptoglobin and Gc types in the Haida Indians. *Vox Sang.* **12**, 151.

MANSA B. (1962) Immunoelectro-phoretic analyses of the Gc types in human sera. *Acta path. microbiol. scand.* **55**, 250.

MAREK Z., BUNDSCHUH G., KERDE C. & GESERICK G. (1963) Untersuchungen über die Anwendbarkeit der menschlichen Gc-Komponenten in der forenischen Serologie. *Arztl. Lab.* **9**, 228.

MELARTIN L. (1965) Studies on the Gc system in Finns and Lapps. *Acta genet.* **15**, 45.

MOULLEC J. (1963) Les groupes Gc étude de 221 donneurs de sang parisiens. *Rev. fr. Clin. Biol.* **8**, 910.

NEEL J.V., SALZANO F.M., JUNQUEIRA P.C., KEITER F. & MAYBURY-LEWIS D. (1964) Studies on the Xavante Indians of the Brazilian Mato Grosso. *Am. J. hum. Genet.* **16**, 52.

NERSTRØM B. (1963) Further investigation on the inheritance of the Gc-system. A Danish mother-child material. *Acta genet.* **13**, 150.

OMOTO K. (1963) Vergleichende Untersuchungen zur Allelenhaüfigkeit des Gc-system bei asiatischen und europäischen populationen. *Medical Dissertation.* München.

PERSSON I. (1963) The Gc-system in Greenland Eskimos. *Acta genet.* **13**, 84.

PERSSON I. & TINGSGAARD P. (1966) Serum types in East Greenland Eskimos. *Acta genet.* **16**, 84.

REINSKOU T. (1965) Distribution of the Gc types in Norway. *Acta genet.* **15**, 33.

REINSKOU T. & KORNSTAD L. (1965) The Gc types of the Norwegian Lapps. *Acta genet.* **15**, 126.

REINSKOU T. & MOHR J. (1962) Inheritance of the Gc-types: 95 Norwegian families with 343 children. *Acta genet.* **12**, 51.

SALZANO F.M. & HIRSCHFELD J. (1965) The dynamics of the Gc polymorphism in a Brazilian population. *Acta genet.* **15**, 116.

SALZANO F.M. & SHREFFLER D.C. (1966) The Gc polymorphism in the Caingang Indians of Brazil. *Acta genet.* **16**, 242.

SCHLESINGER D. (1963) The frequency of Gc groups in the Polish population. *Arch. Immunol. Therap. Exp.* **11**, 615.

SHREFFLER D.C. & STEINBERG A.G. (1967) Further studies on the Xavante Indians. IV. Serum protein groups and the SC_1 trait of saliva in the Simões Lopes and São Marcos Xavantes. *Am. J. hum. Genet.* **19**, 514.

SPEDINI G. (1966) I gruppi sierici 'Gc' nella popolazione Italiana. *Acta genet. med. Roma* **15**, 94.

THOMAS K. & HOFMAN F. (1965) Die Frequenz der Gc-Gruppen in Bezirk Dresden. *Z ärzt. Fortbild.* **59**, 209.

VOGT A., PROKOP O. & SCHLESINGER D. (1963) Die Vererbung der Serumgruppe Gc. *Blut.* **9**, 345.

WALTER H., ARDNT-HAUSER A., BERNHARD W. & HEYDE G. (1964) The frequencies of the Hp, Gc and Gm serum groups in South-West Germany. *Blut* **10**, 225.

WALTER H., BERNHARD W., HASSAN S.T. & BAJATZADEH M. (1966) Untersuchungen über die Verteilung der Hp-, Gc- and Gm-Gruppen in Pakistan. *Humangenetik* **2**, 262.

WALTER H. & PALSSON J. (1964) Zur Häufigkeit der Serumgruppen in Island. *Vox Sang.* **7**, 732.

WENDT G.G. & THEILE U. (1963) Untersuchungen über den Gc-faktor. *Dt. med. Wschr.* **88**, 696.

WIEDERMANN D. (1964) The distribution of serum Gc-types in the population of the Brno region. *Scr. Med. Fac. Med.* **37**, 239.

CHAPTER 5

BETA LIPOPROTEIN ALLOTYPES:
THE Ag AND Lp SYSTEMS

Structure of Beta Lipoprotein 177

Inherited Variations in β Lipo-
protein Structure 178

The Ag System 178
    Serum C de B 178
    Other Ag antisera 178
    Ag phenotypes 179
    Genetic interpretation 180
    Frequencies of Ag phenotypes
    in various populations 181

The Lp System 182

Association of β lipoprotein
concentration and Lp pheno-
type 183

Lp phenotype frequencies in
various populations 184

Factors Influencing the Formation
of Antibodies Against Lp and Ag
Antigens 185

Methods 186

References 187

Normal human serum contains at least four different classes of lipoproteins, separable by ultracentrifugation and electrophoresis: (1) high density ($>$1·065 g/ml) α lipoprotein, (2) low density (1·006–1·064 g/ml) β lipoprotein, (3) very low density (0·95–1·005 g/ml) pre-β or α_2 lipoproteins, and (4) the chylomicra ($<$0·95 g/ml). The latter have a protein content of less than 3 per cent, which may be derived by adsorption.

The apoproteins (i.e. protein moiety) of the first two classes differ in their immunological behavior, total amino acid content and terminal residues. In the very low density class, the protein fraction may represent either a separate structural entity (Gustafson *et al.*, 1966) or a mixture of the α and β proteins (called A and B, respectively) with contaminating proteins such as albumin (Walton & Darke, 1964; Granda & Scanu, 1966). For a review of the structural, functional and metabolic properties of the lipoproteins in health and disease, the reader is referred to the paper of Fredrickson *et al.* (1967).

Beta lipoprotein is of particular interest from a genetical point of view because it is involved in several kinds of inherited variation. These include both hypo and hyperbetalipoproteinemia (Fredrickson *et al.*,

1967) as well as a double β lipoprotein found so far in only one family (Seegars *et al.*, 1965). In addition, there are many minor structural variants detected by serological tests with anti-β lipoprotein antisera. The allotypes thus defined are potentially useful as genetic markers, so they are the subject of this chapter. Further details, as well as a more extensive list of references can be found in the reviews of Allison & Blumberg (1965), Berg (1966, 1968), Hirschfeld (1965, 1968) and Bütler (1967).

STRUCTURE OF BETA LIPOPROTEIN

The β lipoprotein molecules, which are the major cholesterol carriers in plasma, have a very large (about 75 per cent) lipid moiety, consisting of free and esterified cholesterol (43 per cent), phospholipid (22 per cent) and triglyceride (10 per cent) (Bragdon *et al.*, 1956).

Carbohydrates are said to compose about 3 per cent of the molecule (Winzler, 1960). However, since several different enzyme activities—particularly esterase—are associated with the β lipoproteins (Lawrence & Melnick, 1961), it is possible that a portion of the sugar moiety represents small amounts of bound or adsorbed glycoproteins, some of which have enzyme activity. Carbohydrate, in contrast to lipid, is strongly antigenic; thus the sugar moiety needs to be considered in the interpretation of immunological reactions attributed to antibodies against the protein component of β lipoprotein (Allison & Blumberg, 1965).

Structural studies of the protein component have been hampered by aggregation of the molecule during delipidation. However, Granda & Scanu (1966) were able to obtain a lipid-free, water soluble protein by treatment with diethyl ether in the presence of dodecyl sulfate. They then confirmed previous reports of a single N-terminal amino acid (glutamic acid) and tentatively estimated that the molecular weight, assuming a single species of polypeptide chain, was about 70,000. A similar figure of 64,000 was obtained by Shore & Shore (1967). The average molecular weight of intact β lipoprotein is of the order of 2–3 million (Björklund & Katz, 1956). Thus, Granda & Scanu (1966) and Shore & Shore (1967) proposed a molecular model consisting of several protein units held together by lipid bridges, in agreement with a previous suggestion of Margolis & Langdon (1966).

The nature of the bonds linking protein to lipid is uncertain. The fact that few, if any, are covalent (Fisher & Gurin, 1964) is consistent with the fact that there is a constant exchange of lipid between the molecules of plasma lipoprotein and of plasma and tissue lipoprotein.

INHERITED VARIATION IN β LIPOPROTEIN STRUCTURE

Although the biochemical investigations have not revealed heterogeneity of β lipoprotein structure, serological studies have shown that there are genetically controlled differences representing two separate genetic systems. The first, called Ag, has a complexity approaching that of the Gm system of γG globulin; its various types are recognized by the use of human isoimmune precipitating antisera. The second, called Lp, appears to be less complex, and thus more susceptible to genetic analysis. Its phenotypes are demonstrated with heteroprecipitins prepared in various animal species, excluding man.

THE Ag SYSTEM

SERUM C. DE B

Allison & Blumberg (1961) described the serological behavior of an antibody in the serum of a patient (Mr. C. de B.) with a long transfusion history. This antibody, a γG globulin, formed precipitates in agar with over half of the human serum specimens tested. The positively-reacting sera were classified as Ag(a+), and the negatively-reacting sera as Ag(a−) to designate the presence or absence of an AntiGen recognized by the C. de B. antibody. The antigen, subsequently identified as a component of β lipoprotein, was found to be inherited as an autosomal dominant character (Blumberg *et al.*, 1962a,b).

OTHER Ag ANTISERA

The apparent simplicity of the Ag system was found to be illusory when sera from a number of other patients with transfusion histories were found to contain anti-β lipoprotein isoantibodies with specificities differing from that of C. de B. (Blumberg & Riddell, 1963; Bundschuh *et al.*, 1963; Blumberg, 1963, 1964; Blumberg *et al.*, 1964a,b; Hirschfeld & Blombäck, 1964; Hirschfeld *et al.*, 1964, 1966, 1967; Vierucci *et al.*, 1966; Okochi, 1967). Hirschfeld *et al.* (1964) showed that the C. de B. serum contained at least three antibody specificities, which they called anti-Ag(a_1), anti-Ag(x) and anti-Ag(z). Two more apparent Ag antigenic determinants were subsequently identified by antibodies with specificities designated as anti-Ag(y) and anti-Ag(t) (Hirschfeld *et al.*, 1966, 1967). All of the Ag antigens appear to be inherited as autosomal dominants.

Recently, Hirschfeld (1968) has compared the reactions of 28 different human isoprecipitin sera against 462 sera from unrelated individuals. When indicated, extensive absorption tests were performed to determine the number and identity of the antibodies present. Many of the antisera gave poorly reproducible results, and at least seven of them contained a mixture of two or more antibodies separable by absorption. As a further complication, it appeared that three of these seven sera contained 'complex' antibodies reacting with more than one antigenic determinant.

Hirschfeld (1968) concluded that the reactions of the 28 antisera could be interpreted as due to various combinations of five antibody specificities. For example, anti-Ag(x) was found by itself or with weakly-reactive, unidentified precipitins in 17 antisera, and with anti-Ag(t), anti-Ag(a_1) and anti-Ag(a_1x) in one each of three additional sera. Anti-Ag(z) was not found by itself in any antiserum, but it accompanied anti-Ag(a_1), anti-Ag(a_1) plus anti-Ag(x), and anti-Ag(a_1) plus anti-Ag(xz) in one instance each.

Ag PHENOTYPES

Among the 462 donor sera tested, Hirschfeld (1968) found only 13 of 32 theoretically possible phenotypes which would be expected from random combination of the five antigenic determinants. No examples of Ag(x−y−) were found in this or any other population tested. Furthermore, the Ag(x) and Ag(y) antigens appeared to be antithetical, suggesting that they are the products of allelic genes, in agreement with the family studies of Hirschfeld *et al.* (1967), Rittner (1967), Morganti *et al.* (1967) and Okochi (1967). The most common phenotype, found in nearly 28 per cent of specimens, was Ag(x − y + a_1 −

z − t +), closely followed at 24 per cent by Ag(x + y + a$_1$ + z − t +). About 17 per cent were Ag(x − y + a$_1$ + z + t +), 11 per cent Ag(x − y + a$_1$ + z − t +), 7 per cent Ag(x + y + a$_1$ + z + t +), and 5 per cent Ag(x + y − a$_1$ + z − t +). The other seven phenotypes had frequencies ranging from 0·4 to 3·5. Family studies were not included in the report, so associations among the Ag(a$_1$), Ag(z) and Ag(t) antigens could not be defined with certainty. It was noted that no sera of type Ag(a$_1$ − z +) were found, and that the frequency of Ag(z − t −) was considerably lower than would be expected if the two antigens were distributed at random. (See Addenda.)

GENETIC INTERPRETATION

Since it is difficult to obtain clearcut serological reactions with many of the anti-Ag sera, and since only the Ag(x) and Ag(y) antigens have been studied in the same families, it is not yet possible to construct a feasible genetic model to explain the inter-relationships of the various Ag antigens. Nevertheless, it is apparent that the problem character-istic of the Rh, Kell and MNSs blood group systems and the Gm system of the γG heavy chain is also applicable to the Ag system. In other words, whenever two or more identifiable characters belonging to a given genetic system are inherited as a unit, they may represent either the product of a single gene varying at one or more mutable sites or the products of two or more closely linked genes. An excellent discussion of this general problem was presented by Shreffler (1967).

Bundschuh (cited in Hirschfeld, 1968) has shown that the serum from individuals with the phenotype Ag(x + y +) gives reactions of non-identity when tested against anti-Ag(x) and anti-Ag(y) reagents. This finding probably indicates that the two antigenic determinants occupy different β lipoprotein molecules in the tested serum. Since their respective genes, Ag^x and Ag^y are thought to be allelic (Hirsch-feld *et al.*, 1967; Rittner, 1967; Okochi, 1967), the presence of the two antigens on separate molecules in heterozygotes probably indicates that there is no hybrid formation between the polypeptides produced by the two homologous Ag structural genes within a given cell. Alternatively, it is possible (but unlikely) that there is some form of autosomal inactivation analogous to that of the immunoglobulin genes, or that the expression of the Ag antigens is a function of as yet undefined β lipoprotein molecular subclasses analogous to the γG subclasses. If the assumption is correct that intra-cellular dimers are not formed, then it is theoretically possible to distinguish those antigens determined by the Ag locus on one chromosome from those produced by the opposite chromosome, as suggested by Hirschfeld (1968). However, studies of their inheritance would also be required.

Although the Ag antigenic determinants are probably part of the protein moiety of β lipoprotein, it is possible that they represent the lipid, or more likely, the carbohydrate portion. In the latter case, they would be analogous to the blood group (carbohydrate) antigenic determinants, and thus would not be direct gene products. Conceivably, then, the actual gene products of the Ag system are enzymes, capable of mediating the synthesis of Ag antigenic sites. Although unlikely, this hypothesis cannot be ruled out at present, and the genetic mechanisms of the Ag system will remain obscure until further information is available on both the inheritance and the chemical structure of its numerous antigens. (See Addenda.)

FREQUENCIES OF Ag PHENOTYPES IN
VARIOUS POPULATIONS

In the early papers of Allison & Blumberg (1961) and Blumberg et al. (1962a,b), the phenotype of sera reacting with the C. de B. antiserum was recorded as Ag(a+). When further studies showed that this reagent contained antibodies against Ag(a_1), Ag(x) and Ag(z) (Hirschfeld et al., 1964), the previous results required a re-evaluation

TABLE 5.1. Distribution of the Ag^x and Ag^y genes in a small number of populations. (Adapted from Hirschfeld & Okochi, 1967 and Morganti et al., 1967)

Population	Number tested	Gene frequency	
		Ag^x	Ag^y
Sweden	245	0·23	0·77
Switzerland	282	0·24	0·76
Finland	24	0·31	0·69
North Italy	334	0·23	0·77
Thailand	54	0·69	0·31
Japan	1205	0·73	0·27
India	42	0·74	0·26

which has not yet been achieved. Hirschfeld & Okochi (1967) and Morganti *et al.* (1967) have recently presented the results of typing samples from several populations with anti-Ag(x) and anti-Ag(y); these results are summarized in Table 5.1. The potential usefulness of these two Ag antigens as genetic markers is indicated by the high frequency of the Ag^x gene in the Asiatic countries as opposed to a relatively low frequency in Northern Europe. These data also strengthen the idea that the Ag^x and Ag^y genes are truly allelic, since their respective antigens (in contrast, for example, to certain antigens of the Gm system) behave as antithetical characters in several widely separated populations.

THE Lp SYSTEM

Berg (1963) reported that when the serum of rabbits injected with human β lipoprotein was absorbed with the serum of certain human subjects, the absorbed rabbit serum formed a precipitate with some, but not all, human sera. This property of human serum was found to be inherited as an autosomal dominant. The polymorphic system thus detected was called Lp for the LipoProtein antigen, designated Lp(a), which reacted with the anti-Lp(a) antibodies (Berg, 1964b). Subsequently, Bundschuh (1964) and Prokop & Bundschuh (1964) obtained a similar result by immunizing a horse with human β lipoprotein. In addition to the Lp(a) antigen, they were able to demonstrate a second determinant, Lp(x), which, when present, was found to be inherited with Lp(a). Thus, the known phenotypes are Lp(a + x −), Lp(a + x +) and Lp(a − x −) (Bundschuh & Vogt, 1965).

Further studies on the inheritance of Lp(a) confirmed its autosomal mode of inheritance, and showed that it represented a genetic system independent of several other systems, notably including Ag (Berg & Mohr, 1963; Mohr & Berg, 1963; Berg & Wendt, 1964; Gedde-Dahl & Berg, 1965; Berg, 1964a, 1965b, 1966a,b, 1968; Rittner, 1966a). Berg (1964b) found that in serum with the Lp(a+) phenotype, only a portion of the β lipoprotein molecules possess the Lp(a) antigenic determinant. Furthermore, it appears that the antigens of the Lp and Ag systems occupy different β lipoprotein molecules (Berg, 1964a, Rittner, 1966a). Thus, when serum of type Lp(a+), Ag(a+) was treated with anti-Lp(a) serum to remove by precipitation all of the

Lp(a+) molecules, the reactions of the remaining serum with anti-Ag(a) were not apparently affected. Although this finding can be interpreted to indicate that the Ag and Lp systems define different structural classes of lipoproteins (analogous to the γG subclasses), other explanations are also possible. However, until it is determined whether or not the antigenic determinants of both systems are in the protein moiety, further speculation is of no value.

ASSOCIATION OF β LIPOPROTEIN CONCENTRATION AND Lp PHENOTYPE

Newborn infants are capable of synthesizing β lipoprotein (Morganti *et al.* 1967), but the level in plasma is so low that neither Ag nor Lp typing is completely reliable (Blumberg *et al.*, 1962b; Bundschuh *et al.*, 1963; Dürwald *et al.*, 1965). In adults, it has been reported that the plasma of Lp(a+) subjects tends to have a higher β lipoprotein concentration than the plasma of Lp(a−) subjects (Kahlich-Koenner & Weippl, 1965; Reinskou, 1966). However, this finding does not necessarily imply that the Lp phenotype is influenced by the β lipoprotein level in normal adults. As noted by Berg (1968), the results could be due to heteroprecipitins remaining after incomplete absorption of the rabbit anti-Lp(a) serum by human Lp(a−) serum during the preparation of the antiserum. Such heteroprecipitins, unrelated to Lp specificity, might be expected to react with Lp(a−) serum containing large amounts of β lipoprotein. Reinskou (1966) offered as an additional possibility that β lipoprotein molecules carrying the Lp(a) determinant may carry additional determinants related to Lp(a) which are recognized by some anti-Lp(a) antibodies. (See Addenda.)

Jörgensen (1966) compared the Lp phenotypes of healthy blood donors with those of patients having a variety of diagnoses, and found a significantly higher frequency of Lp(a+) among the 143 diabetics tested and lower frequency of Lp(a+) among the 185 patients with hepatic diseases. Similarly, Thomas & Hoffman (1967) found an increase in Lp(a+) among 307 patients with cerebral arteriosclerosis (patients who commonly have high lipoprotein levels). These findings might be due to incomplete absorption of the antisera, since Berg (1966a) and Berg *et al.* (1967) were unable to find a similar correlation in subjects with diabetes or coronary heart disease.

It appears likely that when weak precipitin reactions cannot be ascribed to the effects of a high lipoprotein concentration or to bacterial contamination (see next paragraph), the individual being tested may have an Lp(a) antigen with decreased reactivity. Evidence for inherited variation in antigen reactivity, a characteristic of many of the blood group antigens, was presented by Rittner (1966c).

Lp PHENOTYPE FREQUENCIES IN VARIOUS POPULATIONS

In spite of the difficulties associated with determining Lp phenotypes, most of the reports on their distribution in similar populations are in accord, as shown in Table 5.2. Thus, although the Lp^a gene frequency appears to be very low in certain population isolates, such as the natives of Easter Island and Labrador, there is very little difference among the other tested groups. The rather high frequency of Lp(a+) sera reported by Seidl *et al.* (1966) in residents of Frankfurt may have been due to the formation of non-specific precipitation which has been reported to occur between β lipoprotein and agar (Berg, 1965c).

TABLE 5.2. Distribution of the Lp phenotypes

Population	Number tested	Phenotype frequencies Lp(a+)	Lp(a−)	Lp^a frequency	Reference
Norway	1109	0·35	0·65	0·195	Berg (1968)
United States	126	0·39	0·61	0·218	Berg (1968)
Germany,					
Dresden	416	0·39	0·61	0·217	Berg (1968)
Marburg	301	0·33	0·67	0·179	Berg (1968)
Frankfurt	321	0·51	0·49	0·303	Seidl *et al.* (1966)
Austria,	470	0·35	0·65	0·196	Jarosch (1965)
Vienna	310	0·31	0·69	0·171	Berg *et al.* (1965)
Vienna	377	0·41	0·59	0·235	Speiser & Pausch (1965)
Greece	218	0·27	0·73	0·144	Walter & Yannissis (1967)
U.S., Negroes	242	0·34	0·66	0·189	Berg (1968)
Brazil	104	0·31	0·69	0·168	Berg (1968)
N.W.					
Tanzania	107	0·35	0·65	0·191	Berg (1968)
Easter Island	106	0·09	0·91	0·043	Berg (1968)
Labrador					
(Indians)	234	0·02	0·98	0·009	Berg (1968)

Alternatively, the antiserum could have been contaminated with bacteria such as *Bacillus cereus*, which causes a precipitin reaction when exposed in agar to serum containing anti-bacterial antibodies (Bundschuh *et al.*, 1965).

FACTORS INFLUENCING THE FORMATION OF ANTIBODIES AGAINST Lp AND Ag ANTIGENS

Both anti-Ag and anti-Lp antibodies have been found in the γG globulin fraction of serum (Allison & Blumberg, 1961; Berg, 1965a,b). No examples of human serum with anti-Lp specificity nor of animal precipitins with anti-Ag specificity have been reported. The formation of anti-Ag antibodies is apparently dependent upon a large number of exposures to the antigen. Although the number of blood transfusions required may be as low as ten (Rittner, 1966b), a much larger number, given at intervals over several years, is more likely to stimulate antibody production (Bütler, 1967). There is some evidence that these antibodies may cause reactions to transfusion. (See Addenda.)

Multiple pregnancies are also capable of bringing about Ag immunization. Thus, Dürwald *et al.* (1965) found two examples of anti-Ag precipitins in testing the serum of 157 women with a history of four or more pregnancies. The two immunized women (pregnant six and seven times, respectively) had never been transfused, so their infants probably were the source of the antigen.

Among transfused patients, the diagnosis appears to be an important determinant in anti-Ag formation. Blumberg *et al.* (1964a) found Ag isoprecipitins in 13 of 36 patients with thalassemia who had received 50 or more transfusions, as compared with only three of 34 non-thalassemic patients with similar transfusion histories. Vierucci *et al.* (1966) reported that six of 80 transfused thalassemic children in Italy had Ag antibodies in their sera. The American and Italian investigators subsequently extended their studies in a joint report (Levene *et al.*, 1967), which confirmed the much greater likelihood of immunization in thalassemia. The considerably higher proportion of antibody formers in the affected American children (all of southern European origin) than in the Italian children (30 per cent *vs* 13 per cent) could not be satisfactorily explained. There appeared to be a direct correlation between the number of transfusions and antibody formation in the group studied in America. A similar trend was not apparent in the Italian patients. (See Addenda.)

Berg (1965a,b, 1968) made a thorough study of the factors which determined the likelihood of stimulating anti-Lp antibody formation in rabbits. He reported that the best antisera were obtained when purified β lipoprotein, rather than whole human serum, was injected. He also found it essential to use large amounts of the antigen and to inject it intravenously rather than by other routes.

METHODS

The antigens in both systems can be detected by using one of several modifications of the Ouchterlony immunodiffusion technique (Alison & Blumberg, 1961, 1965; Blumberg & Riddell, 1963; Berg, 1963, 1964b,c; Hirschfeld, 1963; Bundschuh, 1964; Bütler *et al.*, 1967). A microtechnique is most satisfactory, since serum is conserved, and reactions occur fairly rapidly.

The method of Blumberg & Riddell (1963) requires the use of lantern slides ($3\frac{1}{4} \times 4$ inches) and an agar cutter composed of a plastic holder and seven 2 mm metal tubes, six of which are arranged in radial fashion at a 3 mm distance around the seventh (see Fig. 5.1). The slides may be pre-treated either with silicone or with a thin layer of 0·2 per cent agar dried at 56°C. Then ten ml of a 0·9 per cent purified agar or agarose solution in 0·07 M phosphate buffer at pH 7·4 is poured on the slide. When the agar is firm, twenty 'rosette' patterns are cut, so that 120 sera can be tested on one slide. These are placed in the peripheral wells and, after 30 minutes, the central wells are filled with antiserum. The slides are incubated at room temperature in a humid chamber and observed at intervals for 2 days. If desired, pictures can be taken by placing the slide in a photographic enlarger between the light source and a sheet of light sensitive paper. Such a picture, an enlargement of a single rosette, is shown in Fig. 5.2.

For final processing, the slides are soaked in saline for two hours with at least four changes of saline, and dried at 37°C, with overlying filter paper to absorb the moisture, for at least an hour. After a 30 minute fixation in 2 per cent acetic acid, the slides are washed in three or four changes of distilled water for 2 hours and then thoroughly dried overnight at 37°C with overlying filter paper. A fat stain, such

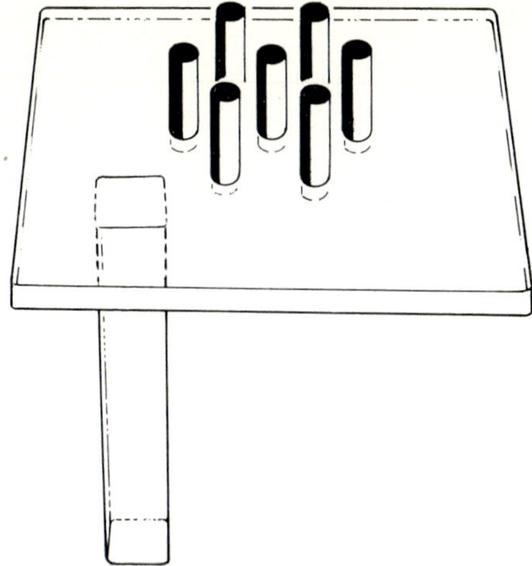

FIGURE 5.1. Drawing of rosette cutter made of a plastic frame with copper tubing as described in the text. The plastic extension at the left serves as a handle.

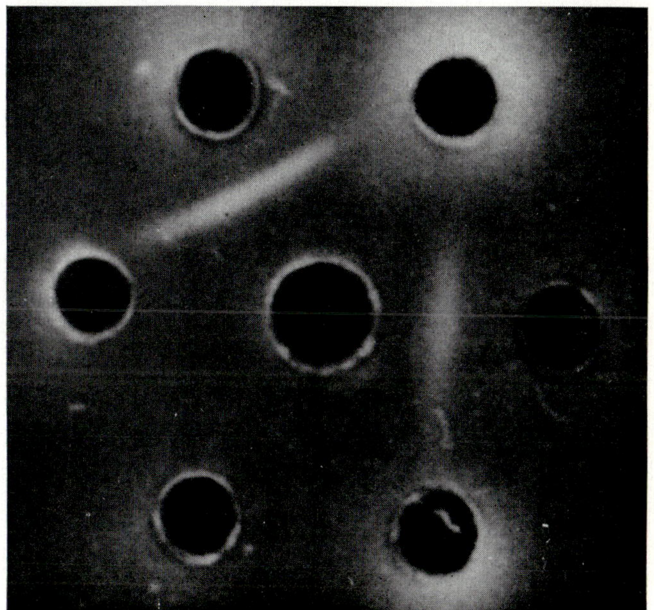

FIGURE 5.2. Enlarged photograph of a single agar rosette, with anti-Ag in the central well, and six different human serum specimens in the peripheral wells. The precipitation bands indicate that two of the six specimens contain the specific antigen.

[*Facing page* 186]

as saturated Sudan Black, is then applied. This stain is prepared by placing one g of dye in one L of 60 per cent ethanol, incubating with occasional stirring at 37°C for 24 hours, filtering and storing at room temperature in a dark bottle. The slides are stained in this solution for 2 hours, washed twice in 50 per cent ethanol and dried at 37°C. If desired, a protein counterstain can also be applied.

It is strongly recommended, particularly when testing for the Lp antigens, that sera be tested within a week or two after the specimens are obtained from the donors, since reactions become weaker after prolonged storage (Berg, 1964b). Known positive and negative controls must always be included, because a number of physical factors affect the rate and amount of precipitate formed.

Berg (1964c,d) recommended that the buffered agar solution should be made with saline, in order to prevent nonspecific electrostatic reactions between albumin and lipoprotein. The addition of sodium azide ($1:10,000$) to the agar solution is also useful to minimize the growth of micro-organisms.

Bütler & Brunner (1966) have described a passive hemagglutination method for detecting the Ag antigens which Bütler *et al.* (1967) found to yield essentially the same results as immunodiffusion. For details, the reader is referred to the original papers.

The antisera for Ag and Lp testing are not commercially available, and both are very difficult to obtain. By far the best source of anti-Ag is a multi-transfused thalassemic patient. However, such patients are rarely over 15 years old, and obtaining a useful quantity of serum is often impossible. Non-thalassemic patients are very unlikely to have useful antisera, in spite of a long history of transfusion. The production of anti-Lp antibodies in animals has been carefully studied by Berg, and his papers should be consulted before Lp immunization is attempted. Although β lipoprotein as such is an exceptionally good antigen, the portion of the molecule which determines Lp specificity is much less likely to stimulate specific antibody formation.

REFERENCES

ALLISON A.C. & BLUMBERG B.S. (1961) An isoprecipitation reaction distinguishing human serum protein types. *Lancet* i, 634.

ALLISON A.C. & BLUMBERG B.S. (1965) Serum lipoprotein allotypes in man, *in* STEINBERG A.G. & BEARN A.G. (eds.) *Progress in Medical Genetics*, vol. 4, p. 176. Grune & Stratton, New York.

BERG K. (1963) A new serum type system in man—the Lp system. *Acta path. microbiol. scand.* **59**, 369.

BERG K. (1964a) Comparative studies on the Lp and Ag serum type systems. *Acta path. microbiol. scand.* **62**, 276.

BERG K. (1964b) Immunological studies of the Lp(a) factor. *Acta path. microbiol. scand.* **62**, 600.

BERG K. (1964c) Studies on the reaction between Lp(a+) human sera and anti-Lp(a) sera from rabbits. *Acta path. microbiol. scand.* **62**, 613.

BERG K. (1964d) Precipitation reactions in agar gel between albumin and β-lipoprotein of human serum. *Acta path. microbiol. scand.* **62**, 287.

BERG K. (1965a) Immunochemical studies of anti-Lp(a) sera. *Acta path. microbiol. scand.* **63**, 142.

BERG K. (1965b) The production of anti-Lp(a) sera in rabbits. *Acta path. microbiol. scand.* **63**, 127.

BERG K. (1965c) Precipitation reactions in agar gel between normal human sera. *Vox Sang,* **10**, 222.

BERG K. (1966a) Serum lipoproteins. Plenary Session, *Proc. 11th Cong. int. Soc. Hematol.* Sydney, p. 260.

BERG K. (1966b) Further studies on the Lp system. *Vox Sang.* **11**, 419.

BERG K. (1968) The Lp system. *Series Haematologica* **I**, 1, 111.

BERG K., AARSETH S., LUNDEVALL J. & REINSKOU T. (1967) Blood groups and genetic serum types in diabetes mellitus. *Diabetologia,* **3**, 30.

BERG K. & MOHR J. (1963) Genetics of the Lp system. *Acta genet.* **13**, 349.

BERG K. & WENDT G.G. (1964) Das Lp System. *Humangenetik* **I**, 24.

BJÖRKLUND R. & KATZ S. (1956) Molecular weights and dimensions of some human serum lipoproteins. *J. Am. chem. Soc.* **78**, 2122.

BLUMBERG B.S. (1963) Iso-antibodies in humans against inherited serum low density beta-lipoproteins. The Ag system. *Ann. N.Y. Acad. Sci.* **103**, 1052.

BLUMBERG B.S. (1964) Polymorphisms of the serum proteins and the development of iso-precipitins in transfused patients. *Bull. N.Y. Acad. Med.* **40**, 377.

BLUMBERG B.S., ALTER H.J., RIDDELL N.M. & ERLANDSON M. (1964a) Multiple antigenic specificities of serum lipoproteins detected with sera of transfused patients. *Vox Sang.* **9**, 128.

BLUMBERG B.S., BERNANKE D. & ALLISON A.C. (1962a) A human lipoprotein polymorphism. *J. clin. Invest.* **41**, 1936.

BLUMBERG B.S., DRAY S. & ROBINSON J.C. (1962b) Antigen polymorphism of a low density beta-lipoprotein. Allotypy in human serum. *Nature* **194**, 656.

BLUMBERG B.S. & RIDDELL N.M. (1963) Inherited antigenic differences in human serum beta lipoproteins. A second antiserum. *J. clin. Invest.* **42**, 867.

BLUMBERG B.S., WORKMAN P.L. & HIRSCHFELD J. (1964b) Gamma-globulin, group specific, and lipoprotein groups in a U.S. white and Negro population. *Nature* **202**, 561.

BRAGDON J.H., HAVEL R.J. & BOYLE E. (1956) Human serum lipoproteins. I. Chemical composition of four fractions. *J. Lab. clin. Med.* **48**, 36.

BUNDSCHUH G.V. (1964) Anti-Lp(a,x) vom Pferd. *Arztl. Lab.* **10**, 309.

BUNDSCHUH G., GESERICK G., MAREK Z. & FÜNFHAUSEN G. (1963) Anti-Ag nach 16 transfusionen-frequenz von Ag in der Berliner Bevölkerung. *Dt. Gesundh Wes.* **18**, 819.

BUNDSCHUH G., KÜNZEL W., GESERICK G. & VOGT A. (1965) Bacillus cereus als Ursache fehlerhafter Präzipitationsbefunde insbesondere beim Ag und Lp Test. *Z. ärztl. Fortbild,* **59**, 217.

BUNDSCHUH G. & VOGT A. (1965) Die Häufigkeit des Merkmals Lp(x) in der Berliner Bevölkerung. *Humangenetik* **1**, 379.

BÜTLER R. (1967) Polymorphisms of the human low-density lipoproteins. *Vox Sang.* **12**, 2.

BÜTLER R. & BRÜNNER E. (1966) A new sensitive method for studying the polymorphisms of the human low density lipoproteins. *Vox Sang.* **11**, 738.

BÜTLER R., BRÜNNER E., VIERUCCI A. & MORGANTI G. (1967) Comparative studies on anti-Ag sera in immunodiffusion and in passive hemagglutination methods. *Vox Sang.* **13**, 327.

DÜRWALD W., LEOPOLD D. & KRÄMER K.H. (1965) The formation of precipitating antibodies after multiple pregnancies. *Vox Sang.* **10**, 94.

FISHER W. & GURIN S. (1964) Structure of lipoproteins: covalently bound fatty acids. *Science* **143**, 362.

FREDRICKSON D.S., LEVY R.I. & LEES R.S. (1967) Fat transport in lipoproteins: an integrated approach to mechanisms and disorders. *New Engl. J. Med.* **276**, 32.

GEDDE-DAHL T. & BERG K. (1965) Linkage in man: the Inv and Lp serum type systems. *Nature* **208**, 1126.

GRANDA J.L. & SCANU A. (1966) Solubilization and properties of the apoproteins of the very low and low-density lipoproteins of human serum. *Biochemistry* **5**, 3301.

GUSTAFSON A., ALAUPOVIC P. & FURMAN R.H. (1966) Studies on the composition and structure of serum lipoproteins: separation and characterization of phospholipid protein residues obtained by partial delipidization of very low density lipoproteins of human serum. *Biochemistry* **3**, 632.

HIRSCHFELD J. (1963) Investigations of a new anti-Ag antiserum with particular reference to the reliability of Ag typing by micro-immuno diffusion tests in agar gel. *Sci. Tools.* **10**, 45.

HIRSCHFELD J. (1965) Human lipoprotein polymorphism. *Proc. 10th Cong. int. Soc. Blood Transf.*, p. 365. S. Karger, Basel.

HIRSCHFELD J. (1968) The Ag system—comparison of different isoprecipitin sera. *Series Haematologica* **i**, 1, 38.

HIRSCHFELD J. & BLOMBÄCK M. (1964) A new anti-Ag serum (L.L.). *Nature* **201**, 1337.

HIRSCHFELD J., BLUMBERG B.S. & ALLISON A.C. (1964) Relationship of human anti-lipoprotein allotypic sera. *Nature* **202**, 706.

HIRSCHFELD J., CONTU L. & BLUMBERG B.S. (1967) Antilipoprotein Nuoro serum (C.P.) and its relation to sera C. de B. and L.L. *Nature* **214**, 495.

HIRSCHFELD J. & OKOCHI K. (1967) Distribution of Ag(x) and Ag(y) antigens in some populations. *Vox Sang.* **13**, 1.

HIRSCHFELD J., UNGER P. & RAMGREN O. (1966) A new anti-Ag serum (serum B.N.) *Nature* **212**, 206.

JAROSCH K. (1965) Ein 'neues' Erbmerkmal des menschlichen Serum Lp^a und seine Phänotypenfrequenz in Oberösterreich. *Wien. med. Wschr.* **115**, 839.

JÖRGENSEN G. (1966) Lp and Krankheit. *Meeting int. Lp Workshop*, Marburg.

KAHLICH-KOENNER D.M. & WEIPPL G. (1965) Lp-Typensystem und β-Lipoprotein-Konzentration. *Humangenetik* **1**, 388.

LAWRENCE S.H. & MELNICK P.J. (1961) Enzymatic activity related to human serum beta-lipoprotein: histochemical, immunoelectrophoretic and quantitative studies. *Proc. Soc. exp. Biol.* **107**, 998.

LEVENE C., BLUMBERG B.S., VIERUCCI A. & RAGAZZINI F. (1967) Incidence of antibodies against β-lipoproteins (Ag system) and the factors influencing isoimmunization in transfused patients in the U.S.A. and Italy. *Lancet* **ii**, 582.

MARGOLIS S. & LANGDON R.G. (1966) Studies on human serum β_1 lipoprotein. III. Enzymatic modification. *J. biol. Chem.* **241**, 485.

MOHR J. & BERG K. (1963) Genetics of the Lp serum types: associations and linkage relations. *Acta genet.* **13**, 343.

MORGANTI G., BEOLCHINI P.E., VIERUCCI A. & BÜTLER R. (1967) Contributions to the genetics of the serum β-lipoproteins in man. I. Frequency, transmission and penetrance of factors Ag(x) and Ag(y). *Humangenetik* **4**, 262.

OKOCHI K. (1967) Serum lipoprotein allotypes Ag(x) and Ag(y) in Japanese. *Vox Sang.* **13**, 319.

PROKOP O. & BUNDSCHUH G. (1964) Anti-Lp(a) vom Pferd. *Z. klin. Chem.* **2**, 193.

REINSKOU T. (1966) Is there any association between Lp type and β lipoprotein concentration? *Meeting int. Lp Workshop*. Marburg.

RITTNER Ch. (1966a) Bestehen Beziehungen zwischen dem Ag und dem Lp-System? *Z. Immun. Forsch.* **130**, 229.

RITTNER Ch. (1966b) Anti-Ag(B-B) nach 10 Transfusionen. *Blut* **12**, 225.

RITTNER Ch. (1966c) Discussion of paper by V. Reinskou. *Meeting int. Lp Workshop*. Marburg.

RITTNER Ch. (1967) Further family data on the inheritance of Ag(x). *Vox Sang.* **12**, 225.

SEEGARS W., HIRSCHHORN K., BURNETT L., ROBSON E. & HARRIS H. (1965) Double beta-lipoprotein: a new genetic variant in man. *Science* **149**, 303.

SEIDL S., FRIEDRICH E. & SPIELMANN W. (1966) The frequency of the Lp(a) factor in Frankfurt/Main. *Vox Sang.* **11**, 730.

SHORE B. & SHORE V. (1967) The protein moiety of human serum β-lipoproteins. *Biochem. biophys. Res. Commun.* **28**, 1003.

SHREFFLER D.C. (1967) Molecular aspects of immunogenetics. *Ann. rev. Genet.* **1**, 163.

SPEISER P. & PAUSCH V. (1965) Das erbliche Serum-β-Lipoprotein-System Lp(a,x). *Ann. paediat.* **205**, 193.

THOMAS K. & HOFFMANN F. (1967) Die Serumgruppen-Systeme Lp(a), Gm(a), Gc und Hp bei Cerebralsklerotikern. *Humangenetik* **4**, 8.

Vierucci A., Morganti G., Varone D. & Borgatti L. (1966) New anti-beta lipoprotein sera in transfused children with thalassemia. *Vox Sang.* 11, 427.

Walter H. & Yannissis C. (1967) Zur Häufigkeit der Serumprotein-Polymorphismen Hp, Gc, Gm, Inv und Lp in Griechenland. *Humangenetik* 4, 130.

Walton K.W. & Darke S.J. (1964) Immunological characteristics of human low-density lipoproteins. *Immunochemistry* 1, 267.

Winzler R.J. (1960) Glycoproteins, *in* Putnam F.W. (ed.) *The Plasma Proteins*, vol. 1, p. 309. Academic Press, New York.

CHAPTER 6

PSEUDOCHOLINESTERASE

The Cholinesterases in Blood 192

Pseudocholinesterase Physiology 193
   Metabolism 193
   Function 194

Structure of Pseudocholinesterase 195
   Heterogeneity of serum cholin-
   esterase 195
   Anionic and esteratic sites 197

Genetics 197
   The E_1 locus 197
     Early studies 197
     Evidence for inherited
     structural variation 198
     Relative amount of atypi-
     cal enzyme in serum 200
     The 'dibucaine number' 201
     Inheritance of the 'atypical',
     dibucaine-resistant enzyme 203
     The fluoride-resistant en-
     zyme variant 203
     The silent allele, E_1^s 204
     Evidence for heterogeneity
     of the E_1^u allele 206
     Nomenclature and charac-
     teristics of the E_1 locus
     phenotypes 206

   Frequencies of the E_1 locus
   genes 207
   Linkage of E_1 and Tf loci 210
The E_2 locus 210
   The C_5 component 210
   Inheritance 211
   Effect of E_1 locus genes on
   C_5 characteristics 211
   Frequency of the C_5+
   phenotype 212
   Other possible E_2 locus
   variants 212

Methods 213
   Screening tests for E_1 variants 213
     Agar diffusion test for I and
     A phenotypes 213
     Rapid tube test for I and A
     phenotypes 214
     The NaCl screening test 215
   Determination of per cent in-
   hibition 216
   Measurement of dibucaine
   number (DN) 216
   Measurement of fluoride num-
   ber (FN) 218
   Detection of E_2 locus variants 218

References 219

THE CHOLINESTERASES IN BLOOD

The term 'cholinesterase' is reserved for those esterases which, although they can hydrolyse esters other than those of choline, are
subject to the inhibitory effects of physostigmine (eserine) as well as
certain organic phosphates such as diisopropyl-fluorophosphate
(DFP) and tetra-ethylpyrophosphate (TEPP). Human blood contains
two kinds of cholinesterase: acetylcholinesterase ('true cholinesterase')
in the red cells, and pseudocholinesterase in the plasma. The red cell
enzyme, designated formally as acetylcholine acetylhydrolase, EC

3.1.1.7, preferentially utilizes acetylcholine as substrate; while the plasma enzyme (acylcholine acylhydrolase, EC 3.1.1.8) is much more active in the hydrolysis of esters with long chain fatty acids. Pseudocholinesterase can be differentiated from acetylcholinesterase by the ability to split benzoylcholine but not acetyl-β-methyl choline. Although acetylcholinesterase plays a very important physiological role at the synapses of cholinergic nerves, its function in the red cell is not currently understood. The function of pseudocholinesterase, which has a wide-spread tissue distribution, is similarly obscure.

This chapter is concerned only with plasma pseudocholinesterase, since it is known to exhibit genetic polymorphism. Several years ago Kalow (1962) noted that the bibliography on cholinesterase exceeded 1300 references; the list by now is considerably longer. Among the many papers reviewing various aspects of this subject are those by Whittaker (1951), Davies & Green (1958), Augustinsson (1960, 1961), Wilson (1960), Kalow (1959a, 1962) and Lehmann & Lidell (1961, 1964, 1966).

PSEUDOCHOLINESTERASE PHYSIOLOGY

METABOLISM

Plasma cholinesterase is probably synthesized in the liver. Studies on normal human liver sections by Gürtner *et al.* (1963) indicated even distribution of the enzyme in the lobules and in the cytoplasm of individual hepatic cells. However, in patients with liver insufficiency, the number of cells containing the enzyme was reduced, and this reduction was reflected in a decreased plasma concentration, as reported earlier by McArdle (1940). Measurement of the plasma level has been used as a diagnostic aid in the differential diagnosis of jaundice (Hunt & Lehmann, 1960), since a decrease does not occur with biliary obstruction in the absence of secondary damage to liver cells.

The enzyme is also present in other tissues, particularly the heart, pancreas and brain. Serum levels are reportedly decreased in several disorders, some of which are not necessarily associated with liver disease, including malnutrition, organophosphate poisoning, uremia and carcinoma of the large bowel (Wetstone *et al.*, 1957, 1960; Waterlow, 1950). Elevated levels have been observed in enphrosis (Kunkel

& Ward, 1947), thyrotoxicosis (Koster & Kisch, 1943); anxiety states (Rose *et al.*, 1965) and obesity (Berry *et al.*, 1954).

Pseudocholinesterase activity is usually measured with either acetylcholine or benzoylcholine as substrate. In the gasometric method of Ammon (1933) and Callaway *et al.* (1951), a unit is that amount of enzyme which releases sufficient acetic acid from acetylcholine to liberate one microliter of carbon dioxide from a bicarbonate buffer in one minute at 37°C. The normal range is about 60–120 units per ml. In the method of Kalow & Lindsay (1955), the hydrolysis of benzoylcholine is followed spectrophotometrically. A conversion factor is used in the calculations, so that the results can be expressed in units as the number of micromoles of acetylcholine hydrolyzed per hour at 37°C. The normal range is about 140–280 per ml (Kalow & Gunn, 1959).

McCance *et al.* (1949) and Lehmann *et al.* (1957) found that newborn infants have somewhat lower values than adults, but by the age of 2 months, the adult level is attained and usually exceeded during the next several years. With advancing age, there is a decline in average values (Kalow & Gunn, 1959; Harris *et al.*, 1960). According to Rider *et al.* (1957), men have higher levels than women, but Simpson (1966) found no effect of sex when corrections were made for age, weight and hematocrit.

The biological half-life of pseudocholinesterase has been determined from the enzyme generation curve after administration of DFP (Neitlich, 1966), and also from the disappearance curve after transfusion of normal plasma in a subject lacking the enzyme (Jenkins *et al.*, 1967). In both studies the half-life was found to be about 10 days.

FUNCTION

The physiological role of pseudocholinesterase is not known, although several hypotheses have been advanced. For example, Clitherow *et al.* (1963) proposed that butyrylcholine, which has an undesirable nicotinic action, is the natural substrate for the enzyme and is prevented from accumulating because of rapid hydrolysis at its site of synthesis in the liver. On the other hand, Funnell & Oliver (1965) have suggested that pseudocholinesterase takes part in a homeostatic mechanism for the control of free choline levels in plasma. However, since those rare individuals without demonstrable pseudocholinesterase have no apparent associated clinical symptoms, it is likely that the enzyme either has no important role or else its function can be taken over by some other metabolic pathway.

Suxamethonium (succinyl dicholine), a muscle-relaxant commonly used in anesthesia, consists of two acetylcholine molecules, joined by their acetate groups; it is rapidly inactivated by pseudocholinesterase (Glick, 1941). The amount of enzyme normally present in plasma is sufficient to destroy 40 mg (the usual dose of suxamethonium) in less than a minute *in vitro* (Kalow, 1959b). However, the pharmacological effect is due to escape of some of the drug into the motor endplate, which it depolarizes. In patients with very low enzyme activity, the endplate is exposed to a much larger amount of the drug, and muscular paralysis is prolonged (Kalow, 1962). The resultant apnea was first described by Bourne *et al.* (1952) and Evans *et al.* (1952), and investigation of this unusual drug response eventually led to the recognition of inherited variation in pseudocholinesterase molecular structure.

STRUCTURE OF
PSEUDOCHOLINESTERASE

Surgenor & Ellis (1954) isolated partially purified cholinesterase from human plasma and characterized it as an α-2 globulin with a molecular weight of about 300,000. Svensmark (1961) showed that the enzyme is a glycoprotein containing several sialic acid residues, and that neuraminidase treatment causes a decrease in electrophoretic mobility without loss of enzyme activity. Haupt *et al.* (1966) obtained a highly purified material with enzyme activity 10,000 times that of an equivalent volume of serum. Its molecular weight was reported to be 348,000, and its content in serum, measured immunologically, about 0·9 mg per 100 ml.

HETEROGENEITY OF SERUM CHOLINESTERASE

The heterogeneity of serum esterases was studied extensively by Augustinsson and his colleagues (reviewed in Augustinsson, 1960, 1961). They found that human serum or plasma contains an arylesterase with rapid electrophoretic mobility and a cholinesterase which migrates as a single slower band on cellulose column electrophoresis. Uriel (1961) also demonstrated serum esterase heterogeneity by immunoelectrophoresis, and Bernsohn *et al.* (1961) found several zones with esterase activity by starch gel electrophoresis.

Harris *et al.* (1962, 1963a) described the appearance and enzymatic behavior of the serum esterases seen after two-dimensional electrophoresis, first on paper, then in starch gel. They found that four components, called C_1, C_2, C_3 and C_4 in order of decreasing mobility, hydrolysed α-naphthyl acetate and were inhibited by physostigmine, DFP and TEPP. An additional diffuse zone of activity migrating with albumin was not similarly inhibited, and it was considered to be the albumin esterase described by Wilde & Kekwick (1962). When β-naphthyl acetate was substituted as substrate, two additional zones of activity not affected by the inhibitors were seen. One, a diffuse and indistinct band with an intermediate migration rate, was thought to be the arylesterase of Augustinsson (1961). The other, associated with β-lipoprotein, also appeared when β-naphthol, instead of its acetate ester, was applied to the gel.

These findings made it apparent that only the four C zones could be designated pseudocholinesterase isozymes. The slowest-moving band, C_4, stained much more intensely than the other three, and it was thus assumed to represent most of the enzyme present in plasma. Figure 6.1 shows a photograph of the starch gel electrophoretic pattern of the four cholinesterase bands. The adjacent drawing of the two-dimensional pattern (Harris *et al.*, 1962) indicates that C_1, C_3 and C_4 have the same mobility on paper electrophoresis, but C_2 is somewhat faster. Thus, at pH 8·6, this component is more negatively charged.

Harris *et al.* (1962) found that the addition of increments of neuraminidase to serum caused a stepwise removal of sialic acid residues and a similar reduction of mobility of all four components. Furthermore, the serum of fetuses and newborn infants regularly contained a heterogeneous component migrating behind C_4, which they ascribed to molecules of cholinesterase lacking their full complement of sialic acid. A similar phenomenon was observed by Parker & Bearn (1962) in the transferrin of fetal serum.

In a subsequent paper, Harris & Robson (1963a) reported that the four regular components of pseudocholinesterase could be separated by passage of serum through a Sephadex G-200 column. They confirmed the fact, predicted from their electrophoretic data, that the relative molecular sizes progressed from C_1 to C_4. Presumably, C_4,

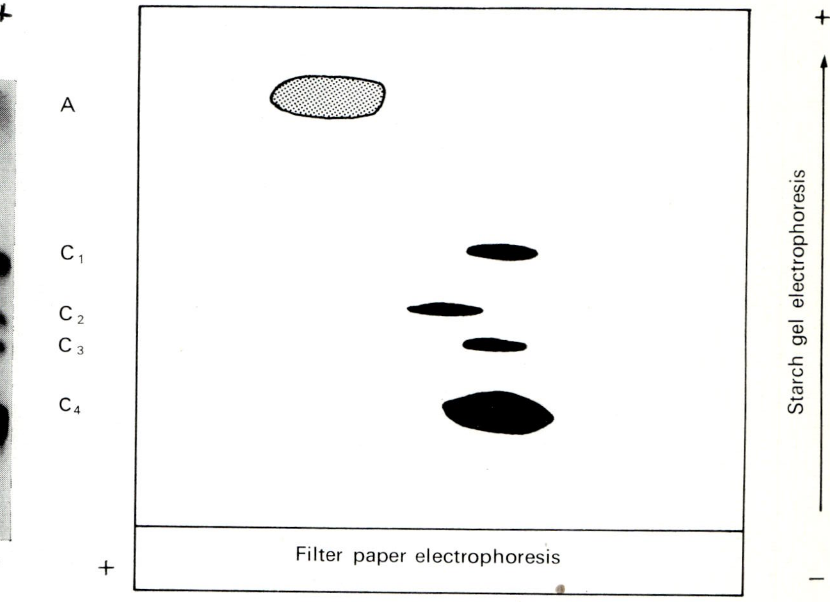

FIGURE 6.1. *Left:* photograph of a starch gel stained for esterase activity after electrophoresis of a normal human serum specimen. The four pseudo-cholinesterase zones (C_1-C_4) and the albumin-esterase (A) are indicated. *Right:* diagram of the same esterase zones as they appear after two- dimensional electrophoresis, first on filter paper and then in starch gel.

the largest in size as well as in quantity, has the molecular weight of 300,000 to 350,000 reported by Surgenor & Ellis (1954) and Haupt *et al.* (1966) for the isolated enzyme. However, it seems somewhat unlikely that the other three components represent a simple polymeric series, since the C_2 component has a different electrical charge (Harris *et al.* 1962) and also appears to lose activity on the addition of mercaptoethanol (author's unpublished observation).

ANIONIC AND ESTERATIC SITES

By analogy with acetylcholinesterase and several other hydrolytic enzymes, pseudocholinesterase has been postulated to have two active centers, each combining with a different part of the substrate molecule (Wilson, 1954). The negatively charged anionic site has an orientation function, combining with a positively charged portion of the substrate, such as the N^+ of choline. The esteratic site initiates hydrolysis by co-valent attachment to the carbonyl group of the substrate ester linkage where cleavage occurs. A peptide containing this active site, localized by its binding of DFP^{32}, has been analyzed by Cohen *et al.* (1959). Its amino acid sequence is very similar to that of a large group of hydrolytic enzymes in which the active site contains a serine residue with Asp or Glu on one side and Gly or Ala on the other (see p. 498).

GENETICS

THE E_1 LOCUS

EARLY STUDIES

Several years ago, Forbat *et al.* (1953) reported that the healthy brother of a patient with a low level of serum cholinesterase activity had a similarly low level, and they suggested the possibility of an inherited trait. Subsequent family studies provided additional support for that assumption (Lehmann & Ryan, 1956; Allott & Thompson, 1956; Kalow, 1956; Lehmann *et al.*, 1958). However, the wide range in serum enzyme activity made genetic interpretation difficult. For example, Kaufman *et al.* (1960) reported the serum cholinesterase levels (expressed in Ammon units) of 93 members of 26 families. In each family, the propositus had less than 36 units of enzyme activity

8

and had come to attention because of suxamethonium-induced apnea in the absence of diseases associated with decreased enzyme. In the parents and children of the propositi, the enzyme level ranged from 26–90 units. They were assumed to be heterozygous, and the propositi homozygous for an autosomal gene determining low enzyme activity. However, since the normal range was considered to be 60–125 units, assignment of a genotype to random specimens was not possible.

This problem was largely resolved by the work of Kalow and his associates, who showed that the enzyme in the serum of individuals with very low activity is a structural variant of the usual enzyme. This conclusion was based on measurements of hydrolytic activity with various substrates and of susceptibility to a number of inhibitors. From these studies, they were able to devise a reliable test for detecting homozygotes and heterozygotes for the variant allele based on differential inhibition.

EVIDENCE FOR INHERITED STRUCTURAL VARIATION

Davies *et al.* (1960) compared the esterase activity of serum containing the 'usual' enzyme with serum containing the 'atypical' enzyme on a series of cholinester substrates. They found that the usual enzyme hydrolysed all of the substrates more rapidly (i.e. the Michaelis constant (K_m) was lower), but the activity ratios between the two kinds of serum varied, indicating a difference in their turnover numbers. Subsequently, Bamford & Harris (1964) showed that with α-naphthyl acetate as substrate, the K_m of the atypical enzyme was actually lower than that of the normal enzyme, although the $V_{\max}$ (μmoles substrate hydrolysed per minute per ml of plasma) of usual enzyme was slightly greater than the $V_{\max}$ of the atypical enzyme (see Table 6.1).

The effect of inhibitors was studied in detail by Kalow & Davies (1958) using benzoyl choline as substrate, and further information was supplied by the studies of Harris & Whittaker (1961, 1962, 1963) and Bamford & Harris (1964). As shown in Table 6.1, the inhibitors can be divided into five groups, based on their effect on the two enzymes. The comparison of inhibition by a given compound is derived from the ratio $I_{50}A : I_{50}U$, in which the expression I_{50} refers to the negative log of the concentration of substance required for 50 per cent inhibition of the atypical (A) and usual (U) enzymes, respectively.

Kalow & Davies (1958) characterized the behavior of the first three groups of inhibitors. The first group consists of organophosphates such as DFP and TEPP, which block the activity of the two enzymes to about the same extent, although inhibition of the atypical enzyme takes place at a slower rate. These compounds react stoichiometrically at the esteratic site of the enzyme molecule, which is probably the same in both 'usual' and 'atypical' esterase.

TABLE 6.I. Behavior of usual (U) and atypical (A) cholinesterases with two different substrates and five classes of inhibitors

	Substrate	
Inhibitors	Benzoylcholine	α-naphthyl acetate
None	U activity 2–5 times A	U activity 1·3 times A
I DFP and TEPP	U inhibition = A inhibition	
II Succinyl choline and Decamethonium	Avidity for A greatly reduced; U/A inhibition = 100	
III Choline, Chlorpromazine, Physostigmine, Dibucaine, Neostigmine, RO2-0683	Differential inhibition of U > A increases with potency of inhibitor	Similar to benzoylcholine
IV Sodium fluoride	U/A inhibition = 5	U/A inhibition < 2
V Sodium chloride	A/U inhibition = 3	

Compounds in the second group can behave either as substrate or inhibitor, depending on concentration; they are the muscle relaxants, succinyl dicholine (suxamethonium) and decamethonium. Both have a greatly decreased avidity for the atypical enzyme, about one-hundredth of that for the usual enzyme. Thus, the occurrence of suxamethonium-induced apnea in patients with the atypical esterase is presumably due to failure of the structurally altered enzyme to hydrolyze, and thus inactivate, the drug.

The third group is composed of physostigmine, some simple quaternary ammonium compounds, and various alkaloids used as local anesthetics, including dibucaine and the neostigmine analogue given the code name RO2-0683 by its manufacturers (Hoffman LaRoche).

These compounds block the action of normal enzyme more than atypical enzyme, and in general, the more potent the inhibitor, the greater the difference in inhibitory capacity. The members of this group contain one and the previous group, two positively charged quaternary nitrogens, which presumably react at the anionic site of the enzyme molecule. It therefore seems likely that the molecular alteration in the atypical enzyme involves the anionic, rather than the esteratic site (Kalow & Davies, 1958). Differentiation of the two enzymes by physico-chemical procedures has been reported by Kalow (1959a) and Liddell *et al.* (1962a).

The fourth kind of inhibitory activity is characteristic of the fluoride ion (Harris & Whittaker, 1961; Bamford & Harris, 1964). With benzoylcholine as substrate, esterase inhibition by sodium fluoride resembles that observed with the third group of inhibitory compounds. However, when α-naphthyl acetate is the substrate, the difference in fluoride inhibition of usual and atypical enzymes is greatly decreased, so that the latter is only slightly less inhibited than the former.

Fluoride is negatively charged, and its mode of action must be quite different from that of the other inhibitors. However, the influence of the substrate—presumably at the site of attachment—is reflected in the similarity of inhibition of the two enzymes by fluoride in the absence of a quaternary nitrogen.

Sodium chloride and (probably) the chloride salts of several other metals, represent the fifth group of inhibitors. Harris & Whittaker (1963) found that 50 per cent inhibition (I_{50}) of the usual enzyme requires 40,000 times more sodium chloride than sodium fluoride, and for the atypical enzyme, the I_{50} of chloride is 2500 times that of fluoride. Also, for any given degree of inhibition, a higher concentration of sodium chloride is required for the usual enzyme than for the atypical enzyme. A slight enhancement of usual enzyme activity by $0 \cdot 1$ M chloride was noted by these authors, and Swift & La Du (1966) subsequently confirmed this effect, using it as the basis for differentiating the two enzymes (see p. 215).

(see p. 215)

RELATIVE AMOUNT OF ATYPICAL ENZYME IN SERUM

Kalow & Davies (1958) showed that DFP, which combines with active sites of esterase molecules on a mole to mole basis, blocks the enzyme in normal and in atypical serum to the same extent. Kalow & Davies (1958) also found that serum from heterozygotes behaves like a mixture of normal and atypical sera. Thus, there is a skew distribution of enzyme activity in response to increments of added RO2-0683, which selectively blocks the normal enzyme. The observed pattern, shown in Fig. 6.2, is consistent with the presence in equimolar proportions of two enzymes with unequal activities. These observations strongly supported the possibility that the number of esterase molecules in the serum of atypical gene homozygotes differs little, if at all, from the number in normal serum.

Confirmatory evidence was supplied by Hodgkin *et al.* (1965), who

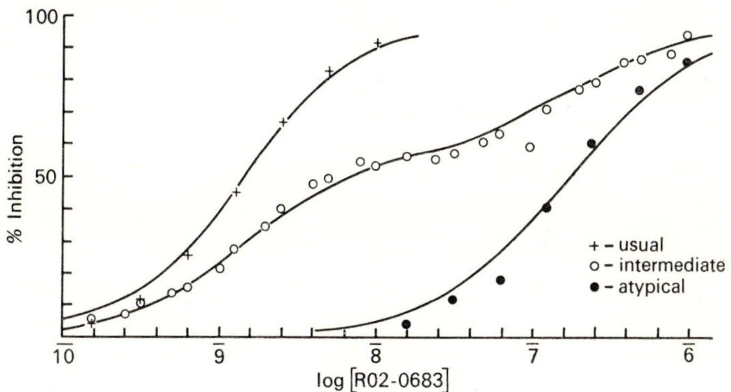

FIGURE 6.2. Inhibition of RO2-0683 of serum cholinesterase of three phenotypes: usual, intermediate and atypical. The skew distribution of points in the intermediate curve is also obtained by mixing serum of the usual and atypical phenotypes (From Kalow & Davies, 1958. Reprinted with permission from *Biochem. Pharmac.* **1**, 183. (Pergamon Press).).

showed that neutralization of anti-human pseudocholinesterase antibodies was achieved with similar volumes of serum from normal and homozygous atypical subjects, even though their esterase activities were considerably different. This cross-reactivity of anti-normal enzyme with the atypical enzyme also showed that the two enzymes have similar molecular structure.

THE 'DIBUCAINE NUMBER'
Kalow & Genest (1957) devised a test to detect the presence of the atypical enzyme based on its relative resistance to inhibition by the local anesthetic dibucaine (cinchocaine or nupercaine). Figure 6.3

shows the inhibitory effect of increasing concentrations of dibucaine on the two enzymes. The maximum difference between the two levels of activity was obtained at a dibucaine concentration of 10^{-5} M Kalow & Genest, 1957). Consequently, this molarity was chosen for differentiating the two enzymes.

In the test, the hydrolysis of benzoylcholine is measured spectro-

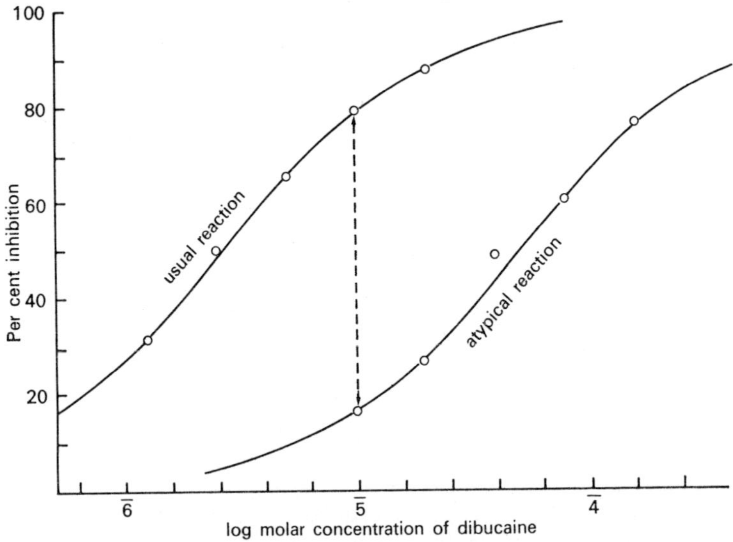

FIGURE 6.3. Inhibition by dibucaine of cholinesterase activity in serum of usual and atypical phenotypes, with 5×10^{-5}M benzoylcholine as substrate. The interrupted line indicates inhibition by 10^{-5}M dibucaine, the concentration chosen for determining dibucaine number, DN. (From Kalow & Genest, 1957. Reprinted with permission of the National Research Council of Canada from the *Can. J. Biochem. Physiol.* **35**, 340.)

photometrically in the absence and in the presence of 10^{-5}M dibucaine. The percentage of inhibition is expressed as the 'dibucaine number' or DN of a given serum (or plasma) specimen. This value, expressing a proportion, is unaffected by the level of cholinesterase activity, and thus is an accurate indicator of the phenotype of newborn infants as well as patients with low levels secondary to disease. Even old specimens of dried blood can be used for this purpose (Lehmann & Davies, 1962).

INHERITANCE OF THE 'ATYPICAL' DIBUCAINE-RESISTANT ENZYME

In tests of large numbers of random serum specimens, Kalow & Staron (1957) and Kalow & Gunn (1959) found a trimodal distribution of DN consistent with the homozygous and heterozygous states of a common or 'usual' gene, E_1^u, and an 'atypical' gene, E_1^a with a frequency of about 0·02. The DN of E_1^u homozygotes ranged from 71–85 with a mean of 79; the DN of E_1^a homozygotes was less than 25; while the DN of heterozygotes had the intermediate range of 50–68 with a mean of 62. Family studies performed in several laboratories verified these findings (Kalow & Staron, 1957; Harris *et al.*, 1960; Bush, 1961; Harris & Whittaker, 1962; Lehmann & Liddell, 1962; Telfer *et al.*, 1964; Goedde *et al.*, 1964; Thompson & Whittaker, 1966). Furthermore, Kalow (1962) and Liddell *et al.* (1963b) obtained postmortem specimens from homozygous and heterozygous subjects, and showed that DN of tissue extracts is similar to that of the serum. Thus, in individuals heterozygous or homozygous for the E_1^a allele, the mutant gene product is found throughout the body.

THE FLUORIDE-RESISTANT ENZYME VARIANT

Harris & Whittaker (1961, 1962) described the inhibition characteristics of sodium fluoride and showed that the fluoride number, FN, could be used in the same way as the DN to differentiate the normal, atypical and intermediate phenotypes, for which the FN was about 57–68, 18–28, and 42–55, respectively. However, in testing large numbers of sera, they found a few that had only a slightly decreased DN but a markedly decreased FN, suggesting another structural variant of pseudocholinesterase. Studies of several families by Harris & Whittaker (1962) indicated that the gene responsible for the altered enzyme is a third allele at the E_1 locus, and it was named E_1^f (Motulsky, 1964). Confirmatory evidence was supplied by Lehmann *et al.* (1963), Liddell *et al.* (1963a), Goedde *et al.* (1964a,b), Griffiths *et al.* (1966), Thompson & Whittaker (1966), Simpson (1967) and Whittaker (1964, 1967). The DN and FN associated with the various phenotypes are given in Table 6.2.

In one of the families reported by Liddell *et al.* (1963a) and Lehmann *et al.* (1963), there were two apparent homozygotes for the E_1^f gene. Both had a moderate reduction in esterase activity and DN. The FN was considerably lower than normal and similar to that found in two

TABLE 6.2. Characteristics of the pseudocholinesterase types

Genotype	Phenotype	Average esterase activity, % of normal	Dibucaine number	Fluoride number	RO2-0683 number	Relative sensitivity to suxa-meth-onium
$E_1^u E_1^u$	U	100	71–85	57–68	>95	Lowest
$E_1^u E_1^s$	U	65	71–85	57–68	>95	Low
$E_1^u E_1^a$	I	78	50–68	42–55	58–76	Low
$E_1^a E_1^a$	A	25	14–25	18–28	<10	High
$E_1^a E_1^s$	A	20	14–25	20–25	<10	High
$E_1^u E_1^f$	UF	80	71–78	50–55	87–95	Low
$E_1^f E_1^f$	F*	60	64–67	32–40	75–86	Intermed.
$E_1^f E_1^s$	F†	60	67	43	–	Probably Intermed.
$E_1^a E_1^f$	IF	60	47–53	30–38	40–60	Intermed.
$E_1^s E_1^s$	S	0	–	–	–	Highest

* Only three examples reported
† Only one example reported

individuals heterozygous for E_1^f and E_1^a. However, differentiation between $E_1^f E_1^f$ and $E_1^f E_1^a$ was achieved by testing the sera for enzyme inhibition by RO2-0683. Kalow and Davies (1958) had shown RO 2-0683 to be particularly effective in differentially inhibiting the normal and dibucaine-resistant variants, and Liddell *et al.* (1963a) found that the fluoride-resistant variant was only slightly less inhibited by this compound than the normal enzyme. Thus, in the $E_1^a E_1^f$ heterozygotes, the 'RO2-0683' number was appreciably lower than in E_1^f homozygotes (see Table 6.2).

THE SILENT ALLELE, E_1^s

In studies on the inheritance of the E_1^a gene, anomalous findings were reported in nine different families (Kalow & Staron, 1957; Harris *et al.*, 1960; Bush, 1961; Liddell *et al.*, 1962b; Simpson & Kalow, 1964). These families contained individuals with the low DN associated with homozygosity for the E_1^a gene, but tests of their parents and/ or children indicated that they were actually heterozygous. Kalow

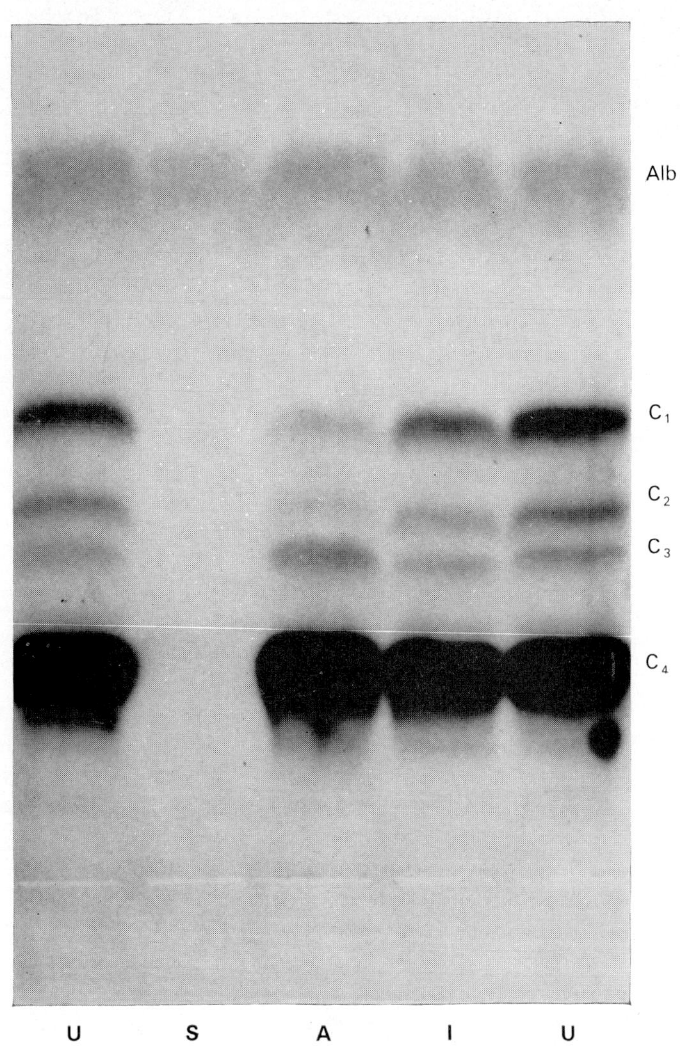

FIGURE 6.4. Photograph of starch gel stained for esterase activity after electrophoresis of five human serum specimens, two of the usual (U) pseudocholinesterase type, and one each of the silent (S), atypical (A) and intermediate (I) types. The albumin-esterase, indicated at the top of the picture, appears in all of the sera, regardless of pseudocholinesterase type.

& Staron (1957) postulated that these individuals might be heterozygous for E_1^a and another, silent allele. However, this hypothesis was dependent upon finding a homozygote for the silent allele, who would presumably have no measurable serum cholinesterase activity.

Such a finding was eventually reported by Hart & Mitchell (1962) and Liddell *et al.* (1962b) in a Greek woman who was brought to their attention because of suxamethonium-induced apnea in the absence of any disease known to be associated with depressed esterase activity. Her mother and two children had low to low normal levels of enzyme activity with normal DN and FN values.

A second case of pseudocholinesterase anenzymia was described by Doenicke *et al.* (1963) in a woman with suxamethonium-induced apnea who had no available family for study. Like the serum, a specimen of liver obtained by biopsy was devoid of enzyme activity.

Several other cases have subsequently been reported (Hodgkin *et al.*, 1965; Dietz *et al.*, 1965; Szeinberg *et al.*, 1965, 1966b; Kattamis *et al.*, 1967 and Jenkins *et al.*, 1967). Family studies were performed in most instances, and they were consistent with the existence of an allele at the E_1 locus which behaves like an amorph in having no recognizeable product. Thus it was named E_1^s (Simpson & Kalow, 1964).

Hodgkin *et al.* (1965) showed that on starch gel electrophoresis, the 'silent' serum of two siblings had none of the pseudocholinesterase components described by Harris *et al.* (1962), although all four bands were clearly visible in the serum of normal subjects as well as E_1^a homozygotes (see Fig. 6.4).

From mixtures of 'silent' and normal sera, Hodgkin *et al.* (1965) were unable to find any evidence of an inhibitor to account for the lack of esterase activity. In agar immunodiffusion and immunoelectrophoresis, the serum of E_1^u and E_1^a homozygotes formed demonstrable bands of precipitation with anti-human pseudocholinesterase serum, but no reactions were observed with the serum of E_1^s homozygotes. Furthermore, apparent lack of cross-reacting material in the 'silent' serum was shown by its failure to neutralize the antiserum to any measurable extent. Similar results were reported in the studies of Szeinberg *et al.* (1966b) and Kattamis *et al.* (1967).

Goedde *et al.* (1965a,b) described two serum specimens in which no enzyme activity could be demonstrated by spectrophotometry, but on electrophoresis, a weakly stained C_4 band was detected. The sera reacted with anti-human pseudocholinesterase to form a precipitate visible on esterase staining, and neutralized the antibodies to about the same extent as normal sera. Presumably this 'nearly silent' variant represents structural alteration with loss of esterase activity but little or no loss in ability to combine with antibodies to normal enzyme. (See Addenda.)

EVIDENCE FOR HETEROGENEITY OF THE E_1^u ALLELE

The wide variation in the serum esterase activity of individuals with the usual type of enzyme suggests that there may be additional genetic influences controlling the amount of enzyme produced. Simpson & Kalow (1963) studied the esterase levels of 107 people from 25 families and of 28 pairs of twins. Analysis of variance with intraclass correlations for parents, sibships and twin pairs provided no evidence for additional genetic factors. However, two later studies (Wetstone *et al.*, 1965; Simpson, 1966) included larger numbers of families, and the results showed a higher intraclass correlation for siblings than for unrelated individuals. Thus, as is the case in several other components of blood (e.g. haptoglobin and G6PD), the quantity as well as the structure of the gene product is an inherited characteristic.

NOMENCLATURE AND CHARACTERISTICS OF THE E_1
LOCUS PHENOTYPES

A nomenclature agreed upon by an international group of investigators was presented by Motulsky (1964); it is currently used in most papers. (An alternative nomenclature suggested by Goedde & Baitsch (1964) was used by some workers for about 2 years). As shown in Table 6.2, the genotypes are composed of various combinations of the E_1^u, E_1^a, E_1^f and E_1^s genes, and the phenotypes of the respective homozygotes are U (usual), A (atypical), F (fluoride) and S (silent). The most common heterozygote, $E_1^u E_1^a$ has the phenotype I (intermediate), while the $E_1^u E_1^f$ heterozygote is called UF, and $E_1^a E_1^f$ is designated IF. Since the E_1^s allele is not expressed, its presence cannot be detected, except in the homozygous state. Thus, the E_1^s heterozygote phenotypes bear the phenotypic name of the expressed gene, as U, A or F for E_1^u, E_1^a and E_1^f respectively.

Differentiation of the phenotypes requires the determination of both dibucaine number and fluoride number. For more precise identification of rare types, the RO 2-0683 number is also required, as previously noted. Table 6.2 (p. 204) lists the genotypes, phenotypes, average figures for the widely ranging esterase activity, and the usual range of DN, FN and per cent inhibition by RO2-0683, as compiled from papers by Motulsky (1964), Harris & Whittaker (1962), Liddell *et al.* (1963a), Simpson (1966) and Whittaker (1967). As indicated, suxamethonium susceptibility is especially associated with the genotypes $E_1^a E_1^a$, $E_1^a E_1^f$, $E_1^f E_1^f$, $E_1^a E_1^s$ and $E_1^s E_1^s$, although Thompson & Whittaker (1966) found that in 78 patients with suxamethonium-induced apnea, there were 25 with $E_1^u E_1^u$ genotype, 10 with $E_1^u E_1^a$ and six with $E_1^u E_1^f$. Thus, it is apparent that factors in addition to pseudocholinesterase activity are involved in physiological susceptibility to suxamethonium.

FREQUENCY OF THE E_1 LOCUS GENES

Most population studies have been performed with methods capable of detecting the E_1^a gene product, but not the E_1^f gene product or the presence of E_1^s (except in homozygotes). The frequency of E_1^a in various populations is shown in Table 6.3. A 3 to 5 per cent frequency of $E_1^u E_1^a$ heterozygotes has been reported in several populations of European origin, including Canadian (Kalow & Gunn, 1959), British, Greek, Portuguese (Kattamis *et al.*, 1962), German (Goedde & Altland, 1963; Altland *et al.*, 1967), Yugoslavian and North American (Motulsky & Morrow, 1968). Similar heterozygote frequencies were found in students from India (Motulsky & Morrow, 1968) in Australian aborigines (Horsfall *et al.*, 1963) Mexican Indians (Lisker *et al.*, 1964) and mixed Brazilians (Simpson & Kalow, 1965). However, in 291 specimens from three South American Indian tribes, no variants were found (Arends *et al.*, 1967), and among 100 specimens from Japanese, there were no E_1^a heterozygotes, although two individuals were said to have the fluoride-resistant variant (Omoto & Goedde, 1965). Studies on other Japanese samples as well as Chinese, Filipinos and Eskimos showed virtual absence of the trait (Altland *et al.*, 1967; Motulsky & Morrow, 1968). Szeinberg *et al.* (1966a) tested 2213 Jews of varied geographic origin and found an overall intermediate (I) phenotype frequency of 4·0 per cent. However, among 381 Jews from Iraq and Iran, the percentage was 9·7, as opposed to 0·7 in 302 Jewish individuals from Morocco, Algiers, Tunis and Libya.

Motulsky & Morrow (1968) found that only 1 per cent of 666 American Negroes tested had the atypical enzyme trait, and of 460 Congolese Negroes, the frequency was even lower, 0·22 per cent. Absence of the trait appears to be characteristic of the pygmies. Motulsky & Morrow

TABLE 6.3. Frequency of the E_1^a allele in various populations

Population	Number tested	Number of heterozygotes	E_1^a frequency*	Reference
European				
British	703	27	0·019	Kattamis *et al.* (1962)
Canadian	2017	74	0·019	Kalow & Gunn (1959)
USA	246	8	0·016	Motulsky & Morrow (1968)
German	8314	264	0·016	Altland *et al.* (1967)
Czechoslovakian	180	12	0·033	Altland *et al.* (1967)
Yugoslavian	248	7	0·016	Motulsky & Morrow (1968)
Greek	561	16	0·014	Motulsky & Morrow (1968)
Greek	360	13	0·018	Kattamis *et al.* (1962)
Portuguese	179	6	0·017	Kattamis *et al.* (1962)
Jewish				
Ashkanazi (Europe)	923	29	0·017	Szeinberg *et al.* (1966)
Non-Ashkanazi: Iraq & Iran	381	37	0·051	Szeinberg *et al.* (1966)
Balkan & Turkey	214	7	0·016	Szeinberg *et al.* (1966)
No. Africa	302	2	0·003	Szeinberg *et al.* (1966)
Yemen	124	7	0·036	Szeinberg *et al.* (1966)

African				
Congolese Negroes	460	1	0·001	Motulsky & Morrow (1968)
Ituri pygmies	125	0	0·000	Motulsky & Morrow (1968)
Babinga pygmies	300	0	0·000	Siniscalco (pers. comm.)
No. American Negroes	666	7	0·005	Motulsky & Morrow (1968)
Arabs (Israel)	110	2	0·009	Szeinberg *et al.* (1966)
Asiatic				
East Indian (Seattle)	98	3	0·015	Motulsky & Morrow (1968)
Japanese (Japan)	371	0	0·000	Altland *et al.* (1967)
Japanese (Seattle)	140	0	0·000	Motulsky & Morrow (1968)
Formosan Chinese	340	1	0·001	Motulsky & Morrow (1968)
Mixed Oriental (Seattle)	426	4	0·005	Motulsky & Morrow (1968)
Thai (Thailand)	723	0	0·000	Altland *et al.* (1967)
The Americas and Oceania				
Eskimo	145	0	0·000	Motulsky & Morrow (1968)
Mexican	377	7	0·009	Lisker *et al.* (1964)
Brazilian (mixed)	2138	60	0·015	Simpson & Kalow (1965)
So. American tribal	291	0	0·000	Arends *et al.* (1967)
Australian	98	1	0·005	Horsfall *et al.* (1963)
Filipino	427	2	0·002	Motulsky & Morrow (1968)

* A small number of sera in a few populations had the A phenotype. Thus, the E_1^a frequency in those populations is slightly higher than would be calculated from the number of heterozygotes.

(1968) found no heterozygotes among 125 Ituri pygmies tested, and they cited the unpublished work of Dr. M. Siniscalco, indicating a similar finding in 300 Babinga pygmies. Since about 30 per cent of the gene pool of Negroes living in the northern cities of the United States is derived from Caucasian admixture (Steinberg *et al.*, 1960), it is likely that the low frequency of the trait in these people reflects an even lower frequency in African Negroes.

The frequency of E_1^f heterozygotes in a mixed Brazilian population was found to be 0·93 per cent by Simpson & Kalow (1965) while a slightly higher frequency was reported in 801 Germans tested by Goedde *et al.* (1964a,b). The E_1^s gene frequency may be even lower than E_1^f, so that homozygotes, with an inherited absence of cholinesterase activity, probably have an incidence of about 1 in 100,000 (Simpson, 1966). (See Addenda).

LINKAGE OF E_1 AND Tf LOCI

Robson *et al.* (1966) noted that in a family they were studying, the E_1^a gene and the transferrin gene, Tf^{B2} appeared to be segregating together. Accordingly, they extended the investigation to include a total of six pedigrees in which there were variants of both cholinesterase and transferrin. One of the pedigrees gave low probability ratios for linkage of the two loci. However, even when this pedigree was included in the calculations, the estimated recombination fraction was 0·16, and the chance of its being outside the range of 0·07–0·29 was less than 1 in 20. When the one pedigree was omitted from the calculations, the estimated recombination fraction was decreased to 0·13. Although additional data are required, it appears nearly certain that the two loci are linked.

THE E_2 LOCUS

THE C_5 COMPONENT

When Harris *et al.* (1962) studied the two-dimensional starch gel electrophoretic pattern of serum esterase, they found that about 5 per cent of sera (13 of 248) had, in addition to the four characteristic bands of cholinesterase activity, a fifth, slower moving band called C_5. In a later paper, Harris *et al.* (1963a) reported that unidimensional starch gel electrophoresis at pH 5 to 6 alters the pattern so that C_1, C_2 and C_3 are undetectable, but C_4 and C_5 are more readily distin-

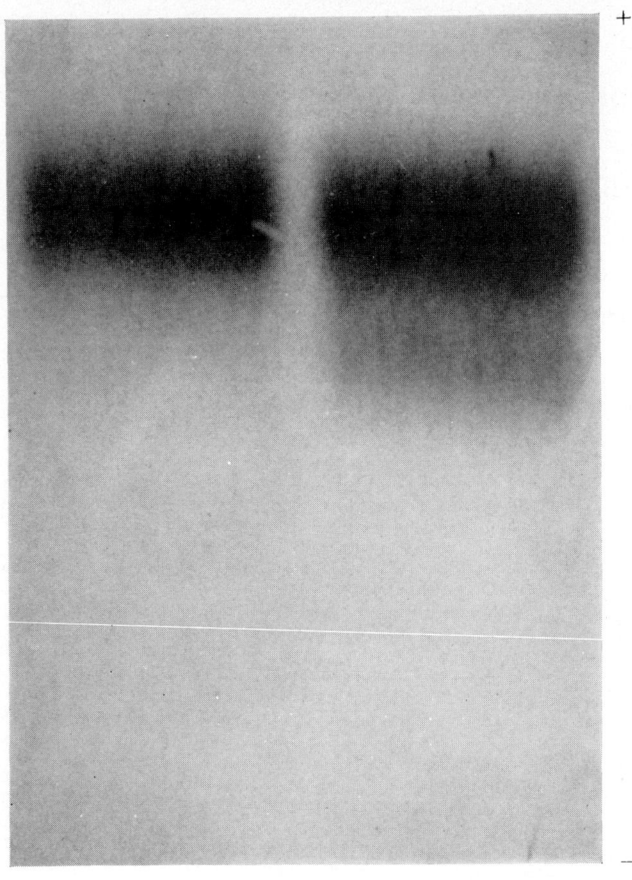

FIGURE 6.5. Starch gel electrophoresis of serum specimens with the phenotypes C$_5$− and C$_5$+. At the acid pH of this starch gel, the C$_1$, C$_2$ and C$_3$ zones do not appear as components separate from C$_4$.

[*Facing page* 211]

guished as individual bands (see Fig. 6.5). This method provided a rapid means of screening large numbers of sera, but two-dimensional tests were thought to be required to confirm the apparent identification of C_5+ sera.

A comparison of C_5+ and C_5- subjects showed no significant difference in the distribution of dibucaine number and fluoride number in the two groups. However, the average level of cholinesterase activity was about 30 per cent higher in C_5+ than in C_5- sera. This difference was considered to be due to activity of the C_5 component, as judged by relative staining intensity of the starch gel pattern.

INHERITANCE

Studies by Harris *et al.* (1963a) of nine British families and 71 families belonging to a highly inbred population from Tristan da Cunha indicated that the C_5 trait is probably inherited as an autosomal dominant. However, in a small number of people apparently carrying the gene, the C_5 component was either very weakly stained or not visible. For example, in four of the 32 $C_5- \times C_5-$ matings, there was at least one C_5+ child. A few exceptions were also noted in a subsequent study of 209 British and Greek families by Robson & Harris (1966) and over a thousand Brazilian families by Ashton & Simpson (1966). In the latter study, there was a higher proportion of C_5+ sera in the male parents (9·3%) than in the female parents (6·9%), and more C_5- children were found than expected.

In spite of these inconsistencies in expression, as well as inability to detect the homozygous variant, the evidence is overwhelming that the C_5+ phenotype is under genetic control. Harris *et al.* (1963b) found two families in which the E_1^u and E_1^a alleles could be shown to assort independently of the genes determining the presence or absence of the C_5 component. Accordingly, the existence of a second locus, E_2 was established, and its two alleles were called E_2^+ and E_2^-. In expressing the phenotypes of both loci, the phenotype of the E_1 locus (U, I, A, etc) is given first, followed by C_5+ or C_5- (Motulsky, 1964).

EFFECT OF E_1 LOCUS GENES ON C_5 CHARACTERISTICS

Harris *et al.* (1963b) examined the effect of the E_1^a and E_1^u genes on the inhibition characteristics of the C_5 esterase component, using RO 2-0683 as the inhibitor in the buffer of starch gels. They found that the C_4 and C_5 zones in UC_5+ serum were almost completely inhibited, while in IC_5+ serum, both zones were only partially inhibited. Thus, the E_1^a gene has the same effect on the inhibition characteristics of the C_5 component as it does on the C_4 component. Since C_4 in $E_1^u E_1^a$ heterozygotes (i.e. those of the I phenotype) consists of both the usual and atypical enzymes, the C_5 component in such individuals may also consist of a mixture of two enzymes, one of which shares a common subunit with the usual enzyme, and the other a common subunit with the atypical enzyme. In other words, it is likely that the C_5 component is determined by the genes of two loci, one of which is shared by the C_4 component (Harris *et al.* 1963b).

Robson & Harris (1966) subsequently determined the E_2 phenotypes in the serum of 68 people with the genotypes $E_1^a E_1^a$ and $E_1^a E_1^s$ (selected by the occurrence of suxamethonium-induced apnea). In no case was the C_5+ phenotype detected with certainty. Further data are needed to determine if the C_5 variant molecule requires the presence of an E_1^u gene for its expression. If this were the case, the behavior of the C_5 component in $E_1^u E_1^a$ heterozygotes would need a more complex explanation.

FREQUENCY OF THE C_5+ PHENOTYPE

Studies on the frequency of the E_2^+ allele are somewhat hampered by its variable expression. Robson & Harris (1966) found the C_5+ phenotype in 10 per cent of the 1941 British people tested, and Ashton & Simpson (1966) reported an incidence of 8 per cent in 2102 mixed Brazilians. Seattle Negroes were tested in both laboratories, 100 by Robson & Harris and 317 by Ashton & Simpson. The percentage of C_5+ was 0·02 and 0·05, respectively. Thus it is likely that the E_2^+ gene, like the E_1^a gene, has a lower frequency in Negroes than in Western Europeans. Robson & Harris (1966) also reported strikingly different frequencies of C_5+ sera in two Greek villages, one being 0·09 and the other 0·29.

OTHER POSSIBLE E_2 LOCUS VARIANTS

Neitlich (1966) reported that the serum of a healthy male subject had three times the mean normal level of pseudocholinesterase activity, and resistance to suxamethonium was similarly elevated. The electrophoretic pattern contained an additional intensely staining component migrating behind the C_4 band. The DN and FN were normal. The same phenotype was found in the proband's mother, female sib and female child, but not in his father, indicating apparent autosomal dominant inheritance. The author postulated that the extreme sensitivity of the proband's enzyme to inhibition by DFP indicated that his serum contained a normal number of cholinesterase molecules, each having three to four times the usual esteratic activity.

Van Ros & Druet (1966) described two unusual electrophoretic patterns in the serum of random Negro males. The first, found in four of 734 subjects tested, contained a component called C_6, which could be differentiated from C_5 by its slower migration rate. The second, found in two subjects, contained two components called C_{7a} and C_{7b} with mobilities slightly faster and slightly slower than C_6. The patterns were apparently not artifacts because they were observed in fresh specimens and presumably in specimens obtained on more than one occasion. Unfortunately, it was not possible to study the inheritance of either of these phenotypes.

METHODS

SCREENING TESTS FOR E_1 VARIANTS

AGAR DIFFUSION TESTS FOR I AND A PHENOTYPES

Harris & Robson (1963b) devised a screening test for detecting the presence of 'atypical' enzyme in either the homozygous state (A phenotype) or heterozygous state (I phenotype). The chemical basis for the test (Gomori, 1953) is the coupling of a diazo reagent to the hydrolysis of α-naphthyl acetate, so that the released α-naphthol forms a colored compound on combining with the diazo reagent. Serum is tested in the presence and absence of the inhibitor, RO 2-0683; the atypical enzyme is detected by its resistance to inhibition, as indicated by the formation of the colored compound.

Agar (1·5 per cent in 0·1 M tris-HCl buffer, pH 7·4) is poured in a 3 mm layer into plastic sandwich boxes or onto glass plates. The agar poured on half of the plates contains RO2-0683, added in a concentration of 10^{-7} M while the agar is liquid. The plates on which the agar does not contain inhibitor are the controls. When the agar has hardened, 4 mm holes are bored at 3 cm intervals. Serum specimens are diluted 1:8 for the plates containing inhibitor and 1:32 for the control plates. The wells are filled, the boxes closed (or plates covered) to prevent evaporation, and incubation is carried out at 37°C overnight.

For color development, the surface of the agar is flooded with a reaction mixture consisting of 100 ml 0·2 M phosphate buffer (pH 7·1), 2 ml 1 per cent α-naphthyl acetate in 50 per cent aqueous acetone, and 20 mg 5-chloro-*o*-toluidine (Fast red TR salt).

After 2 hours at room temperature, the reaction is stopped by pouring off the stain and flooding with a mixture of distilled water, methanol and acetic acid in 50:50:10 proportions. On the plates without added inhibitor, sera within the normal range of enzyme activity appear as brown circular zones about 1 cm in diameter. On the plates with added inhibitor, sera with the U phenotype show little reaction, while sera of A and I type show a brown zone similar to that observed on the control plates. This test does not discriminate between the A and I types, but it is very useful for screening purposes. For example, Simpson & Kalow (1965) compared their results using this method with the spectrophotometric tests on over 6000 serum specimens. Eleven per cent were incorrectly designated as atypical, but only 0·1 per cent were erroneously assigned to the usual phenotype. Thus they concluded that if all screened specimens considered to be atypical are tested for esterase activity and dibucaine number, the accuracy of this screening technique is sufficient to permit its use on large population samples. A similar degree of accuracy was reported by Lee & Robinson (1967).

RAPID TUBE TEST FOR I AND A PHENOTYPES
Morrow & Motulsky (1968) have recently described another screening procedure based on the same chemical principle as the agar diffusion method, but using a series of three test tubes for each serum determination and an incubation period of 1 hour. Preliminary results indicate that this test reliably discriminates between the I and A phenotypes.

The reagent stock solutions include 0·2 M phosphate buffer (pH 7·1), α-naphthyl acetate (0·03 M in 50 per cent aqueous acetone), 10^{-6} M RO 2-0683 (M.W. 393) in the phosphate buffer, and 3 per cent aqueous Duponal (sodium lauryl sulfate).

Just before use, two solutions are prepared. The working substrate contains 1 ml of α-naphthyl acetate stock solution, 20 ml phosphate buffer and 79 ml distilled water. The color reagent consists of 50 mg Fast red TR salt (5-chloro-*o*-toluidine), 15 ml water and 10 ml of 3 per cent Duponal. The latter substance is included to stop the enzyme reaction.

Since red cell esterase interferes with the results, it is not possible to use serum or plasma specimens containing visible hemoglobin. For testing, 0·5 ml of a 1:100 dilution of serum or plasma in buffer is placed in the first two tubes, and 0·5 ml of buffer in the third (reagent blank). The first and third tubes receive 0·5 ml of RO2-0683 solution, and the second tube 0·5 ml buffer. The addition of 3·5 ml of working substrate solution to all three tubes is followed by incubation in a 37°C water bath for 1 hour. The tubes are then removed, and 0·5 ml of color reagent is added. After mixing, the tubes are observed for color development, which occurs in 15 minutes and is stable for an hour.

If the serum contains cholinesterase of the usual type, there is strong inhibition by RO2-0683, so that the color in tube 2, without inhibitor, is deep red-purple, while tube 1, with inhibitor, is faint lavender. The reagent blank, tube 3, should be even more faintly colored than tube 2 unless there has been spontaneous hydrolysis, necessitating the preparation of fresh substrate solution.

If the esterase activity is too low for adequate color development, the test is repeated with a 1:50 serum dilution. Serum of the A phenotype is detectable because the atypical enzyme is only slightly susceptible to inhibition at this concentration of inhibitor. Thus, the first two tubes show similar color development. The serum of heterozygotes has a wide variation in enzyme activity, but since the mixture of enzymes is partially inhibited, tube 1 shows about half the color of tube 2.

Although visual inspection is usually sufficient to determine the phenotype, it is possible to obtain spectrophotometric readings at 555 mμ, expressing per cent inhibition as

$$100\left(1 - \frac{\text{OD tube 1}}{\text{OD tube 2}}\right).$$

Karahasanoglu & Özand (1967) have described a somewhat more difficult screening test which may also distinguish among the three phenotypes. It is based on the differential hydrolysis of butyrylthiocholine, and therefore is not specific for the genetic variants, since the serum of individuals with low levels of the usual enzyme react abnormally.

THE NaCl SCREENING TEST

Swift & La Du (1966) described a method for detecting the E_1^a homozygote based on their observations that 0·1 M NaCl enhances the usual enzyme activity but does not affect the atypical enzyme. Each serum is tested in a pair of test tubes, one containing 0·125 M NaCl, while both tubes contain benzoylcholine as the substrate and phenol red as an indicator of acid production.

The stock reagents include 0·5 per cent phenol red solution, 5 M NaCl, 0·02 M benzoylcholine and 0·1 N NaOH. A substrate-indicator solution is prepared from stock reagents to contain 2·5 mg phenol red per 100 ml and 5×10^{-5} M benzoylcholine. This solution is adjusted to a red-purple color (pH about 7·8) with the 0·1 M NaOH, and 2 ml is placed in each of a series of paired tubes. The second tube of each pair receives 0·05 ml of 5 M NaCl. Before 15 minutes has elapsed (to minimize CO_2 absorption), 0·1 ml of a 1:10 dilution of serum is added to both tubes of a pair, so that the final serum dilution is 1:200 and the molarity of NaCl in the second tube is 0·125. The tubes are corked, inverted, and set side-by-side to compare colors. The tubes must match at the onset, but within a minute, the tube containing NaCl begins to turn color, becoming orange by 5 minutes if the serum enzyme is of the usual or intermediate type. If no perceptible change in color occurs within 15 minutes, the serum presumably contains none of the usual enzyme, and the most likely genotype is $E_1^a E_1^a$.

DETERMINATION OF PERCENT INHIBITION

None of the screening tests outlined above can detect individuals with the UF phenotype, and two are not individually capable of making the distinction between types I and A. Thus, for accurate determination of all types, it is necessary to carry out tests for both dibucaine number (DN) and fluoride number (FN) expressing percent inhibition caused by 10^{-5} M dibucaine and 5×10^{-5} M fluoride, respectively. The original papers describing these techniques should be consulted, but a brief outline of the procedure is presented here.

MEASUREMENT OF DIBUCAINE NUMBER (DN)

In the original description of their spectrophotometric method for measurement of pseudocholinesterase activity, Kalow & Lindsay (1955) employed a Beckman Model DU spectrophotometer with ultraviolet attachment. However, in the subsequent paper describing a method for determining the per cent inhibition by 10^{-5} M dibucaine (i.e. the dibucaine number or DN), Kalow & Genest (1957) used a

Beckman DK 2 recording spectrophotometer with time-drive accessory. Although the latter is preferable, the test can be performed with a manually operated instrument. Temperature regulation in the range of 21–26° is important. With temperature of 35–37°C, the serum dilution is increased to slow the speed of reaction.

Although the absorption peak of the substrate, benzoylcholine, is at 235 mμ, the wavelength is set at 240 mμ to reduce the optical density of diluted serum. At 240 mμ, benzoylcholine has a greater optical density than its reaction products, so the decrease in optical density (ΔOD) is a measure of enzyme activity.

The reagents include M/15 phosphate buffer, pH 7·4 (7·58 g anhydrous Na_2HPO_4 and 1·8 g anhydrous KH_2PO_4 per liter of water), 2×10^{-4} M substrate solution (24·4 mg benzoylcholine in 500 ml buffer) and 4×10^{-5} M inhibitor (13·7 mg dibucaine in one liter of buffer). Serum is also diluted in the buffer, 1 : 50 for working with the recording spectrophotometer and 1 : 100 for hand-operated instruments.

For each test, three cuvettes are used. The first, containing 2 ml of diluted serum plus 2 ml buffer, is the blank for setting the machine at zero, with 240 mμ wavelength, slit width 1·4 and 10 mm light path. The second cuvette contains 1 ml of substrate and 1 ml of inhibitor, quickly mixed with 2 ml of diluted serum at time zero. The optical density is recorded at 20-second intervals for 4–5 minutes. In the third cuvette, buffer is substituted for inhibitor, and the procedure is repeated.

Under the conditions of this assay (i.e. with final serum dilution 1:200, pH 7·5, volume 4 ml, wavelength 240 mμ, light path 10 mm), a decrease in optical density (ΔOD) of 0·66 corresponds to the hydrolysis of 400 mμmole of benzoylcholine in the reaction mixture. Since 20 μl of serum is used, the calculation for number of international units is

$$\frac{400}{0\cdot66} \times \frac{1000}{20} \times \Delta\text{OD per minute}$$

This figure can be corrected to a temperature of 25 or 37°C by using a table recently provided by King (1965), who also described and discussed other methods for measurement of esterase activity.

To determine the DN, it is not necessary to convert the ΔOD into esterase units, since the per cent inhibition can be calculated as

$$\text{DN} = 100\left(1 - \frac{\Delta\text{OD in presence of dibucaine}}{\Delta\text{OD in absence of dibucaine}}\right)$$

MEASUREMENT OF FLUORIDE NUMBER (FN)

The procedure employed by Harris & Whittaker (1961 & 1962) for determining the per cent inhibition by 5×10^{-5} M fluoride (i.e. the fluoride number or FN) is the same as that described above for DN except that 2×10^{-4} M sodium fluoride solution (8·4 mg/liter) is substituted for the dibucaine solution. The ranges in values characteristic of the various phenotypes are given in Table 6.2.

DETECTION OF E_2 LOCUS VARIANTS

Although two-dimensional electrophoresis was used originally for the demonstration of the C_5 variant (Harris *et al.*, 1962), fairly satisfactory results were subsequently obtained with unidimensional starch gel electrophoresis. In the horizontal method described by Harris *et al.* (1963a), the serum is diluted 1:4 and inserted into a gel made from a pH 5·0 buffer containing 0·016 M succinic acid and 0·0184 M tris. The bridge buffer contains 1·6 M NaOH and 0·41 M citric acid. Ashton & Simpson (1966) reported that this buffer system can also be used with the vertical apparatus.

In the horizontal system of Ashton & Simpson (1966) the bridge buffer is prepared from 39·5 g citric acid and 11·1 g anhydrous lithium hydroxide per liter (pH 5·3). The gels are made in a pH 5·3 buffer composed of one volume of bridge buffer and nine volumes of a solution containing 5·5 g tris and 4·2 g succinic acid per liter. Electrophoresis at 4°C is carried out for 16 to 18 hours with a voltage gradient of about 4 volts per cm, or at room temperature at 2 volts per cm.

Zones of enzyme activity can be localized by several different esterase staining procedures, using either α- or β-naphthyl acetate as substrate and a diazo compound such as *o*-dianisidine (Fast blue B), 4-benzoylamino-2:5-dimethyoxyaniline (Fast blue RR) or 5-chloro-*o*-toluidine (Fast red TR). The sliced gel is preincubated for 15 minutes at room temperature in 50 ml of 0·2 M phosphate buffer, pH 7·1 and then placed in a solution containing 25 ml of the buffer, 0·5 ml of 1 per cent α- or β-naphthyl acetate in 50 per cent aqueous acetone and 10–20 mg of the diazo salt.

In the staining method described by Ashton & Simpson (1966), the gel is incubated at 37°C in a freshly prepared mixture of two solutions. The first consists of 20–30 mg α-naphthyl acetate in 10 ml of ethanol or acetone, and 90 ml of 0·9 per cent saline. The second contains 20 to

30 mg Fast blue RR, 10 ml 0·1 N HCl, and 5 ml 1 per cent $CaCl_2$ in 85 ml 0·025 M borate buffer, pH 8·9.

When electrophoresis is performed at pH 5 or 6, only the C_4 component is observed in most sera and, when C_5 is present, it migrates behind C_4, as shown in Fig. 6.5. In order to demonstrate the four common components (as shown in Fig. 6.4), a more alkaline pH is required. Very satisfactory results are obtained with the pH 8·6 discontinuous buffer system of Poulik (1957), in which the gel buffer contains 9·206 g tris (0·076 M) and 1·050 g citric acid (0·005 M) per liter, and the bridge buffer contains 18·55 g boric acid (0·3 M) and 2·40 g NaOH (0·06 M). However, at this pH, C_5 migrates too close to C_4, so that a preliminary electrophoretic separation on filter paper is required for clearcut demonstration (Fig. 6.1). For staining, the preferred substrate is α-naphthyl acetate.

REFERENCES

ALLOTT E.N. & THOMPSON J.C. (1956) The familial incidence of low pseudocholinesterase level. *Lancet* **ii**, 517.

ALTLAND K., EPPLE F. & GOEDDE H.W. (1967) Pseudocholinesterase—variants in Thailand and Japan. *Humangenetik* **4**, 127.

AMMON R. (1933) Die fermentative Spaltung des Acetylcholins. *Pflügers Arch. ges. Physiol.* **233**, 486.

ARENDS T., DAVIES D.A. & LEHMANN H. (1967) Absence of variants of usual serum pseudocholinesterase (acylcholine acylhydrolase) in South American Indians. *Acta genet.* **17**, 13.

ASHTON G.C. & SIMPSON N.E. (1966) C5 types of serum cholinesterase in a Brazilian population. *Am. J. hum. Genet.* **18**, 438.

AUGUSTINSSON K.B. (1960) Butyryl and propionyl-cholinesterases and related types of eserine sensitive esterases, *in* BOYER P.D., LARDY H. & MYRBÄCK K. (eds.) *The Enzymes*, vol. 4, p. 521. Academic Press, New York.

AUGUSTINSSON K.B. (1961) Multiple forms of esterase in vertebrate blood plasma. *Ann. N.Y. Acad. Sci.* **94**, 844.

BAMFORD K.F. & HARRIS H. (1964) Studies on 'usual' and 'atypical' serum cholinesterase using α-naphthyl acetate as substrate. *Ann. hum. Genet.* **27**, 417.

BERNSOHN J., BARRON K.D. & HESS A. (1961) Cholinesterases in serum as demonstrated by starch gel electrophoresis. *Proc. Soc. exp. Biol. Med.* **108**, 71.

BERRY W.T.C., COWIN P.J. & DAVIES D.R. (1954) A relationship between body fat and plasma pseudo-cholinesterase. *Br. J. Nutr.* **8**, 79.

BOURNE J.G., COLLIER H.O.J. & SOMERS G.F. (1952) Succinylcholine (succinoylcholine). Muscle relaxant of short action. *Lancet* **i**, 1225.

BUSH G.H. (1961) Prolonged apnoea due to suxamethonium. *Br. J. Anaesth.* **33**, 454.

CALLAWAY S., DAVIES D.R. & RUTLAND J.P. (1951) Blood cholinesterase levels and range of personal variation in a healthy adult population. *Br. med. J.* **ii**, 812.

CLITHEROW J.W., MITCHARD M. & HARPER N.J. (1963) The possible biological function of pseudocholinesterase. *Nature* **199**, 1000.

COHEN J.A., OOSTERBAAN R.A., JANSZ H.S. & BERENDS F. (1959) The active site of esterases. *J. cell. comp. Physiol.* **54**, suppl. 1, 231.

DAVIES D.R. & GREEN A.L. (1958) The mechanism of hydrolysis by cholinesterase and related enzymes. *Adv. Enzymol.* **20**, 283.

DAVIES R.O., MARTON A.V. & KALOW W. (1960) The action of normal and atypical cholinesterase of human serum upon a series of esters of choline. *Can. J. Biochem. Physiol.* **38**, 545.

DIETZ A.A., LUBRANO T. & RUBENSTEIN H.M. (1965) Four families segregating for the silent gene for serum cholinesterase. *Acta genet.* **15**, 208.

DOENICKE A., GÜRTNER T., KREUTZBERG G., REMES I., SPIESS W. & STEINBEREITHNER K. (1963) Serum cholinesterase anenzymia. *Acta anaesth. scand.* **7**, 59.

EVANS F.T., GRAY P.W.S., LEHMANN H. & SILK E. (1952) Sensitivity to succinylcholine in relation to serum-cholinesterase. *Lancet* **i**, 1229.

FORBAT A., LEHMANN H. & SILK E. (1953) Prolonged apnoea following injection of succinyldicholine. *Lancet* **ii**, 1067.

FUNNELL H.S. & OLIVER W.T. (1965) Proposed physiological function for plasma cholinesterase. *Nature* **208**, 689.

GLICK D. (1941) Some additional observations on the specificity of cholinesterase. *J. biol. Chem.* **137**, 357.

GOEDDE H.W. & ALTLAND K. (1963) Pseudocholinesterase variant in Germany and in Czechoslovakia. *Nature* **198**, 1203.

GOEDDE H.W. & BAITSCH H. (1964) Nomenclature of pseudocholinesterase polymorphism. *Br. med. J.* **ii**, 310.

GOEDDE H.W., FUSS W., GEHRING D. & BAITSCH H. (1964a) Studies on formal genetics of the pseudo-cholinesterase polymorphism; in atypical segregation in a family. *Biochem. Pharmac.* **13**, 603.

GOEDDE H.W., GEHRING D. & HOFFMANN R.A. (1965a) On the problem of the 'silent gene' in pseudocholinesterase polymorphism. *Biochim. biophys. Acta* **107**, 391.

GOEDDE H.W., GEHRING D. & HOFFMANN R.A. (1965b) Biochemische Untersuchungen zur Frage der Existenz eines 'silent gene' im Polymorphismus der Pseudocholinesteren. *Humangenetik.* **1**, 607.

GOEDDE H.W., OMOTO K., RITTER H. & BAITSCH H. (1964b) Zur formalen Genetik der Pseudocholinesterasen. Untersuchung von 408 Familien. *Humangenetik.* **1**, 1.

GOMORI G. (1953) Human esterase. *J. Lab. clin. Med.* **42**, 445.

GÜRTNER T., KREUTZBERG G. & DOENICKE A. (1963) Comparative studies on cholinesterase activity in serum and liver cells. *Acta anaesth. scand.* **7**, 69.

HARRIS H., HOPKINSON D.A. & ROBSON E.B. (1962) Two dimensional electrophoresis of pseudocholinesterase components in human serum. *Nature* **196**, 1296.

HARRIS H., HOPKINSON D.A., ROBSON E.B. & WHITTAKER M. (1963a) Genetical studies on a new variant of serum pseudocholinesterase detected by electro-phoresis. *Ann. hum. Genet.* **26**, 359.

HARRIS H. & ROBSON E.B. (1963a) Fractionation of human serum cholinesterase components by gel filtration. *Biochim. biophys. Acta* **73**, 649.

HARRIS H. & ROBSON E.B. (1963b) Screening tests for the 'atypical' and 'inter-mediate' serum-cholinesterase types. *Lancet* **ii**, 218.

HARRIS H., ROBSON E.B., GLEN-BOTT A.M. & THORNTON J.A. (1963b) Evidence for non-allelism between genes affecting human serum cholinesterase. *Nature* **200**, 1185.

HARRIS H. & WHITTAKER M. (1961) Differential inhibition of human serum cholinesterase with fluoride. Recognition of two new phenotypes. *Nature* **191**, 496.

HARRIS H. & WHITTAKER M. (1962) The serum cholinesterase variants. A study of twenty-two families selected via the 'intermediate' phenotype. *Ann. hum. Genet.* **26**, 59.

HARRIS H. & WHITTAKER M. (1963) Differential inhibition of 'usual' and 'atypical' serum cholinesterase by sodium chloride and sodium fluoride. *Ann. hum. Genet.* **27**, 53.

HARRIS H., WHITTAKER M., LEHMANN H. & SILK E. (1960) The pseudocholinester-ase variants. Esterase levels and dibucaine numbers in families selected through suxamethonium sensitive individuals. *Acta genet.* **10**, 1.

HART S.M. & MITCHELL J.V. (1962) Suxamethonium in the absence of pseudo-cholinesterase. A case report. *Br. J. Anaesth.* **34**, 207.

HAUPT V.H., HEIDE K., ZWISLER O. & SCHWICK H.G. (1966) Isolierung und physikalisch-chemische Characterisierung der Cholinesterase aus Human-serum. *Blut* **14**, 65.

HODGKIN W.E., GIBLETT E.R., LEVINE H., BAUR W. & MOTULSKY A.G. (1965) Complete pseudocholinesterase deficiency: genetic and immunologic con-siderations. *J. clin. Invest.* **44**, 486.

HORSFALL W.R., LEHMANN H. & DAVIES D. (1963) Incidence of pseudocholinester-ase variants in Australian aborigines. *Nature* **199**, 115.

HUNT A.H. & LEHMANN H. (1960) Serum albumin, pseudocholinesterase, and transaminases in the assessment of liver function before and after venous shunt operations. *Gut* **1**, 303.

JENKINS T., BALINSKY T. & PATIENT D.W. (1967) Cholinesterase in plasma: first reported absence in the Bantu; half-life determination. *Science*, **156**, 1748.

KALOW W. (1956) Familial incidence of low pseudocholinesterase level. *Lancet* **ii**, 576.

KALOW W. (1959a) Cholinesterase types, *in* WOLSTENHOLME G.E.W. & O'CONNOR C.M. (eds.) *Ciba Foundation Symposium on Biochemistry of Human Genetics*, p. 39. Little, Brown, Boston.

KALOW W. (1959b) The distribution, destruction and elimination of muscle relaxants. *Anesthesiology*, **20**, 505.

KALOW W. (1962) Heritable factors recognized in man by the use of drugs, *in Pharmacogenetics*, p. 69. W.B. Saunders, Philadelphia.

KALOW W. & DAVIES R.O. (1958) The activity of various esterase inhibitors towards atypical human serum cholinesterase. *Biochem. Pharmac.* **1**, 183.

KALOW W. & GENEST K. (1957) A method for the detection of atypical forms of human serum cholinesterase. Determination of dibucaine numbers. *Canad. J. Biochem. Physiol.* **35**, 339.

KALOW W. & GUNN D.R. (1959) Some statistical data on atypical cholinesterase of human serum. *Ann. hum. Genet.* **23**, 239.

KALOW W. & LINDSAY H.A. (1955) A comparison of optical and manometric methods for the assay of human serum cholinesterase. *Canad. J. Biochem. Physiol.* **33**, 568.

KALOW W. & STARON N. (1957) On distribution and inheritance of atypical forms of human serum cholinesterase, as indicated by dibucaine numbers. *Canad. J. Biochem. Physiol.* **35**, 1305.

KARAHASANOGLU A.M. & ÖZAND P.T. (1967) Rapid screening test for serum cholinesterase. *J. Lab. clin. Med.* **70**, 343.

KATTAMIS C., DAVIES D. & LEHMANN H. (1967) The silent serum cholinesterase gene. *Acta genet.* **17**, 299.

KATTAMIS C., ZANNOS-MARIOLEA L., FRANCO A.P., LIDDELL J., LEHMANN H. & DAVIES D. (1962) Frequency of atypical pseudocholinesterase in British and Mediterranean populations. *Nature* **196**, 599.

KAUFMAN L., LEHMANN H. & SILK E. (1960) Suxamethonium apnoea in an infant. Expression of familial pseudocholinesterase deficiency in three generations. *Br. med. J.* **i**, 166.

KING J. (1965) *Practical Clinical Enzymology*, p. 169. D. van Nostrand Co., Ltd., London.

KOSTER H. & KISCH B. (1943) Procaine esterase activity in human blood serum. A new test for toxic goitre. *Exp. Med. Surg.* **1**, 71.

KUNKEL H.G. & WARD S.M. (1947) Plasma exterase activity in patients with liver disease and nephrotic syndrome. *J. exp. Med.* **86**, 325.

LEE G. & ROBINSON J.C. (1967) Agar diffusion test for serum cholinesterase typing and influence of temperature on dibucaine and fluoride numbers. *J. med. Genet.* **4**, 19.

LEHMANN H., COOK J. & RYAN E. (1957) Pseudocholinesterase in early infancy. *Proc. R. Soc. Med.* **50**, 147.

LEHMANN H. & DAVIES D. (1962) Identification of the pseudocholinesterase type in human blood spots, *in Medicine, Science and the Law*, p. 180. Sweet & Maxwell, London.

LEHMANN H. & DAVIES D. (1963) Incidence of pseudocholinesterase variants in Australian aborigines. *Nature* **199**, 1115.

LEHMANN H. & LIDDELL J. (1961) The cholinesterases, *in* EVANS F.T. & GRAY T.C. (eds.) *Modern Trends in Anaesthesia*, p. 164. Butterworths, London.

LEHMANN H. & LIDDELL J. (1962) Transmission héréditaire de la pseudocholinestérase sérique. *Méd. et Hyg.* **20**, 961.

LEHMANN H. & LIDDELL J. (1964) Genetical variants of human serum cholinesterase, *in* STEINBERG A.G. & BEARN A.G. (eds.) *Progress in Medical Genetics*, vol. 3, p. 75. Grune & Stratton, New York.

Lehmann H. & Liddell J. (1966) Pseudocholinesterase deficiency and some other pharmacogenetic disorders, *in* Stanbury J.B., Wyngaarden J.B. & Fredrickson D.S. (eds.) *Metabolic Basis of Inherited Disease*, 2nd ed., p. 1356. McGraw-Hill, New York.

Lehmann H., Liddell J., Blackwell B., O'Connor D.C. & Daws A.V. (1963) Two further pseudocholinesterase phenotypes as causes of suxamethonium apnoea. *Br. med. J.* i, 1116.

Lehmann H., Patson V. & Ryan E. (1958) The inheritance of an idiopathic low plasma pseudocholinesterase level. *J. clin. Path.* 11, 554.

Lehmann H. & Ryan E. (1956) The familial incidence of low pseudocholinesterase level. *Lancet* ii, 124.

Liddell J., Lehmann H. & Davies D. (1963a) Harris and Whittaker's pseudocholinesterase variant with increased resistance to fluoride. *Acta genet.* 13, 95.

Liddell J., Lehmann H., Davies D. & Sharih A. (1962a) Physical separation of pseudocholinesterase variants in human serum. *Lancet* i, 463.

Liddell J., Lehmann H. & Silk E. (1962b) A 'silent' pseudocholinesterase gene. *Nature* 193, 561.

Liddell J., Newman G.E. & Brown D.F. (1963b) A pseudocholinesterase variant in human tissues. *Nature* 198, 1090.

Lisker R., Del Moral C. & Loria A. (1964) Frequency of the atypical pseudocholinesterase in four Indian (Mexican) tribes. *Nature* 202, 815.

McArdle B. (1940) The serum choline esterase in jaundice and diseases of the liver. *Quart. J. Med.* 33, 107.

McCance R.A., Hutchinson A.O., Dean R.F.A. & Jones P.E.H. (1949) The cholinesterase activity of the serum of newborn animals and of colostrum. *Biochem. J.* 45, 493.

Morrow A.C. & Motulsky A.G. (1968) A rapid screening method for the common atypical pseudocholinesterase variant. *J. Lab. clin. Med.* 71, 350.

Motulsky A.G. (1964) Pharmacogenetics, *in* Steinberg A.G. & Bearn A.G. (eds.) *Progress in Medical Genetics*, vol. 3, p. 49. Grune & Stratton, New York.

Motulsky A.G. & Morrow A. (1968) Atypical cholinesterase gene E_1^a: rarity in Negroes and most orientals. *Science* 159, 202.

Neitlich W. (1966) Increased plasma cholinesterase activity and succinylcholine resistance: a genetic variant. *J. clin. Invest.* 45, 380.

Omoto K. & Goedde H.W. (1965) Pseudocholinesterase variants in Japan. *Nature* 205, 726.

Parker W.C. & Bearn A.G. (1962) Studies on the transferrins of adult serum. cord serum and cerebrospinal fluid: the effect of neuraminidase. *J. exp. Med.* 115, 83.

Poulik M.D. (1957) Starch gel electrophoresis in a discontinuous system of buffers. *Nature* 180, 1477.

Rider J.A., Hodges J.L., Swader S. & Wiggins A.D. (1957) Plasma and red cell cholinesterase in 800 'healthy' blood donors. *J. Lab. clin. Med.* 50, 376.

Robson E.B. & Harris H. (1966) Further data on the incidence and genetics of the serum cholinesterase phenotype C_5+. *Ann. hum. Genet.* 29, 403.

ROBSON E.B., SUTHERLAND I. & HARRIS H. (1966) Evidence for linkage between the transferrin locus (Tf) and the serum cholinesterase locus (E_1) in man. *Ann. hum. Genet.* **29**, 325.

ROSE L., DAVIES D.A. & LEHMANN H. (1965) Serum-pseudocholinesterase in depression with notable anxiety. *Lancet* **ii**, 563.

SIMPSON N.E. (1966) Factors influencing cholinesterase activity in a Brazilian population. *Am. J. hum. Genet.* **18**, 243.

SIMPSON N.E. (1967) A second heterozygote for 'silent' and 'fluoride resistant' genes for serum cholinesterase. *J. med. Genet.* **4**, 264.

SIMPSON N.E. & KALOW W. (1963) Serum cholinesterase levels in families and twins. *Am. J. hum. Genet.* **15**, 280.

SIMPSON N.E. & KALOW W. (1964) The 'silent' gene for serum cholinesterase. *Am. J. hum. Genet.* **16**, 180.

SIMPSON N.E. & KALOW W. (1965) Comparisons of two methods for typing serum cholinesterase and prevalence of its variants in a Brazilian population. *Am. J. hum. Genet.* **17**, 156.

STEINBERG A.G., STAUFFER R. & BOYER S.H. (1960) Evidence for a Gm^{ab} allele in the Gm system of American Negroes. *Nature* **188**, 169.

SURGENOR D.M. & ELLIS D. (1954) Preparation and properties of serum and plasma proteins. Plasma cholinesterase. *J. Am. chem. Soc.* **76**, 6049.

SVENSMARK O. (1961) Human serum cholinesterase as a sialo-protein. *Acta physiol. scand.* **52**, 267.

SWIFT M.R. & LaDu B.N. (1966) A rapid screening test for atypical serum cholinesterase. *Lancet* **i**, 513.

SZEINBERG A., PIPANO S. & OSTFELD E. (1965) A silent pseudocholinesterase gene. Marked suxamethonium sensitivity in a person heterozygous for the atypical and silent pseudocholinesterase genes. *Acta genet.* **15**, 201.

SZEINBERG A., PIPANO S. & OSTFELD E. (1966a) Frequency of atypical pseudo-cholinesterase in different population groups in Israel. *Proc. 2nd Cong. europ. Anaesth.* Copenhagen. *Acta anaesth*, p. 199, suppl. 24.

SZEINBERG A., PIPANO S., OSTFELD E. & EVIATOR L. (1966b) The silent gene for serum pseudocholinesterase. *J. med. Genet.* **3**, 190.

TELFER A.B.M., MacDONALD D.J.F. & DINWOODIE A.J. (1964) Familial sensitivity to suxamethonium due to atypical pseudocholinesterase. *Br. med. J.* **i**, 153.

THOMPSON J.C. & WHITTAKER M. (1966) A study of the pseudocholinesterase in 78 cases of apnoea following suxamethonium. *Acta genet.* **16**, 209.

URIEL J. (1961) Caractérisation des cholinestérases et d'autres estérases carboxyliques après électrophorèse et immunoélectrophorèse en gélose. *Annls Inst. Pasteur.* **101**, 104.

VAN ROS G. & DRUET R. (1966) Uncommon electrophoretic patterns of serum cholinesterase (pseudocholinesterase). *Nature* **212**, 543.

WATERLOW J. (1950) Liver choline-esterase in malnourished infants. *Lancet* **i**, 908.

WETSTONE H.J., HONEYMAN M.S. & McCOMB R.B. (1965) Genetic control of the quantitative activity of a serum enzyme in man. *J. Am. med. Ass.* **192**, 1007.

WETSTONE H.J., LA MOTTA R.V., BELLUCCI A., RENNANT R. & WHITE B.V. (1960) Studies of cholinesterase activity. V. Serum cholinesterase in patients with carcinoma. *Ann. int. Med.* **52**, 102.

WETSTONE J.H., TENNANT R. & WHITE B.V. (1957) Studies of cholinesterase activity. I. Serum cholinesterase, methods and normal values. *Gastroenterology* **33**, 41.

WHITTAKER M. (1964) The pseudocholinesterase variants: esterase levels and increased resistance to fluoride. *Acta genet.* **14**, 281.

WHITTAKER M. (1967) The pseudocholinesterase variants. A study of fourteen families selected via the fluoride resistant phenotype. *Acta genet.* **17**, 1.

WHITTAKER V.P. (1951) Specificity, mode of action and distribution of cholinesterase. *Physiol. Rev.* **31**, 312.

WILDE C.E. & KEKWICK R.G.O. (1962) The aromatic esterase activity of human serum. *Biochem. J.* **83**, 39P.

WILSON I.B. (1954) The active surface of the serum esterase. *J. biol. Chem.* **208**, 123.

WILSON I.B. (1960) Acetylcholinesterase, *in* BOYER P.D., LARDY H. & MYRBÄCK K. (eds.) *The Enzymes*, vol. 4, p. 501. Academic Press, New York.

CHAPTER 7

PLASMA ALKALINE PHOSPHATASE

Intestinal Alkaline Phosphatase in Plasma 227
   Electrophoretic phenotypes..... 227
   Association of phosphatase phenotypes, ABO blood group and secretor status...... 228
   Inheritance of alkaline phosphatase phenotypes 230
   Possible biological significance 231

Placental Alkaline Phosphatase..... 232
Methods 233
   Starch gel electrophoresis......... 233
   Inhibition of the intestinal enzyme with L-phenylalanine 235
   Effect of neuraminidase on electrophoretic mobility...... 235

References...... 235

Electrophoretic heterogeneity of alkaline phosphatase derived from human tissues has been demonstrated with a number of media (Baker & Pellegrino, 1954; Keiding, 1959; Rosenberg, 1959; Estborn, 1959; Kowlessar *et al.*, 1959; Boyer, 1961; Moss & King, 1962; Hodson *et al.*, 1962; Chiandussi *et al.*, 1962; Korner, 1962; Haije & DeJong, 1963; Taswell & Jeffers, 1963). Boyer (1963), using specific anti-enzymes in a method of combined immunoabsorption and electrophoresis, studied the inter- and intra-organ relationships of human alkaline phosphatases. Of the three antigenic classes he distinguished, the first included the enzymes of liver, bone, spleen and kidney. Cross-reactions were observed between the second and third classes, from intestine and placenta, respectively. The isozymes found in human plasma (or serum) are derived from bone, liver and intestine, and, during pregnancy, from the placenta (Schlamowitz, 1958; Schlamowitz & Bodansky, 1959; Gutman, 1959; Boyer, 1961; Nisselbaum *et al.*, 1961; Hodson *et al.*, 1962; Cunningham & Rimer, 1963; Kreischer *et al.*, 1965; Warnock, 1966; Dymling, 1966; Wilding *et al.*, 1966; Posen *et al.*, 1967). An extensive review by Fishman & Ghosh (1967) provides a complete summary of the studies on alkaline phosphatases. The genetically determined variants have been recently reviewed by Beckman (1966, 1968), by Robinson & Goldsmith (1967) and Robson & Harris (1967).

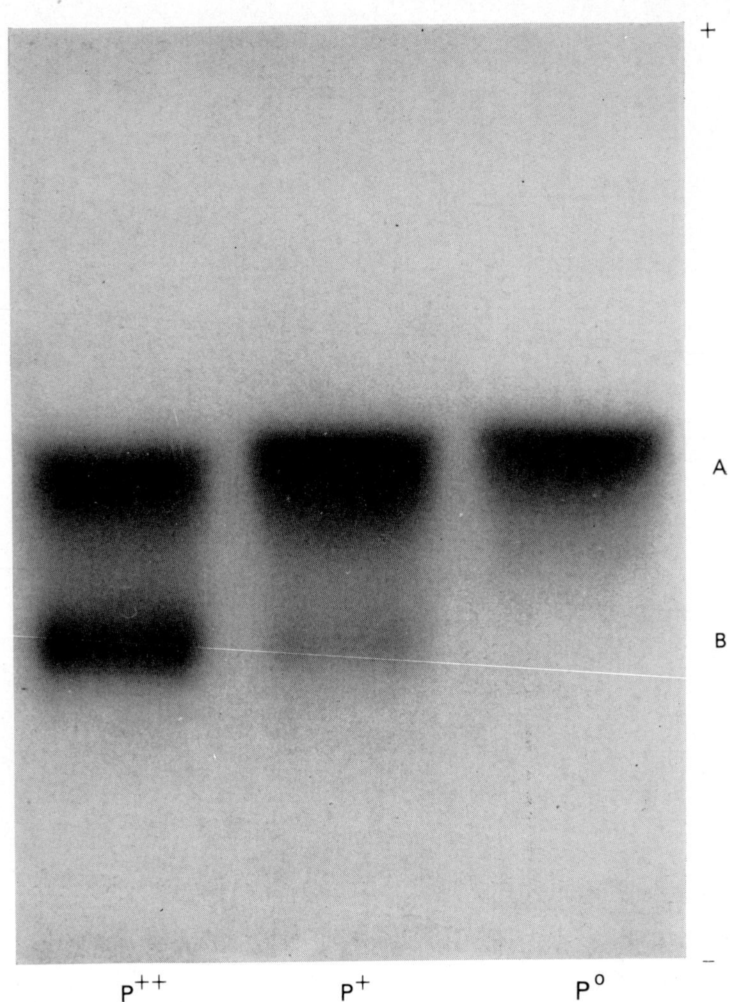

FIGURE 7.1. Photograph of starch gel stained for alkaline phosphatase activity after electrophoresis of three serum specimens of phenotypes p^{++}, p^{+} and p^{0}, indicating the relative amount of slow-moving B zone activity.

INTESTINAL ALKALINE PHOSPHATASE
IN PLASMA

ELECTROPHORETIC PHENOTYPES

On starch gel electrophoresis of normal adult plasma (or serum) some alkaline phosphatase activity may appear in the β-lipoprotein zone, but most of the activity is localized in either a single or a double band migrating near the β-globulin, transferrin. The faster-moving of the two components, zone A, is present in all sera; it apparently is of bone and liver origin, although Fishman & Ghosh (1967) offered an alternative hypothesis (see p. 232). On neuraminidase treatment, its mobility is greatly reduced due to removal of sialic acid residues. The slower-moving zone B, which occurs in about two-thirds of serum specimens, is of intestinal origin. Its mobility is unaffected by neuraminidase, but treatment with L-phenylalanine abolishes its activity (Boyer, 1961; Hodson *et al.*, 1962; Cunningham & Rimer, 1963; Fishman *et al.*, 1963; Arfors *et al.*, 1963a,b; Robinson & Pierce, 1964; Kreisher *et al.*, 1965; Moss, 1965; Shreffler, 1966).

The quantity of the slow component varies considerably from one person to another and also in specimens taken at different times from the same individual (Shreffler, 1965; Langman *et al.*, 1966). Thus, it is necessary to assign a grade to each pattern in order to differentiate weak from strong reactions. Shreffler (1965) used a 1–4 scoring system for this purpose, while Bamford *et al.* (1965) used the phenotypic symbols p^{++}, p^{+} and p^{0} to designate much, little or no demonstrable serum alkaline phosphatase of intestinal origin (see Fig. 7.1). In occasional serum specimens, two or even three slow-moving bands are observed, suggesting a polymorphism of this enzyme. However, the additional bands are probably zone A components partially deficient in sialic acid, since Robinson & Pierce (1964) showed that after neuraminidase treatment, only one of the B bands remains. After this treatment, failure to observe slower-moving zones corresponding to the neuraminidase-sensitive components may be due either to loss of enzyme activity or migration corresponding to that of the altered zone A enzyme (Robinson & Goldsmith, 1967).

Heterogeneity of human intestinal alkaline phosphatase was reported by Moss (1965) and Beckman *et al.* (1966b), who observed two zones of enzyme activity in intestinal extracts, the most intense corresponding to that seen in plasma. Also, Keiding (1966) reported that the alkaline phosphatase in serum is less susceptible to inhibition by L-phenylalanine and has a lower affinity for the substrate, p-nitrophenyl phosphate than the enzyme found in human intestine. However, the serum specimens he tested were from fasting individuals, and diet apparently has a strong influence on the amount of intestinal enzyme present in plasma. This observation was made by Langman *et al.* (1966), who reasoned that since there is a rise in alkaline phosphatase in the thoracic duct lymph after a fatty meal, and this increase is due to the intestinal enzyme (Keiding, 1964; Blomstrand & Werner, 1965), it might be expected that the serum enzyme activity would rise after the ingestion of fat. Accordingly, serum of normal subjects was tested before and after a fatty meal. Some of the subjects initially graded as p^0 or p^+ were subsequently found to have an increase in B zone activity, graded as p^+ and p^{++}, respectively; but none changed from p^{++} or from p^+ to p^0. Further studies by these authors demonstrated a direct correlation of the intestinal alkaline phosphatase concentration in the serum and the intake of fat. This observation is of considerable importance in the interpretation of phosphatase phenotypes determined in family and population studies as well as in patients with various diseases.

ASSOCIATION OF PHOSPHATASE PHENOTYPE, ABO BLOOD GROUP AND SECRETOR STATUS

Arfors *et al.* (1963a,b) found a strong correlation of phenotype p^{++} (type 2 in their nomenclature) with blood groups O and B; while p^0 or p^+ (their type 1) was almost invariably observed in group A subjects. An apparent association with Lewis phenotype was also suggested by these authors. However, it was subsequently shown that this effect is actually due to an association with the Secretor locus, which determines the ability to secrete ABH blood group soluble substances (Beckman, 1964; Shreffler, 1965; Bamford *et al.*, 1965; Evans, 1965; Hope, 1966; Walter *et al.*, 1967; Robinson *et al.*, 1967; Fichtner *et al.*, 1967).

As outlined in detail in Chapter 9, individuals with the red cell phenotype Le(a+) are non-secretors of ABH substances, having the Secretor genotype *se/se*. Among those of Le(a−) phenotype most are Le(a − b +), all of whom secrete blood group specific substance. Only

the small proportion who are Le(a − b −) can have the *se/se* genotype,
and most of them do not. Thus, for practical purposes, the designation
Le(a−) implies the ability to secrete ABH substance, associated with
either *Se/Se* or *Se/se* genotype. Table 7.1, prepared from information
kindly supplied by Professor Harry Harris, presents some data on the
ABO, Lewis and alkaline phosphatase B zone phenotype of normal
subjects presumably not fasting at the time their blood was taken.

TABLE 7.1. The Lewis, ABO and alkaline phosphatase phenotypes of 800 healthy
people tested in the laboratory of Professor Harry Harris. Some of these results
were published in the paper by Bamford, Harris *et al.* (1965)

| Antigens | | Serum alkaline phosphatase (in per cent) | | | Total number |
Lewis	ABO	p^0	p^+	p^{++}	
Le(a−)	O	15	46	39	144
	B	14	50	36	137
	A_1	80	20	0	137
	A_2	73	24	3	79
	A_1B	58	40	2	91
	A_2B	30	44	26	107
Le(a+)	All ABO	94	6	0	105

As shown in the bottom line of Table 7.1, none of the 105 non-
secretors of ABH (i.e. those with Le(a+) red cells) had the p^{++} phosphatase phenotype, and only six were p^+. However, among the ABH
secretors (i.e. phenotype Le(a−)), there was a strong correlation of
ABO groups with phosphatase type. Thus, none of 137 persons with
blood group A_1, and only 3 per cent of the 79 with A_2 were classified as
p^{++}. In subjects typed as A_1 or A_1B, the p^0 type predominated over
p^+, particularly in A_1 subjects. Comparison of A_1 with A_2 showed
only a small difference, but there was a definite shift toward p^+ and
p^{++} in A_2B as compared with A_1B. Among the subjects with blood
groups O and B, a large proportion were p^+ or p^{++} but, in each instance, about 15 per cent had no demonstrable alkaline phosphatase
B zone.

The fat-loading studies of Langman *et al.* (1966) showed that the
phenotype of A secretors was changed from p^0 to p^+ in six of 14

9

instances, but never to p^{++}. Among the non-secretors of ABH substance, only one of 17 tested changed from p^0 to p^+. These studies, as well as the previous reports of Beckman (1964), Shreffler (1965) and Bamford *et al.* (1965) demonstrate that the correlation of p^0 and ABH non-secretion is almost absolute, while the correlation of p^0 with blood type A is high, but less striking. This subject is re-examined in Chapter 9.

INHERITANCE OF ALKALINE
PHOSPHATASE PHENOTYPES

Arfors *et al.* (1963a) tested 89 pairs of presumably monozygotic twins and found concordance of serum alkaline phosphatase phenotypes in all instances. Among dizygotic twins, there was a 35 per cent discordance. Shreffler (1965) also conducted a study of twin pairs, 125 dizygotic and 101 monozygotic, assigning a numerical grade of 0–4 to the starch gel pattern based on the relative intensity of zone B, the slower-moving phosphatase band. An analysis of variance showed an intraclass correlation coefficient of 0·736 for monozygotic twins, more than twice that of the dizygotic twins. A separate analysis restricted to B or O secretors was made on 40 pairs of monozygotic twins and 27 pairs of dizygotic twins. The intraclass correlation coefficients were similar to those obtained in the entire collection of twins. Thus, even when fraternal twins were matched for ABO and secretor status, the correlation of their alkaline phosphatase B zone scores was only about half that found in identical twins. In other words, factors in addition to the genes at the ABO and secretor loci play a role in determining the phosphatase phenotype.

These additional factors are probably genetic in origin. In order to obtain further information on this point, Shreffler (1965) selected 56 families with B or O secretor parents and 121 of their children, who were also B or O secretors. Based on the 0–4 grading system, there was a general correlation of the phosphatase level in the children with that of their parents. The author pointed out that the usefulness of this measurement was diminished by the fact that children in general have lower intestinal phosphatase levels than adults. Also, the effect of fat intake on enzyme levels was not known at that time, so that the diet of these subjects preceding blood sampling was not taken into account.

Beckman (1964) summarized the results of testing 468 unselected

Brazilian parents and their 839 children, the phosphatase types being recorded as 1 (corresponding to p^0 and probably p^+) and 2 (corresponding to p^{++}). These data substantiated the random donor studies which showed a high correlation of p^0 phenotype with ABH non-secretion (regardless of blood type) as well as a strong tendency for individuals of blood group A to possess little or no slow-moving phosphatase. There was also an indication of an additional genetic determinant of alkaline phosphatase phenotype in agreement with the independent study of Shreffler (1965).

Because of the influence of diet and age on the level of intestinal alkaline phosphatase in the serum, it is difficult to determine the population frequency of the gene or genes which, in addition to the genes of the ABO and secretor systems, determine the expression of serum alkaline phosphatase. No additional association has been found between phosphatase phenotype and the genes of any of the known blood groups or serum groups (Beckman, 1964; Shreffler, 1965; Robinson *et al.*, 1967; Walter *et al.*, 1967).

POSSIBLE BIOLOGICAL SIGNIFICANCE

The fact that the bovine J groups and ovine R groups, analogous to the ABO groups of man, are involved in a similar association with serum phosphatase phenotypes, emphasizes the possibility that the relationship plays some important functional role (Gahne, 1963; Rendel & Stormont, 1964; Rendel *et al.*, 1964; Rasmusen, 1965). It has been repeatedly demonstrated by microscopic studies that the major portion of small intestinal alkaline phosphatase activity is localized on the brush border of the absorbing cells of the mucosal villi (Brandes *et al.*, 1956; Clark, 1961; Reale, 1962; Chase, 1963; Watanabe & Fishman, 1964). However, Hugon & Borgers (1966) found that enzyme activity could also be detected in the Golgi zone, in the smooth reticulum cisternae and in the dense bodies, structures which play a role in the absorption of fat and glucose.

Langman *et al.* (1966) postulated that there might be a difference in the physiological behavior of the absorbing cells in subjects of p^0 and p^{++} phenotypes. For example, the amount of phosphatase in the cells or the degree of their permeability might vary according to ABO and secretor status. Another possibility would be a qualitative or quantitative difference in fat absorption. The latter seemed unlikely because

there was no association between the serum turbidity after fat ingestion and the blood type or secretor status. Furthermore, the possibility
of a difference in quantity of mucosal alkaline phosphatase was
diminished by the studies of Shreffler (1966) on human postmortem
specimens of duodenum and jejunum. He was unable to find any
evidence that genes of the ABO and secretor systems exert an effect on
the amount of enzyme produced, and proposed that the difference in
serum levels must be due to increase of an inhibitor or to reduction in
rate of enzyme secretion either from intestinal cells into lymph or from
the thoracic duct lymph into the blood stream. (See Addenda.)

Fishman & Ghosh (1967) have presented evidence that in normal
adults, most of the serum alkaline phosphatase is of intestinal origin,
and that the zone A enzyme is a metabolic product which has lost
its susceptibility to L-phenylalanine treatment, possibly due to
prolonged intestinal absorption. Thus, they proposed that the genetic
effects associated with the blood group and secretor status could
operate at one or more levels affecting rates of absorption and/or
secretion, as well as the chemical environment affecting the enzyme
molecular structure.

There is good evidence that the tendency to develop certain diseases
of the gastro-intestinal tract, particularly duodenal ulcer and gastric
carcinoma, is affected, in some completely unknown manner, by the
individual's ABO and secretor status (Roberts, 1959; Clarke, 1961).
Since these two genetic systems are also associated with the appearance
of intestinal alkaline phosphatase in the blood stream, it is reasonable
to assume that a better understanding of the mechanisms involved
might provide an important clue to the pathogenesis of these diseases.
Preliminary attempts to find an association between alkaline phosphatase B zone and the ABO group of patients with gastro-intestinal disease have not been successful (Fichtner *et al.*, 1967). For
further discussion of ABO-secretor-phosphatase interaction, see
Chapter 9.

PLACENTAL ALKALINE PHOSPHATASE

Boyer (1961) observed differences in the starch gel electrophoretic
pattern of serum from pregnant women which suggested a genetic
polymorphism involving placental alkaline phosphatase. This

enzyme appears in maternal serum between the fourth and seventh month of gestation and is no longer detectable by about five or six weeks after delivery (Boyer, 1961; Robinson *et al.*, 1966a,b). It can be differentiated from the usual alkaline phosphatases in serum by its heat stability (Neale *et al.*, 1965), although, like the zone A isozyme, its electrophoretic mobility decreases after neuraminidase treatment (Kitchener *et al.*, 1965; Robinson *et al.*, 1966a). Recent studies by Robson & Harris (1965, 1967) and by Beckman *et al.* (1966a,b) indicate the existence at an autosomal locus of three common and at least six rare alleles which determine the 15 different phenotypes observed by electrophoresis of human placental extracts. However, since the serum of human non-pregnant subjects does not contain these isozymes, and since the preferred source of material for pheno-typing is placenta rather than blood, this polymorphism will not be discussed further. The interested reader is referred to the paper by Robson & Harris (1967) and to the reviews by Beckman (1966, 1968) and by Robinson & Goldsmith (1967).

METHODS

Demonstrating the A and B zones of serum alkaline phosphatase activity is best achieved by electrophoresis in a medium such as starch gel or acrylamide gel. The former has been used in most of the studies so far reported, but the sharply defined zones characteristic of acryl-amide gel may eventually lead to its preferential use.

STARCH GEL ELECTROPHORESIS

Both horizontal and vertical starch gel systems described in Chapter 2 provide adequate separation of the two isozymes, although there is a tendency for a blurring effect in gels run for long periods, particularly at room temperature. Best results are obtained with a water-cooled apparatus, but electrophoresis in the cold or even at room temperature can also be used. Various techniques have been described by Arfors *et al.* (1963a), Shreffler (1965), Bamford *et al.* (1965), and Fichtner *et al.* (1967).

When the intensity of the B zone is to be graded on a scale of zero to four, the serum is inserted undiluted. For the less sensitive but generally more clear-cut distinction of p^0, p^+ and p^{++}, the serum is diluted 1:3 with water or gel buffer.

A discontinuous buffer system is superior to simple borate or phosphate buffers. Although several authors advocate the system of Poulik (p. 102), the modification described by Ashton & Braden (1961) appears to give better results in a shorter period of time. For this buffer, two solutions are prepared. Solution A contains 1·2 g lithium hydroxide and 11.9 g boric acid per liter of water. Solution B consists of 1·6 g citric acid and 6·3 g Tris (Sigma 7-9 or 121) per liter of water. Solution A is used for the bridge buffer. The gel buffer is made from one part of solution A and nine parts of solution B. (The same buffer system is useful for separating the serum Gc components; see Chapter 4.)

Detection of alkaline phosphatase activity in the sliced gel is dependent upon a coupling system in which the naphthol released from either α- or β-naphthyl phosphate combines with the diazonium salt of one of several dyes, including Fast blue RR (4-benzoylamino-2,5-dimethoxy aniline), Garnet GBC (4-amino-3,1'-dimethyl azo-benzene), Variamine blue B (4-amino-4'-methoxy diphenylamine) or Fast blue VRT (4-amino-diphenylamine).

The procedure found most useful in this laboratory employs a vertical apparatus and the buffer system of Ashton & Braden described above. High voltage electrophoresis in the cold is carried out for 3 hours at a beginning voltage gradient of 7 volts/cm. After about 2 hours, the voltage is increased to 12 volts/cm. After slicing, the gel is incubated at 37°C in the staining solution (modified from Taswell & Jeffers, 1963; see below) for 1–2 hours. It is then washed briefly with water and placed in fixing solution containing methyl alcohol, water and glacial acetic acid in the proportion 50:50:10.

Staining solution

25 ml	1 M Tris
0·25 ml	1 M maleic acid
75 ml	Distilled water
20 mg	α-Naphthyl phosphate (sodium salt)
50 mg	Variamine blue B
60 mg	Magnesium sulfate

INHIBITION OF THE INTESTINAL ENZYME
WITH L-PHENYLALANINE

Fishman & Ghosh (1967) described a postcoupling technique for locating the L-phenylalanine-sensitive intestinal alkaline phosphatase zone by incubating each half of the sliced gel for 2 hours at 37°C with one of two strips of overlaid filter paper. One filter paper is impregnated with the inhibitor (0·005 M L-phenylalanine) plus buffered α-naphthyl phosphate, and the other with the same substrate and 0·005 M D-phenylalanine. Since the latter is not an inhibitor of intestinal alkaline phosphatase, enzyme activity is subsequently revealed by flooding the gel with Fast blue RR salt solution. However, the enzyme zone is not stained by similar treatment of the gel incubated with inhibitor.

EFFECT OF NEURAMINIDASE ON
ELECTROPHORETIC MOBILITY

Demonstration of the differential effect of neuraminidase on the electrophoretic migration rate of the serum alkaline phosphatases (due to removal of sialic acid residues) was described by Robinson & Pierce (1964). In their method, an equal volume of *Vibrio cholerae* neuraminidase, diluted in a solution containing 0·05 M sodium acetate, 0·06 M NaCl, and 0·004 M $CaCl_2$, pH 5·6, is added to the serum to be tested. After incubation for 20 hours at 37°C, electrophoresis is carried out, and the enzyme activity revealed by a dye-coupling technique. Comparison of treated with untreated serum enzyme patterns shows that the migration of the slow-moving B zone enzyme is unaffected, while there is a marked decrease in mobility of the fast-moving component(s).

REFERENCES

ARFORS K.E., BECKMAN L. & LUNDIN L.G. (1963a) Genetic variations of human serum phosphatase. *Acta genet.* **13**, 89.

ARFORS K.E., BECKMAN L. & LUNDIN L.G. (1963b) Further studies on the association between human serum phosphatases and blood groups. *Acta genet.* **13**, 366.

ASHTON G.C. & BRADEN A.W.H. (1961) Serum β-globulin polymorphism in mice. *Aust. J. exp. Biol. med. Sci.* **14**, 248.

BAKER R.W.R. & PELLEGRINO C. (1954) The separation and detection of serum enzymes by paper electrophoresis. *Scand. J. clin. Lab. Invest.* **6**, 94.

BAMFORD K.F., HARRIS H., LUFFMAN J.E., ROBSON E.B. & CLEGHORN T.E. (1965) Serum alkaline-phosphatase and the ABO blood groups. *Lancet* **i**, 530.

BECKMAN L. (1964) Associations between human serum alkaline phosphatases and blood groups. *Acta genet.* **14**, 286.

BECKMAN L. (1966) *Isozyme Variation in Man.* S. Karger, Basel.

BECKMAN L. (1968) Blood groups and serum alkaline phosphatase. *Series Haematologica* **1**, 1, 137.

BECKMAN L., BJÖRLING G. & CHRISTODOULOU C. (1966a) Pregnancy enzymes and placental polymorphism. I. Alkaline phosphatase. *Acta genet.* **16**, 59.

BECKMAN L., BJÖRLING G. & HEIKEN A. (1966b) Human alkaline phosphatases and the factors controlling their appearance in serum. *Acta genet.* **16**, 305.

BLOMSTRAND R. & WERNER B. (1965) Alkaline phosphatase activity in human thoracic duct lymph. *Acta chir. scand.* **129**, 177.

BOYER S.H. (1961) Alkaline phosphatase in human sera and placentae. *Science* **134**, 1002.

BOYER S.H. (1963) Human organ alkaline phosphatases: discrimination by several means including starch gel electrophoresis of antienzyme-enzyme supernatant fluids. *Ann. N.Y. Acad. Sci.* **103**, 938.

BRANDES D., ZETTERQUIST H. & SHELDON H. (1956) Histochemical techniques for electron microscopy: alkaline phosphatase in the intestinal brush border. *Nature* **177**, 382.

CHASE W. (1963) The demonstration of the alkaline phosphatase activity in frozen-dried mouse gut in the electron microscope. *J. Histochem. Cytochem.* **11**, 96.

CHIANDUSSI L., GREENE S.F. & SHERLOCK S. (1962) Serum alkaline phosphatase fractions in hepato-biliary and bone diseases. *Clin. Sci.* **22**, 425.

CLARK S.L. (1961) The localization of alkaline phosphatase in tissues of mice, using the electron microscope. *Am. J. Anat.* **109**, 57.

CLARKE C.A. (1961) Blood groups and disease, *in* STEINBERG A.G. & BEARN A.G. (eds.) *Progress in Medical Genetics*, vol. 1, p. 81. Grune & Stratton, New York.

CUNNINGHAM W.R. & RIMER J.G. (1963) Isozymes of alkaline phosphatase of human serum. *Biochem. J.* **89**, 50P.

DYMLING J.F. (1966) Separation of serum and placental alkaline phosphatase by agarose gel electrophoresis and sephadex chromatography. *Scand. J. clin. Lab. Invest.* **18**, 129.

ESTBORN B. (1959) Visualization of acid and alkaline phosphatase after starch gel electrophoresis of seminal plasma, serum and bile. *Nature* **184**, 1636.

EVANS D.A.P. (1965) Confirmation of association between ABO blood groups and salivary ABH secretor phenotypes and electrophoretic patterns of serum alkaline phosphatase. *J. med. Genet.* **2**, 126.

FICHTNER K., CLEVE H., KRÜPE M. & WENDT G.G. (1967) Die blutgruppenassoziierten Elektrophoresetypen der alkalischen Serumphosphatase bei Gesunden und bei Patienten mit gastrointestinalen Erkrankungen. *Humangenetik* **4**, 244.

FISHMAN W.H. & GHOSH N.K. (1967) Isoenzymes of human alkaline phosphatase *in* BODANSKY O. & STEWART C.P. (eds.) *Advances in Clinical Chemistry*, vol. 10, p. 255. Academic Press, New York.

FISHMAN W.H., GREEN S. & INGLIS N.I. (1963) L-phenylalanine: an organic specific, stereospecific inhibitor of human intestinal alkaline phosphatase. *Nature* **198**, 685.

GAHNE B. (1963) Genetic variation of phosphatase in cattle serum. *Nature* **199**, 305.

GUTMAN A.B. (1959) Serum alkaline phosphatase activity in diseases of the skeletal and hepatobiliary systems. *Am. J. Med.* **27**, 875.

HAIJE W.H. & DeJONG M. (1963) Iso-enzyme patterns of serum alkaline phosphatase in agar-gel electrophoresis and their clinical significance. *Clin. chim. Acta* **8**, 620.

HODSON A.W., LATNER A.L. & RAINE L. (1962) Isozymes of alkaline phosphatase. *Clin. chim. Acta* **7**, 255.

HOPE R.M. (1966) Human serum alkaline phosphatase variants and their association with the ABO blood groups in an Australian sample. *Aust. J. exp. Biol. med. Sci.* **44**, 323.

HUGON J. & BORGERS M. (1966) Ultrastructural localization of alkaline phosphatase activity in the absorbing cells of the duodenum of mouse. *J. Histochem. Cytochem.* **14**, 629.

KEIDING N.R. (1959) Differentiation into three fractions of the serum alkaline phosphatase and the behavior of the fractions in diseases of bone and liver. *Scand. J. clin. Lab. Invest.* **11**, 106.

KEIDING N.R. (1964) The alkaline phosphatase fractions of human lymph. *Clin. Sci.* **26**, 291.

KEIDING N.R. (1966) Intestinal alkaline phosphatase in human lymph and serum. *Scand. J. clin. Lab. Invest.* **18**, 134.

KITCHENER P.N., NEALE F.C., POSEN S. & BRUDENELL-WOODS J. (1965) Alkaline phosphatase in maternal and fetal sera at term and during the puerperium. *Am. J. clin. Path.* **44**, 654.

KORNER N.H. (1962) Distribution of alkaline phosphatase in serum protein fractions. *J. clin. Path.* **15**, 195.

KOWLESSAR O.D., PERT J.H., HAEFFNER L.J. & SLEISENGER M.H. (1959) Localization of 5-nucleotidase and non-specific alkaline phosphatase by starch gel electrophoresis. *Proc. Soc. exp. Biol. Med.* **100**, 191.

KREISCHER J.H., CLOSE V.A. & FISHMAN W.H. (1965) Identification by means of L-phenylalanine inhibition of intestinal alkaline phosphatase components separated by starch gel electrophoresis of serum. *Clin. chim. Acta* **11**, 122.

LANGMAN M.J.S., LEUTHOLD E., ROBSON E.B., HARRIS J., LUFFMAN J.E. & HARRIS H. (1966) Influence of diet on the 'intestinal' component of serum alkaline phosphatase in people of different ABO blood groups and secretor status. *Nature* **212**, 41.

MOSS D.W. (1965) Properties of alkaline-phosphatase fractions in extracts of human small intestine. *Biochem. J.* **94**, 458.

MOSS D.W. & KING E.J. (1962) Properties of alkaline-phosphatase fractions separated by starch-gel electrophoresis. *Biochem. J.* **84**, 192.

NEALE F.C., CLUBB J.S., HOTCHKIS D. & POSEN S. (1965) Heat stability of human placental alkaline phosphatase. *J. clin. Path.* **18**, 359.

NISSELLBAUM J., SCHLAMOWITZ M. & BODANSKY O. (1961) Immunochemical studies of functionally similar enzymes. *Ann. N.Y. Acad. Sci.* **94**, 970.

POSEN S., NEALE F.C., BIRKETT D.J. & BRUDENELL-WOODS J. (1967) Intestinal alkaline phosphatase in human serum. *Amer. J. clin. Path.* **48**, 81.

RASMUSEN B.A. (1965) Inheritance of R-O-i blood groups and alkaline phosphatase polymorphism in sheep. *Genetics* **51**, 767.

REALE E. (1962) Electron microscopic localization of alkaline phosphatase from material prepared with the cryostat microtome. *Exp. cell. Res.* **26**, 210.

RENDEL J., AALUND O., FREEDLAND R.A. & MØLLER F. (1964) The relationship between alkaline phosphatase polymorphism and blood group O in sheep. *Genetics* **50**, 973.

RENDEL J. & STORMONT C. (1964) Variants of ovine alkaline phosphatases and their association with the R-O blood groups. *Proc. Soc. exp. Biol. Med.* **115**, 853.

ROBERTS J.A. FRASER (1959) Some associations between blood groups and disease. *Br. med. Bull.* **15**, 129.

ROBINSON J.C. & GOLDSMITH L.A. (1967) Genetically determined variants of serum alkaline phosphatase: a review. *Vox Sang.* **13**, 289.

ROBINSON J.C., LEVENE C., BLUMBERG B.S. & PIERCE J.E. (1967) Serum alkaline phosphatase types in North American Indians and Negroes. *J. med. Genet.* **4**, 96.

ROBINSON J.C., LONDON W.T. & PIERCE J.E. (1966a) Observations on the origin of pregnancy-associated plasma proteins. *Am. J. Obstet. Gynec.* **96**, 226.

ROBINSON J.C. & PIERCE J.E. (1964) Differential action of neuraminidase on human serum alkaline phosphatases. *Nature* **204**, 472.

ROBINSON J.C., PIERCE J.E. & BLUMBERG B.S. (1966b) The serum alkaline phosphatase of pregnancy. *Am. J. Obstet. Gynec.* **94**, 559.

ROBSON E.B. & HARRIS H. (1965) Genetics of the alkaline phosphatase polymorphism of the human placenta. *Nature* **207**, 1257.

ROBSON E.B. & HARRIS H. (1967) Further studies on the genetics of placental alkaline phosphatase. *Ann. hum. Genet.* **30**, 219.

ROSENBERG I.N. (1959) Zone electrophoretic studies of serum alkaline phosphatase *J. clin. Invest.* **38**, 630.

SCHLAMOWITZ M. (1958) Immunochemical studies on alkaline phosphatase. *Ann. N.Y. Acad. Sci.* **75**, 373.

SCHLAMOWITZ M. & BODANSKY O. (1959) Tissue sources of human serum alkaline phosphatase, as determined by immunochemical procedures. *J. biol. Chem.* **234**, 1433.

SHREFFLER D.C. (1965) Genetic studies of blood group-associated variations in a human serum alkaline phosphatase. *Am. J. hum. Genet.* **17**, 71.

SHREFFLER D.C. (1966) Relationship of alkaline phosphatase levels in intestinal mucosa to ABO and secretor blood group. *Proc. Soc. exp. Biol. Med.* **123**, 423.

TASWELL H.F. & JEFFERS D.M. (1963) Isozymes of serum alkaline phosphatase in hepatobiliary and skeletal disease. *Am. J. clin. Path.* **40**, 349.

WALTER H., NEUMANN S., YANNISSIS C. & STEEGMÜLLER H. (1967) Untersuchungen über die alkalischen Serumphosphatasegruppen. *Humangenetik* **4**, 174.

WARNOCK M.L. (1966) Characterization of tissue and serum alkaline phosphatase. *Clin. chim. Acta* **14**, 156.

WATANABE K. & FISHMAN W.H. (1964) Application of the stereospecific inhibitor L-phenylalanine to the enzymorphology of intestinal alkaline phosphatase. *J. Histochem. Cytochem.* **12**, 252.

WILDING P., STUBRIN M.I., MAJCHER S.J., HAVERBACK B.J. & CHEN D.M. (1966) Serum alkaline phosphatase of intestinal origin. *Enzymol. biol. Clin.* **6**, 248.

CHAPTER 8

OTHER GENETIC MARKERS
IN PLASMA

Albumin 240
Bisalbuminemia (alloalbumin-
emia) 240
An unusual albumin variant 243
Genetic linkage 245
Methods 246

The Protease-Inhibitor (Pi) Sys-
tem: α_1-Antitrypsin 247
Properties of α_1-antitrypsin 247
Phenotypic variants of α_1-anti-
trypsin 248
Frequency of the Pi genes.......... 250
Physiological role of α_1-anti-
trypsin 252

Methods 252
Ceruloplasmin 252
Physical properties of cerulo-
plasmin 253
Genetic variants of cerulo-
plasmin 253
Methods 255
The Xm System (α_2 Macroglobu-
lin) 256
Inheritance of Xm(a).......... 256
Xm gene frequencies.......... 257
Alpha$_1$-Acid Glycoprotein.......... 257
Types I, II and III.......... 258
References.......... 259

There is a growing list of plasma proteins with inherited structural variation. In most instances, the variants are so rare that they have very little use as genetic markers or else their polymorphism is largely confined to a single ethnic group. In a few genetic systems, there are two or more alleles with fairly high frequencies, but either the distinction of the phenotypes is too uncertain or else the reagents necessary for their detection are virtually impossible to obtain. For these reasons, this chapter contains only brief descriptions of several proteins which cannot at present be considered as useful genetic markers except under special circumstances.

ALBUMIN

BISALBUMINEMIA (ALLOALBUMINEMIA)

The occurrence of genetic variation in human albumin was first observed by Scheurlen (1955), who found, in addition to the usual albumin zone, a second albumin with slower electrophoretic mobility.

The familial nature of this rare anomaly was demonstrated by Knedel (1957, 1958) and by Nennsteil & Becht (1957). Subsequently, several more families of European origin with slow-moving albumin variants were described (Earle *et al.*, 1958, 1959; Wuhrmann, 1959; Franglen *et al.*, 1960; Miescher, 1960; Adner & Redfors, 1961; Cooke *et al.*, 1961; Sarcione & Aungst, 1962; Robbins *et al.*, 1963; Efremov & Braend, 1964; Sandor *et al.*, 1965; Drachman *et al.*, 1965; Braend *et al.*, 1965; Ungari & Lopez, 1965; Adams, 1966; Weitkamp *et al.*, 1966). In addition, two fast-moving variants were reported in European families (Wieme, 1960, 1962; Tárnoky & Lestas, 1964). In all instances, the two albumins were present in approximately equal quantities, and the total albumin concentration (when measured) was not elevated. Inheritance of the variants was consistent with autosomal codominance.

Blumberg *et al.* (1968) have suggested substituting the term 'allo-albuminemia' for bisalbuminemia, to indicate (by analogy with the serologically determined allotypes, such as Gm, Inv and Ag) inherited variation. More important, such a term can be applied to homozygosity for a variant allele where the term bisalbuminemia is not appropriate.

Partial chemical analysis of the European slow-moving albumin (so-called type B) performed by Gitlin *et al.* (1961) indicated that it probably represented a single amino acid substitution of lysine for aspartic or glutamic acid. It is reasonable to assume that most or all of the albumin variants represent a series of alleles at the structural locus for albumin. Direct evidence to support this assumption is lacking, but indirect evidence from linkage data (see p. 245) indicates that at least three of these genes are alleles. A few studies suggest that the variants may have different binding constants for certain substances, such as bromphenol blue and thyroxine (Sarcione & Aungst, 1962; Blumberg *et al.*, 1968).

Melartin & Blumberg (1966) found that among certain American Indian tribes, an electrophoretic fast-moving variant called Albumin Naskapi is fairly common. For example, among 151 Naskapi Indians of Quebec, there were 37 carriers of the trait and one apparent homozygote with a single, fast-moving albumin, indicating that the mutant gene has a frequency of about 0·13 in that ethnic group. A similar frequency was found in the nearby Montagnais tribe, which had two homozygotes. In the Athabascan and Tlingit tribes of Alaska, the aberrant gene frequency was lower (0·028 and 0·005), while among the 160 Sioux Indians tested, there were two heterozygotes (gene frequency 0·007). No examples were found in 192 Peruvian Indians, 443 Alaskan Eskimos or 365 Indians from the Canadian Haida tribe.

A fast-moving variant, possibly the same as that reported by Melartin & Blumberg (1966) was independently detected in a Cree Indian family of Alberta, Canada by Bell *et al.* (1967). The frequency of the variant was not given. In both reports, the unusual albumin was inherited as an autosomal co-dominant. Polesky & Rokala (1967) found a similar or identical variant among Ojibwa and Blackfoot Indian tribes of northern Minnesota. The gene frequency was about 0·03.

Melartin *et al.* (1967) subsequently found that ten of 185 serum specimens obtained from Mexican mestizo communities contained a variant called albumin Mexico, which migrated between the usual albumin and the slow-moving variant most frequently observed in Europeans. (See Addenda.)

Weitkamp *et al.* (1967b) compared the electrophoretic mobility of 19 albumin variants obtained from 19 unrelated families: 15 of European origin, 3 from Navajo or Chippewa Indians, and one unknown. These 19 variants could be placed in five classes on the basis of their electrophoretic mobility. Thus, all but one of the slow-moving variants from Europeans had the same mobility, and were designated as 'very slow'. The single exception, which had been described by Sandor *et al.* (1965) had a somewhat more rapid mobility, so it was called 'slow'. Presumably it migrates at a rate similar to that of albumin Mexico described by Melartin *et al.* (1967). However, these two albumin variants were not compared.

Three classes of fast-moving variants were distinguished. One, called 'fast', came from a British family described by Tárnoky & Lestas (1964). The variants described in North American Indians composed the 'faster' class. Finally, there were three examples in the 'very fast' category, including the albumin described by Wieme (1960) and two others supplied by Wieme and Drachmann (unpublished) from a Belgian and Danish family, respectively.

Figure 8.1 is a diagrammatic representation of the various albumins which have been described. In this diagram, the 'slow' variant of Sandor *et al.* (1965) and the Mexican variant of Melartin *et al.* (1967) are considered to have different mobilities on the basis of the individual descriptions given by those two papers as well as that of Weitkamp *et al.* (1967b). Thus, in addition to the commonly observed albumin, there are probably at least six electrophoretic variants which, in

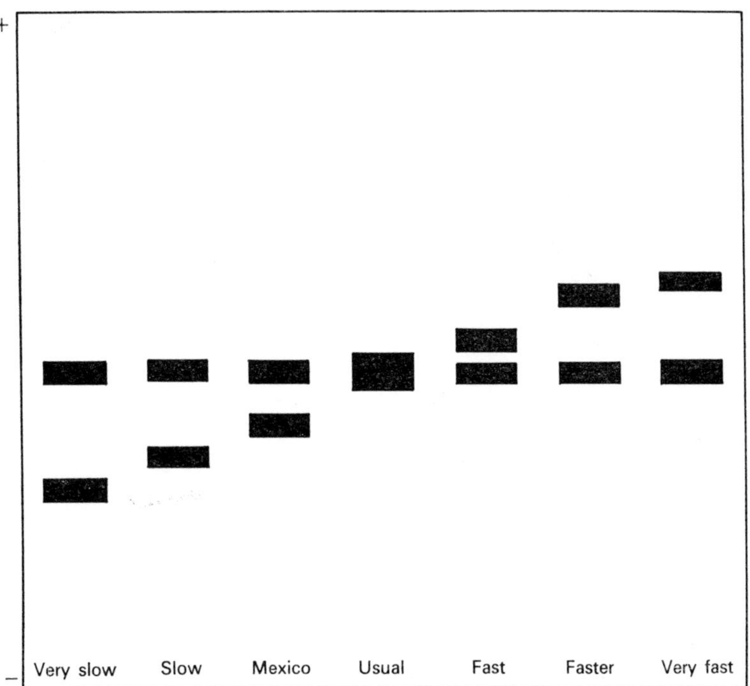

FIGURE 8.1. Diagram of starch gel electrophoretic patterns of three slow-moving and three fast-moving albumin variants represented as bisalbumin-emia in heterozygotes. (Drawn from data in Weitkamp *et al.*, 1967a and Melartin *et al.*, 1967.)

heterozygotes, give the appearance of bisalbuminemia (i.e. two albumin zones with similar staining intensity). A seventh variant, which does not fit into this definition, requires separate consideration.

AN UNUSUAL ALBUMIN VARIANT

Fraser *et al.* (1959) found that the serum of several members of a single family contained a component migrating slightly faster than transferrin on starch gel electrophoresis, and best demonstrated by two-dimensional electrophoresis (i.e. first electrophoresis on filter paper and then, at right angles, in starch gel). Under these circumstances, it was shown to move with albumin on paper electrophoresis but to have a slower mobility when subjected to molecular sieving in starch gel. Its staining intensity was considerably less than that of albumin, and somewhat less than that of transferrin. Figure 8.2, taken from the paper of Fraser *et al.* (1959), illustrates these findings diagrammatically.

Poulik *et al.* (1960) showed that when normal human serum was

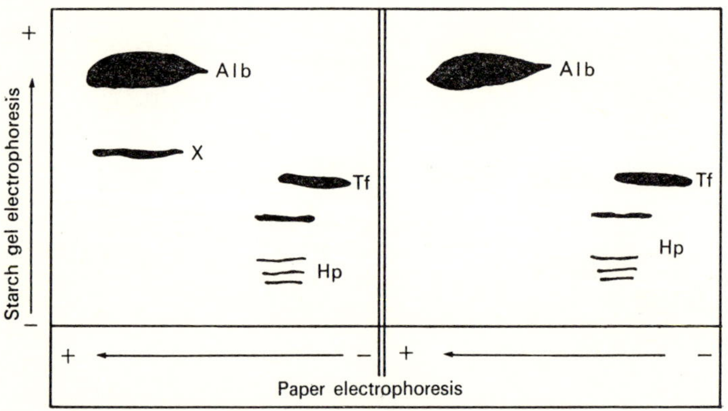

FIGURE 8.2. Diagram of two-dimensional electrophoretic patterns (first on filter paper, then in starch gel) comparing a normal serum specimen (right) with an unusual serum containing an X component (left). Albumin (alb), transferrin (Tf) and haptoglobin (Hp) are shown. (Adapted from Fraser *et al.*, 1959. Reprinted with permission from *Lancet* **i**, 1023).

subjected to two-dimensional electrophoresis, 61 of 64 specimens contained a faint component migrating in the same position as the anomalous protein described by Fraser *et al.* (1959). This component was found to cross-react immunologically with albumin.

The relationship of these two isolated observations was clarified by the studies of Laurell & Niléhn (1966). They found that five of 1550 unrelated orthopedic patients had an unusual albumin pattern on agarose gel electrophoresis which was inherited as an autosomal co-dominant. The albumin, normal in concentration, consisted of a single broad zone extending cathodally to the α_1 region (see Fig. 8.3).

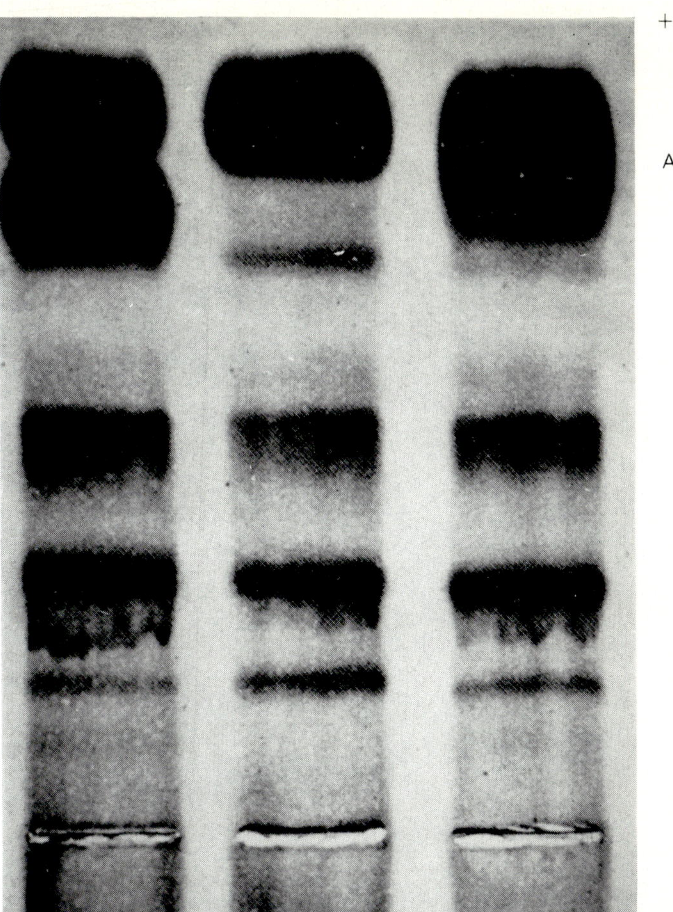

FIGURE 8.3. Photograph of protein-stained agarose electrophoresis of three serum specimens. 1: bisalbuminemia of the 'European' type. 2: normal pattern 3: the unusual albumin variant whose two-dimensional pattern was shown in Fig. 8.2. (From Laurell & Niléhn, 1966. Reprinted with permission from the *J. clin. Invest.* **45**, p. 1935.)

[*Facing page* 244]

In starch gel, however, the albumin did not appear altered, but there was an unusual component similar to that described by Fraser *et al.* (1959) which migrated slightly ahead of transferrin, and had a concentration much lower than that of albumin. The piece of starch gel containing this component was placed in an agarose gel containing anti-albumin, and subsequent electrophoresis into the agarose showed that the unusual protein reacted immunologically as albumin. Examination of normal serum revealed a minute amount of similarly-reacting protein, consistent with the earlier report of Poulik *et al.* (1960). The slow-moving minor component of both normal and unusual sera disappeared after treatment with mercaptoethanol.

Laurell & Niléhn (1966) separated the total albumin of the anomalous serum chromatographically into two components, one consisting of normal albumin and the other of a more positively charged component with about a third of the normal concentration. This finding proved that the anomaly under study was actually bisalbuminemia, as suggested by the electrophoretic pattern in agarose gel, although the electrical charge difference was not sufficient to separate normal and aberrant proteins completely. The authors concluded that the minor, slow-moving component usually seen in normal serum on starch gel electrophoresis is probably dimerized mercaptalbumin, and that the larger amount seen in the aberrant serum is due to a greater tendency of the variant albumin to dimerize. A similar tendency is found in the hemoglobin variant, Hb Porto Alegre (p. 385), in which cysteine replaces a serine residue in the β globin chain.

The possibility that this unusual albumin might be associated with a connective tissue defect was suggested by Laurell & Niléhn, since the anomaly was found in five of 1550 orthopedic cases and in only one of 3200 controls. However, the family evidence presented was insufficient to provide more than slight support for this hypothesis.

GENETIC LINKAGE

Cooke *et al.* (1961) reported that in two families with albumin variants, they were unable to find evidence of linkage with the loci of the 'common' blood group systems or the plasma proteins (unspecified) tested. However, Weitkamp *et al.* (1966) produced good evidence for linkage between the structural loci of albumin and the Gc protein. In the two families they studied, both having the 'very slow' variant, there were 42 informative offspring, of whom only one appeared to be a recombinant. The estimated recombination fraction of 0·015 indicated that the two loci are probably very close together. The authors suggested that this close linkage between the two serum proteins raises the possibility that their genes were derived from a single locus by unequal crossing over and duplication. Biochemical analysis is required to assess this possibility.

Subsequently, Weitkamp *et al.* (1967b) showed that mutant albumin genes occur in coupling with either Gc^1 or Gc^2, as would be expected on theoretical grounds. Moreover, Kaarsalo *et al.* (1967) presented additional evidence for linkage between the two loci from studies of four Naskapi and Montagnais Indian families in whom a fast-moving albumin variant was segregating. The calculated recombination fraction was similar to that reported by Weitkamp *et al.* (1967b) for the slow variant.

Weitkamp *et al.* (1967a) recently reported that they had studied the family with the unusual albumin variant originally described by Fraser *et al.* (1959) (see previous section). The fact that they observed no recombination between this albumin variant and the Gc proteins indicated that the genes determining at least three albumin variants are almost certainly alleles. Such a conclusion is consistent with the assumption of Laurell & Niléhn (1966) that the unusual variant represents a mutation at the albumin structural locus.

METHODS

Variant albumin phenotypes, except for the kind described by Fraser *et al.* (1959) and Laurell & Niléhn (1966), are detectable by most electrophoretic separation methods. With starch gel electrophoresis, either horizontal or vertical, the buffer system used should limit the breadth of the albumin zone. Thus, the discontinuous system of Poulik (p. 102) is preferable to the borate buffer originally described by Smithies (1955). However, the studies of Weitkamp *et al.* (1967b) indicate that two buffer systems are preferable for achieving maximal separation and for comparative studies of two or more variants. In the first system (described by Kueppers & Bearn, 1966a) the cathodal electrode vessel contains 0·125 M sodium acetate and 0·011 M ethylenediaminetetraacetic acid (EDTA), and the anodal vessel contains the same solution at half concentration. The gel buffer is

0·031 M sodium acetate and 0·004 M EDTA. All solutions have a pH of 4·95.

In the second system (described by Welser in a personal communication to Weitkamp *et al.*, 1967b) both electrode vessels contain 0·03 M tris, 0·043 M EDTA and 0·27 M boric acid (pH 6·4), while the gel is made with a 1:6·2 dilution of the bridge buffer. Samples are applied in thin slits, and vertical electrophoresis is carried out for 6 hours at 12 volts/cm in the cold. The sliced gels are stained with amido black, as described in Chapter 2.

Blumberg *et al.* (1968) reported satisfactory separation of albumin bands in the five different variant phenotypes they studied with the discontinuous buffer system described by Ashton and Braden (see p. 234).

THE PROTEASE INHIBITOR (Pi) SYSTEM: α_1-ANTITRYPSIN

The term 'protease inhibitor' is applied to a group of serum proteins which have the property of inhibiting the activity of proteolytic enzymes such as trypsin, plasmin and thrombin. The protease inhibitors which have been subjected to the most investigation (see Ganrot, 1967) are in the α_1 and α_2 fractions. The latter, a macroglobulin, has not been found to exhibit electrophoretic polymorphism. The former, commonly called α_1-antitrypsin, is the subject of this section.

PHYSICAL PROPERTIES OF α_1-ANTITRYPSIN

The protease inhibitor effect of α_1-antitrypsin is due to the formation of an inactive complex with the proteolytic enzyme (Rimon *et al.*, 1966). First given the name of α_1-3·5-glycoprotein for its sedimentation rate (Schultze *et al.*, 1955), this protein was later identified as the major trypsin inhibitor of serum by Schultze *et al.* (1962), who reported a molecular weight of 60,000. Subsequently, Laurell & Eriksson (1963) showed that the α_1 band seen on paper or agar electrophoresis of serum is largely composed of the protease inhibitor. The average normal concentration of α_1-antitrypsin, measured by immunochemical methods, is about 235 mg/100 ml according to Augener (1965) and about 180 mg/100 ml according to Laurell & Eriksson

(1963). A linear increase occurs during pregnancy (Faarvang & Lauritsen, 1963), reaching twice the normal concentration at the time of parturition (Ganrot, 1967). Increased levels have also been noted in response to steroid administration and the stress of surgery (Faarvang & Lauritsen, 1963). In patients with renal homografts, a drop in serum antitrypsin activity is said to be an indicator of graft rejection (Tyler, 1964).

PHENOTYPIC VARIANTS OF α_1-ANTITRYPSIN

Laurell & Eriksson (1963) found that the serum of a small proportion of hospital patients contained very low levels of α_1 anti-trypsin (10 per cent of normal or less), and there was an apparent association of this deficiency and pulmonary disease. Subsequent family studies showed that the deficient state represents homozygosity for a mutant gene, and that heterozygotes have about half normal activity (Eriksson, 1964, 1965; Kueppers *et al.*, 1964; Lopez *et al.*, 1964; Talamo *et al.*, 1966; Ganrot *et al.*, 1967). Laurell (1965a) showed that the deficient serum contained a small amount of protein recognizable immunologically as α_1-antitrypsin, and that it had a migration rate slightly different from that of the normal protein. In heterozygotes both a major (normal) protein and a minor (deficient) protein could be visualized, using the technique of antigen-antibody crossed electrophoresis (Laurell, 1965b). This technique also revealed that each component was heterogeneous, and that the pattern of this heterogeneity differed. Thus, the deficient protein was altered not only in quantity but also in structure.

The existence of other α_1-antitrypsin variants with normal or nearly normal concentration in serum was first reported by Eriksson & Laurell (1963). They described a slow-moving variant subsequently shown to be inherited as an autosomal co-dominant, and its gene was found to be allelic to the gene determining the deficient variant (Axelsson & Laurell, 1965).

Fagerhol & Braend (1965) reported that the electrophoresis of serum in starch gel prepared in a pH 5 buffer permitted the demonstration of several bands with variable patterns in the prealbumin region, indicating the existence of at least five phenotypes. These prealbumin zones were recognized as α_1-antitrypsin by Kueppers & Bearn (1966a), Fagerhol & Braend (1966) and Fagerhol & Laurell (1967). The latter authors reported the results of testing a number of sera with variant phenotypes by starch gel electrophoresis, by agarose gel electrophoresis, and by antigen-antibody crossed electrophoresis. In the latter technique, the initial electrophoresis was performed in the acid starch gel, and a slice of gel containing the prealbumin components was cut out and inserted into a slot cut in agarose gel containing α_1-antitrypsin antiserum. Electrophoresis in the agarose gel, perpendicular to the previous direction, was then carried out. The resulting pattern provided a 'profile' of the α_1-antitrypsin, as shown in Figure 8.4,

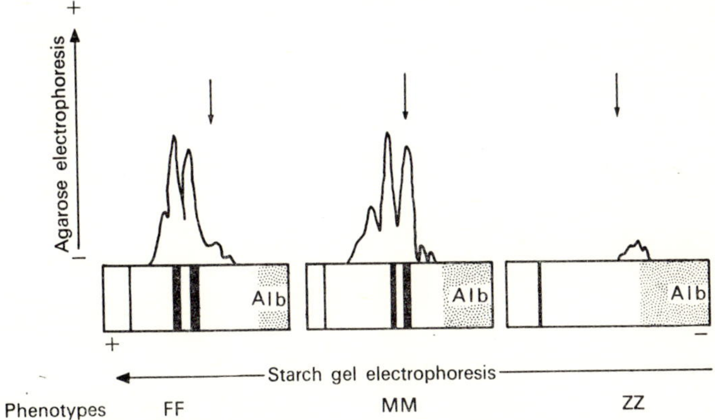

FIGURE 8.4. Diagram of three Pi phenotypes as seen on starch gel and on crossed electrophoresis, in which the agarose contains anti-Pi antibodies. The leading portion of albumin is indicated by Alb. The short arrows point to the same position on each gel, showing the relative migration rates of the variants, FF (fast) and ZZ (slow, deficient) compared with the common phenotype, MM. (Adapted from Fagerhol & Laurell, 1967. Reprinted with permission from *Clinica chim. Acta* **16**, 199.)

adapted from Fagerhol & Laurell (1967). The three phenotypes compared in the drawing are the common type, called MM, a rare type, FF, and the ZZ type, which designates α_1-antitrypsin deficiency. These three types represent homozygosity for the common gene, Pi^M, and the rare genes Pi^F and Pi^Z, respectively. The position of the fourth peak in the MM phenotype is marked with an arrow to show the faster mobility of the F components and the slower mobility of the faint Z components.

According to Fagerhol (1968), there is now evidence for the existence of at least seven alleles at the Pi locus, two having products which migrate faster than that of the common Pi^M gene and four having slower-moving products. In order of decreasing mobility of their respective protein products, they are Pi^F, Pi^I, Pi^M, Pi^S, Pi^V, Pi^X and Pi^Z. An eighth allele, Pi^W, is also postulated. Figure 8.5, from Fagerhol (1968), indicates diagrammatically the appearance of the twelve phenotypes which have been observed in Norwegians.

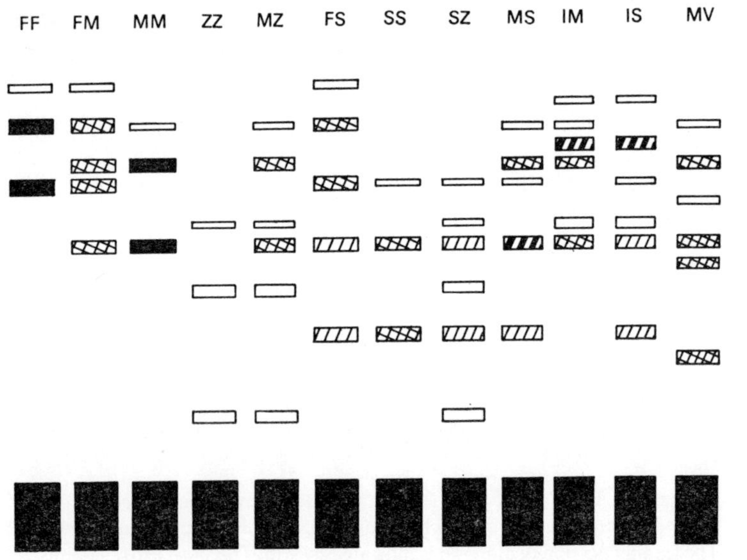

FIGURE 8.5. Diagram of twelve Pi phenotypes, migrating ahead of albumin on starch gel electrophoresis at acid pH. (From Fagerhol, 1968. Reprinted with permission from *Series Haematologica*, **1**, 153.)

FREQUENCY OF THE Pi GENES

Very little information is available on the Pi gene frequencies in various populations because many of the phenotypes are difficult to differentiate. Fagerhol (1968) reported the data presented in Table 8.1, obtained by testing 2803 blood donors and pregnant women in Norway. The common Pi^M gene has a frequency of about 0·95, and three other genes (Pi^S, Pi^Z and Pi^F) each appear to have a frequency somewhat greater than 1 per cent in that population. Further studies using the starch gel acid electrophoretic technique described below are required to determine if other populations exhibit more polymorphism.

TABLE 8.1. Pi gene frequencies in Norwegian, Swedish and American populations (Fagerhol, 1968)

Population	No. of persons	Pi^M	Pi^S	Pi^Z	Pi^F	Pi^I	Pi^V
Norwegian blood donors and pregnant women	2830	0·946	0·023	0·0157	0·0133	0·0012	0·0004
Swedish factory workers, blood donors and patients	3200				0·003		
Adult Swedes	6995			0·024			
White Americans	193			0·010			

PHYSIOLOGICAL ROLE OF α_1-ANTITRYPSIN

It is clear that only a very small proportion of patients with pulmonary emphysema have a deficiency of α_1-antitrypsin (Kueppers *et al.*, 1964; Eriksson, 1965; Talamo *et al.*, 1966). Furthermore, not all people known to have the deficiency develop emphysema. Kueppers & Bearn (1966b) attempted to find a basis for the association of the protein and the disease by testing the inhibitory activity of α_1-antitrypsin against several proteolytic enzymes. Finding that the protein inhibited proteolysis by elastase and also by the enzyme present in human leukocytes, they postulated that a function of the protease inhibitor might be to prevent the digestion of lung tissue during an inflammatory process.

METHODS

According to Fagerhol (1968) it is possible to separate the protease inhibitor variants by horizontal electrophoresis in a 12 per cent starch gel. Two-hundred ml of buffer is made by adding 186 ml distilled water to 14 ml of a stock solution, of which one liter contains 21 g citric acid and 23 g tris. The anodal vessel contains 11·75 g citric acid and 15·7 g $Na_2HPO_4.2H_2O$ per liter. The cathodal vessel contains 9·25 g boric acid and 3 g sodium hydroxide per liter. (The pH was not specified, but these solutions prepared in this laboratory have pH 4·8 (stock buffer) pH 4·5 (anodal buffer) and pH 9·0 (cathodal buffer)). The serum is inserted in filter paper strips, 1 to 2 mm thick. The initial voltage gradient is 8 to 10 volts per cm, and this is increased to about 15 volts/cm when the visible buffer zone has migrated one cm past the insertion (only possible with a water-cooled apparatus). Electrophoresis is stopped when the buffer zone has migrated 10 cm further. The sliced gel is stained with amido black.

CERULOPLASMIN

Ceruloplasmin is an α_2 globulin which contains more than 90 per cent of the plasma copper (Holmberg & Laurell, 1948). The protein is known to have oxidase activity *in vitro*, particularly with *p*-phenylenediamine as substrate (Holmberg & Laurell, 1951; Walaas *et al.*, 1967). Its physiological role has not yet been determined, although a copper storage role has been suggested (Bearn, 1966). The plasma concentration of ceruloplasmin in normal adults is about 35 to 48 mg/100 ml (Aisen *et al.*, 1960; Scheinberg & Sternlieb, 1963) with a half-life of about 4 days (Kekki *et al.*, 1966) while in most patients with Wilson's disease, the level is appreciably lower (Bearn & Kunkel, 1952). A small proportion of Wilson's disease heterozygotes also have low ceruloplasmin levels (Sternlieb *et al.*, 1961), but this measurement is not always a useful indicator of the carrier state, because there is evidence for another kind of familial ceruloplasmin deficiency, unrelated to Wilson's disease or any other obvious clinical abnormality (Cox, 1966).

PHYSICAL PROPERTIES OF CERULOPLASMIN

Ceruloplasmin is a blue colored glycoprotein with a 7 per cent carbohydrate content and a molecular weight of about 160,000 (or 143,000 for the protein moiety) (Kasper & Deutsch, 1963). Each molecule contains eight copper atoms (Holmberg & Laurell, 1951); two pairs in the cuprous, and two pairs in the cupric state (van Gelder & Veldsema, 1966). The heterogeneity of this protein has been recognized for several years (Broman, 1958), and numerous studies have been designed to determine whether the heterogeneity is due to the known instability of the molecule or to inherited alterations in the protein moiety (Morell & Scheinberg, 1960; Poulik, 1962, 1963; Poulik & Bearn, 1962; Kasper & Deutsch, 1963; Poillon & Bearn, 1966). These studies indicate that the protein is usually homogenous but that it consists of subunits, which upon partial dissociation, provide the appearance of heterogeneity. According to the tentative model of Poillon & Bearn (1966), ceruloplasmin may be an octomer with the molecular formula $\alpha_4 \beta_4$, in which α and β are similar subunits with a molecular weight of about 17,000 and a slight difference in net charge.

GENETIC VARIANTS OF CERULOPLASMIN

McAlister *et al.* (1961) described a fast-moving electrophoretic variant of ceruloplasmin inherited as a dominant trait in four generations of a Caucasian family in Seattle. However, in spite of intensive search, no further evidence of inherited variation was produced until the studies of Shreffler and his colleagues.

Shreffler *et al.* (1967) reported the occurrence of five different ceruloplasmin phenotypes among 576 American Negroes tested, and subsequently (Shokeir *et al.*, 1967), extended the number to seven. They found that by subjecting serum to electrophoresis in starch gel with a gel buffer pH of 9·5 and a tank buffer pH of 9·0, clear separations of most of these ceruloplasmin bands could be obtained. From studies of families and twin pairs, they obtained evidence for four alleles at the Cp locus. The fact that only seven of the ten possible

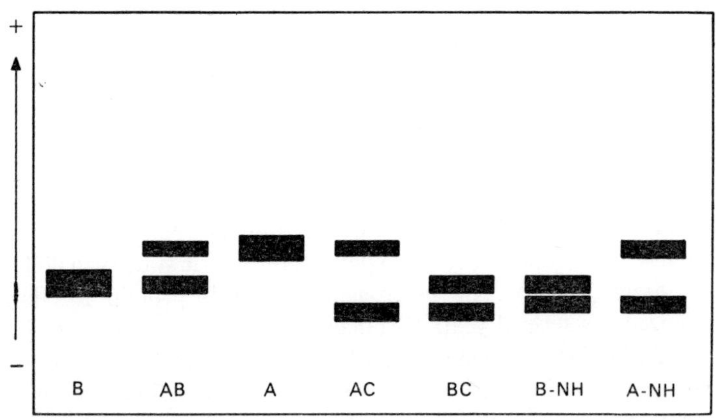

FIGURE 8.6. Diagram of the starch gel electrophoretic patterns of the seven ceruloplasmin phenotypes described by Shreffler *et al.* (1967) and Shokeir *et al.* (1967). The rare Cp phenotype B-NH is very difficult to differentiate from the common Cp B phenotype.

phenotypes were observed is due to the very low frequency of two of the four alleles.

Figure 8.6 depicts diagrammatically the seven phenotypes described by Shreffler *et al.* (1967) and Shokeir *et al.* (1967). The common phenotype, Cp B, represents homozygosity for the gene, Cp^B, which has a frequency of about 0·94 in Negroes and over 0·99 in the non-Negroes tested. The Cp^A gene, with a frequency of 0·05 in Negroes, was also found in the homozygous state, as designated by the phenotype Cp A. The genes Cp^C and Cp^{NH} (NH for New Haven) have very low frequencies, estimated at 0·004, and their gene products have been seen only in combination with Cp^B and Cp^A. The authors pointed out that it is very difficult to resolve the two bands in the Cp B-NH phenotype, so that the Cp^{NH} frequency may be higher than that of Cp^C.

Shreffler *et al.* (1967) compared the ceruloplasmin patterns in their serum specimens with the serum reported to contain an unusual ceruloplasmin (called IF) by McAlister *et al.* (1961). Not only did the Cp 1F protein have a slightly slower mobility than Cp A, but it consisted of a single component. The fact that members of the family had either Cp A or Cp 1F, and none had both ceruloplasmin bands indicated dominant, rather than co-dominant inheritance. Thus the Cp 1F phenotype is not only structurally different from Cp A, but it also appears to have arisen by a different kind of genetic mechanism.

Since the Cp AB phenotype occurs in about 10 per cent of American Negroes, it can be used as a genetic marker in that ethnic group. From preliminary studies, it appears that American Indians and Orientals probably have very low frequencies of the ceruloplasmin variants (Shreffler *et al.*, 1967). (See Addenda.)

METHODS

Shreffler *et al.* (1967) separated the ceruloplasmin variants with a horizontal starch gel electrophoretic system. The gel buffer is 0·016 M boric acid and 0·010 M sodium hydroxide, pH 9·5. The bridge buffer (pH 9·0) is 0·21 M boric acid and 0·085 M sodium hydroxide. Serum samples are diluted one to three and then subjected to electrophoresis at 4°C for 22 hours at 7 volts/cm.

Because of its oxidase activity, ceruloplasmin can be differentiated from other serum proteins by using a suitable substrate. Shreffler *et al.* (1967) used a modification of a staining method described by Owen & Smith (1961). The sliced gel is placed in a solution of 0·1 per cent orthodianisidine and 20 per cent ethanol in 0·04 M acetate buffer at pH 5·5. After incubation at 37°C for about an hour, the ceruloplasmin bands are examined over an X-ray film illuminator or any frosted glass with a light source. For permanent records, the gels are photographed without delay, because the ceruloplasmin zones tend to diffuse after a few hours. The gels can then be stained for the haptoglobin-hemoglobin complex by adding one volume of the ceruloplasmin staining solution, one-half volume of 0·1 M acetate buffer (pH 4·7) and one-fiftieth volume of 3 per cent hydrogen peroxide. Only 5–10 minutes are required for adequate staining.

THE Xm SYSTEM (α_2 MACROGLOBULIN)

A genetic system of great potential importance was described by Berg & Bearn (1966) who found that after absorption, an antiserum prepared by injecting a rabbit with human serum contained an antibody which reacted by immunodiffusion tests against the serum of some, but not all human subjects. To this extent, the system is similar to the Lp lipoprotein system, but the Xm locus is on the X chromosome, and its corresponding antigen is a part of, or is closely associated with, α_2 macroglobulin.

INHERITANCE OF Xm(a)

The antigen revealed by the immune serum was called Xm(a), and its gene, Xm^a. The phenotypes were designated as Xm(a+) and Xm(a−), and the gene allelic to Xm^a was assigned the tentative name Xm, since its product has not yet been demonstrated.

In studies of families, the phenotypes observed in the children were consistent with X-linked inheritance from their parents, except for one Xm(a−) daughter of a mating in which both parents were Xm(a+). Such a finding could be explained on the basis of the Lyon hypothesis if the mother of this girl was heterozygous Xm^a/Xm. Thus, if she transmitted her Xm gene to her daughter, the Xg^a gene inherited from the father might be inactivated in so many cells synthesizing the Xm protein that the Xm(a) antigen would not be detectable. Another possibility could be a delay in development of the Xm(a) antigen.

That the Xm(a+) phenotype may be a fairly late manifestation in girls was postulated by Berg & Bearn (1966) as the result of testing 145 children between the ages of 7 and 19 years. Based on the gene frequencies of the adults from the same geographic source (Easter Island), the 85 boys tested had close to the expected number of Xm(a+) and Xm(a−) phenotypes. However, of the 60 girls tested, 23 were Xm(a+) and 37 Xm(a−), while the expected numbers were 28 and 32, respectively. This small departure from expectation was accentuated in the younger girls, aged 7–10, who had none of the expected five examples of Xm(a+), and in the 10–13 year olds, who had 6 of the 9 expected. The older girls were not deficient in Xm(a+). Whether or not the trend indicated by these findings will be shown in other samples obtained from children remains to be determined.

Xm GENE FREQUENCIES

Four populations were sampled by Berg & Bearn (1966) to determine the Xm gene frequencies. The results, shown in Table 8.2, indicate very little difference among the populations tested. In frequency studies of the X-linked blood group antigen, Xg(a), the early results were similar to those reported for Xm(a), in that only small deviations were noted. It may be that with the exception of relatively isolated population groups, both Xg^a and Xm^a are maintained at fairly constant levels. No linkage studies have been reported, so it is not yet known whether the Xm locus is within mapping distance of any other recognizable character determined by genes on the X chromosome. (See Addenda).

TABLE 8.2. Distribution of Xm phenotypes in four populations (Berg & Bearn, 1966)

Population	Number	Xm(a+) in percent	Xm(a−) in percent	Xm^a gene frequency
Norwegian	100 (M)	23·00	77·00	0·23
	101 (F)	56·44	43·56	0·34
U.S. Whites	57 (M)	26·32	73·68	0·26
	67 (F)	50·75	49·25	0·30
U.S. Negroes	81 (M)	30·86	69·14	0·31
	151 (F)	59·60	40·40	0·36
Easter Island	66 (M)	24·24	75·76	0·24
	80 (F)	47·50	52·50	0·28

ALPHA$_1$-ACID GLYCOPROTEIN
(OROSOMUCOID)

Alpha$_1$-acid glycoprotein, or orosomucoid, has a carbohydrate moiety of over 50 per cent, composed of hexoses (galactose and mannose), hexosamine, a pentose (fucose) and sialic acid. The latter substance comprises over 15 per cent of the molecule, and is responsible for its very low isoelectric point of 2·7 (Schmid, 1953; Winzler, 1958). In the protein moiety, serine is the C-terminal residue (Schmid *et al.*, 1959), and the N-terminal residue is pyroglutamic acid (Ikenaka *et al.*, 1966).

Although the concentration of this protein is elevated in inflammatory and neoplastic diseases, its function remains unknown (Silberberg *et al.*, 1955; Weisman *et al.*, 1961).

Schmid & Binette (1961) and Schmid *et al.* (1962) reported that when purified α_1-acid glycoprotein was submitted to electrophoresis in starch gel at pH values above or below its 2·7 isoelectric point, only a single protein band was seen. However, at pH 2·7, the protein consisted of several bands, varying in number and staining intensity from one specimen to another. Schmid *et al.* (1964) tested the purified protein from serum specimens of two fairly large families and 18 pairs of twins. All specimens had either 5, 6, 7 or 8 bands, and these numbers were concordant among the identical twins. The inheritance patterns were difficult to determine.

TYPES I, II AND III

Tokita & Schmid (1963) found that when most of the sialic acid residues had been removed from the protein, the isoelectric point shifted to 5·2, and when electrophoresis of this material was performed at a pH of 4·8, one of three alternative patterns was observed. These patterns all consisted of one very faint band and two major bands, of which one or the other was decreased in quantity in type I and II; in type III, the two components were about equal. Schmid *et al.* (1965) made sialic acid-free preparations from the sera of 97 random donors and 18 sets of twins from the Boston area and 64 donors living in Japan. Among the American donors, they found 11, 46 and 43 per cent, respectively of types I, II and III. In the Japanese subjects, the percentages were 17, 63 and 20. Thus, if types I and II represented homozygosity and type III, heterozygosity for two different alleles, the Hardy-Weinberg equilibrium would be greatly disturbed in the Japanese group.

Schmid *et al.* (1965) compared the type (that is I, II or III) of the sialic acid-free protein in the 18 sets of twins, with the number of bands (that is 5, 6, 7 or 8) observed in the native protein in each case. No relationship could be seen, although the identical twins were concordant in each kind of electrophoretic pattern. As yet, no family studies on types I, II and III have been reported, so no conclusions can be drawn about their mode of inheritance. A recent paper by Schmid *et al.* (1967) indicates that the difference between the proteins of type I and II may be very minor, possibly a single amino-acid substitution.

Note: Two additional plasma proteins with inherited molecular variation are briefly described in the Addenda to this chapter. They are the clotting factor, fibrinogen, and the complement component, C_3' (β_1c). The gene frequencies in the latter system should make it a useful genetic marker.

REFERENCES

ADAMS M.S. (1966) Genetic diversity in serum albumin. *J. med. Genet.* **3**, 198.

ADNER P.L. & REDFORS A. (1961) A family with two serum-albumin fractions, distinguishable at electrophoresis (bisalbuminemia). *Nord. Med.* **65**, 623.

AISEN P., SCHORR J.B., MORELL A.G., GOLD R.Z. & SCHEINBERG I.H. (1960) A rapid screening test for deficiency of plasma ceruloplasmin and its value in the diagnosis of Wilson's disease. *Am. J. Med.* **28**, 550.

AUGENER W. (1965) Immunanalyse von Glykoproteinen. *XII Colloquium on Protides of Biological Fluids* (ed. PEETERS H.). p. 363. Elsevier, Amsterdam.

AXELSSON U. & LAURELL C.B. (1965) Hereditary variants of serum α_1-antitrypsin. *Am. J. hum. Genet.* **17**, 466.

BEARN A. (1966) Wilson's disease, *in* STANBURY J.B., WYNGAARDEN J.B. & FREDRICKSON D.S. (eds.) *The Metabolic Basis of Inherited Disease*, 2nd edn. p. 761. McGraw-Hill, New York.

BEARN A.G. & KUNKEL H.G. (1952) Biochemical abnormalities in Wilson's disease. *J. clin. Invest.* **31**, 616.

BELL H.E., NICHOLSON S.F. & THOMPSON Z.R. (1967) Bisalbuminemia of the fast type with a homozygote. *Clin. chim. Acta* **15**, 247.

BERG K. & BEARN A.G. (1966) An inherited X-linked serum system in man. The Xm system. *J. exper. Med.* **123**, 379.

BLUMBERG B.S., MARTIN J.R. & MELARTIN L. (1968) Alloalbuminemia. *J. Am. med. Ass.* **203**, 180.

BRAEND M., EFREMOV G., FAGERHOL M.K. & HARTMAN O. (1965) Albumin and transferrin variants in Norwegians. *Hereditas* **53**, 137.

BROMAN L. (1958) Separation and characterization of two ceruloplasmins from human serum. *Nature* **182**, 1655.

COOKE K.B., CLEGHORN T.E. & LOCKEY E. (1961) Two new families with bisalbuminemia: an exploration of possible links with other genetically controlled variants. *Biochem. J.* **81**, 39P.

COX D.W. (1966) Factors influencing serum ceruloplasmin levels in normal individuals. *J. Lab. clin. Med.* **68**, 893.

DRACHMAN O., HARBOE N.M.G., SVENDSEN P.H. & JOHNSON T.S. (1965) Bis- or paralbuminemia. A genetic alteration in plasma albumin. *Dan. med. Bull.* **12**, 74.

EARLE D.P., HUTT M.P., SCHMID K. & GITLIN D. (1958) A unique human serum albumin transmitted genetically. *Trans. Ass. Am. Physns.* **71**, 69.

EARLE D.P., HUTT M.P., SCHMID K. & GITLIN D. (1959) Observations on double albumin: a genetically transmitted serum protein anomaly. *J. clin. Invest* **38**, 1412.

EFREMOV G. & BRAEND M. (1964) Serum albumin: polymorphism in man. *Science* **146**, 1679.

ERIKSSON S. (1964) Pulmonary emphysema and alpha$_1$-antitrypsin deficiency. *Acta med. scand.* **175**, 197.

ERIKSSON S. (1965) Studies on α_1-antitrypsin deficiency. *Acta med. scand.* **177** (supp. 432), 1.

ERIKSSON S. & LAURELL C.B. (1963) A new abnormal serum globulin α_1-antitrypsin. *Acta chem. scand.* **17**, 5150.

FAARVANG H.J. & LAURITSEN O.S. (1963) Increase of trypsin inhibitor in serum during pregnancy. *Nature* **199**, 290.

FAGERHOL M.K. (1968) The Pi system: genetic variants of serum α_1-antitrypsin. *Series Haematologica* I, 1, 153.

FAGERHOL M.K. & BRAEND M. (1965) Serum prealbumin: polymorphism in man. *Science* **149**, 986.

FAGERHOL M.K. & BRAEND M. (1966) Classification of human serum prealbumin after starch gel electrophoresis. *Acta path. microbiol. scand.* **68**, 434.

FAGERHOL M.K. & LAURELL C.B. (1967) The polymorphism of 'prealbumins' and α_1-antitrypsin in human sera. *Clin. chim. Acta* **16**, 199.

FRANGLEN G., MARTIN N.H., HARGREAVES T., SMITH M.J. & WILLIAMS D.I. (1960) Bisalbuminemia, a hereditary albumin abnormality. *Lancet* i, 307.

FRASER G.R., HARRIS H. & ROBSON E.B. (1959) A new genetically determined plasma protein in man. *Lancet* i, 1023.

GANROT P.O. (1967) Studies on serum protease inhibitors with special reference to α_2-macroglobulin. *Acta Univ. lund.* II, No. 2, p. 5.

GANROT P.O., LAURELL C.B. & ERIKSSON S. (1967) Obstructive lung disease and trypsin inhibitors in α_1-antitrypsin deficiency. *Scand. J. clin. Lab. Invest.* **19**, 205.

GELDER B.J. VAN & VELDSEMA A. (1966) Different species of copper in ceruloplasmin. *Biochem. biophys. Acta* **130**, 267.

GITLIN D., SCHMID K., EARLE D.P. & GIVELBER H. (1961) Observations on double albumin. II. A peptide difference between two genetically determined human serum albumins. *J. clin. Invest.* **40**, 820.

HOLMBERG C.G. & LAURELL C.B. (1948) Investigations in serum copper. II. Isolation of the copper containing protein, and a description of some of its properties. *Acta chem. scand.* **2**, 550.

HOLMBERG C.G. & LAURELL C.B. (1951) Oxidase reactions in human plasma caused by ceruloplasmin. *Scand. J. clin. Lab. Invest.* **31**, 103.

IKENAKA T., BAMMERLIN H., KAUFMANN H. & SCHMID K. (1966) The amino-terminal peptide of α_1-acid glycoprotein. *J. biol. Chem.* **241**, 5560.

KAARSALO E., MELARTIN L. & BLUMBERG B.S. (1967) Autosomal linkage between the albumin and Gc loci in humans. *Science* **158**, 123.

KASPER C.B. & DEUTSCH H.F. (1963) Physiochemical studies of human ceruloplasmin. *J. biol. Chem.* **238**, 2325.

KEKKI M., KOSKELO P. & NIKKILÄ E.A. (1966) Turnover of iodine-131-labelled ceruloplasmin in human beings. *Nature* **209**, 1252.

KNEDEL M. (1957) Die Doppel-Albuminämie, eine neue erbliche Proteinanomalie. *Blut* **3**, 129.

KNEDEL M. (1958) Über eine neue vererbte Protein-Anomalie. *Clin. chim. Acta* **3**, 72.

KUEPPERS F. & BEARN A.G. (1966a) Inherited variations of human serum of α_1-antitrypsin. *Science* **154**, 407.

KUEPPERS F. & BEARN A.G. (1966b) A possible experimental approach to the association of hereditary α_1-antitrypsin deficiency and pulmonary emphysema. *Proc. Soc. exp. Biol. Med.* **121**, 1207.

KUEPPERS F., BRISCOE W.A. & BEARN A.G. (1964) Hereditary deficiency of serum α_1-antitrypsin. *Science* **146**, 1678.

LAURELL C.B. (1965a) Electrophoretic microheterogeneity of serum α_1-antitrypsin. *Scand. J. clin. Lab. Invest.* **17**, 1.

LAURELL C.B. (1965b) Antigen-antibody crossed electrophoresis. *Anal. Biochem.* **10**, 358.

LAURELL C.B. & ERIKSSON S. (1963) The electrophoretic α_1-globulin pattern of serum in α_1-antitrypsin deficiency. *Scand. J. clin. Lab. Invest.* **15**, 132.

LAURELL C.B. & NILÉHN J.E. (1966) A new type of inherited serum albumin anomaly. *J. clin. Invest.* **45**, 1935.

LOPEZ V., OETLIKER O., COLOMBO J.P. & BÜTLER R. (1964) Ein Fall von familiärem α_1-Antitrypsin-Mangel. *Helv. paediat. Acta* **19**, 296.

MCALISTER R., MARTIN G.M. & BENDITT E.P. (1961) Evidence for multiple caeruloplasmin components in human serum. *Nature* **190**, 927.

MELARTIN L. & BLUMBERG B.S. (1966) Albumin Naskapi: a new variant of serum albumin. *Science* **153**, 1664.

MELARTIN L., BLUMBERG B.S. & LISKER P. (1967) Albumin Mexico, a new variant of serum albumin. *Nature* **215**, 1288.

MIESCHER F. (1960) Neues Vorkommen der vererbbaren Doppelalbuminämie. *Schweiz. med. Wschr.* **90**, 1273.

MORELL A.G. & SCHEINBERG I.H. (1960) Heterogeneity of human ceruloplasmin. *Science* **131**, 930.

NENNSTIEL H.J. & BECHT T. (1957) Über das erbliche auftreten einer albumin-spaltung im electrophorescdiagramm. *Klin. Wschr.* **35**, 689.

OWEN J.A. & SMITH H. (1961) Detection of ceruloplasmin after zone electrophoresis. *Clin. chim. Acta* **6**, 441.

POILLON W.N. & BEARN A.G. (1966) The molecular structure of human ceruloplasmin. *Biochim. biophys. Acta* **127**, 407.

POLESKY H.F. & ROKALA D.A. (1967) Serum albumin polymorphism in North American Indians. *Nature* **216**, 184.

POULIK M.D. (1962) Electrophoretic and immunological studies on structural sub-units of human ceruloplasmin. *Nature* **194**, 842.

POULIK M.D. (1963) Heterogeneity and structural subunits of human ceruloplasmin. *Proc. 10th Colloquium on Protides of the Biological Fluids* (ed. PEETERS H.) p. 170. Elsevier, Amsterdam.

POULIK M.D. & BEARN A.G. (1962) Heterogeneity of ceruloplasmin. *Clin. chim. Acta* **7**, 374.

10

POULIK M.D., ZUELZER W.W. & MEYER R. (1960) Separation of human serum albumins. *Nature* **188**, 506.

RIMON A., SHAMASH Y. & SHAPIRO B. (1966) The plasmin inhibitor of human plasma. *J. biol. Chem.* **241**, 5102.

ROBBINS J.L., HILL G.A., MARCUS S. & CARLQUIST J.H. (1963) Paralbuminemia: paper and cellulose acetate electrophoresis and preliminary immunoelectrophoretic analysis. *J. Lab. clin. Med.* **62**, 753.

SANDOR G., MARTIN L., PROSIN M., ROUSSEAU A. & MARTIN R. (1965) A new bisalbuminaemic family. *Nature* **208**, 1222.

SARCIONE E.J. & AUNGST C.W. (1962) Studies in bisalbuminemis: binding properties of the two albumins. *Blood* **20**, 156.

SCHEINBERG I.H. & STERNLIEB I. (1963) Wilson's disease and the concentration of ceruloplasmin in serum. *Lancet* **ii**, 1420.

SCHEURLEN P.G. (1955) Über Serumweissenveränderungen beim Diabetes mellitus. *Klin. Wschr.* **33**, 198.

SCHMID K. (1953) Preparation and properties of serum and plasma proteins. XXIX. Separation from human plasma of polysaccharides, peptides and proteins of low molecular weight. Crystallization of an acid glycoprotein. *J. Am. chem. Soc.* **75**, 60.

SCHMID K., BENCZE W.L., NUSSBAUMER T. & WEHRMÜLLER J.O. (1959) Studies on the structure of α_1-acid glycoprotein. *J. biol. Chem.* **234**, 529.

SCHMID K. & BINETTE J.P. (1961) Polymorphism of α_1-acid glycoprotein. *Nature* **190**, 630.

SCHMID K., BINETTE J.P., KAMIYAMA S., PFISTER V. & TAKAHASHI S. (1962) Studies on the structure of α_1-acid glycoprotein. III. Polymorphism of α_1-acid glycoprotein and the partial resolution and characterization of its variants. *Biochemistry* **1**, 959.

SCHMID K., BINETTE J.P., TOKITA K., MOROZ L. & YOSHIZAKI H. (1964) The polymorphic forms of α_1-acid glycoprotein of normal Caucasian individuals. *J. clin. Invest.* **43**, 2347.

SCHMID K., POLIS A., HUNZIKER K., FRICKE R. & YAYOSHI M. (1967) Partial characterization of the sialic acid-free forms of α_1-acid glycoprotein from human plasma. *Biochem. J.* **104**, 361.

SCHMID K., TOKITA K. & YOSHIZAKI H. (1965) The α_1-acid glycoprotein variants of normal caucasian and Japanese individuals. *J. clin. Invest.* **44**, 1394.

SCHULTZE H.E., GÖLLNER I., HEIDE K., SCHÖNENBERGER M. & SCHWICK G. (1955) Zur Kenntnis der α-Globuline des menschlichen Normalserums. *Z. Naturf.* **10b**, 463.

SCHULTZE H.E., HEIDE K. & HAUPT H. (1962) α_1-antitrypsin aus Humanserum. *Klin. Wschr.* **40**, 427.

SHOKEIR M.H., SHREFFLER D.C. & GALL J.C. (1957) Further electrophoretic variation in human ceruloplasmin. *Meeting Am. Soc. hum. Genet.* Toronto.

SHREFFLER D.C., BREWER G.J., GALL J.C. & HONEYMAN M.S. (1967) Electrophoretic variation in human serum ceruloplasmin: a new genetic polymorphism *Biochem. Genet.* **1**, 101.

SILBERBERG S., GOODMAN M., KEFALIDES N.A. & WINZLER R.J. (1955) Immunochemical determination of orosomucoid. *Proc. Soc. exp. Biol. Med.* **90**, 641.

SMITHIES O. (1955) Zone electrophoresis in starch gels: group variations in the serum proteins of normal adult human subjects. *Biochem. J.* **61**, 629.

STERNLIEB I., MORELL A.G., BAUER C.D., COMBER B., DEBOBES-STERNBERG S. & SCHEINBERG I.H. (1961) Detection of the heterozygous carrier of the Wilson's disease gene. *J. clin. Invest.* **40**, 707.

TALAMO R.C., BLENNERHASSETT J.B. & AUSTEN F. (1966) Familial emphysema and alpha$_1$-antitrypsin deficiency. *New Engl. J. Med.* **275**, 1301.

TÁRNOKY A.L. & LESTAS A.N. (1964) A new type of bisalbuminemia. *Clin. chim. Acta* **9**, 551.

TOKITA K. & SCHMID K. (1963) Variants of α_1-acid glycoprotein. *Nature* **200**, 266.

TYLER H.M. (1964) Trypsin inhibitor levels and homograft rejection. *Clin. Sci.* **26**, 315.

UNGARI S. & LOPEZ V. (1965) Doppia albuminemia. *Minerva Pediat.* **17**, 288.

WALAAS E., LOVSTAD R.A. & WALAAS O. (1967) Interaction of dimethyl-p-phenyl-enediamine with ceruloplasmin. *Arch. Biochem. Biophys.* **121**, 480.

WEISMAN S., GOLDSMITH B., WINZLER R. & LEPPER M.H. (1961) Turnover of plasma orosomucoid in man. *J. Lab. clin. Med.* **57**, 7.

WEITKAMP L.R., ROBSON E.B. & SHREFFLER D.C. (1967a) An unusual type of albumin variant: further data on genetic linkage between loci for human serum albumin and the group specific component (Gc). *Meeting Am. Soc. hum. Genet* Toronto, p. 43.

WEITKAMP L.R., RUCKNAGEL D.L. & GERSHOWITZ H. (1966) Genetic linkage between structural loci for albumin and group specific component (Gc). *Am. J. hum. Genet.* **18**, 559.

WEITKAMP L.R., SHREFFLER D.C., ROBBINS J.L., DRACHMANN O., ADNER P.L., WIEME R.J., SIMON N., COOKE K.B., SANDOR G., WUHRMANN F., BRAEND M. & TÁRNOKY A.L. (1967b) An electrophoretic comparison of serum albumin variants from nineteen unrelated families. *Acta genet.* **17**, 399.

WIEME R.J. (1960) On the presence of two albumins in certain normal human sera and its genetic determination. *Clin. chim. Acta* **5**, 443.

WIEME R.J. (1962) On the occurrence of different types in genetically determined bisalbuminemia. *Proc. 9th Colloquium on Protides of the Biological Fluids*, Bruges, p. 222. ed. H. Peeters, Elsevier, Amsterdam.

WINZLER R.J. (1958) Chemistry and biology of mucopolysaccharides, *in* WOHLSTONHOLME G.E.W. & O'CONNOR M. (eds.) *Ciba Symposium on Chemistry and Biology of Mucopolysaccharides*, p. 245. J.A. Churchill, Ltd., London.

WUHRMANN F. (1959) Albumindoppelzacken als vererbbare Bluteiweissanomalie. *Schweiz. med. Wschr.* **89**, 150.

PART 2

GENETIC MARKERS IN

BLOOD CELLS

CHAPTER 9

THE RED CELL ANTIGENS:
BLOOD GROUPS

◈

General Considerations.................. 268
  Blood group antigens 268
  Transglycosylase enzymes as
  blood group gene products..... 270
  The problem of multiple anti-
  genic determinants 271
  Nomenclature 272
  Inheritance of blood group
  antigens..................................... 276
  'Minus-minus' phenotypes 278
The ABO System and its Genetic
Interactions.. 278
  The ABO, Hh, Secretor and
  Lewis systems 278
    A, B and H antigens 278
      Secretion of water-
      soluble A, B and H
      substances 280
      Cellular localization of
      A, B and H substances 280
    The Lewis system 281
    Structural relationships of
    the A, B, H and Lewis
    antigens 284
    Sequential gene activity in
    blood group antigen bio-
    synthesis 288
      Biosynthesis of se-
      creted substances......... 288
      Biosynthesis of red
      cell H, A and B anti-
      gens 290
      Blockage of biosyn-
      thesis: 'Bombay' and
      other rare phenotypes 290
    Changes in phenotype
    associated with disease..... 292
  Interactions of ABO and
  Secretor genes with gene determining plasma alkaline
  phosphatase phenotype 292
    Summary of observations 292

    Suggested mechanisms..... 294
The Ii system and its relation-
ship to ABH antigens.............. 296
  The antigens I and i.............. 296
  Serological association of
  I and ABH.............................. 297
  Possible association of I
  antigen and leukemia.......... 298
Possible Biosynthetic Pathways
in Other Blood Group Systems..... 298
  The P system............................. 298
    Relationship of P, Ii and
    ABH antigens........................ 300
  The MNSs system........................ 301
    Chemical characteristics
    of M and N substances..... 302
    The antigen U........................ 302
  The Rh system............................ 303
    The LW antigen.................... 304
    The Rh_{null} phenotypes..... 305
    Studies on Rh antigen
    structure 309
    Quantitative studies on
    Rh active sites........................ 310
Blood Groups and Genetic Link-
age 311
  Autosomal linkage 311
  X chromosome linkage: Xg^a 312
Blood Groups and Selection.......... 314
  Disease association 314
  Maternal-fetal incompatibil-
  ity 314
  Prezygotic selection.................... 316
Some Blood Group Gene Fre-
quencies 317
Blood Grouping Methods.............. 320
  General principles 321
  The antiglobulin technique..... 322
  Neutralization tests.................... 323
  Pitfalls 323
  Automated blood grouping..... 324
References... 325

The term 'blood groups' could be appropriately applied to all of the polymorphic systems in the blood, including the serum proteins and red cell enzymes, but by convention, the term is reserved for inherited antigens detected on the red cell surface by specific antibodies. The importance of the blood group antigens in blood transfusion and in the pathogenesis of hemolytic disease of the newborn has been recognized for several years (see Mollison, 1967; Allen & Diamond, 1957) but it is only recently that the participation of elaborate gene-interaction sequences in the production of these antigens has been appreciated.

For detailed descriptions of all blood group systems, including their inheritance and applicability to genetic problems, the reader is urged to consult the complete, authoritative and highly-readable text of Race & Sanger (Fourth edition, 1962; Fifth edition, 1968). The purpose of the present chapter is to provide a general introduction to the subject and to discuss some of its aspects as they apply to human genetic polymorphism.

GENERAL CONSIDERATIONS

BLOOD GROUP ANTIGENS

Table 9.1, which was adapted from Race & Sanger (1962), Rosenfield (1967), and Allen (1967), lists the human blood group systems and their respective antigens. The assignment of a newly discovered antigen to a previously unrecognized system is contingent upon the demonstration of independent assortment of its allele from those of the known systems, by studying selected families (Race & Sanger, 1959). Useful information is also derived from testing blood specimens from various racial groups, since some of the antigens are virtually restricted to individuals of a given geographic origin (Mourant, 1954). In some instances, the frequency of an antigen in all populations tested is either so high or so low that finding appropriate families for determining its relationship to the known systems is extremely difficult. The designations 'public' and 'private' have been applied to these antigens (and their genes).

Although by definition, the blood group antigens are red cell reactive sites detected by specific antibodies, certain antigenic

TABLE 9.1. Blood group antigens of human red cells (Adapted from Race & Sanger, 1962; Rosenfield, 1967; Allen, 1967)

System	Positive reactions with a specific antibody	No specific antibody, but characteristic serological reaction
	Antigens detected by	
ABO, H	A_1, B, H, AB (a determinant common to A and B) IH, IA, IB, iH (ABO-Ii interactions)	A_2, A_3
I, i	I, i IP, (I-P interaction), Luke (ABO-P interaction)	I_{int}
P	P_1, P, P^k	
MNSs	M, N, S, s, U, M^g, M_1, M^A, M^e, Tm, Sj, Mi^a, Vw, Mur, Hil, Hu, He, Vr, Vr^a, St^a, Md^a, Mt^a, Ri^a, Cl^a, Ny^a, Sul, Sj, S^B, U^B	M_2, N_2, M^c
Rh, LW	See Table 9.3 and pp. 304–309	D^u, C^u, E^u
Lutheran	Lu^a, Lu^b, Lu^{ab} ? Au^a	
Kell	See Table 9.4	
Lewis	Le^a, Le^b, Le^{ab}, Magard (ABO-Lewis interaction)	
Duffy	Fy^a, Fy^b	
Kidd	Jk^a, Jk^b, Jk^{ab}	
Xg	Xg^a	
Diego	Di^a, Di^b	
Dombrock	Do^a	
Cartwright	Yt^a, Yt^b	
Cost	Cs^a	

Possibly independent systems

Auberger	Au^a (? in Lutheran system)
Sciana	Sm, Bu^a
DBG	Ho, Ho-like, Ot, DBG
Gonzales	Go^a, Go^b (probably in the Rh system)

Not classified into systems:
Frequency very high
At^a, Co^a, Gursha-Chido, Ge, Yussef, Gy^a, Lan, Eldridge, Vel

Frequency very low
Levay, Wr^a, Be^a, By, Sw^a, Good, Biles, Tr^a, Wb, Bp^a, Rd^a, Ls^a, Box, Or, Ht^a, Gf, Wu, Radden, Evans, Reid, Heidl, Jn^a, Torkildsen, Kennedy

determinants, similar to or identical with those of the red cell, are also found in secreted substances. Indeed, in one instance (i.e. the Lewis system), their presence on the red cell is simply due to passive adsorption from the surrounding plasma (Sneath & Sneath, 1955; Mäkelä *et al.*, 1967). Thus, the term 'blood group substance' is applied to any macromolecule on or off the red cell which contains, as a part of its structure, specific antigenic sites recognizable by their reactions with blood group antibodies. Such macromolecules are widely distributed in both the plant and animal kingdoms (Springer, 1967). In the case of the ABH, Lewis, P and MN antigens, the antigenic specificity resides in the carbohydrate moiety of the molecule (Watkins & Morgan, 1952, 1962; Kabat & Leskowitz, 1955; Springer & Ansell, 1958; Morgan & Watkins, 1962; Hotta & Springer, 1965; Uhlenbruch, 1965). Thus, these antigens cannot be direct gene products, because they consist of monosaccharide, rather than amino acid, subunits.

TRANSGLYCOSYLASE ENZYMES AS BLOOD GROUP GENE PRODUCTS

It is reasonable to assume that when the antigenic determinants are carbohydrates, the genetic polymorphism actually involves (protein) enzymes which transfer specific sugar units from the sugar-nucleotides to the end of a growing carbohydrate chain. Such a mechanism has been demonstrated in the synthesis of glycoproteins and glycolipids in the cell walls of bacteria, some of which cross react with antibodies against human blood group substances (Springer *et al.*, 1961, 1964, 1966b; Osborn *et al.*, 1964; Lüderitz *et al.*, 1966; Anderson *et al.*, 1966). Furthermore, enzymes in mammary tissue which catalyse the sequential addition of sugars in the formation of human milk oligosaccharides are known to exist (Grollman *et al.*, 1965). The direct association of one such enzyme, a fucosyl transferase, with secreted blood group substance has been recently established (Shen *et al.*, 1968). Furthermore, radioactive sugars attached to uridine diphosphate (UDP) or guanosine diphosphate (GDP) have been shown to be incorporated into blood group substances by microsomes of mammalian gastric mucosa (Tuppy & Staudenbauer, 1966; Grollman & Marcus, 1966; Ziderman *et al.*, 1967). Nevertheless, the enzymes have not been purified, and thus their structure is completely unknown. As yet, their presence can only be inferred from the results of tests on their end products.

An important unsolved problem associated with the transferase enzyme hypothesis is the nature of the mutational process at a given locus which alters the recognition site of the gene product so that it transfers another monosaccharide with a different structure (Boettcher 1966). However, as will be seen in the discussion of the ABO system, there is only a minor difference in the structure of the A and B antigenic determinants, so that the corresponding genes and their enzyme products may differ at only a single site (Watkins, 1966).

THE PROBLEM OF MULTIPLE ANTIGENIC DETERMINANTS

A growing complication of blood group genetic interpretation, particularly characteristic of the Rh, MNSs and Kell systems, is the inheritance from one parent of two or more antigens belonging to one system. For example, in the Rh system, a child may inherit the antigens, C, D and e from one parent and c, D and E from the other parent. He, in turn, will pass these antigens on to his offspring in precisely the same groups of three. (An apparent example of crossing over within the Rh system has been reported by Steinberg (1965), but such an occurrence must be extremely rare.) According to Fisher's original linked-gene hypothesis (see Race & Sanger, first edition, 1950), this mode of transmission is due to very close linkage of three loci, the genes of which determine the antithetical antigens C or c, D or d and E or e, respectively. On the other hand, a single Rh locus is postulated by Wiener (Wiener & Wexler, 1958) who visualizes a series of alleles, each having its own agglutinogen which in turn, possesses a number of Rh antigenic determinants, or factors (see Fig. 9.1). A similar dilemma in the interpretation of Gm inheritance was discussed in Chapter 1.

These differences in concept cannot be reconciled at present (Shreffler, 1967a,b). A gene (i.e. an allele) is currently considered as a chromosomal segment, the DNA of which determines the structure of a single polypeptide chain. This DNA is itself a polynucleotide chain which thereby contains many mutable sites. Thus, changes due to point mutations, recombination or chromosomal rearrangement, could bring about alterations in two or more sites within a single molecule of the peptide product. For example, in the case of the Gm system of IgG molecules, it is possible to envisage a series of mutational events giving rise to multiple antigenic determinants within a single poly-peptide chain. Since the known blood group antigenic determinants are composed of sugars rather than amino acids, it is difficult to see how the enzyme product of a single gene could bring about multiple differences in carbohydrate molecular conformation. However, in the case of the Rh system, there is evidence that the antigenic specificity resides in the protein moiety of a phospholipid-protein complex (Whittemore *et al.*, 1967; Green, 1965, 1967a,b).

NOMENCLATURE

As in many other human genetic systems, blood group nomenclature is inconsistent. A given antigen and its respective gene are usually given the same name, and, in print, the gene is designated by italics. Several kinds of nomenclature are used. For example, in the ABO system, the three major alleles are named *A*, *B* and *O*, the letter O indicating that there is no recognizable gene product. Alternatively these genes are called I^A, I^B and I^0, in accordance with the use of a single symbol to designate a given locus (Wiener & Wexler, 1958). Several subtypes of A and B are recognized, depending upon their serological reactions. The most important of these subtypes are A_1 and A_2, whose corresponding genes are called A^1 and A^2 (or A_1 and A_2). Red cells of A_2 phenotype (genotype A^2O or A^2A^2) are not recognized directly by a specific antibody, but indirectly, through their agglutinability by anti-A, but not by anti-A_1. Other examples of this kind of antigen are listed in Table 9.1.

In the MNSs system, two alternative antigens are named with successive letters M and N, but the other two are given capital and small letters, S and s. Their respective genes (or gene combinations, depending on the genetic interpretation outlined previously for the Rh system) are *MS*, *Ms*, *NS* and *Ns*. The single letter L (for Landsteiner) has also been used to designate this locus, with the appropriate MNSs superscripts (Wiener & Wexler, 1958). A large number of additional antigens of low frequency have been assigned to alleles at this locus, most of them named according to the first two letters of the name of the original propositus (e.g. Hu for Hunter, He for Henshaw, etc).

Rh nomenclature is particularly confusing (see Table 9.2). According to the theory of R.A. Fisher based on the observations of Race & Sanger (1950), the antithetical antigens and their corresponding genes are C and c, E and e, and D and (d), where d refers not to a specific antigen, but to lack of agglutination by anti-D. These antigens or 'factors' in Wiener's classification are rh′ and hr′; rh″ and hr″, and Rh_0, presented in bold face type (Wiener *et al.*, 1949). There is no equivalent to d, since the factors, shown in Fig. 9.1, are multiple

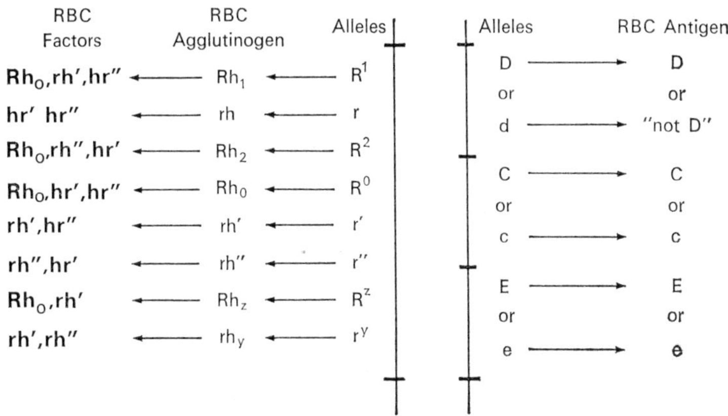

FIGURE 9.1. Diagram representing the Rh multiple-allele theory (*left*) and the linked-gene theory (*right*). In the multiple-allele theory, there is a series of eight alternative genes (alleles), each of which has its own agglutinogen, which is represented by two or more antigenic factors on the red cell. In the linked gene theory, there are three closely-linked genetic sites, each of which has two alternative genes which are expressed by antigens of the same name. A complex of three linked genes in this theory is equivalent to a single gene in the other theory.

serological attributes of the agglutinogens, which are determined by the genes (Wiener, 1966). In referring to the Rh alleles, it is convenient to utilize the names in Table 9.2 proposed by Wiener and subsequently adopted in a slightly different form by Race & Sanger to indicate the Rh gene complex. In recognition of the clinical importance of the potent D antigen, its corresponding gene complexes are designated with an upper case R and an appropriate superscript.

In an entirely different nomenclature system proposed by Rosenfield *et al.* (1962) numbers are employed to designate the various Rh

TABLE 9.2. Comparison of Rh nomenclature according to (a) multiple allele theory and (b) linked gene theory

(a)		(b)	
Allele*	Antigens	Gene complex	Antigens
R^1	Rh_0, rh', hr″	*DCe*	D, C, e
r	hr', hr″	*dce*	c, e
R^2	Rh_0, hr', rh″	*DcE*	D, c, E
R^0	Rh_0, hr', hr″	*Dce*	D, c, e
$r″$	hr', rh″	*dcE*	c, E
r'	rh', hr″	*dCe*	C, e
R^z	Rh_0, rh', rh″	*DCE*	D, C, E
r^y	rh', rh″	*dCE*	C, E

* Listed in order of frequency in western Europe (Race & Sanger, 1962)

antigenic determinants, and these are listed in Table 9.3. Failure to react with a specific antibody is indicated by a minus sign. For example, red cells agglutinated by anti-D, anti-C and anti-e, but not by anti-E

TABLE 9.3. Numerical designation of Rh antigens according to Rosenfield *et al.*, 1962

Antigens	Synonyms in Other Nomenclature Systems		Antigens	Synonyms in Other Nomenclature Systems	
Rh 1	D	Rh_0	Rh 15		Rh^C
2	C	rh'	16		Rh^D
3	E	rh″	17		Hr_0
4	c	hr'	18		Hr
5	e	hr″	19		hr^s
6	ce (f)	hr	20	VS (e^s)	
7	Ce (rh_i)	rh	21	C^G	
8	C^W	rh^{W1}	22	CE	
9	C^X	rh^X	23	D^{Wiel}	
10	V(ce^s)	hr^V	24	E^t	
11	E^W	rh^{W2}	25		LW
12	G	rh^G	26		Deal
13		Rh^A	27	cE	
14		Rh^B	28		hr^H

or anti-c are given the phenotype Rh: 1, 2, −3, −4, 5. Such a designation avoids interpretation at the gene level, and is particularly useful for recording results when certain antibodies are employed which recognize 'compound antigens' (e.g. anti-Rh 6 reacts with ce (f) and anti-Rh 7 reacts with Ce (rh$_i$)). Using this nomenclature, alleles are designated by an italicized R with appropriate numerical superscripts. However, 'shorthand' symbols, such as r'^n or r^{yn} are also employed (Rosenfield *et al.*, 1964a; Morton & Rosenfield, 1967).

Antigens of the Kell system, like those of the Rh system, appear to consist of three antithetical sets: K and k, Kp^a and Kp^b and Js^a and Js^b (Coombs *et al.*, 1946; Levine *et al.*, 1949; Allen & Lewis, 1957; Allen *et al.*, 1958; Giblett, 1958; Walker *et al.*, 1963). The latter two antigens are so-named because the original studies of Js^a by Giblett & Chase (1959) indicated that it probably belonged to a separate blood group system, called Sutter (J and S were the initials of the patient in whom the antibody was found). *Js^b*, the apparent allele of *Js^a*, was subsequently reported by Walker *et al.* (1963), and it was not until the

TABLE 9.4. Numerical designation of Kell antigens according to Allen & Rosenfield (1961) and Rosenfield (1967)

Antigens	Synonyms
K 1	K
2	k
3	Kp^a
4	Kp^b
5	Ku
6	Js^a
7	Js^b
8	Kw
9	'Kell'

work of Stroup *et al.* (1965) and Morten *et al.* (1965) that these alleles were assigned to the Kell locus. The numbers used by Allen & Rosenfield (1961) and by Rosenfield (1967) to designate the antigens of the Kell system are shown in Table 9.4.

The antigens (and their genes) in several other systems are indicated by using letters from the name of the person in whom the antibody, rather than the antigen, was first detected. In addition to Js^a and Js^b, these include Fy^a and Fy^b in the Duffy system (Cutbush & Mollison, 1950; Ikin *et al.*, 1951), Lu^a and Lu^b in the Lutheran system (Callender & Race, 1946; Cutbush & Chanarin, 1956), Jk^a and Jk^b in the Kidd system (Allen *et al.*, 1951; Plaut *et al.*, 1953), Yt^a and Yt^b in the Cartwright system (Eaton *et al.*, 1956; Giles & Metaxas, 1964) and Di^a and Di^b in the Diego system (Layrisse *et al.*, 1955; Thompson *et al.*, 1967). Xg^a is the only X-linked antigen known; the system name is Xg (Mann *et al.*, 1962). Cs^a and Do^a represent single antigens in the separate systems Cs and Dombrock (Giles *et al.*, 1965; Swanson *et al.*, 1965b); while Bu^a (Anderson *et al.*, 1963) and Sm (Schmidt *et al.*, 1962) are apparently antithetical antigens in one blood group system, Sciana, so that they should be appropriately renamed (Lewis *et al.*, 1964, 1967).

INHERITANCE OF BLOOD GROUP ANTIGENS

With the apparent exception of I and i, antigens intrinsic to the red cell (i.e. those which are not acquired by adsorption) are usually inherited as simple Mendelian dominant characters, detectable in single dose. All of the blood group genes except the X-linked Xg^a (Mann *et al.*, 1962) have autosomal loci, but assignment to specific chromosomes has not yet been achieved. Since the 22 autosomes are paired, an individual is either homozygous or heterozygous for the alleles at these loci. In some instances it is possible to predict whether a particular blood group gene is present on only one or on both homologous chromosomes on the basis of dosage effect, i.e. the strength of red cell agglutination by a specific antibody. However, since there is a wide range in normal agglutinability, even within a given specificity, such predictions are often inaccurate. Furthermore, in certain systems, the serological reactivity of an antigen differs in accordance with the presence or absence of another antigen. For example, when the genetic determinants of the Rh antigens D and C are on the same chromosome (i.e. in cis position), the activity of D is weakened. Trans-position effects of C and D can also occur. For example, in some families, red cells of the genotype R^1r' (*CDe/Cde*) react very poorly with anti-D, but in other members of the family, red cells having the same R^1 (*CDe*) gene partnered with r (*cde*), i.e. of genotype R^1r, are strongly agglutinated by the same anti-D (Ceppellini *et al.*, 1955; Chown & Lewis, 1957; Gibbs & Rosenfield, 1966). (*Note:* Although mature erythrocytes do not have chromosomes, it is customary to refer to their genotypes as if the cells still possessed the nucleus present in their precursors.)

When an individual has both antithetical antigens of a given system (such as A and B, K and k, E and e, etc) on his red cells, it is usually safe to assume that he has inherited one antigen from each parent, and thus his genotype can be stated with some degree of confidence. There are, however, exceptional instances in which a child seems to have inherited two antithetical antigens from one parent, such as C and c or E and e (Rosenfield *et al.*, 1960b, 1964a; Morton & Rosenfield, 1967). This phenomenon could occur if, for example, the individual were trisomic for the chromosome carrying that locus, assuming that non-disjunction in the heterozygous parent had occurred during the first meiotic division. However, there is as yet no convincing evidence to support this mechanism in any of the families reported to have anomalous blood group inheritance. Another possibility is blood group mosaicism, in which the individual has two cell populations, each possessing its own set of antigens. This phenomenon is well-known, its major causes being intrauterine grafting of blood-forming tissue between non-identical twins (Dunsford *et al.*, 1953; Booth *et al.*, 1957; Nicholas *et al.*, 1957; van der Hart & van Loghem, 1967) or fertilization of two egg nuclei by two sperm with subsequent development of a single individual (Gartler *et al.*, 1962; Zuelzer *et al.*, 1964; Myhre *et al.*, 1965).

Still a third possibility is unequal crossing over during meiosis in a heterozygote, with the formation of 'hybrid' or 'fusion' genes analogous to the Lepore hemoglobin genes or the Hp^2 genes in the haptoglobin system. Such a mechanism is very difficult to envisage for a blood group system in which the antigenic determinants are carbohydrates. It would require the new gene product, still a transferase enzyme molecule, to have two recognition sites, enabling the enzyme to attach two different monosaccharides to a macromolecular substrate (Grollman, 1967). Morton and his colleagues (1966b) did not speculate about the nature of fine structure changes at the gene level which might be responsible for the coupling as 'bivalents' of two antigenic factors usually found to segregate in repulsion. Analysis of data obtained from a large number of Brazilian Indian families led these authors to conclude that there was evidence for the existence of a *Cc* bivalent in the Rh system. Evidence for bivalence in the ABO system is summarized by Race & Sanger (1968).

'MINUS-MINUS' PHENOTYPES

When an individual possesses only one of two antithetical antigens within a given system, it is common practice to assume that he is a homozygote; indeed, this is usually the case. However, it is hazardous to make a confident statement about the genotype without studying the family, because in nearly every blood group system, a so-called 'minus-minus' phenotype has been reported (reviewed in Race & Sanger, p. 355, 1968). Such phenotypes are characterized by failure of the red cells to react with antibodies recognizing either of the antithetical antigens. In most instances (notably excepting the dominant Lu(a − b −) character), the individual appears to be homozygous for a gene with no recognizable product. A very common example is found in the Duffy system; about 70 per cent of Negroes have the phenotype Fy(a − b −). In matings involving one parent of this phenotype and the other of phenotype Fy(a + b +), the children are either Fy(a − b +) or Fy(a + b −) (Sanger *et al.*, 1955). In testing the blood of such children, it would be fallacious to infer the genotypes Fy^b/Fy^b and Fy^a/Fy^a, since only one gene with a recognizable (antigenic) product is inherited. The nature of the 'silent' gene in this and most of the other systems is unknown (i.e. whether it represents deletion, suppression, inversion, etc or a gene with an as yet unrecognized product). However, in the ABO system, biochemical studies have led to a logical explanation for at least one 'minus-minus' type, the so-called Bombay or O_h phenotype.

THE ABO SYSTEM AND ITS GENETIC INTERACTIONS

THE ABO, Hh, SECRETOR AND LEWIS SYSTEMS

A, B AND H ANTIGENS

A, B and *O* are the three major alleles of the system which bears their name. Disregarding subtypes, the recognizable phenotypes based on reactions with anti-A and anti-B antibodies are A, B, AB and O, representing the genotypes A/A, A/O, B/B, B/O, A/B and O/O. Since the O gene has no known product, its presence can be directly ascertained only in the homozygote, whose red cells are not agglutinated by either anti-A or anti-B. Using routine agglutination tests, it is impossible to differentiate between the genotype A/A and A/O or B/B and B/O in the absence of family studies.

A unique property of the ABO system is the almost invariable presence in the serum of anti-A and/or anti-B when the red cells lack the corresponding antigen. Those rare individuals who do not follow this rule nearly always have some anomaly such as a weak A (or B) subtype, red cell mosaicism or hypogammaglobulinemia.

Cells of group O are strongly agglutinated by anti-H, an antibody which sometimes occurs in the serum of A_1 or A_1B subjects. For laboratory purposes, a better source is the extracts of certain seeds (lectins) such as those of *Ulex europeus* or *Lotus tetragonolobus*. Such extracts can be used to differentiate the A_1 and A_2 subtypes, since A_2 cells have more H antigen and thus react strongly, while A_1 cells react weakly or not at all. There is no specific 'anti-A_2' antibody, and the difference between the A_1 and A_2 antigens is largely quantitative, with a wide range in the number of red cell reactive sites (Cohen & Zuelzer, 1965; Rosenfield & Rubenstein, 1966). Nevertheless, it is possible to prepare specific anti-A_1 antibodies, either by absorbing the serum of a group B person with A_2 cells, or by using a lectin such as that prepared from the seeds of *Dolichos biflorus*. Furthermore, the serum of some A_2 and A_2B individuals contains agglutinins with anti-A_1 specificity.

The red cells of newborn infants who carry an A^1 gene are often found to react weakly or not at all with anti-A_1 of human origin, and are less subject to lysis by hemolytic anti-A antibodies (Witebsky & Engasser, 1949; Crawford *et al.*, 1953a). The A activity of A_2 cells is similarly decreased. In view of the susceptibility of some sugar transferase enzymes in the fetal liver to suppression by maternally produced hormones (Arias *et al.*, 1964), it is conceivable that the 'immaturity' of certain newborn red cell antigens is due to similar suppression of the transferases involved in antigen biosynthesis.

It was first thought that the H antigen was the product of the O gene, and that A_2 cells, which are less strongly agglutinated than A_1 cells by anti-A antibodies, represented the genotype A/O. However, since there is an A^2/B genotype, and since such cells react well with anti-H, this proposal was clearly incorrect. Chemical studies on H substance were not available at that time, but Witebsky & Klendshoj (1941) and later, Morgan & Watkins (1948) suspected that it could be the precursor of A and B. It is now generally accepted that H antigen activity is controlled by a gene, *H*, whose locus is not linked with that of *A*, *B* and *O* (Race & Sanger, 1962). The allele of *H* is called *h*, and it has no known product. Since the *h* gene frequency is very low, individuals with the *h/h* genotype, associated with the 'Bombay' phenotype O_h, are extremely rare (see below).

SECRETION OF WATER SOLUBLE A, B AND H SUBSTANCES
Water soluble substances with A, B and H activity are present in a number of body fluids, particularly saliva, gastric juice and meconium. The ability to secrete substances with these specificities is controlled by a secretor gene, *Se* whose allele, *se* has no known function. Most of the studies on the biochemical basis of blood group specificity have been performed on purified extracts of ovarian cyst fluid. The extensive investigations of Morgan & Watkins (reviewed by Watkins in 1964, 1966, 1967a,b; and by Morgan, 1963, 1964) and by Kabat and his associates (Kabat, 1956; Lloyd *et al.*, 1966a,b, 1967), have shown that the secreted substances are glycoprotein macromolecules. They are more fully described later in this chapter.

Individuals who lack a *Se* gene are often referred to as 'non-secretors', but in fact they do secrete large amounts of glycoprotein which, while lacking ABH activity, cross-reacts with an antiserum to the Type XIV pneumococcal polysaccharide. Furthermore, when such a 'non-secretor' has a Lewis gene (to be discussed), his secretions contain Lewis-active substance. Thus, the function of the *Se* gene is associated with the formation of H, A and B determinants in the water-soluble glycoproteins. Its association with the serum alkaline phosphatase phenotypes was discussed in Chapter 7.

CELLULAR LOCALIZATION OF A, B AND H SUBSTANCES
Studies with fluorescent anti-A, anti-B and anti-H antibodies on the tissues of ABH secretors (Glynn *et al.*, 1957; Szulman, 1960, 1962; Kent, 1964) have shown the presence of ABH specific substances in the salivary glands, in columnar and goblet cells of gastro-intestinal mucosa and in the endocervix of the uterus. In ABH non-secretors (i.e. genotype *se/se*), there is no impairment of ABH activity on the

red cells, indicating independence of this expression from the secretor gene. Furthermore, although ABH activity in non-secretors of ABH is missing from the salivary and intestinal cells typical of the secretor localization, it is readily demonstrated in the deeper layers of the pyloric and intestinal mucosa. Regardless of secretor status, small amounts of specific substances can be demonstrated in nearly all plasma specimens (Høstrup, 1962, 1963) as well as on the leukocytes (Anderson & Walford, 1963) and epidermal cells (Coombs *et al.*, 1956; Swinburne *et al.*, 1961; Yunis & Yunis, 1963). Flory (1966) in a study of buccal cells, detected at least two different H receptor sites, one of which was common to all buccal cells tested, while the other was found only on the cells of secretors.

There is good reason to regard A and B as tissue transplantation antigens, since tissue obtained from ABO incompatible donors has a high probability of rejection (Dausset & Rapaport, 1966; Ceppellini *et al.*, 1966; Kuhns *et al.*, 1966). Also, Szulman (1965, 1966, 1967) has proposed an intriguing theory about the possible function of the blood group substances in organogenesis. This idea was based on his observations that in embryonic life, there is an orderly recession of the ABH antigens from the epithelial cells of most organs coinciding with their morphologic differentiation.

The ABH blood group substances extracted from red cell stroma contain no amino acids, but are composed of glycolipids, in which carbohydrate chains similar to or identical with those of the secreted substances are joined, through sphingosine, to fatty acids (Koscielak, 1963; Watkins, Koscielak & Morgan, 1964). Although red cell glycolipids with A and B specificity have been isolated and purified, it has not yet been possible to obtain a similar preparation with H activity from group O cells, even though such cells react strongly with anti-H antibodies.

THE LEWIS SYSTEM

The main contributions to the currently accepted ideas about Lewis system genetics were made by Grubb (1948, 1951), Watkins & Morgan (1959) and Ceppellini (1955, 1959). In this system there are two allelic genes, *Le* and *le*; the latter has no known function, and is thus analogous to the *O*, *h* and *se* genes. The two Lewis antigens, Le[a] and Le[b], are present on secreted glycoprotein molecules which are only secondarily adsorbed onto the red cell surface. Thus, it is possible to change the phenotype of red cells, for example, from Le(a + b −) to Le(a + b +) and then Le(a − b +) by injecting them into the bloodstream of an individual with the Le(a − b +) phenotype (Sneath & Sneath, 1955; Mollison *et al.*, 1963).

The Le^b antigen does not have a corresponding gene, but appears to be an interaction product of the *Le* and *H* genes. However, since H substance is only found in the secretions when a *Se* gene is present, Le^b is never found in the absence of the *Se* gene.

The *Le* gene is associated with the production of Le^a substance, whether the genotype is *Le/Le* or *Le/le*. When the *H* and *Se* genes are also present, much of the Le^a substance is converted into Le^b, so the secretions of such subjects contain Le^a, Le^b and H substances (as well as A and/or B, depending on the ABO genotype). When a sufficient amount of Le^a substance is present, both antigens can be detected on the red cells. However, (except in infants) Le^b usually predominates, so that the red cell phenotype of people with *H*, *Le* and *Se* genes is Le(a − b +). As would be expected, individuals who lack either the *H* gene (as in the rare Bombay type) or the *Se* gene (as in ABH non-secretors) but who do have the *Le* gene produce only Le^a substance, and their red cell phenotype is Le(a + b −). In the absence of an *Le* gene (i.e. in *le/le* homozygotes), neither Le^a nor Le^b is produced, so the red cells are Le(a − b −) and the secretions contain either ABH substance (in ABH secretors) or no blood group substance (in non-secretors). These findings are summarized in Table 9.5.

During fetal life, the *Le* gene is apparently not very active, so that at birth, red cell reactions with anti-Le^a are very weak, while reactions with anti-Le^b are not observed for several weeks to months (Andresen, 1948; Jordal, 1956; Brendemoen, 1961). In babies who have *H*, *Le* and *Se* genes, it is usually possible to observe a sequence of red cell phenotypes beginning with Le(a + b −), then Le(a + b +) and finally, Le(a − b +) occurring over a variable period of months to years (Cutbush *et al.*, 1956; Lawler & Marshall, 1961).

A phenomenon resembling epistasis is also observed in the Lewis system, in that the plasma of group O and A$_2$ individuals contains more Lewis substance than that of group A$_1$ (Mäkelä & Mäkelä, 1956). Thus O and A$_2$ red cells are more readily agglutinated than A$_1$ cells by certain examples of anti-Le^a and anti-Le^b antibodies (Andresen, 1948). In fact, determination of the red cell Le^b phenotype is notoriously subject to error, and it is generally better to infer the Le(a − b +)

TABLE 9.5. Relationships of phenotypes and genotypes in the (AB)H, Lewis and Secretor Systems (Adapted from Watkins, 1966)

Genotype			Phenotype							
			Secretions			Red Cells				
Hh	Lewis	Secretor	ABH	Le^a	Le^b	ABH	Le^a	Le^b	Lewis Designation	Frequency
H/H or H/h	Le/Le or Le/le	Se/Se or Se/se	+++	+	++	+++	− or ±	++	Le(a − b +)	0·735
H/H or H/h	Le/Le or Le/le	se/se	−	+++	−	+++	+++	−	Le(a + b −)	0·231
H/H or H/h	le/le	Se/Se or Se/se	+++	−	−	+++	−	−	Le(a − b −)	0·028
H/H or H/h	le/le	se/se	−	−	−	+++	−	−	Le(a − b −)	0·006
h/h	Le/Le or Le/le	Se/Se or Se/se	−	+++	−	−	+++	−	Le(a + b −)	Very rare
h/h	Le/Le or Le/le	se/se	−	+++	−	−	+++	−	Le(a + b −)	Very rare
h/h	le/le	Se/Se or Se/se	−	−	−	−	−	−	Le(a − b −)	Very rare
h/h	le/le	se/se	−	−	−	−	−	−	Le(a − b −)	Very rare

Gene Frequencies: Western European (Race & Sanger, 1962)

H nearly 1·000 Le 0·816 Se 0·523
h very rare le 0·184 se 0·477

phenotype from the presence of both Le[a] and H substance in the saliva rather than from the results of testing red cells or saliva with anti-Le[b]. If the saliva is tested for its ability to inhibit anti-Le[b], it is necessary to take into consideration that there are two kinds of anti-Le[b]: anti-Le[bH], which is inhibited by the saliva of all ABH secretors, regardless of Lewis genotype; and anti-Le[bL], which is inhibited only by saliva of ABH secretors who carry the Le gene (Race & Sanger, 1962).

Anti-Lewis antibodies occur more frequently in women during the reproductive period, and they are especially common during pregnancy (Kissmeyer-Nielsen, 1965; author's unpublished data). Conversely, the Le[a] and Le[b] activity on the red cells of pregnant women is decreased (Brendemoen, 1952; Rosenfield et al., 1960a). Hemolytic disease of the newborn due to anti-Lewis antibodies probably never occurs, because not only are the antigens poorly expressed at birth, but the antibodies are nearly always IgM globulin, and therefore do not cross the placental barrier (Polley et al., 1962; Adinolfi et al., 1962).

STRUCTURAL RELATIONSHIPS OF THE
A, B, H, AND LEWIS ANTIGENS

Since it is not yet possible to synthesize molecules with blood group activity, information about the structure of haptenic sites of blood group substance has been derived on the one hand from testing the products of enzymatic degradation and of acid and alkaline hydrolysis for antigenic activity, and on the other hand from measuring the neutralizing effect of various oligosaccharides on specific antibodies. The details of this extensive work formed the basis for the monograph by Kabat (1956) and more recent studies have been reviewed by Schiffman & Marcus (1964), Morgan (1963, 1964), Watkins (1964, 1966, 1967a,b), Szulman (1966) and Lloyd et al. (1966a,b, 1967). The following summary was mainly derived from their papers.

CHEMICAL CHARACTERIZATION OF
SECRETED BLOOD GROUP SUBSTANCES

Ovarian cyst fluid has provided most of the human blood group substances subjected to chemical analysis. These purified secreted substances are glycoproteins with an average molecular weight of about 300,000. A peptide 'backbone' comprises only about 15 per cent of the total molecule; proline, serine, alanine and threonine make up about two-thirds of the amino acids. While the peptide structure is undoubtedly under genetic control, it plays no direct role in determining A, B, H or Lewis serological activity. The carbohydrate moiety is composed of L-fucose (Fuc), D-galactose (Gal), *N*-acetylglucosamine (GNAc), *N*-acetylgalactosamine (GalNAc) and *N*-acetylneuraminic acid (NANA). The latter is a nine carbon sugar with no apparent participation in the activity of these blood group antigens. The sugars are arranged in a large number of fairly short

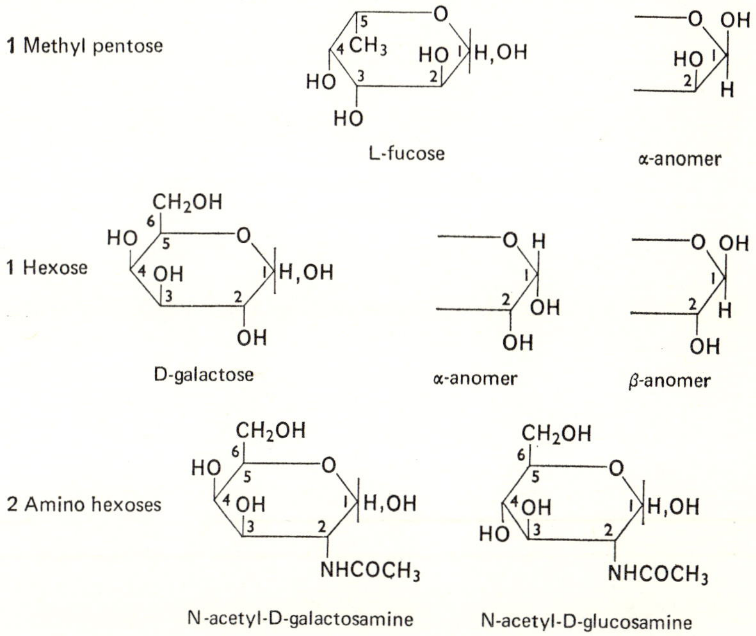

1 Methyl pentose

L-fucose α-anomer

1 Hexose

D-galactose α-anomer β-anomer

2 Amino hexoses

N-acetyl-D-galactosamine N-acetyl-D-glucosamine

FIGURE 9.2. The four sugars (one methyl pentose, one hexose and two amino hexoses) which determine antigenic specificity of A, B, H, Le^a and Le^b blood group substances. The positions of hydrogen and hydroxyl groups in the α and β anomers of D-galactose are also applicable to the anomers of *N*-acetyl-D-galactosamine.

chains attached by covalent linkage to the peptide. The blood group antigenic specificity is associated with the terminal, non-reducing ends of these chains, and involves the four sugars shown in Fig. 9.2. The figure also shows the difference between the α and β anomers of D-galactose (also applicable to GalNAc) and the α configuration of L-fucose. While GNAc does not occupy a terminal position, it is considered to be a part of the complete serological determinant of H, A, B and Lewis specificities.

The secretions of individuals with the genotype *le/le, se/se* have no ABH or Lewis activity, but they do contain glycoproteins of low fucose content which cross react immunologically with Type XIV pneumonococcal polysaccharide. The oligosaccharide chains of this secreted substance have a subterminal GNAc and a terminal Gal on the non-reducing end. In some chains (Type I) these two sugars are joined by $\beta(1\rightarrow3)$ linkage, while in Type II chains they are joined by $\beta(1\rightarrow4)$ linkage (see Fig. 9.3).

In glycoproteins with H activity, the chains are branched, due to the attachment of an α-L-fucosyl residue by $(1\rightarrow2)$ linkage to the terminal Gal of either Type I or Type II chains. In molecules with blood group A activity, the terminal sugar is GalNAc, which is attached by $\alpha(1\rightarrow3)$ linkage to the terminal Gal of the H-active chain. In molecules with blood group B activity, the added sugar is Gal. Thus, the difference between A and B specificity is associated with the chemical substituent on the second carbon of the terminal galactose; namely, a hydroxyl group in B and an *N*-acetylamino group in A (see Fig. 9.2). The fact that the terminal Gal and GalNAc of B and A substance, respectively, provide the only examples of alpha linkage in the unbranched portion of the chain suggests that $\alpha(1\rightarrow3)$ linkage may function as a 'stopper' to the enzyme-mediated growth of the carbohydrate chains (Watkins in Discussion of Grollman, 1967).

Lloyd *et al.* (1966b) observed that while the unbranched trisaccharide, α-D-GalNAc-$(1\rightarrow3)$-β-D-Gal-$(1\rightarrow3)$-GNAc was effective as an inhibitor of A precipitation by anti-A, a similar oligosaccharide containing in addition a terminal α-L-fucosyl branch (like that of the H-specific chain) had much greater inhibiting activity. However, since the H chain itself showed little inhibiting ability, the authors concluded that the fucosyl branch is not a part of A specificity, but rather, it plays a role in stabilizing a preferred conformation of the trisaccharide which represents the actual A antigenic determinant. The potential importance of molecular conformation in determining the specificity not only of oligosaccharides but also of antigens in other chemical classes was emphasized by Kabat (1966).

The trisaccharide, 2'-fucosyllactose, which is α-L-Fuc-$(1\rightarrow2)$$\beta$-D-Gal-$(1\rightarrow4)$-D-Glu, contains the two terminal sugars of the H chain. This trisaccharide is found in the milk of human mothers who are secretors of ABH, but not in the milk of non-secretors (Grollman & Ginsberg, 1967). Since synthesis of the trisaccharide was shown to

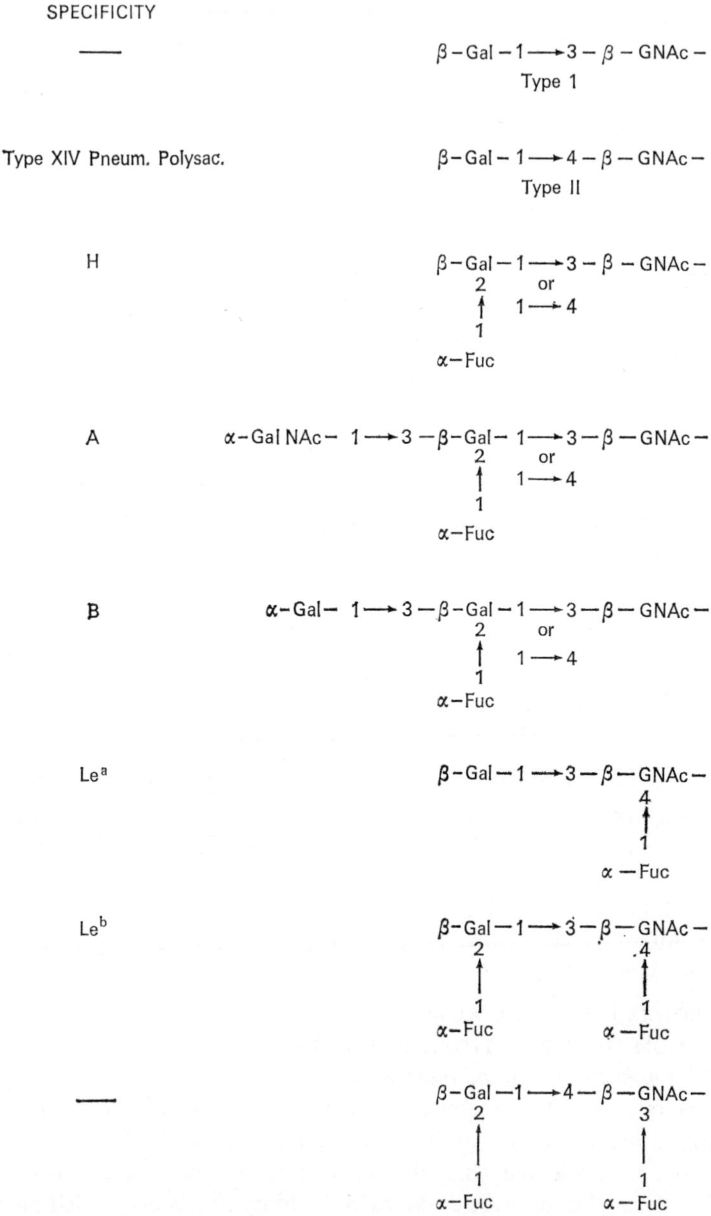

SPECIFICITY

——

$$\beta - \text{Gal} - 1 \longrightarrow 3 - \beta - \text{GNAc} -$$

Type 1

Type XIV Pneum. Polysac.

$$\beta - \text{Gal} - 1 \longrightarrow 4 - \beta - \text{GNAc} -$$

Type II

H

$$\beta - \text{Gal} - 1 \longrightarrow 3 - \beta - \text{GNAc} -$$
$$2 \qquad \text{or}$$
$$\uparrow \quad 1 \longrightarrow 4$$
$$1$$
$$\alpha - \text{Fuc}$$

A

$$\alpha - \text{Gal NAc} - 1 \longrightarrow 3 - \beta - \text{Gal} - 1 \longrightarrow 3 - \beta - \text{GNAc} -$$
$$2 \qquad \text{or}$$
$$\uparrow \quad 1 \longrightarrow 4$$
$$1$$
$$\alpha - \text{Fuc}$$

B

$$\alpha - \text{Gal} - 1 \longrightarrow 3 - \beta - \text{Gal} - 1 \longrightarrow 3 - \beta - \text{GNAc} -$$
$$2 \qquad \text{or}$$
$$\uparrow \quad 1 \longrightarrow 4$$
$$1$$
$$\alpha - \text{Fuc}$$

Le^a

$$\beta - \text{Gal} - 1 \longrightarrow 3 - \beta - \text{GNAc} -$$
$$4$$
$$\uparrow$$
$$1$$
$$\alpha - \text{Fuc}$$

Le^b

$$\beta - \text{Gal} - 1 \longrightarrow 3 - \beta - \text{GNAc} -$$
$$2 \qquad\qquad 4$$
$$\uparrow \qquad\qquad \uparrow$$
$$1 \qquad\qquad 1$$
$$\alpha - \text{Fuc} \qquad \alpha - \text{Fuc}$$

——

$$\beta - \text{Gal} - 1 \longrightarrow 4 - \beta - \text{GNAc} -$$
$$2 \qquad\qquad 3$$
$$\uparrow \qquad\qquad \uparrow$$
$$1 \qquad\qquad 1$$
$$\alpha - \text{Fuc} \qquad \alpha - \text{Fuc}$$

FIGURE 9.3. The terminal and subterminal sugars of the carbohydrate chains associated with ABH and Lewis antigenic specificity. (Adapted from Watkins, 1966 and Lloyd *et al.*, 1967.)

involve the enzymatic transfer of L-fucose from GDP-L-fucose to the galactosyl moiety of lactose (Shen *et al.*, 1968), it is very likely that the responsible fucosyl transferase also acts on the soluble blood group substance. Thus, the *Se* gene appears to have a regulator function, possibly behaving as a 'switch' mechanism for activating the *H* gene in secretory cells to produce this fucosyl transferase (Dr. W.M. Watkins, personal communication).

Figure 9.3 shows that in molecules with Le^a activity, an α-L-fucose residue is attached by (1→4) linkage to the subterminal GNAc. The Le^b determinant contains this fucose as well as the fucose associated with H activity on the terminal Gal (Marr *et al.*, 1967). Only Type I chains possess Lewis activity, because in Type II chains, the fourth carbon of the GNAc residue is already linked to the terminal Gal. In any given macromolecule, both Type I and II chains are present, so that some chains do not have Lewis activity. Oligosaccharides of Type II which have a fucose residue linked (1→3) to GNAc have been isolated. However, such chains have no blood group activity, possibly because of conformational change or steric hindrance (Lloyd *et al.*, 1967).

Enzymatic degradation studies have shown that the presence of A and/or B antigenic activity has a masking effect on H and Le^b antigenic sites. Thus, if blood group substance derived from the secretions of an individual with *H*, *Le*, *Se* and *A* or *B* genes is treated with an enzyme which removes the terminal GalNAc or Gal, the A or B activity is lost, but H activity is intensified (Watkins, 1962; Watkins *et al.*, 1962). Treatment of this H-reactive substance with an enzyme removing the α-L fucose associated with H activity uncovers Le^a reactivity. Finally, enzymatic removal of the α-L fucose associated with Le^a leaves a residual molecule with no blood group activity which cross reacts immunologically with Type XIV pneumococcus polysaccharide (Watkins, 1960).

SEQUENTIAL GENE ACTIVITY IN
BLOOD GROUP ANTIGEN BIOSYNTHESIS

Biosynthesis of secreted substances

The reverse of the sequence described in the previous paragraph was logically applied by Watkins & Morgan (1959) and Watkins (1966) to explain the roles played by the *Le*, *H*, *A* and *B* genes and their respective glycosyl transferase enzymes in the biosynthesis of secreted blood group substances (see Fig. 9.4). Thus, a preformed glycoprotein

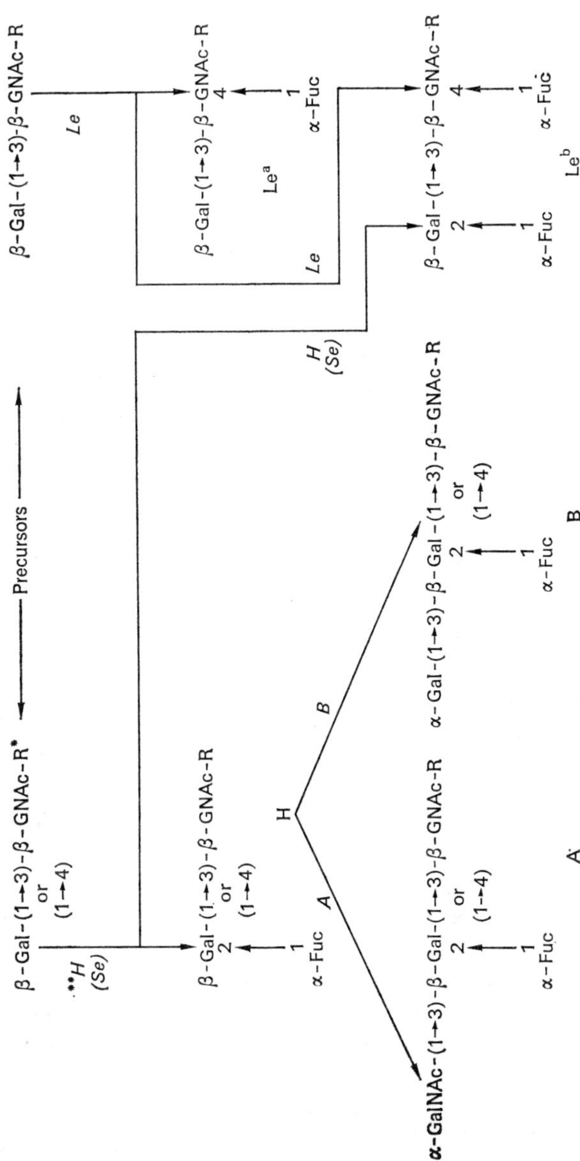

FIGURE 9.4. *R is the rest of the sugar chain, attached to the peptide backbone; **Genes are italicized; antigenic specificity is not.

Sequential steps in the biosynthesis of blood group specific structures mediated by products of the *H* (and *Se*), *Le*, *A* and *B* genes. The *Se* gene is placed in parentheses because it appears to have a regulator function; see text.

molecule, itself the product of an unknown number of biosynthetic steps, acts as a precursor. Transglycosylase enzymes, in stepwise fashion, transfer individual sugar units from sugar-nucleotide complexes to the non-reducing ends of the precursor molecule's multiple carbohydrate chains. The *Le* gene product, which is independent of the *Se* gene, is visualized as adding α-L fucosyl to the subterminal GNAc of Type I chains. In ABH secretors who lack the *Le* gene, the *H* gene product attaches α-L fucosyl to the terminal Gal of both Type I and II chains. When both *H* and *Le* genes are present, addition of fucosyl residues at both sites confers Le^b specificity. Finally, the transferase produced by *A* or *B* attaches either α-GalNAc or α-Gal to the terminal β-Gal of individual chains. This step cannot occur unless the chain has the fucosyl residue provided by the *H* gene product.

It appears that not all of the chains of the glycoprotein molecules undergo the same enzymatic changes. Thus, the secretions of a person with the appropriate genes contain not only A and/or B activity, but also H, Le^a and Le^b activity. Since H and Le^a are masked by A and B (as noted previously), the active H and Le^a determinants of the intact molecules are presumably on different chains, but not necessarily on different molecules.

Biosynthesis of red cell H, A and B antigens
It is assumed that a similar series of biosynthetic steps occurs in the red cell. However, since the *Se* and *Le* genes play no role in determining glycolipid blood group activity, the only genes known to be involved are those at the Hh and ABO loci. Both serological and enzyme inhibition tests have shown that the major determinants of red cell A and B activity are GalNAc and Gal, respectively. However, the isolated glycolipids have not yet been subjected to sequential sugar analysis. Furthermore, attempts to isolate H-active substance from red cells have been unsuccessful. Thus it is probable, but not proven, that the α-Fuc-$(\text{I}\rightarrow 2)\beta$-Gal linkage, which appears to be crucial for the secretion of H (as well as A and B) substances, is necessary for the expression of these antigens on the red cell. If this is not the case, then current theory about the genetic background of the 'Bombay' phenotype would require revision.

Blockage of biosynthesis: 'Bombay' and other rare phenotypes
In the 'Bombay' or O$_h$ phenotype, there is no detectable A, B or H activity on the red cells, and the serum contains anti-A, anti-B and

anti-H. Similarly, the secretions lack A, B and H substances, regardless of whether the individual carries *Se, A* or *B* genes. However, in the presence of an *Le* gene, Le^a activity is readily detected in the secretions and adsorbed on the red cells (Bhende *et al.*, 1952; Hakim *et al.*, 1961; Aloysia *et al.*, 1961; Giles *et al.*, 1963). Levine *et al.* (1955b) and Ceppellini (1959) pointed out that these observations could be explained if the biosynthesis of secreted ABH were blocked at a point beyond the action of the *Le* gene product but before that of the *A* and *B* genes. Watkins & Morgan (1955) suggested that the complete absence of H activity could be due to homozygosity for an inactive allele, *h*. Subsequently, Watkins & Morgan (1959) explained the genetics in biochemical terms, showing that in *hh* homozygotes, the attachment of α-L fucose to the terminal Gal of the oligosaccharide chain could not occur. Thus, there would be a block in the formation of not only H activity, but also A and/or B activity on the red cells as well as in the secretions. Synthesis of Le^a would be unimpaired, but Le^b could not be formed in the absence of an *H* gene.

The rare phenotype A$_m$ is characterized by diminished to absent A, but strong H activity on the red cells, although there is apparently a normal amount of both of these substances in the secretions. Absence of a *Y* gene, necessary for red cell activity, has been suggested as one of the possible underlying mechanisms (Weiner *et al.*, 1957), although in some instances, an allele at the ABO locus appears to be responsible (Salmon *et al.*, 1958a). More difficult to explain is the phenotype described by Gammelgaard (1942) and named A$_m^h$ by Solomon *et al.* (1965) to indicate that the red cells have little or no H antigen but react very weakly with anti-A, while the secretions contain both H and A in normal amounts. The mechanism responsible for suppressing glycolipid H and A formation is unknown, but it appears to be due to homozygosity for a very rare gene which has no effect on the secretions.

In still another rare subtype, A$_x$, the red cells react weakly with Anti-A, and show the expected reciprocal increase in H activity. On the other hand, the secretions, while containing a normal amount of H, cannot be shown by the usual techniques of anti-A inhibition, to contain A substance (André & Salmon, 1957). However, Alter & Rosenfield (1964) absorbed Anti-A with A$_x$ cells and found that the antibody eluted from such cells could be partially inhibited by the addition of the A$_x$ individual's own saliva. Additional evidence for a specific A$_x$ antigen was provided by the work of Yokoyama &

Plocinik (1965). Since family studies indicate that the A_x type does not represent a single genetic phenomenon (Grove-Rasmussen *et al.*, 1952; Glover & Walford, 1958; van Loghem & van der Hart, 1954; Cahan *et al.*, (1957), heterogeneity of the antigenic site(s) is likely.

CHANGES IN PHENOTYPE ASSOCIATED WITH DISEASE
Changes in red cell ABO phenotype associated with certain disease states have received considerable attention in recent years. Most intriguing is the weakening of A (or B) antigenic activity on a variable proportion of red cells, occasionally observed in patients with acute (or, rarely, chronic) leukemia (van Loghem *et al.*, 1957; Stratton *et al.*, 1958; Salmon *et al.*, 1958b; Salmon & Bernard., 1960; Gold *et al.*, 1959; Hoogstraten *et al.*, 1961; Renton *et al.*, 1962; Richards, 1962; van der Hart *et al.*, 1962). There is usually (but not always) little loss of secreted antigen activity, so the phenotype often resembles A_m. In some instances, loss of A from A_1 cells is accompanied by an increase in H activity. Although somatic mutation may be the usual underlying mechanism, there are some cases which are not satisfactorily explained on this basis.

Another phenomenon, acquisition of B antigen activity by A cells (Cameron *et al.*, 1959; Marsh *et al.*, 1959; Giles *et al.*, 1959; Stratton & Renton, 1959; Marsh, 1960) is more easily explained by the adsorption of certain bacterial components, some of which are known to possess B-like antigenic sites (Springer, 1956, 1967; Springer *et al.*, 1961, 1964; Andersen, 1961). The fact that A, but not O cells undergo this change has not yet been explained.

INTERACTIONS OF ABO AND SECRETOR GENES WITH
GENES DETERMINING PLASMA ALKALINE PHOSPHATASE
PHENOTYPES

SUMMARY OF OBSERVATIONS
Chapter 7 described the work of several investigators who showed that when plasma (or serum) specimens are subjected to electrophoresis, either one or two bands with alkaline phosphatase activity are seen in the β globulin region. The faster-moving component, presumably originating in liver and bone, is invariably present; the slower-moving component, originating in the small intestinal mucosa, appears in about two-thirds of random sera. (Additional phosphatase bands of

placental origin are observed during pregnancy; for a review of the placental alkaline phosphatase polymorphism, see Robson & Harris, 1967.)

Langman *et al.* (1966) found that the amount of slow-moving (zone B) phosphatase in serum was influenced by the quantity of fat ingested. However, even when subjects were tested after a fatty meal, the

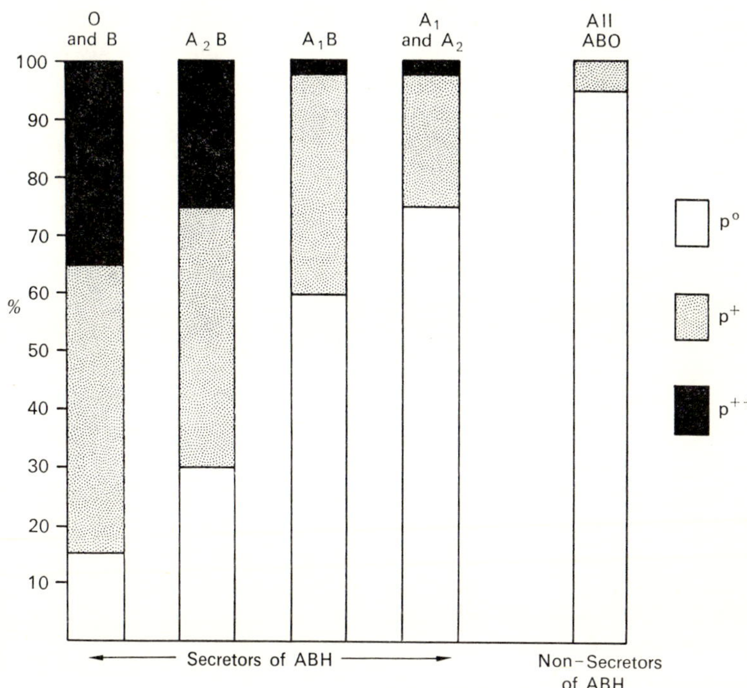

FIGURE 9.5. Association of ABO blood group and ABH secretor status with serum alkaline phosphatase phenotypes.

electrophoretic patterns could be classified as p^{++}, p^+ and p^0 to designate much, little and no demonstrable zone B activity. Furthermore, there was a definite association of alkaline phosphatase phenotype with the individual's ABO blood group and ABH secretor status, confirming the previous observations of Beckman (1964), Shreffler (1965), Bamford *et al.* (1965) and Evans (1965).

Figure 9.5 summarizes in a graph the observations given in Table I

11

of Chapter 7 (p. 229), showing a very strong relationship between ABH non-secretor status and the p^0 phenotype, regardless of the ABO group. There is a definite but less striking association of p^0 phenotype with A_1 and A_2, and among AB individuals, the correlation is considerably higher for A_1B than it is for A_2B. Most, but not all, secretors with O and B types have the slow-moving phosphatase band, and are thus either p^+ or p^{++}.

The fact that even among ABH secretors, there are some B and O people with the p^0 phenotype, and conversely, that there are some people of group A with phosphatase levels graded as p^+ and (rarely) p^{++} indicates that additional factors, probably of genetic origin, are involved in the expression of phosphatase in serum. The family studies of Beckman (1964) and Shreffler (1965) provide strong support for another genetic influence, but the data do not permit further analysis.

As mentioned in Chapter 7, the interaction of these genetic systems in controlling the presence in serum of an intestinal enzyme is particularly intriguing because of the known association of ABO and secretor status with certain diseases of the gastro-intestinal tract, recently reviewed by McConnell (1966). For example, the apparent liability to duodenal ulcer of group O, non-secretors is said to be nearly 2·5 times that of A, B or AB secretors, while liability to carcinoma of the stomach in persons of group A is about 1·3 times that of persons of group O (Clarke, 1961). While there is no lack of hypotheses to account for these associations, none of the postulated mechanisms has so far been convincingly supported by experimental evidence (McConnell, 1966).

SUGGESTED MECHANISMS

Attempts to explain the inter-relationships of the ABO and Secretor genes with the secretion of zone B alkaline phosphatase have not yet been successful. Originally, Arfors, Beckman & Lundin (1963) proposed that the slow-moving enzyme band might represent a complex of the fast-moving (zone A) phosphatase with blood group substances H and B in the serum of secretors. This idea was extended by Robinson & Goldsmith (1967) who proposed that if there were such a complex the fucose added to blood group substance by the *H* gene product (in the presence of an *Se* gene) might be necessary for the complex to be formed, and the terminal GalNAc in A substance would prevent the complex formation. Thus, in A_1 secretors, most of the oligosaccharide chains would have a terminal GalNAc blocking the interaction with phosphatase, while in A_2 secretors, the number of chains with terminal GalNAc would be smaller, but variable, accounting for the larger number of A_2 and A_2B individuals who have the slow-moving phosphatase.

Bamford *et al.* (1965) presented evidence against the theory of complex formation by showing that the total alkaline phosphatase activity in p^+ and p^{++} serum is generally higher than that of p^0 serum. Thus, it would appear that the slow-moving zone B phosphatase is not derived from the fast-moving zone A enzyme. According to the theory of Fishman & Ghosh (1967), both enzymes are of intestinal origin, but the zone A enzyme is a metabolic product of zone B.

Robinson & Goldsmith (1967) pointed out that there need not be a direct reaction between phosphatase and blood group substance, but the carbohydrate moiety of the phosphatase molecule might be modified by the same translycosylases as those involved in the synthesis of ABH substances.

Shreffler (1966) showed that ABO and secretor genetic status is unrelated to the amount of phosphatase in intestinal mucosa and thus it appears likely that the rate of secretion into the lymphatics is influenced by the blood group and secretor genes. Beckman (1968) postulated that a product of the *Se* gene affects the permeability of the secreting cell membrane by providing it with carbohydrate recognition sites. Interaction of such sites with the phosphatase molecule might promote its transfer across the cell membrane. However, if *Se* is a regulator gene, its activation of the *H* gene could bring about an alteration of the cell membrane as well as of the substances synthesized by the cell. Alternatively, there may be no membrane alteration, and the relative rate of secretion may be influenced by the identity of the sugar residues in key positions on the phosphatase molecule. (See Addenda.)

It is possible that some insight into the mechanisms involved might be gained by testing the serum of several individuals with the Bombay (or O_h) type for alkaline phosphatase type, particularly after a fatty meal. If no examples of p^{++} were found, one might conclude that in the absence of the *H* gene product, alkaline phosphatase cannot be secreted, even though the *Se* gene is present. On the other hand, if such individuals do have p^{++} serum, than there can be no relationship between the *H* gene transglycosylase product and the structure of intestinal alkaline phosphatase or the membrane of its synthesizing cell.

THE Ii SYSTEM AND ITS RELATIONSHIP TO
ABH ANTIGENS

THE ANTIGENS I AND i

An antibody called anti-I, which reacted with the red cells of virtually all of 22,000 random blood donors, was first described by Wiener *et al.* (1956) as a potent cold agglutinin in the serum of a patient with severe autoimmune hemolytic anemia. (The letter I was chosen for the antigen to express the individuality of those who lack it.) Subsequently, the studies of Weiner *et al.* (1960), Jenkins *et al.* (1960) and Tippett *et al.* (1960) showed that the antibodies associated with cold auto-agglutinin disease nearly always have anti-I specificity, and that there are two other human sources of anti-I: (1) Auto-agglutinins found in the serum of normal subjects or in patients with certain infectious diseases, such as pneumonia due to Mycoplasma; (2) Iso-agglutinins found in the serum of certain rare 'I-negative' individuals whose red cells react very weakly with potent anti-I.

A wide range in red cell I agglutinability in normal subjects was described by all of these authors, and Tippett *et al.* (1960) noted that most cord blood specimens reacted as weakly as the cells from 'I-negative' adults. Marsh & Jenkins (1960) and Marsh (1961) subsequently found some examples of another cold agglutinin which reacted much more strongly with the red cells of newborn infants and 'I-negative' adults than with usual adult cells, and it was therefore named anti-i. This antibody was first observed in the serum of a few patients with reticulum cell sarcoma and was thought to be of rare occurrence. However Rosenfield *et al.* (1965) and Jenkins *et al.* (1965) found that patients with infectious mononucleosis quite frequently produce antibodies with anti-i specificity. In most instances, the titer is fairly low; however, there have been a few reported cases of autoimmune hemolytic anemia associated with anti-i, presumably as the result of increased i activity on the patients' red cells (Calvo *et al.*, 1965; Troxel *et al.*, 1966).

Marsh (1961) also demonstrated a reciprocal relationship of I and i antigens on the red cells during the first year of life, the gradual 'switch' from i to I being reminiscent of the transition from Le^a to Le^b in the secretions and from Hb F to Hb A in the red cells. This similarity was subsequently recalled by Giblett & Crookston (1964) and by Hillman & Giblett (1965), who found increased i activity on the cells of patients with 'marrow stress' due to some hematological diseases or prolonged blood loss. However, there was no corresponding decrease in I activity or an invariable rise of fetal hemoglobin levels in such patients.

It is probable, but not proven, that normal 'I-negative' adults whose red cells react like those of cord blood are homozygous for a very rare gene which either has a regulator function or else is an amorph, so that there is little or no elaboration of an enzyme conferring recognizable I antigenicity upon the red cell. In some respects this phenotype recalls the persistence of fetal hemoglobin, in which the cells of a homozygote may entirely lack Hb A but are not otherwise abnormal (see Chapter 10). Partial to complete failure of a 'switch' mechanism could be the cause of both genetic anomalies.

There is no evidence that the genes associated with I and i antigenic activity are alleles, although their respective products may take part in a series of biosynthetic enzymatic steps, possible involving genes of the ABO and P system.

SEROLOGICAL ASSOCIATIONS OF I AND ABH

There is some evidence of a serological association of the ABO and Ii systems. For example, Tippett et al. (1960) found that the cells of some group O, 'I-negative' adults react very weakly with anti-H. They also described one example of iso anti-I, formed by a group A$_1$, 'I-negative' person which reacted best with A$_1$ cells and could be converted into a 'pseudo anti-A$_1$' antibody by absorption with group O cells. Subsequently, other authors described antibodies apparently recognizing antigenic determinants formed as joint products of two genetic systems, giving rise to such terms as anti-OI, anti-IH, anti-H(−i), anti-IA and anti-IB (Gold, 1964; Schmidt & McGinniss, 1965; Rosenfield et al. 1964b; Giblett et al., 1965; Issitt, 1967).

Marcus et al. (1963) showed that treatment of red cells with β-galactosidase and glucosaminidase decreased their reactivity with anti-I, and that A$_1$ cells were altered more rapidly and extensively than cells of group O. Similarly, Yokoyama (1965) obtained some evidence from antibody inhibition to support a relationship between A and I antigenic specificity. However, in view of the marked heterogeneity of anti-I specificity and the great disparity in titer between most examples of iso anti-I and auto anti-I, observations relating to one example of the antibody may not apply when other examples are used. Thus Yakulis *et al.* (1966) observed diminished activity of their anti-I sera tested against red cells treated with α-galactosidase but not with β-galactosidase.

POSSIBLE ASSOCIATION OF I ANTIGEN AND LEUKEMIA

McGinniss *et al.* (1964) found decreased to absent reactivity of an iso anti-I against the red cells of patients with a variety of diseases, particularly leukemia; subsequently, Schmidt *et al.* (1965) showed that the same antibody was inhibited by several strains of Mycoplasma isolated from various sources. However, Feizi & Hardisty (1966) could find no weakening of the I antigen in 15 leukemic patients whose red cells were tested with three other examples of anti-I. Also, Feizi & Darrell (1966) failed to demonstrate blocking of the I antigen by *M. pneumoniae in vitro* and *in vivo*, although their use of group O red cells rather than those of group A might have been responsible for this negative result, according to Schmidt & McGinniss (1967). On the other hand, Lewis *et al.* (1961) and Lewis (1962) found that certain examples of auto anti-I from patients with autoimmune hemolytic anemia reacted more strongly than normal with the cells of many patients with hematological disorders, including leukemia; but Ducos *et al.* (1965), using an auto anti-I from a normal I-positive subject found no apparent difference in its reactions with normal red cells and those from leukemia patients.

POSSIBLE BIOSYNTHETIC PATHWAYS IN OTHER BLOOD GROUPS SYSTEMS

THE P SYSTEM

For many years after the discovery of the P system by Landsteiner & Levine (1927), only two human red cell phenotypes, P+ and P−, were recognized. However, it is now known that several antigens are involved in this system, and Fig. 9.6 shows the tentative biosynthetic scheme proposed by Kortekangas *et al.* (1965) to account for the serological reactions they observed. Although there is as yet no biochemical evidence to support this tentative scheme, the serological analogies of the P and ABO systems make it appear likely that step-wise changes in antigenic structure are involved.

According to this scheme, the product of a very common gene called Z initiates the series of reactions by conferring p^k specificity on a precursor substance. If the Z gene is absent, this conversion does not

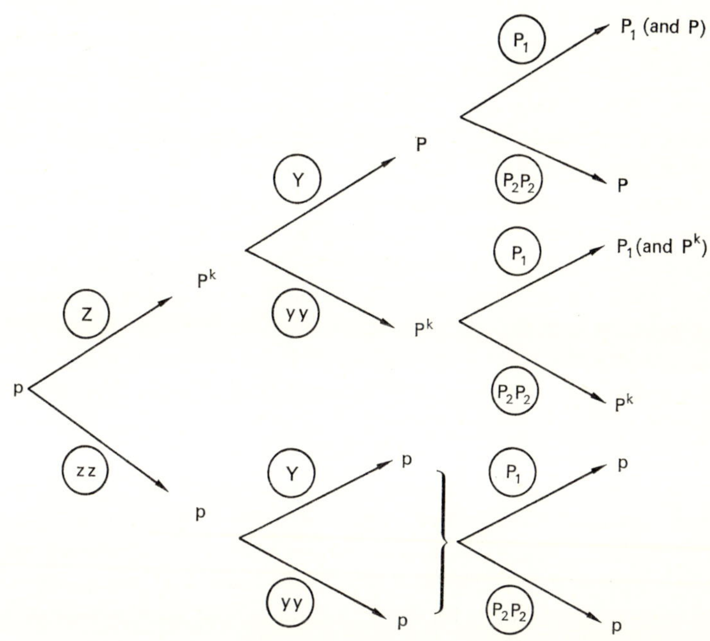

FIGURE 9.6. Tentative scheme for biosynthesis of antigens in the P system, as discussed in the text. Encircled letters indicate genes mediating individual biosynthetic steps. (Adapted from Kortekangas *et al.*, 1965. Reprinted with permission from *Vox Sang.* **10**, 385 (S. Karger, Basel/New York).)

occur, and since no additional biosynthetic steps are possible, the red cells have the p phenotype, analogous to the O_h phenotype in the ABO system. P^k is converted to P under the influence of a second common gene, Y. Conversion of P to P_1 is accomplished by the P_1 gene. P_2, the allele of P_1, has no apparent function, nor does the y gene. Thus in y/y homozygotes, P^k is not converted to P, but P^k can be altered by the P_1 gene product, since some individuals have both P_1 and P^k on their cells (phenotype P_1^k), while others have P^k but no P activity (phenotype P_2^k).

Whether or not this scheme is eventually shown to be correct, it serves an important purpose as a model for biochemists to prove or disprove. Human subjects do not normally secrete P substance, but there is a secreted substance in hydatid cyst fluid which has P_1 specificity (Cameron & Staveley, 1957; Staveley & Cameron, 1958). Morgan & Watkins (1964) and Watkins & Morgan (1964) showed that this substance is a glycoprotein chemically similar to the ABH and Lewis secreted substance. They further showed that the specificity probably resides in the sugar moiety, and more precisely, that α-D-galactosyl units are very likely to be involved.

There is an interesting relationship between the P system and a rare disease, paroxysmal cold hemoglobinuria (PCH). Patients with this disease suffer from hemolytic episodes upon exposure to cold, due to fixation of complement by 'Donath-Landsteiner' antibodies found in the patients' γG globulin (Adinolfi *et al.*, 1962; Hinz, 1963). These antibodies react with virtually all human red cells, except those having the phenotypes p and p^k (Levine *et al.*, 1963a; Worlledge & Rousso, 1965).

RELATIONSHIP OF P, Ii AND ABH ANTIGENS

There are a few clues which suggest that the P system may have a relationship, perhaps in a manner similar to that of the I system, with ABH. Tippett *et al.* (1965) have described an unusual (Luke) antibody with apparent specificity for a determinant of both systems. According to Issitt *et al.* (1968), there are also anti-IP$_1$ antibodies with activity against an antigen jointly determined by genes of the Ii and P systems.

Franks (1966) has shown that various human tissue cells, even after long periods of culture, contain antigens which react with the Donath-Landsteiner antibody as well as with anti-I and anti-i. A similar observation had previously been made with HeLa cell cultures and anti-H by Brand *et al.* (1963) and Chessin *et al.* (1965). Thus it seems well established that antigenic sites determined by genes involved in the structure of H, I and P antigens are intrinsic to other tissues besides the red cell. The previously mentioned A activity of epidermal cells is not retained after prolonged culture, but according to Chessin *et al.* (1965), it may be possible to induce a Gal or GalNAc transferase enzyme by adding blood group precursors to the medium. Further studies will be required to determine whether A and B antigens can actually be used as markers for cultured cells. If not, there are as yet no promising blood group antigens for this purpose, since H, I and P occur on the cells of virtually all human subjects.

THE MNSs SYSTEM

As indicated in Table 9.1, there are many antigens in the MNSs system, and for a full description the reader is referred to Race & Sanger (1968). The common antigen pairs, MS, Ms, NS and Ns are inherited as units, suggesting either that two genetic determinants are associated with one locus or that there are two very closely linked loci. Examples of possible recombination in this system have been reported by Chown *et al.* (1965), Gedde-Dahl *et al.* (1967) and Scholz & Murken (1967).

Tests for the M and N antigens are usually performed with antisera prepared in rabbits by the injection of human red cells. Seed extracts (lectins) and human iso-antibodies are also available, but there is considerable heterogeneity among the reactive sites recognized by these various reagents. For example, anti-M and anti-N of rabbit and human origin recognize a red cell antigenic determinant which is inactivated by viral receptor-destroying enzyme (neuraminidase), but the N-like receptor site recognized by *Vicia graminea* lectin is unaffected by this enzyme. Furthermore, the *Vicia* antigenic site is actually present in both M and N red cells, but the lectin reacts poorly with M cells unless they are treated with neuraminidase (Springer & Ansell, 1958; Mäkelä & Cantell, 1958; Springer *et al.*, 1966a).

The M and N antigens are inherited in a manner consistent with two allelic genes, *M* and *N*, but some evidence indicates that N may be a precursor of M, perhaps analogous to H, as a precursor of A and B. For example, antibodies with anti-N specificity obtained from rabbit, human and *Vicia* cross-react with M antigen, and can be absorbed by human type M cells at a low temperature; but anti-M antibodies do not cross-react with type N red cells (Landsteiner & Levine, 1928; Levine *et al.*, 1955a; Hirsch *et al.*, 1957). Furthermore, N-active haptenic structures reacting with human anti-N have been isolated from M substance prepared from red cells, but the reverse has not been achieved (Nagai & Springer, 1962; Hotta & Springer, 1965). If N is the precursor of M, strong reactions of red cells with both anti-M and anti-N would not represent heterozygosity at an MN locus, but rather heterozygosity for the *M* gene and its 'silent' allele (i.e. a gene with no demonstrable activity). Thus, in the biosynthetic pathway, there would be only partial conversion of N to M.

CHEMICAL CHARACTERISTICS OF M AND N SUBSTANCES
Hotta & Springer (1965) showed that the isolated M and N substances are glycoproteins with constituents similar to those of the ABH and Lewis substances but with a higher content of the sialic acid, NANA (*N*-acetylneuraminic acid), associated with myxovirus receptor activity. More recently, Springer *et al.* (1966a) described the physico-chemical characteristics of N substance prepared from the pooled red cells of 150 type N donors. This apparently homogeneous glycoprotein was found to have a molecular weight of 595,000. It had a 44 per cent peptide content, of which threonine and serine were the most prominent amino acids. In the carbohydrate moiety, galactose was most abundant, followed by sialic acid. Mannose, glucosamine and galactose were present in nearly equal proportions. Both the myxovirus and N receptors were destroyed by neuraminidase, but the *Vicia* N-like receptor activity was enhanced rather than inhibited. From these and other findings, it was inferred that N specificity involves sialylgalactopyranosyl structures, while *Vicia* specificity involves β-galactopyranosyl groupings.

The M antigenic site also appears to contain sialic acid as a part of its structure, but its precise chemical relationship to the N antigenic site remains to be determined. Some observations which may have an indirect bearing on this problem have been made during studies of the rare U-negative phenotype.

THE ANTIGEN U
The existence of the very common antigen, U, was first reported by Wiener *et al.* (1953, 1954) who described an anti-U antibody which reacted with the red cells of all Caucasians and all but 1 per cent of Negroes tested. Wiener *et al.* presumed that the very common *U* gene had an allele, *u*, for which the non-reacting Negroes were homozygous. Greenwalt *et al.* (1954) described a second example of anti-U, and found that the cells which did not react with this antibody also failed to react with either anti-S or anti-s. Since the gene determining this

behavior appeared to be an allele of *S* and *s*, they proposed that it should be renamed S^u.

Anti-U was first thought to behave like anti-Ss, in that it reacted with cells containing either S or s or both. However, the two specificities could not be separated by absorption. Furthermore, some blood specimens were subsequently found (Allen *et al.*, 1960, 1963a; Race & Sanger, 1962) which reacted with anti-U, but failed to react with either anti-S or anti-s. While such U-positive, S- and s-negative cells are very rare, the phenotype U-negative, S- and/or s-positive is even rarer. In fact, it has only been observed in the Rh null, LW negative phenotype (see p. 308), which probably has an associated red cell membrane defect.

Allen *et al.* (1960, 1963a) observed that cells having the phenotype M, U-negative (and thus S- and s-negative) or M, U-positive, but S and s negative, failed to cross react in the usual manner of M cells with anti-N antibodies. These findings were confirmed and expanded by Figur & Rosenfield (1965). However, Uhlenbruck & Krüpe (1965) and Springer *et al.* (1966a) isolated the M-active substance from M, U-negative cells and found that it contained some N activity. Thus, the observed difference in cross reactivity is probably quantitative rather than qualitative, possibly reflecting the near exhaustion of N-reactive substrate required for further conversion to S and s.

Clearly, it is much too early to attempt a reconstruction of the genetic events leading to the production of M and N specificity, but it does appear that the red cell M and N glycoproteins, like the red cell H, A and B glycolipids, depend on carbohydrate structures for their specificity. Nothing is yet known about the chemical structure of S, s and U. It is reasonable but not necessary to suppose that they, too, are carbohydrate-associated antigenic determinants. However, since these antigens are not consistently vulnerable to the enzymes which destroy M and N activity, they are probably not dependent on NANA for their specificity.

THE Rh SYSTEM

The name Rh was given by Landsteiner & Wiener (1940) to an antigen detected on 85 per cent of human red cell specimens by an antibody produced in guinea pigs and rabbits after the injection of *Macacus rhesus* red cells. In the previous year, Levine & Stetson (1939) had detected an antibody in the serum of the mother of an infant severely affected with hemolytic disease of the newborn, and for many years thereafter, it was thought that this human antibody, which also reacted with 85 per cent of human red cell specimens, had the same specificity as that induced by rhesus cells; consequently, it, too, was given the name anti-Rh. When the complex nature of Rh genetics was subsequently revealed, the human antigen was renamed Rh_0 by Wiener & Landsteiner (1943) and D by Race (1944), who credited Fisher with the 'linked-gene' theory underlying that terminology (see p. 273).

The earlier editions of Race & Sanger (1950, 1954) outline the extensive work done by both American and English investigators to determine the genetics of the Rh system, and their later editions (1958, 1962, 1968) show how the problem became even more complicated by the discovery of apparent 'compound antigens', 'deleted' antigens and numerous variants. Their very clear and detailed accounts will not be repeated here. As pointed out on p. 273, the actual nature of the human Rh locus is still undefined, but it is convenient to consider the antigens, C and c, E and e, and D and non-D, as alternative or antithetical expressions of genetic determinants which appear to reside within a single locus or genetic region.

THE LW ANTIGEN

A curious phenomenon was reported by Murray & Clark (1952), who found that guinea pigs, injected with heat extracts of either Rh(D) positive or Rh(D) negative human red cells, produced an antibody which appeared to have anti-D specificity after absorption with Rh negative cells. Several years later, Levine *et al.* (1961a,b) confirmed this observation and showed by blocking experiments that although Rh positive red cells carried the 'D-like' antigen, the reactive sites of D and 'D-like' were different from each other. The designation 'D-like' was subsequently abandoned in favor of 'LW', the initials of Landsteiner and Wiener.

Table 9.6 summarizes the reactions of anti-LW antibodies as described by Levine & Celano (1967). It shows that Rh negative cells have less LW antigen activity than Rh positive cells, and that LW negative cells do not react with anti-LW, even if they are Rh positive.

The fact that the D antigen can be expressed in the absence of LW was first shown by Levine *et al.* (1963b). They found two women with

Rh positive, LW negative red cells whose serum contained isoanti-bodies with anti-LW specificity, reacting strongly with Rh positive cells and weakly with Rh negative cells. Absorption of the serum with Rh negative, LW positive cells removed all antibody activity, and the eluate from the absorbing cells showed the same preferential reactions with Rh positive cells as the unabsorbed serum. Studies of

TABLE 9.6. Reactions of guinea pig anti-LW serum with red cells of various Rh and LW phenotypes (Adapted from Levine & Celano, 1967)

Cells	Anti-LW serum		
	Absorbed once with D−, LW+ cells	Absorbed 2 or 3 times with D−, LW+ cells	Absorbed once with D−, LW− cells
D+, LW+	++++	++	++++
D−, LW+	++	o	++
D+ or D−, LW−	o	o	o

the family of one of these women as well as of an Rh negative, LW negative woman reported by Swanson & Matson (1964) indicated that these LW negative subjects are homozygous for *lw*, the very rare inactive allele of *LW*, and that the genes of the LW system assort independently of the genes in the Rh system.

Swanson *et al.* (1965a) showed that there is a difference in the specificity of anti-LW antibodies produced in human subjects and in animals. They found that red cells of newborn infants reacted more strongly than those of adults when tested with guinea pig anti-LW, but not with human anti-LW. A similar observation had been made by Fisk & Foord (1942) before the difference between Rh and LW antigens was recognized.

THE Rh_{null} PHENOTYPES

Vos *et al.* (1961) found an Australian aborigine with the very rare phenotype now known as Rh_{null}, in which the red cells do not react with any of the antibodies specific for antigens determined by genes at the Rh locus. Levine *et al.* (1962) showed that these Rh_{null} cells were also LW negative. Subsequently, Levine *et al.* (1964, 1965) reported their studies on the family of an American white woman with the Rh_{null}, LW negative phenotype. As shown in Fig. 9.7, the proposita inherited R^1 from one parent and either R^1 or r from the other, but neither gene was expressed on her red cells. She was considered to be homozygous for a very rare X^0r gene, whose allele, X^1r, is necessary for producing a precursor common to Rh and LW. According to this theory, shown in the top half of Fig. 9.9 (p. 308) the production of Rh and LW antigens is independent, unless there is a block in the synthesis of their common precursor.

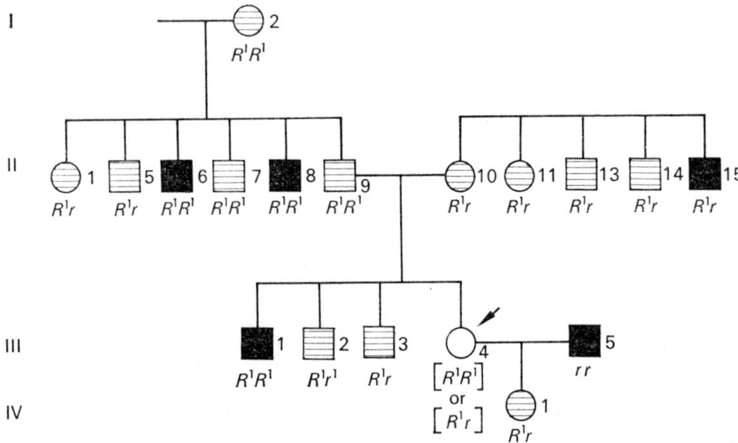

FIGURE 9.7. Apparent Rh genotypes of selected members in the family of a woman with the Rh null phenotype. Differences between weak and strong reactions of Rh antigens are indicated by lined and solid figures, respectively (From Levine *et al.*, 1965. Reprinted by permission from *Transfusion* 5, 492.)

One would expect that in X^1r/X^0r heterozygotes, the decrease in precursor would be reflected in a similar decrease in both Rh and LW antigen reactivity. Levine *et al.* (1965) found an overall decrease in activity of all the Rh antigens, but a decrease in LW was not observed. Swanson *et al.* (1965a) reported similar findings, but Gibbs (1966) and Gibbs & Rosenfield (1966) found that the use of very sensitive serological techniques revealed a correspondence between Rh and LW activity, not only in X^1r/X^0r heterozygotes, but also in randomyl

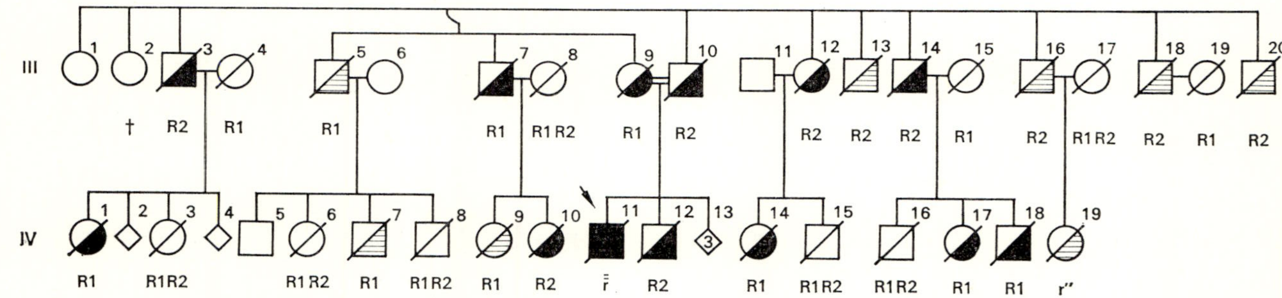

FIGURE 9.8. Rh phenotypes of members in the family of a boy with Rh null phenotype due to homozygosity for an amorph at the Rh locus. There was consanguinity in this pedigree, only part of which is shown here. In obligate carriers, the 'silent' Rh allele is symbolized as solid black; in possible carriers, as hatching. (From Ishimori & Hasekura, 1967. Reprinted with permission from *Transfusion* 7, 84).

chosen subjects. The latter observation is in accord with the previously reported stronger reactions of Rh positive cells with anti-LW.

A complication in interpreting the Rh-LW relationship was introduced by Schmidt & Vos (1967) and Schmidt *et al.* (1967) who reported that in the two Rh_{null}, LW negative women, red cell reactions with

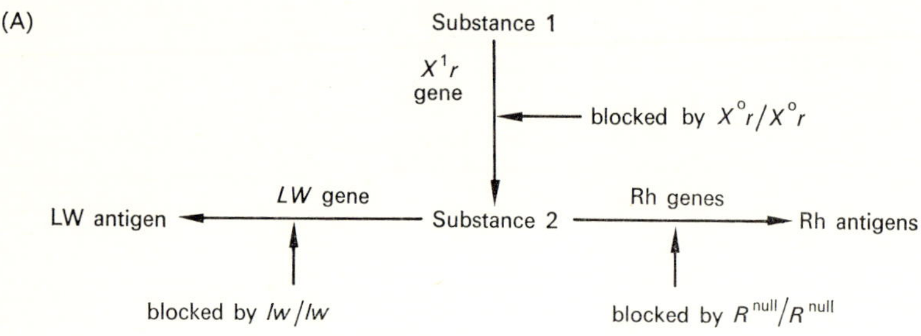

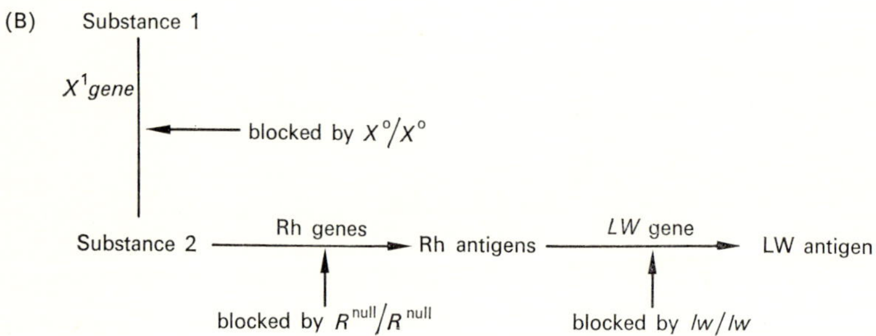

FIGURE 9.9. Alternative genetic pathways in the biosynthesis of Rh and LW antigens. Scheme A is adapted from Race (1965c) and Levine *et al.* (1965). Scheme B was constructed after the report was published on Rh_{null} due to homozygosity for an Rh amorph.

certain other antibodies were distinctly unusual. For example, no reactions were obtained with most examples of anti-U and some examples of anti-s, although the relatives of the American women appeared to have normal expression of S, s and U. There was also an increase in i reactivity, suggesting the possibility of 'marrow-stress' as described by Giblett & Crookston (1964) and Hillman & Giblett (1965). One woman had a mild hemolytic process with 6 per cent reticulocytes. These findings raise the possibility that the entire red cell phenotype has no direct association with the genes at the Rh, LW and MNSs loci, but reflects an inherited anomaly of the red cell membrane due to homzygosity for a gene (called X^0 in scheme B of Fig. 9.9) which interferes with the synthesis of membrane components containing certain blood group antigenic determinants. (See Addenda.)

This possibility was enhanced by the reports of Ishimori & Hasekura (1966, 1967) and Hasekura *et al.* (1967), who described a third example of Rh_{null}, LW negative blood in a healthy Japanese boy with no anomalous S, s or U antigens. An extensive family study (see Fig. 9.8) revealed a different genetic basis, namely, homozygosity for an allele behaving as an amorph at the Rh locus. This gene is tentatively designated as R^{null} in Fig. 9.9. As shown in Fig. 9.8, heterozygotes for this rare allele express the Rh antigens determined by only one of their paired Rh loci. According to the theory that Rh and LW antigens are formed independently from a common precursor (scheme A of Fig. 9.9), an individual homozygous for a 'silent' Rh allele should not be LW negative. An alternative explanation, shown in scheme B of Fig. 9.9, could be that the Rh antigen represents an intermediate step in the biosynthesis of LW. The fact that the cells of some non-human primates (such as the gibbon and the rhesus monkey) react with anti-LW but not with anti-D (Masouredis *et al.*, 1967a) does not contradict this suggestion, because there appears to be another antigen (not LW) on the cells of these animals which cross reacts with all human red cells except those of Rh_{null} phenotype (Boettcher *et al.*, 1965).

STUDIES ON Rh ANTIGEN STRUCTURE

The serological vagaries of the Rh antigens have not yet been explored on a chemical basis. Indeed, the nature of the molecule bearing the antigenic determinants has not been clearly defined. The observations of Dodd *et al.* (1960, 1963, 1964) appeared to implicate NANA in Rh serological specificity, while Boyd & Reeves (1961) reported anti-D partial neutralization by a similar substance, colominic acid. However, Wolff & Springer (1965) could not confirm the neutralizing effect of any sialic acids, colominic acids or gangliosides they tested. The Rh-active substance they prepared from human blood was unaffected by a number of glycosidases, lipases, ribonucleases and (at low concentration) trypsin and papain.

Green (1965, 1967a) from studies of lyophilized red cell ghosts, presented evidence suggesting that the molecule which bears the D antigenic site is a peptide or protein, and that one or more disulfide bonds and one or more sulfhydryl groups may be required for activity. In a further study, Green (1967b) reported that after extraction with butanol, neither the residual red cell membrane nor the extract itself had any D antigen activity. However, when the extract was added back to the extracted membranes, D antigen ativity was restored. Extracts of D-negative cells could also restore activity to extracted D-positive membranes, but the addition of D-positive cell extracts to D-negative extracted membranes had no effect. The restoring factor in butanol extract was found to be a phospholipid. Presumably, the antigenic specificity is on a protein molecule of the membrane which requires the formation of a complex with a phospholipid in order to react with anti-D antibodies.

Whittemore *et al.* (1967) also reported the loss of Rh activity when red cell membranes were dissolved in butanol, and suggested that the behavior of Rh resembles that of another membrane component, cation-activated ATPase, in being dependent on the integrity of a phospholipid-protein complex. It is conceivable that the molecules with Rh activity actually have ATPase activity and thus are involved in ion transport across the cell membrane. If this possibility turns out to be correct, it will provide the first available information about the function of molecules which possess human blood group antigen activity. There is similar speculation that the sheep blood group antigen, M, has ATPase activity (Tosteson, 1963; Shreffler, 1967b).

QUANTITATIVE STUDIES ON Rh REACTIVE SITES

Estimates of the number of red cell Rh reactive sites have been obtained either by determining the uptake of isotope-labelled, purified antibody (Masouredis, 1959; Masouredis *et al.*, 1960; Hughes-Jones *et al.*, 1963a,b; Evans *et al.*, 1963) or by quantitative hemagglutination methods employing an electronic device for measuring free red cells (Gibbs & Becker, 1963; Rosenfield *et al.*, 1964c). Rochna & Hughes-Jones (1965) showed that in unrelated individuals, the number of D reactive sites varied considerably with the phenotype (as would be predicted from results obtained with crude titration methods), but there was also a wide range of activity within each phenotype, resulting in considerable overlap between phenotypes. The overall range was from 10,000–33,000 antigenic sites, the lower numbers being associated with cells having the probable genotype R^1r, and the highest numbers with R^2R^2.

Masouredis *et al.* (1967b) studied 63 members of 11 families selected for a history of maternal-fetal Rh incompatibility. Within most families, it was possible to determine the zygosity of the Rh positive father by comparing the I^{125}-tagged antibody uptake by his red cells with that of his heterozygous offspring. However, the amount of antibody bound to the red cells of unrelated homozygous individuals showed an almost six-fold variation.

BLOOD GROUPS AND GENETIC LINKAGE

AUTOSOMAL LINKAGE

Four examples of autosomal linkage involving a blood group locus are known. These include Lutheran with Secretor (Mohr, 1951; Sanger & Race, 1958; Lawler & Renwick, 1959; Metaxas *et al.*, 1959; Greenwalt, 1961; Cook, 1965); ABO with nail-patella syndrome (Renwick & Lawler, 1955; Jameson *et al.*, 1956; Lawler *et al.*, 1957, 1958; Clarke *et al.*, 1961; Elliott *et al.*, 1963; Renwick & Izatt, 1965; Renwick & Schulze, 1965; Sharma, 1966); Rh with elliptocytosis (Chalmers & Lawler, 1953; Goodall *et al.*, 1953; Lawler & Sandler, 1954; Morton, 1956; Clarke *et al.*, 1960a; Geerdink *et al.*, 1967); and, less well established, Duffy with congenital cataract (Renwick & Lawler, 1963). In addition, there is a possibility of linkage between the ABO and xeroderma pigmentosum loci; however, since over 80 per cent of the 46 patients with xeroderma studied in 34 sibships had group O blood, the apparent genetic linkage may actually be due to predisposition of individuals with this blood group for development of the disease (El-Hefnawi *et al.*, 1965). (See also p. 515.)

An interesting by-product of the search for genetic linkage was the observation by Renwick & Schultze (1965) and by Cook (1965) that the estimated recombination fraction (calculated for ABO-nail patella and Lutheran-Secretor, respectively), was significantly greater for women than for men. In other words, the data indicated that, at least for the autosomal loci involved, more recombination occurs in women than in men. From animal models, Haldane (1922) proposed that there may be greater difficulty of fusion of chromosome pairs in the heterogametic sex, thus restricting the frequency of crossing over in males. However, more recent studies by Dunn & Bennett (1967) failed to detect any consistent differences in male *vs* female recombination frequencies, although there was a general tendency for higher values in females.

A possible association of the Rh and Duffy systems was reported by Jenkins & Marsh (1965), who found a mixture of two red cell populations in the blood of a healthy blood donor. About 70 per cent of the cells were A_1, Fy(a − b +) and Rh negative (rr), while the other 30 per cent were A_1, Fy(a+) , and Rh positive (probably R^1r). Studies of the family showed that in the donor's Rh negative population, there was loss of D and replacement of C by c. The c antigen could not have been inherited in the usual manner, since one parent was apparently homozygous for the R^1 gene (*CDe*). The same parent appeared to be homozygous for the *Fy*ᵃ gene. Thus, the failure of 70 per cent of the donor's cells to express *Fy*ᵃ was in some way associated with the change from R^1 to r (*CDe* to *cde*). Somatic mutation was offered as the most likely explanation for this phenomenon, affecting the serological behavior of C, D and Fyᵃ antigens. This suggestion does not necessarily imply genetic linkage between the Rh and Duffy loci; in fact, other explanations, such as alteration of a common precursor or a cell membrane defect, seem more likely.

Lawler (1964) has reviewed the possible ways in which assignment of a blood group locus to a particular chromosome might be accomplished by family studies of patients with chromosomal aberrations such as satellite enlargement or other kinds of 'marker chromosomes', partial deletion, trisomy, mosaicism or defects secondary to diseases affecting the bone marrow. Many investigations of this kind are in progress, and a few have been reported. For example, the study of Hulten *et al.* (1966) suggested that chromosome 2 might carry the Kidd locus; Price Evans *et al.* (1966) showed that the Kell locus could be on chromosome 21. Crawford *et al.* (1967) reported that chromosome 16 might carry the Duffy locus, but number one seems more likely (Donahue *et al.*, 1968). Much more evidence is needed to determine if these preliminary assignments are correct.

X CHROMOSOME LINKAGE: Xgᵃ

An X-linked antigen, Xgᵃ, was first reported by Mann *et al.* (1962) and has since been studied extensively by Race & Sanger (1968).

It is probable, but not proven, that the locus for *Xg*ᵃ is on the short arm of the X chromosome (Lindsten *et al.*, 1963; Sanger, 1965). It is apparently not within mapping distance of the genes determining hemophilia A or hemophilia B (Davies *et al.*, 1963; Harrison, 1964), deutan or protan color-blindness (Jackson *et al.*, 1964; Renwick & Schulze, 1964), Duchenne muscular dystrophy (Clark *et al.*, 1963; Blyth *et al.*, 1965), congenital hypogammaglobulinemia (Sanger & Race, 1963; Rosen *et al.*, 1965) and Norrie's disease (Warburg *et al.*, 1965). The loci for Xg and G6PD were originally thought to be within mapping distance, but subsequent studies failed to substantiate this finding (Siniscalco *et al.*, 1966; Adam *et al.*, 1966a, 1967).

On the other hand, there is good evidence for linkage of Xg with ichthyosis (Kerr *et al.*, 1964; Adam *et al.*, 1966b; Wells *et al.*, 1966), and fairly conclusive evidence for linkage with angiokeratoma (Optiz *et al.*, 1965) and ocular albinism (Fialkow *et al.*, 1967).

Studies of Xg^a have been very useful in determining the origin of X chromosomes in cases of aneuploidy (Race, 1965a,b; Edwards *et al.*, 1966). For example, of 130 cases of XO Turner's syndrome, about a fourth have been found to inherit their single X from their fathers and the other three-fourths from their mothers. As would be expected from these figures (Fraser, 1963) the Xg^a distribution among XO patients is the same as that of the general male population. On the other hand, Xg^a frequency among patients with XXY Klinefelter's syndrome is about midway between the frequencies of random males and females, thus reflecting the heterogeneity of the parental origin (i.e. the site of non-disjunction). In two Xg(a+) patients with an XXYY karyotype whose fathers were Xg(a+) and mothers Xg(a−), it was possible to infer that non-disjunction had occurred at both meiotic divisions in the paternal gamete, so that an XYY sperm fertilized a normal X-bearing ovum (De la Chapelle *et al.*, 1964; Pfeiffer *et al.*, 1966).

According to the Lyon hypothesis of X chromosome inactivation (see Chapter 12), women who are heterozygous for the Xg^a gene should have Xg mosaicism with a variable proportion of Xg(a+) and Xg(a−) red cells. One study on the agglutinability of such cell samples by anti-Xg^a failed to indicate more than a single population of cells (Gorman *et al.*, 1963). On the other hand, MacDiarmid *et al.* (1967) presented evidence in support of 'Lyonization'. They found that the blood of a woman and two of her four daughters contained two red cell populations, consistent with the diagnosis of hereditary X-linked anemia. When the populations were separated, the normal cells were found to be Xg(a−), while the abnormal cells were Xg(a+). The father was Xg(a−), and the two unaffected girls were Xg(a+). Thus, it was concluded that one of the mother's X chromosomes carried an Xg^a gene and a gene associated with the red cell abnormality, while her other X chromosome, also carrying Xg^a, had the common 'normal' gene at the second locus.

BLOOD GROUPS AND SELECTION

DISEASE ASSOCIATION

The possibility of demonstrating an association between red cell phenotype and the development of disease has intrigued geneticists for many years. Most investigators now agree that blood group O and ABH non-secretor status play a role, as yet unknown, in pre-disposition to duodenal ulcer, and there also appears to be an association of group A with gastric carcinoma and pernicious anemia (see Roberts, 1959; Clarke, 1961; and McConnell, 1966). However, it is unlikely that these particular diseases, which are most prevalent in people beyond the age of reproduction, play a major role in maintaining the marked differences in blood group gene frequencies in the world's populations.

A more likely source of selective pressure is susceptibility to infection with various micro-organisms, and there is some evidence supporting a relationship of blood types with infectious disease. For example, ABH non-secretors of blood group A are said to have an increased susceptibility to rheumatic fever, possibly reflecting decreased resistance to streptococcal infection (Glynn *et al.*, 1956; Clarke *et al.*, 1960b; Glynn & Holborow, 1961; van Rijsewijk & Goslings, 1963). Furthermore, there appears to be an increased frequency of group O among patients infected with influenza A_2 virus (McDonald & Zuckerman, 1962; Potter & Schild, 1967).

MATERNAL-FETAL INCOMPATIBILITY

An obvious agent in selection is maternal-fetal incompatibility, particularly in the ABO and Rh blood group systems. Since mothers of group O have anti-A and anti-B in their serum, and since a variable proportion of these antibodies are IgG globulins capable of crossing the placenta, hemolytic disease of the newborn (HDN) due to anti-A (and anti-B) is not uncommon (Rosenfield, 1955; Kochwa *et al.*, 1961).

Red cell destruction is usually not severe, so that the mortality risk in the full term infant is small; however, a larger fetal loss may occur during early pregnancy. Although the magnitude of this early loss is still an unsettled question, it is thought by some investigators to be high enough to necessitate the existence of one or more strongly compensatory factors operating to maintain the level of *A* and *B* genes observed in different populations (Levene & Rosenfield, 1961; Chung & Morton, 1961).

Hemolytic disease of the newborn not due to ABO incompatibility is usually caused by Rh antibodies, particularly anti-D. Giblett (1964) reported that among 2024 mothers with a positive indirect antiglobulin test, anti-D was found in the serum of 95 per cent. Other antibodies identified were anti-E (2 per cent), anti-c (1 per cent), anti-K (1 per cent) and one to four instances each of anti-Ce, anti-Fy^a, anti-Jk^a, anti-Jk^b, anti-S, anti-s and anti-U. Some of these antibodies were stimulated by transfusion, however, so that among affected infants, the proportion with HDN due to anti-c was slightly higher, and the proportion due to anti-K slightly lower than would have been predicted on the basis of the maternal isosensitizations.

When HDN is due to anti-D, red cell destruction is often severe, so that still-birth in the third trimester occurs in about 12 to 20 per cent of cases (Walker *et al.*, 1957). Various mechanisms have been considered to account for the persistence of polymorphism at the Rh locus (Fisher *et al.*, 1944; Glass, 1950; Li, 1953; Levin, 1967), but the question is by no means settled. An additional complication lies in the fact that HDN due to anti-D occurs more commonly in ABO compatible matings than in ABO incompatible matings. This observation, first made by Levine (1943) has since been confirmed in a large number of reports. The usual explanation is that when the Rh positive red cells of ABO incompatible infants cross the placenta, they are destroyed in the maternal liver, where they rarely stimulate anti-D formation. The complex effect on overall gene frequencies resulting from the interaction of the two blood group systems has been discussed at length by Allison (1955), Cohen (1960) and Stern (1960).

A very important offshoot of Levine's observation was initiated by the work of Stern *et al.* (1961), who found that Rh antibodies could be used to prevent Rh immunization. In their study, 16 Rh negative male subjects were given injections of Rh positive red cells coated with anti-D; none of the subjects subsequently produced anti-D. This work was confirmed and extended by Clarke *et al.* (1963) and by Freda *et al.* (1964), both of whom showed that if anti-D was injected 24–72 hours after the administration of Rh positive cells to Rh negative recipients, protection against anti-D formation was virtually complete. On the basis of these and other observations, the now extensive studies on maternal protection were begun. Of more than a thousand non-immunized Rh negative women given injections of anti-Rh antibodies within 24–72 hours of delivering an ABO compatible, Rh positive child, only six have been shown to develop Rh antibodies, even though many of them have had subsequent pregnancies with an Rh positive fetus (Clarke *et al.*, 1966; Clarke, 1967; Freda *et al.*, 1966, 1967; Schneider & Preisler, 1966; Zipursky & Israels, 1967; Hamilton, 1967; DeWit & Borst-Eilers, 1968). The success of these trials has raised many questions about the actual mechanism of protection, the number of antibody molecules needed to suppress the primary immune response, the possibility of antibody enhancement or of anti-allotype formation, the efficacy of IgG over IgM antibodies, and so forth. Obtaining the answers to these questions should greatly advance the general understanding of immunological mechanisms. (See Addenda.)

PREZYGOTIC SELECTION

Gershowitz *et al.* (1958) detected anti-A and anti-B in the cervical secretions of about a fourth of the women they tested. However, it has not been convincingly demonstrated that these antibodies, particularly in such low titer, could adversely affect the sperm of ABO incompatible males, although the study of Ackerman (1967) was said to support the possibility. Furthermore, although blood group substance is readily demonstrated on the sperm of ABH secretors, it is not clearly established that the ABH activity reported by some (but not by other) investigators on the sperm of non-secretors reflects the gene content of the individual haploid sperm (Gullbring, 1956; Holborow *et al.*, 1960; Levine & Celano, 1961; Shahani & Southam, 1962; Edwards *et al.*, 1964).

Matsunaga & Hiraizumi (1962) found in Japan that matings in which the father was *A/O* or *B/O* resulted in the transmission of more than 50 per cent of O-bearing sperm to the zygote formation of the children. However, their data indicated that this apparent prezygotic selection occurred independently of the maternal environment, and presumably reflected either meiotic drive or sperm competition in the father. Similarly, Hiraizumi (1964) in a later review of Japanese family data, produced no evidence supporting selection at the gametic level. Morton *et al.* (1966a) tested a large number of Brazilian families for genetic markers in 16 polymorphic systems, including ABH secretor status and antigens of seven blood group systems. After examining the data statistically for evidence of natural selection, they concluded that measurable selection was evident only in the ABO system. From the hemolysin titers found in the mothers, they attributed the observed shortage of A_1 children to the postzygotic effect of maternal anti-A antibodies. In two other large population studies, an ABO effect was either absent or minimal (Reed *et al.*, 1964; Peritz, 1967).

SOME BLOOD GROUP GENE
FREQUENCIES

No attempt can be made here to review the very large body of information on the blood group gene frequencies in various populations of the world. Mourant (1954) has written the classical text on the subject and is currently preparing a revised edition. A comprehensive review of gene distribution in the ABO system is also available (Mourant *et al.*, 1958). Race & Sanger (1962) have presented very useful data for all of the blood group systems, especially as it pertains to populations of Western Europe (and thus, to a large proportion of North American inhabitants). Tables 9.7 and 9.8, mainly derived from their figures, also include data on the gene frequencies of American Negroes. In addition to Race & Sanger (1962), the sources used were Giblett *et al.* (1957), Giblett (1961), Rosenfield *et al.* (1960b), Francis & Hatcher (1966), Tippett (1967) and Giles *et al.* (1967). The genes listed in Table 9.7 show considerable quantitative heterogeneity within a given antigenic specificity, but clearcut qualitative heterogeneity, expressed as two or more antigenic determinants, has not been demonstrated. However, in Table 9.8, the column headed 'Gene' refers to the individual genetic determinants within the locus or region of a 'gene complex' which determine two or more antigen sites. In the Rh system, the genes are designated in accordance with the nomenclatures of

Race & Sanger (1962) and Wiener & Wexler (1958), because they are commonly used. For translation into the numerical terminology of

TABLE 9.7. *Frequency of genes which appear to determine a single antigenic site (or none) in 10 blood group systems*

System	Gene	Frequency Caucasian	Frequency Negro
ABO	O	0·660	0·713
	A^1	0·209	0·135
	A^2	0·070	0·038
	B	0·061	0·114
P	P^1	0·540	0·836
	P^2	0·460	0·164
Lutheran	Lu^a	0·039	0·020
	Lu^b	0·961	0·980
Duffy	Fy^a	0·421	0·053
	Fy^b	0·549	0·122
	Fy	0·030	0·825
Kidd	Jk^a	0·486	0·725
	Jk^b	0·514	0·275
Xg^a	Xg^a	0·659	0·550
Cartwright	Yt^a	0·959	
	Yt^b	0·041	
Diego	Di^a	0·001	
	Di^b	0·999	
Auberger	Au^a	0·576	
Dombrock	Do^a	0·420	

Rosenfield *et al.* (1962) (see Table 9.3, p. 274). The numbering method of Allen & Rosenfield (1961) and of Morton *et al.* (1965) is used in the case of the Kell system, however, because there are no comparable

TABLE 9.8. Frequency of 'genes' and 'gene complexes' of three blood group systems in which two or more antigenic determinants are inherited as a unit

System	'Gene'		Frequency Caucasian	Frequency Negro	'Gene Complex'	Frequency Caucasian	Frequency Negro
MNSs	M		0·532	0·495	MS	0·238	0·098
	N		0·468	0·505	Ms	0·305	0·336
	S		0·327	0·171	NS	0·070	0·073
	s		0·673	0·716	Ns	0·386	0·380
	*		0·001	0·113	$M*$ and $N*$	0·001	0·113
Rh	D		0·590	0·713	CDe, R^1	0·421	0·175
	d		0·410	0·287	cde, r	0·389	0·259
	C		0·433	0·179	$cdeV, r^V$		0·021
	c		0·567	0·821	cDE, R^2	0·141	0·107
	E		0·155	0·107	cDe, R^0	0·026	0·292
	e		0·845	0·893	$cDeV, R^{OV}$		0·136
	V		0·001	0·157	cdE, r''	0·012	0·002
					Cde, r'	0·010	0·008
					CDE, R^z	rare	Rare
Kell	K	(K^1)	0·045	0·005	$K(K^{1,-2,-3,4,5,-6,7})$	0·045	0·005
	k	(K^2)	0·954		$k(K^{-1,2,-3,4,5,-6,7})$	0·942	0·851
	Kp^a	(K^3)	0·012		$k^p(K^{-1,2,3,-4,5,-6,7})$	0·012	
	Kp^b	(K^4)	0·988		$k^s(K^{-1,2,-3,4,5,6,-7})$	0·000	0·144
	K^u	(K^5)	0·999		$k^0(K^{-1,-2,-3,-4,-5,-6,-7})$	0·001	
	Js^a	(K^6)	0·000	0·144			
	Js^b	(K^7)	1·000	0·856			

* Gene with no S or s expression; U usually not expressed

terminologies which convey the same amount of information. In each instance, the 'shorthand' version is also indicated.

Considerably greater heterogeneity exists than is noted in the table. First of all, there are many additional antigens (listed in Table 9.1, p. 269) which belong to one of the known systems but, since their antibodies are not readily available, their genes are not mentioned in Tables 9.7 and 9.8. Secondly, there are many variant phenotypes reflecting changes of an undetermined nature in the products of mutant genes. These include such 'quantitative' variants as D^u and E^u, and such 'qualitative' variants as C^w, e^s, M_1, K^w, etc. The Rh antigen, V, which is determined only by genes which also determine c and e (DeNatale *et al.* 1955), has a high frequency in Negroes (see Table 9.8). Similarly, the *Di*^{a} gene, while rarely present in Europeans or Africans, has a variable frequency among Mongoloid populations, particularly in South American (Layrisse, 1958). Finally, in most or all of the systems, there is the so-called minus-minus phenotype which may reflect one of a number of different genetic mechanisms (see p. 278). In most instances, the frequency of the corresponding gene is so low that it would not alter the figures given. However, there are three notable exceptions: *O* in the ABO system, *Fy* in the Duffy system (homozygotes are Fy(a − b −)) and the absence of S or s antigens (indicated in Table 9.8 by an asterisk) in the MNSs system. The symbol *Ss* or *u* could not be used to designate this 'gene', because although most S − s − people are also U−, some are U+ (Francis & Hatcher, 1966; see also p. 302).

BLOOD GROUPING METHODS

There are several manuals and textbooks which provide ample instruction in the art and science of blood grouping (or typing; the terms are often used synonymously, but there is a tendency in some laboratories to talk about grouping cells for ABO and typing them for other antigens. It is difficult for the author to develop any strong prejudices in this matter.) Particularly useful is the most recent edition of Mollison's 'Blood Transfusion in Clinical Medicine' (1967), which contains an up-to-date discussion of various techniques as well as precise instructions for their performance. Another reason for not including details here is that commercially prepared antisera are always accompanied by a brochure which describes the procedure recommended for use with the particular lot of serum. The paragraphs which follow are meant to provide some general information and to point out some of the problems which beset the blood grouper at every stage of the laboratory procedures.

GENERAL PRINCIPLES

Only a few, deceptively simple, methods are used in most blood grouping laboratories. These tests are designed to show that an antigen-antibody reaction has taken place on the red cell surface. When the antibody used is an IgM globulin (see Chapter 1), its reaction with a red cell antigen is often followed by visible agglutination, due to intercellular bonding (i.e. lattice formation). Such 'complete' antibodies, when of sufficient titer and avidity, are simple to use and may require only a short incubation period. Common examples are anti-A, anti-B, anti-M, anti-N, and 'saline-active' anti-D. When obtained commercially, the anti-A and anti-B sera (of human origin) are colored with a distinctive dye to minimize the possibility of mistaken identity. Anti-A_1 is available either as B serum (from which anti-A has been absorbed by A_2 cells) or as a seed extract (lectin), while anti-M and anti-N are usually prepared by injecting animals (especially rabbits) with human red cells and removing the heteroagglutinins by absorption. Other 'saline agglutinins', such as anti-D, are obtained from human subjects immunized by pregnancy, transfusion or deliberate injection of human blood. In most instances, the agglutinating antibodies are found early in the course of the immunization, and they are nearly always replaced by 'incomplete' IgG antibodies within a short period.

IgG antibodies usually do not agglutinate saline or serum-suspended cells, but they may do so after high speed centrifugation, or in the presence of a high-protein medium such as albumin, or when the cells are treated with proteolytic enzymes such as trypsin, papain, bromelin or ficin. The high-protein medium reduces the electrical charges which normally oppose cell aggregation and thus permit the IgG coating the cells to bridge the intercellular gap (Pollack, 1965). The proteases directly reduce the red cell surface charge by removing sialic acid residues (Pollack et al., 1965). There is also an associated increase in antibody uptake (Masouredis, 1962), which is due to a rise in the antigen-antibody association constant (Hughes-Jones *et al.* 1964).

THE ANTIGLOBULIN TECHNIQUE

Some antibodies require the use of the antiglobulin technique, otherwise known as the Coombs test. In this technique, antibodies stimulated in animals against human globulin components form the intercellular bridges by reacting either with the antibodies coating the cells or else with complement components fixed to the cells. Thus, when human incomplete antibodies composed of IgG globulin react with antigens on the cell surface, their presence is detected by exposure to anti-human IgG antibodies. On the other hand, if the human antibodies (IgG or IgM) fix complement to the cells, then the agglutination observed after adding anti-human globulin serum may be due to the reaction of complement with anti-complement antibodies. The two kinds of reaction can be differentiated by mixing IgG with the anti-globulin serum to neutralize the anti-IgG before exposure to the coated cells. If no agglutination occurs, one can conclude that the previously observed agglutination was due to an IgG anti-IgG reaction. Alternatively, the distinction can be made by comparing the reactions of the coated cells with specific antiglobulin sera: anti-IgG and anti-complement.

When the antiglobulin test is performed on red cells which have simply been stored in their own serum at low temperatures, agglutination due to a complement anti-complement reaction may be observed. This phenomenon is due to the presence in nearly all normal human sera of an antibody-like protein with anti-H specificity, which fixes complement to the cells (Dacie, 1950; Crawford *et al.*, 1953b; Adinolfi, 1965a,b). Thus, whenever blood is kept in the refrigerator before being tested for blood group antigens by the antiglobulin technique, the possibility that the cells are already coated with complement should be remembered. Complement components can also be fixed to cells by lowering the ionic strength of the medium (Mollison and Polley, 1964).

In some diseases, the red cells become coated *in vivo* with globulin i.e., the cells give a positive *direct* antiglobulin test. Although the cells of healthy persons rarely take up globulin *in vivo*, such a possibility should be considered in setting up controls for blood grouping by the antiglobulin technique (i.e., the red cells should be tested with the antiglobulin reagent before, as well as after, exposure to the blood-group antibody).

NEUTRALIZATION TESTS

The principle of neutralization is employed when testing body fluids such as saliva, serum, urine, cyst fluids, etc. for the presence of A, B, H or Lewis substances. It is preferable to set up a series of tubes containing the antiserum (with an appropriate amount of antibody activity) and then to add the fluid being tested in a series of dilutions. The highest dilution of fluid (e.g. saliva) which will prevent the agglutination of selected red cells is recorded; it provides a semi-quantitative measure of the blood group substance present.

PITFALLS

Because of the relative simplicity of blood typing tests, there is a tendency to permit *inexperienced personnel* to perform the tests without due regard for the numerous pitfalls that await the unwary. Perhaps the most common source of error is failure to become thoroughly acquainted with the particular antisera being used. There is considerable heterogeneity in the behavior of antibody molecules, even within the same serum specimens, and especially between different lots obtained by bleeding either the same donor or different donors on repeated occasions. Furthermore, some antibodies are very labile, necessitating continuous evaluation of their effectiveness during the time they are used. Antibodies in the IgM fraction are particularly liable to lose activity during storage. Also, most examples of anti-Lewis and anti-Kidd antibodies tend to become inactive because their detection is dependent upon complement fixation, and some of the components of complement rapidly lose activity at temperatures above freezing. Thus, it may be necessary to add fresh serum or better, to perform a two-stage test, first incubating the cells with the antiserum and then, after washing, adding fresh serum to provide complement (Polley & Mollison, 1961). Another maneuver used to increase the effectiveness of such antibodies is the performance of an antiglobulin test on protease-treated cells incubated with antibody (van der Hart & van Loghem, 1953).

A second common source of error is *the inadequate use of controls*. The obvious positive and negative controls (i.e. cells with and without the antigen) are all too frequently omitted. In addition, dosage effect is often forgotten. Some antibodies react strongly with cells from homozygotes but only weakly with cells from heterozygotes; thus, an antigen may not be detected when in single dose. Furthermore, all antisera of human origin, except that from AB individuals, originally contained anti-A and/or anti-B, which was removed by absorption with red cells or neutralization with blood group substances. When this procedure is inadequately performed, weak reactions with A and/or B cells can occur and lead to misinterpretation of results. Thus, negative controls should be selected in such a way that unwanted anti-A and anti-B antibodies are revealed.

Even more confusing is the *presence of other antibodies* in the antiserum, undetected by the 'screening' procedure employed at the time the serum was originally tested. Such unwanted antibodies are usually directed against low-frequency antigens, and they are virtually impossible to exclude altogether, because the cell panel necessary for their detection would have to contain as many rare blood specimens as there are rare antigens. For this reason, it is never possible to guarantee that typing blood with only one example of a given antibody will give completely reliable results. Thus, whenever practicable, a second series of tests with a second antiserum of the same specificity should be included.

AUTOMATED BLOOD GROUPING

Very sensitive and reproducible blood grouping tests can be performed by the use of an auto-analyser (McNeil *et al.*, 1963; Allen *et al.*, 1963b). With the use of proteolytic enzymes and polyvinylpyrolidone (PVP), the number of antibody molecules required to produce 50 per cent agglutination of red cells is reduced by as much as 100-fold (Rosenfield *et al.*, 1964c). In this system, the enzyme-treated red cells are exposed to the antiserum containing PVP. During passage through the auto-analyser tubing, agglutination occurs, and the clumped cells are eliminated. The free cells continue on and are lysed while still in the tubing. The amount of hemoglobin released from these cells is measured and recorded by a photoelectric colorimeter.

While this kind of blood group test is particularly useful for research purposes (Rosenfield & Kochwa, 1964; Rosenfield *et al.*, 1964c), it can also be adapted for routine blood typing in laboratories where large numbers of blood specimens are tested.

REFERENCES

ACKERMAN D.R. (1967) Antibodies of the ABO system and the metabolism of human spermatozoa. *Nature* **213**, 253.

ADAM A., SHEBA CH., SANGER R. & RACE R.R. (1966a) The linkage relation of G-6-PD to Xg. *Am. J. hum. Genet.* **18**, 110.

ADAM A., TIPPETT P., GAVIN J., NOADES J., SANGER R. & RACE R.R. (1967) The linkage relation of Xg to G6PD in Israelis: the evidence of a second series of families. *Ann. hum. Genet.* **30**, 211.

ADAM A., ZIPRKOWSKI L., FEINSTEIN A., SANGER R. & RACE R.R. (1966b) Ichthyosis, Xg blood-groups and protan. *Lancet* **i**, 877.

ADINOLFI M. (1965a) Some serological characteristics of the normal incomplete cold antibody. *Immunology* **9**, 31.

ADINOLFI M. (1965b) Further attempts to characterize the normal incomplete cold antibody. *Immunology* **9**, 365.

ADINOLFI M., POLLEY M.J., HUNTER D.A. & MOLLISON P.L. (1962) Classification of group antibodies as β_2M or gamma globulin. *Immunology* **5**, 566.

ALLEN F.H. (1967) The blood groups and erythroblastosis, *in* REID D.E. (Ed.) *A Textbook of Obstetrics.* 2nd edn., W.B. Saunders, Philadelphia.

ALLEN F.H., CORCORAN P.A. & ELLIS F.R. (1960) Some new observations on the MN system. *Vox Sang.* **5**, 224.

ALLEN F.H. & DIAMOND L.K. (1957) *Erythroblastosis Fetalis.* Little, Brown, Boston.

ALLEN F.H., DIAMOND L.K. & NIEDZIELA B. (1951) A new blood antigen. *Nature* **167**, 482.

ALLEN F.H. & LEWIS S.J. (1957) Kp^a (Penney), a new antigen in the Kell blood group system. *Vox Sang.* **2**, 81.

ALLEN F.H., LEWIS S.J. & FUDENBERG H.H. (1958) Studies of anti-Kp^b, a new antibody in the Kell blood group system. *Vox Sang.* **3**, 1.

ALLEN F.H., MADDEN H.J. & KING R.W. (1963a) The MN gene Mu, which produces M and U but no N, S or s. *Vox Sang.* **8**, 549.

ALLEN F.H. & ROSENFIELD R.E. (1961) Notation for the Kell blood group system. *Transfusion* **1**, 305.

ALLEN F.H., ROSENFIELD R.E. & ADEBAHR M.E. (1963b) Kidd and Duffy blood typing without Coombs serum. Adaptation of the auto-analyzer hemagglutination system. *Vox Sang.* **8**, 698.

ALLISON A.C. (1955) Aspects of polymorphism in man. *Cold Spring Harbor Symposia on Quantitative Biology* **20**, 239.

ALOYSIA M., GELB A.G., FUDENBERG H.H., HAMPER J., TIPPETT P. & RACE R.R. (1961) The expected 'Bombay' groups $O_h^{A_2}$ and $O_h^{A_1}$. *Transfusion* **1**, 212.

ALTER A.A. & ROSENFIELD R.E. (1964) The nature of some subtypes of A. *Blood* **23**, 605.

ANDERSEN J. (1961) Production of B character on human erythrocytes by haemosensitization with purified *Escherichia coli* O 86 substance. *Nature* **190**, 730.

ANDERSON C., HUNTER J., ZIPURSKY A., LEWIS M. & CHOWN B. (1963) An antibody defining a new blood group antigen, Bu^a. *Transfusion* **3**, 30.

12

ANDERSON J., MEADOW P.M., HASKIN M.A. & STROMINGER J.L. (1966) Biosynthesis of the peptidoglycan of bacterial cell walls. *Archs Biochem. Biophys.* **116**, 487.

ANDERSON R.E. & WALFORD R.L. (1963) Direct demonstration of A, B and Rh_0 (D) blood groups antigens on human leucocytes. *Am. J. clin. Path.* **40**, 239.

ANDRE R. & SALMON C. (1957) Étude sérologique comparée de neuf examples non apparentés de groupe sanguin 'A faible'. *Revue Hémat.* **12**, 668.

ANDRESEN P.H. (1948) The blood group system L: A new blood group L_2. A case of epistasy within the blood groups. *Acta path. microbiol. scand.* **25**, 728.

ARFORS K.E., BECKMAN L. & LUNDIN L.G. (1963) Genetic variations of human serum phosphatases. *Acta genet.* **13**, 89.

ARIAS I.M., GARTNER I.M., SEIFTER S. & FURMAN M. (1964) Prolonged neonatal unconjugated hyperbilirubinemia associated with breast feeding and a steroid, pregnane-3 (alpha), 20 (beta) diol, in maternal milk that inhibits glucuronide formation *in vitro. J. clin. Invest.* **43**, 2037.

BAMFORD K.F., HARRIS H., LUFFMAN J.E., ROBSON E.B. & CLEGHORN T.E. (1965) Serum alkaline-phosphatase and the ABO blood groups. *Lancet* **i**, 530.

BECKMAN L. (1964) Associations between human serum alkaline phosphatases and blood groups. *Acta genet.* **14**, 286.

BECKMAN L. (1968) Blood groups and serum alkaline phosphatase. *Series Haematologica* I, 1, 137.

BHENDE Y.M., DESHPANDE C.K., BHATIA H.M., SANGER R., RACE R.R., MORGAN W.T.J. & WATKINS W.M. (1952) A 'new' blood group character related to the ABO system. *Lancet* **i**, 903.

BLYTH H., CARTER C.O., DUBOWITZ V., EMERY A.E.H., GAVIN J., JOHNSTON H.A., MCKUSICK V.A., RACE R.R., SANGER R. & TIPPETT P. (1965) Duchenne's muscular dystrophy and Xg blood groups: a search for linkage. *J. med. Genet.* **2**, 157.

BOETTCHER B. (1966) Modification of Bernstein's multiple allele theory for the inheritance of the ABO blood groups in the light of modern genetical concepts. *Vox Sang.* **11**, 129.

BOETTCHER B., VOS G.H. & HAY J. (1965) The Rh antigens of anthropoid apes in relation to Rh 'deletion' phenotypes. *Am. J. hum. Genet.* **17**, 308.

BOOTH P.B., PLAUT G., JAMES J.D., IKIN E.W., MOORES P., SANGER R. & RACE R.R. (1957) Blood chimaerism in a pair of twins. *Br. med. J.* **i**, 1456.

BOYD W.C. & REEVES E. (1961) Specific inhibition of anti-D antibody by colominic acid. *Nature* **191**, 511.

BRAND G., YUNIS E. & YUNIS J.J. (1963) H antigen in HeLa cells and H antibodies in anti-HeLa sera. *Nature* **200**, 362.

BRENDEMOEN O.J. (1952) Some factors influencing Rh immunization during pregnancy. *Acta path. microbiol. scand.* **31**, 579.

BRENDEMOEN O.J. (1961) Development of the Lewis blood group in the newborn. *Acta path. microbiol. scand.* **52**, 55.

CALVO R., STEIN W., KOCHWA S. & ROSENFIELD R.E. (1965) Acute hemolytic anemia due to anti-i: frequent cold agglutinins in infectious mononucleosis. *J. clin. Invest.* **44**, 1033.

CAHAN A., JACK J.A., SCUDDER J., SARGENT M., SANGER R. & RACE R.R. (1957) A family in which A_x is transmitted through a person of the blood group A_2B. *Vox Sang.* **2**, 8.

CALLENDER S.T. & RACE R.R. (1946) A serological and genetical study of multiple antibodies formed in response to blood transfusion by a patient with lupus erythematosus diffusus. *Ann. Eugen.* **13**, 102.

CAMERON C., GRAHAM F., DUNSFORD I., SICKLES G., MACPHERSON C.R., CAHAN A., SANGER R., & RACE R.R. (1959) Acquisition of a B-like antigen by red blood cells. *Br. med. J.* **ii**, 29.

CAMERON G.L. & STAVELEY J.M. (1957) Blood group P substance in hydatid cyst fluids. *Nature* **179**, 147.

CEPPELLINI R. (1955) On the genetics of secretor and Lewis characters: a family study. *Proc. 5th Cong. int. Soc. Blood Transf.* Paris, p. 207.

CEPPELLINI R. (1959) Physiological genetics of human blood factors, *in* WOLSTEN-HOLME G.E.W. & O'CONNER C.N. (eds.) *Ciba Foundation Symposium on Biochemistry of Human Genetics*, p. 242. Little, Brown, Boston.

CEPPELLINI R., CURTONI E.S., MATTIUZ P.L., LEIGHEB G., VISETTI M. & COLOMBI A. (1966) Survival of test skin grafts in man: effects of genetic relationship and of blood group incompatibility. *Ann. N.Y. Acad. Sci.* **129**, 421.

CEPPELLINI R., DUNN L.C. & TURRI M. (1955) An interaction between alleles at the Rh locus in man which weakens the reactivity of the Rh_0 factor. *Proc. natn. Acad. Sci.* **41**, 283.

CHALMERS J.N.M. & LAWLER S.D. (1953) Data on linkage in man: elliptocytosis and blood groups. I. Families 1 and 2. *Ann. Eugen.* **17**, 267.

CHESSIN L.N., BRAMSON S., KUHNS W.J. & HIRSCHHORN K. (1965) Studies on the A, B, O(H) blood groups on human cells in culture. *Blood*, **25**, 944.

CHOWN B. & LEWIS M. (1957) Occurrence of D^u type of reaction when CDe or cDE is partnered with Cde. *Ann. hum. Genet.* **22**, 58.

CHOWN B., LEWIS M. & KAITA H. (1965) An anomaly of inheritance in the MNSs blood groups. *Am. J. hum. Genet.* **17**, 9.

CHUNG C.P. & MORTON N.E. (1961) Selection at the ABO locus. *Am. J. hum. Genet.* **13**, 9.

CLARK J.I., PUITE R.H., MARCZYNSKI R. & MANN J.D. (1963) Evidence for the absence of detectable linkage between the genes for Duchenne muscular dystrophy and the Xg blood group. *Am. J. hum. Genet.* **15**, 292.

CLARKE C.A. (1961) Blood groups and disease, *in* STEINBERG A.G. & BEARN A.G. (eds.) *Progress in Medical Genetics*, vol. 1, p. 81. Grune & Stratton, New York.

CLARKE C.A. (1967) Prevention of Rh-haemolytic disease. *Br. med. J.* **4**, 7.

CLARKE C.A., DONOHOE W.T.A., FINN R., McCONNELL R.B., SHEPPARD P.M. & NICOL D.S.H. (1960a) Data on linkage in man: ovalocytosis, sickling and the Rhesus blood group complex. *Ann. hum. Genet.* **24**, 283.

CLARKE C.A., DONOHOE W.T.A., McCONNELL R.B., WOODROW J.C., FINN R., KREVANS J.R., KULKE W., LEHANE D. & SHEPPARD P.M. (1963) Further experimental studies on the prevention of Rh haemolytic disease. *Br. med. J.* **i**, 979.

CLARKE C.A., FINN R., LEHANE D., McCONNELL R.B., SHEPPARD P.M. & WOODROW J.C. (1966) Dose of anti-D gamma-globulin in prevention of Rh-haemolytic disease of the newborn. *Br. med. J.* **i**, 213.

Clarke C.A., McConnell R.B. & Sheppard P.M. (1960b) ABO blood groups and secretor character in rheumatic carditis. *Br. med. J.* i, 21.

Clarke C.A., McConnell R.B. & Sheppard P.M. (1961) Data on linkage in man. The nail-patella syndrome, family S. *Ann. hum. Genet.* 25, 25.

Cohen B.H. (1960) ABO-Rh interaction in an Rh-incompatibly mated population. *Am. J. hum. Genet.* 12, 180.

Cohen F. & Zuelzer W.W. (1965) Interrelationship of the various subgroups of the blood group A: study with immunofluorescence. *Transfusion* 5, 223.

Cook P.J.L. (1965) The Lutheran-Secretor recombination fraction in man: a possible sex difference. *Ann. hum. Genet.* 28. 393.

Coombs R.R.A., Bedford B. & Rouillard L.M. (1956) A and B blood group antigens on human epidermal cells demonstrated by mixed agglutination. *Lancet* i, 461.

Coombs R.R.A., Mourant A.E. & Race R.R. (1946) *In vivo* isosensitization of red cells in babies with haemolytic disease. *Lancet* i, 264.

Crawford H., Cutbush M. & Mollison P.L. (1953a) Hemolytic disease of the newborn due to anti-A. *Blood* 8, 620.

Crawford H., Cutbush M. & Mollison P.L. (1953b) Specificity of incomplete 'cold' antibody in human serum. *Lancet* i, 566.

Crawford M.N., Punnett H.H. & Carpenter G.G. (1967) Deletion of the long arm of chromosome 16 and an unexpected Duffy blood group phenotype reveal a possible autosomal linkage. *Nature* 215, 1075.

Cutbush M. & Chanarin I. (1956) The expected blood group antibody, anti-Lu^b. *Nature* 178, 855.

Cutbush M., Giblett E.R. & Mollison P.L. (1956) Demonstration of the phenotype Le(a + b +) in infants and in adults. *Br. J. Haemat.* 2, 210.

Cutbush M. & Mollison P.L. (1950) The Duffy blood group system. *Heredity* 4, 383.

Dacie J.V. (1950) Occurrences in normal human sera of 'incomplete' forms of 'cold' auto-antibodies. *Nature* 166, 36.

Dausset J. & Rapaport F.T. (1966) The role of blood group antigens in human histocompatibility. *Ann. N.Y. Acad. Sci.* 129, 408.

Davies S.H., Gavin J., Goldsmith K.L.G., Graham J.B., Hamper J., Hardisty R.M., Harris J.B., Holman C.A., Ingram G.I.C., Jones T.G., McAfee L.A., McKusick V.A., O'Brien J.R., Race R.R., Sanger R. & Tippett P. (1963) The linkage relations of hemophilia A and hemophilia B (Christmas disease) to the Xg blood group system. *Amer. J. hum. Genet.* 15, 481.

De La Chapelle A., Hortling H., Sanger R. & Race R.R. (1964) Successive non-disjunction at first and second meiotic division of spermatogenesis: evidence of chromosomes and Xg. *Cytogenetics* 3, 334.

DeNatale A., Cahan A., Jack J.A., Race R.R. & Sanger R. (1955) V, a 'new' Rh antigen, common in Negroes, rare in white people. *J. Am. med. Ass.* 159, 247.

DeWit C.D. & Borst-Eilers E. (1968) Failure of anti-D immunoglobulin injection to protect against Rhesus immunization after massive foeto-maternal haemorrhage. Report of 4 cases. *Br. med. J.* i, 152.

Dodd M.C., Bigley N.J. & Geyer V.B. (1960) Specific inhibition of Rh$_0$(D) antibody by sialic acids. *Science* 132, 1398.

DODD M.C., BIGLEY N.J. & GEYER V.B. (1963) Further observations on the serologic specificity of sialic acid for $Rh_0(D)$ antibody. *J. Immunol.* **90**, 518.

DODD M.C., BIGLEY N.J., JOHNSON G.A. & McCLUER R.H. (1964) Chemical aspects of inhibitors of $Rh_0(D)$ antibody. *Nature* **204**, 549.

DONAHUE, R.P., RENWICK J.H., COBOS, L., BORGAONKAR D.S., BIAS W.B. & McKUSICK V.A. (1968) Karyotypic and linkage analysis in two pedigrees with marker chromosomes. *Clin. Res.* **16**, 296.

DUCOS J., RUFFIE J., COLUMBIES P., MARTY Y. & OHAYON E. (1965) I antigen in leukaemic patients. *Nature* **208**, 1329.

DUNN L.C. & BENNETT D. (1967) Sex differences in recombination of linked genes in animals. *Genet. Res.* **9**, 211.

DUNSFORD I., BOWLEY C.C., HUTCHINSON A.M., THOMPSON J.S., SANGER R. & RACE R.R. (1953) A human blood group chimaera. *Br. med. J.* **ii**, 81.

EATON B.R., MORTON J.A., PICKLES M.M. & WHITE K.E. (1956) A new antibody anti-Yt[a], characterizing a blood group of high incidence. *Br. J. Haemat.* **2**, 233.

EDWARDS J.H., FERGUSON-SMITH M.A., FROLAND A., LINDSTEN J., POLANI P.E., RACE R.R. & SANGER R. (1966) The contribution of the Xg blood groups to X chromosome aneuploidy: an interim report. *Meeting Int. Soc. hum. Genet.* Chicago, p. 28.

EDWARDS R.G., FERGUSON L.C. & COOMBS R.R.A. (1964) Blood group antigens in human spermatozoa. *J. Reprod. Fert.* **17**, 153.

EL-HEFNAWI H., SMITH S.M., PENROSE L.S. (1965) Xeroderma pigmentosum—its inheritance and relationship to the ABO blood group system. *Ann. hum. Genet.* **28**, 273.

ELLIOTT G.B., LEWIS M., KAITA H. & CHOWN B. (1963) The blood groups of a further family with nail-patella syndrome. *Am. J. hum. Genet.* **15**, 182.

EVANS D.A.P. (1965) Confirmation of association between ABO blood groups and salivary ABH secretor phenotypes and electrophoretic patterns of serum alkaline phosphatase. *J. med. Genet.* **2**, 126.

EVANS R.S., TURNER E. & BINGHAM M. (1963) Studies of I^{131} tagged Rh antibody of D specificity. *Vox Sang.* **8**, 153.

FEIZI T. & DARRELL J.H. (1966) Failure to demonstrate blocking of I antigen by *Mycoplasma pneumoniae in vitro* and *in vivo*. *Nature* **211**, 1159.

FEIZI T. & HARDISTY R.M. (1966) I antigen in leukemic patients. *Nature* **210**, 1066.

FIALKOW P.J., GIBLETT E.R. & MOTULSKY A.G. (1967) Measurable linkage between ocular albinism and Xg. *Am. J. hum. Genet.* **19**, 63.

FIGUR A.M. & ROSENFIELD R.E. (1965) The crossreaction of anti-N with type M erythrocytes. *Vox Sang.* **10**, 169.

FISHER R.A., RACE R.R. & TAYLOR G.L. (1944) Mutation and the rhesus reaction. *Nature* **153**, 106.

FISHMAN W.H. & GHOSH N.K. (1967) Isoenzymes of human alkaline phosphatase, *in* BODANSKY O. & STEWART C.P. (eds.) *Advances in Clinical Chemistry*, vol. 10. Academic Press, New York.

FISK R.T. & FOORD A.G. (1942) Observations on the Rh agglutinogen of human blood. *Am. J. clin. Path.* **12**, 545.

FLORY L.L. (1966) Differences in the H antigen on human buccal cells from secretor and non-secretor individuals. *Vox Sang.* **11**, 137.

FRANCIS B.J. & HATCHER D.E. (1966) MN blood types. The $S - s - U +$ and the M_1 phenotypes. *Vox Sang.* **11**, 213.

FRANKS D. (1966) Antigenic markers on cultured human cells, I. Ii, Tj^a, Donath-Landsteiner and 'non-specific' autoantigens. *Vox Sang.* **11**, 674.

FRASER G.R. (1963) Parental origin of the sex chromosomes in the XO and XXY karyotypes in man. *Ann. hum. Genet.* **26**, 297.

FREDA V.J., GORMAN J.G. & POLLACK W. (1964) Successful prevention of experimental Rh sensitization in man with an anti-Rh gamma$_2$-globulin antibody preparation: a preliminary report. *Transfusion* **4**, 26.

FREDA V.J., GORMAN J.G. & POLLACK W. (1966) Rh factor: prevention of immunization and clinical trial on mothers. *Science* **151**, 828.

FREDA V.J., GORMAN J.G., POLLACK W., ROBERTSON J.G., JENNINGS E.R. & SULLIVAN J.F. (1967) Prevention of Rh isoimmunization. Progress report of the clinical trial in mothers. *J. Am. med. Ass.* **199**, 390.

GAMMELGAARD A. (1942) On rare weak A antigens (A_3, A_4, A_5 and A_x) in man. *Thesis for Ph.D.*, Univ. of Copenhagen translated by COWAN H., 1964. Walter Reed Army Institute of Research, Washington D.C.

GARTLER S., WAXMAN S.H. & GIBLETT E.R. (1962) An XX/XY hermaphrodite resulting from double fertilization. *Proc. nat. Acad. Sci.* **48**, 332.

GEDDE-DAHL T., GRIMSTAD A.L., GUNDERSON S. & VOGT E. (1967) A probable crossing over or mutation in the MNSs blood group system. *Acta genet.* **17**, 193.

GEERDINK R.A., NIJENHUIS L.E. & HUIZINGA J. (1967) Hereditary elliptocytosis: linkage data in man. *Ann. hum. Genet.* **30**, 363.

GERSHOWITZ H., BEHRMAN S.J. & NEEL J.V. (1958) Hemagglutinins in uterine secretions. *Science* **128**, 719.

GIBBS M.B. (1966) The quantitative relationship of the Rh-like (LW) and D antigens of human erythrocytes. *Nature* **210**, 642.

GIBBS M.B. & BECKER E.L. (1963) Quantitation of haemagglutination by enumeration of free cells by an electronic counter. *Nature* **198**, 90.

GIBBS M.B. & ROSENFIELD R.E. (1966) Immunochemical studies of the Rh system. IV. Hemagglutination assay of antigenic expression regulated by interaction between paired Rh genes. *Transfusion* **6**, 462.

GIBLETT E.R. (1958) Js, a 'new' blood group antigen found in Negroes. *Nature* **181**, 1221.

GIBLETT E.R. (1961) A critique of the theoretical hazard of inter- vs intra-racial transfusion. *Transfusion* **1**, 233.

GIBLETT E.R. (1964) Blood group antibodies causing hemolytic disease of the newborn. *Clin. Obstet. Gynec.* **7**, 1044.

GIBLETT E.R. & CHASE J. (1959) Js^a, a 'new' red cell antigen found in Negroes: evidence for an eleventh blood group system. *Br. J. Haemat.* **5**, 319.

GIBLETT E.R., CHASE J. & MOTULSKY A.G. (1957) Studies on anti-V: a recently discovered Rh antibody. *J. lab. clin. Med.* **49**, 433.

GIBLETT E.R. & CROOKSTON M.C. (1964) Agglutinability of red cells by anti-i in patients with thalassaemia major and other haematological disorders. *Nature* **201**, 1138.

GIBLETT E.R., HILLMAN R.S. & BROOKS L.E. (1965) Transfusion reaction during marrow suppression in a thalassemic patient with a blood group anomaly and an unusual cold agglutinin. *Vox Sang.* **10**, 448.

GILES C.M., HUTH M.C., WILSON T.E., LEWIS H.B.M. & GROVE G.E.B. (1965) Three examples of a new antibody, anti-Cs^a, which reacts with 98% of red cell samples. *Vox Sang.* **10**, 405.

GILES C.M. & METAXAS M.N. (1964) Identification of the predicted blood group antibody, anti-Yt^b. *Nature* **202**, 1122.

GILES C.M., METAXAS-BÜHLER M., ROMANSKI Y. & METAXAS M.N. (1967) Studies on the Yt blood group system. *Vox Sang.* **13**, 171.

GILES C.M., MOURANT A.E. & ATABUDDIN A.H. (1963) A Lewis-negative 'Bombay' blood. *Vox Sang.* **8**, 269.

GILES C.M., MOURANT A.E., PARKIN D.M., HORLEY J.F. & TAPSON K.J. (1959) A weak B antigen, probably acquired. *Br. med. J.* **ii**, 32.

GLASS B. (1950) The action of selection on the principal Rh alleles. *Am. J. hum. Genet.* **2**, 269.

GLOVER S.M. & WALFORD R.L. (1958) A serologic and family study of the rare blood group A$_x$. *Am. J. clin. Path.* **30**, 539.

GLYNN A.A., GLYNN L.E. & HOLBOROW E.J. (1956) The secretor status of rheumatic fever patients. *Lancet* **ii**, 759.

GLYNN L.E. & HOLBOROW E.J. (1961) Relation between blood groups, secretor status and susceptibility to rheumatic fever. *Arthritis Rheumat.* **4**, 203.

GLYNN L.E., HOLBOROW E.J. & JOHNSON G.D. (1957) The distribution of blood group substances in human gastric and duodenal mucosa. *Lancet* **ii**, 1083.

GOLD E.R. (1964) Observations on the specificity of anti-O and anti-A$_1$ sera. *Vox Sang.* **9**, 153.

GOLD E.R., TOVEY G.H., BENNEY W.E. & LEWIS F.J.W. (1959) Changes in the group A antigen in a case of leukaemia. *Nature* **183**, 892.

GOODALL H.B., HENDRY D.W.W., LAWLER S.D. & STEPHEN S.A. (1953) Data on linkage in man: elliptocytosis and blood groups. II. Family 3. *Ann. Eugen.* **17**, 272.

GORMAN J.G., DIRE J., TREACY A.M. & CAHAN A. (1963) The application of Xg^a antiserum to the question of red cell mosaicism in female heterozygotes. *J. Lab. clin. Med.* **61**, 642.

GREEN F.A. (1965) Studies on the Rh (D) antigen. *Vox Sang.* **10**, 32.

GREEN F.A. (1967a) Erythrocyte membrane sulfhydryl groups and Rh antigen activity. *Immunochemistry* **4**, 247.

GREEN F.A. (1967b) Rh antigenicity: an essential component soluble in butanol. *Blood* **30**, 870.

GREENWALT T.J. (1961) Confirmation of linkage between the Lutheran and Secretor genes. *Am. J. hum. Genet.* **13**, 69.

GREENWALT T.J., SASAKI T., SANGER R., SNEATH J. & RACE R.R. (1954) An allele of the S(s) blood group genes. *Proc. nat. Acad. Sci.* **40**, 1126.

GROLLMAN A.P. (1967) Metabolic pathways leading to the biosynthesis of blood group substances, *in* KUHNS W.J. (ed.) *J.A. Hartford Foundation Conference on Blood Groups and Blood Transfusion*, vol. 1, p. 33.

GROLLMAN A.P., HALL C.W. & GINSBURG J. (1965) Biosynthesis of fucosyllactose and other oligosaccharides found in milk. *J. biol. Chem.* **240**, 975.

GROLLMAN A.P. & MARCUS D.M. (1966) Enzymatic incorporation of fucose into blood group H substance. *Biochem. biophys. Res. Commun.* **25**, 542.

GROLLMAN E.F. & GINSBURG V. (1967) Correlation between secretor status and the occurrence of 2'-fucosyllactose in human milk. *Biochem. biophys. Res. Commun.* **28**, 50.

GROVE-RASMUSSEN M., SOUTTER L. & LEVINE P. (1952) A new blood subgroup (A_0) identifiable with group O serum. *Am. J. clin. Path.* **22**, 1157.

GRUBB R. (1948) Correlation between Lewis blood groups and secretor character in man. *Nature* **162**, 933.

GRUBB R. (1951) Observations on the human group system Lewis. *Acta path. microbiol. scand.* **28**, 61.

GULLBRING B. (1957) Investigation on the occurrence of blood group antigens in spermatozoa from man, and serological demonstration of the segregation of characters. *Acta med. scand.* **159**, 169.

HAKIM S.A., VYAS G.N., SANGHVI D.L. & BHATIA H.M. (1961) Eleven cases of 'Bombay' phenotype in six families: suppression of the ABO antigen demonstrated in two families. *Transfusion* **1**, 218.

HALDANE J.B.S. (1922) Sex ratio and unisexual sterility in hybrid animals. *J. Genet.* **12**, 101.

HAMILTON E.G. (1967) Prevention of Rh immunization by injection of anti-D antibody. *Obstet. Gynec.* **30**, 812.

HARRISON J.F. (1964) Haemophilia, Christmas disease and the Xg blood groups. Observations based on the haemophiliacs of Birmingham. *Br. J. Haemat.* **10**, 115.

HASEKURA H., ISHIMORI T. & YAMADA K. (1967) Chromosome studies in the Japanese boy lacking whole Rh blood group antigen and his family members. *Proc. Japan. Acad.* **43**, 249.

HILLMAN R.S. & GIBLETT E.R. (1965) Red cell membrane alteration associated with 'marrow stress'. *J. clin. Invest.* **44**, 1730.

HINZ C.F. (1963) Serologic and physicochemical characterization of Donath-Landsteiner antibodies from six patients. *Blood* **22**, 600.

HIRAIZUMI Y. (1964) Prezygotic selection as a factor in the maintenance of variability. *Cold Spring Harbor Symposia on Quantitative Biology* **29**, 51.

HIRSCH W., MOORES P., SANGER R. & RACE R.R. (1957) Notes on some reactions of human anti-M and anti-N sera. *Br. J. Haemat.* **3**, 134.

HOLBOROW E.J., BROWN P.C., GLYNN L.E., HAWES M.D., GRESHAM G.A., O'BRIEN T.F. & COOMBS R.R.A. (1960) The distribution of the blood group A antigen in human tissues. *Br. J. exp. Path.* **41**, 430.

HOOGSTRATEN B., ROSENFIELD R.E. & WASSERMAN L.R. (1961) Change of ABO blood type in a patient with leukemia. *Transfusion* **1**, 32.

HØSTRUP H. (1962) A and B blood group substances in the serum of normal subjects. *Vox Sang.* **7**, 704.

HØSTRUP H. (1963) A and B blood group substances in the serum of the newborn infant and the foetus. *Vox Sang.* **8**, 557.

HOTTA K. & SPRINGER G.F. (1965) Isolation and partial characterization of blood group N specific haptens from human blood group M and N substances. *Proc. 10th Cong. int. Soc. Blood Transf.* Stockholm, p. 505.

HUGHES-JONES N.C., GARDNER B. & TELFORD R. (1963a) Studies on the reaction between the blood group antibody anti-D and erythrocytes. *Biochem. J.* **88,** 435.

HUGHES-JONES N.C., GARDNER B. & TELFORD R. (1963b) Comparison of various methods of dissociation of anti-D, using ^{131}I labelled antibody. *Vox Sang.* **8,** 531.

HUGHES-JONES N.C., GARDNER B. & TELFORD R. (1964) The effect of ficin on the reaction between anti-D and red cells. *Vox Sang.* **9,** 175.

HULTEN M., LINDSTEN J., PEN-MING L.M., FRACCARO M., MANNINI A., TIEPOLO L., ROBSON E.B., HEIKEN A. & TILLINGER K.G. (1966) Possible localization of the genes for the Kidd blood group on an autosome involved in a reciprocal translocation. *Nature* **211,** 1067.

IKIN E.W., MOURANT A.E., PETTENKOFER H.J. & BLUMENTHAL G. (1951) Discovery of the expected haemagglutinin, anti-Fy^{b}. *Nature* **168,** 1077.

ISHIMORI T. & HASEKURA H. (1966) A case of a Japanese blood with no detectable Rh blood group antigen. Discovery of a homozygote of deleted chromosomes or silent Rh alleles. *Proc. Japan. Acad.* **42,** 658.

ISHIMORI T. & HASEKURA H. (1967) A Japanese with no detectable Rh blood group antigens due to silent Rh alleles or deleted chromosomes. *Transfusion* **7,** 84.

ISSITT P.D. (1967) I blood group system and its relation to other blood group systems. *J. med. lab. Tech.* **24,** 90.

ISSITT P.D., TEGOLI J., JACKSON V., SANDERS C.W. & ALLEN F.H. (1968) Anti-IP$_1$: antibodies that show an association between the I and P blood group systems. *Vox Sang.* **14,** 1.

JACKSON C.E., SYMON W.E. & MANN J.D. (1964) X chromosome mapping of genes for red-green colorblindness and Xg. *Am. J. hum. Genet.* **16,** 403.

JAMESON R.J., LAWLER S.D. & RENWICK J.H. (1956) Nail-patella syndrome: clinical and linkage data on family G. *Ann. hum. Genet.* **20,** 348.

JENKINS W.J., KOSTER H.G., MARSH W.L. & CARTER R.L. (1965) Infectious mononucleosis: an unsuspected source of anti-i. *Br. J. Haemat.* **11,** 480.

JENKINS W.L. & MARSH W.L. (1965) Somatic mutation affecting the Rhesus and Duffy blood group systems. *Transfusion* **5,** 6.

JENKINS W.L., MARSH W.L., NOADES J., TIPPETT P., SANGER R. & RACE R.R. (1960) The I antigen and antibody. *Vox Sang.* **5,** 97.

JORDAL K. (1956) The Lewis blood groups in children. *Acta path. microbiol. scand.* **39,** 399.

KABAT E.A. (1956) *Blood Group Substances.* Academic Press, New York.

KABAT E.A. (1966) The nature of an antigenic determinant. *J. Immunol.* **97,** 1.

KABAT E.A. & LESKOWITZ S. (1955) Immunochemical studies on blood groups. XVII. Structural units involved in blood group A and B specificity. *J. Am. chem. Soc.* **77,** 5159.

KENT S.P. (1964) The demonstration and distribution of water soluble blood group O (H) antigen in tissue sections using a fluorescein labelled extract of *Ulex Europeus* seed. *J. Histochem. Cytochem.* **12,** 591.

KERR C.B., WELLS R.S. & SANGER R. (1964) X-linked ichthyosis and the Xg groups. *Lancet* **ii**, 1369.

KISSMEYER-NIELSEN F. (1965) Irregular blood group antibodies in 200,000 individuals. *Scand. J. Haemat.* **2**, 331.

KOCHWA S., ROSENFIELD R.E., TALLAL L. & WASSERMAN L.R. (1961) Isoagglutinins associated with erythroblastosis. *J. clin. Invest.* **40**, 874.

KORTEKANGAS A.E., KAARSALO E., MELARTIN L., TIPPETT P., GAVIN J., NOADES J., SANGER R. & RACE R.R. (1965) The red cell antigen P^k and the P system: the evidence of three more P^k families. *Vox Sang.* **10**, 385.

KOSCIELAK J. (1963) Blood group A specific glycolipids from human erythrocytes. *Biochim. biophys. Acta.* **78**, 313.

KUHNS W.J., RAPAPORT F.T., LAWRENCE H.S. & CONVERSE J.M. (1966) Relationship of human blood group antigens and antibodies to survival of skin homografts. *Transplantation* **4**, 250.

LANDSTEINER K. & LEVINE P. (1927) Further observations on individual differences of human blood. *Proc. Soc. exp. Biol. Med.* **24**, 941.

LANDSTEINER K. & LEVINE P. (1928) On the inheritance of agglutinogens of human blood demonstrable by immune agglutinins. *J. exp. Med.* **48**, 731.

LANDSTEINER K. & WIENER A.S. (1940) An agglutinable factor in human blood recognized by immune sera for rhesus blood. *Proc. Soc. exp. Biol. Med.* **43**, 223.

LANGMAN M.J.S., LEUTHOLD E., ROBSON E.B., HARRIS J., LUFFMAN J.E. & HARRIS H. (1966) Influence of diet on the 'intestinal' component of serum alkaline phosphatase in people of different ABO blood groups and secretor status. *Nature* **212**, 41.

LAWLER S.D. (1964) Localization of autosomal genes in man. *Hum. Biol.* **36**, 146.

LAWLER S.D. & MARSHALL R. (1961) Lewis and secretor characters in infancy. *Vox Sang.* **6**, 541.

LAWLER S.D. & RENWICK J.H. (1959) Blood groups and genetic linkage. *Br. med. Bull.* **15**, 145.

LAWLER S.D., RENWICK J.H., HAUGE M., MOSBECK J. & WILDERVANCK L.S. (1958) Linkage tests involving the P blood group locus and further data on the ABO: nail-patella linkage. *Ann. hum. Genet.* **22**, 342.

LAWLER S.D., RENWICK J.H. & WILDERVANCK L.S. (1957) Further families showing linkage between the ABO and nail-patella loci, with no evidence of heterogeneity. *Ann. hum. Genet.* **21**, 410.

LAWLER S.D. & SANDLER M. (1954) Data on linkage in man: elliptocytosis and blood groups. IV. Families 5, 6 and 7. *Ann. Eugen.* **18**, 328.

LAYRISSE M. (1958) Anthropological considerations of the Di^a antigen. *Am. J. phys. Anthrop.* **16**, 173.

LAYRISSE M., ARENDS T. & DOMINGUEZ R.S. (1955) Neuvo grupo sanguineo encontrado en descendientes de Indios. *Acta med. venez.* **3**, 132.

LEVENE H. & ROSENFIELD R.E. (1961) ABO incompatibility, *in* STEINBERG A.G. & BEARN A.G. (eds.) *Progress in Medical Genetics*, vol. 1, p. 120. Grune & Stratton, New York.

LEVIN B.R. (1967) The effect of reproductive compensation on the long term maintenance of the Rh polymorphism: the Rh crossroad revisited. *Am. J. hum. Genet.* **19**, 288.

LEVINE P. (1943) Serological factors as possible causes in spontaneous abortions. *J. Hered.* **34**, 71.

LEVINE P. & CELANO M.J. (1961) The question of D (Rh$_0$) antigenic sites on human spermatozoa. *Vox Sang.* **6**, 720.

LEVINE P. & CELANO M.J. (1967) Agglutinating specificity for LW factor in guinea pig and rabbit anti-Rh serums. *Science* **156**, 1744.

LEVINE P., CELANO M.J. & FALKOWSKI F. (1963a) The specificity of the antibody in paroxysmal cold hemoglobinuria. *Transfusion* **3**, 278.

LEVINE P., CELANO M.J., FALKOWSKI F., CHAMBERS J.W., HUNTER O.B. & ENGLISH C.T. (1964) A second example of ---/--- or Rh$_{null}$ blood. *Nature* **204**, 892.

LEVINE P., CELANO M.J., FALKOWSKI F., CHAMBERS J.W., HUNTER O.B. & ENGLISH C.T. (1965) A second example of ---/--- or Rh$_{null}$ blood. *Transfusion* **5**, 492.

LEVINE P., CELANO M., FENICHEL R., POLLACK W. & SINGHER H. (1961a) A 'D-like' antigen in rhesus monkey, human Rh positive and human Rh negative red blood cells. *J. Immunol.* **87**, 747.

LEVINE P., CELANO M., FENICHEL R. & SINGHER H. (1961b) A 'D-like' antigen in rhesus red blood cells and in Rh-positive and Rh-negative red cells. *Science* **133**, 332.

LEVINE P., CELANO M.J., VOS G.H. & MORRISON J. (1962) The first human blood, ---/---, which lacks the D-like antigen. *Nature* **194**, 304.

LEVINE P., CELANO M.J., WALLACE J. & SANGER R. (1963b) A human 'D-like' antibody. *Nature* **198**, 596.

LEVINE P., OTTENSOOSER F., CELANO M.J. & POLLITZER W. (1955a) On reactions of plant anti-N with red cells of chimpanzees and other animals. *Am. J. phys. Anthrop.* **13**, 29.

LEVINE P., ROBINSON E., CELANO M., BRIGGS O. & FALKINBURG L. (1955b) Gene interaction resulting in suppression of blood group substance B. *Blood* **10**, 1100.

LEVINE P. & STETSON R.E. (1939) An unusual case of intragroup agglutination. *J. Am. med. Ass.* **113**, 126.

LEVINE P., WIGOD M., BACKER A.M. & PONDER R. (1949) The Kell-Cellano (K-k) genetic system of human blood factors. *Blood* **4**, 869.

LEWIS M., CHOWN B. & KAITA H. (1967) On the blood group antigens Bu^a and Sm. *Transfusion* **7**, 92.

LEWIS M., CHOWN B., SCHMIDT R.P. & GRIFFITTS J.J. (1964) A possible relationship between the blood group antigens Sm and Bu^a. *Am. J. hum. Genet.* **16**, 254.

LEWIS S.M. (1962) Red cell abnormalities and haemolysis in aplastic anaemia. *Br. J. Haemat.* **8**, 322.

LEWIS S.M., DACIE J.V. & TILLS D. (1961) Comparison of the sensitivity to agglutination and haemolysis by a high-titre cold antibody of the erythrocytes of normal subjects and of patients with a variety of blood diseases including paroxysmal nocturnal haemoglobinuria. *Br. J. Haemat.* **7**, 64.

LI C.C. (1953) Is Rh facing a crossroad? A critique of the compensation effect. *Am. Nat.* **87**, 257.

LINDSTEN J., FRACCARO M., POLANI P.E., HAMERTON J.L., SANGER R. & RACE R.R. (1963) Evidence that the Xg blood group genes are on the short arm of the X chromosome. *Nature* **197**, 648.

LLOYD K.O., BEYCHOK S. & KABAT E.A. (1967) Immunochemical studies on blood groups. XXXVII. The structures of difucosyl and other oligosaccharides produced by alkaline degradation of blood group A, B and H substances. Optical rotatory dispersion and circular dichroism spectra of these oligosaccharides. *Biochemistry* **6**, 1448.

LLOYD K.O., KABAT E.A., LAYUG E.J. & GRUEZO F. (1966a) Immunochemical studies on blood groups. 34. Structures of some oligosaccharides produced by alkaline degradation of blood group A, B and H substances. *Biochemistry* **5**, 1489.

LLOYD K.O., KABAT E.A. & ROSENFIELD R.E. (1966b) Immunochemical studies on blood groups. 35. The activity of fucose-containing oligosaccharides isolated from blood group A, B and H substances by alkaline degradation. *Biochemistry* **5**, 1502.

LÜDERITZ O., STAUB A.M. & WESTPHAL O. (1966) Immunochemistry of O and R antigens of *Salmonella* and related Enterobacteriaceae. *Bact. Rev.* **30**, 192.

MAC DIARMID W.D., LEE G.R., CARTWRIGHT G.E. & WINTROBE M.M. (1967) X-inactivation in an unusual X-linked anemia and the Xg^a blood group. *Clin. Res.* **15**, 132.

MÄKELÄ O. & CANTELL K. (1958) Destruction of M and N blood group receptors of human red cells by some influenza viruses. *Ann. Med. exp. fenn.* **36**, 366.

MÄKELÄ O. & MÄKELÄ P. (1956) Le^b antigen. Studies on its occurrence in red cells, plasma and saliva. *Ann. Med. exp. fenn.* **34**, 157.

MÄKELÄ O., MÄKELÄ P.H. & KORTEKANGAS A. (1967) *In vitro* transformation of the Lewis blood groups of erythrocytes. *Ann. Med. exp. fenn.* **45**, 159.

MANN J.D., CAHAN A., GELB A. G., FISHER N., HAMPER J., TIPPETT P., SANGER R. & RACE R.R. (1962) A sex linked blood group. *Lancet* **i**, 8.

MARCUS D.M., KABAT E.A. & ROSENFIELD R.E. (1963) The action of enzymes from *Clostridium tertium* on the I antigenic determinant of human erythrocytes. *J. exp. Med.* **118**, 175.

MARR A.S., DONALD A.S.R., WATKINS W.M. & MORGAN W.T.J. (1967) Molecular and genetic aspects of human blood group Le^b specificity. *Nature* **215**, 1345.

MARSH W.L. (1960) The pseudo B antigen. A study of its development. *Vox Sang.* **5**, 387.

MARSH W.L. (1961) Anti-i: a cold antibody defining the Ii relationship in human red cells. *Br. J. Haemat.* **7**, 200.

MARSH W.L. & JENKINS W.J. (1960) Anti-i: a new cold autoantibody. *Nature* **188**, 753.

MARSH W.L., JENKINS W.J. & WALTHER W.W. (1959) Pseudo B: an acquired blood group antigen. *Br. med. J.* **ii**, 63.

MASOUREDIS S.P. (1959) Reaction of I^{131} trace labelled human anti-Rh$_0$ (D) with red cells. *J. clin. Invest.* **38**, 279.

MASOUREDIS S.P. (1962) Reaction of I^{131} anti-Rh$_0$ (D) with enzyme treated red cells. *Transfusion* **2**, 363.

MASOUREDIS S.P., CHI C.A. & FERGUSON E. (1960) Relationship between Rh$_0$ (D) genotype and quantity of I^{131} anti-Rh$_0$ (D) bound to red cells. *J. clin. Invest.* **39**, 1450.

MASOUREDIS S.P., DUPUY M.E. & ELLIOT M. (1967a) Distribution of the human Rh$_0$ (D) antigen in the red cells of non-human primates. *J. Immunol.* **98**, 8.

MASOUREDIS S.P., DUPUY M.E. & ELLIOTT M. (1967b) Relationship between Rh$_0$ (D) zygosity and red cell Rh$_0$ (D) antigenic content in family members. *J. clin. Invest.* **46**, 681.

MATSUNAGA E. & HIRAIZUMI Y. (1962) Prezygotic selection in ABO blood groups. *Science* **135**, 432.

MCCONNELL R.B. (1966) *The Genetics of Gastro-Intestinal Disorders.* Oxford University Press, London.

MCDONALD J.C. & ZUCKERMAN A.J. (1962) ABO blood groups and acute respiratory virus disease. *Br. med. J.* **ii**, 89.

MCGINNISS M.H., SCHMIDT P.J. & CARBONE P.P. (1964) Close association of I blood group and disease. *Nature* **202**, 606.

MCNEIL C., HELMICK W.M. & FERRARI A. (1963) A preliminary investigation into automatic blood grouping. *Vox Sang.* **8**, 235.

METAXAS M.N., METAXAS-BÜHLER M., DUNSFORD I. & HOLLÄNDER L. (1959) A further example of anti-Lu^b together with data in support of the Lutheran-Secretor linkage in man. *Vox Sang.* **4**, 298.

MOHR J. (1951) A search for linkage between the Lutheran blood group and other hereditary characters. *Acta path. microbiol. scand.* **28**, 207.

MOLLISON P.L. (1967) *Blood Transfusion in Clinical Medicine*, 4th edn. Blackwell, Oxford.

MOLLISON P.L. & POLLEY M.J. (1964) Uptake of γ-globulin and complement by red cells exposed to serum at low ionic strength. *Nature* **203**, 535.

MOLLISON P.L., POLLEY M.J. & CROME P. (1963) Temporary suppression of Lewis blood group antibodies to permit incompatible transfusion. *Lancet* **i**, 909.

MORGAN W.T.J. (1963) Some observations on the carbohydrate-containing components of human ovarian cyst mucin. *Ann. N.Y. Acad. Sci.* **106**, 177.

MORGAN W.T.J. (1964) Some aspects of immunological specificity in terms of carbohydrate structure. *Bull. Soc. chim. Biol.* **46**, 1627.

MORGAN W.T.J. & WATKINS W.M. (1948) The detection of a product of the blood group O gene and the relationship of the so-called O substance to the agglutinogens A and B. *Brit. J. exp. Path.* **29**, 159.

MORGAN W.T.J. & WATKINS W.M. (1964) Blood group P$_1$ substance. I. Chemical properties. *Proc. 9th Cong. int. Soc. Blood Transf.* 1962. Mexico City, p. 225.

MORTON N.E. (1956) The detection and estimation of linkage between genes for elliptocytosis and the Rh blood type. *Am. J. hum. Genet.* **8**, 80.

MORTON N.E., KRIEGER H. & MI M.P. (1966a) Natural selection on polymorphisms in Northeastern Brazil. *Am. J. hum. Genet.* **18**, 153.

MORTON N.E., KRIEGER H., STEINBERG A.G. & ROSENFIELD R.E. (1965) Genetic evidence confirming the localization of Sutter in the Kell blood group system. *Vox Sang.* **10**, 608.

MORTON N.E., MI M.P. & YASUDA N. (1966b) Bivalent alleles. *Am. J. hum. Genet.* **18**, 233.

MORTON N.E. & ROSENFIELD R.E. (1967) A new Rh allele, $r^{yn}(R^{-1,2,w^3,w^4})$. *Transfusion* **7**, 117.

Mourant A.E. (1954) *The Distribution of the Human Blood Groups*. Blackwell, Oxford.

Mourant A.E., Kopec A.C. & Domaniewska-Sobczak K. (1958) *The ABO Blood Groups: Comprehensive Tables and Maps of World Distribution*. Blackwell, Oxford.

Murray J. & Clark E.C. (1952) Production of anti-Rh in guinea pigs from human erythrocyte extracts. *Nature* 169, 886.

Myhre B.A., Meyer T., Opitz J.M., Race R.R., Sanger R. & Greenwalt T.J. (1965) Two populations of erythrocytes associated with XX/XY mosaicism. *Transfusion* 5, 501.

Nagai Y. & Springer G.F. (1962) Partial hydrolysis of isolated blood group M antigen. *Fed. Proc.* 67, 21.

Nicholas J.W., Jenkins W.J. & Marsh W.L. (1957) Human blood chimaeras. A study of surviving twins. *Br. med. J.* i, 1458.

Opitz J.M., Stiles F.C., Wise D., Race R.R., Sanger R., Von Gemmingen G.R., Kierland R.R., Cross E.G. & De Groot W.P. (1965) The genetics of angiokeratoma corporis diffusum (Fabry's disease) and its linkage relations with the Xg locus. *Am. J. hum. Genet.* 17, 325.

Osborn M.J., Rosen S.M., Rothfield L., Zeleznick L.D. & Horecker B.L. (1964) Lipopolysaccharide of the gram-negative cell wall. *Science* 145, 783.

Peritz E. (1967) A statistical study of intrauterine selection factors related to the ABO system. I. The analysis of data on liveborn children. *Ann. hum. Genet.* 30, 259.

Pfeiffer R.A., Körver G., Sanger R. & Race R.R. (1966) Paternal origin of an XXYY anomaly. *Lancet* i, 1427.

Plaut G., Ikin E.W., Mourant A.E., Sanger R. & Race R.R. (1953) A new blood group antibody, anti-Jk^b. *Nature* 171, 431.

Pollack W. (1965) Some physicochemical aspects of hemagglutination. *Ann. N.Y. Acad. Sci.* 127, 892.

Pollack W., Hager H.J., Reckel R., Toren D.A. & Singher H.O. (1965) A study of the forces involved in the second stage of agglutination. *Transfusion* 5, 158.

Polley M. & Mollison P.L. (1961) The role of complement in the detection of blood group antibodies. Special reference to the antiglobulin test. *Transfusion* 1, 9.

Polley M.J., Mollison P.L. & Soothill J.F. (1962) The role of 19S gamma globulin blood group antibodies in the antiglobulin reaction. *Br. J. Haemat.* 8, 149.

Potter C.W. & Schild G.C. (1967) The incidence of HI antibody to influenza virus A$_2$/Singapore/1/57 in individuals of blood groups A and O. *J. Immunol.* 98, 1320.

Price Evans D.A., Donohoe W.T.A., Bannerman R.M., Mohn J.F. & Lambert R.M. (1966) Blood group gene localization through a study of mongolism. *Ann. hum. Genet.* 30, 49.

Race R.R. (1944) An 'incomplete' antibody in human serum. *Nature* 153, 771.

Race R.R. (1965a) Contributions of blood groups to human genetics. *Proc. R. Soc.* 163, 151.

RACE R.R. (1965b) Identification of the origin of the X chromosome(s) in sex chromosome aneuploidy. *Can. J. Genet. Cytol.* **7**, 214.

RACE R.R. (1965c) Modern concepts of the blood group systems. *Ann. N.Y. Acad. Sci.* **127**, 884.

RACE R.R. & SANGER R. *Blood Groups in Man*, 1st edn., 1950; 2nd edn., 1954; 3rd edn., 1958; 4th edn., 1962; 5th edn., 1968. Blackwell, Oxford.

RACE R.R. & SANGER R. (1959) The inheritance of blood groups. *Br. med. Bull.* **15**, 99.

REED T.E., GERSHOWITZ H., SONI A. & NAPIER J. (1964) A search for natural selection in six blood group systems and ABH secretion. *Am. J. hum. Genet.* **16**, 161.

RENTON P.H., STRATTON F., GUNSON H.H. & HANCOCK J.A. (1962) Red cells of all four ABO groups in a case of leukaemia. *Br. med. J.* **i**, 294.

RENWICK J.H. & IZATT M.M. (1965) Some genetical parameters of the nail-patella locus. *Ann. hum. Genet.* **28**, 369.

RENWICK J.H. & LAWLER S.D. (1955) Linkage between the ABO and nail-patella loci. *Ann. hum. Genet.* **19**, 312.

RENWICK J.H. & LAWLER S.D. (1963) Probable linkage between a congenital cataract locus and the Duffy group locus. *Ann. hum. Genet.* **27**, 67.

RENWICK J.H. & SCHULZE J. (1964) An analysis of some data on the linkage between Xg and colorblindness in man. *Am. J. hum. Genet.* **16**, 410.

RENWICK J.H. & SCHULZE J. (1965) Male and female recombination fractions for the nail-patella: ABO linkage in man. *Ann. hum. Genet.* **28**, 379.

RICHARDS A.G. (1962) Loss of blood group B antigen in chronic lymphatic leukaemia. *Lancet* **ii**, 178.

ROBERTS J.A. FRASER (1959) Some associations between blood groups and disease. *Br. med. Bull.* **15**, 129.

ROBINSON J.C. & GOLDSMITH L.A. (1967) Genetically determined variants of serum alkaline phosphatase: a review. *Vox Sang.* **13**, 289.

ROBSON E.B. & HARRIS H. (1967) Further studies on the genetics of plasma alkaline phosphatase. *Ann. hum. Genet.* **30**, 219.

ROCHNA E. & HUGHES-JONES N.C. (1965) The use of purified ^{125}I-labelled anti-γ globulin in the determination of the number of D antigen sites on red cells of different phenotypes. *Vox Sang.* **10**, 675.

ROSEN F.S., HUTCHISON G.B. & ALLEN F.H. (1965) The Xg blood groups and congenital hypogammaglobulinemia. *Vox Sang.* **10**, 729.

ROSENFIELD R.E. (1955) A-B hemolytic disease of the newborn. Analysis of 1480 cord blood specimens with special reference to the direct anti-globulin test and to the group O mother. *Blood* **10**, 17.

ROSENFIELD R.E. (1967) The current status of some blood group problems. *Seminars in Hematology*, **4**, 133.

ROSENFIELD R.E., ALLEN F.H., SWISHER S.N. & KOCHWA S. (1962) A review of Rh serology and presentation of a new terminology. *Transfusion* **2**, 287.

ROSENFIELD R.E., HABER G.V. & DEGNAN T.G. (1964a) Rh alleles, $^-R^{10,20}$ and new evidence for $R^{2,4}$. *Vox Sang.* **9**, 168.

ROSENFIELD R.E., HABER G.V., KISSMEYER-NIELSEN F., JACK J.A., SANGER R. & RACE R.R. (1960a) Gc, a very common red cell antigen. *Br. J. Haemat.* **6**, 344.

ROSENFIELD R.E., HABER G.V., SCHROEDER R. & BALLARD R. (1960b) Problems in Rh typing as revealed by a single Negro family. *Am. J. hum. Genet.* **12**, 147.

ROSENFIELD R.E. & KOCHWA S. (1964) Immunochemical studies of the Rh system. II. Capacity of antigenic sites of different phenotypes to bind and retain Rh antibodies. *J. Immunol.* **92**, 693.

ROSENFIELD R.E. & RUBENSTEIN P. (1966) Blood groups in immunogenetics, *in Plenary Session Papers*, 11*th Cong. int. Soc. Blood Transf.* Sydney, p. 1.

ROSENFIELD R.E., SCHMIDT P.J., CALVO R.C. & McGINNISS M.H. (1965) Anti-i, a frequent cold agglutinin in infectious mononucleosis. *Vox Sang.* **10**, 631.

ROSENFIELD R.E., SCHROEDER R.S., BALLARD R., VAN DER HART M., MOES M. & VAN LOGHEM J.J. (1964b) Erythrocytic antigenic determinants characteristic of H, I in the presence of H(IH), or H in the absence of i (H[-i]). *Vox Sang.* **9**, 415.

ROSENFIELD R.E., SZYMANSKI I.O. & KOCHWA S. (1964c) Immunochemical studies of the Rh system: III. Quantitative hemagglutination that is relatively independent of source of Rh antigens and antibodies. *Cold Spring Harbor Symposia on Quantitative Biology.* **29**, 427.

SALMON C. & BERNARD J. (1960) Nouvelle observation d'antigène. A 'faible' chez un malade atteint de leucémie aiguë. *Rev. fr. Étud. clin. biol.* **5**, 912.

SALMON C., BORIN P. & ANDRÉ R. (1958a) Le groupe sanguin A_m dans deux générations d'une même famille. *Rev. Hémat.* **13**, 529.

SALMON C., DREYFUS B. & ANDRÉ R. (1958b) Double population de globules différent seulement par l'antigène de groupe ABO, observée chez un malade leucémique. *Revue Hémat.* **13**, 148.

SANGER R. (1965) Genes on the X chromosome. *Can. J. Genet. Cytol.* **7**, 179.

SANGER R. & RACE R.R. (1958) The Lutheran-Secretor linkage in man: support for Mohr's findings. *Heredity* **12**, 513.

SANGER R. & RACE R.R. (1963) The Xg blood groups and familial hypogammaglobulinaemia. *Lancet* **i**, 859.

SANGER R., RACE R.R. & JACK J. (1955) The Duffy blood groups of New York Negroes: the phenotype Fy(a − b −). *Br. J. Haemat.* **1**, 370.

SCHIFFMAN G. & MARCUS D.M. (1964) Chemistry of the ABH blood group substances, *in* MOORE C.V. & BROWN E.B. (eds.) *Progress in Hematology*, vol. 4, p. 97. Grune & Stratton, New York.

SCHMIDT P.J., BARILE M.F. & McGINNISS M.H. (1965) Mycoplasma (pleuropneumonia-like organisms) and blood group I; associations with neoplastic disease. *Nature* **205**, 371.

SCHMIDT P.J., LOSTUMBO M.M., ENGLISH C.T. & HUNTER O.B. (1967) Aberrant U blood group accompanying Rh_{null}. *Transfusion* **7**, 33.

SCHMIDT P.J. & McGINNISS M.H. (1965) Differences between anti-H and anti-OI red cell antibodies. *Vox Sang.* **10**, 109.

SCHMIDT P.J. & McGINNISS M.H. (1967) Cell surfaces, blood groups and microorganisms. *Nature* **214**, 1363.

SCHMIDT P.J. & VOS G.H. (1967) Multiple phenotypic abnormalities associated with Rh_{null} (---/---). *Vox Sang.* **13**, 18.

SCHMIDT R.P., GRIFFITTS J.J. & NORTHMAN F.F. (1962) A new antibody, anti-Sm, reacting with a high incidence antigen. *Transfusion* **2**, 338.

SCHNEIDER J. & PREISLER O. (1966) Prevention of Rh sensitization from fetomaternal microtransfusions. *Obstet. Gynec.* **28**, 615.

SCHOLZ W. & MURKEN J.D. (1967) Beobachtung eines Faktorenaustausches zwischen den Blutgruppen-Genorten MN und Ss. *Humangenetik* **4**, 268.

SHAHANI S. & SOUTHAM A.L. (1962) Immunofluorescent study of the ABO blood group antigens in human spermatozoa. *Am. J. Obstet. Gynec.* **84**, 660.

SHARMA J.C. (1966) Nail-patella syndrome in an Indian family: clinical and linkage data. *Ann. hum. Genet.* **30**, 193.

SHEN L., GROLLMAN E.F. & GINSBURG V. (1968) An enzymatic basis for secretor status and blood group substance specificity in humans. *Proc. nat. Acad. Sci.* **59**, 224.

SHREFFLER D.C. (1965) Genetic studies of blood group-associated variations in a human serum alkaline phosphatase. *Am. J. hum. Genet.* **17**, 71.

SHREFFLER D.C. (1966) Relationship of alkaline phosphatase levels in intestinal mucosa to ABO and secretor blood groups. *Proc. Soc. exp. Biol. Med.* **123**, 423.

SHREFFLER D.C. (1967a) Genetic control of cellular antigens. *Proc. 3rd Cong. int. Soc. Hum. Genet.* Chicago, p. 217.

SHREFFLER D.C. (1967b) Molecular aspects of immunogenetics. *Ann. rev. Genet.* **1**, 163.

SINISCALCO M., FILIPPI G., LATTE B., PIOMELLI S., RATTAZZI M., GAVIN J., SANGER R. & RACE R.R. (1966) Failure to detect linkage between Xg and other X-borne loci in Sardinians. *Ann. hum. Genet.* **29**, 231.

SNEATH J.S. & SNEATH P.H.A. (1955) Transformation of the Lewis groups of human red cells. *Nature* **176**, 172.

SOLOMON J.M., WAGGONER R. & LEYSHON W.C. (1965) A quantitative immuno-genetic study of gene suppression involving A_1 and H antigens of the erythrocyte without affecting secreted blood group substances. The ABH phenotypes A_m^h and O_m^h. *Blood* **25**, 470.

SPRINGER G.F. (1956) Inhibition of blood group agglutinins by substances occurring in plants. *J. Immunol.* **76**, 399.

SPRINGER G.F. (1967) Blood group activity in certain bacteria and plants, *in* KUHNS W.J. (ed.) *J.A. Hartford Foundation Conference on Blood Groups and Blood Transfusion*, vol. 1, p. 137.

SPRINGER G.F. & ANSELL N.J. (1958) Inactivation of human erythrocyte agglutinogens M and N by influenza viruses and receptor-destroying enzyme. *Proc. nat. Acad. Sci.* **44**, 182.

SPRINGER G.C., NAGAI Y. & TEGTMEYER H. (1966a) Isolation and properties of human blood group NN and meconium Vg antigens. *Biochemistry* **5**, 3254.

SPRINGER G.F., NICHOLS J.H. & CALLAHAN H.J. (1964) Galactosidase action on human blood group B-active *Escherichia coli* and ox red cell substances. *Science* **146**, 946.

SPRINGER G.F., WANG E.T., NICHOLS J.H. & SHEAR J.M. (1966b) Relations between bacterial lipopolysaccharide structures and those of human cells. *Ann. N.Y. Acad. Sci.* **133**, 566.

SPRINGER G.F., WILLIAMSON P. & BRANDES W.C. (1961) Blood group activity of gram negative bacteria. *J. exp. Med.* **113**, 1077.

STAVELEY J.M. & CAMERON G.L. (1958) The inhibiting action of hydatid cyst fluid on anti-Tj^a sera. *Vox Sang.* **3**, 114.

STEINBERG A.G. (1965) Evidence for a mutation or crossing-over at the Rh locus. *Vox Sang.* **10**, 721.

STERN C. (1960) *Principles of Human Genetics*, 2nd edn. W.H. Freeman & Co. San Francisco.

STERN K., GOODMAN H.S. & BERGER M. (1961) Experimental isoimmunization in man. *J. Immunol.* **87**, 189.

STRATTON F. & RENTON P.H. (1959) Acquisition of B-like antigen. *Br. med. J.* **ii**, 244.

STRATTON F., RENTON P.H. & HANCOCK J.A. (1958) Red cell agglutinability affected by disease. *Nature* **181**, 62.

STROUP M., MAC ILROY M., WALKER R. & AYDELOTTE J.V. (1965) Evidence that Sutter belongs to the Kell blood group system. *Transfusion* **5**, 309.

SWANSON J. & MATSON G.A. (1964) Third example of a human 'D-like' antibody or anti-LW. *Transfusion* **4**, 257.

SWANSON J., POLESKY H.F. & MATSON G.A. (1965a) The LW antigen of adult and infant erythrocytes. *Vox Sang.* **10**, 560.

SWANSON J., POLESKY H.F., TIPPETT P. & SANGER R. (1965b) A 'new' blood group antigen, Do^a. *Nature* **206**, 313.

SWINBURNE L.M., FRANK B.B. & COOMBS R.R.A. (1961) The A antigen on the buccal epithelial cells of man. *Vox Sang.* **6**, 274.

SZULMAN A.E. (1960) The histological distribution of blood group substances A and B in man. *J. exp. Med.* **111**, 785.

SZULMAN A.E. (1962) The histological distribution of the blood group substances in man as disclosed by immunofluorescence. II. The H antigen and its relation to A and B antigens. *J. exp. Med.* **115**, 977.

SZULMAN A.E. (1965) The ABH antigens in human tissues and secretions during embryonal development. *J. Histochem. Cytochem.* **13**, 752.

SZULMAN A.E. (1966) Chemistry, distribution and function of blood group substances. *Ann. rev. Med.* **17**, 307.

SZULMAN A.E. (1967) The distribution of blood group antigens in tissues, *in* KUHNS W.J. (ed.) *J.A. Hartford Foundation Conference on Blood Groups and Blood Transfusion*, vol. 1, p. 61.

THOMPSON P.R., CHILDERS D.M. & HATCHER D.E. (1967) Anti-Di^b: first and second examples. *Vox Sang.* **13**, 314.

TIPPETT P. (1967) Genetics of the Dombrock blood group system. *J. med. Genet.* **4**, 7.

TIPPETT P., NOADES J., SANGER R., RACE R.R., SAUSAIS L., HOLMAN C.A. & BUTTIMER R.J. (1960) Further studies of the I antigen and antibody. *Vox Sang.* **5**, 107.

TIPPETT P., SANGER R., RACE R.R., SWANSON J. & BUSCH S. (1965) An agglutinin associated with the P and the ABO blood group systems. *Vox Sang.* **10**, 269.

TOSTESON D.C. (1963) Active transport, genetics and cellular evolution. *Fed. Proc.* **22**, 19.

TROXEL D.B., INNELLA F. & COHEN R.J. (1966) Infectious mononucleosis complicated by hemolytic anemia due to anti-i. *Am. J. clin. Path.* **46**, 625.

TUPPY H. & STAUDENBAUER W.L. (1966) Microsomal incorporation of N-acetyl-D-galactosamine into blood group substance. *Nature* **210**, 316.

UHLENBRUCK G. (1965) Immunochemical studies on erythrocyte mucoids: the nature of the M and N specific substances. *Proc. 10th Cong. int. Soc. Blood Transf.* Stockholm, p. 476.

UHLENBRUCK G. & KRÜPE M. (1965) Cryptantigenic N_{Vg} receptor in mucoids from *Mu/Mu* cells. *Vox Sang.* **10**, 326.

VAN DER HART M., VAN DER VEER M. & VAN LOGHEM J.J. (1962) Change of blood group B in a case of leukaemia. *Vox Sang.* **7**, 449.

VAN DER HART & VAN LOGHEM J.J. (1953) A further example of anti-Jk^a. *Vox Sang.* **3**, 261.

VAN DER HART M. & VAN LOGHEM J.J. (1967) Blood group chimerism. *Vox Sang.* **12**, 161.

VAN LOGHEM J.J., DORFMEIER H. & VAN DER HART M. (1957) Two A antigens with abnormal serological properties. *Vox Sang.* **2**, 16.

VAN LOGHEM J.J. & VAN DER HART M. (1954) The weak antigen A_4 occurring in the offspring of group O parents. *Vox Sang.* **4**, 69.

VAN RIJSEWIJK M.G.H. & GOSLINGS N.E.O. (1963) Secretor status of streptococcus pyogenes group A carriers and patients with rheumatic heart disease or acute glomerulonephritis. *Br. med. J.* **ii**, 542.

VOS G.H., VOS D., KIRK R.L. & SANGER R. (1961) A sample of blood with no detectable Rh antigens. *Lancet* **i**, 14.

WALKER R.H., ARGALL C.I., STEANE E.A., SASAKI T.T. & GREENWALT T.J. (1963) Anti-Js^b, the expected antithetical antibody of the Sutter blood group system. *Nature* **197**, 295.

WALKER W., MURRAY S. & RUSSELL J.K. (1957) Stillbirth due to haemolytic disease of the newborn. *J. Obstet. Gynaec. Br. Commonw.* **44**, 573.

WARBURG M., HAUGE M. & SANGER R. (1965) Norrie's disease and the Xg blood group system: linkage data. *Acta genet.* **15**, 103.

WATKINS W.M. (1960) Changes in blood group specificity induced by enzymes. *Bull. Soc. chim. Biol.* **42**, 1599.

WATKINS W.M. (1962) Changes in the specificity of blood group mucopolysaccharides induced by enzymes from *Trichomonas foetus. Immunology* **5**, 245.

WATKINS W.M. (1964) Blood-group substances: their nature and genetics, *in* BISHOP C. & SURGENOR D.M. (eds.) *The Red Blood Cell*, p. 359. Academic Press, New York.

WATKINS W.M. (1966) Blood-group substances. *Science* **152**, 172.

WATKINS W. (1967a) Biochemical genetics of the ABO and Lewis blood group systems, *in* KUHNS W.J. (ed.) *John A. Hartford Foundation Conference on Blood Groups and Blood Transfusion.* vol. 1, p. 7.

WATKINS W.M. (1967b) The possible enzymic basis of the biosynthesis of blood-group substances. *Proc. 3rd Cong. int. Soc. Hum. Genet.* Chicago, p. 171.

WATKINS W.M., KOSCIELAK J. & MORGAN W.T.J. (1964) The relationship between the blood group A and B substances isolated from erythrocytes and from secretions. *Proc. 9th Cong. int. Soc. Blood Tranf.* 1962, Mexico City, p. 213.

WATKINS W.M. & MORGAN W.T.J. (1952) Neutralization of the anti-H agglutinin in eel serum by simple sugars. *Nature* **169**, 825.

WATKINS W.M. & MORGAN W.T.J. (1955) Some observations on the O and H characters of human blood secretions. *Vox Sang.* **5**, 1.

WATKINS W.M. & MORGAN W.T.J. (1959) Possible genetical pathways for the biosynthesis of blood group mucopolysaccharides. *Vox Sang.* **4**, 97.

WATKINS W.M. & MORGAN W.T.J. (1962) Further observations on the inhibition of blood-group specific serological reactions by simple sugars of known structure. *Vox Sang.* **7**, 129.

WATKINS W.M. & MORGAN W.T.J. (1964) Blood-group P_1 substance: II. Immunological properties. *Proc. 9th Cong. int. Soc. Blood Transf.* 1962. Mexico City, p. 230.

WATKINS W.M., ZARNITZ M.L. & KABAT E.A. (1962) Development of H activity by human blood group B substance treated with coffee bean α-galactosidase. *Nature* **195**, 1204.

WEINER W., LEWIS H.B.M., MOORES P., SANGER R. & RACE R.R. (1957) A gene, y, modifying the blood group antigen A. *Vox Sang.* **2**, 25.

WEINER W., SHINTON N.K. & GRAY I.R. (1960) Antibody of blood-group specificity in simple ('cold') haemolytic anaemias. *J. clin. Path.* **13**, 232.

WELLS R.S., JENNINGS M.C., SANGER R. & RACE R.R. (1966) Xg blood groups and ichthyosis. *Lancet* **ii**, 493.

WHITTEMORE N.B., TRABOLD N., WEED R.I., REGA A., REED C.F. & SWISHER S.N. (1967) Relationship of Rh activity to integrity of membrane lipoprotein structure. *Clin. Res.* **15**, 290.

WIENER A.S. (1966) The blood groups. Three fundamental problems—serology, genetics and nomenclature. *Blood* **27**, 110.

WIENER A.S., GORDON E.B. & HANDMAN L. (1949) Heredity of the Rh blood types. *Am. J. hum. Genet.* **1**, 127.

WIENER A.S. & LANDSTEINER K. (1943) Heredity of variants of the Rh type. *Proc. Soc. exp. Biol.* **53**, 167.

WIENER A.S., UNGER L.J. & COHEN L. (1954) Distribution and heredity of blood factor U. *Science* **119**, 734.

WIENER A.S., UNGER L.J., COHEN L. & FELDMAN J. (1956) Type-specific cold auto-antibodies as a cause of acquired hemolytic anemia and hemolytic transfusion reactions: biologic test with bovine red cells. *Ann. Int. Med.* **44**, 221.

WIENER A.S., UNGER L.J. & GORDON E.B. (1953) Fatal hemolytic transfusion reaction caused by sensitization to a new blood factor, U. *J. Am. med. Ass.* **153**, 1444.

WIENER A.S. & WEXLER I.B. (1958) *Heredity of the Blood Groups.* Grune & Stratton, New York.

WITEBSKY E. & ENGASSER L.M. (1949) Blood groups and subgroups of the newborn. I. The A factor of the newborn. *J. Immunol.* **61**, 171.

WITEBSKY E. & KLENDSHOJ N.C. (1941) The isolation of an O specific substance from gastric juice of secretors and carbohydrate-like substances from gastric juice of non-secretors. *J. exp. Med.* **73**, 655.

WOLFF I. & SPRINGER G.F. (1965) Observations on Rh active substances. *Proc. 10th Cong. int. Soc. Blood Transf.* 1964, Stockholm, p. 510.

WORLLEDGE S.M. & ROUSSO C. (1965) Studies on the serology of paroxysmal cold haemoglobinuria (P.C.H.), with special reference to its relationship with the P blood group system. *Vox Sang.* **10**, 293.

YAKULIS V., COSTEA N. & HELLER P. (1966) α-Galactosidase determinants of the I antigen. *Proc. Soc. exp. Biol. Med.* **121**, 812.

YOKOYAMA M. (1965) Close relationship between A and I blood groups. *Nature* **206**, 411.

YOKOYAMA M. & PLOCINIK B. (1965) Serologic and immunochemical characterization of A$_x$ blood. *Vox Sang.* **10**, 149.

YUNIS E. & YUNIS J.J. (1963) Cell antigens and cell specialization. III. On the H antigen receptors of human epithelial cells. *Blood* **22**, 750.

ZIDERMAN D., GOMPERTZ S., SMITH Z.G. & WATKINS W.M. (1967) Glucosyl transferases in mammalian gastric mucosal linings. *Biochem. biophys. Res. Comm.* **29**, 56.

ZIPURSKY A. & ISRAELS L.G. (1967) The pathogenesis and prevention of Rh immunization. *Can. med. Ass. J.* **97**, 1245.

ZUELZER W.W., BEATTIE K.M. & REISMAN L.E. (1964) Generalized unbalanced mosaicism attributable to dispermy and probable fertilization of a polar body. *Am. J. hum. Genet.* **16**, 38.

CHAPTER 10

HEMOGLOBIN

✧

The Normal Hemoglobins — 347
General structure — 347
The allosteric effect: Hemoglobin as an enzyme — 353
Hb A$_2$ — 355
Hb F — 356
Embryonic hemoglobin — 358
Heterogeneity of Hb A and Hb F — 359
The Abnormal Hemoglobins — 360
Hemoglobin homotetramers — 360
The Lepore hemoglobins — 362
Hemoglobins with amino acid substitution or deletion — 363
Hemoglobins causing red cell structural distortion — 370
Sickle hemoglobin: Hb S — 371
Hb C$_{Georgetown}$ — 374
Hb C$_{Harlem}$ — 374
Hb I — 374
Hb C — 375
Hb D$_{Punjab}$ and Hb E — 375
Hemoglobins associated with methemoglobinemia — 375
Hemoglobins with increased oxygen affinity — 377
Hb Chesapeake — 377
Hb J$_{Capetown}$ — 378
Hb Yakima — 378
Hb Rainier — 379
Hb Ypsi — 379
Hemoglobin with decreased oxygen affinity— Hb Kansas — 380
The unstable hemoglobins — 380
Unstable beta chain variants — 380
Hb Zürich — 380
Hb Köln — 381
Hb Genova — 381
Hb Sydney — 382
Hb Hammersmith — 382
Hb Freiburg — 383
Hb Gun Hill — 383
Other beta chain variants associated with molecular instability — 384
Unstable alpha chain variants — 384
Hb Sinai (Sealy) — 384
Hb Bibba — 384
Hb Torino — 385
Hb Porto Alegre: *in vitro* polymerization — 385
Hemoglobin Genetics — 386
Inheritance of structural genes — 386
Synthesis of hemoglobin — 388
Normal synthesis — 388
Synthesis of Hb A — 388
Hemoglobin synthesis during development — 389
Synthesis of structurally abnormal variants — 390
Defects in globin chain synthesis: Thalassemia — 390
Defective synthesis of β or δ chains — 391
β-thalassemia (A$_2$ thalassemia) — 391
$\beta\delta$-thalassemia (F thalassemia) — 393
Hb Lepore syndrome — 393
Hereditary persistence of Hb F — 394
δ-thalassemia — 395
Defective synthesis of α chains — 395
α-Thalassemia — 395
Hb H disease — 397

α-Thalassemia in subjects with β chain abnormalities 398

Genetic mechanisms in thalassemia 399

Evolution of hemoglobin.......... 401

Hemoglobin and balanced polymorphism: Geographic distribution 404

Methods 405

References............................ 406

The study of hemoglobin has made many major contributions toward the development of molecular biology, and so much has been written on hemoglobin structure, function, pathologic physiology and inheritance that even a modest review would occupy a large volume. No attempt can be made in this chapter to cover the early studies which were so important to current concepts in biochemistry, genetics and physiology, nor even to discuss in any detail the more recent work in this field. Many outstanding papers and books reviewing various aspects of the subject have appeared since 1960, and much of this chapter was based on publications by Ingram (1961a, 1963), Jonxis (1961), Rucknagel & Neel (1961), Cullis *et al.* (1962), Bannerman (1961), Schroeder (1963), Baglioni (1963), Huisman (1963), Braunitzer *et al.* (1964), Rossi-Fanelli *et al.* (1964), Jaffé (1964), Motulsky (1964), Allen (1964), Heller (1965, 1966), Itano (1965, 1966), Fessas (1965), Perutz (1965), Huehns (1965), Huehns & Shooter (1965), Weatherall (1965, 1967), Allison (1965), Antonini (1965), Marks (1966), Nathan & Gunn (1966), Gabuzda (1966), Lehmann & Huntsman (1966), Livingstone (1967) and Conley & Charache (1967). (See Addenda.)

THE NORMAL HEMOGLOBINS

GENERAL STRUCTURE

The normal human hemoglobins have the general formula X_2Y_2, where X_2 refers to a pair of α polypeptide chains and Y_2 to a pair of either β, γ, δ or ϵ chains. For example, Hb A, the major hemoglobin in the blood of adults, has the molecular formula $\alpha_2^A \beta_2^A$. The subunits of hemoglobin (like those of enzymes such as G6PD or lactate dehydrogenase, but unlike those of such proteins as haptoglobin and the immunoglobulins) are of similar (but not idential) size, and the forces linking them do not include co-valent bonds. Thus, while reductive cleavage is required to break the disulfide bonds linking the light and heavy chains of haptoglobin or the immunoglobulins, dissociation of the hemoglobin subunits is accomplished more readily (Itano & Singer, 1958). This property of hemoglobin is of considerable importance, from both a physiological and chemical point of view, permitting the dissociation and recombination of subunits under certain conditions of pH and ionic strength.

The sequence of the 141 amino-acid residues in the α chain and of the 146 residues in the non-α chains, has been determined (Konigberg & Hill, 1962; Schroeder, 1963; Braunitzer *et al.*, 1964). Three-dimensional studies of Cullis *et al.* (1962), Muirhead & Perutz (1963), Perutz (1963, 1965) and Perutz *et al.* (1964, 1965) have shown that the individual chains greatly resemble the single chain of sperm whale myoglobin described by Kendrew *et al.* (1960, 1961). The resemblance is due in small part to homologous amino acids, but in much larger part to the common feature of polar residue exclusion from the interior of the globin chains. Nearly all of the charged side groups of the amino acids are directed toward the exterior, or toward the water-filled central cavity of the hemoglobin molecule. Those surface residues which have large non-polar side chains tend to bury themselves in crevices in order to avoid contact with water. Gene mutations resulting in replacement of polar by non-polar (or vice versa) amino acids at the external surface do not necessarily affect the tertiary structure, but replacement of a non-polar by a polar residue at an interior site may disrupt the structural integrity of the molecule (Perutz *et al.*, 1965).

Figure 10.1 is a diagrammatic representation adapted from Schroeder (1963) and Heller (1966), showing the β chain pulled apart to avoid overlapping segments but retaining the helices, designated by the letters A through H, which alternate with the non-helical portions. This figure is supplemented by the data in Fig. 10.2, which compares the amino acid sequences of the α and β chains and indicates their corresponding three-dimensional positions as assigned by Perutz (1965).

The α chain has a similar structure, except that the D helix is missing (i.e. a so-called 'Braunitzer gap'). The α chain also lacks a reactive sulfhydryl (SH) group, so that the two reactive sulfhydryls of the intact molecule belong to the cysteine residues at position 93 of the β chains. Inactive SH groups are located at 104α, in a cavity between the B and G helices, and at 112β, which lies in an area of contact between α and β chains.

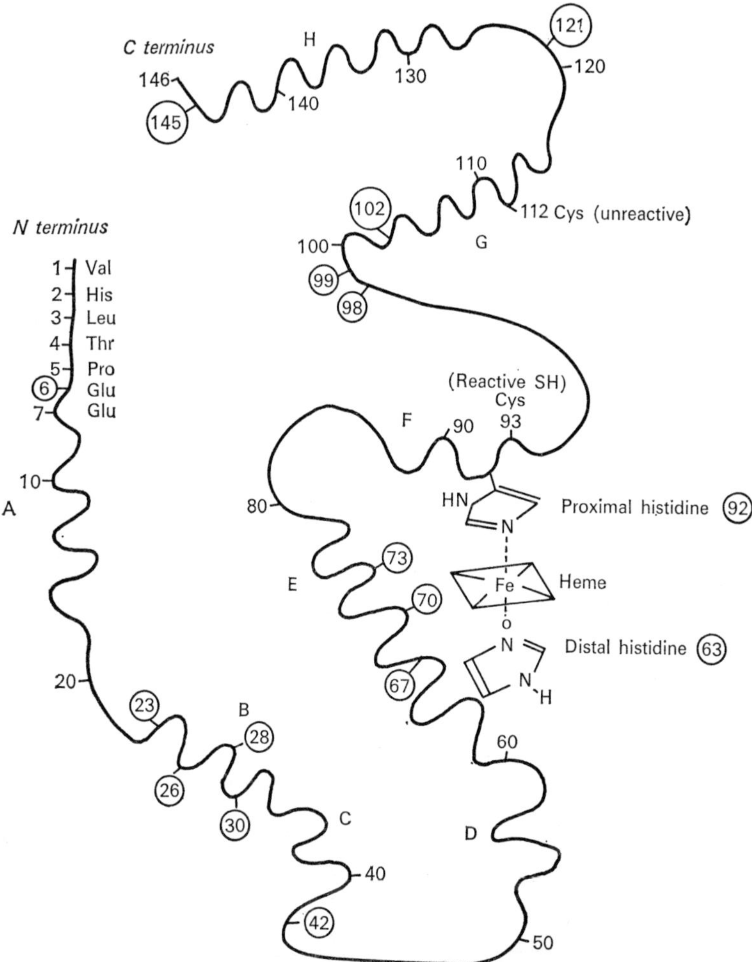

FIGURE 10.1. Schematic structure of the hemoglobin beta chain, pulled apart but still retaining its helices, A through H. The position of every tenth amino acid is numbered. The encircled numbers refer to positions on the β chain in which amino acid substitutions due to gene mutations are known to alter the function of hemoglobin. The positions of the reactive and unreactive sulfhydryl groups, the proximal and distal histidines, and the location of heme between the E and F helices are also indicated. In the three-dimensional model, there is close spatial contact between the B and E helices near the heme prosthetic group. (Fig. suggested by drawings in Schroeder, 1963 and Heller, 1966a).

Chain	Residues	Region	1	2	3	4	5	6	7	8	9	10	11	12	13	14	15	16	17	18	19	20	21	22	23	24
α	1-2	NA		Val	Leu																					
β	1-3		Val	His	Leu																					
α	3-18	A	Ser	Pro	Ala	Asp	Lys	Thr	Asn	10 Val	Lys	Ala	Ala	Try	Gly	Lys	Val	Gly								
β	4-18		Thr	Pro	Glu	Glu	Lys	Ser	10 Ala	Val	Thr	Ala	Leu	Try	Gly	Lys	Val									
α	19	AB	Ala																							
β	—		—																							
α	20-35	B	20 His	Ala	Gly	Glu	Tyr	Gly	Ala	Glu	Ala	Leu	30 Glu	Arg	Met	Phe	Leu	Ser								
β	19-34		Asn	20 Val	Asp	Glu	Val	Gly	Gly	Glu	Ala	Leu	Gly	30 Arg	Leu	Leu	Val	Val								
α	36-42	C	Phe	Pro	Thr	Thr	40 Lys	Thr	Tyr																	
β	35-41		Tyr	Pro	Try	Thr	Gln	40 Arg	Phe																	
α	43-49	CD	Phe	Pro	His	Phe	Asp		Leu	Ser																
β	42-49		Phe	Glu	Ser	Phe	Gly	Asp	Leu	Ser																
α	50-51	D						50 His	Gly																	
β	50-56		50 Thr	Pro	Asp	Ala	Val	Met	Gly																	
α	52-71	E	Ser	Ala	Gln	Val	Lys	Gly	His	Gly	60 Lys	Lys	Val	Ala	Asp	Ala	Leu	Thr	Asn	Ala	70 Val	Ala				
β	57-76		Asn	Pro	Lys	60 Val	Lys	Ala	His	Gly	Lys	Lys	Val	Leu	Gly	70 Ala	Phe	Ser	Asp	Gly	Leu	Ala				
α	72-79	EF	His	Val	Asp	Asp	Met	Pro	Asn	Ala																
β	77-84		His	Leu	Asp	80 Asn	Leu	Lys	Gly	Thr																
α	80-88	F	80 Leu	Ser	Ala	Leu	Ser	Asp	Leu	His	Ala															
β	85-93		Phe	Ala	Thr	Leu	Ser	90 Glu	Leu	His	Cys															
α	89-93	FG	His	90 Lys	Leu	Arg	Val																			
β	94-98		Asp	Lys	Leu	His	Val																			
α	94-112	G	Asp	Pro	Val	Asn	Phe	Lys	100 Leu	Leu	Ser	His	Cys	Leu	Leu	Val	Thr	Leu	110 Ala	Ala	His					
β	99-117		Asp	100 Pro	Glu	Asn	Phe	Arg	Leu	Leu	Glu	Asn	Val	110 Leu	Val	Cys	Val	Leu	Ala	His	His					
α	113-117	GH	Leu	Pro	Ala	Glu	Phe																			
β	118-122		Phe	Gly	120 Lys	Glu	Phe																			
α	118-141	H	Thr	Pro	120 Ala	Val	His	Ala	Ser	Leu	Asp	Lys	Phe	Leu	130 Ala	Ser	Val	Ser	Thr	Val	Leu	Thr	Ser	Lys	140 Tyr	Arg
β	123-146		Thr	Pro	Pro	Val	Gln	Ala	Ala	130 Tyr	Gln	Lys	Val	Val	Ala	Gly	Val	Ala	Asn	140 Ala	Leu	Ala	His	Lys	Tyr	His

Each of the globin chains has a heme prosthetic group (Fig. 10.3), which is located in a crypt between the E and F helices (see Fig. 10.1). The fifth coordination position of the heme iron is linked directly with the imidazole of a histidine residue on the F helix in position 92 of the β chain or 87 of the α chain; these are the so-called proximal histidine

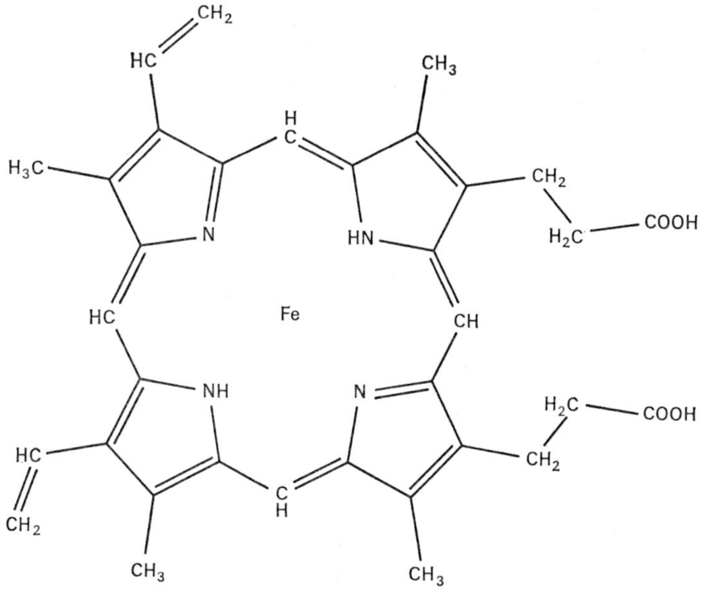

FIGURE 10.3. The prosthetic group of hemoglobin, iron protoporphyrin IX.

residues. On the E helix, there is a distal histidine at 58α and 63β, respectively, which points toward the sixth coordination position of heme iron, and is involved in an unknown way with the reversible bonding of molecular oxygen. This problem has been discussed extensively as, for example, in Hemes and Hemoproteins, a symposium edited by Chance *et al.* (1966).

The two propionic acid side groups of heme are hydrogen-bonded

FIGURE 10.2. The amino acid sequences of the α and β chains of human hemoglobin (*numbered on the left*), arranged according to their three-dimensional positions (*numbered at the top of the figure*) as assigned by Perutz (1965). Every tenth amino acid in each chain is also numbered according to linear sequence.

to amino acid side groups, and (in addition to the two histidines) each polypeptide has several amino acids whose side groups are located within van der Waals contact distance of the porphyrin ring. Most of these are non-polar residues not in contact with the molecular surface (Perutz, 1965; Watson, 1966). (See Addenda).

The four polypeptide chains in the intact molecule contain a total of 574 amino acids and four hemes, with a calculated molecular weight of 64,450. However, by various measurements, the figure is closer to 66,000. The molecule is spheroidal, with dimensions of 64 Å × 55 Å ×

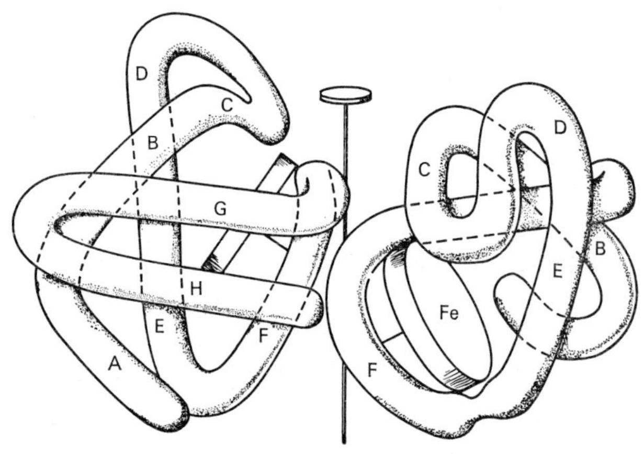

FIGURE 10.4. Diagram of the two beta chains of oxyhemoglobin, showing their orientation about the two-fold symmetry axis. (From Muirhead & Perutz, 1963. Reprinted with permission from *Nature* **199**, 633.)

50 Å. The chains are arranged tetrahedrally about an axis of two-fold symmetry (see Cullis *et al.*, 1962, for pictures of the molecular model). Figure 10.4, taken from Muirhead & Perutz (1963), shows the positioning of the two β chains and their hemes about an axis of symmetry. The α chains form a similar pair, and thus, like members of each pair are so arranged that they could replace each other's position by rotation of 180 degrees about the axis. If all four chains were identical, then rotation through either of the two perpendicular planes would also bring any two chains into congruence, since the molecule would be perfectly symmetrical. However, the difference

in α and β chain structure distorts the symmetry, so that 'pseudo-axes' pass through the molecule at the two perpendicular planes which relate the α and β chains: α^1 to β^1 and α^2 to β^2 in one plane, and α^1 to β^2 and α^2 to β^1 in the other plane.

Although each chain has some contact with the other three chains, unlike chains (α and β) are in much closer contact than like chains (α and α or β and β), suggesting that when the molecule is split, symmetrical dissociation into $\alpha\beta$ half-molecules is more likely to occur than asymmetrical dissociation into α_2 and β_2 (Perutz, 1965). Schroeder (1963) pointed out that two kinds of symmetrical dissociation are possible, since the distance between one set of unlike chains, α^1—β^1 and α^2—β^2, is 30·4 Å (measured between their respective iron atoms), while the distance between the other set of unlike chains, α^1—β^2 and α^2—β^1, is 25·2 Å (Perutz, 1965). Recent studies of Rosemeyer & Huehns (1967) showed that in the initial dissociation of hemoglobin by p-chloromercuribenzoate and other reagents, the (identical) dimers $\alpha^1\beta^1$ and $\alpha^2\beta^2$ are formed (indicating breakage of the α^1—β^2 and α^2—β^1 bonds). It is likely that physiological dissociation also involves these two dimers.

THE ALLOSTERIC EFFECT: HEMOGLOBIN AS AN ENZYME

On oxygenation, the distance between the two β chains (but not the α chains) is decreased in a kind of 'molecular respiratory movement'. However, even though the heme groups of these chains move closer together by about 7 Å, the hemes are still separated by at least 25 Å (Muirhead & Perutz, 1963). This finding raised serious doubts about the concept of direct heme-heme interaction as the basis for the sigmoid shape of the oxygen dissociation curve. In fact, the concept was no longer tenable when it was found that neither the β nor the γ chain tetramers show any evidence of heme-heme interaction (see p. 360), and also, that there is equivalent strength of binding at all four sites in normal hemoglobin (Wyman, 1964; Alben & Caughey, 1966).

Muirhead & Perutz (1963) and Wyman (1963) pointed out that the sigmoid shape of the oxygen dissociation curve is best explained by considering hemoglobin as an enzyme with allosteric properties. In other words, the combination of the enzyme with its substrate (i.e.

hemoglobin with oxygen) at one site on the molecule causes changes in configuration which facilitate the combination of enzyme and substrate at other sites on the molecule. Monod *et al.* (1965) showed that such allosteric effects could be best accounted for by a model in which the protein has two conformational states of widely different affinity for the substrate.

Sizable evidence suggests that the functional unit of hemoglobin is not the intact molecule, but the $\alpha\beta$ dimer (Rossi-Fanelli *et al.*, 1964; Riggs, 1964; Antonini *et al.*, 1965, 1966; Benesch *et al.*, 1965; Antonini, 1967; Rosemeyer & Huehns, 1967). According to Benesch *et al.* (1965, 1966), the normal process of oxygenation requires a

(a)　　Dissociation of high affinity tetramer:

$$\begin{matrix}(\alpha\,\beta)^{\circ}\\(\beta\,\alpha)^{\circ}\end{matrix} \longrightarrow (\alpha\,\beta)^{\circ} \quad + \quad (\alpha\,\beta)^{\circ}$$

(b)　　Exchange of high-affinity for low-affinity dimer:

$$\begin{matrix}\alpha\,\beta\\\beta\,\alpha\end{matrix} \quad + \quad (\alpha\,\beta)^{\circ} \longrightarrow \begin{matrix}(\alpha\,\beta)^{*}\\(\beta\,\alpha)^{\circ}\end{matrix} \quad + \quad \alpha\,\beta$$

(c)　　Oxygenation of intermediate tetramer to high-affinity tetramer:

$$\begin{matrix}(\alpha\,\beta)^{*}\\(\beta\,\alpha)^{\circ}\end{matrix} \longrightarrow \begin{matrix}(\alpha\,\beta)^{\circ}\\(\beta\,\alpha)^{\circ}\end{matrix}$$

FIGURE 10.5. Successive oxygenation of the two $\alpha\beta$ dimers of the hemoglobin molecule. (Adapted from Benesch *et al.*, 1966.)

dissociation of the tetramer into dimers, followed by an exchange of oxygenated dimers in one conformation state with unoxygenated tetramers in another conformation state.

For this mechanism to operate, it is necessary to postulate the three tetramers shown in Fig. 10.5. One is composed of two unoxygenated $\alpha\beta$ dimers with a low affinity for oxygen; the second consists of two oxygenated dimers, $(\alpha\beta)^0$ in a high-affinity conformational state; and the third, intermediate, form is composed of one $(\alpha\beta)^0$ and one $\alpha\beta^*$ dimer. The asterisk is used by Benesch *et al.* (1966) to indicate that the $\alpha\beta^*$ dimer is not yet oxygenated, but it has the high-affinity conformational state of $(\alpha\beta)^0$. The existence of such an intermediate 'ligand conformation without ligand' had been shown by Gibson (1959), using high energy light flash dissociation of carbonmonoxyhemo-globin.

From experiments with ferro- and ferrihemoglobin, Benesch *et al.* (1965, 1966) demonstrated indirectly an exchange of subunits within Hb A and also between Hb A and Hb S molecules, the latter indicating that in the red cells of individuals heterozygous for the Hb S (β chain abnormality) gene, there are not only the expected tetramers, $\alpha_2^A\beta_2^A$ and $\alpha_2^A\beta_2^S$, but also the mixed tetramer, $\alpha_2^A\beta_1^A\beta_1^S$.

The possibility of such a 'mixed molecule' had been considered previously by a number of workers, and Guidotti *et al.* (1963) pointed out that if all three tetramers were in rapid equilibrium with their $\alpha\beta$ dimers, separation methods such as electrophoresis and chromato-graphy, by disturbing the equilibrium in favor of $\alpha_2^A\beta_2^A$ and $\alpha_2^A\beta_2^S$, would make it impossible to detect the mixed tetramer. According to Benesch *et al.* (1966), it might be possible to isolate such a tetramer under conditions in which subunit exchange is suppressed, that is, when the oxygen dissociation curve lacks the sigmoid shape indicative of 'heme-heme interaction'. It may therefore be pertinent that Ruck-nagel *et al.* (1967) have described a hemoglobin with increased oxygen affinity, Hb Ypsi, which appears to form separable $\alpha_2^A\beta_1^A\beta_1^{\text{Ypsi}}$ tetramers when mixed with Hb A (see p. 379). Other hemoglobins with in-creased oxygen affinity have not been reported to show this kind of separation, but it is conceivable that they might do so under certain conditions of electrophoresis or chromatography.

Hb A$_2$

While most of the hemoglobin in normal human subjects past infancy is Hb A, Hb A$_2$ is also present, but in a concentration of only 1·5 to 3 per cent of the total hemoglobin content (Kunkel *et al.*, 1957). Its molecular formula is $\alpha_2^A\delta_2$, indicating that the two α chains are identical with those of Hb A, but that the other two (identical) chains are sufficiently different from β chains to warrant the designation δ (Ingram & Stretton, 1961). In fact, the δ chain has the same number of amino-acid residues as the β chain, including the two cysteines, and there are only ten differences in their sequence (Hill & Kraus, 1963; Jones, 1964). These differences are listed in Table 10.1, which indicates the site of amino-acid substitution in both δ and γ chains, as compared with the β chain.

The studies of Eddison *et al.* (1964) indicate that Hb A_2 has the same oxygen affinity and equilibrium curve as Hb A. Somewhat different

TABLE 10.1. Amino acid differences of the γ and δ chains as compared with the β chain. Adapted from Schroeder (1963) and Ingram & Stretton (1962a,b)

β Chain	γ Chain	δ Chain	3 D	β Chain	γ Chain	δ Chain	3 D
1 Val	Gly		NA1	71 Phe	Leu		E15
3 Leu	Phe		NA3	72 Ser	Gly		E16
5 Pro	Glu		A2	74 Gly	Ala		E18
7 Glu	Asp		A4	75 Leu	Ileu		E19
9 Ser	Ala	Thr	A6	76 Ala	Lys		E20
10 Ala	Thr		A7	80 Asn	Asp		EF4
11 Val	Ileu		A8	86 Ala		Ser	F2
12 Thr		Asn	A9	87 Thr	Gln	Gln	F3
13 Ala	Ser		A10	104 Arg	Lys		G6
21 Asp	Glu		B3	112 Cys	Thr		G14
22 Glu	Asp	Ala	B4	116 His	Ileu	Arg	G18
23 Val	Ala		B5	117 His		Asn	G19
27 Ala	Thr		B9	125 Pro	Glu	Gln	H3
43 Glu	Asp		CD2	126 Val		Met	H4
47 Asp	Asn		CD7	129 Ala	Ser		H7
50 Thr	Ser	Ser	D1	130 Tyr	Try		H8
51 Pro	Ala		D2	133 Val	Met		H11
52 Asp	Ser		D3	135 Ala	Thr		H13
54 Val	Ileu		D5	139 Asn	Ser		H17
69 Gly	Thr		E13	142 Ala	Ser		H20
70 Ala	Ser		E14	143 His	Ser		H21
				144 Lys	Arg		H22

results were reported by Huisman *et al.* (1962). On electrophoresis, Hb A_2 has a much slower migration rate than Hb A and is readily separable by chromatographic techniques.

Hb F

Fetal hemoglobin, or Hb F, has a concentration of less than 0·5 per cent after the first few years of life in normal subjects, but it is the major hemoglobin component during fetal development. Like Hb A and A$_2$, Hb F contains two α^A chains, but a different pair of chains, called γ, completes the tetramer, so the molecular formula is $\alpha_2^A\gamma_2^F$. The γ chain differs from the β chain in 39 amino-acid residues (Schroeder, 1963), although the total number of residues is the same (see Table 10.1). The reactive SH group at γ93 is the same as on the β chain, but the cysteine at position 112 is replaced by threonine. Also, while the β chain contains no isoleucine, this amino acid is present in the γ chain at positions 11, 54, 75 and 116 (A8, D5, E19 and G18).

Allen *et al.* (1953) showed that although suspensions of fetal red cells have a much higher oxygen affinity than adult red cells, the difference disappears when hemolysates of the same blood are similarly examined. This property of fetal cells is unrelated to their Hb F content, since the cells of adults with the benign condition, hereditary persistence of fetal hemoglobin (p. 394) have normal oxygen dissociation curves (Schruefer *et al.*, 1962; Nechtman & Huisman, 1964). The fetal cell membrane, which has both structural and functional differences from that of normal adult blood, is probably responsible for the oxygen avidity (Zipursky, 1965). Another explanation could be a difference in intracellular organic phosphate. Benesch & Benesch (1967) found ATP and 2,3-diphosphoglycerate (and inositol hexaphosphate in avian cells) to be very effective in shifting the oxygen dissociation curve to the right (i.e. decreasing the oxygen affinity of Hb). Presumably this effect is due to the binding of organic phosphate to deoxyhemoglobin, with alteration in the allosterism of hemoglobin (Benesch *et al.*, 1968). Although the ATP level is elevated in fetal cells (Gross *et al.*, 1963), 2,3-DPG appears to be less readily synthesized, at least during *in vitro* incubation (Greenwalt & Ayers, 1960; Zipursky *et al.*, 1960). During early development, chickens undergo a rapid rise in red cell inositol hexaphosphate, during which the oxygen affinity shows a rapid decline (Benesch *et al.*, 1968). Conceivably, a similar mechanism may operate during the first months of human life.

Hb F is strikingly resistant to alkaline and acid denaturation. This property is used for its quantitation in blood specimens (Singer *et al.*, 1951; Betke *et al.*, 1959) and in individual red cells (Kleihauer *et al.*, 1957; Shepherd *et al.*, 1962). Hb F and Hb A are readily distinguished at pH 6·2 in agar gel (Robinson *et al.*, 1957). Cellulose acetate and starch gel may also be used with appropriate buffer systems at alkaline pH (Chernoff & Pettit, 1964; Chernoff *et al.*, 1965;

13

Briere *et al.*, 1965). Preparations of pure Hb F are best made by column chromatography (Allen *et al.*, 1958).

Although Hb F is only a trace component of the red cells of normal adults, an increased proportion is often encountered in certain acquired hematological diseases such as hypoplastic anemia, pernicious anemia and leukemia; in some of the hemoglobinopathies, such as thalassemia and sickle cell anemia; and in the hereditary persistence of Hb F.

EMBRYONIC HEMOGLOBINS

The studies of Huehns *et al.* (1961, 1964) showed that there are at least two hemoglobins in the red cells of human embryos which do not normally persist throughout the fetal period. An embryonic Hb ϵ chain, not yet subjected to fine structure analysis, may be elaborated before α chain production begins. It has been identified in Hb Gower-1, which may be ϵ_4, and in Hb Gower-2, which is $\alpha_2\epsilon_2$. Both of these hemoglobins have markedly slower electrophoretic migration rates than Hb A.

Kaltsoya *et al.* (1966) reported that another hemoglobin in embryos, originally thought to be Hb A, could be readily differentiated from it by agar electrophoresis at acid pH. The nature of this A-like component was not clarified, but the authors suggested that it probably had no β chain content, since they could not detect any β chain synthesis in embryos with a crown-rump length of less than 100 mm. The possibility of an $\epsilon_2\gamma_2$ tetramer lacked precedent, since the analog, $\beta_2\gamma_2$, had never been detected. However, Capp *et al.* (1967) described a hemoglobin composed of two different non-α chains, one of which was identified as γ, and the other as an unknown, possibly ϵ. This hemoglobin, called Hb Portland 1, migrated slightly faster than Hb A in starch gel at alkaline pH. It was found in a Chinese infant with multiple congenital anomalies associated with a complex karyotype, showing mosaicism involving trisomy 16 and deletion of the short arm of a 17–18 chromosome. Subsequently, Hecht *et al.* (1967) reported that by subjecting hemolysates to IRC-50 ion exchange chromatography, they detected the presence of Hb Portland 1 in five other infants with chromosomal abnormalities. In two normal infants, the hemoglobin was present in smaller proportions. The persistence of ϵ chain synthesis after birth was originally described in five patients with D

trisomy (Huehns *et al.*, 1964) and it is probable that the presence of
Hb Portland 1 also represents prolonged ϵ chain production. Alternat-
ively, still another embryonic hemoglobin chain might be present in the
Hb Portland 1 molecule.

HETEROGENEITY OF Hb A AND Hb F

In normal adult and cord blood specimens, the existence of minor
hemoglobins has been demonstrated both by electrophoresis (Kunkel
& Wallenius, 1955; Robinson *et al.*, 1957) and by chromatography
(Schnek & Schroeder, 1961; Huisman & Dozy, 1962, 1965; Huisman
et al., 1966; Atassi, 1964). Conflicting terminology has been used to
characterize these minor components, some of which apparently
occur *in vivo*, while others are associated with red cell aging *in vitro*.
Hb A_{Ic}, of biological origin, was reported to have the formula
$\alpha_2^A\beta_1^A\beta_1^X$, where the β^X chain differs from β^A in that its N-terminal
amino group is blocked by an aldehyde or a ketone with a molecular
weight of about 280 (Holmquist & Schroeder, 1964, 1966). Hb A_{Id},
which is a mixed disulfide of Hb A and glutathione residues (Huisman
et al., 1966), is a part of the mixture of molecular species called Hb A_3,
which has an electrophoretic migration slightly faster than Hb A.
Mixed disulfides of Hb F and of Hb A_2 also provide additional
components, particularly in stored blood. In fetal blood, the so-called
F_I component was reported to have the molecular formula $\alpha_2^A\gamma^F\gamma^X$,
where γ^X differs from γ^F by substitution of an acetyl group for the
terminal amino group (Schroeder *et al.* 1962). Thus it can be designated as γ^{AcF}.

Huehns & Shooter (1966) pointed out that molecules with the
mixed tetrameric structure proposed for Hb A_{Ic} and Hb F_I should not
be demonstrable unless their dissociation properties are such that the
tetramers are not in rapid equilibrium with their subunit dimers. By
dissociation and recombination experiments, as well as by self-hybri-
dization of Hb F_I molecules, they showed that Hb F_I is in equilibrium with its subunits, and that no γ^F chains are present in the
molecule. However, since Hb F_I appears to contain two kinds of γ
chains, still another structural alteration, possibly a blocked N-
terminal amino group, is apparently characteristic of the γ chain
accompanying γ^{AcF} in Hb F_I.

According to Marti *et al.* (1967), patients with severe intravascular hemolysis have a fast-moving hemoglobin in their plasma and urine, produced by the removal of the C terminal arginine of the α chain, presumably through the action of a plasma carboxypeptidase. They named this acquired component Hb Koelliker, with the molecular formula $\alpha_2^{\text{minus 141 Arg}} \beta_2^{A}$.

THE ABNORMAL HEMOGLOBINS

HEMOGLOBIN HOMOTETRAMERS

Hemoglobins composed of only one kind of normal polypeptide chain (Jones *et al.*, 1959; Hunt & Lehmann, 1959; Dance & Huehns, 1962; Huehns *et al.*, 1964) do not appear in normal blood except in the embryo, which contains both Hb Gower-1 (ϵ_4) and Hb Barts (γ_4). The β chain tetramer, Hb H, is not found in the blood of infants or adults, unless there is depressed α chain synthesis. The α chain apparently does not form tetramers, at least *in vivo*, but precipitated α^{A}, as well as α^{A} monomers and dimers have been isolated from the red cells of patients whose β chain synthesis is suppressed (Fessas & Loukopoulos, 1964; Fessas *et al.*, 1966; Huehns & Modell, 1967).

The oxygen equilibrium curves of Hb H and Hb Barts are similar to that of myoglobin, that is, the sigmoid shape and Bohr effect are lost and oxygen affinity is increased (Benesch *et al.*, 1961; Horton *et al.*, 1962); thus, they are useless in oxygen transport. The loss of allosteric effect in these homotetramers is further shown by the fact that they do not undergo a change in molecular configuration associated with oxygenation such as that observed in Hb A. Instead, the β_4 molecule resembles that of normal deoxygenated Hb, with the characteristic increase in distance between two of the β chains, and no change in position of the other two β chains which occupy the region normally held by α chains (Perutz & Mazzerella, 1963). Inability of these two homotetramers (β_4 and γ_4) to combine with haptoglobin is another striking difference from both normal and abnormal hemoglobins (Nagel & Ranney, 1964), a property they share with deoxygenated Hb A (Nagel *et al.*, 1965; Chiancone *et al.*, 1966). The most likely explanation for both deviations from normal behavior is resistance to subunit dissociation, which is necessary for normal molecular function. Furthermore, since the isolated α and β chain monomers have physiological properties similar to those of the homotetramers, it appears that a dimer consisting of two different chains is required for functional activity (Antonini *et al.*, 1965).

Hb H is very unstable, and thus its presence in more than trace quantities may in itself be deleterious to red cells. First described by Rigas *et al.* (1955) and Gouttas *et al.* (1955), Hb H has been the object of intense investigation, summarized by Gabuzda (1966). The occurrence of Hb H is most often associated with α thalassemia, but it has also been reported in patients with leukemia (White *et al.*, 1960; Beaven *et al.*, 1963). When blood containing Hb H is incubated with brilliant cresyl blue, the red cells subsequently contain characteristic small inclusion bodies having a speckled appearance. These so-called BCB inclusion bodies are apparently an artifact produced by precipitation with the dye, because in hemolysates of dye-treated cells, the Hb H remains in the stromal fraction and cannot be detected in the supernatant solution (Fessas, 1959). On the other hand, preformed inclusion bodies consisting of denatured Hb H are demonstrable in bone marrow red cell precursors and, after splenectomy, in peripheral blood specimens, indicating that the spleen preferentially removes the Hb H, either by 'pitting' it from the cells or by removing the intact cells from circulation (Rigas & Koler, 1961a). This assumption is supported by the finding that patients synthesize Hb H at least as rapidly as Hb A, but the latter is by far the predominant Hb in peripheral blood cells (Gabuzda *et al.*, 1965). The intracellular denaturation of Hb H is probably associated with methemoglobin formation, since Hb H is highly susceptible to oxidation (Rigas & Koler, 1961b). The fact that all eight of the sulfhydryl groups of Hb H (2 for each chain) are reactive may be an important factor in determining the physical properties of the molecule (Benesch & Benesch, 1964; Gabudza, 1966). Gabuzda *et al.* (1967) found that Hb H has a low redox potential, and that binding of glutathione (GSH) to Hb H greatly exceeded its binding to Hb A during oxidation with 2,6-dichlorobenzenoneindophenol. This increased binding was thought to be associated with the reduction in GSH level which occurs *in vivo* in cells containing Hb H.

Hb H is readily demonstrated by electrophoresis, since it has an unusually rapid migration rate. However, the instability of the protein requires that fresh specimens be used for electrophoresis and for column chromatography (Horton *et al.*, 1962; Weatherall, 1965).

THE LEPORE HEMOGLOBINS

In the Lepore hemoglobins, the α chains are normal, but the non-alpha chains consist of the N-terminal portion of the δ chain joined to the C-terminal portion of the β chain, with the total number of amino acids the same as in the β chain or δ chain (Baglioni, 1962a). Hemoglobins of the Lepore type have been found in the Mediterranean area, in New Guinea, and in a New York Negro family. The Italian Lepore hemoglobin is called Hb Lepore_Boston (or Lepore_Washington) (Gerald & Diamond, 1958; Gerald *et al.* (1961; Baglioni, 1962a, Pearson &

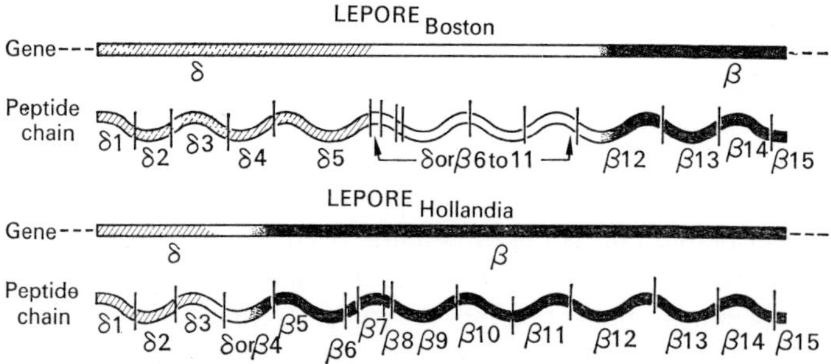

FIGURE 10.6. Diagram of two different Lepore hemoglobin genes and their corresponding peptide chains. Each Lepore gene contains a portion of the Hbδ gene (cross-hatched) and a portion of the Hbβ gene (solid). The empty areas may be derived from either gene. The vertical lines indicate sites of cleavage by trypsin. (From Baglioni, 1962a. Reprinted with permission from the *Proc. natn. Acad. Sci.* **48**, 1880.)

McFarland, 1962), and has the same structure as Hb Lepore_Augusta (Labie *et al.*, 1966). The Greek Lepore hemoglobin was called Hb Pylos by Fessas *et al.* (1962); its fine structure has not yet been reported. Hb Lepore_Hollandia, which occurs in New Guinea natives (Neeb *et al.*, 1961), differs from the Italian Lepore hemoglobin in that its abnormal chain contains a shorter δ chain contribution and a longer segment of β chain (Barnabas & Muller, 1962; Baglioni, 1962a). Hb Lepore_Cyprus, found in Cypriots of Turkish origin (Beaven *et al.*, 1964) and Hb Lepore_The Bronx, detected in a New York Negro family (Ranney & Jacobs, 1964) have not been differentiated from the Italian (and Greek) Lepore hemoglobins. Clinically, patients who are heterozygous for the Lepore hemoglobin genes resemble patients with β thalassemia minor or major, respectively (see p. 391).

Baglioni (1962a, 1963) suggested that the most likely genetic event leading to formation of the chains shown in Fig. 10.6, which contain components of both β and δ chains, was displaced synapsis during meiosis, with unequal crossing-over between homologous portions of the genes coding for β and δ chains, respectively (see p. 564). The same kind of genetic event was postulated by Smithies *et al.* (1962) to account for Hp^2 gene variability in the haptoglobin system (see p. 81). Implicit in this mechanism is very close linkage of the β and δ chain loci on the same chromosome, and this linkage has been established (Reviewed by Weatherall, 1965).

HEMOGLOBINS WITH AMINO ACID SUBSTITUTIONS OR DELETIONS

Most of the abnormal hemoglobins appear to be products of point mutation with a single amino-acid substitution in the α, β, γ or δ peptide chains. The resultant change in the whole hemoglobin molecule may be so benign that no physiological effect is detectable and even the rate of synthesis remains normal. However, in most instances, there is a variable degree of curtailment in synthesis even though the affected individual remains hematologically normal (Itano, 1966). Furthermore, changes in the physical properties of some molecular variants are sufficient to bring about measurable changes in the red cell. When the defect is severe, the heterozygote is clinically recognizable and the homozygous state may be lethal. When the defect is moderate, the heterozygote may be detected by special tests, but only the homozygote has clinical symptoms under usual environmental conditions. A large number of the apparently 'benign' abnormal hemoglobins have so far been detected only in the heterozygous state, so that the possible deleterious effects of homozygosity are not known. However, in some instances, these variants have been observed in combination with thalassemia, which depresses the synthesis of either the α or β chain product of the homologous locus. Under these circumstances, the ability of the variant gene to compensate for the loss of its homologous gene product provides an indication of the degree of its impairment.

Detection of a structurally abnormal hemoglobin is usually

TABLE 10.2. Abnormal hemoglobins with amino acid substitution or deletion at one or more sites

	Third dimensional position	Position in chain	Amino acid in α^A chain	Substituted amino acid	Migration rate relative to Hb A	Unusual characteristics	References
α chain variants:							
J-Toronto	A-3	5	Ala	Asp	Fast		Crookston *et al.* (1965)
J-Paris	A-10	12	Ala	Asp	Fast		Rosa (1966)
J-Oxford	A-13	15	Gly	Asp	Fast		Liddell *et al.* (1964)
I	A-14	16	Lys	Glu	Fast		Beale & Lehmann (1965)
J-Medellin	B-3	22	Gly	Asp	Fast		Gottlieb *et al.* (1964)
Memphis	B-4	23	Glu	Gln	Slow		Kraus *et al.* (1966)
Audhali	B-4	23	Glu	Val	?		Cited in Wintrobe (1967)
G-Honolulu	B-11	30	Glu	Gln	Slow		Swenson *et al.* (1962)
Torino	CD-1	43	Phe	Val	Same	Unstable	Beretta *et al.* (1968)
L-Ferrara-Beilinson	CD-5	'47	Asp	Gly	Slow		Baglioni & Colombo (1964) De Vries *et al.* (1963)
Sinai-Sealy	CD-5	47	Asp	His	Slow	Unstable	Charache *et al.* (1967), Schneider *et al.* (1967)

Russ	D-7	51	Gly	Arg	Slow		Reynolds & Huisman (1967)
Shimonoseki	E-3	54	Gln	Arg	Slow		Hanada & Rucknagel (1963)
Mexico	E-3	54	Gln	Glu	Fast		Jones et al. (1963)
Norfolk	E-6	57	Gly	Asp	Fast		Baglioni (1962c)
M-Boston	E-7	58	His	Tyr	Same	Methemoglobinemia	Gerald & Efron (1961)
G-Philadelphia	E-17	68	Asn	Lys	Slow		Baglioni & Ingram (1961)
Ube-2	E-17	68	Asn	Asp	Fast		Miyaji et al. (1967)
Stanleyville II	EF-7	78	Asn	Lys	?		Cited in Wintrobe (1967)
G-Norfolk	F-6	85	Asp	Asn	Slow		Cited in Wintrobe (1967)
M-Iwate	F-8	87	His	Tyr	Same	Methemoglobinemia	Miyaji et al. (1963)
J-Broussais	FG-2	90	Lys	Asn	Fast		Cited in Wintrobe (1967)
J-Capetown	FG-4	92	Arg	Gln	Fast	High O_2 affinity	Botha et al. (1966)
Chesapeake	FG-4	92	Arg	Leu	Fast	High O_2 affinity	Clegg et al. (1966)
J-Tongariki	GH-3	115	Ala	Asp	Fast		Gajdusek et al. (1967)
O-Indonesia	GH-4	116	Glu	Lys	Slow		Baglioni & Lehmann (1962)
Bibba	H-19	136	Leu	Pro	Slow	Unstable	Kleihauer et al. (1968)
Koelliker	H-24	141	Arg	None	Fast	*In vitro* change	Marti et al. (1967)
G-Baltimore		T IX	Asp	His	Slow		Rieder & Naughton (1965)
Q		T IX			Slow		Schroeder (1963)

TABLE 10.2—*continued*

	Third dimensional position	Position in chain	Amino acid in β chain	Substituted amino acid	Migration rate relative to Hb A	Unusual characteristics	References
β Chain Variants:							
Tokuchi	NA-2	2	His	Tyr	Slow		Shibata *et al.* (1963a)
S	A-3	6	Glu	Val	Slow	Cell sickling	Ingram (1961a)
C-Harlem*	A-3	6	Glu	Val	Slow	Cell sickling	Bookchin *et al.* (1967)
C	A-3	6	Glu	Lys	Slow	Cell rigidity	Hunt & Ingram (1960)
C-Georgetown*	A-3	6	Glu	Val	Slow	Cell sickling	Gerald & Rath (1966)
G-San Jose	A-4	7	Glu	Gly	Slow		Hill *et al.* (1960)
Siriraj	A-4	7	Glu	Lys	Slow		Tuchinda *et al.* (1965)
Porto Alegre	A-6	9	Ser	Cys	Same	Polymer forms *in vitro*	Bonaventura & Riggs (1967)
J-Baltimore	A-13	16	Gly	Asp	Fast		Baglioni & Weatherall (1963)
D-Bushman	A-13	16	Gly	Arg	Slow		Wade *et al.* (1967)
G-Coushatta	B-4	22	Glu	Ala	Slow		Bowman *et al.* (1967)
Freiburg	B-5	23	Val	None	Slow	Increased O_2 affinity	Jones *et al.* (1966c)
E	B-8	26	Glu	Lys	Slow	Moderate anemia in homozygotes	Hunt & Ingram (1961)
Genova	B-10	28	Leu	Pro	Same	Unstable	Sansone *et al.* (1967)
Tacoma	B-12	30	Arg	Ser	?	Increased O_2 affinity	Jones *et al.* (unpubl.)

Hammersmith	CD-1	42	Phe	Ser	Same	Unstable	Dacie *et al.* (1967)
G-Galveston	CD-2	43	Glu	Ala	Slow		Bowman *et al.* (1964)
K-Ibadan	CD-5	46	Gly	Glu	Fast		Allan *et al.* (1965)
G-Copenhagen	CD-6	47	Asp	Asn	Slow		Sick *et al.* (1967)
J-Bangkok	D-7	56	Gly	Asp	Fast		Clegg *et al.* (1966)
Hikari	E-5	61	Lys	Asn	Fast		Shibata *et al.* (1964)
N-Seattle	E-5	61	Lys	Glu	Fast		Jones *et al.* (in press 1968)
M-Saskatoon	E-7	63	His	Tyr	Same	Methemoglobinemia	Gerald & Efron (1961)
Zürich	E-7	63	His	Arg	Slow	Unstable	Muller & Kingma (1961)
Sydney	E-11	67	Val	Ala	Slow	Unstable	Carrell *et al.* (1967)
M-Milwaukee	E-11	67	Val	Glu	Same	Methemoglobinemia	Gerald & Efron (1961)
J-Cambridge	E-13	69	Gly	Asp	Fast		Sick *et al.* (1967)
Seattle	E-14	70 (or 76)	Ala	Glu	Fast	Unstable	Huehns & Shooter (1965)
C-Harlem*	E-17	73	Asp	Asn	Slow	Cell sickling	Bookchin *et al.* (1967)
J-Iran	EF-1	77	His	Asp	Fast		Rahbar *et al.* (1967)
G-Accra	EF-3	79	Asp	Asn	Slow		Lehmann *et al.* (1964)
D-Ibadan	F-3	87	Thr	Lys	Slow		Watson-Williams *et al.* (1965)
Agenogi	F-6	90	Glu	Lys	Slow		Miyaji *et al.* (1966)
M-Hyde Park	F-8	92	His	Tyr	Same	Methemoglobinemia	Heller *et al.* (1966a)
Gun Hill	F-9, FG 1–4	93–97	See text	None	Slow	Unstable	Bradley *et al.* (1967)
Oak Ridge	FG-1	94	Asp	Asn	?		Cited in Wintrobe (1967)
N-Baltimore	FG-2	95	Lys	Glu	Fast		Clegg *et al.* (1965)
Köln	FG-5	98	Val	Met	Slow	Unstable	Carrell *et al.* (1966)

* These hemoglobins have 2 amino acid substitutions. Both are known for Hb C-Harlem.

TABLE 10.2—*continued*

	Third dimensional position	Position in chain	Amino acid in β chain	Substituted amino acid	Migration rate relative to Hb A	Unusual characteristics	References
Kempsey	G-1	99	Asp	Asn	Slow	Increased O$_2$ affinity	Jones *et al.* (1967)
Yakima	G-1	99	Asp	His	Sl. slow	Increased O$_2$ affinity	Osgood *et al.* (1967)
Kansas	G-4	102	Asn	Thr	Same	Decreased O$_2$ affinity	Riggs & Bonaventura (1967)
New York	G-15	113	Val	Glu	Sl. fast		Ranney *et al.* (1967)
Hijiyama	GH-3	120	Lys	Glu	Fast		Miyaji *et al.* (1968a)
D-Punjab	GH-4	121	Glu	Gln	Slow	Moderate anemia in homozygote	Baglioni (1962b)
O-Arabia	GH-4	121	Glu	Lys	Slow		Baglioni & Lehmann (1962)
Hofu	H-4	126	Val	Glu	Fast		Miyaji *et al.* (1968b)
K-Woolwich	H-10	132	Lys	Gln	Fast		Allan *et al.* (1965)
Hope	H-14	136	Gly	Asp	Fast		Minnich *et al.* (1965)
Kenwood	H-21	143	His	Asp	Fast		Bayrakci *et al.* (1964)
Rainier	H-23	145	Tyr	His	Same	Increased O$_2$ affinity; Alkali resistant	Stamatoyannopoulos *et al.* (1968)
J-Rambam	E-13 or 18	69 or 74	Gly	Asp			Salomon *et al.* (1964)
R (Durham I)		T I					Chernoff & Liu (1961)

	Third dimensional position	Position in chain	Amino acid in γ or δ chain	Substituted amino acid	Migration rate relative to Hb A_2	References
γ Chain Variants:						
F-Texas I	A-2	5	Glu	Lys	Slow	Jenkins *et al.* (1967)
F-Texas II	A-3	6	Glu	Lys	Slow	Schneider & Jones (1965)
F-Houston			Glu	Ala	Slow	Schneider *et al.* (1966b)
F-Warren						Huisman *et al.* (1965)
F-Roma					Fast	Silvestroni & Bianco (1963)
F-Hull	GH-4	121	Glu	Lys	Slow	Sacker *et al.* (1967)
δ Chain Variants:						
Sphakiá	NA-2	2	His	Arg	Slow	Jones *et al.* (1966a)
A_2' or B_2	A-13	16	Gly	Arg	Slow	Jones *et al.* (1965)
Flatbush	B-4	22	Ala	Glu	Fast	Jones *et al.* (1966b)

dependent on a difference in its isoelectric point demonstrated by electrophoresis. Such variants were originally designated by a letter, but more recently, the geographic origin is used. When two or more hemoglobins have the same electrophoretic mobility but a different amino-acid substitution, they are assigned the same letter, but a different geographic name. A molecular formula is assigned when the precise site of amino-acid substitution is known. For example, Hb C is $\alpha_2^A\beta_2^{6\,Glu\rightarrow Lys}$, meaning that lysine is substituted for glutamic acid at the sixth position of the β chain. The charge differences responsible for variations in electrophoretic migration rate often represent substitutions of charged for uncharged amino-acid residues, or vice versa. However, in some instances, the mobility of the hemoglobin does not reflect the change expected from the substituted amino acid. For example, in Hb Köln, one neutral amino acid is replaced by another (Carrell *et al.*, 1966), and the separated α and β globin chains have the same mobility as α^A and β^A (R.V. Jones *et al.*, 1967). However, the intact variant molecule has a slower mobility than Hb A. On the other hand, Hb New York represents the substitution of neutral valine by the negatively charged glutamic acid, but its migration rate is only slightly faster than that of Hb A (Ranney *et al.*, 1967). Hb Freiburg provides still another example. In this instance, loss of a neutral valine without any substituting residue has resulted in a difference in net electrical charge (Jones *et al.*, 1966c). The explanation for these anomalies must lie in the third dimensional position of the amino acid, and also in the effect of adjacent residues.

Table 10.2 presents a list of the abnormal hemoglobins and their individual amino-acid substitutions in order of linear sequence and position within the helical and non-helical portions of the globin chains. The migration rate relative to that of Hb A at alkaline pH is also given, as well as unusual characteristics of structure and/or function. Since the list is long, synonyms have been omitted.

Among the hemoglobins listed in Table 10.2, most have been found unaccompanied by any hematological disorder, during population surveys. However, those abnormal hemoglobins associated with altered red cell structure and/or function have provided a great deal of information about the effects on the hemoglobin molecule of certain amino acid substitutions, the locations of which are shown in Figs. 10.1 and 10.7. These hemoglobins are briefly considered in the next section. (See also the Addenda.)

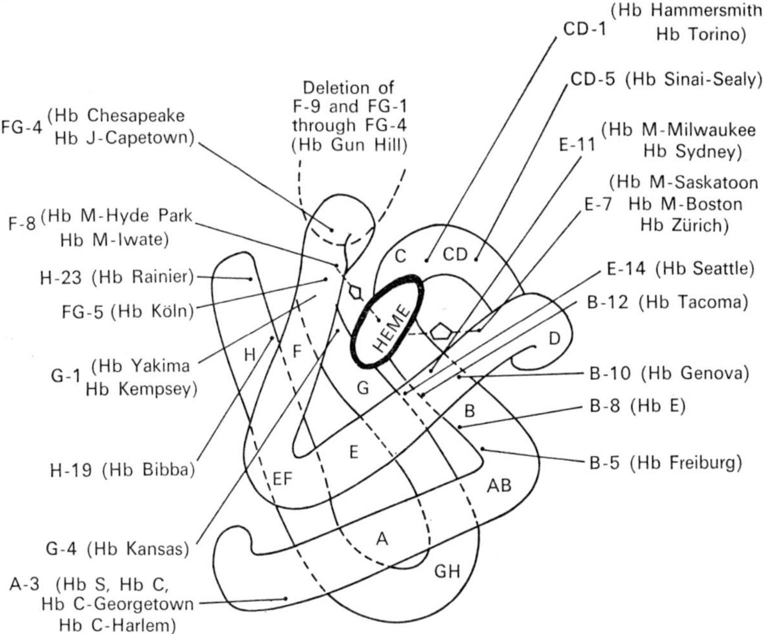

FIGURE 10.7. Sites of amino acid substitution in the globin chains of hemoglobin associated with altered molecular function, as indicated in Table 10.2. The beta chain is depicted, but alpha chain variants are also included in the figure. (See also Addenda)

HEMOGLOBINS CAUSING RED CELL STRUCTURAL DISTORTION
Changes in red cell physical structure which are sufficient to cause clinical symptoms in heterozygotes are characteristic of only a small number of abnormal hemoglobins. In most instances, heterozygotes can be shown to be affected, but to a much lesser extent than homozygotes; cell viability is usually normal. The precise mechanisms responsible for cell destruction in homozygotes are not completely understood, but progress in this difficult area is rapid, particularly in the understanding of red cell sickling.

Sickle hemoglobin: Hb S
Hemoglobin S was found by Pauling *et al.* (1949) to have an isoelectric point different from that of Hb A; from this and other findings, they deduced the now classical concept of molecular disease. Hunt &

Ingram (1959) subsequently showed that the chemical difference in
Hb S is the substitution of valine for glutamic acid at the sixth position
of the β chain, giving the molecular formula $\alpha_2^A\beta_2^S$; more specifically,

FIGURE IO.8. Diagram of the seven amino acids at the N-terminus of the
β^S chain, showing hydrophobic bonding between valines at positions 1 and
6 and a hydrogen bond from the carbonyl group of valine to the amino
group of threonine at positions 1 and 4, respectively. (From Murayama *et
al.*, 1965. Reprinted with permission from *Biochim. biophys. Acta*, **94**, 194.)

$\alpha_2^A\beta_2^{6\,Glu\rightarrow Val}$. The relationship between this amino-acid substitution
and the cell sickling tendency under reduced oxygen tension has
received considerable attention, reviewed by Harris (1959). The low
solubility of deoxygenated Hb S, and the spindle-shaped configuration
of its crystals were reported some years ago (Perutz & Mitchison,

1950; Harris, 1950). However, demonstrating the intracellular crystallization of deoxygenated Hb S in sickled red cells proved difficult, and was only recently achieved (Stetson, 1966).

The mechanism which is probably responsible for this alteration in three-dimensional structure of Hb S was deduced from steric considerations by Murayama (1965, 1966) and Murayama *et al.* (1965). In this model, shown in Fig. 10.8, a hydrophobic bond forms between the N-terminal valine and the (substituted) valine in the sixth position of the same β chain. This intramolecular bonding causes a change in surface configuration which is stabilized by hydrogen bonding between the carbonyl group of the first valine and the amino group of the threonine at position 4. Thus, a stabilized, cyclic structure projecting from the molecular surface is visualized as a 'key' which fits into a complementary site of an α chain belonging to an adjacent Hb S molecule. In this manner, the molecules undergo 'head to tail' stacking with formation of filaments which then may twist together into hollow cables (or 'microtubules'), forming the basis for the crystalline structure. Failure of sickling to occur in oxygenated Hb S is explained by the fact that the β chains move toward each other on oxygenation, thus increasing the distance between the β and α chains of adjacent molecules. Since the 'key' would then be out of the 'lock', stacking would no longer be possible.

The degree of cellular sickling is influenced not only by the concentration of Hb S in the cell (Singer & Singer, 1953) but also by the identity of the accompanying hemoglobin. Thus, when Hb A and Hb S are mixed *in vitro*, part of the Hb A is co-precipitated with the less soluble Hb S when exposed to buffers of high molarity (Itano, 1953). However in a mixture of Hb S with Hb F, only the Hb S is precipitated (Heller, 1965). Similarly, intact cells are less susceptible to sickling when they contain similar concentrations of Hb A and Hb S (Charache & Conley, 1964). In people who are heterozygous for the genes determining Hb S and the hereditary persistence of Hb F, despite a red cell Hb S concentration as high as 70 per cent, the uniform distribution of Hb F is sufficient to prevent a hemolytic process (Conley *et al.*, 1963). However, in patients who are homozygous for the Hb S gene, there is considerable difference in the distribution of Hb F from one cell to another. Thus the cells containing little or no Hb F can become permanently sickled in spite of re-oxygenation, and are subject to the most rapid destruction (Shen *et al.*, 1949; Singer &

Chernoff, 1952). Recent studies suggest that in such cells, the membrane is 'permanently sickled' but the Hb S is not permanently crystallized (Bertles *et al.*, 1967).

Some abnormal hemoglobins interact with Hb S to increase molecular aggregation and cellular sickling tendency, as compared with a similar concentration of Hb A. This interaction has been noted particularly with Hb D_{Punjab} and Hb C (Charache & Conley, 1964).

Hb C$_{Georgetown}$

Another Hb which sickles when deoxygenated was first reported by Pierce *et al.* (1963). Investigation by Gerald & Rath (1966) showed that Hb C$_{Georgetown}$ has the same $\beta^{6 Glu \to Val}$ substitution as Hb S, as well as a second β chain amino-acid substitution, involving one of the 'core' residues, which occupy positions 83 through 120.

Hb C$_{Harlem}$

A third sickling Hb was described by Bookchin *et al.* (1967, 1968). Like Hb C$_{Georgetown}$, Hb C$_{Harlem}$ contains two amino-acid substitutions, one of which is the same as that found in Hb S ($\beta^{6 Val}$). In the second substitution, asparagine replaces aspartic acid at 73β. (Thus Hb C$_{Harlem}$ is listed twice in Table 10.2.) The electrophoretic migration rate is slightly faster than that of Hb C, and there are minor differences in sickling and gelation properties, as compared with Hb S. The authors pointed out that the two sites of substitution are spatially related, since 73 β, which is the seventeenth amino acid of the E helix, has a third-dimensional orientation near the N-terminus of the chain. According to Miyaji *et al.* (1967), substitution of aspartic acid for asparagine at the homologous E 17 position of the α chain (68 α) characterizes Hb Ube-2, which has no associated hematological abnormalities. Another 'benign' hemoglobin with a 68 α substitution is Hb G$_{Philadelphia}$, which has the formula $\alpha_2^{68\,Lys}\beta_2^A$ (Baglioni & Ingram, 1961). For a description of another hemoglobin with the same 73β substitution as in Hb C$_{Harlem}$, see the Addenda.

Hb I

This hemoglobin has the molecular formula $\alpha_2^{16\,Lys \to Glu}\beta_2^A$ (Beale & Lehmann, 1965). Atwater *et al.* (1960) noted that the red cells of a patient heterozygous for the Hb I and α thalassemia genes contained 70 per cent Hb I, and that they tended to sickle on exposure to sodium metabisulfite. However, the abnormal hemoglobin possessed none of the gelling properties of Hb S, and the authors concluded that they were observing a property of the cell membrane, rather than of the hemoglobin itself. In a subsequent paper, Schneider *et al.* (1966a) implied that the sickling could have been an artifact due to a high concentration of metabisulfite.

Hb C

Like Hb S, Hb C has an amino-acid substitution at 6 β, but the replacement of glutamic acid by lysine, rather than by valine, is accompanied by a completely different physico-chemical effect (Hunt & Ingram, 1960). Hb C does not cause cell sickling, but the molecule is less soluble than Hb A, possibly because of electrostatic interactions between positively charged β 6 side groups and negatively charged groups on adjacent molecules (Charache *et al.*, 1967). Patients homozygous for the Hb C gene have a large number of target cells and microspherocytes in peripheral blood films, and they usually have splenic enlargement, but the degree of red cell destruction is generally moderate (Smith & Krevans, 1959). Diggs & Bell (1965) called attention to rod-like inclusion bodies in the red cells of such patients, and Charache *et al.* (1967) reported that reduced plasticity prevents the passage of the cells through Millipore filters easily traversed by normal cells. According to Murphy (1967), this reduced plasticity is due to an increased binding of cell water by the abnormal hemoglobin, with a proportionate decrease in the amount of water available for dispersion of hemoglobin and dilution of cations within the cell. Charache *et al.* (1967) postulated that the intracellular Hb C may be in a precrystalline state due to an abnormally high MCHC (mean corpuscular hemoglobin concentration) and molecular packing.

Hb D$_{Punjab}$ and Hb E

A mild hemolytic syndrome was reported by Bird & Lehmann (1956) in individuals homozygous for the Hb D$_{Punjab}$ gene and by Lehmann *et al.* (1956) in Hb E gene homozygotes. Their respective molecular formulae are $\alpha_2^A \beta_2^{121\ Glu \to Gln}$ and $\alpha_2^A \beta_2^{26\ Glu \to Lys}$. (Baglioni, 1962b; Hunt & Ingram, 1961). In the latter, hemolysis may be related to alteration in the $\alpha'\beta'$ interchain binding.

HEMOGLOBINS ASSOCIATED WITH METHEMOGLOBINEMIA

Patients who are heterozygous for the Hb M genes have cyanosis because the amino-acid substitution in the Hb M favors methemoglobin (ferri-hemoglobin) formation in spite of a normal quantity of the enzyme, methemoglobin reductase (NADH diaphorase). First described by Hörlein and Weber (1948), this form of inherited methemoglobinemia has never been reported in the homozygous state, presumably because it would be incompatible with life.

The abnormality leading to methemoglobinemia may occur in either the α or the β chain (see Table 10.3). Thus, cyanosis is observed in newborn infants with the α chain variants, but in those with β chain variants, there is a delay in the appearance of cyanosis until β chain synthesis has reached a sufficient level (Betke, 1962).

TABLE 10.3. Amino acid substitutions in the M hemoglobins

Hb M	Synonyms	Amino acid replacement	Reference
		Proximal Histidine	
M_{Iwate}	Kankakee	$\alpha^{87\,His\rightarrow Tyr}$	Miyaji *et al.* (1963)
$M_{Hyde\,Park}$		$\beta^{92\,His\rightarrow Tyr}$	Heller *et al.* (1966a)
		Distal Histidine	
M_{Boston}	Osaka	$\alpha^{58\,His\rightarrow Tyr}$	Gerald & Efron (1961)
	Gothenburg		
$M_{Saskatoon}$	Emory	$\beta^{63\,His\rightarrow Tyr}$	Gerald & Efron (1961)
	Kurume		
	Radom		
	Chicago		
		Near a Distal Histidine	
$M_{Milwaukee-1}$		$\beta^{67\,Val\rightarrow Glu}$	Gerald & Efron (1961)

The amino-acid substitution of the M hemoglobins occurs either at the site of the heme-linked proximal histidine (87 α and 92 β), the distal histidine (58 α and 63 β) or in the vicinity of the distal histidine (e.g. 67 β). For spatial orientation of these sites, see Figs. 10.1 and 10.7. In the four known instances involving a histidine replacement with resulting methemoglobinemia, tyrosine is the substituted amino acid (Note: replacement of distal histidine by arginine at 63 β is found in Hb Zürich, an unstable hemoglobin unaccompanied by cyanosis; see p. 380.) When a distal histidine is replaced ($M_{Saskatoon}$ and M_{Boston}), the phenolic group of tyrosine is thought to form an internal complex with the sixth coordination position of heme iron. This second bonding of iron to a peptide site stabilizes it in the ferric form (Gerald & George, 1959). Similarly, in Hb M$_{\text{Milwaukee-1}}$, the substitution of glutamic acid for valine occurs at 67 β, which is one helical turn removed from 63 β, and interaction occurs between the carboxyl group of the glutamic acid and the ferric iron of heme (Pisciotta *et al.*, 1959; Gerald & Efron, 1961).

In Hb M$_{\text{Iwate}}$ and Hb M$_{\text{Hyde Park}}$, a proximal histidine is replaced, and the mechanism responsible for the methemoglobinemia is more difficult to explain (Heller, 1965, 1966; Heller *et al.*, 1966a). The functions of the distal and proximal histidine may be reversed, so that the distal histidine becomes the bond to heme iron. According to Caughey (1967), the substitution of tyrosine for histidine provides an electron source in addition to that provided by the oxidation of ferrous to ferric iron, so that rapid two-electron reductions of oxygen would be expected to occur. Caughey also pointed out that in Hb Zürich, the arginine replacing histidine at 63 β would be expected to form a ligand with iron, but failure of this hemoglobin to cause cyanosis may be due to the fact that arginine, like histidine, is a poorer electron source than tyrosine.

In studies on oxygen saturation of Hb M$_{\text{Hyde Park}}$, Heller *et al.* (1966a) found that although the aberrant hemoglobin accounted for 40 per cent of the patient's hemoglobin, only 20 per cent of the total hemoglobin was incapable of combining with oxygen. This and studies on the other Hbs M showed that the presence of a non-functional heme group on one chain of an $\alpha\beta$ dimer does not prevent the heme group on the other chain from functioning (Heller *et al.*, 1962; Gibson *et al.*, 1966), although there may be marked changes in the oxygen dissociation curve (Hayashi *et al.*, 1966).

HEMOGLOBINS WITH INCREASED OXYGEN AFFINITY

Hb Chesapeake

Charache *et al.* (1966) and Clegg *et al.* (1966) found a moderate increase in the hematocrit of carriers of a new variant, Hb Chesapeake. From their studies, extended by Nagel *et al.* (1967), this hemoglobin was shown to have an increased oxygen affinity and considerable loss of the sigmoid shape of the oxygenation curve. There was no alteration in red cell MCV or MCHC, and the authors concluded that the number of cells was increased, probably through erythropoietin stimulation, to compensate for the decreased amount of oxygen delivered per cell to the tissues. The electrophoretic mobility of Hb Chesapeake was faster than Hb A but slower than Hb Barts, N or I. No inclusion bodies were observed.

Structural studies revealed the replacement of arginine at position 92 α by leucine (molecular formula $\alpha_2^{92\,\text{Leu}}\,\beta_2^{A}$). The authors pointed out that since the proximal histidine of the α chain, linked to the heme group, is only five amino-acid residues away from 92 α, the change in charge might affect the avidity of heme iron for electrons donated by oxygen. Furthermore, 92 α has the spatial position of FG 4, on the non-helical portion of the α chain between F and G helix (Fig. 10.1). This position is in the contact α^1–β^2 (by the designation of Perutz, 1965), which is probably broken during dissociation (Rosemeyer & Huehns, 1967). Thus, a spatial change resulting from the substitution of a non-polar group at this site might have a profound effect on molecular dissociation, and thus on the oxygenation process. Nagel *et al.* (1967) demonstrated that Hb Chesapeake does not undergo the same conformational change with deoxygenation as that observed in normal hemoglobin. However, unlike the molecules with very high oxygen affinity (e.g. Hb H, HpHb, Hb α^A), the Hb Chesapeake molecule has some 'heme-heme interaction' and a normal Bohr effect.

Hb J$_{Capetown}$

A hemoglobin with the molecular formula $\alpha_2^{92\,\text{Arg}\to\text{Gln}}\,\beta_2^{A}$ was discovered by Botha *et al.* (1966). Because the glutamine substitution occurred at the same 92 α position as the leucine substitution in Hb Chesapeake, oxygen binding studies were performed by Lines & McIntosh (1967). Their results indicated that Hb J$_{Capetown}$ behaves like Hb Chesapeake in having a steep and left-shifted oxygen dissociation curve. Again, the alteration in the region of α-β contact was invoked as the cause of the abnormality. The authors further pointed out that these two hemoglobins are the only known examples of mammalian hemoglobin in which arginine at the 92 α position is replaced by another amino acid.

Hb Yakima

Osgood *et al.* (1967) found Hb Yakima in three members of a family who also had increased hematocrits but were otherwise normal. The oxygen affinity of the abnormal hemoglobin was greatly increased

(Novy *et al.*, 1967). On electrophoresis, Hb Yakima moved just behind Hb A, and made up about 40 per cent of the total hemoglobin. Replacement of the usual aspartic acid residue at 99 β by a histidyl residue was found (molecular formula $\alpha_2^A\beta_2^{99\,His}$). R.T. Jones *et al.* (1967) pointed out that position 99 β is located at G1, just below α F64 (92 α) where α^1 and β^2 are in contact. Thus, there is further evidence that amino acid replacement in the F–G region can disturb the spatial rearrangements which occur during oxygenation and deoxygenation. These authors also mentioned another hemoglobin variant with similar properties called Hb Kempsey, in which the aspartic acid at 99 β is replaced by asparagine.

Hb Rainier

Hb Rainier was also found in association with familial erythremia (Stamatoyannopoulos *et al.*, 1968). The molecular formula in this case is $\alpha_2^A\beta_2^{145\,Tyr\rightarrow His}$, representing the only known hemoglobin chain in which position H 23 is not occupied by tyrosine. Perutz *et al.* (1965) listed the H 23 tyrosine as one of nine invariant amino acids in the polypeptide chains of hemoglobins from several different species. It probably plays an important role in determining three-dimensional structure, linking the G and H helices through an internal hydrogen bond with the α carbonyl of valine at position FG 5, the FG corner. H 23 is an internal site, and substitution of the imidazole group of histidine for the phenolic side group of tyrosine might be expected to bring about a conformational change altering chain interaction in favor of the oxyhemoglobin quaternary structure. The nature of this alteration is not clear, nor is it known why Hb Rainier has a marked resistance to alkaline denaturation. (However, see Addenda.)

Hb Ypsi

Rucknagel *et al.* (1967) described Hb Ypsi, which has an as yet undetermined abnormality in the β chain. This hemoglobin also occurred in association with an increased hematocrit, but on electrophoresis of the hemolysate, four components were detected in addition to Hb A and A$_2$. The two major components were thought to have the structure $\alpha_2^A\beta_2^{Ypsi}$ and $\alpha_2^A\beta_1^A\beta_1^{Ypsi}$, while the minor components were presumed to be $\alpha_2^A\beta_1^{Ypsi}\gamma_1$ and $\alpha_2^A\beta_1^{Ypsi}\delta_1$. Amino-acid analysis of the abnormal β chain was not reported. However, it appeared that the altered chain was responsible not only for increasing the oxygen affinity, but also for changing the equilibrium between $\alpha_2^A\beta_2^A$, $\alpha_2^A\beta_2^{Ypsi}$ and their molecular hybrid, which is not demonstrable in other heterozygous states (see pp. 355 and 387).

HEMOGLOBIN WITH DECREASED OXYGEN AFFINITY: Hb KANSAS

Hb Kansas, first described by Reissmann *et al.* (1961) was found in association with familial cyanosis. This unusual hemoglobin has an abnormal absorption spectrum and an oxygen affinity much lower than normal. The molecular formula, determined by Riggs & Bonaventura (1967) is $\alpha_2^A\beta_2^{102\,Asn\to Thr}$. Position 102 β is on the G helix, which normally makes an α^1–β^2 contact with the G helix of the α chain at the center of the molecule. Breakage of this bond would favor formation of $\alpha\beta$ dimers, altering O_2 affinity.

THE UNSTABLE HEMOGLOBINS

All but three of the known examples of unstable hemoglobin have β chain structural abnormalities, perhaps because α chain instability would deter normal fetal development. Alternatively, unstable α chain variants may have escaped detection because the Hb_α^A gene on the opposite chromosome is capable of nearly full compensation, as is evident from the phenotype of α-thalassemia trait (see p. 395). Most of these rare hemoglobins have an increased but variable tendency to autoxidation and Heinz body formation, and are associated (in the heterozygous state) with congenital non-spherocytic hemolytic anemia.

It is likely that most cases of congenital 'Heinz body anemia' are caused by an unstable hemoglobin. The occurrence in these patients' red cells of free α (or β) chains is consistent with an increased rate of β (or α) chain degradation and possibly, a decreased rate of abnormal chain synthesis. The presence of inclusion bodies is often difficult to detect unless the patient has undergone splenectomy. Although some of the unstable hemoglobins have electrophoretic migration rates which differ from that of Hb A (see Table 10.2), their separated globin chains migrate at the same rate as normal globin chains. Thus, the amino acid substitution in the abnormal chain is not directly responsible for the altered charge of the intact molecule (Carrell *et al.*, 1967).

Unstable beta chain variants

Hb Zürich is a particularly interesting molecule because, as indicated by its molecular formula, $\alpha_2^A\beta_2^{63\,Arg}$, the distal histidine of the

β chain is replaced by arginine (Muller & Kingma, 1961). However, although the rate of methemoglobin formation in red cells containing Hb Zürich is considerably higher than that of normal cells when incubated at 37°C, significant amounts of methemoglobin occur *in vivo* only after the ingestion of oxidant drugs, such as those inducing red cell lysis in G6PD deficiency. Under these circumstances, people with a gene determining Hb Zürich undergo severe hemolytic episodes, and large inclusion bodies are demonstrable in their red cells (Frick *et al.*, 1962; Bachmann & Marti, 1962). As in Hb H disease, there appears to be preferential removal of this abnormal Hb from the red cells, since the rates of synthesis of Hb A and Hb Zürich are comparable, but the concentration of Hb A is higher in the peripheral blood (Rieder *et al.*, 1965).

Hb Köln has been described in two families (Pribilla *et al.* 1965; Hutchison *et al.*, 1964; R.V. Jones *et al.*, 1967). All affected members had mild hemolysis and red cell inclusion bodies of variable degree. Incubation of their blood was associated with increased methemoglobin formation. The structural abnormality was identified as the substitution of methionine for valine at position 98 β, so the molecular formula is $\alpha_2^A \beta_2^{98\,Met}$ (Carrell *et al.*, 1966). R.V. Jones *et al.* (1967) postulated that the introduction of a physically larger amino acid into the heme-containing pocket formed by the E and F helices may permit the entry of water into the position vacated by oxygen upon deoxygenation. Thus, ferrihemoglobin (i.e. methemoglobin) formation might be facilitated.

The mechanism of red cell destruction as well as the excretion of heme pigment in this form of Heinz body anemia has been extensively explored by Jacob *et al.* (1967). They reported that the substitution of methionine for valine at 98 β is responsible for altering the behavior of both the cysteine (containing the reactive SH group) at position 93 β and the heme-binding histidine at 92 β. Thus, both cellular and membrane thiols bind the 93 β sulfhydryls as mixed disulfides. With loss of avidity for heme by the proximal histidine at 92 β, heme is lost, and Heinz bodies are formed by the heme-deficient globin. These bodies attach to, and thus deplete, SH groups of the membrane; this depletion of sulfhydryls causes membrane hyperpermeability, splenic entrapment and osmotic destruction.

Hb Genova (Sansone *et al.*, 1967) has the same electrophoretic mobility as Hb A but is separable by mild heat precipitation. The proposita and her daughter had severe hemolytic anemia associated with Heinz bodies in many of their red cells. The structural formula, $\alpha_2^A \beta_2^{28\,\text{Leu}\rightarrow\text{Pro}}$, represented the first known hemoglobin gene mutation resulting in the substitution of proline for leucine. The second, *Hb Bibba*, has a similar substitution in the α chain (see p. 384). The authors pointed out that the presence of proline at the tenth position on the B helix would be expected to disrupt the helix and change its direction. The resulting distortion of the β chain could alter its interaction with the α chain and thus make the molecule unstable.

Hb Sydney (Carrell *et al.*, 1967) was also found in association with hemolytic anemia and red cell inclusion bodies. The site of amino-acid substitution is the same as that of Hb M Milwaukee. In spite of its slow rate of electrophoretic migration, Hb Sydney has a β chain substitution of one neutral amino acid (valine) by another (alanine): $\alpha_2^A \beta_2^{67\,\text{Val}\rightarrow\text{Ala}}$. As in Hb Köln, the replaced residue (valine at 98 β in Hb Köln; valine at 67 β in Hb Sydney) forms a hydrophobic bond with heme, and the substituted amino acids (methionine and alanine, respectively) are also non-polar and uncharged. Carrell *et al.* pointed out that this kind of uncharged amino-acid substitution, which is characteristic of most of the unstable hemoglobins, involves amino-acid residues located at internal positions, at which replacement by a charged amino acid would totally disrupt the molecule (Perutz *et al.*, 1965). Presumably, then, the degree of molecular stability, although decreased, is sufficient for viability if the substitution involves a residue differing in dimension but not in electrical charge.

Two exceptions were cited: Hb Zürich (63 β His $\rightarrow$ Arg), which is unstable only on exposure to oxidants, and Hb M Milwaukee (67 β Val $\rightarrow$ Glu). The latter is not unstable, although its amino-acid replacement occurs at the same position as in Hb Sydney. However, more stability is imparted to the Hb M Milwaukee molecule because the acidic group of the substituted glutamic acid is bonded to the heme iron, and its side chain forms the same hydrophobic bond with the porphyrin ring as that formed by valine.

Hb Hammersmith was first reported by Grimes & Meisler (1962) and Grimes *et al.* (1964) as a heat-precipitable hemoglobin in a child with severe Heinz body anemia. A second case in an unrelated child was found by Dacie *et al.* (1967), and the abnormal molecule was characterized chemically as $\alpha_2^A \beta_2^{42\,\text{Phe}\rightarrow\text{Ser}}$. The replaced phenylalanine at position CD 1 is one of the nine invariant residues described by

Perutz *et al.* (1965). According to Dacie *et al.* (1967), this residue, by forming strong van der Waals bonding with heme, is a major stabilizing force within the molecule. Thus, its replacement by the non-charged but polar serine causes more severe instability than the replacement of heme-binding valine residues in Hb Köln and Hb Sydney. The fact that neither of the parents of the affected children had the abnormal hemoglobin is consistent with the severity of the anemia. Thus, the chances of survival and reproduction are so small that one might expect Hb Hammersmith to be limited to patients in whom the molecule is the result of a new mutation (Dacie *et al.*, 1967).

Hb Freiburg, associated with moderate methemoglobinemia, cyanosis and hemolysis, is characterized chemically by a deleted rather than a substituted amino acid at 23 β, where the usual valine residue is missing (Jones *et al.*, 1966c). This Hb was found, along with Hb A, in the proposita and in two of her three children, but not in either of her parents. The authors postulated that this first known example of a deletion among the abnormal hemoglobins probably arose as a result of unequal crossing over at the normal β chain locus during meiosis in one of the proposita's parents (see p. 562).

Hb Gun Hill (Bradley & Rieder, 1966; Bradley *et al.*, 1967) is an unstable hemoglobin with an even more remarkable β chain abnormality: five amino acids are deleted in the heme-binding region. Thus, only the two α chains have heme prosthetic groups, and the molecule has a high oxygen affinity, with absence of 'heme-heme interaction'. Three of the missing amino acids are cysteine, aspartic acid and lysine, at positions 93, 94 and 95 β. The other two, leucine and histidine, could represent loss from the β chain of the residues at positions 91 and 92 or 95 and 96 or 92 and 96, since the Leu-His combination occurs at both ends of the Cys-Asp-Lys tripeptide. In any case, the resultant sequence is consistent with deletion of the terminal residue (F 9) of the F helix and the first four residues of the FG corner (FG 1–FG 4). Thus, although the abnormal chain has a proximal histidine at 93 β, it is peptide-bonded to valine, rather than to cysteine, probably altering the spatial orientation of the histidine. The binding of heme is probably blocked further by loss of the lysine at FG 2 (which normally forms a salt bridge with a propionic acid side chain of heme) and of leucine at FG 3, thought to make van der Waals contact with a CH group of the heme ring. Although the primary structure of the Gun Hill abnormal β chain is considerably altered by the loss of five consecutive amino acids, the tertiary structure is sufficiently intact to permit combination with normal α chains (Bradley *et al.*, 1967).

Other beta chain variants associated with molecular instability. According to Huehns & Shooter (1965), *Hb Seattle* is less stable than Hb A, but inclusion bodies are not observed unless the cells are incubated *in vitro*. The abnormality of this Hb is also in the β chain, a lysine residue being replaced by glutamic acid probably at position 70, although 76 is also possible. The hemolytic process associated with *Hb St. Mary's* is more severe, and there is a greater tendency of the Hb molecules to polymerize. The β chain abnormality presumably resides in the core. Milder hemolysis is associated with *Hb Tacoma*, $\alpha_2^A \beta_2^{12\,\mathrm{Arg}\to\mathrm{Ser}}$ (Jones *et al.*, unpublished). *Hb Ube-1* (Shibata *et al.*, 1963b) was found in a Japanese subject with severe Heinz-body anemia. Loss of SH activity at 93 β was the only molecular abnormality noted. (For more unstable variants, see Addenda.)

Unstable alpha chain variants

Hb Sinai (Sealy). Two preliminary reports describing the same hemoglobin in two different parts of the United States recently appeared at the same time (Charache & Mondzac, 1967; Schneider *et al.*, 1967). This α-chain variant has the molecular formula $\alpha_2^{47\,\mathrm{Asp}\to\mathrm{His}}\beta_2^A$, the amino-acid substitution occurring at the same position as in Hb L-Ferrara (Beilinson), in which glycine is the substituted amino acid (Baglioni & Colombo, 1964; DeVries *et al.*, 1963). No hematological abnormality was found in association with Hb Sealy by Schneider *et al.* although they did find that the abnormal hemoglobin accounted for only 14–18 per cent of the total hemoglobin. Charache & Mondzac found a similar low percentage in the family with Hb Sinai. In addition, they reported moderate elevations in reticulocyte count and a slightly shortened survival of Cr^{51}-labelled red cells. An increase in heat lability at 60–65°C was also noted.

Hb Bibba. Severe anemia, increased reticulocytes and many Heinz bodies were found in the blood of a 27-year-old orphaned woman with a history of early splenectomy who was described by Kleihauer *et al.* (1968). In fresh blood, there was 12·3 per cent methemoglobin, compared with 2·6 per cent in a normal control. Electrophoresis revealed in addition to Hb A, a slow-moving hemoglobin subsequently called Hb Bibba, which was found to have the molecular formula $\alpha_2^{\text{Leu}\rightarrow\text{Pro}}\beta_2^A$. The authors suggested that the substitution of proline for leucine in the C-terminal part of the H helix probably caused a reversal in direction of the helix and an alteration in the interaction of α and β chains similar to that postulated for the β chain variant, Hb Genova. (See Addenda for two more Leu→Pro variants.)

Hb Torino. An alpha-chain variant analogous to the beta-chain variant, Hb Hammersmith, was found in an Italian man with severe Heinz body anemia by Beretta *et al.* (1968). An abnormal hemoglobin (Hb Torino) was separated by heat precipitation and found to have the molecular formula $\alpha_2^{43\,\text{Phe}\rightarrow\text{Val}}\beta_2^A$. Thus, the phenylalanine at position CD_1 was substituted by valine, rather than by serine, as in Hb Hammersmith. The authors suggested that the more severe anemia in the beta chain variant might be related to the fact that serine resembles phenylalanine less than valine does. However, another important differences lies in the 8 per cent level of α chain variant in the patient with Hb Torino as opposed to 30 per cent of β chain variant in the patient with Hb Hammersmith. The increased ability of the normal α chain gene to compensate for a faulty gene on the homologous chromosome was mentioned earlier.

Hb Porto Alegre: in vitro polymerization

Hb Porto Alegre has the structural formula $\alpha_2^A\beta_2^{9\,\text{Ser}\rightarrow\text{Cys}}$ (Bonaventura & Riggs, 1967). This hemoglobin is not associated with any clinical abnormality, even in the homozygous state, but its unusual physico-chemical properties justify its inclusion in the group of unstable hemoglobins. According to Tondo *et al.* (1963), the migration of Hb Porto Alegre is markedly retarded on starch gel electrophoresis as compared with electrophoresis on filter paper or starch block. Furthermore, its sedimentation rate is higher than that of normal hemoglobin. These findings taken together reflect enlargement due to polymerization. Since fresh specimens show a nearly normal pattern, and the slow-moving component increases on aging or on oxidation by ferricyanide, Hb Porto Alegre molecules probably exist as normal tetramers *in vivo*. However, the substitution of cysteine for serine at position 6 of the β chain A helix, while not altering the number of ionizable groups, provides an externally directed SH group which can react with the corresponding residue of another molecule to form polymers of various sizes (Bonaventura & Riggs, 1967).

HEMOGLOBIN GENETICS

INHERITANCE OF STRUCTURAL GENES

As would be expected from the previous section, each of the Hb polypeptide chains, α, β, γ, δ and ϵ, has a corresponding chromosomal locus, named Hb_α, Hb_β, Hb_γ, Hb_δ and Hb_ϵ. Weatherall (1965) and Huehns & Shooter (1965) have reviewed the evidence regarding linkage between these various loci. It is well established that the genetic loci controlling α and β chain structure are not closely linked, and they may be on different chromosomes. However, a considerable

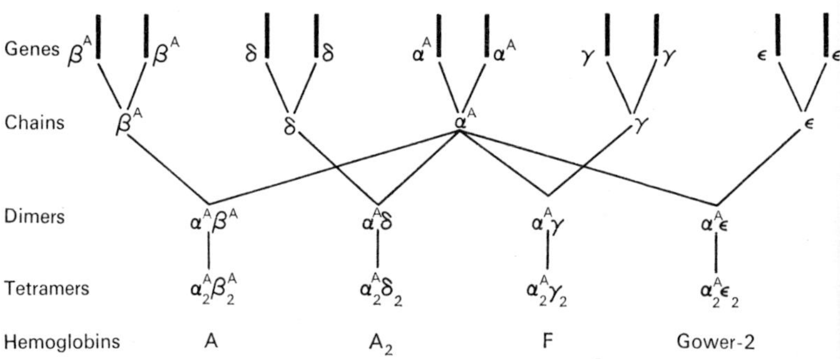

FIGURE 10.9. Diagram showing the five pairs of genes controlling the synthesis of globin chains in human hemoglobin. (Adapted from Huehns & Shooter, 1965. Reprinted with permission from the *J. med. Genet.* **2**, 48.)

amount of evidence indicates that there is very close linkage between the β and δ chain loci. In man, there are no useful data regarding the possible linkage of β and δ to the γ chain locus, although such linkage has been postulated (Rucknagel & Neel, 1961; Nance, 1963). (See Addenda.) Nothing is known about the ϵ chain locus except that Hb Gower-2 is persistently elevated in infants with D_1 trisomy, an association as yet unexplained (Huehns *et al.*, 1964; Capp *et al.*, 1967).

The genes associated with the chains of Hb A, α^A and β^A, are named Hb_α^A and Hb_β^A. Normal individuals are homozygous for these two genes and thus they produce $\alpha_2^A\beta_2^A$. In addition, the Hb_γ^F and Hb_δ^{A2} gene products also form tetramers with the α^A chain: Hb F ($\alpha_2^A\gamma_2^F$) and

Hb A_2 ($\alpha_2^A \delta_2^{A_2}$). In early fetal life, the locus controlling ϵ chain structure is active, so that Hb Gower-2 ($\alpha_2 \epsilon_2$), Hb Gower-1 (ϵ_4) and possibly $\epsilon_2 \gamma_2$ are produced (see Fig. 10.9).

In an individual heterozygous for Hb_β^A and an abnormal gene such as Hb_β^S, a fourth Hb—in this case Hb S ($\alpha_2^A \beta_2^S$)—is also formed, by the association of $\alpha^A \beta^S$ dimers. Theoretically, hybrid molecules, composed of the two dimers, $\alpha^A \beta^A$ and $\alpha^A \beta^S$, could be formed. Failure to observe such $\alpha_2^A \beta_1^A \beta_1^S$ tetramers by conventional separation techniques is probably due to the fact that they are in rapid equilibrium with $\alpha_2^A \beta_2^A$ and $\alpha_2^A \beta_2^S$ (Guidotti *et al.*, 1963). Nevertheless, their existence has been postulated on theoretical grounds (Hill, 1964) and from oxygen-binding measurements (Benesch *et al.*, 1966). The subject was discussed on p. 355.

Individuals who are heterozygous at the Hb_γ or Hb_δ loci also have an additional hemoglobin, such as $\alpha_2^A \gamma_2^{Roma}$ (Silvestroni & Bianco, 1963) or $\alpha_2^A \delta_2^{B_2}$ (Ceppellini, 1959). (For evidence of more than one normal γ chain locus, see Addenda.)

Since all of the normal hemoglobins (A, A_2, F and Gower-2) contain α chains, it follows that individuals who are heterozygous at the Hb_α locus will have not only the usual Hbs A, A_2 and F, but also three additional hemoglobins containing the new α chain, such as $\alpha_2^I \beta_2^A$ (Hb I), $\alpha_2^I \delta_2$ and $\alpha_2^I \gamma_2$. The fetal homolog is usually seen only in infants (Minnich *et al.*, 1962) and the δ-containing homolog only in later life, because δ chain production is curtailed *in utero*, while there is very little γ chain production after birth. A useful preliminary guide in differentiating α from β chain variants in non-cord blood specimens is the presence of two faintly-staining bands in the Hb A_2 region on starch gel electrophoresis.

In an individual homozygous for an abnormal β gene such as Hb_β^S, there can be no Hb A production, because there are no β^A genes. Consequently, the hemoglobins observed in this instance are S, A_2 and F. Due to decreased synthesis of the abnormal β chains, there is a compensatory increase in γ chain production. Thus, the concentration of Hb S is low, relative to normal, while that of Hb F is elevated.

As would be expected, individuals who have different mutant genes, such as Hb_β^S and Hb_β^C occupying the same locus on homologous chromosomes, produce both variants; in this case Hb S ($\alpha_2^A \beta_2^S$) and Hb C ($\alpha_2^A \beta_2^C$), but no Hb A. However, in individuals who are heterozygous at both the α and β loci, a large number of Hb species are formed. For example, in the patient described by Raper *et al.* (1960), heterozygosity for Hb_α^G and Hb_α^A at one locus and Hb_β^C and Hb_β^A at the other locus produced four major Hb species: Hb A, Hb C, Hb G and a mixed molecule containing both abnormal chains, $\alpha_2^G\beta_2^C$. In addition, such individuals at birth have normal Hb F and a second fetal Hb consisting of $\alpha_2^G\gamma_2$. Subsequently they form both normal Hb A_2 and its abnormal counterpart $\alpha_2^G\delta_2$.

SYNTHESIS OF HEMOGLOBIN

NORMAL SYNTHESIS

Synthesis of Hb A

As with other proteins, the synthesis of a globin chain begins at the N-terminus and proceeds to the C-terminus by successive addition of amino acids coded in the mRNA template as it moves along the polysome (Dintzis, 1961). According to Warner *et al.* (1963) each polysome of reticulocytes contains from four to six ribosomes. Winslow & Ingram (1966) have shown that the sequential addition of amino acids does not proceed at a uniform rate, both α and β chain synthesis being slowed down at about position 90. Since this position is at or near the attachment of heme to the proximal histidyl, it is possible that the slowing of synthesis is associated with heme incorporation while the globin chain is still on the ribosome. However, according to Baglioni (1966), the polysomes of reticulocytes incubated with Fe^{59} take up some radioactivity, but the nascent globin chains on the polysomes are not labeled by the isotope. Thus it is likely that the attachment of heme to the globin chains occurs after their synthesis is complete. Also, Heywood & Hillman (1967) have shown that normal cells contain a small amount of free $\alpha\beta$ globin which probably represents an intermediate phase between the release of globin dimer and the attachment of heme.

The studies of Baglioni & Colombo (1964) and Weatherall *et al.* (1965) indicate that under normal conditions the rate of synthesis of individual α chains equals or exceeds that of β chains. On completion, the α chains tend to remain attached to the ribosomes until released as the $\alpha\beta$ dimer. Thus, β chain synthesis is probably the rate-limiting step in normal globin production. Winterhalter & Huehns (1964) have shown that a human globin preparation consisting of $\alpha\beta$ dimers readily forms $\alpha_2\beta_2$ tetramers upon the addition of heme. A similar observation on globin synthesized by a cell-free rabbit reticulocyte preparation was made by Zucker & Schulman (1967). It is reasonable to assume that *in vivo*, a similar series of events may take place in the normal marrow, i.e., β chains released from their polysomes combine with completed α chains and the resulting $\alpha\beta$ dimers then incorporate their individual heme moieties and condense to form the tetramer. However, further data are required before this hypothesis can be regarded as established.

Hemoglobin synthesis during development
During early embryonic life, the synthesis of β chains is completely suppressed, so that the α chains produced combine with either ϵ or γ

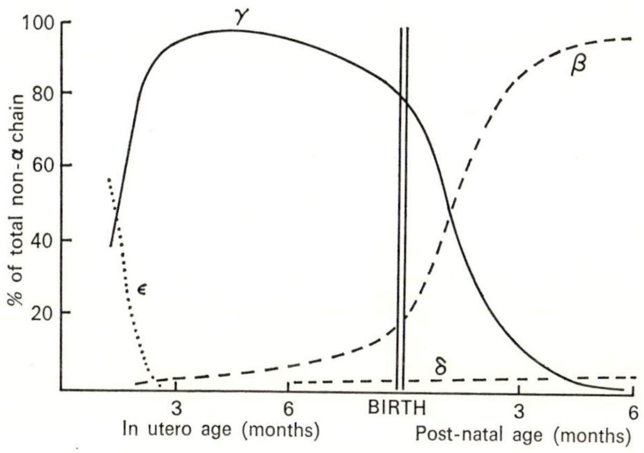

FIGURE 10.10. Changes in production of non-alpha globin chains during early human development. (Adapted from Huehns & Shooter, 1965. Reprinted with permission from the *J. med. Genet.* **2**, 48.)

chains, both of which are present at a very early stage (see Fig. 10.10). Presumably, there is an imbalance of chain production at that time, because both ϵ_4 and γ_4 tetramers are also normally present (Huehns *et al.*, 1964). The production of ϵ chains is usually 'turned off' early in fetal development, and β chain synthesis begins, increasing gradually until the time of birth, when it rapidly overtakes and exceeds γ chain

14

synthesis, the latter decreasing to a low level within a few months. Thus, in most infants, the Hb F concentration is about 15 per cent of the total Hb at four months of age. However, the decline in γ chain production is not so rapid thereafter, so that Hb F may be elevated above the normal adult value of about 0·5 per cent for three to five years (Beaven *et al.*, 1960; Zipursky *et al.*, 1962; Raper, 1963). Alpha chains combine preferentially with β over γ chains, so that traces of γ_4 but not β_4, are often detectable in the newborn (Huehns *et al.*, 1964). Delta chain synthesis in the fetus begins at or about the same time as β chain synthesis, and the increase in Hb A concentration is accompanied by a rise in Hb A_2 (Kunkel *et al.*, 1957). Winslow & Ingram (1966) have shown that in normal bone marrow, the rate of δ chain synthesis is considerably slower than that of the α or β chain, consistent with the small amount of Hb A_2 normally present in peripheral blood.

Synthesis of structurally abnormal variants
According to the 'structure-rate' hypothesis of Itano (1957, 1966) the rate of amino-acid assemblage on the polysome may be decreased at the site of amino-acid substitution in variant globin chain synthesis. In general, the proportion of an abnormal hemoglobin in Hb_α heterozygotes is lower than in Hb_β heterozygotes. In both instances, the normal gene can compensate for the chain deficit. Thus, unless there is an associated hemolytic process, as in the unstable hemoglobins or in thalassemia, the total hemoglobin level of the blood is normal.

DEFECTS IN GLOBIN CHAIN SYNTHESIS: THALASSEMIA
The term 'thalassemia' is applied to a group of inherited disorders in which there is a variable decrease in net synthesis of a particular globin chain without an associated change in the structure of that chain. In the Hb Lepore syndromes, the abnormality is associated with a structural alteration, but the clinical and hematological results resemble β-thalassemia. Also included are the various forms of hereditary persistence of Hb F production, which represents a suppression of β and δ chain synthesis fully compensated by γ chain production. Table 10.4 briefly summarizes the major features of β-thalassemia and related conditions.

TABLE 10.4. Beta thalassemia and related syndromes

Thalassemia type	Heterozygous (trait)	Homozygous
β-thalassemia with little or no synthesis of β chains by chromosome with abnormal gene	Hematologically very variable; usually, normal Hct and increased RBC. Hb A_2 increased; Hb F normal or variably increased	Severe hypochromic microcytic anemia; little or no Hb A; very high proportion of Hb F
β-thalassemia with variably reduced β chain synthesis by chromosome with abnormal gene	Same as above	Severe anemia; high proportion of Hb F; Hb A proportion variable.
βδ-thalassemia (F-thalassemia)	Hb A_2 normal or reduced; Hb F 5–30% distributed unevenly	Moderate anemia; no Hb A or Hb A_2; only Hb F present.
Lepore thalassemia	Hb A_2 level normal; Hb F mod. increased; about 10% Hb Lepore	Severe anemia; no Hb A or Hb A_2; only Hb F and Hb Lepore present

Defective synthesis of β and/or δ chains

β-Thalassemia (A_2-thalassemia) is characterized in the heterozygote by variable depression in β chain synthesis, a moderate increase in Hb A_2 (up to as high as 6 or 7 per cent of the total hemoglobin) and, in about half the cases, a slight increase in Hb F (1–5 per cent). The MCH is decreased, and the red cells show slight to moderate morphological abnormalities such as microcytosis and target cells. The degree of severity is consistent within families, but the wide variability among different families and geographic regions suggests a series of β thalassemia genes associated with different degrees of suppressed β chain synthesis (Fessas, 1965; Weatherall, 1967).

Evidence for such heterogeneity is best obtained by studies of families carrying a β thalassemia gene and a structural gene variant at the Hb_β locus. For example, in a family described by Heller *et al.* (1966b), the individuals doubly heterozygous for genes determining

β thalassemia and either Hb S or Hb C had about one-third Hb A and two-thirds Hb S or C, but no increase in Hb F. However, in other reported families, such heterozygotes have either little Hb A or (less often) none at all, and high levels of Hb F. In these cases, the Hb F is not evenly distributed among the red cells, indicating a compensatory increase in γ chain production in some, but not in other, cell clones.

In patients who are homozygous for β thalassemia, anemia is usually very severe (thalassemia major or Cooley's anemia) and death occurs during childhood, even with adequate supportive medical care. The blood films of these patients show very marked anisochromia, anisocytosis, poikilocytosis, and variable numbers of nucleated red cells. The bone marrow is greatly hypertrophied and extremely active, but since most of the erythropoiesis is ineffective, the cells actually released into the circulating blood represent only a fraction of those produced and destroyed in the marrow (Sturgeon & Finch, 1957). In the usual case, the Hb F concentration of the peripheral blood is about 40–60 per cent, but the individual cell content of Hb F varies considerably, as does the cell viability (Kaplan & Zuelzer, 1950; Shepherd *et al.*, 1962). Furthermore, the turnover of Hb A is considerably more rapid than that of Hb F, and there seems little doubt that the cells most liable to destruction are those containing little or no Hb F (Gabuzda *et al.*, 1962, 1963, 1964).

Fessas (1963) described the appearance of inclusion bodies in a fairly large proportion of marrow cells but in only rare cells of the peripheral blood of β-thalassemia patients. Loukopoulos & Fessas (1965) and Nathan & Gunn (1966) showed that these inclusion bodies occur in cells destined for rapid destruction. The identity of these inclusions as precipitated excess α chains was proposed by Fessas (1963); and Fessas *et al.* (1966) demonstrated that the peptide fingerprint of inclusion body protein greatly resembled that of Hb α^A.

Studies of hemoglobin synthesis in homozygous β-thalassemia, using radioactive amino acids, have been reported by Heywood *et al.* (1965), Weatherall (1965) Bank & Marks (1966a), Bargellesi *et al.* (1967) and Huehns & Modell (1967). All have shown that α chains are synthesized at several times the rate of β chains, suggesting that α chains are freely released from the ribosomes. Huehns & Modell (1967) separated the free, labelled α chains by column chromatography. The fact that these chains had a very high specific activity compared with those found in Hb A suggested that in thalassemia, free α chains are intermediate in hemoglobin synthesis.

The production of γ chains with Hb F formation in the red cells of β-thalassemia patients is thus a life-saving mechanism, since uncombined α chains, being highly unstable, precipitate as inclusion bodies and bring about cell destruction either in the marrow or in the spleen. Why it is possible for only a proportion of the red cell precursors to produce the necessary complement of γ chains is unknown. That the elevation of Hb A_2 found in the heterozygote is not a characteristic of the homozygote is probably due, in part, to the fact that cells with the highest proportion of Hb A_2 have the lowest proportion of Hb F, and thus are subject to more rapid removal (Loukopoulos & Fessas, 1965).

The β-thalassemia described by Schokker *et al.* (1966) appears to be a separate entity from the more common β-thalassemias because the heterozygotes have higher levels of Hb F (5–14 per cent) and Hb A_2 (5–8 per cent), and the homozygotes have only moderate anemia. In this state, Hb F accounts for 98 per cent of the hemoglobin, and Hb A_2 for the other 2 per cent.

$\beta\delta$-Thalassemia (F-thalassemia) is characterized by depression of both β and δ chain synthesis, so that heterozygotes have no elevation of Hb A_2, and the level of Hb F, distributed unevenly among the cells, ranges from about 5–30 per cent of the total hemoglobin (Zuelzer *et al.*, 1961; Fessas, 1962, 1965; Gabuzda *et al.*, 1964; Fraser *et al.*, 1964; Comings & Motulsky, 1966). The only individual known to be homozygous for the $\beta\delta$ thalassemia gene has relatively moderate anemia (Hb 9 gm%), and her red cells contain only Hb F, with no trace of either Hb A or Hb A_2 (Brancati & Baglioni, 1966). In patients heterozygous for both β-thalassemia and $\beta\delta$-thalassemia genes, anemia is moderately severe. The hemoglobin consists of Hb F (90 per cent or more) and either normal or decreased amounts of Hb A_2. In some cases, Hb A is not demonstrable.

Hb Lepore syndrome is included with the thalassemias because it resembles β-thalassemia both clinically and hematologically, although the Lepore hemoglobins are structurally abnormal. The various Hb Lepore genes, discussed on p. 362, probably arose from unequal crossing over between homologous regions of the β and δ genes (Baglioni, 1962a). In heterozygotes, the blood contains about 10 per cent Hb Lepore, a normal amount of Hb A_2, and a moderate increase in Hb F. Homozygotes are severely anemic, and there is no synthesis of β or δ chains, so that only the Lepore hemoglobin and Hb F are present (summarized by Weatherall, 1967).

In both Hb Lepore$_{Boston}$ and Hb Lepore$_{Hollandia}$, the non-alpha chain has the N-terminal sequence of the δ chain and the C-terminal sequence of the β chain, but the site of crossing over differs (Baglioni, 1962a, 1963). Winslow & Ingram (1966) have demonstrated that δ chain synthesis is considerably slower than β chain synthesis, and it has been postulated that the presumed slow rate of Lepore chain synthesis may reflect the presence of a control mechanism in the N terminus of the δ chain (Smithies, 1964).

TABLE 10.5. Hereditary persistence of fetal hemoglobin

Type	Heterozygous	Combined with β thalassemia trait
African type (no β or δ chain synthesis by chromosome with abnormal gene)	No hematological abnormality; 20–30% Hb F, evenly distributed; Hb A$_2$ slightly decreased	Mild anemia; 60–70% Hb F; Hb A$_2$ normal or elevated.
Greek type (Reduced β and δ chain synthesis by chromosome with abnormal gene)	No hematological abnormality; 10–20% Hb F, evenly distributed; Hb A$_2$ level normal	Anemia more severe than above; 20–30% Hb F; Hb A$_2$ slightly increased.
Swiss type	No hematological abnormality 1–3% Hb F, unevenly distributed; Hb A$_2$ level normal	Unknown

Hereditary persistence of Hb F differs from thalassemia in that there is no hematological abnormality, but there is depression to a variable degree of β and δ chain synthesis, fully compensated by γ chain synthesis. The details of this condition were summarised by Motulsky (1964), Fessas & Stamatoyannopoulos (1964) and Weatherall (1967), who described the African Negro, Greek and Swiss types. Table 10.5

was adapted from their papers. The Negro and Greek types are similar in that heterozygotes have an evenly distributed increase in Hb F and a decrease in Hb A_2, but both alterations are more marked in the African Negro type. Furthermore, when the respective genes are partnered by a β thalassemia gene, the Hb F level is much higher in the African type heterozygote, but the severity of the syndrome is greater in Greeks than in Africans. The protective effect of the 'high Hb F' gene in individuals carrying a gene for Hb S on the homologous chromosome was mentioned on p. 373.

The African gene is associated with complete suppression of β and δ chain synthesis, so that the one known homozygote has neither Hb A nor Hb A_2. His red cells show some minor morphological changes, but the hematocrit is normal (Wheeler & Krevans, 1961). Although no homozygote for the Greek gene has been identified, it seems likely that such an individual would not have complete absence of Hb A and A_2.

The Swiss variant bears no resemblance to the other two forms of persistent Hb F, first because the levels of Hb F are only slightly elevated, and second, because the cellular distribution of Hb F is not homogeneous. According to Weatherall (1967), population surveys in Great Britain and Switzerland suggest that this condition occurs in about 1 per cent of people.

δ-*Thalassemia* was first invoked by Fessas & Stamatoyannopoulos (1962) to account for the absence of Hb A_2 in a patient who otherwise had the characteristics of $\beta\delta$ thalassemia trait. Fraser *et al.* (1964) described two Greek families in which some members had typical β-thalassemia trait, others had the hematological trait without elevated Hb A_2, and still others without the trait had low Hb A_2 levels. These findings were interpreted to indicate the existence of a gene associated with selective depression of δ chain synthesis. Better evidence was provided by Thompson *et al.* (1965), who studied a Negro family in which one parent had hereditary persistence of Hb F and the other had a low Hb A_2 level; their child, who inherited the high Hb F gene, had no Hb A_2. Another explanation for the absence of Hb A_2 could be a mutation of the δ chain gene which resulted in a variant Hb A_2 migrating with Hb A (Baglioni, 1964).

Defective synthesis of α chains

α-*Thalassemia*. Complete or nearly complete absence of α chains is incompatible with life because α chains are required for all of the normal hemoglobins and their structural variants. The homotetramers, β_4 and γ_4, have no utility for oxygen exchange since they have a high oxygen affinity and lack the allosteric effects reflected in the sigmoid shape of the normal oxygen dissociation curve. Lie Injo *et al.* (1962) described several stillborn infants with severe hydrops fetalis without blood group incompatibility whose blood was found to contain Hb Barts (γ_4) in a concentration as high as 80 per cent. Hb A was either present in small amounts or absent, and Hb H (β_4) was usually detectable. Failure to observe any Hb F ($\alpha_2\gamma_2$) is consistent with the known preference of α chains for β over γ chains (Huehns *et al.*, 1960).

It seems very likely that these infants were homozygous for an α-thalassemia gene, and it should have been possible to confirm the diagnosis by finding α-thalassemia trait in both parents. Some minor abnormalities of red cell morphology were in fact noted. However, this problem of detecting the heterozygous state is characteristic of α-thalassemia, since there is no elevation of Hb A_2 or Hb F, and red cell changes of microcytosis and target formation may be minimal. Presumably, the normal Hb α gene on the homologous chromosome is able to provide nearly complete compensation. One way to recognize α-thalassemia trait is to find increased levels of Hb Barts (γ_4) in the blood of newborn infants. In older subjects, an analogous rise in Hb H (β_4) is often not observed, although if considerable care is taken to avoid precipitation of this highly unstable molecule, traces of it may be observed on starch gel electrophoresis. Alternatively, after incubation of blood with brilliant cresyl blue, the so-called BCB inclusion bodies may be seen in occasional red cells (Malamos *et al.*, 1962; Weatherall, 1963).

Another way to detect the α-thalassemia gene in a family is by observing its effects on the hemoglobin patterns of individuals who are also carrying a gene for an α chain structural abnormality. Alpha chain variants have very low frequencies, so such combinations are rare. Dormandy *et al.* (1961) found that a child who inherited Hb Q (an α chain variant) from his father had no Hb A at all. His mother, the α-thalassemia carrier, had only mild anemia and slight red cell morphological abnormalities. Similar cases of Hb Q-α thalassemia were described by Lie Injo *et al.* (1966). In these cases, the α-thalassemia gene had no demonstrable α chain product. However, studies in other families have revealed the existence of α-thalassemia gene heterogeneity. For example, Atwater *et al.* (1960) described a woman heterozygous for Hb I and α-thalassemia whose cells contained 70 per cent Hb I ($\alpha_2^I\beta_2^A$) and 30 per cent Hb A, but no Hb H. Nevertheless, her newborn infant had a high level of Hb Barts, accounting for 10 per cent or more of the total hemoglobin. The existence of even milder forms of α-thalassemia has been deduced from studies of families with so-called Hb H disease.

Hb H disease has been called α-thalassemia intermedia by Huehns (1965). It was first described in a Chinese family by Rigas *et al.* (1955) and in a Greek family by Gouttas *et al.* (1955) as a thalassemia-like syndrome of moderate severity associated with the presence of 5–25 per cent Hb H. The latter is demonstrable both on starch gel electrophoresis and as inclusion bodies after incubation of blood with brilliant cresyl blue. Small amounts of Hb Barts are often present, as well as another component first thought to consist entirely of δ chains (Dance *et al.*, 1963), but now considered to have been the mixed tetramer, $\beta_2\delta_2$ (Rosemeyer & Huehns, 1967). Pootrakul *et al.* (1967) have shown that in the cord blood of infants who are later found to have Hb H disease, Hb Barts has a concentration of 25–35 per cent.

Based on family studies such as those reported by Wasi *et al.* (1964), which were later extended by Pootrakul *et al.* (1967), Huehns (1962, 1965) proposed that Hb H disease is caused by the interaction of two allelic α-thalassemia genes, one of which is the same as that causing fetal death in the homozygous state. The second, much milder α-thalassemia gene is very difficult to detect, because adults who carry the trait have no hematological abnormality, while in infants, the trait is only recognizable by the presence of 1–2 per cent Hb Barts. According to both Weatherall (1967) and Huehns (personal communication), Na-Nakorn and his colleagues in Bangkok have unpublished data indicating the existence of such a gene with fairly high frequency in Thailand. The phenotype of individuals homozygous for this gene is not yet known. It may resemble either Hb H disease or α-thalassemia trait. A summary of the phenotypes associated with the genes for severe and mild forms of α-thalassemia is presented in Table 10.6.

Clegg & Weatherall (1967) showed that the rate of α chain synthesis is considerably slower than β chain synthesis in Hb H disease. They also found that the β chains of Hb H can exchange reversibly with those of Hb A in the intact red cell. The nature of the defective synthesis (i.e. whether it involves alterations at the DNA, RNA or ribosomal level) is not known.

TABLE 10.6. Alpha thalassemia syndromes

Thalassemia type	Findings at birth	Findings in adult
α-thalassemia, severe (little or no α chain synthesis by chromosome with the abnormal gene)	*Heterozygote*: mild red cell changes; Hb Barts 5–15%. *Homozygote*: erythroblastosis fetalis and death; only Hb Barts & Hb H present	*Heterozygote*: not anemic; mild red cell changes; trace Hb Barts; Hb H inclusion bodies in occasional red cells.
α-thalassemia, mild (moderate reduction of α chain synthesis by chromosome with abnormal gene)	*Heterozygote*: 1–2% Hb Barts; no red cell abnormalities. *Homozygote*: not yet defined	*Heterozygote*: not detectable *Homozygote*: not yet defined; may resemble α-thalassemia trait or Hb H disease
Hemoglobin H disease (heterozygous for severe and mild α-thalassemia genes)	Moderate anemia and red cell abnormalities; 25–35% Hb Barts	Moderate anemia and red cell abnormalities; frequent Hb H inclusion bodies; Hb H 5 to 30%; Hb A_2 decreased

α-Thalassemia in subjects with β chain abnormalities. Fessas (1965) summarized the data obtained from families in which one or more members were heterozygous for both α- and β-thalassemia. These individuals showed the mild features of thalassemia trait, as expected from lack of interaction between the two non-allelic genes; the only diagnostic criteria were an elevated Hb A_2 level and rare red cells with BCB inclusion bodies. Fessas also pointed out that when α-thalassemia is combined with hereditary persistence of Hb F (Greek type), the syndrome may be difficult to distinguish from βδ thalassemia trait, since both conditions have slight changes in red cell morphology and an associated increase in Hb F, but not Hb A_2.

When the α-thalassemia gene is inherited along with a β chain structural variant, the level of the abnormal hemoglobin (e.g. Hb S or Hb C) is lower in comparison with individuals who have only the β chain variant trait (Weatherall, 1965). This observation was also made by Tuchinda *et al.* (1964), who found in Thailand that the ratio of Hb E ($\alpha_2^A\beta_2^E$) to Hb A is about 35:65 in Hb E trait and about 25:75 when the Hb E trait is combined with α-thalassemia. Furthermore, when an individual has Hb H disease with Hb E trait, the ratio is reduced to 15:85. Thus, α^A chains bind β^A chains in preference to structurally abnormal chains.

Genetic mechanisms in thalassemia

The β chain structural gene and the β-thalassemia gene are either alleles or their loci are very closely linked. The structural β and δ gene loci are sufficiently close together on the same chromosome that crossing-over is very rare. Somewhat less closely linked are the β-thalassemia and δ chain structural loci. According to one suggested sequence for these loci on the autosome, the determinant for the δ chain precedes that of the β structural gene, which is, in turn, adjacent to the locus for at least one of the β-thalassemia genes (summarized in Weatherall, 1967).

In the thalassemias, heme, as well as globin, is synthesized at a reduced rate (Bannerman, 1961). However, the defect in heme synthesis is secondary, probably reflecting a negative feedback effect by heme on the enzyme δ-aminolevulinic acid synthetase, which catalyses the condensation of glycine with succinyl-coenzyme A in the protoporphyrin biosynthetic pathway (Karibian & London, 1965; Gallo, 1967). The studies of Marks & Burka (1964), Heywood *et al.* (1965), Weatherall *et al.* (1965) and Bank & Marks (1966a,b) have shown that in β-thalassemia, there is a defective capacity for incorporation of amino acids in the synthesis of β chains by normal ribosomes. Thus, a pre-ribosomal abnormality is implicated. This abnormality is unlikely to involve the general ability to form transfer RNA (i.e. sRNA or soluble RNA), because there is no evidence for defective synthesis of non-globin proteins in thalassemia. Thus, decreased synthesis of the β chain must be assigned either to the DNA structural gene (with or without a resultant defect in mRNA) or to a closely linked operator.

An appropriate model employing the regulator and operon concept can be constructed to fit almost any condition in which there is an inherited difference in the quantity of protein synthesized (Neel,

1961; Motulsky, 1962). However, such models are not subject to verification by the currently available experimental techniques, so this approach is entirely speculative. Nevertheless, the hereditary persistence of Hb F seems to be especially suited to a mechanism of this kind, in which a regulator gene, analogous to the operator gene of Jacob & Monod (1961) controls the 'switch' from the synthesis of γ chains to the production of β and δ.

The sequence of bases in DNA cannot be determined as yet (McCarthy, 1967) so there is no direct way of testing for an alteration in DNA, such as that postulated by Itano (1965). This hypothesis is based on the known degeneracy of the genetic code (i.e. the ability of more than one base triplet to code for a given amino acid). Thus, if a point mutation substituted a base which did not alter the code, a change would occur in the sRNA employed in the transfer of the same amino acid to the nascent peptide chain. If that sRNA were not readily available, peptide synthesis would be retarded.

Abnormalities in messenger RNA have also been suggested as the basis for defective synthesis of globin chains. According to Ingram (1964), blockage of ribosomal sites by abnormal messenger could make those sites unavailable for polypeptide synthesis. Alternatively, instability of messenger RNA, as demonstrated in the 'amber' and 'ochre' mutants of certain phage types (reviewed by Stretton, 1965) could limit the number of globin chains synthesized (Baglioni & Colombo, 1964).

Nance (1963) and Smithies (1964) have proposed mechanisms of chromosomal rearrangement due to unequal crossing-over, in which the gene product would not be readily recognized as abnormal, but would interfere with normal synthesis. In Nance's model, fusion genes at the δ-β-γ locus could cause the production of defective mRNA, competing with normal mRNA at the ribosomal level. (There is recent evidence for linkage of the γ chain locus with the linked β and δ loci; see Addenda.) Smithies (1964) listed the electrical charge differences and calculated the expected frequencies of all possible gene duplications and deficiencies which could arise from homologous but unequal crossing-over at the δ-β locus. Among the resulting 'Lepore' hemoglobins, some would have the same charge as Hb A and might not be differentiated from it by conventional techniques. The basis for the resulting thalassemia syndrome could only be revealed by examining the amino acid sequence of the affected hemoglobin chain. This attractive hypothesis could account for some cases of thalassemia, but the thalassemia-derived hemoglobins so far subjected to chemical analysis have not been distinguished from normal.

According to Brancati & Baglioni (1966) the most likely explanation to account for the findings in $\beta\delta$ thalassemia is a partial or complete deletion of the β and δ structural genes. Proof for this hypothesis will require more sophisticated laboratory techniques than are presently available.

From this very brief survey, it is clear that postulated mechanisms for inherited variability in the quantitative expression of the hemoglobin structural genes are in great need of experimental verification. In view of the applicability of such investigations to the general problem of heterogeneity in protein synthesis, studies in this difficult field are of particular importance.

EVOLUTION OF HEMOGLOBIN

The possibility of reconstructing evolutionary events in protein synthesis, based on amino acid sequence and the underlying genetic code, has intrigued a number of investigators. Excellent reviews pertinent to hemoglobin evolution will be found in the books of Anfinsen (1959), Ingram (1963) and Jukes (1966), and in the papers of Zuckerkandl & Pauling (1962), Braunitzer *et al.* (1964), Epstein & Motulsky (1965), Margoliash & Smith (1965) and Dixon (1966).

Since proteins are not preserved in fossilized remains, the only feasible approach to a study of hemoglobin evolution is to make structural comparisons between related polypeptides within a given species (e.g. the α, β, δ and γ hemoglobin chains in man) and within a wide range of species extending through the phylogenic series.

Determining the size of the ancestral molecule which may have been the precursor of cytochrome, as well as of myoglobin and hemoglobin, is extremely difficult, because if the large molecules are assumed to have arisen from the smaller by a process of repeated duplication, the genes in each species studied would be expected to have undergone point mutations, as well. Nevertheless, Cantor & Jukes (1966) have presented evidence that a primordial heptapeptide may have been the common precursor.

More solid evidence is provided by comparing the globin chains of myoglobin and hemoglobin. Such comparisons led Ingram (1961b,

1963) to propose that the α chain gene arose by duplication of, and then by translocation away from, a gene for an early myoglobin-like molecule with a molecular weight of about 17,000 (see Fig. 10.11). The initial gene evolved into the present-day myoglobin molecule by mutation, while the α chain gene precursor diverged so that its product took on the function of hemoglobin. The next step, according to Epstein & Motulsky (1965), was duplication of the α chain gene at some stage of its evolution to give rise to the ϵ chain.

Briehl (1963) reported that in the lamprey eel, there is only one

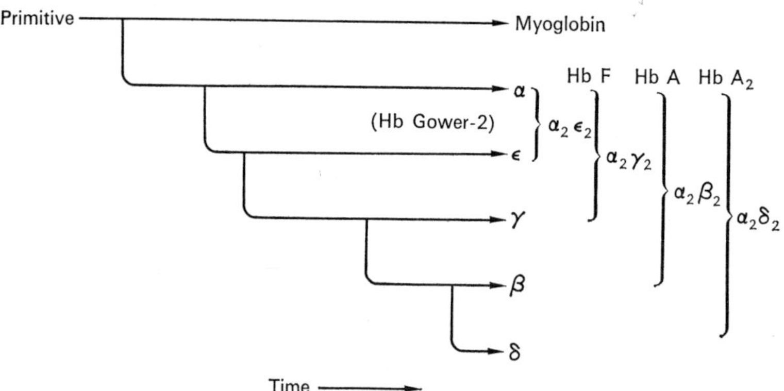

FIGURE 10.11. Evolution of hemoglobin. The horizontal lines represent evolutionary changes in the individual globin chain genes after stepwise duplication, beginning with a primitive myoglobin precursor at left of top line. (Adapted from Epstein & Motulsky, 1965. Reprinted with permission of Grune & Stratton, Inc. from *Progr. med. Genet.* **4**, 85.)

kind of polypeptide chain, which aggregates into dimers and tetramers in the deoxygenated form, but exists as a monomer in the oxygenated form. Presumably, aggregation and de-aggregation is responsible for the sigmoid shape of the oxygen dissociation curve found in lamprey hemoglobin. This process of reversible aggregation cannot be carried out by the hemoglobins of vertebrates at the phylogenetic level of the fishes and above, as is exemplified by the behavior of both the isolated α chain of human hemoglobins and the β and γ chain tetramers (see p. 360). Thus, it appears to be necessary for the hemoglobin of the higher vertebrates to contain two different chains in order to maintain the allosteric effect of heme-heme interaction. If, as Ingram (1963) has proposed, the γ, β and δ chains evolved (in that order) after the α chain ancestor was established, it seems likely that the primitive hemoglobin, consisting of an α-like chain, may have functioned in the manner of the present-day lamprey hemoglobin.

According to Margoliash & Smith (1965), among the cytochrome c molecules from different animal species, about half of the amino-acid residues are invariant (i.e. they invariably occur in the same position on the chain). One would reasonably expect that the hemoglobin polypeptide chains might have a similar proportion of residues in common, particularly since there is a strong resemblance in tertiary structure. Accordingly, Perutz *et al.* (1965) examined the published amino acid sequences of two myoglobins and of normal hemoglobin chains from nine different vertebrate species. They could find only nine amino acids invariably occupying the same site. These included five with heme contacts: the distal histidine at E 7 (see Figs. 10.1 & 10.2), the proximal histidine at F 8, threonine at C 4, phenyl-alanine at CD 1 and leucine at F 4. In addition, they reported an invariant glycine at B 6 (known to make a contact with E 8), proline at C 2 (forming the BC corner), tyrosine at H 23 (forming an internal hydrogen bond in the FG corner) and lysine at H 10, having no obvious function.

It appeared, then, that the common feature responsible for the characteristic tertiary structure was not usually limited to specific amino acids but rather to a variety of amino acids sharing a similar property. Perutz *et al.* (1965) found that maintenance of the helices was associated with the occurrence of regularly repeating interior non-polar residues. (Note: considered as non-polar were not only Gly, Ala, Val, Leu, Phe, Cys and Met, but also Try and Tyr, since their large rigid side chains seek a non-polar environment). Thus, although nearly all of the codons determining the amino acids of hemoglobin have undergone one or more changes in the vertebrate species, such changes have been limited by the characteristics of the resultant amino acid (particularly its polarity) necessary for maintaining the structure of the molecule.

A general expression of this principle was derived by Epstein (1964, 1967), who assigned factors for polarity and for size to the twenty amino acids and derived equations so that each amino acid could be compared with every other amino acid by an index or coefficient of difference. These indices were then calculated for the amino acid exchange in several pairs of homologous proteins (such as the Hb α and β or β and γ chains, the ribonuclease molecules of cows and rats, and two varieties of Tobacco mosaic virus). In all instances, the index of difference was considerably lower than could be accounted for by random mutations, so it was concluded that natural selection has acted during protein evolution to favor amino-acid substitutions compatible with retention of the original molecular conformation.

HEMOGLOBIN AND BALANCED POLYMORPHISM:
GEOGRAPHIC DISTRIBUTION

As early as 1949, Haldane suggested that the gene for thalassemia, a nearly lethal condition in the homozygous state, might be maintained in certain populations because of its protective effect against malaria. Subsequently, Allison (1954) proposed that sickle trait might have a selective advantage in malarious areas, and showed that the incidence of *Plasmodium falciparum* parasitemia was significantly lower among Ganda children with sickle trait than in children without the trait. Further studies by a number of workers on the degree of parasitism among other African children as well as on the distribution of sickle trait in various geographic areas were reviewed by Allison (1964, 1965). The results leave no doubt that in malarious regions, a state of balanced polymorphism exists, in which the harmful effect of the Hb_β^S gene is balanced by the selective advantage it confers against falciparum malaria.

Hb C has also been said to have a protective effect against malaria by Thompson (1962), based on his work in Accra. However, Edington & Laing (1957) found no evidence to support this contention among inhabitants of northern Ghana. Such a selective advantage would probably be difficult to prove, because the homozygous condition is much milder for Hb C than for Hb S. Thus, the Hb_β^C gene would be expected to survive well even if the protective effect against malaria were very small.

The geographic distribution of the β thalassemia genes (Chernoff, 1959) corresponds in many areas, notably excepting central Africa, with the incidence of malaria. However, there is currently no good evidence to indicate a selective relationship between the two disorders (Weatherall, 1965). The α thalassemia genes occur with fairly high frequencies in certain areas of China, Africa and Greece (Weatherall,

1967). Again, no relationship with malaria, based on selection, is evident.

The sickle gene has its highest frequencies in northwestern and central Africa, but it also occurs in the Mediterranean regions as well as in southern Arabia and India. Hb C is found almost exclusively in western Africa, except for parts of America, the West Indies, etc, where large numbers of Africans have settled. On the other hand, Hb E is rarely found in any region except Southeastern Asia, where its frequency exceeds ten percent in some tribal groups. One of the D hemoglobins, D_{Punjab}, was reported to have a frequency of about 2 per cent in the Sikhs of Punjab (Bird & Lehmann, 1956).

These four abnormal hemoglobins, among the many which have been described, are the only variants with a frequency of more than one per cent (aside from isolated tribes) in any known population. Since selective advantage for sickle trait is an established fact in malarious regions, it is tempting to believe that the other variants are maintained for similar reasons.

Further discussion of this subject is outside the scope of this chapter, and the reader is referred to Huehns & Shooter (1965), Livingstone (1967) and to the symposium *Abnormal Hemoglobins in Africa* edited by Jonxis (1965) for discussions by workers active in the field.

METHODS

The identification of hemoglobin structural variants has reached a level of sophistication much higher than that of any other human polymorphic system. Since adequate amounts of globin are readily obtainable, and since it is possible to characterize a given molecular species on the basis of a single amino acid substitution at a given point in the sequence of a peptide, nothing less than this precision is now acceptable. Furthermore, a thorough investigation of the physical properties of the variant hemoglobin is also required, particularly if there is an associated alteration in red cell morphology, viability or function as an oxygen carrier.

An adequate description of all of the techniques involved in the physico-chemical and hematological characterization of aberrant hemoglobins is far beyond the scope of this book. The particular interests of the reader must serve as a guide for appropriate references, many of which have been mentioned in this chapter. Huisman (1963), Watson-Williams & Weatherall (1965), Lehmann & Huntsman (1966) and Hutchison (1967) provide fairly extensive coverage, while Weatherall (1965) describes the methods used in differentiating the various thalassemia syndromes. Instructions for performing hematological measurements are provided by several texts, particularly the Sixth Edition of *Clinical Hematology*, by Wintrobe (1967).

REFERENCES

ALBEN J.O. & CAUGHEY W.S. (1966) Carbonyl stretching frequencies and carbon monoxide binding in red blood cells, hemoglobins and heme, *in* CHANCE B., ESTABROOK R. W. & YONETANI T. (eds.) *Hemes and Hemoproteins*, p. 139. Academic Press, New York.

ALLAN N., BEALE D., IRVINE D. & LEHMANN H. (1965) Three haemoglobins K: Woolwich, an abnormal, and Cameroon and Ibadan, two unusual variants of human haemoglobin A. *Nature* **208**, 658.

ALLEN D.W. (1964) Hemoglobin metabolism within the red cell, *in* BISHOP C. & SURGENOR D.M. (eds.) *The Red Blood Cell*, p. 309. Academic Press, New York.

ALLEN D.W., SCHROEDER W.A. & BALOG J. (1958) Observations on the chromatographic heterogeneity of normal adult and fetal human hemoglobin: a study of the effects of crystallization and chromatography on the heterogeneity and isoleucine content. *J. Am. chem. Soc.* **80**, 1628.

ALLEN D.W., WYMAN J. & SMITH C.A. (1953) The oxygen equilibrium of fetal and adult human hemoglobin. *J. biol. Chem.* **203**, 81.

ALLISON A.C. (1954) Protection afforded by sickle cell trait against subtertian malarial infection. *Br. med. J.* **i**, 290.

ALLISON A.C. (1964) Polymorphism and natural selection in human populations. *Cold Spring Harbor Symposia on Quantitative Biology*, **29**, 137.

ALLISON A.C. (1965) Population genetics of abnormal haemoglobins and glucose-6-phosphate dehydrogenase deficiency, *in* JONXIS J.H.P. (ed.) *Abnormal Haemoglobins in Africa*, p. 365. F.A. Davis, Philadelphia.

ANFINSEN C.B. (1959) *The Molecular Basis of Evolution*. John Wiley & Sons, New York.

ANTONINI E. (1965) Interrelationship between structure and function in hemoglobin and myoglobin. *Physiol. Rev.* **45**, 123.

ANTONINI E. (1967) Hemoglobin and its reactions with ligands. *Science* **158**, 1417.

ANTONINI E., BRUNORI M., WYMAN J. & NOBLE R.W. (1966) Preparation and kinetic properties of intermediates in the reaction of hemoglobin with ligands. *J. biol. Chem.* **241**, 3236.

ANTONINI E., BUCCI E., FRONTICELLI C., WYMAN J. & ROSSI-FANELLI A. (1965) The properties and interactions of the isolated α and β chains of human haemoglobin. III. Observations on the equilibria and kinetics of the reactions with gases. *J. molec. Biol.* **12**, 375.

ATASSI M.Z. (1964) Chemical studies on haemoglobins A_1 and A_0. *Biochem. J.* **93**, 189.

ATWATER J., SCHWARTZ I.R., ERSLEV A.J., MONTGOMERY T.L. & TOCANTINS L.M. (1960) Sickling of erythrocytes in a patient with thalassemia-hemoglobin I disease. *New Engl. J. Med.* **263**, 1215.

BACHMANN F. & MARTI H.R. (1962) Hemoglobin Zürich. II. Physicochemical properties of abnormal hemoglobin. *Blood* **20**, 272.

BAGLIONI C. (1962a) The fusion of two peptide chains in hemoglobin Lepore and its interpretation as a genetic deletion. *Proc. nat. Acad. Sci.* **48**, 1880.

BAGLIONI C. (1962b) Abnormal human haemoglobins. VIII. Chemical studies on haemoglobin D. *Biochim. biophys. Acta* **59**, 437.

BAGLIONI C. (1962c) A chemical study of hemoglobin Norfolk. *J. biol. Chem.* **237**, 69.

BAGLIONI C. (1963) Correlations between genetics and chemistry of human hemoglobins, *in* TAYLOR J.H. (ed.) *Molecular Genetics*, p. 405. Academic Press, New York.

BAGLIONI C. (1964) Discussion of paper by A.G. Motulsky. *Cold Spring Harbor Symposia on Quantitative Biology* **29**, 412.

BAGLIONI C. (1966) Chromosomal and cytoplasmic regulation of haemoglobin synthesis. *Plenary Session 11th Cong. int. Soc. Hemat.* Sydney, p. 413.

BAGLIONI C. & COLOMBO B. (1964) Control of hemoglobin synthesis. *Cold Spring Harbor Symposia on Quantitative Biology*, **29**, 347.

BAGLIONI C. & INGRAM V. (1961) Abnormal human haemoglobins. V. Chemical investigation of haemoglobins A, G, C and X from one individual. *Biochim. biophys. Acta* **48**, 253.

BAGLIONI C. & LEHMANN H. (1962) Chemical heterogeneity of haemoglobin O. *Nature* **196**, 229.

BAGLIONI C. & WEATHERALL D.J. (1963) Abnormal human hemoglobins. IX. Chemistry of hemoglobin J-Baltimore. *Biochim. biophys. Acta* **78**, 637.

BANK A. & MARKS P.A. (1966a) Excess α chain synthesis relative to β chain synthesis in thalassaemia major and minor. *Nature* **212**, 1198.

BANK A. & MARKS P.A. (1966b) Protein synthesis in cell-free human reticulocyte system; ribosome function in thalassemia. *J. clin. Invest.* **45**, 330.

BANNERMAN R.M. (1961) *Thalassemia: A Survey of Some Aspects.* Grune & Stratton, New York.

BARGELLESI A., PONTREMOLI S. & CONCONI F. (1967) Absence of β-globin and excess α-globin synthesis in homozygous β-thalassaemia. *Eur. J. Biochem.* **1**, 73.

BARNABAS J. & MULLER C.J. (1962) Haemoglobin—Lepore$_{Hollandia}$. *Nature* **194**, 931.

BAYRAKCI C., JOSEPHSON A., SINGER L., HELLER P. & COLEMAN R.D. (1964) Hb Kenwood, a new fast hemoglobin. *Proc. 10th Cong. int. Soc. Hemat.* Stockholm.

BEALE D. & LEHMANN H. (1965) Abnormal haemoglobins and the genetic code. *Nature* **207**, 259.

BEAVEN G.H., ELLIS M.L. & WHITE J.C. (1960) Study on human foetal haemo-globin. II. Foetal haemoglobin levels of healthy children and adults and in certain haematological disorders. *Br. J. Haemat.* **6**, 201.

BEAVEN G.H., STEVENS B.L., DANCE N. & WHITE J.C. (1963) Occurrence of haemoglobin H in leukaemia. *Nature* **199**, 1297.

BEAVEN G.H., GRATZER W.B., STEVENS B.L., SHOOTER E.M., WHITE J.C., ELLIS M.J. & GILLESPIE J.E.O'N. (1964) An abnormal haemoglobin (Lepore/Cyprus) resembling haemoglobin Lepore and its interaction with thalassaemia. *Br. J. Haemat.* **10**, 159.

BENESCH R. (1966) The molecular origin of the control mechanisms in haemo-globin. *Plenary Session, 11th Cong. int. Soc. Blood Transf.* Sydney, p. 163.

BENESCH R. & BENESCH R.E. (1964) Properties of haemoglobin H and their significance in relation to function of haemoglobin. *Nature* **202**, 773.

BENESCH R. & BENESCH R.E. (1967) The effect of organic phosphates from the human erythrocyte on the allosteric properties of hemoglobin. *Biochem. biophys. Res. Commun.* **26**, 162.

BENESCH R.E., BENESCH R. & MACDUFF G. (1965) Subunit exchange and ligand binding: a new hypothesis for the mechanism of oxygenation of hemoglobin. *Proc. natn. Acad. Sci.* **54**, 535.

BENESCH R., BENESCH R.E. & TYUMA I. (1966) Subunit exchange and ligand bind-ing. II. The mechanism of the allosteric effect in hemoglobin. *Proc. natn. Acad. Sci.* **56**, 1268.

BENESCH R., BENESCH R.E. & YU C.I. (1968) Reciprocal binding of oxygen and diphosphoglycerate by human hemoglobin. *Proc. natn. Acad. Sci.* **59**, 526.

BENESCH R.E., RANNEY H.M., BENESCH R. & SMITH G.M. (1961) The chemistry of the Bohr effect. II. Some properties of hemoglobin H. *J. biol. Chem.* **236**, 2926.

BERETTA A., PRATO V., GALLO E. & LEHMANN H. (1968) Haemoglobin Torino—$\alpha 43$ (CD1) phenylalanine $\rightarrow$ valine. *Nature* **217**, 1016.

BERTLES J.F., ROTH E. & ANKU V. (1967) The hemoglobin of irreversibly sickled erythrocytes. *J. clin. Invest.* **46**, 1037.

BETKE K. (1962) Hämoglobin M. Typen und ihre Differenzierung (Übersicht), *in* LEHMANN H. & BETKE K. (eds.) *Haemoglobin Colloquium.* p. 39. Thieme Verlag, Stuttgart.

BETKE K., MARTI H.R. & SCHLICHT I. (1959) Estimation of small percentages of foetal haemoglobin. *Nature* **184**, 1877.

BIRD G.W.G. & LEHMANN H. (1956) Haemoglobin D in India. *Br. med. J.* **i**, 514.

BONAVENTURA J. & RIGGS A. (1967) Polymerization of hemoglobins of mouse and man: structural basis. *Science* **158**, 800.

BOOKCHIN R.M., DAVIES R.P. & RANNEY H.M. (1968) Clinical features of hemo-globin C$_{Harlem}$, a new sickling hemoglobin variant. *Ann. intern. Med.* **68**, 8.

BOOKCHIN R.M., NAGEL R.L. & RANNEY H.M. (1967) Structure and properties of hemoglobin C$_{Harlem}$, a human hemoglobin variant with amino acid substitu-tions in 2 residues of the β-polypeptide chain. *J. biol. Chem.* **242**, 248.

BOTHA M.C., BEALE D., ISAACS W.A. & LEHMANN H. (1966) Haemoglobin J Cape Town—$\alpha_2^{92\,\text{Arg}\rightarrow\text{Glun}} \beta_2$. *Nature* **212**, 792.

BOWMAN B.H., BARNETT D.R. & HITE R. (1967) Hemoglobin G$_{Coushatta}$: a beta variant with a delta-like substitution. *Biochem. biophys. Res. Commun.* **26**, 466.

BOWMAN B.H., OLIVER C.P., BARNETT D.R., CUNNINGHAM J.E. & SCHNEIDER R.G. (1964) Chemical characterization of three hemoglobins G. *Blood* **23**, 193.

BRADLEY T.B. & RIEDER R.F. (1966) Hemoglobin Gun Hill, a β chain abnormality associated with a hemolytic state. *Blood* **28**, 975.

BRADLEY T.B., WOHL R.C. & RIEDER R.F. (1967) Hemoglobin Gun Hill: deletion of five amino acid residues and impaired heme-globin binding. *Science* **157**, 1581.

BRANCATI C. & BAGLIONI C. (1966) Homozygous $\beta\delta$ thalassaemia ($\beta\delta$ microcythaemia). *Nature* **212**, 262.

BRAUNITZER G., HILSE K., RUDOLPH V. & HILSCHMANN N. (1964) The hemoglobins, *in* ANFINSEN C.B., ANSON M.L., EDSALL J.T. & RICHARDS F.M. (eds.) *Advances in Protein Chemistry*, vol. 19, p. 1. Academic Press, New York.

BRIEHL R.W. (1963) The relation between the oxygen equilibrium and the aggregation of subunits in Lamprey hemoglobin. *J. biol. Chem.* **238**, 2361.

BRIERE R.O., GOLIAS T. & BATSAKIS J.G. (1965) Rapid qualitative and quantitative hemoglobin fractionation. *Am. J. clin. Path.* **44**, 695.

CANTOR C.R. & JUKES T.H. (1966) The repetition of homologous sequences in the polypeptide chains of certain cytochromes and globins. *Proc. natn. Acad. Sci.* **56**, 177.

CAPP G.L., RIGAS D.A. & JONES R.T. (1967) Hemoglobin Portland 1: a new human hemoglobin unique in structure. *Science* **157**, 65.

CARRELL R.W., LEHMANN H. & HUTCHISON H.E. (1966) Haemoglobin Köln (β-98 valine $\rightarrow$ methionine): an unstable protein causing inclusion-body anaemia. *Nature* **210**, 915.

CARRELL R.W., LEHMANN H., LORKIN P.A., RAIK E. & HUNTER E. (1967) Haemoglobin Sydney: β67 (E 11) Valine $\rightarrow$ Alanine: an emerging pattern of unstable haemoglobins. *Nature* **215**, 626.

CAUGHEY W.S. (1967) Porphyrin proteins and enzymes. *A. Rev. Biochem.* **36**, 611.

CEPPELLINI R. (1959) Discussion of paper by J.A. Hunt and V.M. Ingram, *in* WOLSTENHOLME G.E.W. & O'CONNOR C.M. (eds.) *Ciba Foundation Symposium on Biochemistry of Human Genetics*, p. 134. Little, Brown, Boston.

CHANCE B., ESTABROOK R.W. & YONETANI T. (1966) *Hemes and Hemoproteins.* Academic Press, New York.

CHARACHE S. & CONLEY C.L. (1964) Rate of sickling of red cells during deoxygenation of blood from persons with various sickling disorders. *Blood* **24**, 25.

CHARACHE S., CONLEY C.L., WAUGH D.F., UGORETZ R.J. & SPURRELL J.R. (1967) Pathogenesis of hemolytic anemia in homozygous hemoglobin C disease. *J. clin. Invest.* **46**, 1795.

CHARACHE S. & MONDZAC A.M. (1967) Hemoglobin Sinai ($\alpha_2^{47\,his}\,\beta_2$): an unstable hemoglobin causing occult hemolysis. *Blood* **30**, 879.

CHARACHE S., WEATHERALL D.J. & CLEGG J.B. (1966) Polycythemia associated with a hemoglobinopathy. *J. clin. Invest.* **45**, 813.

CHERNOFF A.I. (1959) The distribution of the thalassemia gene: a historical review. *Blood* **14**, 899.

CHERNOFF A.I. & LIU J.C. (1961) The amino acid composition of hemoglobin. II. Analytical techniques. *Blood* **17**, 54.

CHERNOFF A.I. & PETTIT N.M. (1964) Some notes on the starch gel electrophoresis of hemoglobins. *J. Lab. clin. Med.* **63**, 290.

CHERNOFF A.I., PETTIT N. & NORTHROP J. (1965) The amino acid composition of hemoglobin. V. The preparation of purified hemoglobin fractions by chromatography on cellulose exchangers and their identification by starch gel electrophoresis using Tris-borate-EDTA buffer. *Blood* **25**, 646.

CHIANCONE E., WITTENBERG J.B., WITTENBERG B.A., ANTONINI E. & WYMAN J. (1966) The combination of haptoglobin 2-2 with the oxy and deoxy forms of human hemoglobin before and after digestion of carboxypeptidase A, and with isolated α chains. *Biochim. biophys. Acta* **117**, 379.

CLEGG J.B., NAUGHTON M.A. & WEATHERALL D.J. (1965) An improved method for the characterization of human haemoglobin mutants: identification of $\alpha_2\beta_2^{95\ Glu}$ haemoglobin N (Baltimore). *Nature* **207**, 945.

CLEGG J.B., NAUGHTON M.A. & WEATHERALL D.J. (1966) Abnormal human haemoglobins. Separation and characterization of the α and β chains by chromatography, and the determination of two new variants, Hb Chesapeake and Hb J (Bangkok). *J. molec. Biol.* **19**, 91.

CLEGG J.B. & WEATHERALL D.J. (1967) Haemoglobin synthesis in α-thalassemia (Haemoglobin H disease). *Nature* **215**, 1241.

COMINGS D.E. & MOTULSKY A.G. (1966) Absence of *cis* delta chain synthesis in ($\delta\beta$) thalassemia (F-thalassemia). *Blood* **28**, 54.

CONLEY C.L. & CHARACHE S. (1967) Mechanisms by which some abnormal hemoglobins produce clinical manifestations. *Seminars in Hematology*, **4**, 53.

CONLEY C.L., WEATHERALL D.J., RICHARDSON S.N., SHEPARD M.K. & CHARACHE S. (1963) Hereditary persistence of fetal hemoglobin: a study of 79 affected persons in 15 Negro families in Baltimore. *Blood* **21**, 261.

CROOKSTON J.H., BEALE D., IRVINE D. & LEHMANN H. (1965) A new haemoglobin, J Toronto (α 5 alanine → aspartic acid). *Nature* **208**, 1059.

CULLIS A.F., MUIRHEAD H., PERUTZ M.F., GROSSMAN M.G. & NORTH A.C.T. (1962) The structure of hemoglobin. IX. A three-dimensional Fourier synthesis at 5·5 Å resolution: description of the structure. *Proc. R. Soc. A.* **265**, 161.

DACIE J.V., SHINTON N.K., GAFFNEY P.J., CARRELL R.W. & LEHMANN H. (1967) Haemoglobin Hammersmith (β42 (CD1) phe → ser). *Nature* **216**, 663.

DANCE N. & HUEHNS E.R. (1962) A haemoglobin containing only δ chains. *Biochem. biophys. Res. Commun.* **7**, 444.

DANCE N., HUEHNS E.R. & BEAVEN G.H. (1963) The abnormal haemoglobins in haemoglobin H disease. *Biochem. J.* **87**, 240.

DEVRIES A., JOSHUA H., LEHMANN H., HILL R.L. & FELLOWS R.E. (1963) The first observation of an abnormal haemoglobin in a Jewish family: haemoglobin Beilinson. *Br. J. Haemat.* **9**, 484.

DIGGS L.W. & BELL A. (1965) Intraerythrocytic hemoglobin crystals in sickle cell-hemoglobin C disease. *Blood* **25**, 218.

DINTZIS H.M. (1961) Assembly of peptide chains of hemoglobin. *Proc. natn. Acad. Sci.* **47**, 247.

DIXON G.H. (1966) Mechanisms of protein evolution. *Essays in Biochemistry* **2**, 148.

DORMANDY K.M., LOCK S.P. & LEHMANN H. (1961) Haemoglobin Q-alpha-thalassemia. *Br. med. J.* **i**, 1582.

EDDISON G.G., BRIEHL R.W. & RANNEY H.M. (1964) Oxygen equilibria of hemoglobin A_2 and hemoglobin Lepore. *J. clin. Invest.* **43**, 2323.

EDINGTON G.M. & LAING W.N. (1957) Relationship between haemoglobin C and S and malaria in Ghana. *Br. med. J.* **ii**, 143.

EPSTEIN C.J. (1964) Relation of protein evolution to tertiary structure. *Nature* **203**, 1350.

EPSTEIN C.J. (1967) Non-randomness of amino acid changes in the evolution of homologous proteins. *Nature* **215**, 355.

EPSTEIN C.J. & MOTULSKY A.G. (1965) Evolutionary origins of human proteins, *in* STEINBERG A.G. & BEARN A.G. (eds.) *Progress in Medical Genetics*, vol. 4, p. 85. Grune & Stratton, New York.

FESSAS P. (1959) Thalassaemia and alterations in haemoglobin pattern, *in* JONXIS J.H.P. & DELAFRESNAYE J.F. (eds.) *Abnormal Haemoglobins—A Symposium*, p. 134. Blackwell, Oxford.

FESSAS P. (1962) The beta-chain thalassaemias, *in* LEHMANN H. & BETKE K. (eds.) *Haemoglobin Colloquium*, p. 90. Thieme Verlag, Stuttgart.

FESSAS P. (1963) Inclusions of hemoglobin in erythroblasts and erythrocytes of thalassemia. *Blood* **21**, 21.

FESSAS P. (1965) Forms of thalassemia, *in* JONXIS J.H.P. (ed.) *Abnormal Haemoglobins in Africa*, p. 71. F.A. Davis, Philadelphia.

FESSAS P. & LOUKOPOULOS D. (1964) Alpha-chain of human hemoglobin: occurrence *in vivo*. *Science* **143**, 590.

FESSAS P., LOUKOPOULOS D. & KALTSOYA D. (1966) Peptide analysis of the inclusions of erythroid cells in β-thalassemia. *Biochim. biophys. Acta* **124**, 430.

FESSAS P. & STAMATOYANNOPOULOS G. (1962) Absence of haemoglobin A_2 in an adult. *Nature* **195**, 1215.

FESSAS P. & STAMATOYANNOPOULOS G. (1964) Hereditary persistence of fetal hemoglobin. A study and a comparison. *Blood* **24**, 223.

FESSAS P., STAMATOYANNOPOULOS G. & KARAKLIS A. (1962) Hemoglobin 'Pylos': study of a hemoglobinopathy resembling thalassemia in the heterozygous and double heterozygous state. *Blood* **19**, 1.

FRASER G.R., STAMATOYANNOPOULOS G., KATTAMIS C., LOUKOPOULOS D., DEFARANAS B., KITSOS C., ZANNOS-MARIOLEA L., CHOREMIS C., FESSAS P. & MOTULSKY A.G. (1964) Thalassemias, abnormal hemoglobins and glucose-6-phosphate dehydrogenase deficiency in the Arta area of Greece: diagnostic and genetic aspects of complete village studies. *Ann. N.Y. Acad. Sci.* **119**, 415.

FRICK P.G., HITZIG W.H. & BETKE K. (1962) Hemoglobin Zürich. I. A new hemoglobin anomaly associated with acute hemolytic episodes with inclusion bodies after sulfonamide therapy. *Blood* **20**, 261.

GABUZDA T.G. (1966) Hemoglobin H and the red cell. *Blood* **27**, 568.

GABUZDA T.G., LAFORET M.T. & GARDNER F.H. (1967) Oxidative precipitation of hemoglobin H and its relation to reduced glutathione. *J. Lab. clin. Med.* **70**, 581.

GABUZDA T.G., NATHAN D.G. & GARDNER F.H. (1962) Comparative metabolism of haemoglobins A and F in thalassaemia. *Nature* **196**, 781.

GABUZDA T.G., NATHAN D.G., GARDNER F.H. & LIMAURO A. (1963) The turnover of hemoglobins A, F and A_2 in the peripheral blood of three patients with thalassemia. *J. clin. Invest.* **42**, 1678.

GABUZDA T.G., NATHAN D.G. & GARDNER F.H. (1964) Thalassemia trait. Genetic combinations of increased fetal and A_2 hemoglobins. *New Engl. J. Med.* **270**, 1212.

GABUZDA T.G., NATHAN D.G. & GARDNER F.H. (1965) The metabolism of the individual C^{14}-labeled hemoglobins in patients with H-thalassemia, with observations on radiochromate binding to the hemoglobins during red cell survival. *J. clin. Invest.* **44**, 315.

GAJDUSEK D.C., GUIART J., KIRK R.L., CARRELL R.W., IRVINE D., KYNOCH P.A.M. & LEHMANN H. (1967) Haemoglobin J Tongariki (α115 alanine → aspartic acid): the first new haemoglobin variant found in a Pacific (Melanesian) population. *J. med. Genet.* **4**, 1.

GALLO R.C. (1967) The inhibitory effect of heme on heme formation *in vivo*: possible mechanism for the regulation of hemoglobin synthesis. *J. clin. Invest.* **46**, 124.

GERALD P.S. & DIAMOND L.K. (1958) A new hereditary hemoglobinopathy (the Lepore trait) and its interaction with thalassemia trait. *Blood* **13**, 835.

GERALD P. & EFRON M.L. (1961) Chemical studies of several variants of Hb M. *Proc. natn. Acad. Sci.* **47**, 1758.

GERALD P.S., EFRON M.L. & DIAMOND L.K. (1961) A human mutation (the Lepore hemoglobinopathy) possibly involving two cistrons. *Am. J. Dis. Childh.* **102**, 514.

GERALD P.S. & GEORGE P. (1959) Second spectroscopically abnormal methemoglobin associated with hereditary cyanosis. *Science* **129**, 393.

GERALD P. & RATH C.E. (1966) Hb $C_{Georgetown}$; first abnormal hemoglobin due to two different mutations in the same gene. *J. clin. Invest.* **45**, 1012.

GIBSON Q.H. (1959) The photochemical formation of a quickly reacting form of haemoglobin. *Biochem. J.* **71**, 293.

GIBSON Q.H., HELLER P. & YAKULIS V. (1966) The rate of reaction of carbon monoxide with hemoglobins M. *J. biol. Chem.* **241**, 1650.

GOTTLIEB A.J., RESTREPO A. & ITANO H.A. (1964) Hb J-Medellin. Chemical and genetic study. *Fed. Proc.* **23**, 172.

GOUTTAS A., FESSAS P., TSEVRENIS H. & XEFTERI E. (1955) Description d'une nouvelle variété d'anémia hémolytique congénitale (Etude hématologique, electrophorétique, et génétique). *Sang.* **26**, 911.

GREENWALT T.J. & AYERS V.E. (1960) Phosphate partition in the erythrocytes of normal newborn infants and infants with erythroblastosis fetalis. II. Quantitative paper chromatography. *Blood* **15**, 698.

GRIMES A.J. & MEISLER A. (1962) Possible cause of Heinz bodies in congenital Heinz-body anaemia. *Nature* **194**, 190.

GRIMES A.J., MEISLER A. & DACIE J.V. (1964) Congenital Heinz-body anaemia. Further evidence on the cause of Heinz-body production in red cells. *Br. J. Haemat.* **10**, 281.

GROSS R.T., SCHROEDER E.A.R. & BRAUNSTEIN S.A. (1963) Energy metabolism in the erythrocytes of premature infants compared to full term infants and adults. *Blood* **21**, 755.

GUIDOTTI G., KONIGSBERG W. & CRAIG L.C. (1963) On the dissociation of normal adult human hemoglobin. *Proc. natn. Acad. Sci.* **50**, 774.

HALDANE J.B.S. (1949) Disease and evolution. *La Ricerca Scientif.* suppl. **19**, 68.

HANADA M. & RUCKNAGEL D.L. (1963) The abnormality of the primary structure of hemoglobin Shimonoseki. *Biochem. biophys. Res. Commun.* **11**, 229.

HARRIS J.W. (1950) Studies on the destruction of red blood cells. *Proc. Soc. exp. Biol. Med.* **75**, 197.

HARRIS J.W. (1959) The role of physical and chemical factors in the sickling phenomenon, *in* TOCANTINS L.M. (ed.) *Progress in Hematology*, vol. 2, p. 47. Grune & Stratton, New York.

HAYASHI N., MOTOKAWA Y. & KIKUCHI G. (1966) Studies on relationships between structure and function of hemoglobin M_{Iwate}. *J. biol. Chem.* **241**, 79.

HECHT F., JONES R.T. & KOLER R.D. (1967) Newborn infants with Hb Portland 1, an indicator of α-chain deficiency. *Ann. hum. Genet.* **31**, 215.

HELLER P. (1965) The molecular basis of the pathogenicity of abnormal hemoglobins—some recent developments. *Blood* **25**, 110.

HELLER P. (1966) Hemoglobinopathic dysfunction of the red cell. *Am. J. Med.* **41**, 799.

HELLER P., COLEMAN R.D. & YAKULIS V. (1966a) Structural studies of Haemoglobin $M_{Hyde\ Park}$. *Plenary Session 11th Congr. int. Soc. Blood Transf.* Sydney, p. 184.

HELLER P., WEINSTEIN H., YAKULIS V. & ROSENTHAL I. (1962) Hemoglobin $M_{Kankakee}$, a new variant of Hemoglobin M. *Blood* **20**, 287.

HELLER P., YAKULIS V.J., ROSENZWEIG A.I., ABILDGAARD C.F. & RUCKNAGEL D.L. (1966b) Mild homozygous beta-thalassemia. Further evidence for the heterogeneity of beta-thalassemia genes. *Ann. intern. Med.* **64**, 52.

HEYWOOD J.D. & HILLMAN R.S. (1967) Free $\alpha\beta$ globin; a possible intermediate in hemoglobin synthesis. *Clin. Res.* **15**, 279.

HEYWOOD J.D., KARON M. & WEISSMAN S. (1965) Asymmetrical incorporation of amino acids in the alpha and beta chains of hemoglobin synthesized by thalassemic reticulocytes. *J. Lab. clin. Med.* **66**, 476.

HILL R.J. (1964) Discussion of paper by H.A. Itano, E.A. Robinson and A.J. Gottlieb, *Brookhaven Symp. Biol.* **17**, 201.

HILL R.J. & KRAUS A.P. (1963) Studies on the amino acid sequence of Hb A_2. *Fed. Proc.* **22**, 597.

HILL R.L., SWENSON R.T. & SCHWARTZ H.C. (1960) Characterization of a chemical abnormality in hemoglobin G. *J. biol. Chem.* **235**, 3182.

HOLMQUIST W.R. & SCHROEDER W.A. (1964) Properties and partial characterization of adult human hemoglobin A_{Ic}. *Biochim. biophys. Acta* **82**, 639.

HOLMQUIST W.R. & SCHROEDER W.A. (1966) A new N-terminal blocking group involving a Schiff base in hemoglobin A_{Ic}. *Biochemistry* **5**, 2489.

HÖRLEIN H. & WEBER H. (1948) Über chronische familiäre Methämoglobinämie und eine neue Modifikation des Methämoglobins. *Dsch. med. Wschr.* **73**, 476.

HORTON B.F., THOMPSON R.B., DOZY A.M., NECHTMAN C.M., NICHOLS E. & HUISMAN T.H.J. (1962) Inhomogeneity of hemoglobin. VI. The minor hemoglobin components of cord blood. *Blood* **20**, 302.

HUEHNS E.R. (1962) Haemoglobin H disease. Clinical and experimental studies. *M.D. Thesis.* University of London.

HUEHNS E.R. (1965) Thalassemia. *Postgrad. Med. J.* **41**, 718.

HUEHNS E.R., DANCE N., BEAVEN G.H., HECHT F. & MOTULSKY A.G. (1964) Human embryonic hemoglobins. *Cold Spring Harbor Symposia on Quantitative Biology* **29**, 327.

HUEHNS E.R., FLYNN F.V., BUTLER E.A. & BEAVEN G.H. (1961) Two new haemoglobin variants in a very young human embryo. *Nature* **189**, 496.

HUEHNS E.R., FLYNN F.V., BUTLER E.A. & SHOOTER E.M. (1960) The occurrence of haemoglobin 'Barts' in conjunction with haemoglobin H. *Br. J. Haemat.* **6**, 388.

HUEHNS E.R. & MODELL C.B. (1967) Haemoglobin synthesis in thalassaemia. *Trans. R. Soc. trop. Med. Hyg.* **61**, 157.

HUEHNS E.R. & SHOOTER E.M. (1965) Review article: human haemoglobins. *J. med. Genet.* **2**, 1.

HUEHNS E.R. & SHOOTER E.M. (1966) The properties and reactions of haemoglobin F_1 and their bearing on the dissociation equilibrium of haemoglobin. *Biochem. J.* **101**, 852.

HUISMAN T.H.J. (1963) Normal and abnormal human hemoglobins, *in* SOBOTKA H. & STEWART C.P. (eds.) *Advances in Clinical Chemistry*, vol. 6, p. 231. Academic Press, New York.

HUISMAN T.H.J. & DOZY A.M. (1962) Studies on the heterogeneity of hemoglobin. V. Binding of hemoglobin with oxidized glutathione. *J. Lab. clin. Med.* **60**, 302.

HUISMAN T.H.J. & DOZY A.M. (1965) Studies on the heterogeneity of hemoglobin. IX. The use of tris (hydroxymethyl) aminomethane-HCl buffers in the anion-exchange chromatography of hemoglobins. *J. Chromat.* **19**, 160.

HUISMAN T.H.J., DOZY A.M., HORTON B.F. & NECHTMAN C.M. (1966) Studies on the heterogeneity of hemoglobin. X. The nature of various minor hemoglobin components produced in human red blood cell hemolysates on aging. *J. Lab. clin. Med.* **67**, 355.

HUISMAN T.H.J., DOZY A.M., HORTON B.E. & WILSON J.B. (1965) A fetal hemoglobin with abnormal γ-polypeptide chains: hemoglobin Warren. *Blood* **26**, 668.

HUISMAN T.H.J., DOZY A.M., NECHTMAN C. & THOMPSON R.B. (1962) Oxygen equilibrium of haemoglobin A_2 and its variant, haemoglobin A_2' (or B_2) *Nature* **195**, 1109.

HUNT J.A. & INGRAM V.M. (1959) A terminal peptide sequence of human haemoglobin. *Nature* **184**, 640.

HUNT J.A. & INGRAM V.M. (1960) Abnormal human haemoglobins. IV. The chemical difference between normal haemoglobin and haemoglobin C. *Biochim. biophys. Acta* **42**, 409.

HUNT J.A. & INGRAM V.M. (1961) Abnormal human haemoglobins. VI. The chemical difference between haemoglobins A and E. *Biochim. biophys. Acta* **49**, 520.

HUNT J.A. & LEHMANN H. (1959) Abnormal human haemoglobins. Haemoglobin 'Bart's': a foetal haemoglobin without α chains. *Nature* **184**, 872.

HUTCHISON H.E. (1967) *Introduction to the Haemoglobinopathies and the Methods used for their Recognition.* Edward Arnold, London.

HUTCHISON H.E., PINKERTON P. H., WATERS P.,DOUGLAS A.S., LEHMANN H. & BEALE D. (1964) Hereditary Heinz-body anaemia, thrombocytopenia and haemoglobinopathy (Hb Köln) in a Glasgow family. *Br. med. J.* **ii**, 1099.

INGRAM V.M. (1961a) *Hemoglobin and Its Abnormalities.* C.C. Thomas, Springfield.

INGRAM V.M. (1961b) Gene evolution and the haemoglobins. *Nature* **189**, 704.

INGRAM V.M. (1963) *The Hemoglobins in Genetics and Evolution.* Columbia Univ. Press, New York.

INGRAM V.M. (1964) A molecular model for thalassemia. *Ann. N.Y. Acad. Sci.* **119**, 485.

INGRAM V.M. & STRETTON A.O.W. (1961) Human haemoglobin A_2: chemistry, genetics and evolution. *Nature* **190**, 1079.

INGRAM V.M. & STRETTON A.O.W. (1962a) Human haemoglobin A_2. I. Comparison of haemoglobin A_2 and A. *Biochim. biophys. Acta* **62**, 456.

INGRAM V.M. & STRETTON A.O.W. (1962b) Human haemoglobin A_2. II. The chemistry of some peptides peculiar to haemoglobin A_2. *Biochim. biophys. Acta* **63**, 20.

ITANO H.A. (1953) Solubilities of naturally occurring mixtures of human hemoglobin. *Archs Biochem.* **47**, 148.

ITANO H.A. (1957) The human hemoglobins: their properties and genetic control, *in* ANFINSEN C.B., ANSON M.L., BAILEY K. & EDSALL J.T. (eds.) *Advances in Protein Chemistry*, vol. 12, p. 215. Academic Press, New York.

ITANO H.A. (1965) The synthesis and structure of normal and abnormal hemoglobins, *in* JONXIS J.H.P. (ed.) *Abnormal Haemoglobins in Africa*, p. 3. F.A. Davis, Philadelphia.

ITANO H.A. (1966) Genetic regulation of peptide synthesis in hemoglobins. *J. Cell. Physiol.* **67**, suppl. 1, 65.

JACOB F. & MONOD J. (1961) Genetic regulatory mechanisms in the synthesis of proteins. *J. molec. Biol.* **3**, 318.

JACOB H.S., BRAIN M.C. & DACIE J.V. (1967) Blockade of membrane and globin thiols in the pathogensis of congenital Heinz body hemolytic anemia (CHBHA) with dipyrroluria. *J. clin. Invest.* **46**, 1073.

JAFFE E.R. (1964) Metabolic processes involved in the formation and reduction of methemoglobin in human erythrocytes, *in* BISHOP C. & SURGENOR D.M. (eds.) *The Red Blood Cell*, p. 397. Academic Press, New York.

JENKINS G.C., BEALE D., BLACK A.J., HUNTSMAN G.R. & LEHMANN H. (1967) Haemoglobin F-Texas I ($\alpha_2\gamma_2^{5Glu\rightarrow Lys}$): a variant of haemoglobin F. *Br. J. Haemat.* **13**, 252.

JONES R.T. (1964) Structural studies of aminoethylated hemoglobins by automatic peptide chromatography. *Cold Spring Harbor Symposia on Quantitative Biology* **29**, 297.

JONES R.T., BRIMHALL B., HUEHNS E.R. & BARNICOT N.A. (1966a) Hemoglobin Sphakiá: a delta chain variant of hemoglobin A_2 from Crete. *Science* **151**, 1406.

JONES R.T., BRIMHALL B. & HUISMAN T.H.J. (1966b) Chemical studies of hemoglobin A_2-δ-Flatbush $\alpha_2\delta_2^{22\,Glu}$. *Clin. Res.* **14**, 168.

JONES R.T., BRIMHALL B., HUISMAN T.H.J., KLEIHAUER E. & BETKE K. (1966c)

Hemoglobin Freiburg: abnormal hemoglobin due to deletion of a single amino acid residue. *Science* **154**, 1024.

JONES R.T., KOLER R.D. & LISKER R. (1963) The chemical structure of hemoglobin Mexico determined by automatic peptide chromatography and subunit hybridization. *Clin. Res.* **11**, 105.

JONES R.T., OSGOOD E.E., BRIMHALL B. & KOLER R.D. (1967) Hemoglobin Yakima. I. Clinical and biochemical studies. *J. clin. Invest.* **46**, 1840.

JONES R.T., SCHROEDER W.A., BALOG J.E. & VINOGRAD J.R. (1959) Gross structure of hemoglobin H. *J. Am. chem. Soc.* **81**, 3161.

JONES R.T., WESTENDORP F. & HUISMAN T.H.J. (1965) Structural characterization of Hb A_2' (B_2): $\alpha_2\delta_2^{16\,Arg}$. *Paper presented Amer. Soc. hum. Genet. meeting.*

JONES R.V., GRIMES A.J. CARRELL R.W. & LEHMANN H. (1967) Köln haemoglobinopathy. Further data and a comparison with other hereditary Heinz body anaemias. *Br. J. Haemat.* **13**, 394.

JONXIS J.H.P. (1961) Hemoglobinopathies. *Ann. Rev. Med.* **14**, 297.

JONXIS J.H.P. (1965) *Abnormal Haemoglobins in Africa, a C.I.O.M.S. Symposium.* F.A. Davis, Philadelphia.

JUKES T.H. (1966) *Molecules and Evolution.* Columbia Univ. Press, New York.

KALTSOYA A., FESSAS P. & STAVROPOULOS A. (1966) Hemoglobins of early human embryonic development. *Science* **153**, 1417.

KAPLAN E. & ZUELZER W.W. (1950) Erythrocyte survival in childhood. II. Studies in Mediterranean anemia. *J. Lab. clin. Med.* **36**, 517.

KARIBIAN D. & LONDON I.M. (1965) Control of heme synthesis by feedback inhibition. *Biochem. biophys. Res. Commun.* **18**, 243.

KENDREW J.C., DICKERSON R.E., STRANDBERG B.E., HART R.G., DAVIES D R., PHILLIPS D.C. & SHORE V.C. (1960) Structure of myoglobin. A three-dimensional Fourier synthesis at 2 Å resolution. *Nature* **185**, 422.

KENDREW J.C., WATSON H.C., STRANDBERG B.E., DICKERSON R.E., PHILLIPS D.C. & SHORE V.C. (1961) The amino acid sequence of sperm whale myoglobin. *Nature* **190**, 666.

KLEIHAUER E., BRAUN H. & BETKE K. (1957) Demonstration von Fetalem Hämoglobin in den Erythrocyten eines Blut ausstrichs. *Klin. Wschr.* **35**, 637.

KLEINAUER E.F., REYNOLDS C.A., DOZY A.M., WILSON J.B., MOORES R.R., BERENSON M.P., WRIGHT C.S. & HUISMAN T.H.J. (1968) Hemoglobin$_{Bibba}$ or $\alpha_2^{136\,Pro}\beta_2$, an unstable α chain abnormal hemoglobin. *Biochim. biophys. Acta* **154**, 220.

KONIGSBERG W. & HILL R.J. (1962) The structure of human hemoglobin. III. The sequence of amino acids in the tryptic peptides of the α chain. *J. biol. Chem.* **237**, 2547.

KRAUS A.P., MIYAJI, T., IUCHI I. & KRAUS L.M. (1966) A new variant of sickle cell anemia with clinically mild symptoms due to an α chain variant of hemoglobin ($\alpha^{23\,Glu\,NH_2}$). *J. Lab. clin. Med.* **66**, 886.

KUNKEL H.G., CEPPELLINI R., MULLER-EBERHARD U. & WOLF J. (1957) Observations on the minor basic hemoglobin component in the blood of normal individuals and patients with thalassemia. *J. clin. Invest.* **36**, 1615.

KUNKEL H.G. & WALLENIUS G. (1955) New hemoglobin in normal adult-blood. *Science* **122**, 288.

LABIE D., SCHROEDER W.A. & HUISMAN T.H.J. (1966) The amino acid sequence of the δ-β chains of hemoglobin Lepore$_{Augusta}$ = Lepore$_{Washington}$. *Biochim. biophys. Acta* **127**, 428.

LEHMANN H., BEALE D. & BOI-DOKU F.S. (1964) Haemoglobin G$_{Accra}$. *Nature* **203**, 363.

LEHMANN H. & HUNTSMAN R.G. (1966) *Man's Haemoglobins*. J.B. Lippincott, Philadelphia.

LEHMANN H., STORY P. & THEIN H. (1956) Haemoglobin E in Burmese. Two cases of haemoglobin E disease. *Br. med. J.* **i**, 544.

LIDDELL J., BROWN D., BEALE D., LEHMANN H. & HUNTSMAN R.G. (1964) A new haemoglobin—J$_\alpha$ Oxford found during a survey of an English population. *Nature* **204**, 269.

LIE-INJO, LUAN ENG (1962) Alpha-chain thalassemia and hydrops fetalis in Malaya: report of five cases. *Blood* **20**, 581.

LIE-INJO, LUAN ENG, PILLAY R.P. & THURAISINGHAM V. (1966) Further cases of Hb-Q-H disease (Hb-Q-α thalassemia). *Blood* **28**, 830.

LINES J.G. & MCINTOSH R. (1967) Oxygen binding by haemoglobin J$_{Capetown}$ ($\alpha^{92\,Arg\rightarrow Gln}$). *Nature* **215**, 297.

LIVINGSTONE F.B. (1967) *Abnormal Hemoglobins in Human Populations*. Aldine, Chicago.

LOUKOPOULOS K. & FESSAS P. (1965) The distribution of hemoglobin types in thalassemic erythrocytes. *J. clin. Invest.* **44**, 231.

MALAMOS B., FESSAS P. & STAMATOYANNOPOULOS G. (1962) Types of thalassaemia trait carriers as revealed by study of their incidence in Greece. *Br. J. Haemat.* **8**, 5.

MARGOLIASH E. & SMITH E.L. (1965) Structural and functional aspects of cytochrome c in relation to evolution, *in* BRYSON V. & VOGEL H.J. (eds.) *Evolving Genes and Proteins*. p. 221. Academic Press, New York.

MARKS P.A. (1966) Thalassemia syndromes. Biochemical, genetic and clinical aspects. *New Engl. J. Med.* **275**, 1363.

MARKS P.A. & BURKA E.R. (1964) Hemoglobin synthesis in human reticulocytes: a defect in globin formation in thalassemia major. *Ann. N.Y. Acad. Sci.* **119**, 513.

MARTI H.R., BEALE D. & LEHMANN H. (1967) Haemoglobin Koelliker: a new acquired haemoglobin appearing after severe haemolysis: $\alpha_2^{minus\,141\,Arg}\,\beta_2$. *Acta haemat.* **37**, 174.

MCCARTHY B.J. (1967) Arrangement of base sequences in deoxyribose nucleic acid. *Bact. Rev.* **31**, 215.

MINNICH V., CORDONNIER J.K., WILLIAMS W.J. & MOORE C.V. (1962) Alpha, beta and gamma hemoglobin polypeptide chains during the neonatal period with description of a fetal form of hemoglobin D$_{\alpha\,St.\,Louis}$. *Blood* **19**, 137.

MINNICH V., HILL R.H., KHURI P.D. & ANDERSON M.E. (1965) Hemoglobin Hope: a beta chain variant. *Blood* **25**, 830.

MIYAJI T., IUCHI I., SHIBATA S., TAKEDA I. & TAMURA A. (1963) Possible amino acid substitution in the α chain (α87 Tyr) of Hb M-Iwate. *Acta haemat. Japan* **26**, 538.

MIYAJI T., IUCHI I., YAMAMOTO K., OHBA Y. & SHIBATA S. (1967) Amino acid substitution of hemoglobin Ube-2 ($\alpha_2^{68\,\text{Asp}}\ \beta_2$): an example of successful application of partial hydrolysis of peptide with 5% acetic acid. *Clin. chim. Acta* **16**, 347.

MIYAJI T., OHBA Y., YAMAMOTO K., SHIBATA S., IUCHI I. & HAMILTON H.B. (1968a) Hemoglobin Hijiyama: a new fast-moving hemoglobin in a Japanese family. *Science* **159**, 204.

MIYAJI T., OHBA Y., YAMAMOTO K., SHIBATA S., IUCHI I. & TAKENAKA M. (1968b) A Japanese haemoglobin variant. *Nature* **217**, 89.

MIYAJI T., SUZUKI H., OHBA Y. & SHIBATA S. (1966) Hemoglobin Agenogi ($\alpha_2\beta_2^{90\,\text{Lys}}$), a slow moving hemoglobin of a Japanese family resembling Hb E. *Clin. chim. Acta* **14**, 624.

MONOD J., WYMAN J. & CHANGEUX J.P. (1965) On the nature of allosteric transitions: a plausible model. *J. molec. Biol.* **12**, 88.

MOTULSKY A.G. (1962) Controller genes in synthesis of human haemoglobin. *Nature* **194**, 607.

MOTULSKY A.G. (1964) Current concepts of the genetics of the thalassemias. *Cold Spring Harbor Symposia on Quantitative Biology* **29**, 399.

MUIRHEAD H. & PERUTZ M.F. (1963) Structure of haemoglobin. A three-dimensional Fourier synthesis of reduced human haemoglobin at 5·5 Å resolution. *Nature* **199**, 633.

MULLER C.J. & KINGMA S. (1961) Haemoglobin Zürich, $\alpha_2\beta_2^{63\,\text{Arg}}$. *Biochim. biophys. Acta* **50**, 595.

MURAYAMA M. (1965) Orientation of sickled erythrocytes in a magnetic field. *Nature* **206**, 420.

MURAYAMA M. (1966) Molecular mechanism of red cell 'sickling'. *Science* **153**, 145.

MURAYAMA M., OLSON R.A. & JENNINGS W.H. (1965) Molecular orientation in horse hemoglobin crystals and sickled erythrocytes. *Biochim. biophys. Acta* **94**, 194.

MURPHY J.R. (1967) Hb CC disease: an abnormality in cell water. *J. clin. Invest.* **46**, 1099.

NAGEL R.L., GIBSON Q.H. & CHARACHE S. (1967) Relation between structure and function in Hemoglobin Chesapeake. *Biochemistry* **6**, 2395.

NAGEL R.L. & RANNEY H.M. (1964) Haptoglobin binding capacity of certain abnormal hemoglobins. *Science* **144**, 1014.

NAGEL R.L., ROTHMAN M.C., BRADLEY T.B. & RANNEY H.M. (1965) Comparative haptoglobin binding properties of oxyhemoglobin and deoxyhemoglobin. *J. biol. Chem.* **240**, PC4543.

NANCE W.E. (1963) Genetic control of hemoglobin synthesis. *Science* **141**, 123.

NATHAN D.G. & GUNN R.B. (1966) Thalassemia: the consequences of unbalanced hemoglobin synthesis. *Am. J. Med.* **41**, 815.

NECHTMAN C.M. & HUISMAN T.H.J. (1964) Comparative studies of oxygen equilibria of human adult and cord red cell hemolysates and suspensions. *Clin. chim. Acta* **10**, 165.

NEEB H., BEIBOER J.L., JONXIS J.H.P., KAARS-SIJPENSTEIJN J.A. & MULLER C.J. (1961) Homozygous Lepore haemoglobin disease appearing as thalassaemia major in two Papuan siblings. *Trop. geogr. Med.* **13**, 207.

NEEL J.V. (1961) The hemoglobin genes: a remarkable example of the clustering of related genetic functions on a single mammalian chromosome. *Blood* **18**. 769.

Novy M.J., Edwards M.J. & Metcalfe J. (1967) Hemoglobin Yakima. II. High blood oxygen affinity associated with compensatory erythrocytosis and normal hemodynamics. *J. clin. Invest.* **46**, 1848.

Osgood E.E., Jones R.T., Brimhall B. & Koler R.D. (1967) Hemoglobin Yakima: clinical and biochemical studies. *Clin. Res.* **15**, 134.

Pauling L., Itano H.A., Singer S.J. & Wells I.C. (1949) Sickle-cell anemia, a molecular disease. *Science* **110**, 543.

Pearson H.A. & McFarland W. (1962) Erythrokinetics in thalassemia. II. Studies in Lepore trait and hemoglobin H disease. *J. Lab. clin. Med.* **59**, 147.

Perutz M.F. (1963) X-ray analysis of hemoglobin. *Science* **140**, 863.

Perutz M.F. (1965) Structure and function of haemoglobin. I. A tentative atomic model of horse oxyhaemoglobin. *J. molec. Biol.* **13**, 646.

Perutz M.F., Bolton W., Diamond R., Muirhead H. & Watson H.C. (1964) Structure of haemoglobin. *Nature* **203**, 687.

Perutz M.F., Kendrew J.C. & Watson H.C. (1965) Structure and function of haemoglobin. II. Some relations between polypeptide chain configuration and amino acid sequence. *J. molec. Biol.* **13**, 669.

Perutz M.F. & Mazzarella L. (1963) A preliminary X-ray analysis of haemoglobin H. *Nature* **199**, 639.

Perutz M.F. & Mitchison J.M. (1950) State of haemoglobin in sickle-cell anaemia. *Nature* **166**, 677.

Pierce L.E., Rath C.E. & McCoy K. (1963) A new hemoglobin variant with sickling properties. *New Engl. J. Med.* **268**, 862.

Pisciotta A.V., Ebbe S.N. & Hinz J.E. (1959) Clinical and laboratory features of two variants of methemoglobin M disease. *J. Lab. clin. Med.* **54**, 73.

Pootrakul S., Wasi P. & Na-Nakorn S. (1967) Studies on haemoglobin Barts (Hb-γ_4) in Thailand: the incidence and the mechanism of occurrence in cord blood. *Ann. hum. Genet.* **31**, 149.

Pribilla W., Klesse P., Betke K., Lehmann H. & Beale D. (1965) Hämoglobin-Köln-Krankheit: Familiäre hypochrome hämolytische Anämie mit Hämoglobinanomalie. *Klin. Wschr.* **43**, 1049.

Rahbar S., Beale D., Isaacs W.A. & Lehmann H. (1967) Abnormal haemoglobins in Iran. Observation of a new variant: haemoglobin J Iran ($\alpha_2\beta_2^{77\,\text{His}\rightarrow\text{Asp}}$), *Br. med. J.* **i**, 674.

Ranney H.M. & Jacobs A.S. (1964) Simultaneous occurrence of haemoglobins C and Lepore in an Afro-American. *Nature* **204**, 163.

Ranney H.M., Jacobs A.S. & Nagel R.L. (1967) Hemoglobin New York. *Nature* **213**, 876.

Raper A.B. (1963) The occurrence of foetal erythropoiesis after infancy. *Archs Dis. Childh.* **38**, 553.

Raper A.B., Gammack D.B., Huehns E.R. & Shooter E.M. (1960) Four haemoglobins in one individual. *Br. med. J.* **ii**, 1257.

Reissmann K.R., Ruth W.E. & Nomura T. (1961) A human hemoglobin with lowered oxygen affinity and impaired heme-heme interactions. *J. clin. Invest.* **40**, 1826.

REYNOLDS C.A. & HUISMAN T.H.J. (1966) Hemoglobin Russ or $\alpha_2^{51\,Arg}\beta_2$. *Biochim. biophys. Acta* **130**, 541.

RIEDER R.F. & NAUGHTON M.A. (1965) Hemoglobin G$_{Baltimore}$: a new abnormal hemoglobin and an additional individual with four hemoglobins. *Bull. Johns Hopkins Hosp.* **116**, 17.

RIEDER R.F., ZINKHAM W.H. & HOLTZMAN N.A. (1965) Hemoglobin Zürich. Clinical, chemical and kinetic studies. *Am. J. Med.* **39**, 4.

RIGAS D.A. & KOLER R.D. (1961a) Decreased erythrocyte survival in hemoglobin H disease as a result of the abnormal properties of hemoglobin H: the benefit of splenectomy. *Blood* **18**, 1.

RIGAS D.A. & KOLER R.D. (1961b) Erythrocyte enzymes and reduced glutathione (GSH) in hemoglobin H disease: relation to cell age and denaturation of hemoglobin H. *J. Lab. clin. Med.* **58**, 417.

RIGAS D.A., KOLER R.D. & OSGOOD E.E. (1955) New hemoglobin possessing a higher electrophoretic mobility than normal adult hemoglobin. *Science* **121**, 372.

RIGGS A. (1964) The relation between structure and function in hemoglobins. *Can. J. Biochem. Physiol.* **42**, 763.

RIGGS A. & BONAVENTURA J. (1967) Properties of hemoglobin Kansas: $\alpha_2\beta_2^{102\,Thr}$. *Fed. Proc.* **26**, 673.

ROBINSON A.R., ROBSON M., HARRISON A.P. & ZUELZER W.W. (1957) A new technique for differentiation of hemoglobin. *J. Lab. clin. Med.* **50**, 745.

ROSA J., LABIE D., MALEKNIA N. & BLUM N. (1966) Une nouvelle hémoglobine abnormale: l'hémoglobine J$_{\alpha Paris}$ 12 ala $\rightarrow$ asp. *Nouv. Revue fr. Hémat.* **6**, 423.

ROSEMEYER M.A. & HUEHNS E.R. (1967) On the mechanism of the dissociation of haemoglobin. *J. molec. Biol.* **25**, 253.

ROSSI-FANELLI A., ANTONINI E. & CAPUTO A. (1964) Hemoglobin and myoglobin, *in* ANFINSEN C.B., ANSON M.L., EDSALL J.T. & RICHARDS F.M. (eds.) *Advances in Protein Chemistry*, vol. 19, p. 73. Academic Press, New York.

RUCKNAGEL D.L., GLYNN K.P. & SMITH J.R. (1967) Hemoglobin Ypsi, characterized by increased oxygen affinity, abnormal polymerization and erythremia. *Clin. Res.* **15**, 270.

RUCKNAGEL D.L. & NEEL J.V. (1961) The hemoglobinopathies, *in* STEINBERG A.G. & Bearn, A.G. (ed.) *Progress in Medical Genetics*, vol. 1, p. 158. Grune & Stratton, New York.

SACKER L.S., BEALE D., BLACK A.J., HUNTSMAN R.G., LEHMANN H. & LORKIN P.A. (1967) Haemoglobin F Hull (γ 121 glutamic acid $\rightarrow$ lysine), homologous with haemoglobins O Arab and O Indonesia. *Br. med. J.* **3**, 531.

SALOMON H., TATARSKI I., DANCE N., HUEHNS E.R. & SHOOTER E.M. (1964) A new haemoglobin variant found in a Bedouin tribe: haemoglobin 'Rambam'. *Proc. int. Cong. Pac. Asiatic Path. Soc.*, Haifa.

SANSONE G., CARRELL R.W. & LEHMANN H. (1967) Haemoglobin Genova: β28 (B 10) Leucine $\rightarrow$ Proline. *Nature* **214**, 877.

SCHNEIDER R.G., ALPERIN J.B., BEALE D. & LEHMANN H. (1966a) Hemoglobin I in an American Negro family: structural and hematologic studies. *J. Lab. clin. Med.* **68**, 940.

SCHNEIDER R.G. & JONES R.T. (1965) Hemoglobin F$_{Texas}$: a gamma chain variant. *Science* **148**, 240.

Schneider R.G., Jones R.T. & Suzuki K. (1966b) Hemoglobin $F_{Houston}$: a fetal variant. *Blood* 27, 670.

Schneider R.G., Ueda S., Alperin J.B., Brimhall B. & Jones R.T. (1967) Hemoglobin Sealy ($\alpha_2^{47\,His}$), a new variant in a Jewish family. *Program Amer. Soc. hum. Genet. Meeting.* Toronto, p. 42.

Schnek A.G. & Schroeder W.A. (1961) The relation between the minor components of whole normal adult hemoglobin as isolated by chromatography and starch block electrophoresis. *J. Am. chem. Soc.* 83, 1472.

Schokker R.C., Went L.N. & Bok J. (1966) A new genetic variant of β thalassaemia. *Nature* 209, 44.

Schroeder W.A. (1963) The hemoglobins. *Ann. rev. Biochem.* 32, 301.

Schroeder W.A., Cua J.T., Matsuda G. & Fenniger W.D. (1962) Hemoglobin F_1, an acetyl-containing hemoglobin. *Biochim. biophys. Acta* 63, 532.

Schruefer J.J.P., Heller C.J., Battaglia F.C. & Hellegers A.E. (1962) Independence of whole blood and haemoglobin solution oxygen dissociation curves from haemoglobin type. *Nature* 196, 550.

Shen S.C., Fleming E.M. & Castle W.B. (1949) Studies on the destruction of red blood cells. V. Their experimental production *in vitro*. *Blood* 4, 498.

Shepherd M.K., Weatherall D.J. & Conley C.L. (1962) Semi-quantitative estimation of distribution of fetal hemoglobin in red cell populations. *Bull. Johns Hopkins Hosp.* 110, 293.

Shibata S., Iuchi I., Miyaji T. & Takeda I. (1963a) Hemoglobinopathy in Japan. *Bull. Yamaguchi Med. Sch.* 10. 1.

Shibata S., Iuchi I., Miyaji T., Ueda S. & Takeda I. (1963b) Hemolytic disease associated with the production of abnormal hemoglobin and intraerythrocytic Heinz bodies. *Acta haemat. Japan* 26, 164.

Shibata S., Miyaji T., Iuchi I., Ueda S. & Takeda I. (1964) Hemoglobin Hikari ($\alpha_2^A \beta_2^{61\,Asp\,NH_2}$), a fast moving hemoglobin found in two unrelated Japanese families. *Clin. chim. Acta* 10, 101.

Sick K., Beale D., Irvine D., Lehmann H., Goodall P.T. & MacDougall S. (1967) Haemoglobin $G_{Copenhagen}$ and haemoglobin $J_{Cambridge}$; two new β-chain variants of haemoglobin A. *Biochim. biophys. Acta* 140, 231.

Silvestroni E. & Bianco I. (1963) A new variant of human fetal hemoglobin: Hb F_{Roma}. *Blood* 22, 545.

Singer K. & Chernoff A.I. (1952) Studies on abnormal hemoglobins. III. The interrelationship of the S (sickle-cell) hemoglobin and type F (alkali-resistant) hemoglobin in sickle-cell anemia. *Blood* 7, 47.

Singer K., Chernoff A.I. & Singer L. (1951) Studies on abnormal hemoglobins. I. Their demonstration in sickle-cell anemia and other hematologic disorders by means of alkali denaturation. *Blood* 6, 413.

Singer K. & Singer L. (1953) Studies on abnormal hemoglobins. VIII. The gelling phenomenon of sickle-cell hemoglobin: its biologic and diagnostic significance. *Blood* 8, 1008.

Smith E.W. & Krevans J.R. (1959) Clinical manifestations of hemoglobin C disorders. *Bull. Johns Hopkins Hosp.* 104, 17.

Smithies O. (1964) Chromosomal rearrangements and protein structure. *Cold Spring Harbor Symposia on Quantitative Biology* 29, 309.

15

SMITHIES O., CONNELL G.E. & DIXON G.W. (1962) Chromosomal rearrangements and evolution of haptoglobin genes. *Nature* **196**, 232.

STAMATOYANNOPOULOS G., YOSHIDA A., ADAMSON J. & HEINENBERG S. (1968) Hemoglobin Rainier ($\beta^{145\ \text{Tryosine}\rightarrow\text{Histidine}}$): alkali-resistant hemoglobin with increased oxygen affinity. *Science* **159**, 741.

STETSON C.A. (1966) The state of hemoglobin in sickled erythrocytes. *J. exp. Med.* **123**, 341.

STRETTON A.O.W. (1965) The genetic code. *Br. med. Bull.* **21**, 229.

STURGEON P. & FINCH C.A. (1957) Erythrokinetics in Cooley's anemia. *Blood* **12**, 64.

SWENSON R.T., HILL R.L., LEHMANN H. & JIM R.T.S. (1962) A chemical abnormality in hemoglobin G from Chinese individuals. *J. biol. Chem.* **237**, 1517.

THOMPSON G.R. (1962) Significance of haemoglobins S and C in Ghana. *Br. med. J.* **i**, 682.

THOMPSON R.B., WARRINGTON R., ODOM J. & BELL W.N. (1965) Interaction between genes for delta thalassemia and hereditary persistence of fetal hemoglobin. *Acta genet.* **15**, 190.

TONDO C.V., SALZANO F.M. & RUCKNAGEL D.L. (1963) Hemoglobin Porto Alegre, a possible polymer of normal hemoglobin in a Caucasian Brazilian family. *Am. J. hum. Genet.* **15**, 265.

TUCHINDA S., BEALE D. & LEHMANN H. (1965) A new haemoglobin in a Thai family. A cause of haemoglobin Siriraj-β thalassaemia. *Br. med. J.* **1**, 1583.

TUCHINDA S., RUCKNAGEL D.L., MINNICH V., BOONYAPRAKOB U., BALANKURA K. & SUVANTEE V. (1964) The coexistence of the genes for hemoglobin E and α thalassemia in Thais, with resultant suppression of hemoglobin E synthesis. *Am. J. hum. Genet.* **16**, 311.

WADE P.T., JENKINS T. & HUEHNS E.R. (1967) Haemoglobin variant in a Bushman: Haemoglobin Dβ-Bushman, $\alpha_2\beta_2^{16\ \text{Gly}\rightarrow\text{Arg}}$. *Nature* **216**, 688.

WARNER J.R., KNOPF P.M. & RICH A. (1963) A multiple ribosomal structure in protein synthesis. *Proc. natn. Acad. Sci.* **49**, 122.

WASI P., NA-NAKORN S. & SVINGDUMRONG A. (1964) Haemoglobin H disease in Thailand: a genetical study. *Nature* **204**, 907.

WATSON H.C. (1966) Structural interactions of heme with protein, *in* CHANCE B., ESTABROOK R. & YONETANI T. (eds.) *Hemes and Hemoproteins*. Academic Press, New York.

WATSON-WILLIAMS E.J., BEALE D., IRVINE D. & LEHMANN H. (1965) A new haemoglobin, D Ibadan (β-87 threonine $\rightarrow$ lysine), producing no sickle-cell-haemoglobin D disease with haemoglobin S. *Nature* **205**, 1273.

WATSON-WILLIAMS E.J. & WEATHERALL D.J. (1965) The laboratory characterization of human haemoglobin variants, *in* JONXIS J.H.P. (ed.) *Abnormal Haemoglobins in Africa*, p. 99. F.A. Davis, Philadelphia.

WEATHERALL D.J. (1963) Abnormal haemoglobins in the neonatal period and their relationship to thalassaemia. *Br. J. Haemat.* **9**, 265.

WEATHERALL D.J. (1965) *The Thalassaemia Syndromes*. Blackwell, Oxford.

WEATHERALL D.J. (1967) The thalassemias, *in* STEINBERG A.G. & BEARN A.G. (eds.) *Progress in Medical Genetics*, vol. 5, p. 8. Grune & Stratton, New York.

WEATHERALL D.J., CLEGG J.B. & NAUGHTON M.A. (1965) Globin synthesis in thalassaemia: an *in vitro* study. *Nature* **208**, 1061.

WHEELER J.T. & KREVANS J.R. (1961) The homozygous state of persistent fetal hemoglobin and the interaction of persistent fetal hemoglobin with thalassemia. *Bull. Johns Hopkins Hosp.* **110**, 217.

WHITE J.C., ELLIS M., COLEMAN P.N., BEAVEN G.H., GRATZER W.B., SHOOTER E.M. & SKINNER E.R. (1960) An unstable haemoglobin associated with some cases of leukaemia. *Br. J. Haemat.* **6**, 171.

WINSLOW R.M. & INGRAM V.M. (1966) Peptide chain synthesis of human hemoglobins A and A$_2$. *J. biol. Chem.* **241**, 1144.

WINTERHALTER K.H. & HUEHNS E.R. (1964) Preparation, properties and specific recombination of $\alpha\beta$-globin subunits. *J. biol. Chem.* **239**, 3699.

WINTROBE M.M. (1967) *Clinical Hematology*, 6th edn. Lea & Febiger, Philadelphia.

WYMAN J. (1963) Allosteric effects in hemoglobins. *Cold Spring Harbor Symposia on Quantitative Biology.* **28**, 483.

WYMAN J. (1964) Linked functions and reciprocal effects in hemoglobin: a second look, *in* ANFINSEN C.B., ANSON M.L., EDSALL J.T. & RICHARDS F.M. (eds.) *Advances in Protein Chemistry*, vol. 19, p. 223. Academic Press, New York.

ZIPURSKY A. (1965) The erythrocytes of the newborn infant. *Seminars in Hematology.* **2**, 167.

ZIPURSKY A., LaRue T. & ISRAELS L.G. (1960) The *in vitro* metabolism of erythrocytes from newborn infants. *Can. J. Biochem. Physiol.* **38**, 727.

ZIPURSKY A., NEELANDS P.J., POLLOCK J., CHOWN B. & ISRAELS L.G. (1962) The distribution of fetal hemoglobin in the blood of normal children and adults. *Pediatrics* **30**, 262.

ZUCKER W.V. & SCHULMAN H.M. (1967) The synthesis of globin dimers by a reticulocyte cell-free system. *Biochim. biophys. Acta* **138**, 400.

ZUCKERKANDL E. & PAULING L. (1962) Molecular disease, evolution and genic heterogeneity, *in* KASHA M. & PULLMAN B. (eds.) *Horizons in Biochemistry*, p. 189. Academic Press, New York.

ZUELZER W.W., ROBINSON A.R. & BOOKER C.R. (1961) Reciprocal relationship of hemoglobins A$_2$ and F in beta chain thalassemias; key to genetic control of hemoglobin F. *Blood* **17**, 393.

CHAPTER 11

RED CELL ACID PHOSPHATASE

Early Studies	424	Physico-Chemical Properties of Acid Phosphatase Isozymes	432
Red Cell Acid Phosphatase Phenotypes	425	Acid Phosphatase Gene Frequencies	434
Common types	426	Methods	435
Rare types	427	Electrophoresis using a tris-succinic-citrate buffer system	438
Quantitative Studies	428	Electrophoresis using phosphate buffer	438
Relationship of phenotype to enzyme activity	428	Electrophoresis using formate buffer	439
Increased phosphatase activity in megaloblastic anemia	431	Staining	439
Phosphatase activity and G6PD deficiency	431	References	440

EARLY STUDIES

The existence in red cells of phosphomonoesterase activity in a pH range of 4·8 to 6·1 was apparently first noted by Behrendt (1943), who subsequently reported (Behrendt, 1949) that this enzyme was much more active than serum acid phosphatase with phenylphosphate as the substrate. A similar observation was made by King *et al.* (1945) and by Abul-Fadl & King (1949) who concluded that the red cell enzyme differs from that of the prostate, both in substrate affinity and susceptibility to inhibitors; the red cell phosphatase activity being subject to inhibition by formalin, but not by D(+) tartrate. The studies of Tsuboi & Hudson (1953–56) further characterized the enzyme as a phosphotransferase, taking up phosphorus from the substrate and transferring it to acceptor molecules, such as glycerol or methanol. However, the precise functional role of the enzyme in red cells was (and is) unknown. When purified, red cell acid phosphatase was found to be readily inactivated unless protected by sulfhydryl-containing substances such as mercaptoethanol. The systematic name for this

enzyme is orthophosphate monoester phosphohydrolase (3.1.3.2).
As in the case of so many enzymes, this designation also applies to
other enzymes with similar, but far from identical properties.

Valentine *et al.* (1961) measured the acid phosphatase activity of
red cells obtained from healthy subjects and patients with various
diseases. In the assay, disodium *p*-nitrophenylphosphate at pH 5·7
was used as the substrate. They found that enzyme activity was almost
entirely inhibited by cupric salts and by *p*-chloromercuribenzoate
(PCMB), the latter indicating the involvement of sulfhydryl groups in
the enzymatic function. In 26 normal subjects, the activity (expressed
as mg of *p*-nitrophenol released by 10^{10} red cells per hour) ranged from
9–35, with a mean of 22 ± 8. The activity in the six patients tested with
megaloblastic anemia was considerably higher, with a range of 42–55,
and a mean of 47 ± 4. Only part of the increase could be attributed to
the larger size of the cells, and since these patients were in relapse,
their reticulocyte counts were low. In other hematological diseases,
there appeared to be an elevation in enzyme activity related to
the degree of reticulocytosis, but no distinctive patterns were
observed.

Some evidence for heterogeneity of red cell acid phosphatase,
based on column chromatography, was presented by Angeletti &
Gayle (1962). However, the fact that the enzyme exists in several
heritable forms was first reported by Hopkinson *et al.* (1963). Subsequently, a number of studies have been performed to determine the
physicochemical nature of the variants, their mode of inheritance
and their frequency.

RED CELL ACID PHOSPHATASE
PHENOTYPES

In their original report, Hopkinson *et al.* (1963) described five distinct
red cell acid phosphatase patterns from electrophoresis of hemolysates
in starch gels, using as substrate phenolphthalein diphosphate. No
activity was noted when the more common phosphatase substrates,
α- and β-naphthyl phosphate, were used. Furthermore, differences in
the phenotypes were best demonstrated with an unusual buffer
system, containing tris-succinic acid in the gel and citric acid-sodium
hydroxide in the bridge buffer, at pH 6·0.

COMMON TYPES

The common phenotypes are shown in the first five electrophoretic patterns of Fig. 11.1, designated as A, BA, B, CA and CB. Each consists of two or more isozymes in fresh specimens; in stored specimens, one or more additional faster-moving components are observed.

From family studies, Hopkinson *et al.* (1963) demonstrated that the inheritance of the phenotypes can be explained by the existence of three allelic autosomal genes, which they called P^a, P^b and P^c. Thus, the phenotypes of the three heterozygous genotypes, P^b/P^a, P^c/P^a and P^c/P^b, are BA, CA and CB, respectively. The homozygotes, P^a/P^a and P^b/P^b have the phenotypes A and B. The third homozygote, P^c/P^c (phenotype C) was not found during the initial studies because of the relatively low frequency of the P^c gene. It was subsequently described by Lai *et al.* (1964), and is shown as the sixth pattern of Fig. 11.1. Lai & Kwa (1968) have recently proposed a revision of the locus name from P to PHs (for *h*uman *s*pecies). However, there appears to be little precedent for defining a locus on the basis of the animal species, and the need for this revision is not apparent.

Family studies performed in other laboratories have confirmed the mode of inheritance proposed by Hopkinson *et al.* (1963) for these three alleles (Giblett & Scott, 1965; Fuhrmann & Lichte, 1966; Karp & Sutton, 1967; Speiser & Pausch, 1967). Attempts to establish linkage between the acid phosphatase locus and that of the blood group antigens and serum groups have not yet been successful (Conneally *et al.*, 1965).

With the buffer system described by Hopkinson *et al.* (1963), the homozygous types all consist of two components. The isozymes of type A are of about equal activity, and migrate more rapidly than those of the other homozygous types. The isozymes of types B and C have the same electrophoretic mobility, but they differ markedly in their relative proportions. The heterozygous types have the general appearance expected from a mixture of the isozymes of their corresponding homozygous types. When a phosphate buffer is used at a similar pH (6·2), the isozymes representing the P^b and P^c genes have about the same migration rate, relative to each other, as in the citrate-succinate buffer. However, the P^a gene product is considerably slower (Karp & Sutton, 1967). Luffman & Harris (1968) tested the effect of adding a

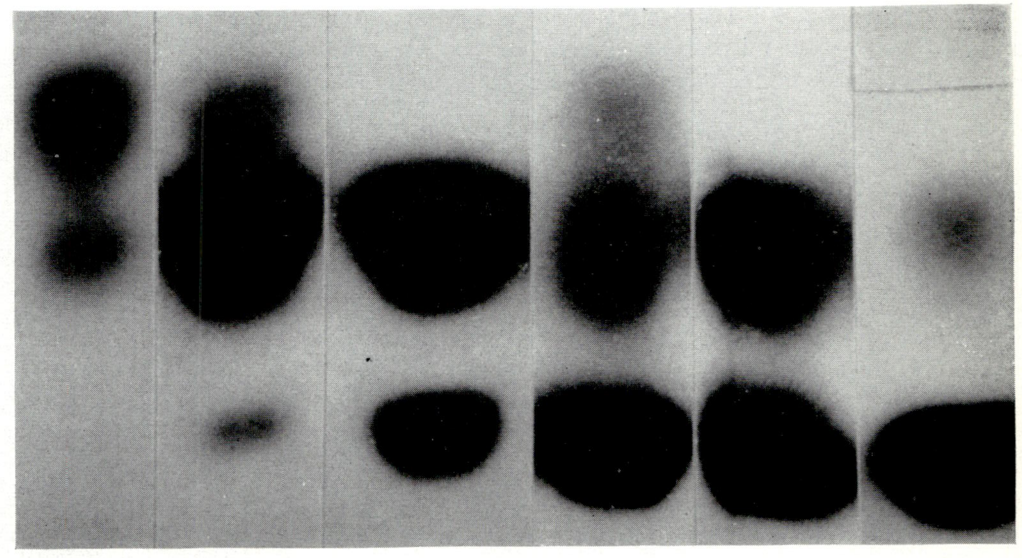

FIGURE 11.1. Starch gel electrophoretic patterns of red cell hemolysates with the six phenotypes representing homozygosity and heterozygosity for the three common genes at the acid phosphatase locus, P^a, P^b and P^c.

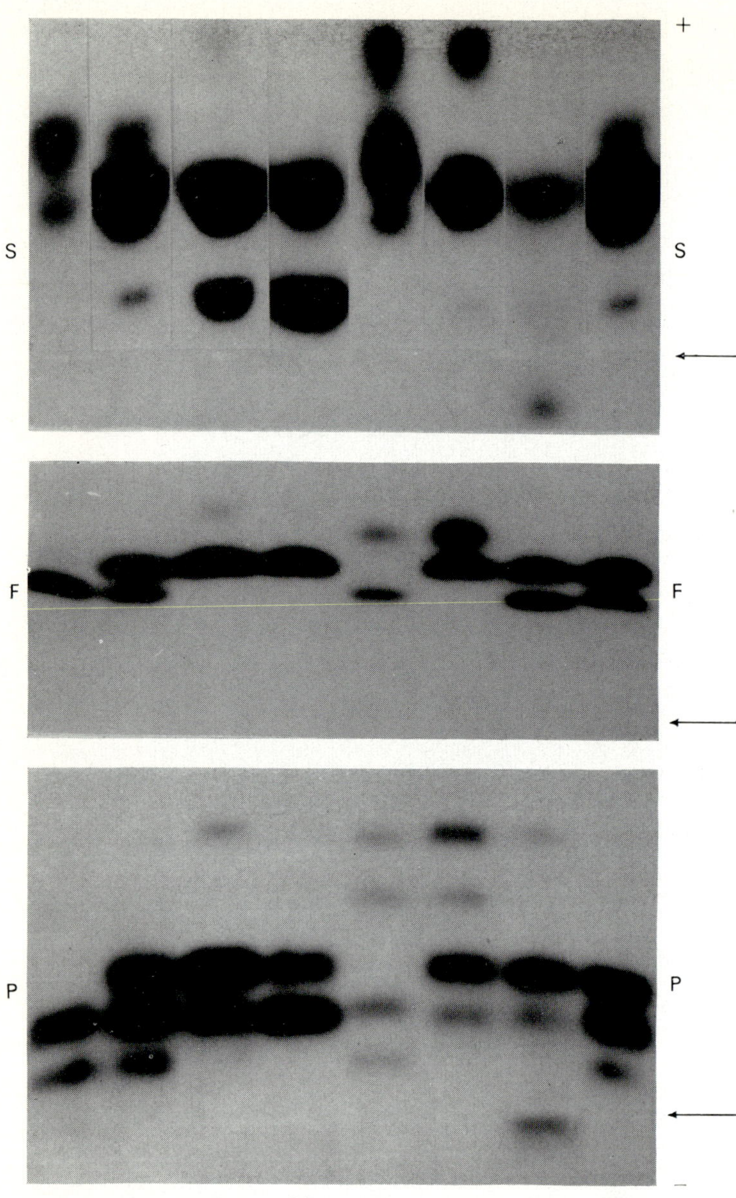

number of different substances to a standard phosphate buffer system, and found that four main electrophoretic patterns were observed:

Pattern I: similar to that obtained without additives to the phosphate system. These substances included monocarboxylic acids and dicarboxylic acids with two- and three-carbon chains.

Pattern II: mobility of the fast A component increased over the fast B component; slow A and B about the same. These substances included EDTA, phthalate, isophthalate and terephthalate.

Pattern III: same as that observed with original citrate-succinate system. Substances behaving similarly to citrate included isocitrate, aconitate and propane-1:2:3 carboxylic acid, all having in common a three-carbon chain with three carboxyl groups. A similar result was obtained with tricarboxylic acid derivatives of the benzene ring.

Pattern IV: mobility of the fast A component similar to that in pattern III, but the slow A component faster than the fast B component. This change was associated with the addition of benzene ring derivatives with more than three carboxyl groups.

RARE TYPES

The existence of two rare genes, P^r and P^d, was subsequently demonstrated by using other buffer systems for electrophoresis. Thus, Giblett & Scott (1965) reported that an unusual phenotype, RA, was readily detectable with a formate buffer system at pH 5·0, although in the succinate-tris-citrate system, RA was not always clearly different from A. A study of the proposita's family indicated that the RA phenotype represented heterozygosity for the gene P^a and a fourth allele P^r. The RB phenotype, like RA, was found in a Negro blood donor by Giblett (1967), and tests of the family showed the genotype to be P^r/P^b. The patterns of these two phenotypes in various buffer systems are shown in Fig. 11.2.

Karp & Sutton (1967) also found the RA and RB phenotypes among Negroes. However, they used a phosphate buffer at pH 6·2, which proved to be superior to the formate buffer described by Giblett & Scott (1965). From family studies, particularly one large pedigree, they confirmed the mode of inheritance of the P^r allele. They also found an additional phenotype, BD, which appeared to represent heterozygosity for the P^b gene with a new, fifth allele, P^d. In this electrophoretic pattern, the major component has a faster migration rate toward the cathode than in any other type (see Fig. 11.2).

Family members were not available for Karp & Sutton (1967) to confirm the mode of inheritance of the BD phenotype. However, Giblett & Scott (unpublished) subsequently observed the pattern in a blood specimen sent to them from Thailand by Dr. Dasnayanee Chandanayingyong for a study on Thai blood genetics which she was conducting with Dr. T.J. Greenwalt (Chandanayingyong *et al.*, 1967). Dr. Chandanayingyong was able to obtain blood specimens from the parents and five sibs of the propositus. Although the phenotypic pattern of the father was not sufficiently intense to show up clearly on a photograph, its appearance was consistent with that expected of the phenotype DA. The mother had the typical B phenotype. Five of the six children had typical BA patterns, so that apparently only the propositus inherited his father's P^d gene. While the evidence provided by this family is not contradictory, further studies will be required to determine with certainty that P^d is an allele in the acid phosphatase system. A further example of the DB phenotype was found in a Seattle Negro donor, but his family was unavailable.

QUANTITATIVE STUDIES

RELATION OF PHENOTYPE TO ENZYME ACTIVITY

Spencer *et al.* (1964) and Hopkinson *et al.* (1964) measured the acid phosphatase activity in hemolysates of blood obtained from several individuals of the five common types. Their findings are given in Fig. 11.3, which shows the distribution of enzyme activity levels among the various types, expressed as μM p-nitrophenol liberated per g hemoglobin per half-hour. Although there is intra-type variability, it is clear that the enzyme activity of the P^b gene product is higher than that of P^a. By calculating the mean level of activity in the different phenotypes, Spencer *et al.* (1964) deduced that the *average* activity attributable to each of the three genes was 61, 94 and 120 units, respectively, for P^a, P^b and P^c, representing an approximate 2:3:4 ratio. A similar ratio was subsequently reported by Shinoda (1967) in studies of blood specimens obtained from random Japanese of various phenotypes.

Modiano *et al.* (1967) determined the red cell acid phosphatase

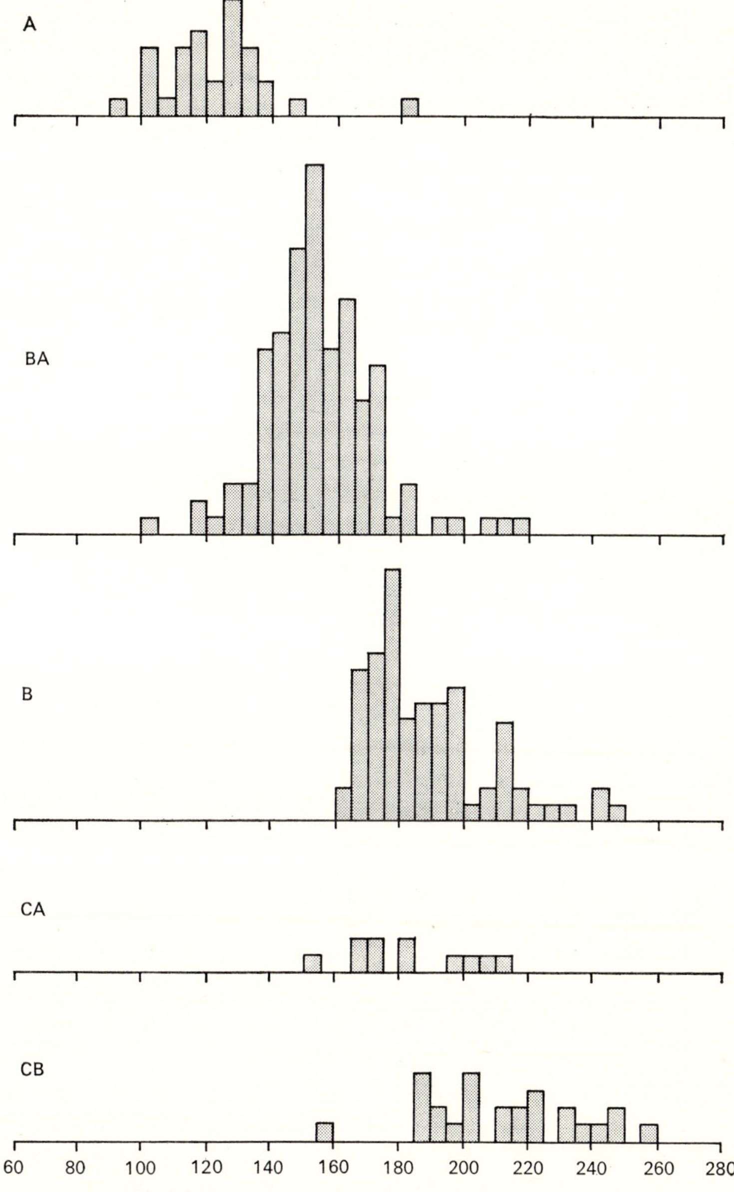

FIGURE 11.3. Distribution of acid phosphatase activities in hemolysates from individuals of five different phenotypes, determined with p-nitrophenyl phosphate as substrate. (From Spencer *et al.*, 1964. Reprinted with permission from *Nature* **201**, 299.)

phenotypes and enzyme activity of 167 healthy inhabitants of lowland Sardinia, an area in which malaria was once highly endemic. They found that the calculated phosphatase activity per gene was virtually the same as that reported by Spencer *et al.* (1964) in the case of the P^a gene, but slightly lower for P^b and P^c. Since the gene frequencies and enzyme activities did not vary significantly from that of inhabitants of

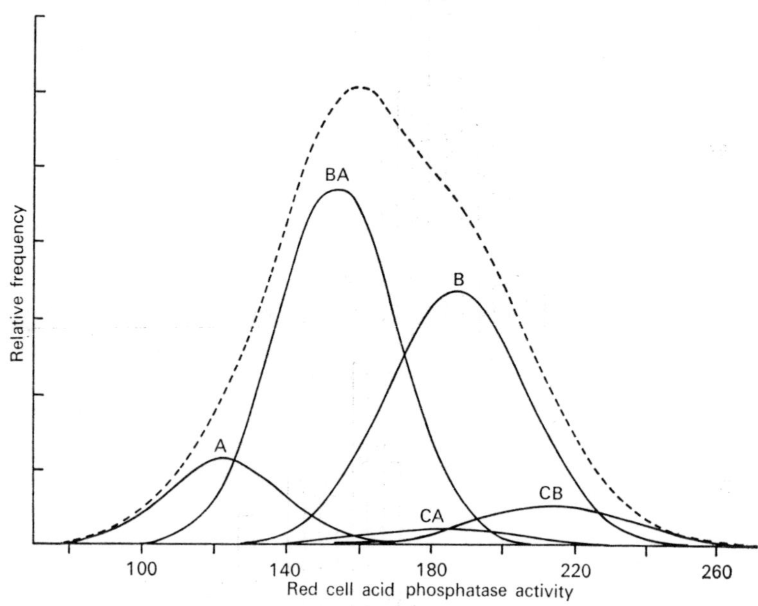

FIGURE 11.4. Unimodal distribution curve of red cell acid phosphatase activity in a random population composed of several distinct phenotypes with overlapping activities. (From Harris, 1966. Reprinted with permission from *Proc. R. Soc.*, B. **164**, 298.)

regions historically free of malaria, they concluded that malaria did not have any measurable selective effect on phosphatase gene frequencies or their quantitative expression.

Harris (1966) showed that if the relative frequencies of the acid phosphatase phenotypes in a random population sample are plotted against enzyme activity, a continuous unimodal distribution curve is obtained, as shown in Fig. 11.4. Thus, a unimodal curve cannot be used as evidence against a polymorphic system, since it may simply represent the summation of a series of different phenotypes with separate but overlapping distribution.

INCREASED PHOSPHATASE ACTIVITY
IN MEGALOBLASTIC ANEMIA

Hopkinson *et al.* (1964) confirmed the finding of Valentine *et al.* (1961) that red cell acid phosphatase activity is elevated in megaloblastic anemia, and were unable to relate this rise to a particular enzyme phenotype. Thus, among 27 patients, there were 2 type A, 12 type BA, 9 type B, 2 type CA and 2 type CB. Furthermore, the differences in activity associated with the individual phosphatase genes in normal subjects were also observed in the patients with megaloblastic anemia.

PHOSPHATASE ACTIVITY AND
G6PD DEFICIENCY

Oski *et al.* (1963) reported that in all of five Caucasian subjects with a deficiency of glucose-6-phosphate dehydrogenase (G6PD), the level of acid phosphatase was more than two standard deviations below the mean level measured in normal males. No such deficiency was observed among Negroes with G6PD deficiency, a condition known to be milder than that observed in Caucasians (see Chapter 12). The authors also found that normal red cells incubated without glucose for 24 hours showed a steady decline in phosphatase activity which did not begin until the glutathione level had dropped to about 10 per cent of its original level. Incubation with acetylphenylhydrazine also produced a decline in activity of both enzymes. Bottini & Modiano (1964, 1966) showed that treatment of normal red cells with oxidized glutathione (GSSG) induced both a fall in phosphatase activity and an alteration in electrophoretic migration rate. They suggested that the low phosphatase levels observed in Caucasian patients with G6PD deficiency may reflect the increased amount of GSSG present in their red cells.

Choremis *et al.* (1964) tested 32 Greek patients with G6PD deficiency, 24 of whom had favism in acute hemolytic crisis. Four of the eight with no history of favism had decreased acid phosphatase activity. More striking, however, was the reduction of phosphatase in 21 of the 24 patients with favism, as well as in their mothers. From this study, as well as that of Oski *et al.* (1963); it appears very likely that there is an association, possibly mediated by GSH depletion, between decreased red cell acid phosphatase activity and at least one variety of G6PD deficiency in the Mediterranean region. Both studies also indicate that the level of acid phosphatase in reticulocytes is probably not appreciably different from that of more mature red cells.

Schettini *et al.* (1965) were unable to find a similar correlation between red cell G6PD and acid phosphatase activity in Sardinians, and the problem was re-examined by Modiano *et al.* (1968). These authors reasoned that the co-existence of thalassemia might distort the results, because higher than normal enzyme levels are obtained with cells having a low MCH, when activity is expressed per gram of hemoglobin. This distortion should disappear when the activity is related to the number of red cells.

They studied 276 Sardinians, of whom 119 had G6PD deficiency, 36 had β-thalassemia trait, 42 had both traits, and 79 had neither. A statistical analysis of the laboratory data indicated that even in the absence of thalassemia, G6PD deficiency was associated with about a 10 per cent *increase* in acid phosphatase activity.

Since the G6PD deficiency of Chinese is apparently at least as severe as that in Italy, it was of interest to examine the association between G6PD and acid phosphatase levels among Chinese. Lu *et al.* (1967) tested 14 newborn Chinese infants with G6PD deficiency and found no difference in their levels of phosphatase activity as compared with 23 normal Chinese infants.

It may be that the heterogeneity in the studies of G6PD and acid phosphatase interaction is due to differences in the G6PD enzyme of various geographic areas. At present, the problem is unresolved.

PHYSICO-CHEMICAL PROPERTIES OF
ACID PHOSPHATASE ISOZYMES

E.M. Scott (1966) made purified enzyme preparations from red cells of acid phosphatase homozygous types A and B, using ammonium sulfate precipitation and DEAE cellulose column chromatography. He found that the relative activity of the two isozyme preparations was virtually the same on a molar basis when tested with a number of substrates. These substrates, in decreasing order of activity at pH

6·05 and 30°C, were phenolphthalein diphosphate, p-nitrophenyl phosphate, riboflavin 5′-phosphate and phenyl phosphate. Much less activity was measured with α-glycerophosphate, ribose-5-phosphate and 3-phosphoglycerate as substrates, while no activity was detected with adenosine 5′-phosphate, β-glycerophosphate, glucose 1-phosphate and glucose 6-phosphate.

Minor differences in the kinetic behavior of A and B enzymes were noted by Scott (1966). For example, the rate of B enzyme activity was slightly less inhibited by inorganic phosphate and less dependent on substrate concentration than that of the A enzyme. It was concluded that these small kinetic differences could not be used to explain the large differences in assays of their activity in red cells. The latter could thus be due to either decreased synthesis or increased lability of the P^a gene product relative to the P^b gene product.

Fenton & Richardson (1967) used DEAE-Sephadex chromatography for separating the acid phosphatase isozymes from red cells of types B, BA, CA and CB. They found that all four types contained three separable fractions with enzyme activity varying with the phenotype. The fractions were named E_s, E_i and E_f for slow, intermediate and fast electrophoretic mobility; the highest E_s activity was found in the CA phenotype, the highest E_i in phenotype B, and the highest E_f in BA and B. The pH optima for these three fractions were reported as 4·75, 5·25 and 5·75, respectively, using p-nitrophenyl phosphate as the substrate. A difference in susceptibility to inhibitors was also noted. For example, oxalate and EDTA strongly inhibited E_s, but not E_i or E_f, while E_i, but not E_s or E_f, resisted the inhibiting effects of sodium fluoride.

Hopkinson & Harris (1967) separated the isozymes of the five common phenotypes by DEAE column chromatography and observed the same heterogeneity in the eluates as that predicted from their electrophoretic patterns. Assuming that the retention of the enzyme on DEAE at pH 8 was due primarily to ion-exchange processes, then the relative isoelectric points of the various components, reflected by their elution order from the column, were slow C zone and slow B zone > slow A zone > fast A zone and fast B zone. These observations indicated that the electrophoretic heterogeneity of the homozygous types is due to net differences in charge between the slow and fast zones.

Luffman & Harris (1967) studied the characteristics of the five common phenotypes in fresh hemolysates. They found that loss of activity after heating was consistently more rapid in types A, BA and B than in types CA and CB. The faster component in each homozygous type became weaker and disappeared first, while the slower component showed an initial increase in intensity, suggesting that the fast component may be converted by heat into a substance similar to, or identical with, the normal slow component. This finding could be taken as evidence for a difference in conformation rather than in primary structure of the two components.

In agreement with Scott (1966), Luffman & Harris (1967) found no differences in the pattern of substrate specificity among the various phenotypes. Also, no differences in phosphotransferase activity were detected when methanol, glycerol, propanol and ethanol were used as acceptors of phosphate from p-nitrophenyl phosphate. From gel-filtration studies, using Biogel P-60 and appropriate markers, it appeared that the molecular weight of all of the acid phosphatase isozymes is surprisingly low, possibly in the 7000 to 10,000 range.

Luffman & Harris (1967) were able to locate the two peaks of acid phosphatase activity previously reported from Sephadex G75 chromatography by Georgatsos (1965). One of these peaks (second in order of elution), corresponds to the activity represented in the red cell acid phosphatase isozymes; it does not require Mg^{++} ions for activity. The earlier peak is Mg^{++} dependent, has a higher pH optimum and a rapid migration rate toward the anode at pH 7·4. Presumably this enzyme is entirely distinct from the phosphatase which takes part in the described polymorphism.

In a later paper, Luffman & Harris (1968) examined the effect on the phenotypic pattern of adding various substances to the gel buffer. From their results, summarized on p. 427, they concluded that the P^a gene product, but not the P^b or P^c gene products, contains an amino acid substitution which is subject to alteration in the presence of certain organic acids (e.g. citric acid) which consist of a three-carbon chain with three carboxyl groups.

ACID PHOSPHATASE
GENE FREQUENCIES

With the exception of a few population isolates, it appears that the P^b gene is generally more common than P^a, and P^c is relatively infrequent. Table 11.1 contains a summary of the available data from a number of different areas. It shows that western Europeans have P^a gene frequencies in the 0·26–0·37 range, while P^c frequencies do not exceed 10 per cent among any of those people tested.

Negroes appear to have a somewhat lower frequency of P^a, and the P^c gene is quite uncommon. The P^r gene, not yet found in other racial groups, has a frequency similar to that of P^c, i.e. about 0·015, in American Negroes. However, in Mozambique, the P^r frequency is higher, and P^c lower.

In most of the specimens from Asian inhabitants, the frequency of P^c is also low, while the P^a gene occurs with a variable frequency ranging from 0·20–0·40. Among Japanese, the gene distribution found in Hokkaido (including a high proportion of Ainus) by Giblett & Scott (1965) was virtually identical to that obtained by Shinoda (1967) in the Tokyo area. However, when a Seattle Oriental population (presumed to be mostly of Japanese origin) was tested by Giblett & Scott (1965 and unpublished), a much lower frequency (0·195) was found for P^a than in the Tokyo and Hokkaido populations (0·392 and 0·399). This difference is probably due to genetic drift, although the resemblance of acid phosphatase frequencies in the Seattle Orientals to those of the Chinese reported by Lai & Kwa (1968) raises the question of a higher than expected Chinese admixture.

Like the Negroes and Orientals, the aboriginal inhabitants of North and South America, as well as of New Guinea, have a low frequency of P^c. The Athabascan Indians of Alaska were found by E.M. Scott et al. (1966) to have the highest known frequency of P^a (0·67), and in the Eskimos and Aleuts, the frequency of P^a was also high: 0·56 and 0·48, respectively. Among the Indians of Florida, Mexico and Brazil, the P^a gene frequency is relatively low (0·19–0·26), while in New Guinea there is considerable heterogeneity. (See also the Addenda.)

METHODS

Starch gel has so far proven to be the best medium for separating the components of the acid phosphatase phenotypes. In this laboratory, three different buffer systems have been used, and they are briefly described below. In routine screening, it is preferable to use the phosphate buffer system (modified from Karp & Sutton, 1967) and to check those specimens which have an apparent CA, CB or C phenotype

TABLE 11.1. Geographic distribution of genes in the acid phosphatase system

Population	Number Tested	Gene frequencies				References
		P^a	P^b	P^c	Other	
European						
England (London)	367	0·36	0·60	0·04		Hopkinson *et al.* (1964)
Australia (Sydney)	260	0·33	0·64	0·03		Lai (196)
Germany (Berlin)	1188	0·37	0·57	0·06		Radam & Strauch (1966)
(Heidelberg)	401	0·33	0·63	0·04		Fuhrmann & Lichte (1966)
(Freiburg)	171	0·27	0·72	0·01		Goedde *et al.* (1966)
Austria (Vienna)	410	0·37	0·57	0·06		Speiser & Pausch (1967)
Norway	65	0·26	0·74	0		Radam & Strauch (1966)
Italy (Rome & Sardinia)	782	0·27	0·65	0·08		Modiano *et al.* (1967)
Tristan da Cunha (Island)	140	0·09	0·91	0		Hopkinson *et al.* (1964)
USA (Seattle)	193	0·39	0·55	0·06		Giblett & Scott (1965)
Negro						
Mozambique	317	0·16	0·81	0·001	$P^r = 0·033$	Giblett *et al.* (unpubl.)
USA (Texas)	294	0·21	0·76	0·015	$P^r = 0·012$ $P^d = 0·003$	Karp & Sutton (1967)
(Seattle)	429	0·25	0·72	0·014	$P^r = 0·015$ $P^d = 0·001$	Giblett & Scott (unpubl.)
(Ann Arbor)	224	0·17	0·83	0·004		Brewer *et al.* (1967)
Brazil (mixed)	369	0·20	0·77	0·03		Lai *et al.* (1965)

Oriental

Japanese (Tokyo)	488	0·39	0·59	0·02	Shinoda (1967)
(Hokkaido)	178	0·40	0·60	0	Giblett & Scott (1965)
Chinese (Singapore)	620	0·22	0·78	0	Lai & Kwa (1968)
Mixed (Seattle)	221	0·20	0·80	0·002	Giblett & Scott (unpubl.)
Malayan (Singapore)	260	0·34	0·66	0·002	Lai & Kwa (1968)
Thai	446	0·29	0·71	0·001 $P^d = 0·001$	Giblett & Scott (unpubl.)
Indian (Singapore)	116	0·25	0·75	0	Lai & Kwa (1968)

Aboriginal

Eskimos	264	0·56	0·44	0	Scott *et al.* (1966)
Aleuts	43	0·48	0·51	0·01	Scott *et al.* (1966)
New Guinea (Kundiawa)	130	0·26	0·74	0	Lai (1966)
(Oksapmin)	119	0·38	0·61	0·01	Lai (1966)
(Maprik)	54	0·29	0·71	0	Lai (1966)
(Trobriand)	484	0·20	0·79	0·01	Lai (1966)

Indians

Athabascan (Alaska)	118	0·67	0·33	0	Scott *et al.* (1966)
Seminole (Florida)	345	0·25	0·75	0·005	Tashian *et al.* (1967)
Hausteco (Mexico)	224	0·22	0·78	0	Lisker & Giblett (1967)
Huichol (Mexico)	71	0·26	0·73	0·01	Lisker & Giblett (1967)
Xavante (Brazil)	376	0·19	0·81	0	Tashian *et al.* (1967)

with the tris-succinic-citrate system of Hopkinson *et al.* (1964). If the latter system is used for screening, some of the rare variants such as RA and RB may not be recognized. The formate buffer of Giblett & Scott (1965) is not routinely used, but it is helpful for confirming the RA and RB variants and for characterizing the electrophoretic behavior of any new variants which may be found.

ELECTROPHORESIS USING A TRIS-SUCCINATE-CITRATE BUFFER SYSTEM

The gel buffer is 0·0026 M succinic acid and 0·0046 M tris pH 6·0. The bridge buffer is 0·41 M citric acid and 1·18 M sodium hydroxide, pH 6·0. Hemolysates are prepared from blood collected into ACD solution and kept for as long as 3 weeks in the refrigerator. Hemolysates should be run within 24 hours of preparation or within 72 hours if frozen after the addition of 2-mercaptoethanol (1 μl per ml). According to Karp & Sutton (1967), longer storage can be achieved by preparing acetone powders by the method of Georgatsos (1965).

For preparing fresh hemolysates, washed packed red cells are lysed with an equal volume of distilled water, extracted with a half-volume of toluene and centrifuged until free of stroma. Thick filter paper (Whatman No. 17) is used for inserting the samples, and horizontal electrophoresis is carried out at 4°C for 16 hours at 6–7 volts/cm. During electrophoresis, the gel becomes constricted at the anodal end, about 8 cm from the origin. (If the constriction is severe, the flow of current is interrupted). At the end of the run, the enzyme is usually located between the origin and the constricted region.

ELECTROPHORESIS USING PHOSPHATE BUFFER

The bridge buffer is 0·1 M, pH 7·0. It contains $Na_2HPO_4.7H_2O$ (17·2 g/l) and $NaH_2PO_4.H_2O$ (4·94 g/l). For the gels, the bridge buffer is diluted 1:20 and adjusted to pH 7·0 with 0·01 M NaH_2PO_4. In the method of Karp & Sutton (1967), the bridge buffer is 0·393 M phosphate pH 6·05, and the gel buffer is a 1:80 dilution, pH 6·2. Both systems give satisfactory results.

Hemolysates are prepared as described above and are placed in the slits of a gel for vertical electrophoresis at 4°C for 16 hours at 5 volts/cm.

ELECTROPHORESIS USING FORMATE BUFFER

The bridge buffer is 0·2 M formic acid and 0·23 M sodium hydroxide, pH 5·0. This buffer is diluted 1:10 for the gels. The hemolysates are inserted and vertical electrophoresis is performed with the electrodes reversed (cathode at the bottom, insertion near the top) at 4°C, beginning at 2·5 volts/cm and a current of 40 ma. The current requires adjustment to 40 ma frequently during the first hour or two of the 18-hour run.

STAINING

(Method of Hopkinson et al., 1964)

The gels are sliced in the usual manner in this laboratory. (Karp & Sutton (1967) recommend a fairly low starch concentration in the gels, which are stained intact without slicing.) The substrate consists of the trisodium salt of phenolphthalein diphosphate (M.W. 574), 143 mg dissolved in 50 ml of 0·05 M citrate buffer, pH 6·0 (citric acid 10·51 g/l, NaOH 5·38 g/l). The pH is readjusted to pH 6·0 with 0·41 M citric acid after the phenolphthalein salt is dissolved. (For older specimens separated with phosphate buffer at pH 7, the stain pH is lowered to 5·0.) This solution is poured over the cut surface of the gel, which is incubated in a covered plastic box at 37°C for 3 hours. After the solution is poured off, about 2 ml of concentrated ammonium hydroxide is placed in the box with the gel and the lid is replaced. The areas of liberated phenolphthalein are revealed as bluish red zones. As shown in Fig. 11.2, the distribution of the zones in relation to each other and to the point of insertion varies considerably, depending on the buffer system used.

Since the color fades and cannot be preserved with any method yet devised in this laboratory, it is necessary to take pictures of the gels before the patterns are distorted by diffusion or lost through fading (i.e. within 20–30 minutes). Use of a green filter enhances the contrast between the red zones and the white background.

REFERENCES

ABUL-FADL M.A.M. & KING E.J. (1949) Properties of the acid phosphatases of erythrocytes and of the human prostate gland. *Biochem. J.* **45**, 51.

ANGELETTI P.U. & GAYLE R. (1962) Chromatography of red cell hemolysates. *Blood* **20**, 51.

BEHRENDT H. (1943) Phosphatase activity of human erythrocytes. *Proc. Soc. exp. Biol. Med.* **54**, 268.

BEHRENDT H. (1949) Phosphatases in blood of man. Values in whole blood, plasma, cytolysates and erythrocyte suspensions. *Am. J. clin. Path.* **19**, 167.

BOTTINI E. & MODIANO G. (1964) Effect of oxidized glutathione on human red cell acid phosphatase. *Biochem. biophys. Res. Commun.* **17**, 260.

BOTTINI E. & MODIANO G. (1966) On the effect of oxidized glutathione and acetyl-phenylhydrazine on red cell acid phosphatase. *Blood* **27**, 281.

BREWER G.J., BOWBEER D.R. & TASHIAN R.E. (1967) The electrophoretic pheno-types of red cell phosphoglucomutase, adenylate kinase, and acid phosphatase in the American Negro. *Acta genet.* **17**, 97.

CHANDANAYINGYONG D., SASAKI T.T. & GREENWALT T.J. (1967) Blood groups of the Thais. *Transfusion* **7**, 269.

CHOREMIS C., KATTAMIS C. & ZANNOS-MARIOLEA L. (1964) Erythrocyte acid phosphomonoesterase in glucose-6-phosphate dehydrogenase deficient Greeks. *Lancet* **i**, 108.

CONNEALLY P.M., NEVO S., CLEGHORN T.E., HARRIS H., HOPKINSON D.A., ROBSON E.B., SPENCER N. & STEINBERG A.G. (1965) Linkage studies with the human red blood cell acid phosphatase locus and blood group, Hp, Gm and Inv loci. *Am. J. hum. Genet.* **17**, 109.

FENTON M.R. & RICHARDSON K.E. (1967) Isolation of three acid phosphatase isozymes from human erythrocytes. *Archs Biochem. Biophys.* **120**, 332.

FUHRMANN W. & LICHTE K.H. (1966) Human red cell acid phosphatase poly-morphism. *Humangenetik* **3**, 121.

GEORGATSOS J.G. (1965) Acid phosphatase of human erythrocytes. *Archs Biochem. Biophys.* **110**, 354.

GIBLETT E.R. (1967) Variant phenotypes: haptoglobin, transferrin and red cell enzymes, *in* GREENWALT T.J. (ed.) *Advances in Immunogenetics*, p. 99. J.B. Lippincott, Philadelphia.

GIBLETT E.R. & SCOTT N.M. (1965) Red cell acid phosphatase: racial distribution and report of a new phenotype. *Am. J. hum. Genet.* **17**, 425.

GOEDDE H.W., RITTER H., CALLSEN U. & FLOCK H. (1966) Untersuchungen zum Polymorphismus der sauren Erythrocytenphosphatasen. *Humangenetik* **3**, 113.

HARRIS H. (1966) Enzyme polymorphisms in man. *Proc. R. Soc. B.* **164**, 298.

HOPKINSON D.A. & HARRIS H. (1967) Column chromatography of human red cell acid phosphatase. *Ann. hum. Genet.* **31**, 29.

HOPKINSON D.A., SPENCER N. & HARRIS H. (1963) Red cell acid phosphatase variants: a new human polymorphism. *Nature* **199**, 969.

HOPKINSON D.A., SPENCER N. & HARRIS H. (1964) Genetical studies on human red cell acid phosphatase. *Am. J. hum. Genet.* **16**, 141.

KARP G.W. & SUTTON H.E. (1967) Some new phenotypes of human red cell acid phosphatase. *Am. J. hum. Genet.* **19**, 54.

KING E.J., WOOD E.J. & DELORY G.E. (1945) Acid phosphatase of the red cells. *Biochem. J.* **39**, XXIV.

LAI L.Y.C. (1966) Hereditary red cell acid phosphatase types in Australian white and New Guinea native populations. *Acta genet.* **16**, 313.

LAI L.Y.C. & KWA S.B. (1968) Red cell acid phosphatase types in some populations of south-east Asia. *Acta genet.* **18**, 45.

LAI L., NEVO S. & STEINBERG A.G. (1964) Acid phosphatases of human red cells: predicted phenotype conforms to a genetic hypothesis. *Science* **145**, 1187.

LU TSUNG-CHO, WEI H. & BLACKWELL R.Q. (1967) Erythrocyte acid phosphomonoesterase activity in newly born Chinese deficient in glucose-6-phosphate dehydrogenase. *Nature* **213**, 707.

LUFFMAN J.E. & HARRIS H. (1967) A comparison of some properties of human red cell acid phosphatase in different phenotypes. *Ann. hum. Genet.* **30**, 387.

LUFFMAN J.E. & HARRIS H. (1968). Personal Communication.

MODIANO G., FILIPPI G., BRUNELLI F., FRATTAROLI W. & SINISCALCO M. (1967) Studies on red cell acid phosphatases in Sardinia and Rome. Absence of correlation with past malarial morbidity. *Acta genet.* **17**, 17.

MODIANO G., FILIPPI G., BRUNELLI F. & SINISCALCO M. (1968) Red cell phosphatase activity in carriers of beta-thalassaemia and/or glucose-6-phosphate dehydrogenase deficiency. *Acta genet.* (in press).

OSKI F.A., SHAHIDI N.T. & DIAMOND L.K. (1963) Erythrocyte acid phosphomonoesterase and glucose-6-phosphate dehydrogenase deficiency in Caucasians. *Science* **139**, 409.

RADAM G. & STRAUCH H. (1966) Populationsgenetik der sauren Erythrocytenphosphatase. *Humangenetik* **2**, 378.

SCHETTINI F., MELONI T., MELA C. & FANCIULLI G. (1965) Red cell acid phosphatase in normal and G6PD-deficient Sardinian subjects. *Acta haemat.* **33**, 230.

SCOTT E.M. (1966) Kinetic comparison of genetically different acid phosphatases of human erythrocytes. *J. biol. Chem.* **241**, 3049.

SCOTT E.M., DUNCAN I.W., EKSTRAND V. & WRIGHT R.C. (1966) Frequency of polymorphic types of red cell enzymes and serum factors in Alaskan Eskimos and Indians. *Am. J. hum. Genet.* **18**, 408.

SHINODA T. (1967) Red cell acid phosphatase types in a Japanese population. *Jap. J. hum. Genet.* **11**, 252.

SPEISER P. & PAUSCH V. (1967) The distribution of the red cell acid phosphatase variants in Vienna. *Vox Sang.* **13**, 12.

SPENCER N., HOPKINSON D.A. & HARRIS H. (1964) Quantitative differences and gene dosage in the human red cell acid phosphatase polymorphism. *Nature* **201**, 299.

TASHIAN R.E., BREWER G.J., LEHMANN H., DAVIES D.A. & RUCKNAGEL D.L. (1967) Further studies on the Xavante Indians. V. Genetic variability in some serum and erythrocyte enzymes, hemoglobin, and the urinary excretion of β-amino-isobutyric acid. *Am. J. hum. Genet.* **19**, 524.

Tsuboi K.K. & Hudson P.B. (1953) Acid phosphatase. I. Human red cell phosphomonoesterase; general properties. *Archs Biochem. Biophys.* **43**, 339.

Tsuboi K.K. & Hudson P.B. (1954) Acid phosphatase. II. Purification of human red cell phosphomonoesterase. *Archs Biochem. Biophys.* **53**, 341.

Tsuboi K.K. & Hudson P.B. (1955) Acid phosphatase. V. The nature of inactivation and stabilization of purified human red cell phosphomonoesterase. *Archs Biochem. Biophys.* **55**, 206.

Tsuboi K.K. & Hudson P.B. (1956) Acid phosphatase. VI. Kinetic properties of purified yeast and red cell phosphomonoesterase. *Archs Biochem. Biophys.* **61**, 197.

Valentine W.N., Tanaka K.R. & Fredricks R.E. (1961) Erythrocyte acid phosphatase in health and disease. *Am. J. clin. Path.* **36**, 328.

CHAPTER 12

GLUCOSE-6-PHOSPHATE
DEHYDROGENASE

Physiology of G6PD 444
    Glycolysis in the red cell 444
    Tissue distribution of G6PD 447

Molecular Structure of G6PD 448

Structural Variants of G6PD 449
    G6PD in Negroes 450
        G6PD deficiency 450
        The B, A and A− enzyme
        types 452
    G6PD deficiency in non-
    Negroes 453
    G6PD variants with altered
    electrophoretic mobility 459
    G6PD deficiency with con-
    genital nonspherocytic hemo-
    lytic disease 459

G6PD Genetics 460
    Lyon hypothesis 460
        Demonstration of G6PD
        cellular mosaicism 460
        G6PD as an indicator of
        the embryonic X inactiv-
        ation event 463
    Linkage studies 463
    Heterogeneity of G6PD gene
    expression 464
    G6PD and natural selection 465

Methods 465
    Brilliant cresyl blue dye
    screening test 466
    Measurement of G6PD en-
    zymatic activity 466
    Starch gel electrophoresis 467

References 469

Glucose-6-phosphate dehydrogenase (G6PD) has the systematic name D-glucose 6-phosphate: NADP oxidoreductase, and the Enzyme Commission number 1.1.1.49. The study of this enzyme in man has a fascinating history, which was reviewed a few years ago by Tarlov *et al.* (1962) and more recently, by Beutler (1965, 1966a) and Motulsky (1965). Interest in the physiology and biochemistry of G6PD and other glycolytic enzymes was greatly stimulated when the association of G6PD deficiency with the hemolytic reactions of 'primaquine sensitivity' (Hockwald *et al.*, 1952) was reported by Carson *et al.* (1956). Much has been learned about the relationship of glucose metabolism and the maintenance of red cell viability, as well as about the role of various drugs in precipitating hemolytic episodes in individuals with decreased enzyme activity. Recent reviews of this complex subject were written by J. Harris (1963), Marks (1964, 1967), Jandl (1965), Desforges (1965), Beutler (1965, 1966a,b) and Zinkham

(1967). In addition, there are two excellent series of papers on red cell metabolic defects: one was published in the November 1966 issue of the *American Journal of Medicine* (vol. 41), and the other was published as a book edited by Beutler (1968).

The fact that both 'primaquine sensitivity' and susceptibility to *Vicea faba* beans (favism) are hereditable disorders was suspected for some time before the biochemical lesion was identified. X-linkage was established by Childs *et al.* (1958), Gross *et al.* (1958), Siniscalco *et al.* (1960) and Larizza (1961). In addition, with the introduction of more effective methods for separating and analysing proteins, it became apparent that G6PD exists in a variety of inherited forms (Boyer *et al.*, 1962; Kirkman & Hendrickson, 1963). Furthermore, it was determined on the basis of population sampling that about 100 million people in certain areas of the world have á deficiency of G6PD activity (Carson & Tarlov, 1962). The recent studies of Yoshida *et al.* (1967a,b) indicate that most cases of enzyme deficiency are probably associated with molecular structural variation. The geographic distribution of the affected individuals has prompted considerable speculation about the possibility of selective advantage of G6PD deficiency in malarious regions (Motulsky, 1960; Allison, 1961; Siniscalco *et al.*, 1961, 1966). Recent papers reviewing various aspects of G6PD genetics were written by Kirkman (1963), Kirkman *et al.* (1964a), Nance (1964), Davidson (1964), Livingstone (1964, 1967), Childs (1965) Motulsky (1965), Beutler (1965, 1966a), Carson & Frischer (1966), Motulsky & Stamatoyannopoulos (1966) and Yoshida *et al.* (1967b). In addition, a report prepared by a WHO scientific group which met in December, 1966, is available (WHO Report, 1967).

PHYSIOLOGY OF G6PD

GLYCOLYSIS IN THE RED CELL

During maturation, the red cell loses its nucleus, and the ribosomes, mitochondria and Golgi apparatus disappear from the cytoplasm. Thus, the ability to synthesize proteins or to obtain energy through the Krebs cycle is lost. By utilizing some of its remaining enzymes, the cell can make simple compounds, such as the tripeptide of glutamic acid, cysteine and glycine, known as glutathione or GSH (Dimant *et al.*, 1955; Szeinberg *et al.*, 1959; Miller & Horiuchi, 1962, 1965). The energy required for this and other functions, such as maintenance of structure and the ion concentration gradient across the cell mem-

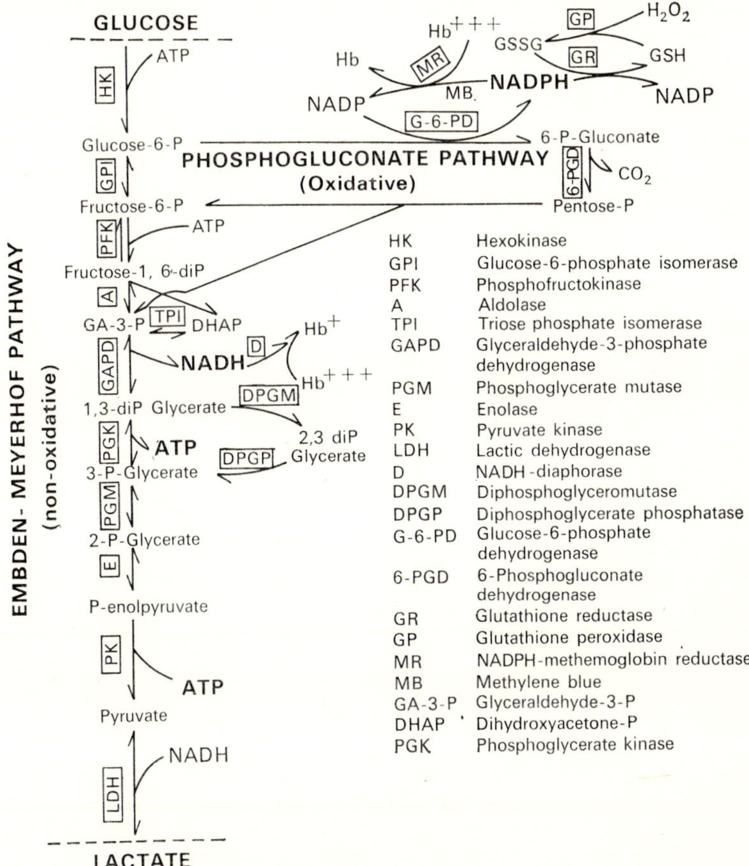

FIGURE 12.1. Glycolysis in the red cell. Enzymes are encircled. Abbreviations are explained on the right. (From Simon *et al.*, 1966.)

brane, is derived from the high energy compound, adenosine triphosphate (ATP), which is generated by the anaerobic Embden-Meyerhof glycolytic pathway (J. Harris, 1963). As shown in Fig. 12.1, energy for the reduction of methemoglobin is also provided by this pathway through the generation of reduced nicotinamide adenine dinucleotide (NADH: formerly DPNH), from which electrons are transferred to methemoglobin (under normal conditions) by the NADH diaphorase, methemoglobin reductase (Scott, 1960; Scott & McGraw, 1962).

Normally, about 90 per cent of red cell glycolysis occurs through the anaerobic pathway, and only 10 per cent through the aerobic pentose phosphate or hexosemonophosphate (HMP) shunt pathway (Murphy, 1960). This pathway begins with the only two reactions available to a cell deprived of its mitochondria for the generation of reduced nicotinamide adenine dinucleotide phosphate (NADPH; formerly TPNH). The first reaction, oxidation of glucose-6-phosphate (G6P) to 6 phosphogluconate (6PG) is catalyzed by G6PD. Since the second oxidation step, catalyzed by 6-phosphogluconate dehydrogenase (6PGD) cannot occur in the absence of the first, a deficiency of G6PD is associated with decreased NADPH production from both sources.

Rose & O'Connell (1964) showed that the level of G6P in red cells plays a role in controlling the uptake of glucose, and thus in determining the activity of both glycolytic pathways. According to Avigad (1966) the concentration of ATP relative to G6P may be involved in controlling the rate of G6P oxidation by the HMP shunt pathway. His experiments, performed in a non-cellular environment, indicated that ATP and other nucleoside triphosphates inhibit G6PD activity at low levels of the substrate, G6P. However, the importance of such a mechanism for controlling the flow of intermediates to synthetic pathways largely unavailable to the mature red cell is questionable.

Another kind of mechanism, which is known to regulate shunt pathway activity in red cells, involves the reversible oxygenation of glutathione. As shown in Fig. 12.1, the enzyme, glutathione reductase (GR) catalyzes the reduction of oxidized glutathione (GSSG) to GSH by transfer of electrons from its co-enzyme, NADPH, which is itself oxidized to NADP. Jacob & Jandl (1966a) reasoned that the state of glutathione oxidation must play an important part in regulating the rate of aerobic glycolysis. In keeping with previous observations by Mills (1957, 1959), Cohen & Hochstein (1961), Hill *et al.* (1964) and Jacob *et al.* (1965), they demonstrated that low levels of peroxide compounds are not inactivated by catalase, and their oxidation of GSH to GSSG is catalyzed by another enzyme, GSH peroxidase

(GSP). Thus, shunt pathway activity is stimulated when peroxide-generating substances oxidize GSH to GSSG, the reversal of which requires oxidation of NADPH to NADP.

A variety of exogenous agents, such as primaquine, acetylphenyl-hydrazine, nitrofurantoin, α naphthol, methylene blue, Nile blue and fava bean extracts, as well as such physiological substances as cysteine, pyruvate and ascorbic acid stimulate the shunt pathway (Desforges *et al.*, 1960; Szeinberg & Marks, 1961). Presumably, some of these substances, such as ascorbic acid, generate hydrogen peroxide by reacting with oxyhemoglobin (Cohen & Hochstein, 1964), and thus exert their stimulating effect by oxidizing GSH. Other substances, such as acetylphenylhydrazine and methylene blue, catalyze the direct transfer of electrons from NADPH to molecular oxygen, so that the GSH peroxidase mechanism is bypassed (Jacob & Jandl, 1966a).

As indicated in Fig. 12.1, the shunt pathway can also bring about the reduction of methemoglobin by an NADPH-diaphorase described by Huennekens *et al.* (1957). However, this mechanism is not normally operative, and can only be demonstrated when the red cells are exposed to a redox catalyst such as methylene blue, in the presence of glucose (Jaffé, 1963).

Red cells deficient in G6PD are limited in their ability to generate NADPH, so they are unable to respond normally to increased demands on the shunt pathway. Thus, on exposure to the stimulating substances previously listed, they cannot maintain their GSH levels (Beutler *et al.*, 1957a,b; Beutler, 1959; Walker & Bowman, 1960). Similarly, the reduction of methemoglobin is greatly curtailed in the presence of redox catalysts and glucose; however, methemoglobin-emia is not a feature of G6PD deficiency, since the NADH-linked methemoglobin reduction system is intact (Carson, 1960; Brewer *et al.* 1962; Jaffé, 1963).

TISSUE DISTRIBUTION OF G6PD

G6PD has been demonstrated in white cells and platelets; in the liver, spleen, adrenal cortex, skin, lens and lactating mammary glands; and in saliva (Glock & McLean, 1954; Marks & Gross, 1959; Ramot *et al.*, 1959a,b; 1960; Zinkham, 1960; Wurzel *et al.*, 1961; Gartler *et al.*, 1962 and Chan *et al.*, 1965). Bonsignore *et al.* (1966a,b) reported that G6PD isolated from white cells differed in stability, electrophoretic mobility and Michaelis constant from the red cell enzyme. However, Yoshida *et al.* (1967b) showed that when white cell preparations were treated by DEAE cellulose column chromatography, the differences were no longer demonstrable.

In nucleated cells, enzymes are synthesized throughout the period of cell viability. However, mature red cells cannot replenish their enzymes, many of which (including G6PD) decline in activity during red cell aging. Thus, the older the red cell, the lower the G6PD activity (Marks *et al.*, 1958). An overall decrease in the levels of red cell G6PD is found in patients with hypothyroidism and hypopituitarism (Kidson & Gajdusek, 1962a; Root *et al.*, 1967), while increased levels occur in hyperthyroidism (Pearson & Druyan, 1961; Baikie, 1965). Although determinations of red cell G6PD level can provide a sensitive index of change in a patient's metabolic status, the mechanism underlying the observed changes is not clearly defined (Root *et al.*, 1967).

Changes in the G6PD level of other tissues have been reported to accompany starvation and refeeding (Glock & McClean, 1955), and the G6PD levels of red cells and probably of other tissues, increases in proportion to the degree of obesity (Lopez & Krehl, 1967). Carson *et al.* (1963) found that in subjects with G6PD deficiency given tracer doses of glucose-1-C^{14}, the amount of $C^{14}O_2$ expired (due to decarboxylation of G-1-C^{14} via the shunt pathway) is decreased. In addition, however, these authors found a decrease in the amount of expired CO_2 derived from G-6-C^{14}, indicating a general depression of glycolysis.

Abnormal responses to oral glucose tolerance tests in Negro subjects with G6PD deficiency were reported by Chanmugan & Frumin (1964), but not in G6PD deficient Iraqi Jewish children (Nitzan & Josefsberg, 1966). Eppes *et al.* (1966) were unable to detect abnormalities of oral glucose tolerance in enzyme-deficient Negroes, but they did find that during cortisone-modified glucose tolerance tests, blood glucose levels were elevated above those of normal subjects. None of these authors found any increase in the frequency of G6PD deficiency among patients with diabetes mellitus.

MOLECULAR STRUCTURE OF G6PD

In common with most tissue enzymes, G6PD is present in very low concentration, so it is extremely difficult to obtain a sufficient amount of purified material for chemical analysis. In addition, the presence of its coenzyme, NADP, is necessary for both molecular stability and enzyme activity (Kirkman & Hendrickson, 1962; Chung & Langdon, 1963b). Inactivation due to NADP removal occurs very readily when the enzyme is incubated with red cell stroma, which contains NADPase activity (Carson *et al.* 1959).

In earlier studies performed on enzyme preparations of relatively low purity, molecular weight measurements were conflicting (Kirkman 1962; Chung & Langdon, 1963a; Balinsky & Bernstein, 1963). More recently, Yoshida (1966, 1967a) reported that highly purified G6PD isolated from normal human red cells has a molecular weight of about 240,000. He found that the molecule contains 110–115 lysine residues and 125–130 arginine residues, and the peptide fingerprint has about 40 spots. He concluded that the molecule is a hexamer, composed of six similar or identical subunits, each with a molecular weight of 40,000 and a content of about 19 lysine and 22 arginine residues. Removal of NADP causes dissociation of the hexamer into a nearly inactive product about one-half the size of the original molecule.

The findings of Yoshida (1966, 1967a) are highly suggestive of a single structural gene locus. However, since Chung & Langdon (1963a) reported two N-terminal amino acids, alanine and tryosine, in their G6PD preparation, it is conceivable that there are two sets of very similar subunits governed by two structural loci. The fact that several zones of enzyme activity can be observed on starch gel electrophoresis under certain conditions would be consistent with more than one locus (Boyer *et al.*, 1962). However, such heterogeneity is more likely to be due to metastable states of the same molecule, particularly when insufficient NADP is present (Kirkman *et al.*, 1964a).

STRUCTURAL VARIANTS OF G6PD

Until recently, there was considerable question about the genetic defect which underlies the common types of G6PD deficiency, since no difference was observed in the electrophoretic rate of the enzyme from cells of affected individuals as compared with cells from subjects with normal enzyme levels. Thus G6PD deficiency was tentatively considered to be in the same category as the thalassemia syndromes in that the gene product appeared to be reduced in quantity rather than altered in structure. However, suspicions to the contrary were aroused by reports that samples of the enzyme from certain rare subjects with G6PD deficiency were found to vary both in electrophoretic mobility and enzymatic properties from normal (Porter *et al.*, 1964; Kirkman *et al.*, 1964a). Subsequent studies of Luzzatto & Allan (1965), Luzzatto & Okoye (1967), Yoshida *et al.* (1967a,b) and Yoshida (1967a,b) have clearly demonstrated that alterations in molecular structure are actually responsible for the aberrant behavior in most, and perhaps all, cases of G6PD deficiency. The precise differences in amino acid sequence have yet to be determined, because liters of blood (or several livers or placentas) from enzyme-deficient subjects are needed to obtain sufficient purified enzyme for fine structure analysis. When available, such data will be of considerable interest, because the G6PD variants, like those of hemoglobin, have different molecular characteristics which are reflected in the syndrome of the affected individual.

It is convenient to discuss the G6PD variants on the basis of their geographic distribution as well as the severity of their effects on the red cell. Population surveys have most commonly employed quantitative screening devices for detecting G6PD deficiency without further characterization of the deficient enzyme. Thus, the degree of G6PD molecular heterogeneity within certain populations, particularly in the Mediterranean region and in Asia, is not known. Nevertheless, some pertinent data are available.

G6PD IN NEGROES

G6PD DEFICIENCY

The military use of the 8-amino-quinolines as anti-malarial drugs led to the recognition of 'primaquine sensitive' hemolytic anemia occurring almost entirely in Negroes of the American forces (Hockwald *et al.*, 1952). In the typical clinical course (Dern *et al.* 1954), the patient's hematocrit began to descend within two or three days of primaquine administration, and sometimes led to severe acute hemolytic episodes with hemoglobinuria. Even when the drug was continued, the patients generally recovered, and this self-limiting course was found to be due to the fact that only the older red cells were susceptible to hemolysis (Beutler *et al.*, 1954). The only remarkable morphological characteristic of the red cells was the early appearance of Heinz bodies in the susceptible cells, and these disappeared as anemia progressed.

In vitro studies of red cells from drug-sensitive patients showed them to be very susceptible to Heinz body formation upon exposure to acetylphenylhydrazine (Beutler *et al.*, 1955a). Similarly, they had limited ability to convert methemoglobin to oxyhemoglobin after stimulation of the shunt pathway with methylene blue (Brewer *et al.*, 1960, 1962). Furthermore, their glutathione levels could be rapidly depleted by acetylphenylhydrazine treatment (Beutler, 1957; Beutler *et al.*, 1957a,b). Laboratory tests based on measurements of glutathione stability and methemoglobin reduction were widely adopted for detecting drug-sensitive individuals. Simpler screening tests are now available (see WHO Report, 1967).

Carson *et al.* (1956) showed that the underlying defect was red cell G6PD deficiency. However, subsequent attempts to define the precise sequence which leads to red cell destruction have produced a large body of literature with many conflicting opinions. Most workers agree that the role of reduced glutathione (GSH) in maintaining the thiol groups of the cell and its enzymes in the reduced state is critical for normal red cell viability (Benesch & Benesch, 1954; Mills, 1960; Rapoport & Scheuch, 1960; Jandl *et al.*, 1960; Allen & Jandl, 1961; Jacob & Jandl, 1962, 1966a; Cohen & Hochstein, 1961; Scheuch *et al.*, 1961; Eldjarn & Bremer, 1962; Kosower *et al.*, 1964; Beutler, 1965; Harley, 1965; Shrago, 1966; Carson & Frischer, 1966; Marks, 1967). In patients with hereditary inability to synthesize the glutathione tripeptide, a compensated hemolytic process is characteristic (Oort *et al.*, 1961; Prins *et al.*, 1966; Waller & Gerok, 1964; Boivin & Galand, 1965).

Under normal circumstances, Negroes with G6PD deficiency have red cell GSH levels somewhat lower than normal (Beutler *et al.*, 1955b) and an increase in oxidized glutathione (Srivastava & Beutler, 1968). A sharp decline in GSH follows administration of primaquine (Flanagan *et al.*, 1958). Similarly, the red cell viability is only slightly decreased prior to administration of the drug (Brewer *et al.*, 1961), and acute hemolysis does not occur until GSH is nearly exhausted (Beutler *et al.*, 1957a,b). The studies of Jacob & Jandl (1962) suggest that this cell destruction is associated with loss of membrane sulfhydryl groups and the formation of hemoglobin denaturation products which appear as Heinz bodies. Formation of mixed disulfides between denatured hemoglobin and the remaining SH groups of the membrane may further damage the cell and render it susceptible to removal by the reticuloendothelial system.

In addition to exogenous substances which stress the shunt pathway, certain endogenous substances with similar effects on the red cell may be produced during disease states, particularly bacterial and viral infections, diabetic acidosis, and hepatitis, both acute and chronic (Burka *et al.*, 1966; Salen *et al.*, 1966; Motulsky & Stamatoyannopoulos, 1966).

The hereditary nature of primaquine sensitivity was suspected for several years, but the studies of Childs *et al.* (1958) as well as Gross *et al.* (1958) established it as an X-linked trait. About 10–14 per cent of American Negro males and about 2 per cent of females were found to have marked glutathione instability, corresponding to G6PD levels about 8–15 per cent of normal. The enzyme levels of normal males and normal females were similar to each other, so it was evident that dosage compensation was operating in the females. The female heterozygotes presented a difficult problem, in that although most of them were found to have intermediate enzyme levels, a few were in the normal range, and a few had marked depletion. A major factor underlying these findings is cellular mosaicism, which is due to random inactivation of the X chromosome in females during early embryonic development (see p. 460 *et seq.*).

THE B, A AND A− ENZYME TYPES

Starch gel electrophoretic studies of red cell and white cell lysates by Boyer *et al.* (1962) and Kirkman & Hendrickson (1963) revealed that non enzyme-deficient American Negro males have one or the other of two G6PD types, A and B (see Fig. 12.2). The B phenotype, about four times more common than A, appears to be the same as that observed in normal individuals of other racial groups. Type A, which has a faster electrophoretic mobility, is largely confined to individuals of African Negro ancestry. As would be expected from X-linked inheritance, the AB phenotype is observed only among females, since males are hemizygous.

Yoshida (1966, 1967a,b) isolated the normal type B enzyme and the Negro A variant from human red cells and found that their tryptic peptide fingerprints were identical, with the exception of a single peptide, called I in the A enzyme and II in the B enzyme. Amino acid

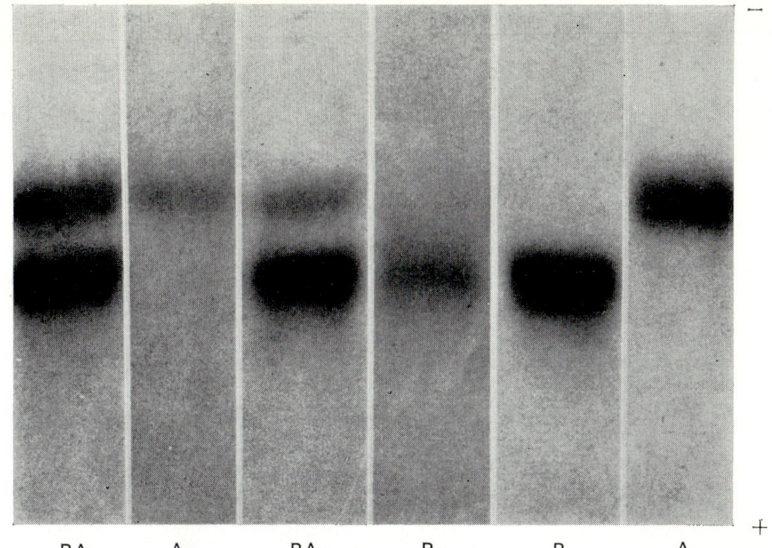

FIGURE 12.2. Starch gel electrophoretic patterns of six different G6PD phenotypes in red cell hemolysates.

analysis of these two peptides revealed a single amino acid substitution, asparagine in the B enzyme being replaced by aspartic acid in the Negro variant; this represents a single step transition in the gene DNA.

Although weaker in staining intensity, the G6PD in lysates of red and white cells from enzyme-deficient Negroes was found to have the same electrophoretic migration rate as the A type enzyme (Boyer *et al.*, 1962; Kirkman & Hendrickson, 1963), so it was given the name A— (see Fig. 12.2). Furthermore, since the studies of Kirkman & Crowell (1963) and of Marks & Tsutsui (1963) indicated no difference in the kinetic or immunological properties of A and A— enzymes, it was concluded that the A— phenotype might represent impaired synthesis or increased catabolism of the A enzyme. However, Luzzatto & Allan (1965) observed differences in the heat stability and eluate patterns from DEAE-Sephadex column chromatography when A and A—were compared. Subsequently, Yoshida *et al.* (1967a) separated the two enzymes by chromatography on a carboxymethyl-Sephadex column and found by quantitative immunological neutralization tests that the A— enzyme had about the same catalytic and serological activity as the A and B enzymes. In addition, the activity of the A— variant in reticulocytes was similar to the A and B activity of young cells obtained from normal individuals, but a marked disparity was found in older cell populations, A— activity being only 10–15 per cent of A or B activity. It was therefore concluded that the underlying defect in this kind of G6PD deficiency is a structural gene mutation, the product of which is normal in quantity but much more susceptible to degradation during cell aging. Thus, type A— G6PD may be analogous to some of the abnormal hemoglobins which, because of amino acid substitution, are less stable than normal hemoglobin molecules.

G6PD DEFICIENCY IN NON-NEGROES

Deficiency of G6PD is prevalent in a number of populations of the Mediterranean area, including Sicily, Sardinia and Italy (Sansone *et al.*, 1958; Siniscalco *et al.*, 1961; Siniscalco, 1963a; Brunetti *et al.*, 1965; Bonsignore *et al.*, 1966a), Greece (Zannos-Mariolea & Kattamis, 1961; Choremis *et al.*, 1962; Allison *et al.*, 1963; Stamatoyannopoulos & Fessas, 1964; Stamatoyannopoulos *et al.*, 1964, 1966b), Yugoslavia (Fraser *et al.*, 1966), Turkey (Say *et al.*, 1965), Cyprus (Plato *et al.*,

16

1964), Iran (Walter & Bowman, 1959), Egypt (Ragab *et al.*, 1966), Lebanon (Taleb *et al.*, 1964), Saudi Arabia (Gelpi, 1965), Kuwait (Shaker *et al.*, 1966) and Israel (Szeinberg *et al.*, 1960; Sheba *et al.*, 1962; Szeinberg, 1963). At least one region of Europe (Hungary) not on the Mediterranean Sea also has the deficiency (Walter *et al.*, 1968).

Another large area in which the frequency of G6PD deficiency is variably elevated includes China (Kirkman *et al.*, 1964a; Wong *et al.*, 1965; McCurdy *et al.*, 1966), Taiwan (Lee *et al.*, 1963; Motulsky *et al.*, 1965), India (Vella, 1959; Baxi *et al.*, 1961), Malay (Vella, 1961; Lie-Injo & Chin, 1964; Lie-Injo & Ti, 1964; Lie-Injo *et al.*, 1966), Thailand (Kruatachue *et al.*, 1962; Flatz & Sringam, 1963, 1964), Indonesia (Lie-Injo & Poey Oey, 1964), the Philippines (Motulsky *et al.*, 1964), Papua, New Guinea and New Britain (Kidson, 1961; Kidson & Gajdusek, 1962a,b; Kidson & Gorman, 1962; Giles *et al.*, 1967).

In contrast to the Negro type of enzyme deficiency, which is largely manifested in the red cells (Marks & Gross, 1959; Marks *et al.*, 1959), the 'non-Negro' deficiency is also expressed in the leukocytes (Ramot *et al.*, 1959a), as well as a number of other tissues including the liver (Brunetti *et al.*, 1960; Chan *et al.*, 1965), intestinal mucosa (Panizon, 1961) and skin (Gartler *et al.*, 1962). Furthermore, the red cell G6PD activity is lower than in the Negro, ranging from about two to seven per cent of normal (Kirkman *et al.*, 1964c). There is less difference in enzyme activity between young and old red cells than there is in the Negro deficiency (Bonsignore *et al.*, 1964), and cells of all ages are destroyed when primaquine is ingested (Salvidio *et al.*, 1967).

Other red cell enzymes, such as pyrophosphatase (Brunetti *et al.*, 1962) and phosphomonoesterase (acid phosphatase) are reportedly decreased in the Mediterranean type of deficiency, and can be further decreased by incubation, probably because of GSH depletion (Oski *et al.*, 1963; see Chapter 11).

A list of oxidant drugs associated with the acute hemolysis of G6PD deficiency, recently compiled by a WHO committee, is reproduced in Table 12.1. Enzyme deficient Negroes are less susceptible to these agents than non-Negroes (Tarlov *et al.*, 1962; Pannacciulli *et al.*, 1965; Carson & Frischer, 1966). This difference in susceptibility is especially pronounced in the case of chloramphenicol, quinine and quinidine (Larizza *et al.*, 1960) and the bean, *Vicea faba*. Thus, favism is common in the Mediterranean area but virtually unknown among

Negroes. Similarly, Negro infants with G6PD deficiency have little or no increased frequency of jaundice (Wolff *et al.*, 1967), but moderate to severe neonatal bilirubinemia unrelated to hemoglobinopathy or blood group incompatibility has been reported in G6PD deficient Chinese (Smith & Vella, 1960; Yue & Strickland, 1965; Lu *et al.*, 1966), Thais (Flatz *et al.*, 1963), Italians (Harley & Robin, 1962), Sardinians (Panizon, 1960), Greeks (Doxiadis *et al.*, 1961) and Israelis (Sheba *et al.*, 1962).

TABLE 12.1. Association of Drugs with Hemolysis in G6PD Deficiency (Adapted from WHO Report No. 366, 1967)

Drugs capable of producing clinically significant hemolysis

Acetanilid	Nitrofurazone
Phenylhydrazine	Nitrofurantoin
Sulfanilamide	Furazolidone
Sulfacetamide	Furaltodone
Sulfapyridine	Quinidine
Sulfamethoxypyridazine	Primaquine
Salicylazosulfapyridine	Pamaquine
Thiazosulfone	Pentaquine
Diaminodiphenylsulfone	Quinocide
Trinitrotoluene	Naphthalene
	Neosalvarsan
	Fava beans

Drugs capable of producing hemolysis when accompanied by other predisposing factors, particularly infection

Aniline	Benemid		Dimercaprol
Para-aminophenol	Sulfadiazine		Quinine
Para-hydroxyacetanilid	Sulfamerazine		Chloroquine
Acetophetidin	Sulfisoxazole		Quinacrine
Para-aminobenzoic acid	Sulfathiazole		Deraprim
Acetylsalicylic acid	Sulfoxone		Menadione Na bisulfate
Pronestyl	Chloramphenicol		
Diphenylhydramine	Nitrite	infants	Menadione Na diphosphate
Tripelennamine	Methylene blue		
Antistine	Ascorbic acid		Vitamin K_1

The existence of additional predisposing factors, apparently genetic in origin, has been suggested by observations of a marked familial tendency, both in favism (Stamatoyannopoulos *et al.*, 1966a)

TABLE 12.2. G6PD Variants (Adapted from WHO Report)

I G6PD Types with normal or nearly normal enzyme activity

Gd Type	Population	Enzyme activity % of normal	Electr. mobility % of B type	Km G6P (μM)	Km NADP (μM)	2dG6P Utiliza-tion	Heat stability	pH Optima	References	Remarks
B	All tested	100	100	50–78	2·9–4·4	<4	Normal	8, truncate		
Madison	Norwegian	100	70 (TEB) 90 (TRIS)						Nance & Uchida (1964)	
A	Negro	88	110	Normal	Normal	<4	Normal	8, truncate	Boyer *et al.* (1962) Kirkman & Hendrickson (1963)	Single amino acid substit. Yoshida 1967)
Baltimore-Austin II	Negro	75	90	68	3·1	<4	Normal		Porter *et al.* (1964) Long *et al.* (1965)	
Ibadan Austin I	Negro	72	80	62–72	3·3	<4	Normal		Porter *et al.* (1964) Long *et al.* (1965)	

II G6PD types with increased enzyme activity

Gd Type	Population	Enzyme activity % of normal	Electr. mobility % of B type	References
Hartford	Negro	200	100	WHO Report (1967)
Hektoen	German or Am. Indian	400	100	Dern (1966)

III G6PD types with 50% or less enzyme activity, but no anemia unless shunt pathway stimulated

Gd Type	Population	Enzyme activity % of normal	Electr. mobility % of B type	Km G6P (μM)	Km NADP (μM)	2dG6P Utiliza-tion	Heat stability	pH Optima	References
Barbieri	Italian	50	135	Incr	Incr		Normal		Marks *et al.* (1962)
Kerala	Asiatic Indian	50	75 (TEB) 90 (TRIS)	23	1·5	7·4	Normal	Biphasic	Azevedo *et al.* (1965)

Tel Hashomer	Sephardic Jew	35	60	30–40		<4		sl. biphasic	Ramot & Brok (1964)	
Columbus	Negro	35	100	Normal	Normal	<4			Pinto *et al.* (1966)	
Athens	Greek	25	98	16–19	2·5–6·5	10–15	Normal	8·5; sl. biphasic	Stamatoyannopoulos *et al.* (1967b)	
Kabyle	Algerian	14–36	104 (pH 8) 101 (pH 7)			<4	Normal	Truncate	Kaplan *et al.* (1967)	
Seattle	Welsh-Scots	8–21	90	15–25	2·4–2·8	7–11	Normal	sl. biphasic	Kirkman *et al.* (1965b)	
Loyola	Irish-German	15	90						Shows *et al.* (1964)	Probably same as Seattle
West Bengal	Asiatic Indian	9	90 (TEB) 94 (TRIS)	31	6·6	<4	Normal	Truncate	Azevedo *et al.* (1965)	
A—	Negro	8–20	110	Normal	Normal	<4	Normal	Truncate	Boyer *et al.* (1962) Kirkman *et al.* (1964a)	
Canton	So. Chinese	4–24	105	20–36	2·0–2·4	4–5	sl. reduced	Biphasic	McCurdy *et al.* (1966)	
B—	Mediterran. Asiatic	0–7	100	19–26	1·2–1·6	23–37	Labile	Biphasic	Ramot *et al.* (1964) Kirkman *et al.* (1964c)	Low specific activity
Markham	New Guinea	0·5–10	105	4·3–6·3		162–222	Labile	Biphasic	Kirkman *et al.* (*in* Beutler, 1968)	Utilizes NAD NADP; and Gal-6-P G6P

IV G6PD Types with low activity, associated with unremitting hemolysis

Oklahoma	W. Europe	4–10	100	127–200	20	<4	Labile	Narrow peak	Kirkman & Riley (1961)	
Chicago	W. Europe	9–26	100	58–76	3·1–3·7	<4	V. Labile	Truncate	Kirkman *et al.* (1964b)	
Eyssen	W. Europe	0	90				Labile		Boyer *et al.* (1962)	
Ohio	Italian	2–16	110	sl. incr.	sl. incr.	<4	Labile		Pinto *et al.* (1966)	
Tübingen	Mixed European	0·3	100	decr.	decr.				Helge & Borner (1966	
Berlin	German	0–1	100						Waller *et al.* (1966)	
Albuquerque	English	1	100	115	11	0	V. labile	8·5	Beutler *et al.* (1968)	
Duarte	W. Europe	8	100	58	5	5·4	V. labile	7·0	Beutler *et al.* (1968)	

and newborn jaundice (Fessas *et al.*, 1962; Doxiadis *et al.*, 1964). Furthermore, in Israel, there appears to be little greater risk of neonatal jaundice in Jewish tribes with a high incidence of enzyme deficiency than among those with a low frequency (Sheba *et al.*, 1962; Szeinberg & Goldschmidt, 1963; Szeinberg *et al.*, 1963).

The electrophoretic mobility of G6PD in most enzyme deficient individuals of the Mediterranean area is the same as the B enzyme of normal Caucasians and Negroes. However, Kirkman *et al.* (1964c) showed that it has greater lability, a bimodal pH optimum and a lower Michaelis constant for G6P and NADP, as well as the unique ability to oxidize 2-deoxyglucose-6-phosphate at a relatively rapid rate. In addition, Yoshida *et al.* (1967b) reported the separation of this so-called B— enzyme from the normal B enzyme by chromatography in a calcium phosphate gel column. They also determined that the specific activity of B—, unlike A—, is only about one-third that of the normal enzyme. Thus, there can be no question that the B and B— types represent structurally different forms. (See Addenda.)

Diversity in the red cell G6PD enzymatic behavior observed in Greek children by Kirkman *et al.* (1965a), as well as the wide range in red cell enzyme levels reported by Stamatoyannopoulos *et al.* (1964) in male subjects from different regions of Greece suggest that variation exists within the category of B— phenotypes, possibly reflecting changes in amino acid sequence not detectable by electrophoresis. Furthermore, since the structural gene of the B— enzyme occurs in diverse geographical regions, its origin by independent mutation on at least two occasions is likely. Evidence favoring that assumption was provided by Adam (1961), who showed that in Israel, the allele associated with the B— type of G6PD rarely occurs on the same X chromosome as the allele for color blindness, while in Sardinia (Siniscalco *et al.*, 1960), the two genes are usually in coupling phase.

Linkage with color blindness has also been used to show that the genes determining A— and B— phenotypes are probably alleles (see p. 463) and, since the normal A and B enzymes differ from each other by a single amino acid (Yoshida, 1967a), it seems almost certain that the structural genes of all four types are alleles at a single locus on the X chromosome. This locus has been designated Gd, so that the phenotypic and genotypic symbols are Gd and *Gd*, respectively. For example, a typical genotype in a Negro female could be Gd^A/Gd^B (WHO Report, 1967).

G6PD VARIANTS WITH ALTERED
ELECTROPHORETIC MOBILITY

Examples of X-linked G6PD variants with electrophoretic migration rates either faster or slower than types A or B have been noted, often in surveys or by chance, in a number of individuals from a variety of ethnic groups, including western Europeans. All but one, the Canton phenotype found in southern Chinese (McCurdy *et al.*, 1966) are thought to be very uncommon, although the extent of certain variants, such as the Athens type (Stamatoyannopoulos *et al.*, 1967b) or the 'Markham' type (Kirkman *et al.*, *in* Beutler, 1968) found in New Guinea may have fairly high frequencies in their respective areas.

The variants and their properties are listed in Table 12.2, adapted from a recent WHO report (1967). In general, there is an associated decrease in enzyme activity, usually quite pronounced. However, the two slow variants found in Negroes, Baltimore-Austin and Ibadan-Austin, have only a small decrease in activity (Porter *et al.*, 1964; Long *et al.*, 1965), and the Madison type described by Nance & Uchida (1964) has no apparent loss of activity. Two variants, 'Hektoen' (Dern, 1966) and 'Hartford' (WHO Report, 1967) reportedly have activity considerably higher than normal. (See also Addenda).

Individuals with the variants so far described are hematologically normal, since their red cells have a normal or only slightly decreased life span (Brewer *et al.*, 1961; Bernini *et al.*, 1964) unless they are exposed to one or more of a variety of substances which stress the HMP pathway. About 100 million of the earth's inhabitants have this kind of G6PD deficiency.

G6PD DEFICIENCY WITH CONGENITAL
NONSPHEROCYTIC HEMOLYTIC DISEASE (CNHD)

A very much smaller group of males, usually of Caucasian ancestry, have G6PD deficiency associated with a marked decrease in red cell viability, chronically elevated reticulocyte levels and often, anemia (Zinkham & Lenhard, 1959). Whenever adequate family data have been available, the enzyme defect has been shown to be X-linked. As with the milder forms of G6PD deficiency, the enzyme isolated from patients with this disease has shown variable physico-chemical characteristics, which are shown in Table 12.2.

Kirkman *et al.* (1964b) pointed out that in comparing the enzyme levels of such patients with those of other individuals, the difference in red cell age must be taken into account. For example, in the Chicago type of enzyme deficiency, the G6PD level is often higher than that found in G6PD deficient Mediterranean subjects with normal cell survival. However, the red cell age distribution in the latter group is very wide, while the former have mainly young cells with somewhat higher enzyme levels. Other factors include differences in thermostability and artifacts arising from the conditions of assay (Beutler, 1965).

While complete absence of G6PD has been reported in occasional cases of CNHD, most enzyme deficient Caucasians, according to Kirkman *et al.* (1964a) are found to have over 3 per cent of normal activity if precautions are taken to exclude heavy metal contaminants, to prepare hemolysates at 0°C and to add small amounts of NADP and 2-mercaptoethanol.

G6PD GENETICS

Because the production of G6PD is X-linked, studies of this enzyme are uniquely useful for testing the Lyon hypothesis as it applies to man, and also for determining the relative position of some of the genes on the X chromosome. Furthermore, quantitation of enzyme activity can supply data concerning the heterogeneity of alleles at the Gd locus, the possible effects of control mechanisms, and the theoretical selective advantages which account for the prevalence of G6PD deficiency in so many of the world's populations.

LYON HYPOTHESIS

DEMONSTRATION OF G6PD CELLULAR MOSAICISM

At the time of Lyon's original suggestion (Lyon, 1961) that phenotypic expression of coat color in female mice is due to cell mosaicism following random inactivation of the X chromosome in early embryonic development, Beutler *et al.* (1962) were formulating a similar hypothesis for human genetics. They found that the red cells of women heterozygous for G6PD deficiency behaved in the same way as a mixture of normal and enzyme-depleted cells when tested for glutathione stability as well as methemoglobin reduction. In a subsequent study, Beutler & Baluda (1964) were able to separate G6PD deficient cells from non-deficient cells in the blood of a female heterozygote who also had sickle trait. They first converted the hemoglobin in her cells to methemoglobin with sodium nitrite and then treated the cells with Nile-blue sulphate. Under this stress to the shunt pathway, the cells with normal enzyme content, but not the deficient cells, reduced their methemoglobin to oxyhemoglobin. Since Hb S does not undergo sickling when oxidized to methemoglobin, the cells with normal G6PD were detectable by their tendency to sickle upon exposure to low oxygen tension. Subsequent millipore filtration trapped the sickled cells, permitting the collection of a high proportion of unsickled methemoglobin-containing cells with very low content of G6PD.

Sansone *et al.* (1963) as well as Tönz & Rossi (1964) utilized a simpler technique to show that when red cells of heterozygous females were treated with nitrite, incubated with Nile-blue sulphate or methylene blue and then subjected to cyanmethemoglobin elution and subsequent staining (method of Kleihauer & Betke, 1963), two red cell populations could be distinguished on the basis of oxyhemoglobin content. Their conclusions were challenged by Papyannopoulou & Stamatoyanno-poulos (1964) who obtained a similar mosaic effect with the cells of mildly deficient Greek males, in whom the younger red cells contained sufficient enzyme to bring about methemoglobin reduction. However, Stamatoyannopoulos *et al.* (1966c, 1967a) subsequently showed that at identical enzyme levels, the methemoglobin reduction tests of cells from male hemizygotes were considerably less abnormal than those of female heterozygotes. Furthermore, in the latter cells, enzyme mosaicism was demonstrable regardless of cell age.

Further evidence to substantiate participation of the G6PD locus in 'Lyonization' of the X chromosome was supplied by Grumbach *et al.* (1962), Davidson *et al.* (1963a) and H. Harris *et al.* (1963). In these three studies, 28 out of 29 subjects whose karyotypes showed either three or four X chromosomes had red cell G6PD activity within the normal range. Similar results were obtained in 20 cases of XO Turner's syndrome, 19 with Kleinfelter's syndrome and four XY subjects with testicular feminization.

Direct proof of cellular mosaicism in heterozygous females was obtained by Davidson *et al.* (1963b). Previous studies by Gartler *et al.* (1962) had shown that in human skin taken from normal and G6PD deficient Caucasian males and maintained in continuous tissue culture for over four months, the enzyme activity remained relatively constant, and accurately reflected their gene-controlled differences. Davidson *et al.* (1963b) reasoned that from the Lyon hypothesis, clones of single cells obtained from skin cultures of Caucasian females heterozygous for G6PD deficiency should contain only one operative Gd locus and thus should have either normal or greatly reduced enzyme activity, but not an intermediate level. In four clones each of tissue cultured from two such women, the enzyme levels were indeed either normal or very low. Furthermore, when extracts of clones from six Negro females of phenotype AB were subjected to starch gel electrophoresis, they contained either type A or type B enzyme, never both types. Similar results were later obtained in cell cultures of certain tumors indicating their origin from a single cell (Gartler & Linder, 1964; Linder & Gartler, 1965a; Gartler *et al.*, 1966).

Another way of using G6PD to demonstrate the validity of the Lyon hypothesis was described by Yoshida *et al.* (1967c). They showed that *in vitro* hybridization of human G6PD of types A and B could be accomplished by dissociation of a mixture of the two molecular species with mild acid hydrolysis followed by reactivation in NADP solution buffered at pH 6·8. The hybrid thus formed is not observed in lysates of red cells taken from Negro women heterozygous for Gd^A and Gd^B. In contrast, individuals heterozygous at *autosomal* loci determining such enzymes as lactate dehydrogenase or 6-phosphogluconate dehydrogenase, regularly produce hybrid enzymes *in vivo*. Thus, it appears very likely that in the female, only one of the two Gd alleles is functioning in any single cell.

Brewer *et al.* (1967) measured the enzyme levels in eight sets of monozygous and nine sets of dizygous Negro female twins, all of whom were heterozygous for G6PD deficiency. They found significantly less within-pair variation in the monozygous twins. Similar results were obtained with 17 additional Negro twin girls who were not G6PD deficient. The fact that monozygotic twins showed the least difference in their enzyme levels was interpreted to mean that there are genetic influences on quantitative G6PD expression which are superimposed on random X inactivation and on the Gd structural alleles.

G6PD AS AN INDICATOR OF THE EMBRYONIC X-INACTIVATION EVENT

In order to obtain information about the stage of embryonic development at which X chromosome inactivation occurs, Linder & Gartler (1965b) biopsied the skin of Negro females with the AB type of G6PD and measured the extent of A and B variegation in cultures of skin of known dimensions. In the absence of selective factors, one would expect that the smaller the number of cells in the embryo at the time of inactivation, the greater would be the area of skin containing descendents of an inactivated cell (i.e. the 'patch size'). From the difference in A:B enzyme activity, measured by starch gel electrophoresis, of several epithelial samples taken from the same subject, they calculated the average patch size in five of six AB heterozygotes to be less than 0·3 cm^2, corresponding to an embryonic stage of about 70,000 cells at the time of inactivation. However, from the considerable variation observed between the mean A:B activity of different skin samples, they concluded that this estimate was much too high, and that the discrepancy might be due to mixing of cells during growth and/or rates of growth of the two cell types in response to selective factors.

Nance (1964) approached the problem in a somewhat different way by comparing the ratios of A:B enzyme activity after starch gel electrophoresis of red cell hemolysates from 30 normal women whose heterozygosity was confirmed by their sons' phenotypes. In this case one would expect that the extent of variation in A:B ratio of the 30 red cell specimens would vary inversely with the size of the embryo at X chromosome inactivation. From calculations based on these data, Nance tentatively concluded that inactivation occurs before the 32-cell stage of embryonic development, when erythroblast differentiation has not yet occurred.

LINKAGE STUDIES

Studies of families belonging to several different racial groups have established that there is close linkage on the X chromosome between the loci for color-blindness and Gd^{B-}. In 16 of 19 Jewish families studied in Israel by Adam *et al.* (1963) the genes were in repulsion. However, in most of the 67 Sardinian families reported by Siniscalco *et al.* (1964) they were in coupling phase, with a recombination frequency of less than 0·04 for deutan colorblindness (57 families) and a similar figure for protan colorblindness (10 families). Porter *et al.* (1962)

found that in seven American Negro families, Gd^{A-} was coupled with the deutan defect in two families and with the protan defect in two others; but in three families with the deutan defect the genes were in repulsion. Deviations from genetic equilibrium in the two Mediterranean populations are probably best explained by drift rather than by selection or by a short duration of the mutant in the gene pool.

Boyer & Graham (1965) studied three Negro kindreds in which the genes for G6PD deficiency and hemophilia A were segregating. Out of 17 opportunities for recombination between the respective genes, none was observed. Their median estimate from the 95 and 99 per cent confidence limits yielded a recombination fraction of only 0·038. This study and a similar one by Siniscalco in southern Europeans (cited by Motulsky, 1968) show that the Gd locus and the hemophilia A locus are very closely linked on the X chromosome.

Studies by Adam *et al.* (1963) of Israeli families and by Fraser *et al.* (1964) of Greek families initially suggested that the Xg^a gene is also within mapping distance of the Gd locus. However, data from a large number of Sardinian families studied by Siniscalco *et al.* (1966b) indicated a fairly high recombination frequency. Further studies by Adam *et al.* (1966) disclosed that their earlier results had, by chance, provided evidence for linkage but, with accumulated data, the apparent relationship disappeared.

HETEROGENEITY OF Gd GENE EXPRESSION

If a single genetic mechanism were responsible for the deficiency of G6PD activity encountered within a given racial group, one would expect that there would be approximate uniformity of G6PD levels among the affected individuals living in a similar environment. Davidson *et al.* (1964) measured the red cell enzyme activity associated with various genotypes in Negro and Sardinian populations as well as in individual families. As anticipated, the Sardinians had lower enzyme levels than the Negroes, but there was considerable variability within the two groups. On the other hand, the enzyme levels of sib pairs within families showed a highly significant correlation which did not appear to be due to environmental or technical influences. They concluded that the quantitative heterogeneity must be genetically controlled, possibly by a system of genetic modifiers but, more likely, by a series of multiple alleles.

G6PD AND NATURAL SELECTION

During the past few years, considerable evidence has accumulated to support the contention of Motulsky (1960) that falciparum malaria confers a selective advantage on G6PD deficiency as well as sickle hemoglobin trait. This subject has been extensively reviewed by Livingstone (1964, 1967), Motulsky (1964, 1965) and Siniscalco *et al.* (1966a). The proposed selective mechanism is based on the requirement of malarial parasites for the oxidative shunt pathway in the parasitized red cell. Thus, individuals with G6PD deficiency presumably have a lower malaria mortality because of decreased parasite proliferation.

Studies of various populations throughout the world have shown, in general, a convincing correlation among the frequencies of sickle trait, G6PD deficiency and malaria. Some exceptions have been noted, particularly by Kidson & Gorman (1962), and these have been used as an argument against the basic hypothesis. However, Allison (1963) pointed out that the only convincing discrepancy would be the finding of frequent G6PD deficiency within populations known to have lived in non-malarious regions for many generations. According to Motulsky (1965), a remarkable feature of the enzyme defect is its high frequency in so many malarious areas, rather than its occasional absence from populations infested with malaria.

METHODS

A recent WHO Report (1967) includes directions for the performance of several standardized tests for detecting G6PD deficiency and for measuring the enzyme activity, electrophoretic migration rate and other physico-chemical properties. This report is inexpensive (1 dollar or 5 shillings) and readily available throughout the world. In the United States, it is sold by Columbia University Press at 136 South Broadway, Irvington-on-Hudson, New York, 10533. In Canada, the source is The Queen's Printer in Ottawa. In the United Kingdom, copies may be purchased at H.M. Stationery Office in London, Edinburgh, Cardiff, Belfast, Manchester, Birmingham and Bristol.

Only three laboratory procedures are briefly outlined here, since they are used in the author's laboratory. For full details and further tests, consult the WHO Report (1967).

BRILLIANT CRESYL BLUE DYE SCREENING TEST
(Motulsky & Campbell-Kraut, 1961)

Brilliant cresyl blue (BCB) both stimulates the activity of the shunt pathway and acts as an indicator of NADPH generation, changing color from blue to colorless. Only certain lots of dye are useful for this purpose. The National Aniline product gives satisfactory results; others may not. The time required for decolorization is a measure of the amount of G6PD activity.

Reagents include the sodium salt of glucose-6-phosphate (825 mg/100 ml), NADP (50 mg/100 ml), BCB (32 mg/100 ml) and tris buffer pH 8·5 (8·96 gm tris in 97 ml water and 3 ml conc. HCl).

The test is performed by adding 0·02 ml of blood (capillary blood or blood stored at 4°C in ACD solution) to 1·0 ml distilled water. To this lysate is added 0·65 ml of a freshly prepared mixture of stock solutions in the proportions of 2 parts G6PD, 2 parts NADP, 5 parts BCB and 4 parts tris buffer. The tube is mixed by rolling between the palms, and a few drops of mineral oil are layered on the top to prevent entry of air.

During incubation in a 37°C water bath, the tubes are observed at 5–10-minute intervals after an initial period of 40 minutes. Decolorization from blue to the pink color of hemoglobin is normally complete in 65 minutes. About 100–150 minutes are required in the Negro type of enzyme deficiency, and more than 3 hours in the Mediterranean type.

Several other simple screening tests have been described, such as the modification of the methemoglobin-reduction test of Brewer *et al.* (1962) by Bowman *et al.* (1965) and Knutsen & Brewer (1966), the spot tests of Fairbanks & Beutler (1962) and Beutler (1966c) and the ascorbate test of Jacob & Jandl (1966b).

MEASUREMENT OF G6PD ENZYMATIC ACTIVITY
(WHO Report, 1967)

Blood collected in ACD solution and stored up to 3 or 4 days at 4°C is satisfactory for assay. It is centrifuged and washed twice in saline to remove plasma and buffy coat. Thirty volumes of a hemolyzing solution containing 10 μM NADP, 7 mM 2-mercaptoethanol and 2·7 mM sodium EDTA, pH 7·0, are mixed with one volume of packed cells for

5–10 minutes. The tube is then centrifuged in the cold at 2000 g for 30 minutes, and the hemoglobin content of the solution is measured.

To 0·050 ml of hemolysate in each of two cuvettes with a 1 cm light path, is added 0·85 ml of a mixture (stored at −20°C) containing NADP (0·2 mM), tris-HCl, pH 8 (0·1 M), and $MgCl_2$ (0·01 M). The reaction is started at 25°C by the addition of 0·10 ml of 6 mM G6P, while water is added to the blank. The optical density at 340 mμ is recorded at 2–3 minute intervals for about 20 minutes. The rate of change becomes linear after a few minutes, and this slope is used for the calculation.

G6PD Activity (IU) per g Hb =

$$\frac{(\Delta OD \text{ per minute}) \, 10^5}{6 \cdot 22 \text{ (Hb conc. in g/100 ml) } (\mu l \text{ enzyme/ml reaction mixt.})}$$

One I.U. is that amount of G6PD which reduces one μM of NADP (absorbance of 6·22) per minute. Its expression per gram of hemoglobin is acceptable in normal subjects but in patients with a high proportion of young cells or with a low MCH, expressing the enzyme activity per cell is more meaningful for comparative purposes. Although some 6PGD activity is also measured in this assay, conditions are such that the interference is minimal. The normal level is about 13 I.U. per gram Hb, with a standard deviation of about 3, although these figures should be determined independently in any laboratory using this procedure.

STARCH GEL ELECTROPHORESIS

Several electrophoretic media and buffer systems can be used for separating the G6PD variants. For starch gels, the WHO Committee (1967) recommended either a tris/HCl or a tris, EDTA, borate system. In our experience, superior results are obtained with a phosphate buffer and a method similar to that described for 6PGD by Fildes & Parr (1963). The gel is made from 0·01 M phosphate buffer, pH 7·0, and 2·5 mg NADP is added per 500 ml of gel just before degassing. The bridge buffer is 0·1 M phosphate buffer, pH 7·0, which also contains 3·5 mg NADP in the cathode vessel. The hemolysate is prepared by mixing one volume of red cells with one volume of water and one-half volume of toluene. Thorough removal of stroma is accomplished by centrifuging and removing the clear red hemolysate three times.

Vertical or horizontal electrophoresis is performed in a cold room or with cooling plates at 4–5 volts/cm for 18 hours. The gel is sliced in the usual manner, and strips of excess gel are used to form a boundary 2–3 mm high about an area inclusive of the 10 cm anodal to the site of insertion. Small amounts of agar effectively seal the corners of the 'box' so constructed.

The staining of G6PD is based on reduction of the tetrazolium salt,

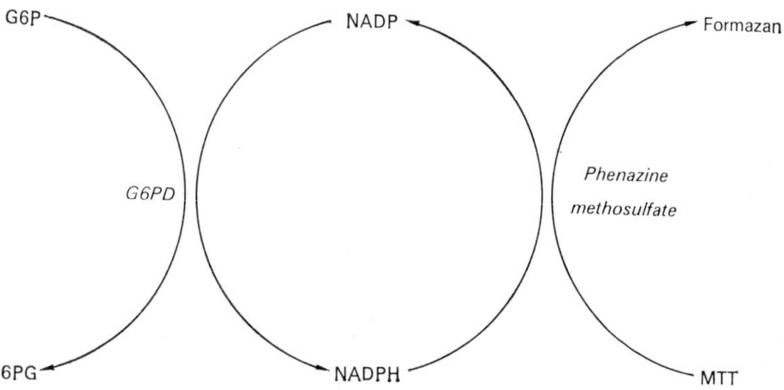

FIGURE 12.3. Principle of staining method employed for detection of G 6 PD activity, with the reduction of the tetrazolium salt, MTT, to a dark blue insoluble formazan.

MTT, to an insoluble dark blue formazan dye, mediated by phenazine methosulfate when NADPH is generated from NADP at the site of G6PD activity. This general principle, shown in Fig. 12.3, is also employed for the detection of several other enzymes described in the following chapters.

The stain consists of 12 mg disodium G6P, 2·4 mg MTT, 2·4 mg $MgCl_2.6H_2O$ and 0·5 mg phenazine methosulfate dissolved in 6 ml 0·2 M tris-HCl buffer, pH 8·0 (1·6 g tris in 0·55 ml conc. HCl diluted to 100 ml). This solution is added to 6 ml of 2 per cent melted aqueous agar at about 60°C; the final pH is about 8·0. The agar-substrate mixture is poured without delay into the boxed area of the starch gel and incubation is carried out in the dark at 37°C for 30–45 minutes. The zones of G6PD activity appear as shown in Fig. 12.2, p. 452.

REFERENCES

ADAM A. (1961) Linkage between deficiency of G6PD and colour blindness. *Nature* **189**, 686.

ADAM A., SHEBA C., SANGER R. & RACE R.R. (1966) The linkage relation of G6PD to Xg. *Am. J. hum. Genet.* **18**, 110.

ADAM A., SHEBA C., SANGER R., RACE R.R., TIPPETT P., HAMPER J., GAVIN J. & FINNEY D.J. (1963) Data for X-mapping calculations. Israeli families tested for Xg, G6PD and for colour vision. *Ann. hum. Genet.* **26**, 187.

ALLEN D.W. & JANDL J.H. (1961) Oxidative hemolysis and precipitation of hemoglobin. II. Role of thiols in oxidant drug action. *J. clin. Invest.* **40**, 454.

ALLISON A.C. (1961) Genetic factors in resistance to malaria. *Ann. N.Y. Acad. Sci.* **91**, 710.

ALLISON A.C. (1963) Malaria and G6PD deficiency. *Nature* **197**, 609.

ALLISON A.C., ASKONAS B.A., BARNICOT N.A., BLUMBERG B.S. & KRIMBAS C. (1963) Deficiency of G6PD in Greek populations. *Ann. hum. Genet.* **26**, 237.

AVIGAD G. (1966) Inhibition of glucose-6-phosphate dehydrogenase by adenosine 5′-triphosphate. *Proc. natn. Acad. Sci.* **56**, 1543.

AZEVEDO E., MORROW A.C., KIRKMAN H.N. & MOTULSKY A.G. (1965) Glucose-6-phosphate dehydrogenase mutations in Asiatic Indians. *Meeting Am. Soc. hum. Genet.* Seattle.

BAIKIE A.G. (1965) Glucose-6-phosphate dehydrogenase activity and an osmotic abnormality of erythrocytes in thyrotoxicosis. *Lancet* i, 86.

BALINSKY D. & BERNSTEIN R.E. (1963) The purification and properties of glucose-6-phosphate dehydrogenase from human erythrocytes. *Biochim. biophys. Acta.* **67**, 313.

BAXI A.J., BALAKRISHNAN V. & SANGHVI L.D. (1961) Deficiency of glucose-6-phosphate dehydrogenase: observations on a sample from Bombay. *Current Sci.* **30**, 16.

BENESCH R.E. & BENESCH R. (1954) Relation between erythrocyte integrity and sulfhydryl groups. *Archs Biochem.* **48**, 38.

BERNINI L., LATTE B., SINISCALCO M., PIOMELLI S., SPADA U., ADINOLFI M. & MOLLISON P.L. (1964) Survival of Cr51 labelled red cells in subjects with thalassaemia trait or G6PD deficiency or both abnormalities. *Br. J. Haemat.* **10**, 171.

BEUTLER E. (1967) The glutathione instability of drug-sensitive red cells. A new method for the *in vitro* detection of drug sensitivity. *J. Lab. clin. Med.* **49**, 84.

BEUTLER E. (1959) The hemolytic effect of primaquine and related compounds: a review. *Blood* **14**, 103.

BEUTLER E. (1965) G6PD deficiency and nonspherocytic congenital hemolytic anemia. *Seminars in Haematology* **2**, 91.

BEUTLER E. (1966a) Glucose 6-phosphate dehydrogenase deficiency, *in* STANBURY J.B., WYNGAARDEN J.B. & FREDRICKSON D.S. (eds.) *The Metabolic Basis of Inherited Disease*, 2nd edn., p. 1060. McGraw-Hill, New York.

BEUTLER E. (1966b) Abnormalities of glycolysis (HMP shunt). *Plenary Session Prog. 11th Cong. int. Soc. Blood Transf*, p. 122. Sydney.

BEUTLER E. (1966c) A series of new screening procedures for pyruvate kinase deficiency, glucose-6-phosphate dehydrogenase deficiency and glutathione reductase deficiency. *Blood* 4, 553.

BEUTLER E. (ed.) (1968) *Hereditary Disorders of Erythrocyte Metabolism*. Grune & Stratton, New York.

BEUTLER E. & BALUDA M.C. (1964) The separation of G6PD deficient erythrocytes from the blood of heterozygotes for G6PD deficiency. *Lancet* i, 189.

BEUTLER E., DERN R.J. & ALVING A.S. (1954) The hemolytic effect of primaquine. IV. The relationship of cell age to hemolysis. *J. Lab. clin. Med.* 44, 439.

BEUTLER E., DERN R.J. & ALVING A.S. (1955a) The hemolytic effect of primaquine. VI. An *in vitro* test for sensitivity of erythrocytes to primaquine. *J. Lab. clin. Med.* 45, 40.

BEUTLER E., DERN R.H. & ALVING A.S. (1957a) The glutathione instability of drug sensitive red cells. A new method for the *in vitro* detection of drug sensitivity. *J. Lab. clin. Med.* 49, 84.

BEUTLER E., DERN R.J., FLANAGAN C.L. & ALVING A.S. (1955b) The hemolytic effect of primaquine. VII. Biochemical studies of drug sensitive erythrocytes. *J. Lab. clin. Med.* 45, 286.

BEUTLER E., MATHAI C.K. & SMITH J.E. (1968) Biochemical variants of glucose-6-phosphate dehydrogenase giving rise to congenital nonspherocytic hemolytic disease. *Blood* 31, 131.

BEUTLER E., ROBSON M.J. & BUTTENWIESER E. (1957b) The mechanism of glutathione destruction and protection in drug-sensitive and non-sensitive erythrocytes: *in vitro* studies. *J. clin. Invest.* 36, 617.

BEUTLER E., YEH M. & FAIRBANKS V.F. (1962) The normal human female as a mosaic of X-chromosome activity: studies using the gene for G6PD deficiency as a marker. *Proc. natn. Acad. Sci.* 48, 9.

BOIVIN P. & GALAND C. (1965) La synthèse du glutathion au cours de l'anémie hémolytique congénitale avec déficit en glutathion réduit. *Nouv. Revue fr. Hémat.* 5, 707.

BONSIGNORE A., FORNAINI G., FANTONI A., LEONCINI G. & SEGNI P. (1964) Relationship between age and enzymatic activities in human erythrocytes from normal and fava bean sensitive subjects. *J. clin. Invest.* 43, 834.

BONSIGNORE A., FORNAINI G., LEONCINI G. & FANTONI A. (1966a) Electrophoretic heterogeneity of erythrocyte and leucocyte glucose-6-phosphate dehydrogenase in Italians from various ethnic groups. *Nature* 211, 876.

BONSIGNORE A., FORNAINI G., LEONCINI G., FANTONI A. & SEGNI P. (1966b) Characterization of leucocyte glucose-6-phosphate dehydrogenase in Sardinian mutants. *J. clin. Invest.* 45, 1865.

BOWMAN J.E., CARSON P.E., FRISCHER H., KAHN M. & AJMAR F.A. (1965) A capillary tube, Nile blue methemoglobin reduction for glucose-6-phosphate dehydrogenase deficiency. *Proc. 10th Cong. int. Soc. Blood Transf.*, p. 592. Stockholm.

BOYER S.H. & GRAHAM J.B. (1965) Linkage between the X chromosome loci for glucose-6-phosphate dehydrogenase electrophoretic variation and hemophilia A. *Am. J. hum. Genet.* **17**, 320.

BOYER S.H., PORTER I.H. & WEILBACHER R.G. (1962) Electrophoretic heterogeneity of G6PD and its relationship to enzyme deficiency in man. *Proc. natn. Acad. Sci.* **48**, 1868.

BREWER G.J., GALL J.C., HONEYMAN M., GERSHOWITZ H., SHREFFLER D.C., DERN R.J. & HAMES C. (1967) Inheritance of quantitative expression of erythrocyte glucose-6-phosphate dehydrogenase activity in the Negro—a twin study. *Biochem. Genet.* **1**, 41.

BREWER G.J., TARLOV A.R. & ALVING A.S. (1960) Methemoglobin reduction test. A new simple, *in vitro* test for identifying primaquine-sensitivity. *Bull. Wld. Hlth. Org.* **22**, 633.

BREWER G.J., TARLOV A.R. & ALVING A.S. (1962) The methemoglobin reduction test for primaquine-type sensitivity of erythrocytes. *J. Am. med. Ass.* **180**, 386.

BREWER G.J., TARLOV A.R. & KELLERMEYER R.W. (1961) The hemolytic effect of primaquine. XII. Shortened erythrocyte life span in primaquine-sensitive male Negroes in the absence of drug administration. *J. Lab. clin. Med.* **58**, 217.

BRUNETTI P., GRIGNANI F. & ERNISLI G. (1962) Behavior of the erythrocyte pyrophosphatase activity in the enzyme-deficient hemolytic anemias. I. Quantitative modifications of the enzyme. II. A new test for the detection of the enzyme defect. *Acta haemat.* **27**, 146.

BRUNETTI P., PARNA A., NENCI G., MIGLIORINI E. & BRUNELLI B. (1965) The incidence of G6PD deficiency in Central Italy. *Haematologica* **50**, 203.

BRUNETTI P., ROSSETTI R. & BROCCHIA G. (1960) Nouve acquisizioni in tema di bioenzimologia del favismo itteroemoglobinurico: Nata III. L'attivita glucosio-6-fosfato-deidrogenasica del parenchima epatico. *Rass. Fisiopat. clin. terap.* **32**, 338.

BURKA E.R., WEAVER Z. & MARKS P.A. (1966) Clinical spectrum of hemolytic anemia associated with glucose-6-phosphate dehydrogenase deficiency. *Ann. intern. Med.* **64**, 817.

CARSON P.E. (1960) Glucose-6-phosphate dehydrogenase deficiency in hemolytic anemia. *Fed. Proc.* **19**, 995.

CARSON P.E., FLANAGAN C.L., ICKES C.E. & ALVING A.S. (1956) Enzymatic deficiency in primaquine-sensitive erythrocytes. *Science* **124**, 484.

CARSON P.E. & FRISCHER H. (1966) Glucose-6-phosphate dehydrogenase deficiency and related disorders of the pentose phosphate pathway. *Am. J. Med.* **41**, 744.

CARSON P.E., OKITA G.T., FRISCHER H., HIRASA J., LONG W.K. & BREWER G.J. (1963) Patterns of hemolytic susceptibility and metabolism. *Proc. 9th Cong. Europ. Soc. Haemat.*, p. 655. Karger, New York.

CARSON P.E., SCHRIER S.L. & KELLERMEYER R.W. (1959) Mechanism of inactivation of glucose-6-phosphate dehydrogenase in human erythrocytes. *Nature* **184**, 1292.

CARSON P.E. & TARLOV A.R. (1962) Biochemistry of hemolysis. *Ann. Rev. Med.* **13**, 105.

CHAN T.K., TODD D. & WONG C.C. (1965) Tissue enzyme levels in erythrocyte glucose-6-phosphate dehydrogenase deficiency. *J. Lab. clin. Med.* **66**, 937.

CHANMUGAM D. & FRUMIN A.M. (1964) Abnormal glucose tolerance response in erythrocyte G6PD deficiency. *New Engl. J. Med.* **271**, 1202.

CHILDS B. (1965) Genetic origin of some sex differences among human beings. *Pediatrics* **35**, 798.

CHILDS B., ZINKHAM W., BROWNE E.A., KIMBRO E.L. & TORBET J.V. (1958) A genetic study of a defect in glutathione metabolism of the erythrocyte. *Bull. Johns Hopkins Hosp.* **102**, 21.

CHOREMIS C., ZANNOS-MARIOLEA L. & KATTAMIS M.D.C. (1962) Frequency of G6PD deficiency in certain highly malarious areas of Greece. *Lancet* **i**, 17.

CHUNG A.E. & LANGDON R.G. (1963a) Human erythrocyte glucose-6-phosphate dehydrogenase. I. Isolation and properties of the enzyme. *J. biol. Chem.* **238**, 2309.

CHUNG A.E. & LANGDON R.E. (1963b) II. Enzyme-coenzyme interrelationship. *J. biol. Chem.* **238**, 2317.

COHEN G. & HOCHSTEIN P. (1961) G6PD and detoxification of hydrogen peroxide in human erythrocytes. *Science* **134**, 1756.

COHEN G. & HOCHSTEIN P. (1964) Generation of hydrogen peroxide in erythrocytes by hemolytic agents. *Biochem. J.* **3**, 895.

DAVIDSON R.G. (1964) The Lyon hypothesis. *J. Pediat.* **65**, 765.

DAVIDSON R.G., CHILDS B. & SINISCALCO M. (1964) Genetic variations in the quantitative control of erythrocyte G6PD activity. *Ann. hum. Genet.* **28**, 61.

DAVIDSON R.G., MIGEON B.R., BORDEN M. & CHILDS B. (1963a) Dosage compensation in the regulation of erythrocyte G6PD activity. *Bull. Johns Hopkins Hosp.* **112**, 318.

DAVIDSON R.G., NITOWSKY H.M. & CHILDS B. (1963b) Demonstration of two populations of cells in the human female heterozygous for G6PD variants. *Proc. natn. Acad. Sci.* **50**, 481.

DERN R.J. (1966) A new hereditary quantitative variant of glucose-6-phosphate dehydrogenase characterized by a marked increase in enzyme activity. *J. Lab. clin. Med.* **68**, 560.

DERN R.J., BEUTLER E. & ALVING A.S. (1954) The hemolytic effect of primaquine. II. The natural course of the hemolytic anemia and the mechanism of its self-limited character. *J. Lab. clin. Med.* **44**, 171.

DESFORGES J.F. (1965) Erythrocyte metabolism in hemolysis. *New Engl. J. Med.* **273**, 1310.

DESFORGES J.F., KALAW E. & GILCHRIST P. (1960) Inhibition of G6PD by hemolysis inducing drugs. *J. Lab. clin. Med.* **55**, 757.

DIMANT E., LANDBERG E. & LONDON I.M. (1955) The metabolic behavior of reduced glutathione in human and avian erythrocytes. *J. biol. Chem.* **213**, 769.

DOXIADIS S.A., FESSAS P. & VALAES T. (1961) G-6-PD deficiency. A new aetiological factor of severe neonatal jaundice. *Lancet* **i**, 297.

DOXIADIS S.A., VALAES T., KARAKLIS A. & STAVRAKAKIS D. (1964) Risk of severe jaundice in glucose-6-phosphate dehydrogenase deficiency of the newborn. *Lancet* **ii**, 1210.

ELDJARN L. & BREMER J. (1962) The inhibitory effect at the hexokinase level of disulphides in glucose metabolism in human erythrocytes. *Biochem. J.* **84**, 286.

EPPES R.B., BREWER G.J., DEGOWN R.L., MCNAMARA J.V., FLANAGAN C.L., SCHRIER S.L., TARLOV A.R., POWELL R.D. & CARSON P.E. (1966) Oral glucose tolerance in Negro men deficient in glucose-6-phosphate dehydrogenase. *New Engl. J. Med.* **275**, 855.

FAIRBANKS V.F. & BEUTLER E. (1962) A simple method for detection of erythrocyte G6PD deficiency (G6PD spot test). *Blood* **20**, 591.

FESSAS P., DOXIADIS S.A. & VALAES T. (1962) Neonatal jaundice in G6PD deficient infants. *Br. med. J.* **ii**, 1359.

FLANAGAN C.L., SCHRIER S.L., CARSON P.E. & ALVING A.S. (1958) The hemolytic effect of primaquine. VIII. The effect of drug administration on parameters of primaquine sensitivity. *J. Lab. clin. Med.* **51**, 600.

FLATZ G. & SRINGAM S. (1963) Malaria and G6PD deficiency in Thailand. *Lancet* **ii**, 1248.

FLATZ G. & SRINGAM S. (1964) G6PD deficiency in different ethnic groups in Thailand. *Ann. hum. Genet.* **27**, 315.

FLATZ G., SRINGAM S., PREMYOTHIN C., PENBHARKKUL S., KETUSINGH R. & CHULAJATA R. (1963) G6PD deficiency and neonatal jaundice. *Archs Dis. Childh.* **38**, 566.

FRASER G.R., DEFARANAS B., KATTAMIS C.A., RACE R.R., SANGER R. & STAMA-TOYANNOPOULOS G. (1964) G6PD, colour vision and Xg blood groups in Greece: linkage and population data. *Ann. hum. Genet.* **27**, 395.

FRASER G.R., GRUNWALD P. & STAMATOYANNOPOULOS G. (1966) Glucose-6-phosphate dehydrogenase (G6PD) deficiency, abnormal haemoglobins and thalassaemia in Yugoslavia. *J. med. Genet.* **3**, 1.

GARTLER S.M., GANDINI E. & CEPPELLINI R. (1962) G6PD deficient mutant in human cell culture. *Nature* **193**, 602.

GARTLER S.M. & LINDER D. (1964) Selection in mammalian mosaic cell populations. *Cold Spring Harbor Symposia on Quantitative Biology* **29**, 253.

GARTLER S.M., ZIPROWSKI L., KRAKOWSKI A., EZRA R., SZEINBERG A. & ADAM A. (1966) Glucose-6-phosphate dehydrogenase mosaicism as a tracer in the study of hereditary multiple trichoepithelioma. *Am. J. hum. Genet.* **18**, 282.

GELPI A.P. (1965) Glucose-6-phosphate dehydrogenase deficiency in Saudi Arabia: a survey. *Blood* **25**, 486.

GILES E., CURTAIN C.C. & BAUMGARTEN A. (1967) Distribution of β-thalassemia trait and erythrocyte glucose-6-phosphate dehydrogenase deficiency in the Markham River Valley of New Guinea. *Am. J. phys. Anthrop.* **27**, 83.

GLOCK G.E. & MCLEAN P. (1954) Levels of enzymes of the direct oxidative pathways of carbohydrate metabolism in tissues and tumors. *Biochem. J.* **56**, 171.

GLOCK G.E. & MCLEAN P. (1955) A preliminary investigation of the hormonal control of the hexose-monophosphate oxidative pathway. *Biochem. J.* **61**, 390.

GROSS R.T., HURWITZ R.E. & MARKS P.A. (1958) An hereditary enzymatic defect in erythrocyte metabolism: G6PD deficiency. *J. clin. Invest.* **37**, 1176.

GRUMBACH M.M., MARKS P.A. & MORISHIMA A. (1962) Erythrocyte G6PD activity and X-chromosome polysomy. *Lancet* **i**, 1330.

HARLEY J.D. (1965) Role of reduced glutathione in human erythrocytes. *Nature* **206**, 1054.

HARLEY J.D. & ROBIN H. (1962) 'Late' neonatal jaundice in infants with G6PD deficient erythrocytes. *Australasian Ann. Med.* **11**, 148.

HARRIS H., HOPKINSON D.A., SPENCER N., COURT BROWN W.M. & MANTLE D. (1963) Red cell G6PD activity in individuals with abnormal numbers of X chromosomes. *Ann. hum. Genet.* **27**, 59.

HARRIS J.W. (1963) *The Red Cell. Production, Metabolism, Destruction: Normal and Abnormal.* Harvard Univ. Press, Cambridge.

HELGE H. & BORNER K. (1966) Kongenitale nichtschärozytäre hämolytische Anämie, Katarakt und Glucose-6-phosphat-Dehydrogenase-Mangel. *Dsch. med. Wschr.* **91**, 1584.

HILL A.S., HAUT A., CARTWRIGHT G.E. & WINTROBE M.M. (1964) The role of nonhemoglobin proteins and reduced glutathione in the protection of hemoglobin from oxidation *in vitro. J. clin. Invest.* **43**, 17.

HOCKWALD R.S., ARNOLD J., CLAYMAN C.B. & ALVING A.S. (1952) Status of primaquine. IV. Toxicity of primaquine in Negroes. *J. Am. med. Ass.* **149**, 1568.

HUENNEKENS F.M., CAFFREY R.W., BASFORD R.E. & GABRIO B.W. (1957) Erythrocyte metabolism. IV. Isolation and properties of methemoglobin reductase. *J. biol. Chem.* **227**, 261.

JACOB H.S., INGBAR S.H., JANDL J.H. & BELL S.C. (1965) Oxidative hemolysis and erythrocyte metabolism in hereditary acatalasia. *J. clin. Invest.* **44**, 1187.

JACOB H.S. & JANDL J.H. (1962) Effects of sulfhydryl inhibition on red blood cells. I. Mechanisms of hemolysis. *J. clin. Invest.* **41**, 779.

JACOB H.S. & JANDL J.H. (1966a) Effects of sulfhydryl inhibition on red blood cells. II. Glutathione in the regulation of the hexose monophosphate pathway. *J. biol. Chem.* **241**, 4243.

JACOB H.S. & JANDL J.H. (1966b) A simple visual screening test for glucose-6-phosphate dehydrogenase deficiency employing ascorbate and cyanide. *New Engl. J. Med.* **274**, 1162.

JAFFÉ E.R. (1963) The reduction of methemoglobin in erythrocytes of a patient with congenital methemoglobinemia, subjects with erythrocyte G6PD deficiency and normal individuals. *Blood* **21**, 561.

JANDL J.H. (1965) Leaky red cells. *Blood* **26**, 367.

JANDL J.H., ENGLE L.E. & ALLEN D.W. (1960) Oxidative hemolysis and precipitation of hemoglobin. I. Heinz body anemias as an acceleration of red cell aging. *J. clin. Invest.* **39**, 1818.

KAPLAN J.C., ROSA R., SERINGE P. & HEOFFEL J.C. (1967) Le polymorphisme génétique de la glucose-6-phosphate déshydrogénase érythrocytaire chez l'homme. II. Étude d'une nouvelle variété à activité diminuée: le type 'Kabyle' *Enzym. biol. clin.* **8**, 332.

KIDSON C. (1961) Erythrocyte G6PD deficiency in New Guinea and New Britain. *Nature* **190**, 1120.

KIDSON C. & GAJDUSEK D.C. (1962a) Congenital defects of the central nervous system associated with hyperendemic goiter in a neolithic highland society of

Netherlands New Guinea. II. Glucose-6-phosphate dehydrogenase in the Mulia population. *Pediatrics* **29**, 364.

KIDSON C. & GAJDUSEK D.C. (1962b) Glucose-6-phosphate dehydrogenase deficiency in Micronesian peoples. *Aust. J. Sci.* **25**, 61.

KIDSON C. & GORMAN J.G. (1962) A challenge to the concept of selection by malaria in G6PD deficiency. *Nature* **196**, 49.

KIRKMAN H.N. (1962) Electrophoretic differences of human erythrocytic G6PD. *Am. J. Dis. Child.* **104**, 566.

KIRKMAN H.N. (1963) Genetic control of human enzymes. *Pediat. clin. N. America.* **10**, 299.

KIRKMAN H.N. & CROWELL B.B. (1963) Molecular deficiency of G6PD in primaquine sensitivity. *Nature* **197**, 286.

KIRKMAN H.N., DOXIADIS S.A., VALAES T., TASSOPOULOS N. & BRINSON A.G. (1965a) Diverse characteristics of G6PD from Greek children. *J. Lab. clin. Med.* **65**, 212.

KIRKMAN H.N. & HENDRICKSON E.M. (1962) Glucose-6-phosphate dehydrogenase from human erythrocytes. II. Subactive states of the enzyme from normal persons. *J. biol. Chem.* **237**, 2371.

KIRKMAN H.N. & HENDRICKSON E.M. (1963) Sex-linked electrophoretic difference in G6PD. *Am. J. hum. Genet.* **15**, 241.

KIRKMAN H.N., McCURDY P.R. & NAIMAN J.L. (1964a) Functionally abnormal G6PD's. *Cold Spring Harbor Symposia on Quantitative Biology* **29**, 391.

KIRKMAN H.N. & RILEY H.D. (1961) Congenital non-spherocytic hemolytic anemia. *Am. J. Dis. Child.* **102**, 313.

KIRKMAN H.N., ROSENTHAL I.M., SIMON E.R., CARSON P.E. & BRINSON A.G. (1964b) 'Chicago 1' variant of G6PD in congenital hemolytic disease. *J. Lab. clin. Med.* **63**, 715.

KIRKMAN H.N., SCHETTINI F. & PICKARD B.M. (1964c) Mediterranean variant of G6PD. *J. Lab. clin. Med.* **63**, 726.

KIRKMAN H.N., SIMON E.R. & PICKARD B.M. (1965b) Seattle variant of glucose-6-phosphate dehydrogenase. *J. Lab. clin. Med.* **66**, 834.

KLEIHAUER E. & BETKE K. (1963) Elution procedure for the demonstration of methaemoglobin in red cells of human blood smears. *Nature* **199**, 1196.

KNUTSEN C.A. & BREWER G.J. (1966) The micro-methemoglobin reduction test for glucose-6-phosphate dehydrogenase deficiency. A simple field screening procedure. *Am. J. clin. Path.* **45**, 82.

KOSOWER N.S., VANDERHOFF G.A. & LONDON I.M. (1964) Hexokinase activity in normal and G6PD deficient erythrocytes. *Nature* **201**, 684.

KRUATRACHUE M., CHAROENLARP P., CHONGSUPHAJAI-SIDDHI T. & HARNINASUTA C. (1962) Erythrocyte G6PD and malaria in Thailand. *Lancet* **ii**, 1183.

LARIZZA P. (1961) Enzymopenic hemolytic anemias. *Folia haemat.* **6**, 19.

LARIZZA P., BRUNETTI P. & GRIGNANI F. (1960) Anemie emolitiche enzimopeniche. *Haematologica* **45**, 1.

LEE T.C., SHIH L.Y., HUANG P.C., LIN C.C., BLACKWELL B.N., BLACKWELL R.Q. & HSIA D.Y.Y. (1963) G6PD deficiency in Taiwan. *Am. J. hum. Genet.* **15**, 126.

LIE-INJO L.E. & CHIN J. (1964) Abnormal haemoglobin and G6PD deficiency in Malayan aborigines. *Nature* **204**, 291.

LIE-INJO L.E., PILLAY R.P. & VIRIK H.K. (1966) Haemolysis due to glucose-6-phosphate dehydrogenase deficiency in Malaya. *Trans. R. Soc. trop. Med. Hyg.* **60**, 262.

LIE-INJO L.E. & POEY-OEY H.G. (1964) G6-PD deficiency in Indonesia. *Nature* **204**, 88.

LIE-INJO L.E. & TI T.S. (1964) Glucose-6-phosphate dehydrogenase deficiency in Malayans. *Trans. R. Soc. trop. Med. Hyg.* **58**, 500.

LINDER D. & GARTLER S.M. (1965a) Glucose-6-phosphate dehydrogenase mosaicism: utilization as a cell marker in the study of leiomyomas. *Science* **150**, 67.

LINDER D. & GARTLER S.M. (1965b) Distribution of G6PD electrophoretic variants in different tissues of heterozygotes. *Am. J. hum. Genet.* **17**, 212.

LIVINGSTONE F.B. (1964) Aspects of the population dynamics of the abnormal hemoglobin and G6PD deficiency genes. *Am. J. hum. Genet.* **16**, 435.

LIVINGSTONE F.B. (1967) *Abnormal Hemoglobins in Human Populations.* Aldine, Chicago.

LONG W.K., KIRKMAN H.N. & SUTTON H.E. (1965) Electrophoretically slow variants of G6PD from red cells of Negroes. *J. Lab. clin. Med.* **65**, 81.

LOPEZ–S. A. & KREHL W.A. (1967) A possible interrelation between glucose-6-phosphate dehydrogenase and dehydroepiandrosterone in obesity. *Lancet* **ii**, 485.

LU T.C., WEI H. & BLACKWELL R.Q. (1966) Increased incidence of severe hyperbilirubinemia among new-born Chinese infants with G-6-PD deficiency. *Pediatrics* **37**, 994.

LUZZATTO L. & ALLAN N.C. (1965) Different properties of glucose-6-phosphate dehydrogenase from human erythrocytes with normal and abnormal enzyme levels. *Biochem. biophys. Res. Commun.* **21**, 547.

LUZZATTO L. & OKOYE V.C.N. (1967) Resolution of genetic variants of human erythrocyte glucose-6-phosphate dehydrogenase by thin layer chromatography. *Biochem. biophys. Res. Commun.* **29**, 705.

LYON M.F. (1961) Gene action in the X-chromosome of the mouse (*mus musculus* L). *Nature* **190**, 372.

MARKS P.A. (1964) Glucose-6-phosphate dehydrogenase: its properties and role in mature erythrocytes, *in* BISHOP C. & SURGENOR D.M. (eds.) *The Red Blood Cell*, p. 211. Academic Press, New York.

MARKS P.A. (1967) Glucose-6-phosphate dehydrogenase in mature erythrocytes. *Am. J. clin. Path.* **47**, 287.

MARKS P.A., BANKS J. & GROSS R.T. (1959) Glucose-6-phosphate dehydrogenase thermostability in leucocytes of Negroes and Caucasians with erythrocyte deficiency of the enzyme. *Biochem. biophys. Res. Commun.* **1**, 199.

MARKS P.A., BANKS J. & GROSS R.T. (1962) Genetic heterogeneity of G6PD deficiency. *Nature* **194**, 454.

MARKS P.A. & GROSS R.T. (1959) Erythrocyte G6PD deficiency: evidence of differences between Negroes and Caucasians with respect to this genetically determined trait. *J. clin. Invest.* **38**, 2253.

MARKS P.A., JOHNSON A.B. & HIRSCHBERG E. (1958) Effect of age on the enzyme activity in erythrocytes. *Proc. natn. Acad. Sci.* **44**, 529.

MARKS P.A. & TSUTSUI E.A. (1963) Human glucose-6-P dehydrogenase: studies on the relation between antigenecity and catalytic activity; the role of TPN. *Ann. N.Y. Acad. Sci.* **103**, 902.

McCURDY P.R., KIRKMAN H.N., NAIMAN J.L., JIM R.T.S. & PICKARD B.M. (1966) A Chinese variant of glucose-6-phosphate dehydrogenase. *J. Lab. clin. Med.* **67**, 374.

MILLER A. & HORIUCHI M. (1962) Erythrocyte glutathione. I. *In vitro* incorporation of radioactive amino acid precursors into the glutathione of human erythrocytes. *J. Lab. clin. Med.* **60**, 756.

MILLER A. & HORIUCHI M. (1965) Erythrocyte glutathione. II. The effect of acetylphenylhydrazine and primaquine on the entry and incorporation of C^{14} glycine into glutathione. *J. Lab. clin. Med.* **66**, 84.

MILLS G.C. (1957) Hemoglobin catabolism. I. Glutathione peroxidase, an erythrocyte enzyme which protects hemoglobin from oxidative breakdown. *J. biol. Chem.* **229**, 189.

MILLS G.C. (1959) The purification and properties of glutathione peroxidase of erythrocytes. *J. biol. Chem.* **234**, 502.

MILLS G.C. (1960) Glutathione peroxidase and the destruction of hydrogen peroxide in animal tissues. *Archs Biochem.* **86**, 1.

MOTULSKY A.G. (1960) Metabolic polymorphisms and the role of infectious diseases in human evolution. *Hum. Biol.* **32**, 28.

MOTULSKY A.G. (1961) Glucose-6-phosphate dehydrogenase deficiency, haemolytic disease of the newborn and malaria. *Lancet* **i**, 1168.

MOTULSKY A.G. (1964) Hereditary red cell traits and malaria. *Am. J. trop. Med. Hyg.* **13**, 147.

MOTULSKY A.G. (1965) Theoretical and clinical problems of G6PD deficiency. Its occurrence in Africans and its combination with hemoglobinopathy, *in* JONXIS J.H.P. (ed.) *Abnormal Hemoglobins in Africa*, p. 143. Davis, Philadelphia.

MOTULSKY A.G. (1968) Contributions of hereditary disorders of red cell metabolism to human genetics, *in* BEUTLER E. (ed.) *Hereditary Disorders of Erythrocyte Metabolism*. p. 303, Grune & Stratton, New York.

MOTULSKY A.G. & CAMPBELL-KRAUT J.M. (1960) Population genetics of G6PD deficiency of the red cell, *in* BLUMBERG B.S. (ed.) *Proc. Conf. Genet. Polymorphism and Geogr. Variations in Disease*, p. 159. Grune & Stratton, New York.

MOTULSKY A.G., LEE T.C. & FRASER G.R. (1965) G6PD deficiency, thalassemia and abnormal haemoglobin in Taiwan. *J. med. Genet.* **2**, 18.

MOTULSKY A.G. & STAMATOYANNOPOULOS G. (1966) Clinical implications of glucose-6-phosphate dehydrogenase deficiency. *Ann. intern. Med.* **65**, 1329.

MOTULSKY A.G., STRANSKY E. & FRASER G.R. (1964) G6PD deficiency, thalassemia and abnormal haemoglobins in the Philippines. *J. med. Genet.* **1**, 102.

MURPHY J.R. (1960) Erythrocyte metabolism. II. Glucose metabolism and pathways. *J. Lab. clin. Med.* **55**, 286.

NANCE W.E. (1964) Genetics tests with a sex-linked marker: G6PD. *Cold Spring Harbor Symposia on Quantitative Biology* **29**, 415.

NANCE W.E. & UCHIDA I. (1964) Turner's syndrome, twinning and an unusual variant of G6PD. *Am. J. hum. Genet.* **16**, 380.

NITZAN M. & JOSEFSBERG Z. (1966) Glucose tolerance in G-6-PD deficient children. *Am. J. Dis. Child.* 111, 406.

OORT M., LOOS J.A. & PRINS H.K. (1961) Hereditary absence of reduced glutathione in the erythrocytes—a new clinical and biochemical entity. *Vox Sang.* 6, 370.

OSKI F.A., SHAHIDI N.T. & DIAMOND L.K. (1963) Erythrocyte acid phosphomonoesterase and G6PD deficiency in Caucasians. *Science* 139, 409.

PANIZON F. (1960) Erythrocyte enzyme deficiency in unexplained kernicterus. *Lancet* 2, 1093.

PANIZON F. (1961) Studio dell'attuvuta enzimatica della mucosa digiunale in soggetti con glucoso-6-fosfato deidrogenasi-penia eritrocitaria. *Studi Sassar.* 39, 210.

PANNACCIULLI I., TIZIANELLO A., AJMAR F. & SALVIDIO E. (1965) The course of experimentally induced hemolytic anemia in a primaquine-sensitive Caucasian. A case study. *Blood* 25, 92.

PAPAYANNOPOULOU T. & STAMATOYANNOPOULOS G. (1964) Pseudo-mosaicism in males with mild G6PD deficiency. *Lancet* ii, 1215.

PEARSON H.A. & DRUYAN R. (1961) Erythrocyte G6PD activity related to thyroid activity. *J. Lab. clin. Med.* 57, 343.

PINTO P.V.C., NEWTON W.A. & RICHARDSON K.E. (1966) Evidence for four types of erythrocyte glucose-6-phosphate dehydrogenase from G-6-PD deficient human subjects. *J. clin. Invest.* 45, 823.

PLATO C.C., RUCKNAGEL D.L. & GERSHOWITZ H. (1964) Studies on the distribution of glucose-6-phosphate dehydrogenase deficiency, thalassemia and other genetic traits in the coastal and mountain villages of Cyprus. *Am. J. hum. Genet.* 16, 267.

PORTER I.H., BOYER S.H., WATSON-WILLIAMS E.J., ADAM A., SZEINBERG A. & SINISCALCO M. (1964) Variations of G6PD in different populations. *Lancet* i, 895.

PORTER I.H., SCHULZE J. & McKUSICK V.A. (1962) Genetical linkage between the loci for G6PD deficiency and colour-blindness in American Negroes. *Ann. hum. Genet.* 26, 107.

PRINS H.K., OORT M., LOOS J.A., ZURCHER C. & BECKERS T. (1966) Congenital nonspherocytic hemolytic anemia, associated with glutathione deficiency of the erythrocytes. *Blood* 27, 145.

RAGAB A.H., EL-ALFI O.S. & ABBOUD M.A. (1966) Incidence of glucose-6-phosphate dehydrogenase deficiency in Egypt. *Am. J. hum. Genet.* 18, 21.

RAMOT B., BAUMINGER S., BROK F., GAFNI D. & SHWARTZ J. (1964) Characterization of glucose-6-phosphate dehydrogenase in Jewish mutants. *J. Lab. clin. Med.* 64, 895.

RAMOT B. & BROK F. (1964) A new G6PD mutant (Tel-Hashomer mutant). *Ann. hum. Genet.* 28, 167.

RAMOT B., FISHER S., SZEINBERG A., ADAM A., SHEBA C. & GAFNI D. (1959a) A study of subjects with erythrocyte G6PD deficiency. II. Investigation of leukocyte enzymes. *J. clin. Invest.* 38, 2234.

RAMOT B., SHEBA C., ADAM A. & ASHKENASI I. (1960) Erythrocyte G6PD deficient subjects: enzyme level in saliva. *Nature* 185, 931.

RAMOT B., SZEINBERG A., ADAM A., SHEBA C. & GAFNI D. (1959b) A study of subjects with erythrocyte G6PD deficiency: investigation of platelet enzymes. *J. clin. Invest.* **38**, 1659.

RAPOPORT S. & SCHEUCH D. (1960) Glutathione stability and pyrophosphatase activity in reticulocytes; direct evidence for the importance of glutathione for the enzyme status in intact cells. *Nature* **186**, 967.

ROOT A.W., OSKI F.A., BONGIOVANNI A.M. & EBERLEIN W.R. (1967) Erythrocyte glucose-6-phosphate dehydrogenase activity in children with hypothyroidism and hypopituitarism. *J. pediat.* **70**, 369.

ROSE I.A. & O'CONNELL E.L. (1964) The role of glucose-6-phosphate in regulation of glucose metabolism in human erythrocytes. *J. biol. Chem.* **239**, 12.

SALEN G., GOLDSTEIN F., HAURANI F.I. & WIRTS C.W. (1966) Acute hemolytic anemia complicating viral hepatitis in patients with glucose-6-phosphate dehydrogenase deficiency. *Ann. intern. Med.* **65**, 1210.

SALVIDIO E., PANNACCIULLI I., TIZIANELLO A. & AJMAR F. (1967) Nature of hemolytic crises and fate of G6PD deficient, drug-damaged erythrocytes in Sardinians. *New Engl. J. Med.* **276**, 1339.

SANSONE G., PIGA A.M. & SEGNI G. (1958) *Il Favismo*. Minerva, Turin.

SANSONE G., RASORE-QUARTINO A. & VENEZIANO G. (1963) Demonstration on blood smears of a double erythrocytic population in females heterozygous for G6PD deficiency. *Pathologica* **55**, 371.

SAY B., OZAND P., BERKEL I. & CEVIK N. (1965) Erythrocyte G6PD deficiency in Turkey. *Acta paediat. scand.* **54**, 319.

SCHEUCH D., KAHRIG C., OCKEL E., WAGENKNECHT C. & RAPOPORT S.M. (1961) Role of glutathione and of a self-stabilizing chain of SH enzymes and substrates in the metabolic regulation of erythrocytes. *Nature* **190**, 631.

SCOTT E.M. (1960) The relation of diaphorase of human erythrocytes to inheritance of methemoglobinemia. *J. clin. Invest.* **39**, 1176.

SCOTT E.M. & McGRAW J.C. (1962) Purification and properties of diphosphopyridine nucleotide diaphorase of human erythrocytes. *J. biol. Chem.* **237**, 249.

SHAKER Y., ONSI A. & AZIZ R. (1966) The frequency of glucose-6-phosphate dehydrogenase deficiency in the newborn and adults in Kuwait. *Am. J. hum. Genet.* **18**, 609.

SHEBA C., SZEINBERG A., RAMOT B., ADAM A. & ASHKENAZI I. (1962) Epidemiologic surveys of deleterious genes in different population groups in Israel. *Am. J. Pub. Hlth.* **52**, 1101.

SHOWS T.B., TASHIAN R.E. & BREWER G.J. (1964) Erythrocyte G6PD in Caucasians: a new inherited variant. *Science* **145**, 1056.

SHRAGO E. (1966) The effect of primaquine and acetylphenylhydrazine on the enzymatic and nonenzymatic oxidation of reduced pyridine nucleotides. *J. Lab. clin. Med.* **68**, 1.

SIMON E.R., GIBLETT E.R. & FINCH C.A. (1966) *Red Cell Manual*. Univ. of Washington Press, Seattle.

SINISCALCO M. (1963) Field and laboratory studies on favism and thalassemia in Sardinia, *in* GOLDSCHMIDT E. (ed.) *Genetics of Migrant and Isolate Populations*, p. 72, Williams & Wilkins, Baltimore.

SINISCALCO M., BERNINI L., FILIPPI G., LATTE B., MEERA KHAN P., PIOMELLI S. &

RATTAZZI M. (1966a) Population genetics of hemoglobin variants, thalassemia and G-6-PD deficiency, with particular reference to the malaria hypothesis. *Wld. Hlth. Org. Bull.* **34**, 379.

SINISCALCO M., BERNINI L., LATTE B. & MOTULSKY A.G. (1961) Favism and thalassemia in Sardinia and their relationship to malaria. *Nature* **190**, 1179.

SINISCALCO M., FILIPPI G. & LATTE B. (1964) Recombination between Protan and Deutan genes: data on their respective positions in respect to the G6PD locus. *Nature* **204**, 1062.

SINISCALCO M., FILIPPI G., LATTE B., PIOMELLI S., RATTAZZI M., GAVIN J., SANGER R. & RACE R.R. (1966b) Failure to detect linkage between the Xg and other X-borne loci in Sardinians. *Ann. hum. Genet.* **29**, 231.

SINISCALCO M., MOTULSKY A.G., LATTE B., BERNINI L. & MONTALENTI G. (1960) Indagini genetiche sulla predisposizione al favismo. II. Dati familiari: associazione genica con il daltonismo. *Accad. naz. dei. lincei.* VIII. 28.

SMITH G.D. & VELLA F. (1960) Erythrocyte enzyme deficiency in unexplained kernicterus. *Lancet* **i**, 1133.

SRIVASTAVA S.K. & BEUTLER E. (1968) Elevated levels of oxidized glutathione (GSSG) in G-6-PD deficient red cells. *Clin. Res.* **16**, 125.

STAMATOYANNOPOULOS G. & FESSAS P. (1964) Thalassaemia, glucose-6-phosphate dehydrogenase deficiency, sickling and malarial endemicity in Greece: a study of five areas. *Br. med. J.* **i**, 875.

STAMATOYANNOPOULOS G., FRASER G.R., MOTULSKY A.G., FESSAS P., AKRIVAKIS A. & PAPAYANNOPOULOU TH. (1966a) On the familial predisposition to favism. *Am. J. hum. Genet.* **18**, 253.

STAMATOYANNOPOULOS G., PANAYOTOPOULOS A. & MOTULSKY A.G. (1966b) The distribution of glucose-6-phosphate dehydrogenase deficiency in Greece. *Am. J. hum. Genet.* **18**, 296.

STAMATOYANNOPOULOS G., PANAYOTOPOULOS A. & PAPAYANNOPOULOU T. (1964) Mild G6PD deficiency in Greek males. *Lancet* **ii**, 932.

STAMATOYANNOPOULOS G., PAPAYANNOPOULOU T., BACOPOULOS C. & MOTULSKY A.G. (1966c) Methemoglobin reduction and the inactive X hypothesis. *Am. J. hum. Genet.* **18**, 417.

STAMATOYANNOPOULOS G., PAPAYANNOPOULOU T., BACOPOULOS C. & MOTULSKY A.G. (1967a) Detection of glucose-6-phosphate dehydrogenase deficient heterozygotes. *Blood* **29**, 87.

STAMATOYANNOPOULOS G., YOSHIDA A., BACOPOULOS C. & MOTULSKY A.G. (1967b) Athens variant of glucose-6-phosphate dehydrogenase. *Science* **157**, 831.

SZEINBERG A. (1963) G6PD deficiency among Jews, *in* GOLDSCHMIDT E. (ed.) *Genetics of Migrant and Isolate Populations*, p. 69. Williams & Wilkins, Baltimore.

SZEINBERG A., ADAM A., RAMOT B., SHEBA C. & MYERS F. (1959) The incorporation of isotopically labelled glycine into the glutathione of erythrocytes with glucose-6-phosphate dehydrogenase deficiency. *Biochim. biophys. Acta* **36**, 65.

SZEINBERG A. & GOLDSCHMIDT E. (1963) Some remarks on neonatal risk in Jewish communities with high frequencies of G6PD deficiency, *in* GOLDSCHMIDT E.

(ed.) *Genetics of Migrant and Isolate Populations*, p. 90. Williams & Wilkins, Baltimore.

SZEINBERG A., KELLERMAN J., ADAM A., SHEBA C. & RAMOT B. (1960) Hemolytic jaundice following aspirin administration to a patient with deficiency of glucose-6-phosphate dehydrogenase in erythrocytes. *Acta haemat.* **23**, 58.

SZEINBERG A. & MARKS P.A. (1961) Substances stimulating glucose catabolism by the oxidative reactions of the pentose phosphate pathway in human erythrocytes. *J. clin. Invest.* **40**, 914.

SZEINBERG A., OLIVER M., SCHMIDT R., ADAM A. & SHEBA C. (1963) Glucose-6-phosphate dehydrogenase deficiency and haemolytic disease of the newborn in Israel. *Archs Dis. Childh.* **38**, 23.

TALEB N., LOISELET J., GHORRA F. & SFEIR H. (1964) Sur la déficience en glucose-6-phosphate dehydrogénase dans les populations autochtones du Liban. *C.R. Acad. Sci.* **258**, 5749.

TARLOV A.R., BREWER G.J., CARSON P.E. & ALVING A.S. (1962) Primaquine sensitivity; G6PD deficiency: an inborn error of metabolism of medical and biological significance. *Archs intern. Med.* **109**, 209.

TÖNZ O. & ROSSI E. (1964) Morphological demonstration of two red cell populations in human females heterozygous for G6PD deficiency. *Nature* **202**, 606.

VELLA F. (1959) Favism in Asia. *Med. J. Australia* **2**, 196.

VELLA F. (1961) The incidence of erythrocyte glucose-6-phosphate dehydrogenase deficiency in Singapore. *Experientia* **17**, 181.

WALKER D.G. & BOWMAN J.E. (1959) Glutathione stability of the erythrocytes in Iranians. *Nature* **184**, 1325.

WALKER D.G. & BOWMAN J.E. (1960) *In vitro* effect of Vicia faba extracts upon reduced glutathione of erythrocytes. *Proc. Soc. exp. Biol. Med.* **103**, 476.

WALLER H.D. & GEROK W. (1964) Schwere strahleninduzierte Hämolyse bei hereditärem Mangel an reduziertem Glutathion in Blutzellen. *Klin. Wschr.* **42**, 948.

WALLER H.D., LOHR G.W. & GAYER J. (1966) Hereditäre nichtsphärocytäre hämolytische Anämie durch Glucose-6-phosphatdehydrogenase-Mangel. *Klin. Wschr.* **44**, 122.

WALTER H., NEUMANN S. & NEMESKÉRI J. (1968) Investigations on the occurrence of glucose-6-phosphate dehydrogenase deficiency in Hungary. *Acta genet.* **18**, 1.

WHO (1967) Standardization of procedures for the study of glucose-6-phosphate dehydrogenase. *Wld. Hlth. Org. Tech. Rep. Ser.* No. 366.

WOLFF J.A., GROSSMAN B.H. & PAYA K. (1967) Neonatal serum bilirubin and glucose-6-phosphate dehydrogenase. *Am. J. Dis. Child.* **113**, 251.

WONG P.W.K., SHIH L.Y. & HSIA D.Y.Y. (1965) Characterization of glucose-6-phosphate dehydrogenase among Chinese. *Nature* **208**, 1323.

WURZEL H., McCREARY T., BAKER L. & GUMERMAN L. (1961) G6PD activity in platelets. *Blood* **17**, 314.

YOSHIDA A. (1966) Glucose-6-phosphate dehydrogenase of human erythrocytes. I. Purification and characterization of normal (B+) enzyme. *J. biol. Chem.* **241**, 4966.

YOSHIDA A. (1967a) A single amino acid substitution (asparagine to aspartic acid) between normal (B+) and the common Negro variant (A+) of human glucose-6-phosphate dehydrogenase. *Proc. natn. Acad. Sci.* **57**, 835.

YOSHIDA A. (1967b) Human glucose-6-phosphate dehydrogenase: purification and characterization of Negro type variant (A+) and comparison with normal (B+). *Biochem. Genet.* **1**, 81.

YOSHIDA A., STAMATOYANNOPOULOS G. & MOTULSKY A.G. (1967a) Negro variant of glucose-6-phosphate dehydrogenase deficiency (A−) in man. *Science* **155**, 97.

YOSHIDA A., STAMATOYANNOPOULOS G. & MOTULSKY A.G. (1967b) Biochemical genetics of glucose-6-phosphate dehydrogenase. *Ann. N. Y. Acad. Sci.* (in press).

YOSHIDA A., STEINMANN L. & HARBERT P. (1967c) *In vitro* hybridization of normal and variant human glucose-6-phosphate dehydrogenase. *Nature* **216**, 275.

YUE P.C.K. & STRICKLAND M. (1965) Glucose-6-phosphate dehydrogenase deficiency and neonatal jaundice in Chinese male infants in Hong Kong. *Lancet* **i**, 350.

ZANNOS-MARIOLEA L. & KATTAMIS C. (1961) G-6-PD deficiency in Greece. *Blood* **18**, 34.

ZINKHAM W.H. (1960) Enzyme studies on lenses from persons with primaquine sensitive erythrocytes. *Am. J. Dis. Child.* **100**, 525.

ZINKHAM W.H. (1967) The selective hemolytic action of drugs: clinical and mechanistic considerations. *J. pediat.* **70**, 200.

ZINKHAM W.H. & LENHARD R.E. (1959) Metabolic abnormalities of erythrocytes from patients with congenital non-spherocytic hemolytic anemia. *J. pediat.* **55**, 319.

CHAPTER 13

6 PHOSPHOGLUCONATE
DEHYDROGENASE

Inherited Variation of 6PGD in Man 485
   Electrophoretic variants 485
      Phenotypes I, II and III..... 485
      Phenotypes IV, V and VI 487
         The Richmond variant (IV) 487
         The Hackney variant (V) 487
         The Friendship variant (VI) 488
   Inherited 6PGD deficiency..... 489
      Ilford and Newham variants 489
      Whitechapel and Dalston variants 489
      Other phenotypes.............. 490

Geographic Distribution of 6PGD Phenotypes.......... 490
Methods 491
   Measurement of 6PGD activity 491
   Differentiation of 6PGD phenotypes by starch gel electrophoresis 494

References.......... 495

In the previous chapter, the sequence of reactions in the HMP shunt pathway leading to the production of ribulose 5-phosphate (R5P) was condensed in order to focus attention on the initial oxidative step by G6PD and its association with glutathione reduction through NADPH. Between the first and second oxidative steps, there is an intermediate reaction, the hydrolysis of 6-phospho-δ gluconolactone to 6-phosphogluconate (6PG), which may be catalyzed by a lactonase. The oxidative decarboxylation of 6PG to R5P is catalyzed by 6-phosphogluconate dehydrogenase (6PGD), generating a second molecule of NADPH. The systematic name for this enzyme is D-glucose-1-phosphate phosphotransferase (decarboxylating), E.C. number 1.1.1.44.

Kazazian (1966) found that the molecular weight of 6PGD prepared from human red cells as well as from homogenates of *Drosophila* was about 80,000 and concluded that in both instances, the enzyme is a dimer containing subunits of about 40,000 molecular weight. The studies of Kazazian *et al.* (1965) and Young (1966) showed that in

TABLE 13.1. 6PGD phenotypes

Phenotype number	Trivial name	Presumed genotype	Activity (% of normal)	Stability
Electrophoretic Variants:				
I	Usual phenotype	$PGD^A\ PGD^A$	100	Normal
II	Common variant	$PGD^A\ PGD^C$	80–100	Intermediate
III	Canning variant	$PGD^C\ PGD^C$	70–90	Low
IV	Richmond variant	$PGD^A\ PGD^R$	100	Intermediate or normal
V	Hackney variant	$PGD^A\ PGD^H$	100	Normal
VI	Friendship variant	$PGD^A\ PGD^F$	100	Intermediate
Low Activity Variants:				
	Ilford variant	$PGD^A\ PGD^O$	50–60	Normal
	Newham variant	$PGD^C\ PGD^O$	40–50	Intermediate
	Whitechapel variant	$PGD^W\ PGD^W$	1–5	? very low
	Dalston variant	$PGD^A\ PGD^W$	60–80	Intermediate

Drosophila, the genetic locus for 6PGD, like that for G6PD, is on the X chromosome. However, abundant evidence indicates that in man as in pigeons, rats, mice, rabbits, cats, pigs and monkeys, the locus (called PGD) for this enzyme is autosomal (Fildes & Parr, 1963; Parr, 1966; Shaw, 1965; Parr & Fitch, 1964, 1967; Young, 1966; Dern *et al.*, 1966; Bowman *et al.*, 1966; Thuline *et al.*, 1967; Davidson, 1967; Personal communication from R. Saison).

INHERITED VARIATION OF 6PGD IN MAN

This enzyme has been found to vary in both electrophoretic pattern and catalytic activity, so that it is convenient to consider the variants under two different headings. The nomenclature for the phenotypes has undergone considerable change. With a few minor exceptions, the names proposed by Parr & Fitch (1967) and Davidson (1967) are used in this chapter.

Table 13.1 summarizes the trivial names, presumed genotypes, relative enzymatic properties and stability characteristics of the six variants which can be differentiated by their electrophoretic patterns. These types are designated by Roman numerals, according to Parr & Fitch (1967). In addition, there are four variants not assigned a Roman numeral since they are not recognizable on the basis of electrophoretic pattern, but rather by a variable decrease in enzyme activity. Sources for the data in Table 13.1 are given in the following paragraphs.

ELECTROPHORETIC VARIANTS

Fildes & Parr (1963) first described the appearance of red cell 6PGD on starch gel electrophoresis and reported the occurrence of an inherited variant. These findings were subsequently enlarged by Parr (1966), who described five different electrophoretic patterns. A sixth was recently described by Davidson (1967).

PHENOTYPES I, II AND III
The electrophoretic patterns of phenotypes I, II and III in hemolysates and white cell extracts are shown in Fig. 13.1. The pattern of the most common type consists of a single band, named a, which migrates

17

toward the anode at pH 7 (Fildes & Parr, 1963). On prolonged incubation, additional, faster-moving components appear in the pattern of red cell lysates (Fildes & Parr, 1964; Bowman *et al.*, 1966; Giblett, 1967), but not of white cells and other tissues (Parr, 1966; Davidson, 1967). Furthermore, when NADP is added to the starch gel, one or more additional cathodal bands appear, particularly on prolonged incubation during staining. Parr & Fitch (1967) called this the Type I or Usual phenotype, and found it in over 95 per cent of a London population sample.

A second pattern, phenotype II or the common variant, was found in 4–5 per cent of subjects tested (Parr & Fitch, 1964; Parr, 1966). It consists of band a and two more cathodal bands, b and c. As with phenotype I, additional anodal components are observed in the red cell pattern when incubation is prolonged. Furthermore, bands a and b are of about equal intensity in the hemolysate pattern, but in white cells, the b band predominates, a and c being about equal. In hemolysates, band c is weak; it originally escaped detection (Parr, 1966; Parr & Fitch, 1967; Davidson, 1967).

Phenotype III has a very low frequency. Its pattern in red cells also consists of bands a, b and c, with band a as a minor component (Parr, 1966; Bowman *et al.*, 1966). In other tissues, only band c is observed (Parr & Fitch, 1967; Davidson, 1967, Carter *et al.*, 1968). Parr found this variant in the Canning Town district of London, so it was originally named the Canning phenotype.

Family studies showed that phenotypes I and III represent homozygosity for the autosomal allelic genes, PGD^A and PGD^C, respectively, while phenotype II is the heterozygous pattern (Parr, 1966; Bowman *et al.*, 1966; Parr & Fitch, 1967; Carter *et al.*, 1968). It is currently assumed that the enzyme exists as a dimer, so that in the two homozygous types, the gene product is a subunit, (i.e. S^A or S^C), which dimerizes to form either the a band ($S^A S^A$) or the c band ($S^C S^C$). In the heterozygote, both bands a and c are present, as well as the b band, which is thought to be a mixed dimer, $S^A S^C$ (Parr, 1966; Parr & Fitch, 1967).

The nature of the heterogeneity observed in red cell (but not white cell) lysate patterns of type III has not been clarified. Bowman *et al.* (1966) noted that additional *cathodal* bands appeared in hemolysate patterns of all three types when NADP was incorporated into the gel. Carter *et al.* (1968) also reported the appearance of cathodal bands

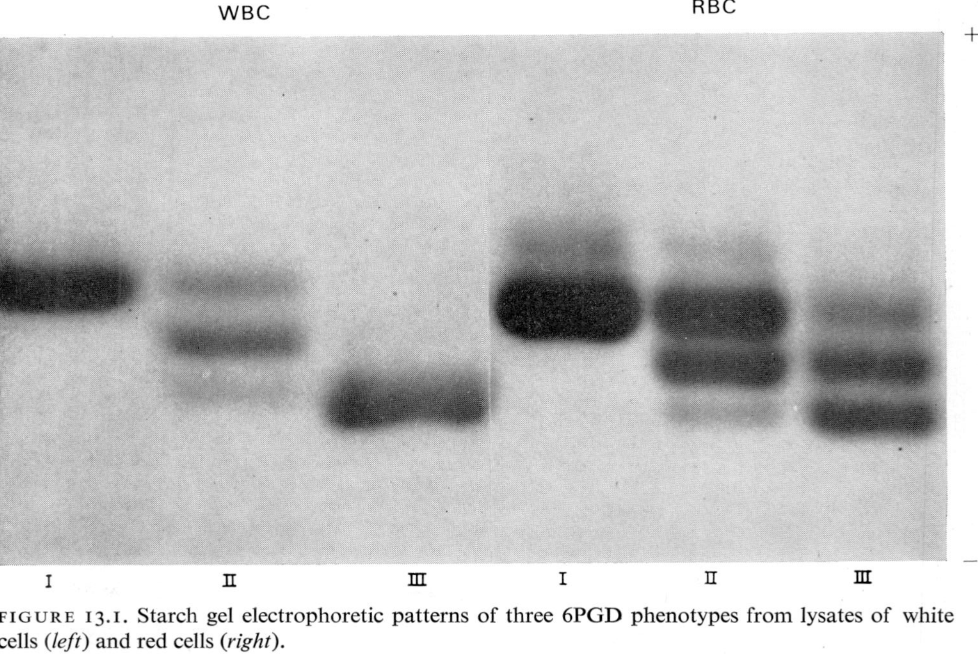

FIGURE 13.1. Starch gel electrophoretic patterns of three 6PGD phenotypes from lysates of white cells (*left*) and red cells (*right*).

associated with NADP incorporation as well as with prolonged storage of hemolysates. Carson *et al.* (1966) found that partial inactivation of 6PGD occurred when hemolysates were incubated with stroma in the presence of NADP. This phenomenon was investigated further by Ajmar *et al.* (1968), using hemolysates containing 6PGD of type I. They found that a product of the reaction between NADP and stromal NADPase (2'-phosphoadenosine diphosphate ribose, or P-ADPR) modified both the structure and activity of 6PGD. The extent of electrophoretic pattern modification was dependent on the time period of hemolysate incubation, with the progressive appearance of slower-moving bands accompanied by disappearance of the major (anodal) component. Presumably, the heterogeneity of type III 6PGD in electrophoretic patterns of hemolysates also reflects the interaction of this form of the enzyme with some substance(s) present in red cells but not in other tissues.

Parr (1966) found that the enzyme activity of phenotype II was about 80–100 per cent of normal (i.e. of type I) while that of type III was 70–90 per cent. Carson *et al.* (1966) similarly reported that the c component of types II and III was more labile than the a component. According to Carter *et al.* (1966), treatment of the three enzyme types with urea decreased the activity of type I by 37 per cent, of type II by 70 per cent and of type III, by 95 per cent. Even more striking differentiation was achieved by iodoacetate treatment, reducing activity by 20 per cent, 50 per cent and 80 per cent, respectively.

PHENOTYPES IV, V AND VI

The 'Richmond' variant, type IV, was observed in two presumably unrelated subjects among 4558 tested by Parr (1966) and in four members of an American family by Davidson (1967). The corresponding genotype appears to be PGD^APGD^R. The phenotypic pattern, illustrated diagrammatically in Fig. 13.2, consists of three bands, the most cathodal of which corresponds to the usual a band, while the intermediate band predominates. Davidson (1967) found no difference in the red cell and white cell pattern. The enzyme activity was in the normal range, and treatment with urea or iodoacetate decreased activity to about the same extent as that observed with type III. According to Carter *et al.* (1968), type IV enzyme behaves like type I when treated with these substances.

The 'Hackney' variant, phenotype V, described by Parr (1966) was found in two of 4558 samples tested. This phenotype apparently represents heterozygosity for PGD^A and PGD^H. It consists of three bands with mobility only slightly slower than bands a, b and c. The intensity of band b in this pattern differentiates it from phenotype II.

The 'Friendship' variant is actually the ninth phenotype described in the PGD system, but since three of the phenotypes (designated VI, VII and VIII by Parr, 1966) represent reduction in enzyme activity rather than altered electrophoretic mobility, they are excluded from

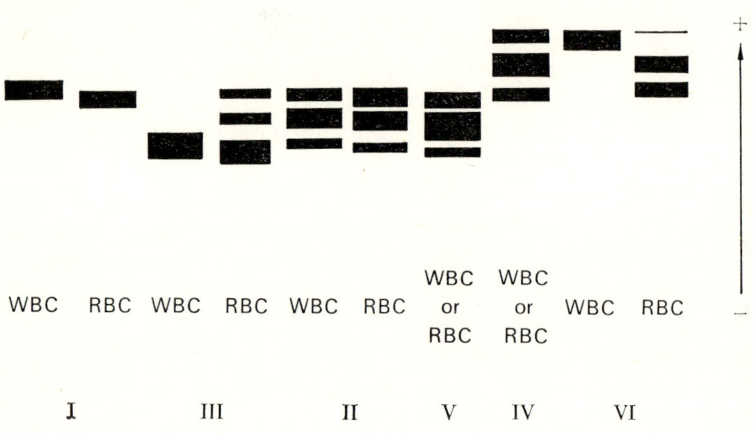

FIGURE 13.2. Diagram of electrophoretic patterns of six different 6PGD phenotypes from red cells and white cells. The 'Hackney' (V) and 'Richmond' (IV) patterns of red cells and white cells do not differ. (Based on Carter *et al.*, 1968.)

the numerical series in this chapter. Thus, the designation phenotype VI is here assigned to the rare Friendship variant reported by Davidson (1967). The gene in this instance can be called PGD^F, since the pedigree indicates that a fifth allele at the PGD locus is segregating in the family of the propositus. The electrophoretic pattern of the red cell enzyme somewhat resembles that of phenotype IV, but there is a notable difference in the white cell patterns, since in phenotype VI, there is only one component, which has a faster migration rate than the normal a band. However, Davidson (1967) found that addition of NADP to the starch gel caused an alteration in the type VI patterns of both red cell and white cell 6PGD, converting them to a single band with the same mobility as the a band of phenotype I. Enzyme activity of this variant was found to be in the normal range, and it was decreased by urea or iodoacetate to about the same extent as phenotype II.

INHERITED 6PGD DEFICIENCY

Parr & Fitch (1964) and Brewer & Dern (1964) found that among hematologically normal subjects, a small proportion (probably less than one per cent) have about 50–60 per cent of normal red cell 6PGD activity. Brewer & Dern (1964) determined that with this level of activity, the methemoglobin reduction, glutathione stability and dye decolorization tests used for detecting G6PD deficiency were normal, even though some decrease in NADP reduction would be expected.

ILFORD AND NEWHAM VARIANTS

The studies of Parr & Fitch (1964, 1967) suggested that the rare, so-called 'Ilford' variant which has about 50 per cent of normal activity, can be explained by the inheritance of a 'silent' PGD^0 gene partnered with a PGD^A gene, resulting in an electrophoretic pattern indistinguishable from phenotype I, except for the decrease in activity. In one family they found two brothers with 40–50 per cent of normal enzyme activity, in whom the pattern was the same as phenotype III. This 'Newham' variant could be explained by heterozygosity for PGD^C and PGD^0.

WHITECHAPEL AND DALSTON VARIANTS

In another family, Parr & Fitch (1967) found two siblings, a man and a woman, with the 'Whitechapel' phenotype, characterized by enzyme levels of only 1–5 per cent of normal. In both instances, the G6PD level was higher than normal, suggesting a younger age red cell population, due to decreased cell survival. Although it was stated that neither of these subjects had clinical manifestations of a hemolytic process, the hematocrit, reticulocyte levels and cell indices were not reported.

Some of the members of the Whitechapel family, including the granddaughter of the woman with marked deficiency, had 6PGD levels about three-fourths normal. Their phenotype was designated as the 'Dalston' variant. None was found to be in the middle range characteristic of individuals heterozygous for PGD^0. Furthermore, some evidence for instability of the Dalston variant enzyme was obtained. The authors proposed that the Whitechapel variant represents homozygosity for a PGD^W gene, the product of which is highly unstable in the absence of the PGD^A gene. However, in the PGD^APGD^W heterozygotes (i.e. those with the Dalston variant), formation of the dimer from subunits S^W and S^A results in a fairly stable enzyme. Thus, instead of the 50 per cent enzyme activity seen in the Ilford and Newham variants (both of which are heterozygous for PGD^0), there is a higher level (about 75 per cent) in the PGD^WPGD^A heterozygotes.

OTHER PHENOTYPES

Dern *et al.* (1966) found four families in which the half normal enzyme levels of some members suggested the presence of a gene behaving like the proposed amorph, PGD^0. However, in three other families, a more complicated genetic interpretation was required, particularly since one individual with low enzyme activity was reported to have phenotype II. Their data suggested that genes at other loci may play a regulator role in controlling the amount of PGD gene product.

Two instances of 6PGD partial deficiency have been reported to be associated with hemolytic anemia. Scialom *et al.* (1965) described a young man in whom the red cell enzyme level was 60–80 per cent of normal, G6PD activity was elevated (in accordance with the young cell population), and glutathione stability was normal. The author suggested that the moderate deficiency of 6PGD was probably not directly related to the hemolytic process. The patient described by Lausecker *et al.* (1965) was a newborn infant with decreased 6PGD and GSH. However, NADH and ATP levels were also low, suggesting a defect involving the anaerobic as well as the oxidative glycolytic pathway.

GEOGRAPHIC DISTRIBUTION
OF 6PGD PHENOTYPES

Only a small number of populations have been tested for their 6PGD phenotypes by starch gel electrophoresis of red cell hemolysates. The data currently available are listed in Table 13.2 along with some previously unpublished data from this laboratory. The table indicates that the PGD^A gene is prevalent among all populations tested. However

PGD^C, which has a frequency of about 0·03 among Europeans, is found in higher frequencies among African Negroes and at least some Orientals, such as the Chinese, northern Japanese and Thais. The highest frequency of *PGD*^{C} yet reported is 0·247 in 95 natives of Bhutan. Although more data are needed, the reports of Arends *et al.* (1967) and Tashian *et al.* (1967) show that *PGD*^{C} may have a very low frequency in South American Indians.

Some heterogeneity has been noted among Negroes. For example, a *PGD*^{C} frequency of 0·152 was found in a sample of 200 South African natives (Gordon *et al.*, 1967), and in 318 natives of Mozambique the frequency was 0·091 (Giblett *et al.*, in preparation). Nigerians and Ugandans had a frequency of 0·057 (Parr, 1966), while frequencies among American Negroes have ranged from 0·030 in Chicago (Dern *et al.*, 1966) to 0·054 in Seattle (Giblett & Scott, unpublished).

It appears from the limited data that *PGD*^{C} probably is more common in Asia and Africa than in Europe and America; further information is required to determine if these trends are meaningful.

METHODS

MEASUREMENT OF 6PGD ACTIVITY
(Parr & Fitch, 1964)

A 1:50 hemolysate is prepared by pipetting 0·1 ml of blood into 4·9 ml of water. Ten minutes later, the tube is centrifuged at 2000 g for 10 minutes, and the clear hemolysate is tested without delay. One-half ml of hemolysate is added (in a cuvette with 1 cm path length) to a mixture of 1·0 ml 0·3 M tris buffer (pH 8·0), 0·3 ml of 0·1 M MgCl$_2$, 0·1 ml 18 mM sodium phosphogluconate and 1·4 ml water. A second cuvette for measuring the combined 6PGD and G6PD activities contains the same mixture plus 0·1 ml of 18 mM sodium glucose 6-phosphate. The reactions are started by adding 0·1 ml of 6 mM NADP. Optical density readings are made at 1 minute intervals at 340 mμ for 10 minutes against a blank containing 0·5 ml of hemolysate, 1·0 ml of buffer and 1·5 ml of water. Hemoglobin is measured against a water blank at 540 mμ. The 6PGD activity in international units per gram of hemoglobin is then calculated:

$$6\text{PGD IU per gram Hb} = \frac{137\cdot9\ (\Delta\text{OD at 340 m}\mu)}{\Delta\text{OD at 540 m}\mu}$$

TABLE 13.2. Population distributed of 6PGD variants: Phenotype and gene frequencies

| Population | Number | Phenotype frequency | | | Gene frequency | | References |
		I	II	III	PGD^A	PGD^C	
European							
England	4558	0·959	0·041	0·001	0·979	0·021	Parr (1966)
Greece	128	0·922	0·078	—	0·961	0·039	Carter *et al.* (1968)
So. Africa	200	0·035	0·065	—	0·968	0·032	Gordon *et al.* (1967)
Chicago	600	0·924	0·074	0·002	0·961	0·039	Bowman *et al.* (1966)
Buffalo	1377	0·953	0·045	0·002	0·976	0·024	Davidson (1967)
Seattle	647	0·965	0·034	0·001	0·981	0·019	Giblett & Scott (unpubl.)
Jews							
Habbanite	499	0·860	0·132	0·008	0·926	0·074	Carter *et al.* (1968)
Negro							
Nigeria & Uganda	209	0·885	0·115	—	0·943	0·057	Parr (1966)
So. Africa Bantus	200	0·715	0·265	0·020	0·848	0·152	Gordon *et al.* (1967)
So. Africa Cape Colored	200	0·910	0·090	—	0·955	0·045	Gordon *et al.* (1967)
Mozambique	318	0·821	0·176	0·003	0·919	0·091	Giblett *et al.* (in preparation)
Chicago	296	0·939	0·061	—	0·970	0·030	Dern *et al.* (1966)
Chicago	416	0·926	0·074	—	0·961	0·039	Bowman *et al.* (1966)

Buffalo	1226	0·930	0·068	0·002	0·964	0·036	Davidson (1967)
Seattle	506	0·893	0·103	0·004	0·946	0·054	Giblett & Scott (unpubl.)
Asiatic							
Iran	322	0·944	0·056	—	0·972	0·028	Bowman & Ranaghy (1967)
Malay	100	0·950	0·050	—	0·975	0·025	Gordon *et al.* (1967)
China	228	0·873	0·123	0·004	0·934	0·066	Shih *et al.* (1968)
Japan							
(Hokkaido)	180	0·860	0·140	—	0·931	0·069	Giblett & Scott (unpubl.)
Thailand	441	0·859	0·141	—	0·930	0·070	Giblett & Scott (unpubl.)
Bhutan	95	0·547	0·411	0·042	0·753	0·247	Carter *et al.* (1968)
Seattle (mixed)	176	0·909	0·085	0·006	0·952	0·048	Giblett & Scott (unpubl.)
Indians							
Venezuela	283	0·997	0·003	—	0·998	0·002	Arends *et al.* (1967)
Brazil	181	1·000	—	—	1·000	—	Tashian *et al.* (1967)
Mexico							
(Yucatan)	85	1·000	—	—	1·000	—	Bowman *et al.* (1966)

The G6PD activity is calculated by subtracting the 6PGD activity from the combined activity of G6PD plus 6PGD. Temperature correction to 25°C is made from the assumption that the activity of both enzymes increases linearly from 74 per cent at 20°C to 126 per cent at 30°C. Expression of 6PGD activity per unit of G6PD activity is often desirable, particularly when the cell population contains a high proportion of young cells or the MCH is low, as in thalassemia or iron deficiency. Since the activities of both enzymes are elevated in reticulocytes and decrease with age, expressing the activity of either enzyme in terms of hemoglobin can be misleading (see Chapter 12). In normal subjects, 6PGD activity per gram of Hb is about 3 units, and the 6PGD:G6PD ratio is about 0·8.

DIFFERENTIATION OF 6PGD PHENOTYPES
BY STARCH GEL ELECTROPHORESIS

(Fildes & Parr, 1963)

Electrophoresis in starch gels has so far been used for determining 6PGD phenotypes, but other media can probably be employed. Good results are obtained with several buffers. In this laboratory, the buffer system and starch gel preparation described for G6PD (see Chapter 12) is also used for 6PGD, so that one half of the gel can be stained for one enzyme and the second half for the other enzyme. If G6PD staining is not required, the NADP is omitted from the buffer and gel, since the addition of NADP is associated with the appearance of additional cathodal bands in the 6PGD zymogram.

Staining is carried out in the same manner as for G6PD except that the substrate for 6PGD is 6 phosphogluconate (12·0 mg in 6 ml of tris-HCl buffer solution containing NADP, $MgCl_2$, MTT and PMS, as described in Chapter 12). This solution, mixed with 6 ml of molten agar, is poured over the gel, and the bands of enzyme activity develop at 37°C in the dark for about 30 minutes.

Fitch & Parr (1966) have introduced a method of applying only 1 ml of reaction mixture (without agar) to the gel, using a small paint brush. Development at room temperature is said to require 30–40 minutes. Since this procedure is both simple and economical of reagents, it would appear to be preferable to the agar overlay.

REFERENCES

AJMAR F., SCHARRER B., HASHIMOTO F. & CARSON P.E. (1968) Interrelation of stromal NAD(P)ase and human erythrocytic 6-phosphgoluconate dehydrogenase. *Proc. natn. Acad. Sci.* **59**, 538.

ARENDS T., BREWER G., CHAGNON N., GALLANGO M.L., GERSHOWITZ H., LAYRISSE M., NEEL J., SHREFFLER D., TASHIAN R. & WEITKAMP L. (1967) Intratribal genetic differentiation among the Yanomama Indians of southern Venezuela. *Proc. natn. Acad. Sci.* **57**, 1252.

BOWMAN J.E., CARSON P.E., FRISCHER H. & DEGARAY A.L. (1966) Genetics of starch-gel electrophoretic variants of human 6-phosphogluconic dehydrogenase: population and family studies in the United States and in Mexico. *Nature* **210**, 811.

BOWMAN J.E. & RONAGHY H. (1967) Hemoglobin, glucose-6-phosphate dehydrogenase, phosphogluconate dehydrogenase and adenylate kinase polymorphism in Moslems in Iran. *Am. J. phys. Anthrop.* **27**, 119.

BREWER G.J. & DERN R.J. (1964) A new inherited enzymatic deficiency of human erythrocytes: 6-phosphogluconate dehydrogenase deficiency. *Am. J. hum. Genet.* **16**, 472.

CARSON P.E., AJMAR F., HASHIMOTO F. & BOWMAN J.E. (1966) Electrophoretic demonstration of stromal effects on haemolysate glucose-6-phosphate dehydrogenase and 6-phosphogluconic dehydrogenase. *Nature* **210**, 813.

CARTER N.D., FILDES R.A., FITCH L.I. & PARR C.W. (1968) Genetically determined electrophoretic variations of human phosphogluconate dehydrogenase. *Acta genet.* **18**, 109.

CARTER N.D., GOULD S.R., PARR C.W. & WALTER P.H. (1966) Differential inhibition of human red cell phosphogluconate-dehydrogenase variants. *Biochem. J.* **99**, 17.

DAVIDSON R.G. (1967) Electrophoretic variants of human 6-phosphogluconate dehydrogenase: population and family studies and description of a new variant. *Ann. hum. Genet.* **30**, 355.

DERN R.J., BREWER G.J., TASHIAN R.E. & SHOWS T.B. (1966) Hereditary variation of erythrocytic 6-phosphogluconate dehydrogenase. *J. Lab. clin. Med.* **67**, 255.

FILDES R.A. & PARR C.W. (1963) Human red cell phosphogluconate dehydrogenase. *Nature* **200**, 890.

FILDES R.A. & PARR C.W. (1964) Various forms of human erythrocytic 6-phosphogluconate dehydrogenase. *Proc. 6th Cong. int. Biochem.* p. 229.

FITCH L.I. & PARR C.W. (1966) Development of zymograms by the paint brush technique. *Biochem. J.* **99**, 20P.

GIBLETT E.R. (1967) Variant phenotypes: haptoglobin, transferrin and red cell enzymes, *in* GREENWALT T.J. (ed.) *Advances in Immunogenetics*, p. 99. J.B. Lippincott, Philadelphia.

GORDON H., KERAAN M.M. & VOOIJS M. (1967) Variant of 6-phosphogluconate dehydrogenase within a community. *Nature* **214**, 466.

KAZAZIAN H.H. (1966) Molecular size studies on 6-phosphogluconate dehydrogenase. *Nature* **212**, 197.

KAZAZIAN H.H., YOUNG W.J. & CHILDS B. (1965) X-linked 6-phosphogluconate dehydrogenase in *Drosophila:* subunit associations. *Science* **150**, 1601.

LAUSECKER C., HEIDT P., FISCHER D., HARTLEYB H. & LOHR G.W. (1965) Anémie hémolytique constitutionnelle avec déficit en 6-phosphogluconate déshydrogénase. *Archs fr. pédiat.* **21**, 789.

PARR C.W. (1966) Erythrocyte phosphogluconate dehydrogenase polymorphism. *Nature* **210**, 487.

PARR C.W. & FITCH L.I. (1964) Hereditary partial deficiency of human erythrocyte phosphogluconate dehydrogenase. *Biochem. J.* **93**, 28c.

PARR C.W. & FITCH L.I. (1967) Inherited quantitative variations of human phosphogluconate dehydrogenase. *Ann. hum. Genet.* **30**, 339.

SCIALOM C., NAJEAN Y. & BERNARD J. (1966) Anémie hémolytique congénitale non sphérocytaire avec déficit incomplet en 6-phosphogluconate déshydrogénase. *Nouv. Revue fr. Hémat.* **6**, 452.

SHAW C.R. (1965) Electrophoretic variations in enzymes. *Science* **149**, 936.

SHIH L.Y., HSIA D.Y.Y., BOWMAN J.E., SHIH S.C. & SHIH P.L. (1968) The electrophoretic phenotypes of red cell 6-phosphogluconate dehydrogenase and adenylate kinase in Chinese populations. *Am. J. hum. Genet.* **20**, 474.

TASHIAN R.E., BREWER G.J., LEHMANN H., DAVIES D.A. & RUCKNAGEL D.L. (1967) Further studies on the Xavante Indians. V. Genetic variability in some serum and erythrocyte enzymes, hemoglobin, and the urinary excretion of β-aminoisobutyric acid. *Am. J. hum. Genet.* **19**, 524.

THULINE H.C., MORROW A.C., NORBY D.E. & MOTULSKY A.G. (1967) Autosomal phosphogluconic dehydrogenase polymorphism in the cat. *Science* **157**, 431.

YOUNG W.J. (1966) X-linked electrophoretic variation in 6-phosphogluconate dehydrogenase in *Drosophila melanogaster. J. Hered.* **57**, 58.

CHAPTER 14

PHOSPHOGLUCOMUTASE

Properties of PGM	497	The PGM₃ locus		504
The PGM Phenotypes	499	Distribution of PGM iso-		
Isozymes of the PGM₁ locus		zymes in blood cells		505
genes	499	Geographic Distribution of PGM		
Common types	499	Genes in Various Populations		505
Rare PGM₁ locus variants	500	Methods		508
Linkage studies	501	Determination of PGM types		
Isozymes of the PGM₂ locus		by starch gel electrophoresis		508
genes	501	References		510

Properties of PGM ... 497
The PGM Phenotypes ... 499
Isozymes of the PGM$_1$ locus genes ... 499
Common types ... 499
Rare PGM$_1$ locus variants ... 500
Linkage studies ... 501
Isozymes of the PGM$_2$ locus genes ... 501
The PGM$_3$ locus ... 504
Distribution of PGM isozymes in blood cells ... 505
Geographic Distribution of PGM Genes in Various Populations ... 505
Methods ... 508
Determination of PGM types by starch gel electrophoresis ... 508
References ... 510

Phosphoglucomutase (α-D-glucose-1,6 diphosphate: α-D-glucose-1 phosphate phosphotransferase, EC 2.7.5.1) is an important and ubiquitous enzyme which reversibly catalyzes the transfer of phosphate from the first to the sixth position of glucose. Najjar & Pullman (1954) showed that phosphoglucomutase (PGM) exists in phospho and dephospho forms, and concluded that the reaction occurs in two steps:

Glucose-1-phosphate + phosphoenzyme $\rightleftharpoons$

glucose-1,6 diphosphate + dephospho-enzyme $\rightleftharpoons$

glucose-6-phosphate + phosphoenzyme

It appears that free glucose-1,6 diphosphate (G 1,6 diP) may not be an obligatory intermediate in this reaction; although its presence is necessary for PGM activity, its apparent role is to prevent the depletion of phosphate from the phosphoenzyme (Ray & Roscelli, 1964; Gounaris *et al.*, 1967).

PROPERTIES OF PGM

The general properties of PGM were reviewed in 1962 by Najjar (Najjar, 1962). Since that date, further studies have added to the extensive literature on this subject. For example, Bocchini *et al.*

(1967) found that PGM isolated from rabbit muscle has six cysteine residues per estimated 74,000 molecular weight, and the titration rate of dephospho-PGM with p-hydroxymercuribenzoate is much faster than that of phospho-PGM, indicating a greater accessibility of two of the sulfhydryl groups. These authors felt that the changes in molecular conformation reflected by the titration data were associated with the state of phosphorylation rather than to an induced fit to an enzyme-substrate complex.

Studies on the structure of PGM from various bacterial and animal sources (summarized by Handler *et al.*, 1965) indicate that the active center of the molecule consists of a pentapeptide, in which serine is the actual site of phosphate attachment (Bocchini *et al.*, 1964). Thus, PGM belongs to a family of 'serine enzymes' in which the prototype of the active center is:

$$---\left\{\begin{matrix} \text{Asp} & & \text{Gly} \\ \text{or} & \text{Ser} & \text{or} \\ \text{Glu} & & \text{Ala} \end{matrix}\right\}---$$

This family includes the bacterial enzymes, alkaline phosphatase and proteinase (Schwartz *et al.*, 1963) as well as the mammalian enzymes, trypsin, chymotrypsin, thrombin, elastase, pseudocholinesterase and liver aliesterase (Dixon *et al.*, 1958; Cohen *et al.*, 1959; Walsh & Neurath, 1964; Hartley, 1964; Smillie & Hartley, 1964). It is postulated that these enzymes may have arisen by gene duplication from a common precursor (summarized by Epstein & Motulsky, 1965).

Joshi *et al.* (1967) compared the column chromatography elution patterns of purified PGM obtained from human, rabbit, rat, flounder, potato, yeast and *E. coli*. In most preparations, there were two major peaks of enzyme activity. In some instances, a third (and rarely, a fourth) minor peak was identified as the dephospho form of a major peak. The major enzymes (I and II) of rabbit muscle were shown to have differences in amino acid composition and in peptide fingerprints. The authors tentatively concluded that each of these two enzymes, having a molecular weight of about 62,000, represents the phenotypic expression of related but independent genes.

The existence of PGM genetic polymorphism in human subjects was first demonstrated by Spencer *et al.* (1964). Subsequently Hopkinson & Harris (1965, 1966) showed that two different PGM genetic

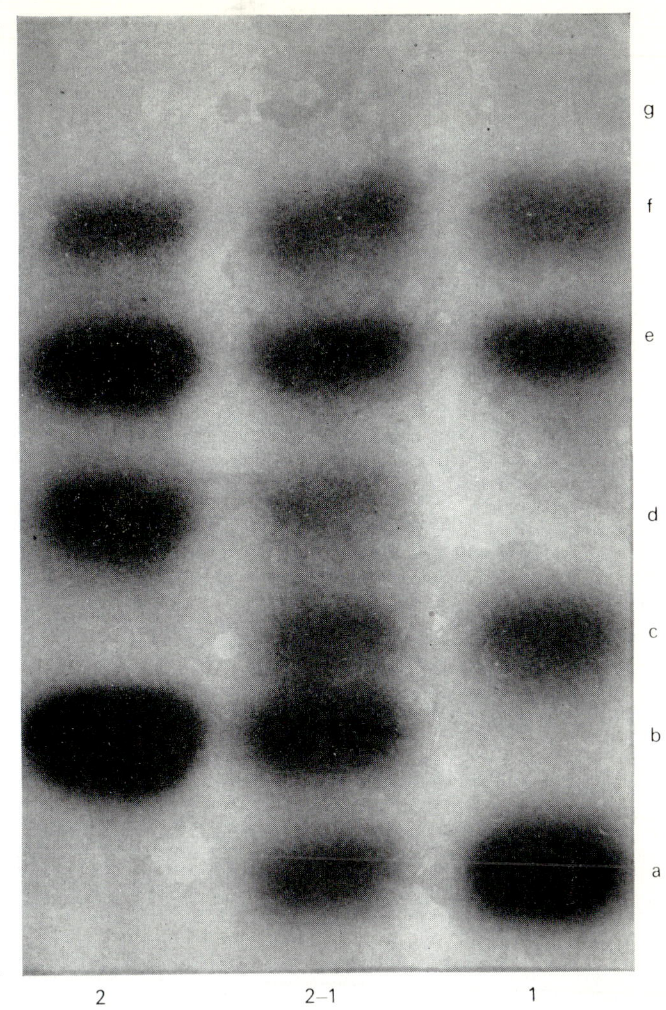

FIGURE 14.1. Starch gel electrophoretic patterns of the three common PGM types. The appearance of components e, f (and g) does not vary; these isozymes are controlled by genes at a second PGM locus.

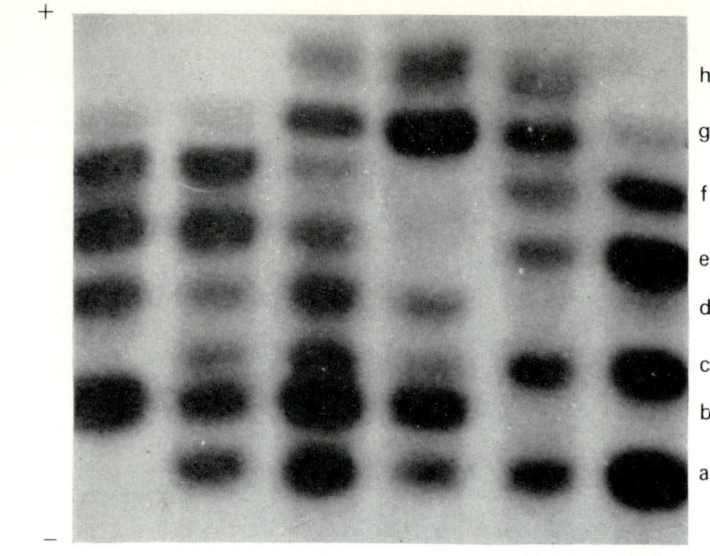

FIGURE 14.3. Starch gel electrophoretic patterns of six different PGM phenotypes reflecting gene variation at the PGM_2 locus, as well as the PGM_1 locus. Two examples of the 'Atkinson' type are shown (patterns 3 and 5), and there is one example of apparent homozygosity for the Atkinson (PGM_2^2) allele, in which bands e and f are missing (pattern 4). This is the Mozambique variant described in the text.

[*Facing page* 499]

systems are demonstrable in human blood and several other tissues. The products of a third PGM locus are represented in extracts of placenta (Harris *et al.* 1968).

THE PGM PHENOTYPES

ISOZYMES OF THE PGM_1 LOCUS GENES

COMMON TYPES

Spencer *et al.* (1964) found that when human red cell lysates are separated by starch gel electrophoresis and stained for PGM activity, three distinct phenotypic patterns are observed. As shown in Fig. 14.1, there are seven PGM isozyme bands, labelled a through g. The three fastest moving bands (e, f and g) are found in all three phenotypes although the g band is usually weak or absent. In PGM 1, zones a and c are present; in PGM 2, zones b and d are present. All four zones are seen in the PGM 2-1 pattern, which has the appearance of a mixture of PGM 1 and PGM 2.

From studies of 338 random adults, Spencer *et al.* determined the phenotype frequencies of PGM 1, 2 and 2-1 to be 0·55, 0·07 and 0·38, respectively. They then tested blood obtained from members of 133 families, and found that the results could be explained by the segregation of two autosomal allelic genes, PGM_1^1 and PGM_1^2. Thus, the phenotypes PGM 1 and 2 represent homozygosity, and PGM 2-1 represents heterozygosity for these two alleles. (Since subsequent studies showed that zones e, f and g are determined by alleles at a second locus, PGM_2, it is necessary to designate the alleles with the appropriate subscript, as well as a superscript.)

Spencer *et al.* (1964) proposed that the allele PGM_1^1 determines the structure of a polypeptide which is common to both the a and c isozymes, while the b and d isozymes share a common polypeptide determined by the allele PGM_1^2. A difference in peptide structure, possibly single amino-acid substitution, might account for the differences in mobility between isozyme a in PGM 1 and isozyme b in PGM 2, as well as c and d of the two phenotypes. However, the relationship of a and c or b and d was (and is) more difficult to ascertain. The suggestion that they may represent the phospho and dephospho forms of the enzyme has not yet been explored directly, although the previously mentioned findings of Joshi *et al.* (1967) are consistent with that possibility.

RARE PGM₁ LOCUS VARIANTS

Hopkinson & Harris (1965, 1966) subsequently described nine additional phenotypes whose isozyme patterns and mode of inheritance indicated heterozygosity for either PGM_1^1 or PGM_1^2 and one of five different rare alleles at the same locus. These phenotypes, shown diagrammatically in Fig. 14.2, were detected during the testing of over

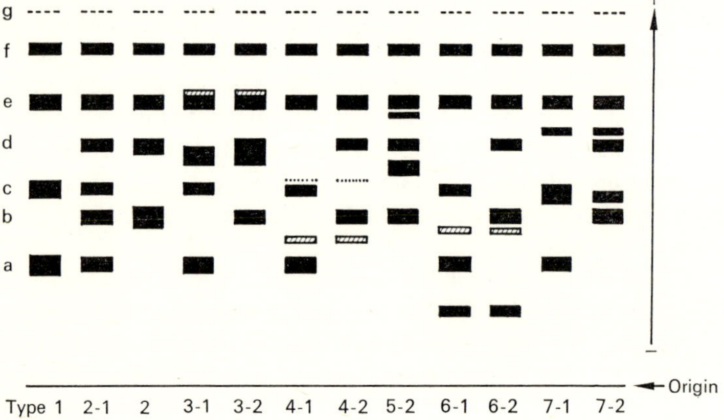

FIGURE 14.2. Diagram of starch gel electrophoretic patterns of the three common types and nine variants representing rare mutants at the PGM₁ locus. (From Hopkinson & Harris, 1966. Reprinted with permission from the *Ann. hum. Genet.* **30**, 167).

3,000 people. With the exception of phenotypes 6-1 and 7-1 (each found in two unrelated individuals and their families), these unusual phenotypes were confined to single families. For example, the eight individuals with PGM 3-1 and two with PGM 3-2 occurred in the same family. A tenth phenotype (PGM 8-1) was later found by Harris *et al.* (1968). (See Fig. 14.6, p. 504.)

These studies indicated that in addition to the common alleles, PGM_1^1 and PGM_1^2, there are at least six rare alleles at this locus: $PGM_1^3, PGM_1^4, PGM_1^5, PGM_1^6. PGM_1^7$ and PGM_1^8. In each instance, the rare gene product appears to consist of a pair of isozymes analogous to a and c or b and d, with the difference in mobility of the slow and fast product about the same in all types. This mobility difference is difficult to determine from Fig. 14.2, because the zones of enzyme activity in some phenotypes, such as 3-1, 3-2, 5-2 and 6-2, are either very close to, or overlap, the common zones. Figure 14.6 (p. 504) shows diagrammatically the individual components determined by the eight postulated alleles at the PGM_1 locus. As suggested by Hopkinson & Harris (1966), such a series of zones could be explained by a series of single amino-acid substitutions. Again, the consistent pattern of two isozymes assumed to be products of a single allele remains unexplained.

LINKAGE STUDIES

Combined data from the studies of Hopkinson & Harris (1966), Robson *et al.* (1966), Tippett (1967) and Edwards & Wingham (1967) revealed no close linkage between the PGM_1 locus and the loci for ABO, MNSs, Rh, K, Le, Lu, Fy, Jk, Dombrock, Hp, Tf, 6PGD, acid phosphatase or placental alkaline phosphatase. Many of these findings were confirmed by Gedde-Dahl & Monn (1967). However, these authors found that in ten informative matings, the calculated recombination fraction between the PGM_1 locus and PTC locus (i.e. the locus determining ability to taste phenylthiocarbamide) was 0.09, with 90 per cent confidence limits of 0·03–0·24. The authors pointed out that these findings are highly suggestive of linkage, but additional data are required for confirmation.

ISOZYMES OF THE PGM_2 LOCUS GENES

Hopkinson & Harris (1965) found that while the electrophoretic mobility of zones e and f is not affected by variation at the PGM_1 locus, there are some phenotypes in which these zones are altered. In two unusual phenotypes, PGM 1 Atkinson and PGM 2-1 Atkinson, the PGM_1 locus isozymes are those characteristic of PGM 1 and PGM 2-1, but the g zone (usually weak or absent) is intensified, while zones e and f are less intense than usual. Also, a faster zone, h, is observed. Examples of these patterns are shown in Fig. 14.3.

Hopkinson & Harris found these unusual electrophoretic patterns in the Atkinson (Negro) family, whose pedigree is given in Fig. 14.4.

They suggested that PGM 1 At. and 2-1 At. represent heterozygosity for two structural genes at the PGM_2 locus, which is not closely linked to the PGM_1 locus. Thus, in the pedigree, the non-identical twins, III-2 and III-3, both inherited their father's PGM_1^1 gene; but while III-2 received the father's 'Atkinson' allele (PGM_2^2), III-3 received his PGM_2^1 allele.

Other examples of these Atkinson types were found by Lie-Injo (1966), Brewer *et al.* (1967) and Giblett (1967). In all cases, the subjects were Negroes. An apparent example of an individual homozygous for

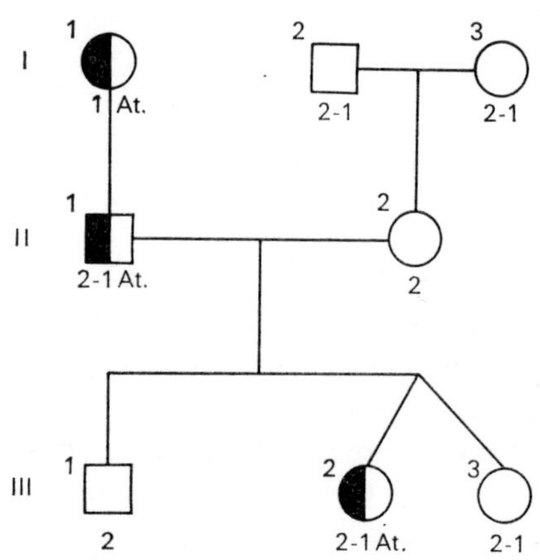

FIGURE 14.4. Pedigree of a family with the Atkinson (At.) phenotype, showing independent assortment of genes at the PGM_1 and PGM_2 loci. (Adapted from Hopkinson & Harris, 1966. Reprinted with permission from the *Ann. hum. Genet.* **30**, 167).

the PGM_2^2 gene was found by Giblett *et al.* (in preparation) in a Mozambique male with no available relatives. His presumed genotype is PGM_1^2/PGM_1^1; PGM_2^2/PGM_2^2. The electrophoretic pattern of this Mozambique phenotype, compared with the Atkinson type in Fig. 14.3, shows the expected accentuation of zones g and h, and the apparent absence of zones e and f.

The 'Palmer' phenotype, shown diagrammatically in Fig. 14.5, probably represents another mutation at the PGM_2 locus (Hopkinson & Harris, 1966). Components e, f and g are reduced in staining intensity, and each is accompanied by a slightly faster component of about the same intensity, suggesting heterozygosity for the common PGM_2^1 allele with a rare PGM_2^3 allele. Phenotypes suggesting the existence of two more PGM_2 alleles, PGM_2^4 and PGM_2^5, were recently found by Harris & Hopkinson (personal communication). Their respective isozyme patterns are shown in Fig. 14.6. (See also Addenda.)

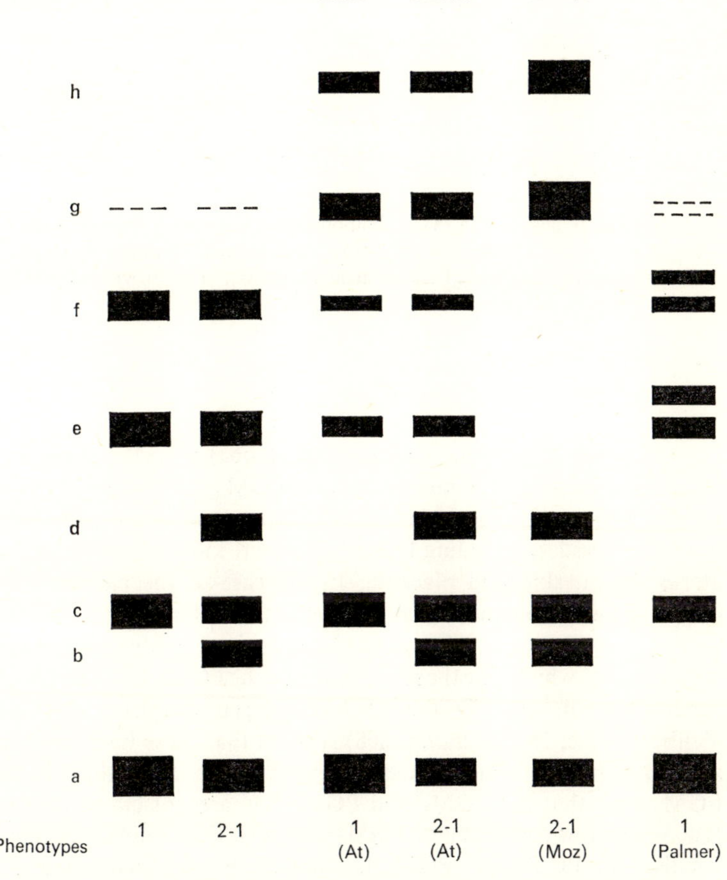

FIGURE 14.5. Diagram of starch gel electrophoretic patterns of the common PGM phenotypes 1 and 2-1 compared with uncommon phenotypes which reflect variation at the PGM₂ locus: Atkinson (At.), Mozambique (Moz.) and Palmer. The Mozambique phenotype represents apparent homozygosity for the Atkinson gene, PGM_2^2.

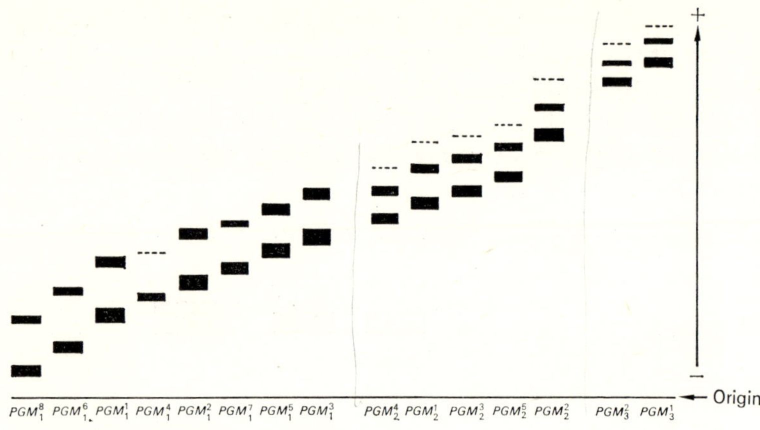

FIGURE 14.6. Diagram of PGM components which are determined by the eight postulated alleles at locus PGM_1, five postulated alleles at locus PGM_2 and two postulated alleles at locus PGM_3. (Fig. provided by H. Harris & D.A. Hopkinson; see Addenda.)

THE PGM_3 LOCUS

In their original studies, Spencer *et al.* (1964) reported that other tissues besides red cells could be used for PGM typing, since all of the isozyme zones (a-g) observed in red cell lysates were present in extracts of various tissues, including leukocytes, liver, kidney, muscle, heart, uterus, brain, skin and placenta. In the case of the placenta, the phenotype was the same as that observed in the infant, rather than the mother. Furthermore, the placenta contained additional faster-moving zones which, like those representing the PGM_1 and PGM_2 loci, occurred in pairs (see Fig. 14.6). From studies of the placenta of non-identical twins, Harris *et al.* (1968) showed that these faster-moving zones are products of at least two allelic genes at a third locus, PGM_3, and that the PGM_1 and PGM_3 loci are not closely linked. Due to the low frequency of PGM_2 variants, it was not determined whether the PGM_2 and PGM_3 loci are genetically linked (see Addenda).

The PGM_3 isozymes are fairly distinct in hemolysates of blood containing a high proportion of reticulocytes (Dr. J. Detter, personal communication). They can also be seen in concentrates of other

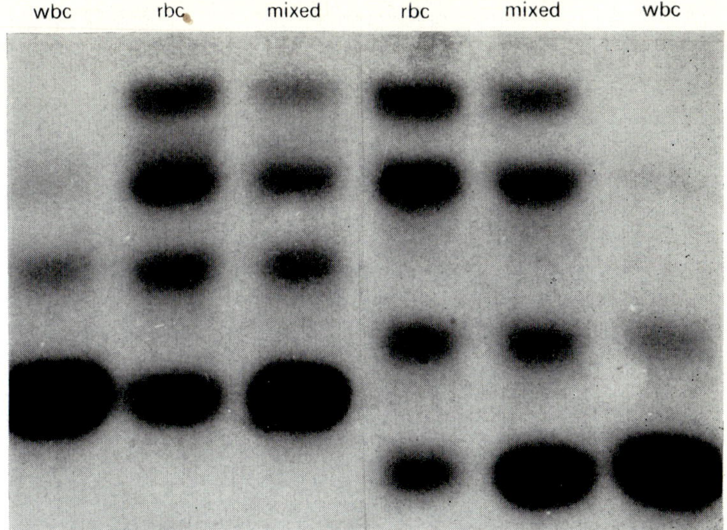

FIGURE 14.7. Starch gel electrophoretic patterns of phenotypes PGM 2 and PGM 1 in lysates of white cells (WBC), red cells (RBC) and mixtures of red and white cells. The PGM_2^1 gene products are only faintly visible in the white cells, while they equal or exceed the PGM_1^1 and PGM_1^2 gene products in the red cells.

tissue extracts, but since the zones are somewhat blurred and difficult to read accurately, placenta remains the only satisfactory source for examining the products of the PGM_3 genes.

DISTRIBUTION OF THE PGM ISOZYMES
IN BLOOD CELLS

Detter, Anderson & Giblett (in preparation) noted that in the patterns of red cell lysates, the intensity of the PGM_2 locus isozymes (e, f and g) was as great or greater than that of the PGM_1 locus isozymes (a, b, c and d). This observation was confirmed when the staining intensity of the individual bands was measured spectrophotometrically after electrophoresis on cellulose acetate. On the other hand, the patterns presented in the paper of Spencer *et al.* (1964) and Hopkinson & Harris (1965, 1966) showed an opposite trend. This difference was apparently resolved by the results of studies on red cell and white cell extracts, shown in Fig. 14.7. In the white cells, the activity of zones a and b predominates over c and d, while the e and f zones are very faint. In the red cells, zones a and b are only slightly stronger than c and d, while e and f equal or exceed the activity of the slower zones. Thus, the phenotypic pattern of a particular specimen varies with the proportion of white cells included in the preparation. When most of the white cells are removed, there is a relative increase in intensity of the faster-moving isozymes (i.e. the pattern is more representative of the red cells).

A similar observation was made by Lie-Injo *et al.* (1968) during studies of patients with leukemia. They suggested that the quantitative distribution of the PGM isozymes in various tissues may have some practical significance analogous to the difference in distribution of the lactate dehydrogenase isozymes.

GEOGRAPHIC DISTRIBUTION OF PGM
GENES IN VARIOUS POPULATIONS

A rather small number of populations have been tested for their PGM phenotypes. The available data, listed in Table 14.1, show that the PGM_1^1 gene frequency is appreciably higher than that of PGM_1^2 in all but one of the populations. The exception, an isolate of Habbanite

TABLE 14.1. Distribution of the common PGM_1 genes and the unusual PGM phenotypes in various populations

| Population | Number tested | Unusual phenotypes | | Common PGM_1 gene frequencies | | References |
		No.	Type	PGM_1^1	PGM_1^2	
Caucasoid						
England	2115	1 each	3-1, 4-2, 5-2, 7-1, 7-2	0·764	0·235	Hopkinson & Harris (1966)
San Francisco	271			0·777	0·223	Lie-Injo (1966)
Seattle	508			0·752	0·248	Giblett & Scott (unpubl.)
Greece	88			0·693	0·307	Hopkinson & Harris (1966)
Cyprus (Turkish)	243			0·698	0·300	Hopkinson & Harris (1966)
Sardinia	633			0·762	0·238	Modiano *et al.* (1967)
Iceland	129			0·818	0·182	Mourant & Tills (1967)
Habbanite Jews	222			0·430	0·570	Mourant & Tills (1967)
Iraqi Jews	69			0·674	0·326	Hopkinson & Harris (1966)
Negro						
		1	1 At.			
Nigeria (Yoruba)	153	1	2-1 At.	0·758	0·239	Hopkinson & Harris (1966)
		1	6-1			
		13	1 At.			
Mozambique	318	5	2-1 At.	0·781	0·219	Giblett *et al.* (in prep.)
		1	1 Moz.			

S. Africa (Bantus)	99	5	1 At.	0·792	0·202	Hopkinson & Harris (1966)
		1	6-2			
England	103	2	1 At.	0·786	0·214	Hopkinson & Harris (1966)
		1	2-1 At.			
San Francisco	284	2	1 At.	0·805	0·195	Lie-Injo (1966)
		2	1 At.			
Seattle	654	4	2-1 At.	0·809	0·191	Giblett & Scott (unpubl.)
		1	1 Palmer			
Ann Arbor	202	1	1 At.	0·841	0·169	Brewer *et al.* (1967)
Oriental						
		4	6-1			
Chinese (misc.)	417	1	6-2	0·750	0·242	Lie-Injo *et al.* (1968)
		2	7-1			
Thailand	503	1	7-1	0·730	0·270	Giblett & Scott (unpubl.)
Japan (Ainu)	182			0·900	0·100	Giblett (1967)
Seattle (misc.)	212			0·776	0·224	Giblett & Scott (unpubl.)
Indians and Eskimos						
Alaska: Eskimos	299			0·824	0·176	Scott *et al.* (1966)
Aleuts	53			0·858	0·142	Scott *et al.* (1966)
Athabascans	127			0·894	0·106	Scott *et al.* (1966)
Mexico: Huasteco	233			0·805	0·195	Lisker & Giblett (1967)
Cora	100			0·890	0·110	Lisker & Giblett (1967)
Huichol	72			0·833	0·167	Lisker & Giblett (1967)
Venezuela:						
10 villages	338			0·938	0·062	Arends *et al.* (1967)

Jews in Southern Arabia, was found to have a PGM_1^2 frequency of 0·570 (Mourant & Tills, 1967). The lowest frequencies of PGM_1^2 were also found in tribal groups: 0·100 among 182 Ainus in northern Japan (Giblett, 1967); 0·106 among 127 Athabascan Indians in Alaska; and 0·110 among 100 Mexican Indians of the Cora tribe (Lisker & Giblett, 1967).

All of the Negro populations samples contained one or more examples of Atkinson phenotypes, indicating the presence of the PGM_2^2 gene with a frequency as high as 0·03 in Mozambique (Giblett *et al.*, in preparation), but 0·01 or less among Negroes living in England and America.

Since some of the rare phenotypes associated with mutation at the PGM_1 locus are difficult to identify without making careful comparisons with other specimens of known phenotypes, it is possible that their frequencies may be slightly higher than Table 14.1 indicates. However, it is unlikely that an appreciable number have escaped detection. They may eventually be found to be more prevalent in certain regions of Asia than in other parts of the world, as suggested by the data of Lie-Injo *et al.* (1968). However, much larger numbers of people must be tested before this possibility can be substantiated.

METHODS

DETERMINATION OF PGM TYPES BY STARCH GEL ELECTROPHORESIS

(Spencer *et al.*, 1964)

Although electrophoresis in agar or cellulose acetate can be used for determining PGM phenotypes, most studies have been performed with starch gel. Hemolysates are prepared from washed, packed red cells lysed with an equal volume of water and extracted with one-half volume of toluene, leaving a stroma-free aqueous extract. The pH 7·4 bridge buffer contains tris (12·1 g), maleic acid (11·6 g), disodium EDTA (3·73 g), $MgCl_2.6H_2O$ (2·03 g), NaOH (5·2 g) and distilled water to make 1 liter. A 1 to 10 dilution of this buffer is used for making the gels.

Horizontal or vertical electrophoresis at 4°C is carried out overnight.

Voltage adjustments are made during the first hour to maintain 5–6 volts/cm.

The gel is sliced, and the bottom half is used for staining. The PGM$_1$ and PGM$_2$ locus isozymes are found within a distance of about 10 cm toward the anode from the point of insertion. The PGM$_3$ isozymes have a faster migration rate, making it necessary to stain the entire length of the gel when placental extracts are inserted.

Detection of enzyme activity is based on the reaction shown in Fig. 14.8. Only a small amount of G1,6 diP is required for this reaction to occur, and it is usually present as a contaminant in G1P, unless the

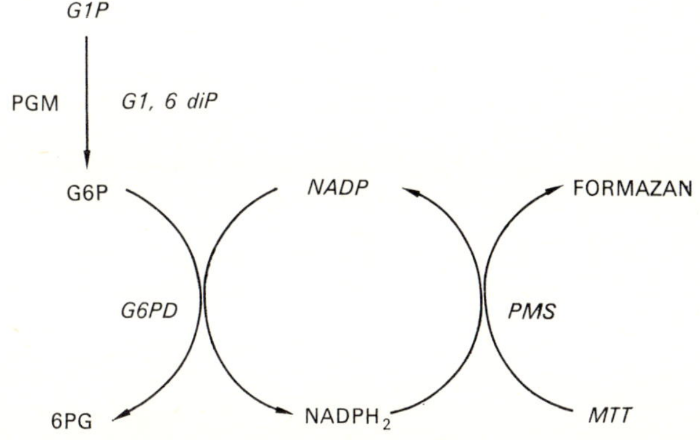

FIGURE 14.8. Chemical reactions employed in staining phosphoglucomutase isozymes; see text. (From Spencer *et al.*, 1964. Reprinted with permission from *Nature* **201**, 299.)

preparation is highly purified. As shown in the figure, G1P is converted to G6P at the sites of PGM activity. The G6P is then oxidized by G6PD, with the concomitant reduction of its coenzyme, NADP. The reduction of MTT to insoluble dark blue formazan is mediated by phenazine methosulfate.

The reaction mixture is applied as an agar overlay, made by mixing 6 ml of 2 per cent molten agar (50°C) with 6 ml of tris-HCl buffer, pH 8·0, containing the dipotassium salt of G1P (20·5 mg), MgCl$_2$.6H$_2$O (11·4 mg) NADP (1·1 mg) PMS (1·2 mg) and MTT (1·2 mg). After incubation in the dark at 37°C for an hour, the PGM isozymes are visible as dark blue bands.

REFERENCES

ARENDS T., BREWER G., CHAGNON N., GALLANGO M.L., GERSHOWITZ H., LAY-RISSE M., NEEL J., SHREFFLER D., TASHIAN R. & WEITKAMP L. (1967) Intra-tribal genetic differentiation among the Yanomama Indians of southern Venezuela. *Proc. natn. Acad. Sci.* **57**, 1252.

BOCCHINI V., ALIOTO M.R. & NAJJAR V.A. (1967) The sulfhdryl groups of rabbit muscle phosphoglucomutase. *Biochemistry* **6**, 313.

BOCCHINI V., HARSHMAN S. & NAJJAR V.A. (1964) The activation and active site of phosphoglucomutase. *Fed. Proc.* **23**, 423.

BREWER G.J., BOWBEER D.R. & TASHIAN R.E. (1967) The electrophoretic phenotypes of red cell phosphoglucomutase, adenylate kinase and acid phosphatase in the American Negro. *Acta genet.* **17**, 97.

COHEN J.A., OOSTERBAAN R.A., JANSZ H.S. & BERENDS F. (1959) The active site of esterases. *J. cell. comp. Physiol.* **54**, suppl. 1, 231.

DIXON G.H., KAUFFMAN D. & NEURATH H. (1958) Amino acid sequence in the region of diisopropylphosphoryl binding in diisopropylphosphoryl-trypsin. *J. biol. Chem.* **233**, 1373.

EDWARDS J.H. & WINGHAM J. (1967) Data on linkage between the locus determining alkaline phosphatase (P1) and other markers. *Ann. hum. Genet.* **30**, 233.

EPSTEIN C.J. & MOTULSKY A.G. (1965) Evolutionary origins of human proteins, *in* STEINBERG A.G. & BEARN A.G. (eds.) *Progress in Medical Genetics*, vol. 4, p. 85. Grune & Stratton, New York.

GEDDE-DAHL T. & MONN E. (1967) Linkage relations of the phosphoglucomutase PGM_1 locus in man. *Acta genet.* **17**, 482.

GIBLETT E.R. (1967) Variant phenotypes: haptoglobin, transferrin and red cell enzymes, *in* GREENWALT T.J. (ed.) *Advances in Immunogenetics*, p. 99. J.B. Lippincott, Philadelphia.

GOUNARIS A.D., HORTON H.R. & KOSHLAND D.E. (1967) The phosphogluco-mutase reactions: investigation of enzyme dephosphorylation during its reaction with substrate. *Biochim. biophys. Acta* **132**, 41.

HANDLER P., HASHIMOTO T.A., JOSHI J.G., DOUGHERTY H., HANBUSA K. & DELRIO C. (1965) Phosphoglucomutase: evolution of an enzyme. *Israel J. med. Sci.* **1**, 1173.

HARRIS H., HOPKINSON D.A., LUFFMAN J.E. & RAPLEY S. (1968) Electrophoretic variation in erythrocyte enzymes, *in Genetically Determined Abnormalities of Red Cell Metabolism*, City of Hope Symposium Series, vol. 1, p. 1. Grune & Stratton, New York.

HARTLEY B.S. (1964) Amino acid sequence of bovine chymotrypsinogen. *Nature* **201**, 1284.

HOPKINSON D.A. & HARRIS H. (1965) Evidence for a second 'structural' locus determining human phosphoglucomutase. *Nature* **208**, 410.

HOPKINSON D.A. & HARRIS H. (1966) Rare phosphoglucomutase phenotypes. *Ann. hum. Genet.* **30**, 167.

JOSHI J.G., HOOPER J., KUWAKI T., SWANSON J.R., SUKURADH, T. & HANDLER P.

(1967) Phosphoglucomutase. V. Multiple forms of phosphoglucomutase. *Proc. natn. Acad. Sci.* **57**, 1482.

LIE-INJO L.E. (1966) Phosphoglucomutase polymorphism: an unusual type in Negroes. *Nature* **210**, 1183.

LIE-INJO L.E., LOPEZ C.G. & POEY-OEY-HOEY G. (1968) Erythrocyte and leucocyte phosphoglucomutase in Chinese. *Am. J. hum. Genet.* **20**, 101.

LISKER R. & GIBLETT E.R. (1967) Studies on several genetic hematological traits of Mexicans. XI. Red cell acid phosphatase and phosphoglucomutase in the three Indian groups. *Am. J. hum. Genet.* **19**, 174.

MODIANO G., SCOZZARI R., GIGLIANI F., FILIPPI G. & LATTE B. (1967) Studies on red cell phosphoglucomutase and adenylate kinase polymorphisms in Sardinia. *Acc. Naz. Linc.* **42**, 906.

MOURANT A.E. & TILLS D. (1967) Phosphoglucomutase frequencies in Habbanite Jews and Icelanders. *Nature* **214**, 810.

NAJJAR V.A. (1962) Phosphoglucomutase, *in* BOYER P.D., LARDY H. & MYRBÄCK K. (eds.) *The Enzymes*, vol. 6, p. 161. Academic Press, New York.

RAY W.J. & ROSCELLI G.A. (1964) A kinetic study of the phosphoglucomutase pathway. *J. biol. Chem.* **239**, 1228.

ROBSON E.B., SUTHERLAND I. & HARRIS H. (1966) Evidence for linkage between the transferrin locus (Tf) and the serum cholinesterase locus (E_1) in man. *Ann. hum. Genet.* **29**, 325.

SCHWARTZ J.H., CRESTFIELD A.M. & LIPMAN F. (1963) The amino acid sequence of a tetradecapeptide containing the reactive serine in *E. coli* alkaline phosphatase. *Proc. natn. Acad. Sci.* **49**, 722.

SCOTT E.M., DUNCAN I.W., EKSTRAND V. & WRIGHT R.C. (1966) Frequency of polymorphic types of red cell enzymes and serum factors in Alaskan Eskimos and Indians. *Am. J. hum. Genet.* **18**, 408.

SMILLIE L.B. & HARTLEY B.S. (1964) Histidine sequences in the active centers of some 'serine' enzymes. *J. Molec. Biol.* **10**, 183.

SPENCER N., HOPKINSON D.A. & HARRIS H. (1964) Phosphoglucomutase polymorphism in man. *Nature* **204**, 742.

TIPPETT P. (1967) Genetics of the Dombrock blood group system. *J. med. Genet.* **4**, 7.

WALSH K.A. & NEURATH H. (1964) Trypsinogen and chymotrypsinogen as homologous proteins. *Proc. natn. Acad. Sci.* **52**, 884.

CHAPTER 15

ADENYLATE KINASE

Electrophoretic Variants of AK 512

Inheritance of AK Phenotypes 514
    Linkage studies 515

Geographic Distribution of AK
Genes 515

Methods 517
    Demonstration of AK pheno-
    types by starch gel electro-
    phoresis 517

References 518

Adenylate kinase (AK) or ATP: AMP phosphotransferase (EC 2.7.4.3) reversibly catalyzes the reaction:

$$2\,ADP \rightleftharpoons ATP + AMP$$

This enzyme is found in many tissues of the body, especially in red cells and in muscle, where it is also referred to as myokinase. The physico-chemical and catalytic properties of AK were reviewed by Noda (1962), while the activity of the enzyme in red cells was the topic of papers by Kashket & Denstedt (1958), Tatibana *et al.* (1958), Cerletti & Bucci (1960) and Levin & Beutler (1967).

The remarkable stability of AK to heat and extremes of pH sets it apart from most other enzymes. Its molecular weight is about 21,000 (rabbit muscle) to 24,000 (human red cells). On electrophoresis, the isozyme patterns of rabbit muscle and red cell AK have the same appearance (Fildes & Harris, 1966), and the patterns of human hemolysates and various tissue extracts are very similar (Dr. J. Detter, personal communication).

ELECTROPHORETIC VARIANTS OF AK

Fildes & Harris (1966) reported that AK in red cell lysates could be demonstrated after starch gel electrophoresis by coupling the forward reaction with the conversion of hexokinase to glucose-6-phosphate (G6P) or the reverse reaction with the conversion of phosphoenolpyruvate to pyruvate. The former method was found most satisfactory for routine testing.

The three AK electrophoretic isozyme patterns described by Fildes & Harris (1966) are shown diagrammatically in Fig. 15.1. The most common phenotype, AK 1, consists of two major components, of

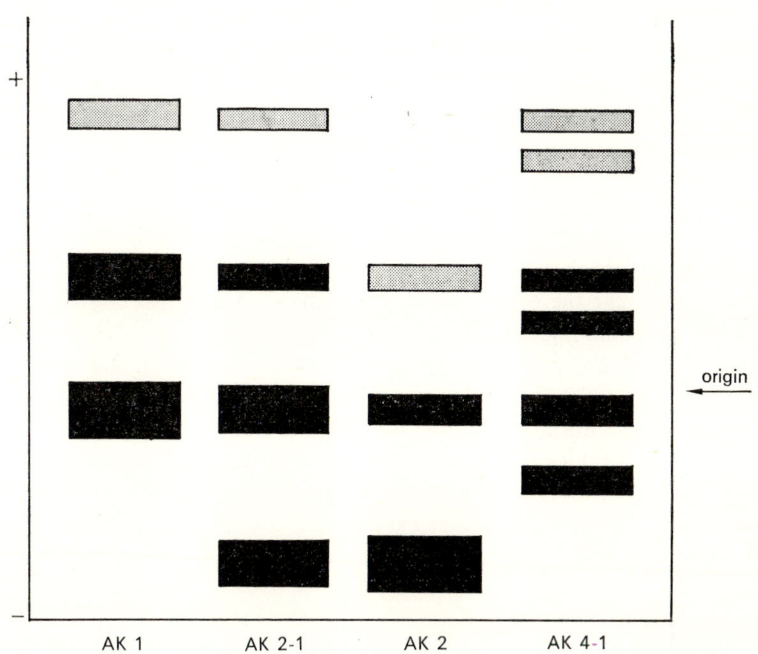

FIGURE 15.1. Diagram of starch gel electrophoretic patterns of four AK phenotypes. The fastest-moving zones have much less enzyme activity than the other zones, and they are often difficult to discern clearly. However, in AK 4-1, they appear to be doubled, as shown. (Adapted from Rapley *et al.*, 1967. Reprinted with permission from *Ann. hum. Genet.* **31**, 237.)

which the most intense remains near the origin (pH 7·0), and the less active zone migrates toward the anode (i.e. is negatively charged). A third, faster-moving component stains poorly and tends to be diffuse. The AK 2-1 pattern consists of the same three components, but with relatively less activity, and a fourth zone which migrates toward the cathode. AK 2 resembles AK 2-1 except that the cathodal zone is more intense than the zone near the origin, and the most anodal band is absent.

A fourth (rare) phenotype, AK 4-1, was subsequently found by Rapley *et al.* (1967) and independently, by Giblett & Scott (unpubl.). It probably corresponds with the AK 4-1 mentioned by Bowman *et al.* (1967). Its pattern, shown diagrammatically in Fig. 15.1, is also presented in a photograph (Fig. 15.2) which compares AK 4-1 with AK 1 and AK 2-1. The fastest-moving (anodal) bands are not shown in Fig. 15.2 because they are too faint. In the AK 4-1 pattern, the components characteristic of AK 1 occur in pairs, of which one band has the usual position, and the other has a slower migration rate. Thus, the gene mutation has produced a charge alteration in all three of the AK 1 components.

Bowman *et al.* (1967) described their results with a different buffer system, in which the AK 1 pattern contains only a single band, migrating near the origin. With this method, AK 2 also contains a single, cathodal band, while AK 2-1 has both components. They also described a rare phenotype called 3-1, which contained the AK 1 component and a second zone of activity migrating toward the anode. These authors did not observe the additional components described by Fildes & Harris (1966), and felt that the multiple bands seen with their system after prolonged incubation represented G6PD or 6PGD. However, a similar explanation cannot be applied to the additional zones which appear when the buffer of Fildes & Harris is used.

INHERITANCE OF AK PHENOTYPES

Fildes & Harris (1966) studied 54 families in which one parent had the phenotype AK 1 and the other, AK 2-1. Among the 136 children, 72 were AK 1 and 64 were AK 2-1. In 62 additional families with both parents AK 1, all 141 children were also AK 1. When the families of two individuals with the (rare) AK 2 phenotype were studied, the results were consistent with the segregation of a pair of autosomal allelic genes, AK^1 and AK^2. Similar results were reported from further studies by Rapley *et al.* (1967). Thus, individuals with the common AK 1 phenotype are homozygous for the AK^1 gene, while those with the rare AK 2 phenotype are homozygous for AK^2. Heterozygotes hav-

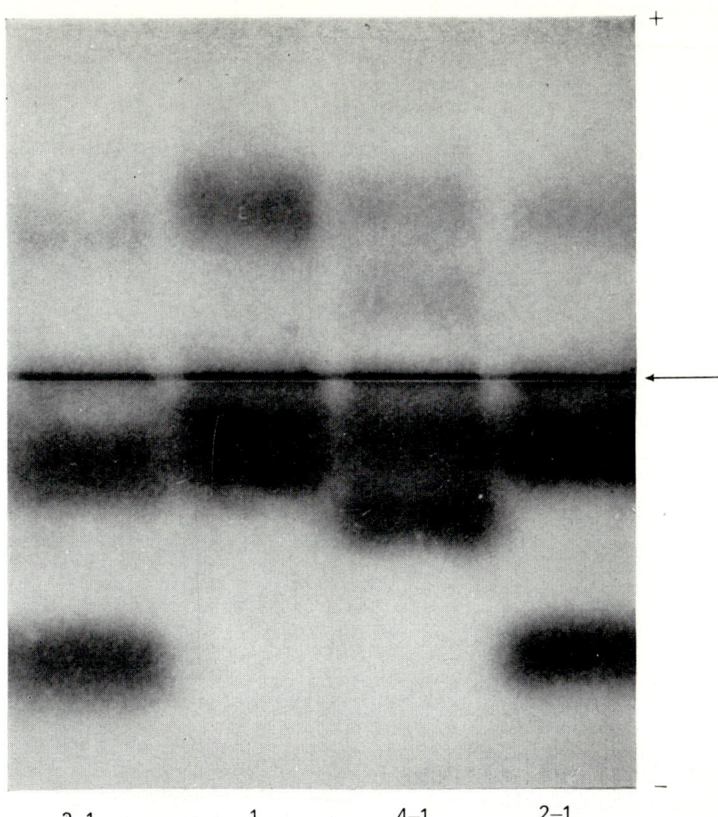

FIGURE 15.2. Photograph of starch gel electrophoretic patterns at pH 7·0 of three AK phenotypes in four hemolysates. The fastest-moving components shown diagrammatically in Fig. 15.1 are not seen in this photograph. The arrow points to the origin.

ing the genotype AK^2/AK^1 occur with a frequency of about 9 per cent in the British population tested by Fildes & Harris (1966) and Rapley *et al.* (1967). The AK 4-1 phenotype represents heterozygosity for the common AK^1 gene with the rare allele, AK^4 (Rapley *et al.*, 1967).

LINKAGE STUDIES

Rapley *et al.* (1967) carried out tests for linkage of the AK genetic system with a number of blood group, serum protein and red cell enzyme systems. Close linkage with some systems, such as Rh, MNSs, P, Secretor, Duffy, haptoglobin, and acid phosphatase was ruled out. A suggestion of linkage with the ABO system was obtained from 36 informative sibships, which provided preliminary evidence for a recombination value of about 0·2. Rapley *et al.* (1967) are pursuing this lead to obtain more information about the possible linkage of AK and ABO.

GEOGRAPHIC DISTRIBUTION
OF AK GENES

Although the AK polymorphism was described fairly recently, several studies on the distribution of its genes in various populations have been performed, as shown in Table 15.1. The frequency of AK^2 in European populations so far tested does not exceed 0·05. Both the Lapps of Finland and the inhabitants of Sardinia appear to have decreased frequencies of AK^2.

The natives of Africa and Central America show very little polymorphism at the AK locus. The presence of AK^2 in American Negroes probably represents gene flow from non-Negroes (Bowman *et al.*, 1967). Among Asiatic populations, the AK^2 frequency varies considerably. All but one of the Chinese and Japanese (the latter in Seattle) tested have the AK 1 phenotype. However, the highest frequencies of AK^2 in any group tested were found in people from India and Pakistan presently living in England (Rapley *et al.*, 1967), while the Thais, Malayans, Iranians and Iraqi Jews have frequencies resembling those of Europeans.

TABLE 15.1. Distribution of Adenylate Kinase Types in Various Populations

Population	Number	AK phenotypes			AK^2 gene frequency	Reference
		1	2–1	2		
European						
England	1887	1720	1652		0·045	Rapley *et al.* (1967)
Finland	77	71	6	0	0·039	Rapley *et al.* (1967)
Finnish Lapps	307	304	3	0	0·005	Rapley *et al.* (1967)
Rome	738	686	52	0	0·035	Modiano *et al.* (1967)
Sardinia	1033	1004	28	1	0·015	Modiano *et al.* (1967)
South Africa	100	92	7	1	0·045	Gordon *et al.* (1966)
Chicago	1314	1193	118	3	0·047	Bowman *et al.* (1967)
Seattle	172	163	9	0	0·031	Giblett & Scott (unpubl.)
Ann Arbor	254	240	14	0	0·028	Brewer *et al.* (1967)
Negro						
Mozambique	318	316	2	0	0·003	Giblett *et al.* (in prep.)
Nigeria	153	153	0	0	0·000	Rapley *et al.* (1967)
Ghana and Nigeria	800	800	0	0	0·000	Bowman *et al.* (1967)
S. Africa	100	98	2	0	0·010	Gordon *et al.* (1966)
Babinga pygmies	300	300	0	0	0·000	Modiano *et al.* (1967)
West Indies	85	84	1	0	0·006	Rapley *et al.* (1967)
Chicago	1062	1049	13	0	0·006	Bowman *et al.* (1967)
Ann Arbor	139	135	4	0	0·014	Brewer *et al.* (1967)
Seattle	223	220	3	0	0·007	Giblett & Scott (unpubl.)

TABLE 15.1—*continued*

Population	Number	AK phenotypes			AK^2 gene frequency	Reference
		I	2–I	2		
Asiatic						
China and Taiwan	227	226	1	0	0·002	Shih *et al.* (1968)
Thailand	201	192	9	0	0·022	Giblett & Scott (unpubl.)
Malaya (S. Africa)	100	93	7	0	0·035	Gordon *et al.* (1966)
India (Engl.)	132	107	24	1	0·098	Rapley *et al.* (1967)
Pakistan (Engl.)	54	40	14	0	0·130	Rapley *et al.* (1967)
Mixed oriental (Seattle)	146	146	0	0	0·000	Giblett & Scott (unpubl.)
Iran Moslems	322	290	32	0	0·050	Bowman & Ronaghy (1967)
Iraqi Jews	139	126	13	0	0·047	Rapley *et al.* (1967)
American Indian						
Lacandon (Mayan)	150	150	0	0	0·000	Bowman *et al.* (1967)
Yanomama (Venezuela)	?	?	0	0	0·000	Arends *et al.* (1967)

METHODS

DEMONSTRATION OF AK PHENOTYPES
BY STARCH GEL ELECTROPHORESIS

(Fildes & Harris, 1966)

Hemolysates are prepared by the method described in the previous chapters and inserted into gels for either horizontal or vertical electrophoresis. A histidine-citrate discontinuous buffer system (pH 7·0) is used. The 0·005 M histidine gel buffer provides a medium for the

18

growth of micro-organisms, so it is best stored as a stock solution with ten times the concentration used for preparing the gel. This stock solution contains DL histidine (10·485 g) and NaOH (1·85 g) in each liter of water. The bridge buffer (also pH 7·0) is 0·41 M citric acid and 1·24 M sodium hydroxide. (One liter contains 86·2 g citric acid and 49·7 g NaOH.) Electrophoresis is carried out at 4°C overnight at 4 volts/cm, or at 9 volts/cm with cooling plates for 4 hours. The gel is sliced and stained with an agar overlay.

The chemical reaction responsible for reducing MTT to formazan at the sites of AK activity is shown in Fig. 15.3. The enzymatic conversion of ADP to ATP provides the phosphate required for conversion of glucose to G6P by hexokinase. Then G6P is oxidized

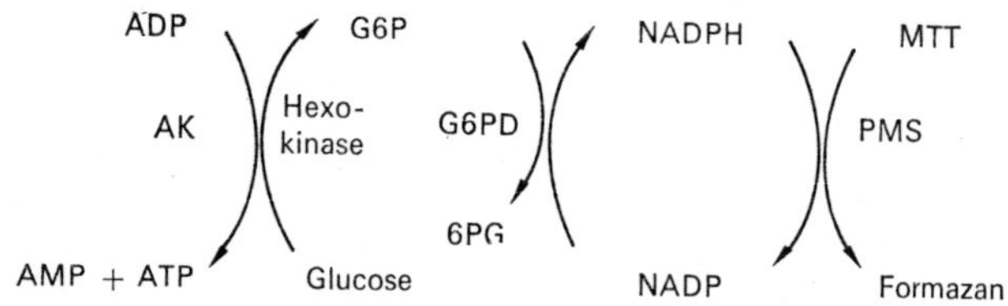

FIGURE 15.3. Chemical reactions employed in the stain for adenylate kinase activity. (See text.)

by G6PD, and with the concomitant reduction of NADP to NADPH, MTT is reduced to dark blue formazan in the presence of PMS.

The agar overlay is prepared by mixing 11 ml of 2 per cent molten agar (50°C) with 11 ml of 0·2 M tris-HCl buffer (pH 8·0) containing glucose (36·6 mg), $MgCl_2.6H_2O$ (74 mg), ADP (9·7 mg), NADP (4·6 mg), MTT (2·1 mg), PMS (2·1 mg), G6PD (0·72 units) and hexokinase (1·4 units). The zones of AK activity develop during incubation at 37°C in the dark for 30–45 minutes.

REFERENCES

ARENDS T., BREWER G., CHAGNON N., GALLANGO M.L., GERSHOWITZ H., LAYRISSE M., NEEL J., SHREFFLER D., TASHIAN R. & WEITKAMP L. (1967) Intratribal differentiation among the Yanomama Indians of southern Venezuela. *Proc. natn. Acad. Sci.* **57**, 1252.

BOWMAN J.E., FRISCHER H., AJMAR F., CARSON P.E. & GOWER M.K. (1967)

Population, family and biochemical investigation of human adenylate kinase polymorphism. *Nature* **214**, 1156.

BOWMAN J.E. & RONAGHY H. (1967) Hemoglobin, glucose-6-phosphate dehydrogenase, phosphogluconate dehydrogenase and adenylate kinase polymorphism in Moslems in Iran. *Am. J. phys. Anthrop.* **27**, 119.

BREWER G.J., BOWBEER D.R. & TASHIAN R.E. (1967) The electrophoretic phenotypes of red cell phosphoglucomutase, adenylate kinase, and acid phosphatase in the American Negro. *Acta genet.* **17**, 97.

CERLETTI P. & BUCCI E. (1960) Adenylate kinase of mammalian erythrocytes. *Biochim. biophys. Acta* **38**, 45.

FILDES R.A. & HARRIS H. (1966) Genetically determined variation of adenylate kinase in man. *Nature* **209**, 261.

GORDON H., VOOIJS M. & KERRAN M.M. (1966) Genetical variation in some human red cell enzymes: an interracial study. *S. Afr. med. J.* **40**, 1031.

KASHKET S. & DENSTEDT O.F. (1958) The metabolism of the erythrocyte. XV. Adenylate kinase of the erythrocyte. *Can. J. Biochem. Physiol.* **36**, 1057.

LEVIN E. & BEUTLER E. (1967) Human erythrocyte adenylate kinase. *Haematologica* **1**, 19.

MODIANO G., SCOZZARI R., GIGLIANI F., FILIPPI G. & LATTE B. (1967) Studies on red cell phosphoglucomutase and adenylate kinase polymorphisms in Sardinia. *Acc. Naz. Linc.* **42**, 906.

NODA L. (1962) Nucleoside triphosphate-nucleoside monophosphokinase, *in* BOYER P.D., LARDY H. & MYRBÄCK K. (eds.) *The Enzymes*, vol. 6, p. 139. Academic Press, New York.

RAPLEY S., ROBSON E.B., HARRIS H. & SMITH S.M. (1967) Data on the incidence, segregation and linkage relations of the adenylate kinase (AK) polymorphism. *Ann. hum. Genet.* **31**, 237.

SHIH L.Y., HSIA D.Y.Y., BOWMAN J.E., SHIH S.C. & SHIH P.L. (1968) The electrophoretic phenotypes of red cell 6-phosphogluconate dehydrogenase and adenylate kinase in Chinese populations. *Am. J. hum. Genet.* **20**, 474.

TATIBANA M., NAKAO M. & YOSHIKAWA H. (1958) Adenylate kinase in human erythrocytes. *J. Biochem.* (Tokyo) **45**, 1037.

CHAPTER 16

OTHER ENZYME MARKERS IN BLOOD CELLS

Lactate Dehydrogenase.................... 521
   Subunit structure....................... 521
   Genetic variants 522
   Methods 523
Malate Dehydrogenase.................... 524
   Cytoplasmic MDH 524
   Mitochondrial MDH 525
   Methods 527
Phosphohexose Isomerase.............. 527
   Variants found at random..... 527
   Variants associated with enzyme deficiency....................... 528
   Methods 529
Carbonic Anhydrase....................... 529
   Genetic variation....................... 530
   Quantitative variation.............. 531
   Methods 533
Red Cell Esterases........................... 533
   Isozyme classification.............. 533
   Variants.................................... 535
   Methods 535
Peptidases 535
   Isozyme characteristics 535

Variants.. 537
Methods 539
Glutathione Reductase.................... 540
   Heterogeneity 540
   Methods 541
Methemoglobin Reductase: NADH and NADPH Diaphorases 541
   Evidence for heterogeneity..... 542
   Methods 543
Catalase 543
   Molecular variation.................. 544
   Methods 545
Galactose-1-Phosphate Uridyl Transferase 546
   Enzyme variability in galactosemia 546
   Methods 547
'Indophenol Oxidase'...................... 547
   The achromatic regions in tetrazolium-stained gels......... 547
   A variant phenotype.................. 548
   Methods 549
References...................................... 549

The cellular enzymes described in the previous chapters are sufficiently polymorphic in several populations to be useful as genetic markers. This chapter contains briefer descriptions of enzymes whose electrophoretic variants are either very infrequent or too difficult to differentiate with certainty. In one instance, the peptidases, the polymorphism has been described too recently to provide sufficient data for a separate chapter, although the peptidases exhibit considerably greater variation than that observed among the other enzymes included in this chapter.

LACTATE DEHYDROGENASE

The function of lactate dehydrogenase (L-lactate: NAD oxidoreductase, EC 1.1.1.27) is to catalyze the reversible conversion of pyruvate to lactate. Studies on the lactate dehydrogenase (LDH) isozymes derived from human and animal sources have provided much important information about such subjects as subunit hybridization, tissue isozyme localization under conditions of embryogenesis and disease, and molecular evolution. However, since LDH electrophoretic variation is rare in man, the details of these extensive investigations are not included here. The reader is referred to such papers as those by Markert & Møller (1959), Kaplan et al. (1960), Allen (1961), Markert & Ursprung (1962), Cahn et al. (1962), Markert & Appella (1963), Vesell & Philip (1963), Wilson et al. (1964), Kaplan (1964), Vesell & Brody (1964), Salthe et al. (1965), and Chilson et al. (1965). Vesell (1965a) has written an excellent review.

SUBUNIT STRUCTURE

On electrophoresis, LDH is separated into five isozymes, each of which is a tetramer composed of the electrophoretically distinct subunits A and/or B. The molecular weight of each subunit is about 34,000, and by random combination, they form the series of isozymes called LDH-1 through LDH-5 with the structure of BBBB, BBBA, BBAA BAAA and AAAA, respectively. LDH-1, composed of four B subunits, has the fastest electrophoretic mobility; LDH-5, with four A subunits, is the slowest-moving isozyme. When the two subunits are present in equal proportions, the quantitative ratio of the five isozymes is 1:4:6:4:1 (Appella & Markert, 1961; Markert, 1963). However, the proportion of subunits varies in different tissues. For example, in heart and kidney, the B subunit predominates, so LDH-1 and LDH-2 are present in greatest quantity. Conversely, skeletal muscle and liver largely produce the A subunit, so most of their enzyme activity is in LDH-4 and LDH-5 (Vesell & Bearn, 1962a,b). Some workers prefer the term M (for muscle) instead of A, and H (for heart) instead of B.

The LDH electrophoretic patterns of normal red cells and white cells are shown in Fig. 16.1. The high proportion of LDH-1 and the virtual lack of LDH-5 in red cells (also characteristic of platelets) is only found after the nucleus has been extruded. Thus, in nucleated red cells and reticulocytes, the pattern resembles that of white cells (Vesell & Bearn, 1962; Starkweather *et al.*, 1965). In hemolysates, LDH-2 and LDH-3 tend to appear as one major and one or more minor components when electrophoresis is performed for more than a few hours. Neither the number nor the pattern of the minor components is constant for a given specimen, and it seems likely that this kind of heterogeneity represents chemical changes not directly influenced by the LDH structural genes.

GENETIC VARIANTS

Boyer *et al.* (1963) found that in the starch gel electrophoretic pattern of a Nigerian blood specimen, LDH-1 consisted of five different components, while LDH-2, LDH-3 and LDH-4 consisted of four, three and two components, respectively. Unlike the variable minor bands described above, these components were constant on repeated electrophoretic determinations, and a second blood specimen behaved similarly. Boyer *et al.* (1963) concluded from this observation that the donor of the blood was probably heterozygous at the structural gene locus determining the B subunit, and that the multiple components observed represented random association as tetramers of the three subunits, A, B, and B′. Although LDH-5 was not demonstrable in the hemolysate, it would have been expected to consist of a single tetramer, AAAA. The two LDH-4 components were thought to represent AAAB and AAAB′. Similar hybrid tetramers would then account for the increasing numbers of LDH-3, -2 and -1 isozymes observed, the latter consisting of BBBB, BBBB′, BBB′B′, BB′B′B′ and B′B′B′B′.

Nance *et al.* (1963) described a variant pattern in a Brazilian family, in which the propositus and three of his children appeared to be heterozygous at the autosomal gene locus determining the A subunit. Since neither LDH-5 nor LDH-4 was visible in their electrophoretic system, they inferred the genetic variability on the basis of a doubling of LDH-2 and what appeared to be only doubling of LDH-3. They concluded that the three gene products, A, A′ and B, did not associate at random. However, Kraus & Neely (1964), in a study of three different A subunit variants, reported results consistent with random

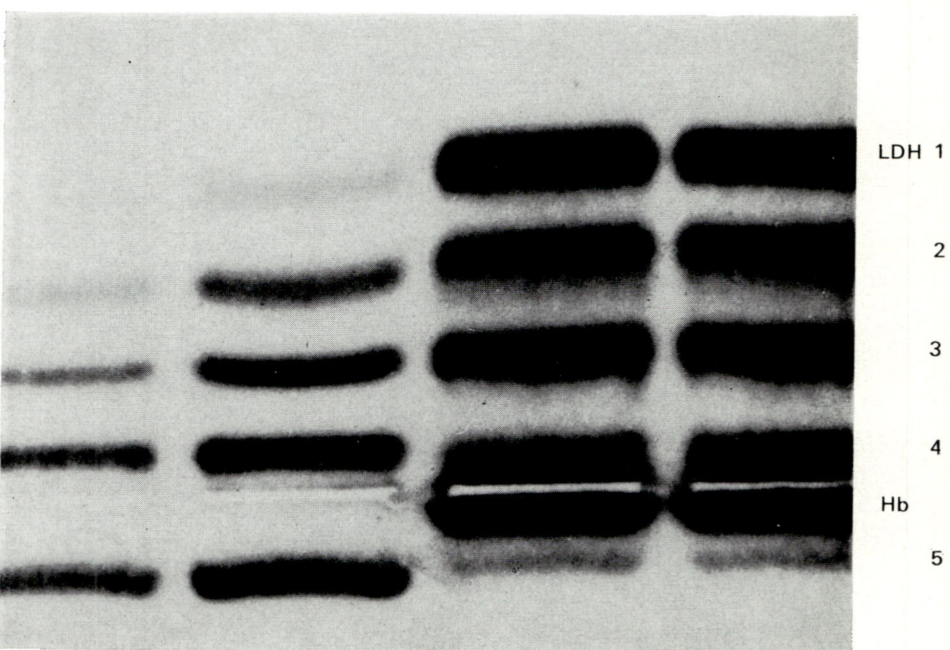

FIGURE 16.1. Starch gel electrophoretic patterns of LDH isozymes in lysates of white cells (left) and red cells (right). The presence of hemoglobin (Hb) makes the activity of LDH-4 appear much greater than it actually is in red cells.

hybrid formation, and these results were confirmed by Davidson *et al.* (1965) and Vesell (1965b,c).

The incidence of LDH phenotypic variation is very low in all of the Caucasian populations so far tested. Among Negroes, the overall incidence may be as high as 1 per cent (Kraus & Neely, 1964; Vesell, 1965b, Davison *et al.*, 1965), but since at least four A subunit variants and probably two B subunit variants have been found in this ethnic group, the individual mutant gene frequencies appear to be less than 0·003. Studies in individual families have shown no close linkage of the A or B LDH loci with the loci of the ABO, MNS, Duffy, Kell, Hp and Gm systems (Nance *et al.*, 1963; Kraus & Neely, 1964; Vesell, 1965a).

METHODS

The LDH genetic variants can be demonstrated by cellulose acetate and agar electrophoresis, but the preferred medium appears to be starch gel. Either a horizontal or vertical system may be used, and the buffer system is not critical. For example, Davidson *et al.* (1965) used horizontal electrophoresis in a pH 7·0 phosphate buffer (gel buffer 0·01 M, bridge buffer 0·2 M) at 5 volts/cm for 16 hours at 4°C or at 10 volts/cm with water-cooled apparatus. Kraus & Neely (1964) used vertical electrophoresis at 4°C for 14–16 hours at 8 volts/cm with a tris-EDTA buffer at pH 8·6. The gels were prepared with a 1:20 dilution of a solution containing 0·9 M tris, 0·02 M EDTA and 0·5 M boric acid. This buffer was diluted 1:5 for the anodic bridge vessel and 1:7 for the cathodic bridge vessel.

The isozymes are readily stained by one of a number of methods which vary only slightly in their results. The one used in the author's laboratory was taken from the paper by Davidson *et al.* (1965), and is a modification of the method described by Dewey & Conklin (1960). It depends on the PMS-mediated reduction of MTT to formazan, induced by the transfer of electrons from the NADH formed when lactate is oxidized to pyruvate. The gels are incubated in the dark for 1–2 hours at 37°C in a reaction mixture containing 0·25 ml 10 per cent lactic acid, 5 mg NAD, 5 mg MTT tetrazolium and 5 mg phenazine methosulfate (PMS) in 50 ml of 0·1 M tris buffer at pH 8·0. The insoluble dark blue formazan indicates the sites of LDH activity.

MALATE DEHYDROGENASE

Malate dehydrogenase (L-malate: NAD oxidoreductase, EC 1.1.1.37) catalyzes the reversible conversion of malate to oxaloacetate. There are two forms of malate dehydrogenase (MDH), which differ in their chemical and physical properties. One is found in the cytoplasm and the other is tightly bound to the mitochondria (Christie & Judah, 1953; Shrago, 1965). Evidence for genetic variation in both of these enzymes has been presented by Davidson & Cortner (1967a,b).

CYTOPLASMIC MDH

Davidson & Cortner (1967a) studied the starch gel electrophoretic patterns at pH 7·0 of hemolysates prepared from the red cells of 1440

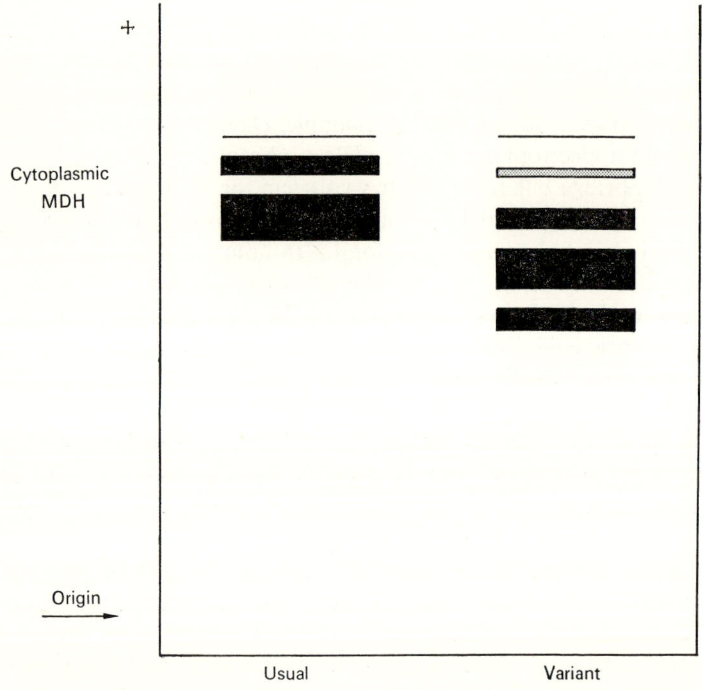

FIGURE 16.2. Diagram of starch gel electrophoretic patterns of cytoplasmic malate dehydrogenase in red cell lysates. A variant pattern is shown on the right. (Drawn from a photograph in Davidson & Cortner, 1967a.)

white and 1470 Negro subjects. They found that in all but one, the MDH pattern consisted of a major anodal band and two minor bands migrating slightly faster. The one variant pattern observed consisted of three major bands, one band corresponding to that of the usual pattern, and the other two with slower mobility, as shown diagrammatically in Fig. 16.2. The minor bands appeared the same as those of the usual phenotype. The variant was inherited as an autosomal co-dominant. The slowest-moving component of the variant and the major component of the usual phenotype were obtained by concentration of starch block eluates followed by column chromatography. When these isozymes were subjected to reversible inactivation and recombination, an intermediate band was observed, suggesting that the enzyme exists as a dimer. Thus, in the variant pattern, the fastest and slowest bands appear to be dimers of two allelic gene products, while the intermediate band is a mixed dimer of the two gene products.

MITOCHONDRIAL MDH

Extracts of various tissues, including white cells and placentae, were studied by Davidson & Cortner (1967b) to determine the electrophoretic pattern of the mitochondrial MDH. They found that the usual pattern consists of five bands, four of which migrate toward the cathode at pH 7·0, while the fastest migrates just anodal to the origin. The major band is most cathodal; the remaining bands have progressively less activity. In five unrelated individuals of 523 white subjects tested, they found an electrophoretic variant of mitochondrial MDH (see Fig. 16.3). It consisted, in all instances, of three major and two minor bands. The major bands had somewhat more activity, but the same mobility as bands 2 and 1 of the usual phenotype. Family studies showed autosomal co-dominant inheritance. The cytoplasmic MDH pattern of the mitochondrial MDH heterozygotes was unaffected.

Davidson & Cortner (1967b) pointed out that their studies demonstrated for the first time genetic variation in a human mitochondria enzyme. Furthermore, its autosomal inheritance indicated that the mitochondrial MDH in man does not belong to a separate genetic system transmitted by cytoplasmic inheritance through the maternal parent, such as has been demonstrated for mitochondrial enzymes in Drosophila and other insects (Sager, 1964). Cortner & Davidson

(1967) have recently demonstrated genetic variation in another mito-chondrial enzyme, glutamic oxaloacetic acid transaminase. This enzyme, like mitochondrial MDH, is inherited as an autosomal

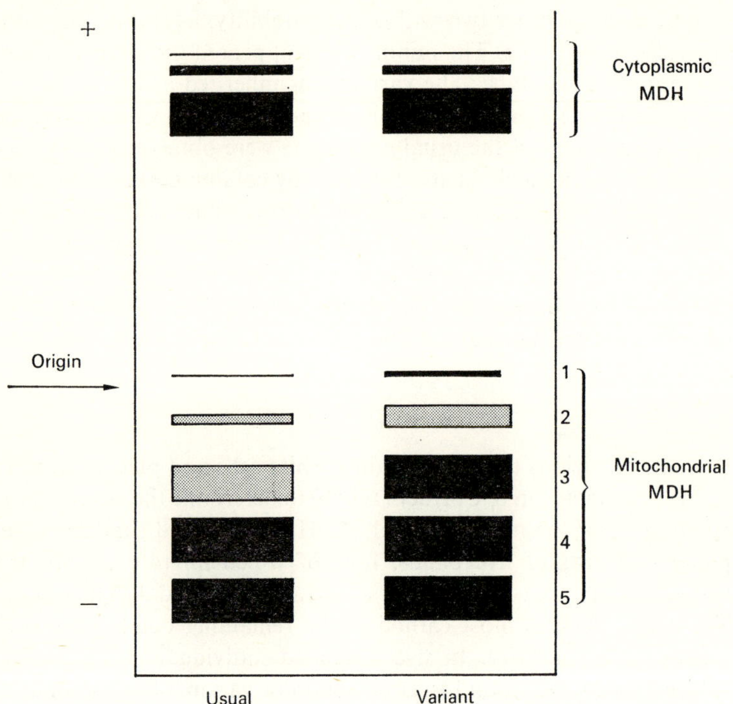

FIGURE 16.3. Diagram of starch gel electrophoretic patterns of anodally-migrating cytoplasmic malate dehydrogenase (MDH) and cathodally-migrating mitochondrial MDH from white cell lysates. A variant mito-chondrial MDH pattern is on the right. (Drawn from a photograph in Davidson & Cortner, 1967b.)

codominant, and thus, like MDH, is not controlled by a separate genetic system of the mitochondrion.

In studies of mitochondrial MDH of the chicken, Kitto *et al.* (1966) provided evidence suggesting that the multiple components of this enzyme represent conformational changes rather than subunit hybrids, such as those associated with the LDH subunits. Conformational variation may be responsible for the minor components observed in nearly all of the enzymes studied by electrophoresis. On the other hand, they may also be due to formation of complexes with various chemical substances, minor degradation in an *in vitro* environment, or any of a number of other non-genetic factors.

METHODS

To demonstrate red cell MDH, Davidson & Cortner (1967a) use horizontal starch gel electrophoresis of hemolysates in a pH 7·0 phosphate buffer, 0·01 M in the gel and 0·2 M in the vessels. Electrophoresis is carried out in a water-cooled apparatus at 10 volts/cm for 5 hours or in a cold room at 4 volts/cm for 16 hours. They prefer a vertical system for the mitochondrial enzyme, using a pH 7·0 phosphate-citric acid buffer, as described by Thorne *et al.* (1963). Leucocyte extracts are prepared by removing the buffy-coat layer from sedimented whole blood, washing briefly in distilled water to lyse the red cells, and then washing in saline. After resuspension in an equal volume of distilled water, the cells are subjected to sonication for 1 minute.

After the gels are sliced, they are placed in a developing solution consisting of 725 mg DL-malic acid, 5 mg NAD, 5 mg PMS and 5 mg MTT in 50 ml of 0·5 M tris-HCl buffer at pH 8·6. This stain reveals the positions of both the cytoplasmic and mitochondrial enzymes after incubation at 37°C. The principle is the same as that for the LDH stain, the malic acid being oxidized to oxaloacetic acid, with reduction of NAD to NADH and reduction of MTT to formazan.

PHOSPHOHEXOSE ISOMERASE

VARIANTS FOUND AT RANDOM

Phosphohexose isomerase (PHI), also known as phosphoglucose isomerase, glucose phosphate isomerase, or D-glucose-6-phosphate ketol isomerase (EC 5.3.1.9) catalyzes the reversible conversion of glucose-6-phosphate to fructose-6-phosphate. Detter *et al.* (1968) examined the hemolysates of nearly 3400 unrelated individuals from several different populations and found in addition to the usual pattern, called PHI 1, eight variant phenotypes, designated PHI 2-1, 3-1, 4-1, 5-1, 6-1, 7-1, 8-1 and 9-1. Autosomal codominant inheritance was demonstrated for PHI 3-1, 5-1, 6-1 and 9-1, but families were not available to study the inheritance of phenotypes 2-1, 4-1, 7-1 and 8-1.

The PHI phenotypes are pictured in Fig. 16.4, which shows that the usual (PHI 1) pattern consists of three components, one major and two minor, migrating toward the cathode at pH 8·0; the major band is most cathodal. The variant phenotypes contain three major components, one of which corresponds to the main component of PHI 1. The triplet pattern of the variants suggests a dimer structure for the enzyme. Thus, it is assumed that with heterozygosity at the PHI locus, two subunits, one usual and one variant, are produced. In the case of PHI 2-1, 3-1, 4-1, 5-1 and 6-1, the variant subunit is more negatively charged than the usual subunit, so its dimer has a more anodal migration. However, in PHI 8-1 and 9-1 (and probably, 7-1), the variant subunit is more positively charged. In all instances but PHI 7-1, there is a band of intermediate mobility, representing the hybrid dimer. The enzyme activity was determined for three variants (PHI 3-1, 5-1 and 6-1). All were within the normal range of $13·6 \pm 1·9$ μmoles NADPH per minute per 10^{10} red cells.

PHI 5-1 differs from the other types, in that the isozyme pattern of fresh hemolysates is poorly defined, but in older lysates, the sharply demarcated zones shown in Fig. 16.4 are seen. This pattern can be reverted to the pattern of fresh lysate by adding 2-mercaptoethanol, suggesting that the alteration on storage probably involves disulfide bonding with a substance such as oxidized glutathione.

All of these variant isozyme patterns are rare in the populations sampled, with the exception of PHI 3-1, which was found in about one per cent of 574 Asiatic Indians tested. Of the total of 3397 people from various populations tested, only 20 were found to have a variant; eleven of these were PHI 3-1, of which six occurred in the Asiatic Indians.

VARIANTS ASSOCIATED WITH
ENZYME DEFICIENCY

Baughan *et al.* (1968) studied a patient with congenital non-spherocytic hemolytic anemia in whom both the red cells and white cells were markedly deficient in PHI activity. In both parents, enzyme activity was about half normal. Detter *et al.* (1968) found that the phenotype of the mother and several of her relatives, including a sib of the propositus, was similar to that of PHI 9-1. However, the phenotype of the father

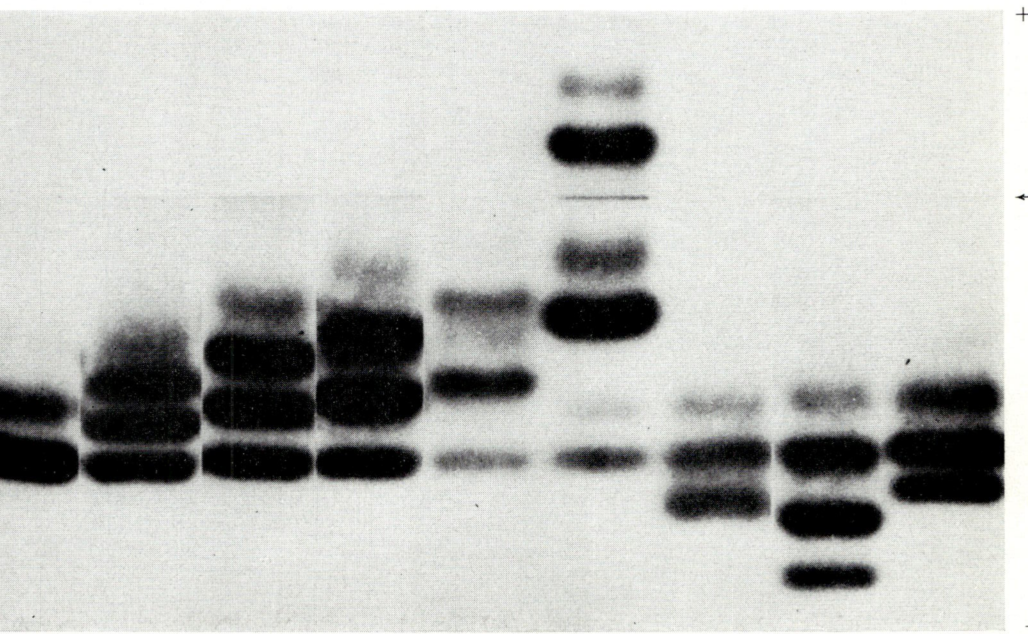

FIGURE 16.4. Starch gel electrophoretic patterns of nine phosphohexose isomerase (PHI) phenotypes. The usual phenotype, PHI 1, is on the left. The arrow points to the origin.

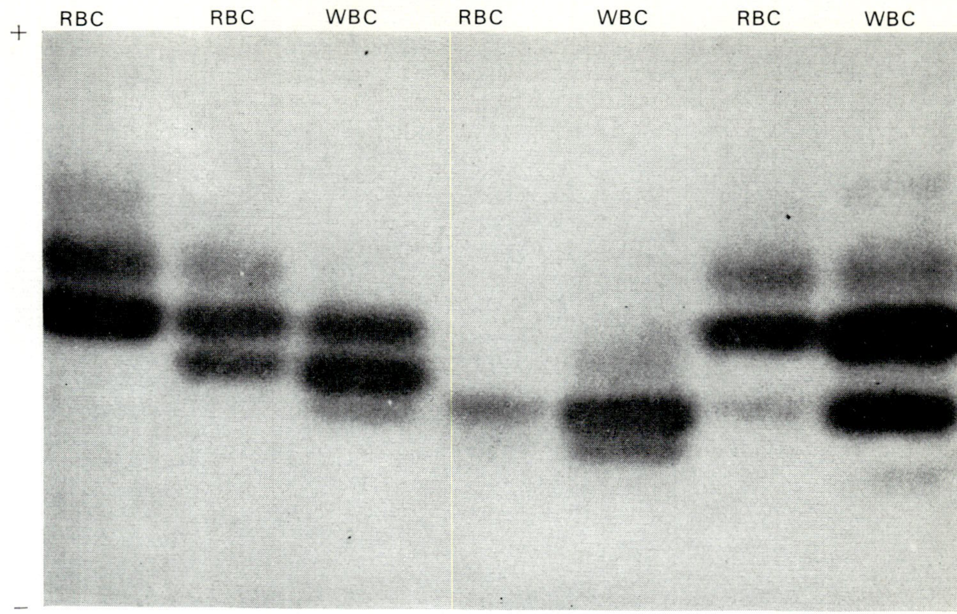

FIGURE 16.5. Starch gel electrophoretic patterns of PHI phenotypes in lysates of red cells (RBC) and white cells (WBC) of a child with PHI deficiency (9–10), his mother (9–1) and his father (10–1). The pattern of the usual phenotype, PHI 1, is included for comparison. The arrow at the top of the figure points to the origin.

and some of his relatives was distinctly different, so it was called PHI 10-1. The pattern of the propositus, shown in Fig. 16.5, contained none of the bands characteristic of the usual phenotype, and since it apparently represented heterozygosity for the two abnormal PHI genes of his parents, the phenotype was called PHI 9-10. The decrease in enzyme activity of the two gene products may reflect molecular instability, relative catalytic inactivity, a slower rate of synthesis or a combination of these factors. (See Addenda.)

METHODS

Starch gel electrophorsis as performed by Detter *et al.* (1968) at pH 8·0, uses a bridge buffer of 0·25 M tris and 0·057 M citric acid and a gel buffer of 0·017 M tris and 0·0032 M citric acid. After 18 hours electrophoresis at 6 volts/cm in the cold, the gel is sliced and stained with a 1 per cent agar overlay containing $9·2 \times 10^{-4}$ M fructose-6-phosphate, 1·1 mg NADP, 1·2 mg MTT, 0·6 mg PMS and 1 unit of G6PD in 12 ml of 0·03 M tris buffer at pH 8·1. The blue formazan deposition at the site of PHI activity is dependent upon the conversion of F6P to G6P by the enzyme, with subsequent conversion of G6P to 6PG by G6PD, and resultant reduction of MTT to formazan, mediated by PMS.

CARBONIC ANHYDRASE

Carbonic anhydrase (carbonate hydro-lyase, EC 4.2.1.1) is second only to hemoglobin as the major protein component of red cells. The enzyme catalyzes the reversible conversion of carbon dioxide and water to carbonic acid. Thus, in red cells, it provides an important mechanism for the release of carbon dioxide as blood passes through the pulmonary capillaries. There are two different molecular forms of red cell carbonic anhydrase, designated CA I and CA II by Tashian (1965) and B and C, respectively by Nyman & Lindskog (1964), Rickli *et al.* (1964) and Armstrong *et al.* (1966). A third, minor component, designated A by Rickli *et al.* (1964) has the same amino-acid composition and tryptic peptide pattern as CA I, and probably represents a small surface charge difference which is not genetically controlled (Nyman & Lindskog, 1964; Laurent *et al.*, 1965).

The properties of the two molecular forms were reviewed by Armstrong *et al.* (1966). CA I has about one-twentieth the specific activity but a concentration about five times that of CA II. Both enzymes have a molecular weight of about 28,000, but they differ in isoelectric point and amino-acid composition, although they each consist of a single polypeptide chain containing one cysteine residue and one zinc atom per molecule. CA I has a more anodal migration rate than CA II at alkaline pH. Both have relatively low esterase activity, which permits their visualization on starch gel electrophoresis using various phenyl and naphthyl acetates as substrates. Both the esterase and carbonic anhydrase activities are strongly inhibited by sulfonamides such as acetazolamide, but not by diisopropylfluorophosphate. Studies on the amino acid sequence of CA I have been reported by Nyman *et al.* (1966) and Laurent *et al.* (1967). Catalytic properties have recently been studied by Verpoorte *et al.* (1967) and Whitney *et al.* (1967a,b).

GENETIC VARIATION

Tashian (1961) found that when hemolysates of human red cells were subjected to starch gel electrophoresis and stained by dye-coupling with α-napthol esters as substrates, a large series of bands with esterase activity was observed. Shaw *et al.* (1962) subsequently reported the occurrence of a rare, slow-moving band which they identified as a slow variant of CA I. It was found in only one of 2638 random specimens tested. This variant, subsequently named CA Ib, was shown to be inherited as an autosomal codominant.

Tashian *et al.* (1963) reported a second slow-moving variant, CA Ic, also inherited as an autosomal codominant in four different families inhabiting the islands of Guam and Saipan. In a subsequent study, this variant, now called CA Ic$_{Guam}$, was subjected to chemical analysis (Tashian *et al.*, 1966). The results indicated the probable substitution of a glycine residue for arginine at the eighth position from the C-terminal end of the polypeptide chain. Lie-Injo (1967) has recently reported the occurrence of this variant in two of 120 Filipinos living in San Francisco.

A third variant, CA Id$_{Michigan}$ was found in a Negro woman and her daughter by Tashian (1965) and was later studied for its physico-chemical properties by Shows (1967). He found this enzyme more thermolabile than the normal enzyme, CA Ia, and determined that the two enzymes had similar Michaelis constants, pH profiles and rates of inhibition by acetazolamide. Chemical analysis indicated that in the variant molecule there is a substitution of lysine for threonine.

The relative mobilities of these three rare variants, compared with the pattern of the usual enzyme, CA Ia, is diagrammatically shown in Fig. 16.6. No CA II variants have been described, and this isozyme provides a useful reference point for comparing relative mobilities of

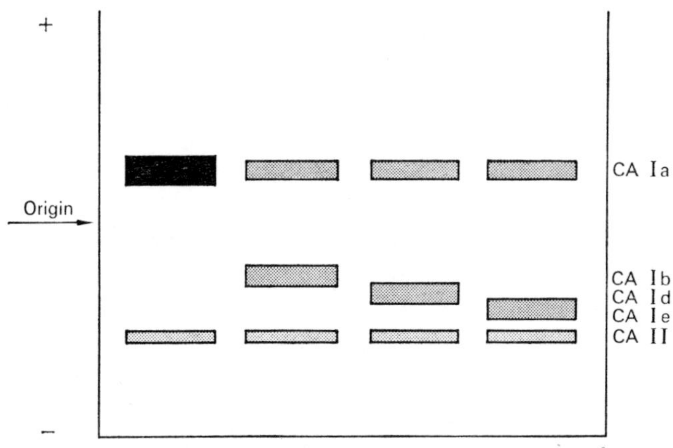

FIGURE 16.6. Diagram of starch gel electrophoretic patterns of red cell carbonic anhydrases CA I and CA II. The usual pattern is shown on the left: the three known variants of CA I are on the right. (Adapted from Tashian, 1965. Reprinted with permission from the *Am. J. hum. Genet.* **17**, 257.)

the CA I variants. All of the carbonic anhydrases are distinguishable from the other red cell enzymes with esterase activity by their slower mobility, and their susceptibility to inactivation by sulfonamides.

QUANTITATIVE VARIATION

In hemolysates of normal cord blood, the carbonic anhydrase activity is greatly reduced, so that neither CA I nor CA II is readily visualized on electrophoresis of hemolysates (Haut *et al.*, 1962). However, in adults, carbonic anhydrase deficiency is very rare. Rieder & Weatherall (1964) reported that in a study of about 1400 blood specimens from adults, the only carbonic anhydrase abnormality noted was the virtual absence of CA I, observed in a Negro woman with thalassemia minor. Her total carbonic anhydrase level was normal, presumably indicating a compensatory increase in CA II production. Neither of her parents nor her three daughters had a similar decrease in CA I activity, so it could not be determined whether the defect was inherited as a recessive trait or had a non-genetic origin.

Lie-Injo & Tarail (1966) subsequently described a Greek male with persistence of fetal hemoglobin and a thalassemia-like syndrome in whom the activity of both CA I and CA II was decreased. No elevation of Hb F or decreased activity of carbonic anhydrase was found in his two children, two brothers and two maternal uncles.

The unusual nature of this case prompted Lie-Injo *et al.* (1967) to study the levels of red cell carbonic anhydrase in various disease states. They found a decreased level of the enzyme in a Negro woman with thyrotoxicosis, sickle trait, and a high level of fetal hemoglobin. After treatment of her thyrotoxic state, the enzyme level increased to normal, and her Hb F fell from 19 to 6 per cent. Tests were then performed on twelve other patients with thyrotoxicosis, many of whom were being treated. In nine, the carbonic anhydrase level was lower than normal; in two, the Hb F was moderately increased. Furthermore, the authors stated that from a personal communication, they learned that the patient described by Rieder & Weatherall (1964) had hyperthyroidism.

These findings suggest that the excess of thyroid hormone associated with thyrotoxicosis may suppress the production of carbonic anhydrase or bring about an increase in its catabolic rate with or without a compensatory increase in synthesis. The elevated levels of fetal hemoglobin, observed in only three cases, may not be directly related to the underlying disease, but additional evidence is required. It does appear that there is no evidence at present for hereditary deficiency of carbonic anhydrase activity due to variation at the structural gene loci for CA I and CA II. In the case of the structural variants, CA Ic$_{Guam}$ and CA Id$_{Michigan}$ the ratio of variant to normal enzyme activity in the heterozygotes (whose total enzyme is normal) is approximately 1 : 2 (Tashian *et al.*, 1966; Shows, 1967), indicating some increase in production by the normal allele to compensate for the decreased amount of variant. Homozygotes for these rare variants have not been seen.

METHODS

In the method of Tashian (1961, 1965) and Tashian & Shaw (1962), red cell hemolysates are prepared by lysing the cells with one volume of distilled water and extracting with 0·4 volume of toluene, using adequate centrifugation to remove all traces of stroma. Electrophoresis is performed at 4°C and 8 volts/cm for 18 hours in 0·02 M borate, 0·0046 M NaOH gel buffer, pH 8·6, and 0·3 M borate, 0·05 M NaOH bridge buffer, pH 8·0, containing 0·03 M NaCl. After slicing, the gels are placed in an incubation mixture containing 2 ml of 1 per cent α- or β-naphthyl acetate in acetone, 10 ml of 0·4 M tris buffer, pH 7·0, and 100 mg Blue RR salt (diazotized 4′-amino-2′,5′-dimethoxy-benzanilide) in 100 ml distilled water. The zones of esterase activity are colored blue-red, due to the release of α- or β-naphthol from the acetate, with subsequent coupling of napthol with the diazo salt to form a colored complex. Although a large number of zones with esterase activity are stained, the carbonic anhydrases can be differentiated by their slow mobility, CA Ia being anodal, and CA II cathodal to the origin.

RED CELL ESTERASES

ISOZYME CLASSIFICATION

Tashian (1965) classified the various 'nonspecific' esterases of primate erythrocytes into three main groups, depending upon their substrate preference for the α- and β-naphthol esters of butyrate, propionate and acetate. Thus, he could differentiate butyrylesterase (B esterase), propionylesterase (P esterase) and acetylesterases (C esterase, A esterase and carbonic anhydrase). The appearance of these enzymes in human hemolysates, which lack the P esterase, is shown diagrammatically in Fig. 16.7. The carbonic anhydrases and the very faint C esterase are not included in the figure. The B esterase appears as a sharply defined band which acts as a marker for the other bands. This enzyme, which chemically resembles the arylesterase of plasma, is inhibited by diisopropylfluorophosphate (DFP) but not by physostigmine (eserine); it is activated by p-chloromercuribenzoate (PCMB). The C esterase appears as a series of poorly defined, rapidly migrating bands which are inactivated by PCMB but not by DFP. The A esterase is subdivided into three groups, A_1, A_2, and A_3, each appearing as

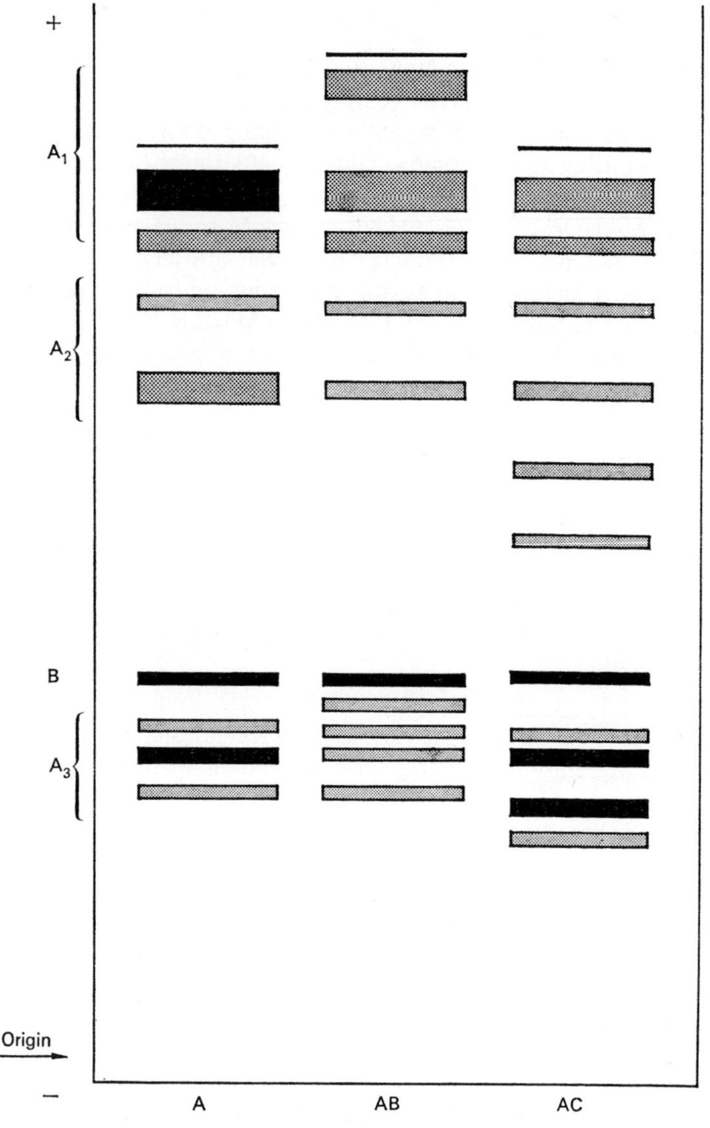

FIGURE 16.7. Diagram of starch gel electrophoretic patterns of red cell esterases. The variants, AB and AC, differ from the usual type A in all three sets of esterase bands, A_1, A_2 and A_3. (Adapted from Tashian, 1965. Reprinted with permission from the *Am. J. hum. Genet.* **17**, 257.)

doublets or triplets which vary slightly in susceptibility to inhibitors such as iodoacetate and urea (Tashian, 1965).

VARIANTS

In a survey of 4100 human hemolysates, Tashian (1961, 1965) and Tashian & Shaw (1962) reported a total of three instances of variation in the A bands, two designated as type AB and one as AC. As shown in Fig. 16.7, members of all three groups of A esterase isozymes are altered in these phenotypes; thus Tashian (1965) suggested that they all share a common polypeptide subunit. The AB pattern was found in one of 2600 samples from Caucasian donors. The AC variant and another example of AB were detected in one of 600 samples from Negro donors. The autosomal codominant inheritance of type AB was shown by Tashian & Shaw (1962).

METHODS

The method described for distinguishing the carbonic anhydrase variants is applicable to the A esterase variants, as well. Tashian & Shaw (1962) recommended the use of two gels, one buffered at pH 8·4 and the other at 8·6 to ensure that all components are visualized. Omission of NaCl from the bridge is sometimes desirable, since the salt alters the migration rate of the B esterase, which may overlie some of the A esterase bands.

PEPTIDASES

ISOZYME CHARACTERISTICS

Lewis & Harris (1967) have recently described the starch gel electrophoretic behavior of several peptidases in red cells and other tissues, and have presented evidence for inherited variation. Using the twelve dipeptides and four tripeptides listed in Table 16.1 as substrates, they were able to differentiate five different peptidase patterns in all hemolysates tested. These zones of peptidase activity, called A, B, C, D and E, are shown diagrammatically in Fig. 16.8.

All of the dipeptides except Leu-Pro were hydrolysed by peptidase A. Peptidase B reacted best with the tripeptide, Leu-Gly-Gly; appreciable activity was also observed with Phe-Tyr and Phe-Leu, while there was weak activity with dipeptides Lys-Leu and Pro-Phe as well as with the three other tripeptides. Like peptidase A, peptidase C hydrolysed most of the dipeptides, but there were marked differences in the preference of the two enzymes. Only the dipeptide Leu-Pro provided a suitable substrate for peptidase D, while peptidase E could be detected by four dipeptides and leucyl-β-naphthylamide, but all of its reactions were weak.

TABLE 16.1. Relative activity (on a scale of 3) of the five red cell peptidases with dipeptides, tripeptides and Leucyl-β-naphthylamide. Adapted from Lewis & Harris (1967)

Peptide	Peptidase				
	A	B	C	D	E
Val-Leu	3	—	—	—	—
Gly-Leu	3	—	I	—	—
Leu-Gly	3	—	I	—	—
Gly-Phe	2	—	I	—	—
Leu-Ala	2	—	3	—	—
Leu-Leu	2	—	2	—	I
Gly-Trp	2	—	2	—	—
Lys-Leu	I	I	3	—	I
Pro-Phe	2	I	2	—	—
Phe-Leu	2	2	3	—	I
Phe-Tyr	2	2	3	—	I
Leu-Pro	—	—	—	2	—
Leu-Gly-Gly	—	3	—	—	—
Leu-Gly-Phe	—	I	—	—	—
Tyr-Tyr-Tyr	—	I	—	—	—
Leu-Leu-Leu	—	I	—	—	—
Leu-βNAmide	—	—	—	—	I

Peptidases C, D and E were observed to undergo changes in migration with aging of the hemolysate or incubation with oxidized glutathione. The pattern of aged samples could be restored to normal by adding mercaptoethanol. Similar changes have been noted with the phosphohexose isomerase variant, PHI 5-1 (see p. 528), suggesting that this enzyme variant, as well as the three peptidases, may have sulfhydryl groups available for disulfide bond formation with the oxidized glutathione present in hemolysates.

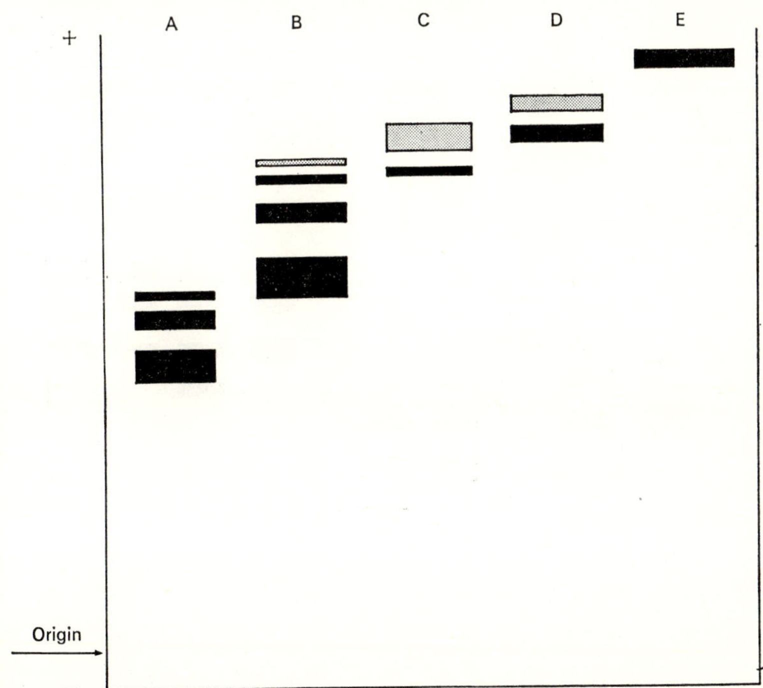

FIGURE 16.8. Diagram of starch gel electrophoretic patterns of the five different peptidases, A through E, commonly seen in red cell lysates. (Adapted from Lewis & Harris, 1967. Reprinted with permission from *Nature* **215**, 351.)

VARIANTS

Peptidases A and B were found by Lewis & Harris (1967) to exhibit genetic variation. The authors referred to the common types of peptidase A and peptidase B as Pep A 1 and Pep B 1, respectively. Variant types of peptidase A were designated as Pep A 2-1, Pep A 2, Pep A 3-1 and Pep A 4-1. The variants of peptidase B were called Pep B 2-1, Pep B 3-1 and Pep B 4-1. These phenotypes are shown diagrammatically in Fig. 16.9. (See also the Addenda.)

Over 2000 Europeans and 550 Negroes were tested for their peptidase A and B types. The results, shown in Table 16.2, indicate a 13 per cent incidence of Pep A 2-1 and 1·3 per cent incidence of Pep A 2 among the Negroes. Neither type occurred in the Europeans. Pep A 3-1 was found in one European and one Negro, while the three examples of Pep A 4-1 were confined to Europeans. Nine people with one of the variant peptidase B types were found; all were European.

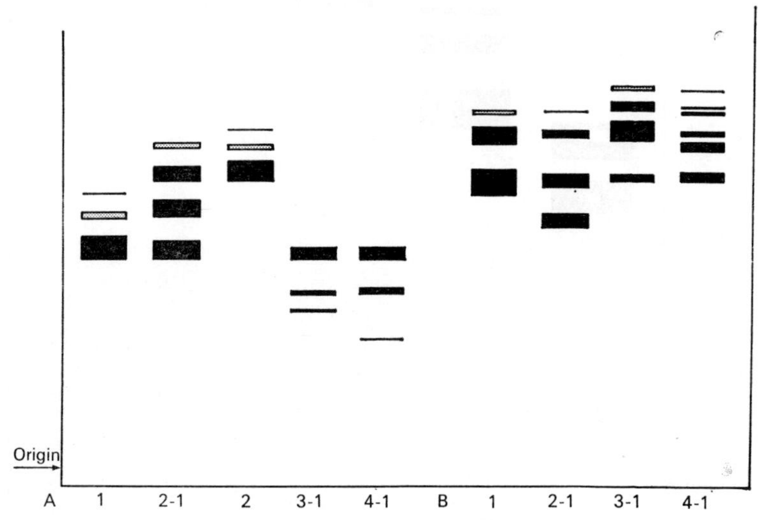

FIGURE 16.9. Diagram of starch gel electrophoretic patterns of five different peptidase A phenotypes and four different peptidase B phenotypes. (Adapted from Lewis & Harris, 1967. Reprinted with permission from *Nature* **215**, 351.)

Thirty families with at least one parent Pep A 2-1 or Pep A 2 were tested. These and other data strongly support the likelihood that individuals with Pep A 2-1 are heterozygous, and individuals with Pep A 2 are homozygous for a variant allele at an autosomal locus determining peptidase A structure. Evidence for two more alleles at that locus was obtained from family studies of individuals with the phenotypes Pep A 3-1 and Pep A 4-1. The authors concluded that the major components in each of the Pep A phenotypes represent dimers of subunits, so that the three components of the variant types are composed of one mixed and two unmixed dimers, similar to those postulated for the variants of other enzymes, such as 6PGD and PHI.

TABLE 16.2. Incidence of Peptidase A and B phenotypes among Europeans and three (combined) Negro populations resident in England, Nigeria and South Africa. Adapted from Lewis & Harris (1967)

Peptidase phenotypes		Europeans	Negroes
Peptidase A	1	2279	468
	2-1	—	72
	2	—	8
	3-1	1	1
	4-1	3	—
Total		2283	549
Peptidase B	1	2188	549
	2-1	6	—
	3-1	2	—
	4-1	1	—
Total		2197	549

Autosomal codominant inheritance was also found in family studies of the peptidase B variants. The common type, Pep B 1, resembled the pattern characteristic of PHI 1 in having one major and two minor components. However, the three heterozygote patterns (Fig. 16.9) did not give the appearance of mixed and unmixed hybrids. They appeared to represent a mixture of two overlapping series of bands, each series representing a single gene product. Thus, the three Pep B 1 bands were demonstrable in all of the heterozygote patterns, the intensity varying with the amount of overlapping isozyme. Each variant contained an additional slow or fast-moving band, representing the non-overlapping product of the variant allele.

METHODS

According to the method of Lewis & Harris (1967), hemolysates are prepared either by sonication of washed cells or the addition of two volumes of water. Starch gel electrophoresis is performed in the cold for 18 hours at 5 volts/cm, with pH 7·4 buffer (bridge buffer 0·1 M tris/maleate; gel buffer 0·005 M tris/maleate).

The staining procedure is dependent upon a series of three chemical reactions:

$$(1)\ \text{Peptide} + H_2O \xrightarrow{\text{Peptidase}} \text{L-amino acid}$$

$$(2)\ \text{L-amino acid} + O_2 \xrightarrow{\text{L-amino acid oxidase}} \text{Keto acid} + NH_3 + H_2O_2$$

$$(3)\ H_2O_2 + o\text{-dianisidine} \xrightarrow{\text{Peroxidase}} \text{oxidized dianisidine}$$
$$\text{(dark brown color)}$$

A freshly prepared reaction mixture is poured over the surface of the cut gel. This mixture is made from 25 ml of 2 per cent aqueous agar at 55°C to which is added 25 ml 0·2 M phosphate-HCl buffer, pH 7·5, containing 20 mg peptide, 5 mg crude *Crotalus adamanteus* venom (Sigma) as the source of L-amino oxidase, 10 mg horseradish peroxidase POD II (Boehringer), 5 mg *o*-dianisidine hydrochloride and 0·5 ml 0·1 M manganese chloride. After incubation for an hour at 37°C, the zones of dark brown oxidized dianisidine indicate the sites of peptidase activity.

GLUTATHIONE REDUCTASE

Glutathione reductase or NAD(P)H$_2$: glutathione oxidoreductase, EC 1.6.4.2, catalyses the reduction of oxidized glutathione (GSSG) by NADPH (or NADH) to reduced glutathione (GSH). Deficiency of glutathione reductase is associated with a hemolytic process of varying severity (Carson *et al.*, 1961; Löhr & Waller, 1962; Waller *et al.*, 1965).

HETEROGENEITY

Inherited structural variation of the enzyme was reported by Long (1967), who found that in 46 of 196 hemolysates from Negroes the electrophoretic pattern of GSR was different from that of the remaining 150. In six, there was a single band migrating more rapidly than the usual enzyme; in 40, there was a wide zone of enzyme activity overlapping the region of the fast and slow bands.

Long (1967) found from family studies that the wide zone represented heterozygosity, and the fast zone, homozygosity for an autosomal co-dominant allele with a frequency in Negroes of about 0·13. Enzyme assays indicated that the variant had a somewhat greater activity than the usual slower-moving enzyme. Furthermore, among

28 Negro patients with primary gout, 15 appeared to be heterozygous and eight homozygous for the variant allele, a proportion much higher than that observed in the general Negro population.

In the staining procedure described by Long (1967) the zones of activity were seen as white spots against a dark blue background representing regions where NADPH was oxidized to NADP by GSR. Subsequently, Kaplan & Beutler (1968a) described a different method for detecting GSR activity. They confirmed the occurrence of a fast-moving variant in Negroes, and also showed that the GSR of both white cells and red cells in a given individual has the same electrophoretic mobility, regardless of whether the hydrogen donor is NADPH or NADH.

METHODS

Since the staining method of Kaplan & Beutler (1968a) provides a dark blue pattern of enzyme activity against a white background, it is currently preferred. These authors based their stain on the fact that the GSH formed from reduction of GSSG rapidly reduces 2,6 dichlorophenol indophenol (DCIP), which then reacts non-enzymatically with the tetrazolium salt, MTT, to form a dark blue formazan precipitate.

Hemolysates are prepared from washed red cells by freezing and thawing three times after the addition of two volumes of cold distilled water. Stroma is discarded after centrifugation for 30 minutes at 31,000 g. Horizontal electrophoresis at 4°C is performed at 20–25 V/cm for 4 hours. The bridge buffer consists of 0·1 M tris-HCl at pH 9·6, containing 4·5 mM EDTA. This buffer diluted 1:10 is used in preparing the starch gel.

The sliced gel is stained in the dark at room temperature for 2 hours in a freshly prepared solution containing 3·4 mM GSSG, 1·3 mM NADPH, 1·2 mM MTT and 0·004 mM DCIP in a 0·25 M tris-HCl buffer at pH 8·4.

METHEMOGLOBIN REDUCTASE:
NADH AND NADPH DIAPHORASES

In normal red cells, the level of methemoglobin is maintained at a concentration of less than 1 per cent of the total hemoglobin. The two enzymes which catalyze the reduction of methemoglobin are NADPH diaphorase (Huennekens *et al.*, 1957, 1958) and NADH diaphorase (Scott, 1960; Scott & McGraw, 1962). A deficiency of the latter enzyme, which occurs in individuals homozygous for an abnormal gene, results in methemoglobinemia (Scott & Griffith, 1959; Jaffé, 1963; Jaffé & Heller, 1964; Jaffé *et al.*, 1966).

EVIDENCE FOR HETEROGENEITY

The existence of variation at the structural gene locus for NADH diaphorase can be inferred from the studies of West *et al.* (1967) and of Kaplan & Beutler (1967). Of three patients with congenital methemoglobinemia studied by Kaplan & Beutler, two patients had about 10 per cent of the normal level of red cell NADH-diaphorase. When the hemolysates were subjected to starch gel electrophoresis, the enzyme could not be visualized in one case, even in concentrated extracts. But in the second case, it appeared as a very weak band migrating at the same rate as the normal enzyme. In the third patient, whose enzyme activity was 25 per cent of normal, electrophoresis revealed no activity at the usual site of NADH diaphorase, but a weak, faster band was observed. The hemolysate electrophoretic pattern of this patient's daughter, who had 55 per cent of normal enzyme activity, contained two bands, one with mobility of the normal enzyme and one with faster mobility identical to that of her mother. The aberrant enzyme was called the California variant. These authors also tested 200 normal hemolysates without finding any evidence for variation in either NADH or NADPH diaphorases. As expected, the NADPH diaphorase pattern was normal in all three patients with NADH diaphorase deficiency.

West *et al.* (1967) prepared a ten-times concentrated extract from normal red cells and the cells of a patient with mild methemoglobinemia due to NADH diaphorase deficiency. With their staining method, the normal enzyme on starch gel electrophoresis appeared as two faint bands. The patient had only one very faint band with intermediate mobility, while his mother had all three bands. Presumably, the variant enzyme in this case was not the same as that described by Kaplan & Beutler (1968b), suggesting that several structural variants may exist.

METHODS

In the starch gel electrophoretic method described by Kaplan & Beutler (1968b), washed red cells are hemolyzed in one volume of distilled water, frozen and thawed three times and centrifuged for 30 minutes at 31,000 g. The bridge buffer consists of 0·1 M tris and 4·5 mM EDTA adjusted to pH 9·3 with HCl. This buffer is diluted 1:10 for preparation of the gel. Horizontal electrophoresis is performed at 4°C for 4 hours at 20–25 volts/cm.

Staining is based on the use of 2,6 dichlorophenol indophenol (DCIP) as the electron acceptor from NADH (or NADPH) in the presence of the NADH (or NADPH) diaphorase, followed by non-enzymatic reduction of MTT to insoluble blue formazan by the reduced DCIP. The freshly prepared reaction mixture is composed of 1·3 mM NADH (or NADPH), 1·2 mM MTT and 0·06 mM DCIP in a 0·25 M tris-HCl buffer, pH 8·4. The gels are developed in this solution at room temperature for at least 3 hours.

When NADH is included in the staining solution, its diaphorase appears behind the hemoglobin band. With NADPH, a fairly weak, broad zone of activity, migrating far ahead of Hb A is observed.

The staining mixture of West *et al.* (1967) also included DCIP and NADH, but no tetrazolium dye was employed. They used an overlay technique, in which a piece of Whatman 3 MM chromatography paper was dipped into a filtered solution of concentrated DCIP containing 50 mg of NADH per 100 ml of 1 M tris buffer, pH 7·5, with 0·3 mM EDTA. This paper was placed on the cut gel surface and covered with a thin plastic sheet to exclude air. On subsequent removal of the paper, the diaphorase activity could be seen as decolorized zones migrating behind hemoglobin. (Their electrophoresis was performed in Tris-borate-EDTA buffer at pH 8·6.) A somewhat similar method has also been described by Brewer *et al.* (1967).

CATALASE

Catalase, or $H_2O:H_2O_2$ oxidoreductase (EC 1.11.1.6), is a heme-containing enzyme which very rapidly catalyzes the breakdown of hydrogen peroxide to water and oxygen. Homozygosity for a very rare autosomal gene results in acatalasemia or, more correctly, acatalasia, since the enzyme, normally present in many body tissues, is not found in the tissues of these homozygotes (Takahara, 1952; Takahara *et al.*, 1952, 1960; Aebi *et al.*, 1963). The absence of accelerated hemolysis in acatalasia is due to the fact that peroxides, either generated by metabolism or induced by drugs, are dissipated by the action of another enzyme, glutathione peroxidase (Jacob *et al.*, 1965).

MOLECULAR VARIATION

According to Thorup *et al.* (1964) catalase in normal subjects resembles hemoglobin and many other enzymes in that it demonstrates chromatographic heterogeneity which is probably the result of chemical alterations occurring over a period of time after the molecule is synthesized. However, the existence of molecular variation due to rare mutations at the structural gene locus can also be inferred from two lines of evidence.

Aebi & Cantz (1966) showed that normal red cells could be differentiated from acatalatic cells when blood films were stained by an elution method devised by Kleihauer & Betke (1963) to detect cells containing methemoglobin. (This procedure involves the exposure of cyanide-treated blood films to a mixture of ethanol, citric acid and hydrogen peroxide, followed by staining with hematoxylin and erythrosine. Cells containing oxyhemoglobin are stained red, while those containing methemoglobin appear as ghosts. Cells devoid of catalase behave like methemoglobin-containing cells, presumably because they cannot prevent the conversion of oxy- to methemoglobin in the presence of hydrogen peroxide.)

In blood films of patients with acatalasia, Aebi & Cantz (1966) found that about 1 per cent of the red cells were stained red, and thus contained a fairly normal level of catalase. Furthermore, a direct relation between catalase and red cell age was demonstrated, reticulocytes having about 300 times the activity of mature red cells. They concluded that these patients produce a structurally altered catalase which, due to its marked instability, can only be demonstrated in young red cells. This suggestion was confirmed by Matsubara *et al.* (1967) who demonstrated that the catalase isolated from the reticulocytes of acatalatic subjects was less stable than the normal enzyme and had different chromatographic and electrophoretic behavior.

Baur (1963) found that the red cells of six people from three generations of a Caucasian family had an electrophoretically fast-moving catalase which merged with the normally migrating protein and appeared to be due to heterozygosity for a very rare gene at the catalase structural gene locus. In this instance, there was no decrease in enzyme activity, indicating that the mutant gene was not the same as that (or those) associated with acatalasia. (See Addenda.)

METHODS

Both starch gel and acrylamide gel electrophoresis have been used for demonstrating catalase activity in human hemolysates (Thorup *et al.*, 1961; Baur, 1963; Baumgarten, 1963; Kleihauer & Brandt, 1964; Tudhope, 1965). Three staining techniques have been devised. The first (Thorup *et al.*, 1961) is a negative stain in which the enzyme rapidly breaks down hydrogen peroxide. Thus at the site of enzyme activity, there is no hydrogen peroxide to convert iodide to iodine, so that no blue coloration of the starch gel occurs when the gel is immersed in a solution of sodium thiosulfate and hydrogen peroxide followed by potassium iodide. However, with this method, the unstained region of enzyme activity is usually not sharply defined, and if a variant is present, it is difficult to detect (Baur, 1963).

In the second staining method (Kleihauer & Brandt, 1964) a small amount of hemoglobin is incorporated into the gel. After electrophoresis of hemolysates, the gel is stained with *o*-dianisidine and hydrogen peroxide. The hemoglobin from the hemolysates is dark brown, and the hemoglobin in the gel is light brown to violet. The site of catalase activity is revealed as a white band against the slightly dark background. Again, the diffuse appearance is a disadvantage.

The third stain, preferred in the author's laboratory, is based on the development of a green color at the site of enzyme activity when the detergent, Teepol, reacts with *o*-dianisidine. In this method, described by Tudhope (1965), washed red cells are twice frozen and thawed, and the hemolysate diluted with one-half volume of water. A half volume of carbon tetrachloride is added, and after shaking, the aqueous layer is removed and centrifuged at 4°C and 55,000 g for 60 minutes. The hemoglobin content is adjusted to 15 g per 100 ml with water. Electrophoresis is performed in the tris-citrate-borate buffer of Haut *et al.*, (1962) in which the gel buffer is 0·074 M tris adjusted to pH 9·4 with

10 per cent citric acid, and the bridge buffer is 0·1 M boric acid adjusted to pH 9·4 with 0·2 N NaOH.

An *o*-dianisidine solution is prepared by dissolving 0·3 g of the dye in 100 ml methanol. Acetate buffer, pH 4·6, is made by mixing 1·5 M acetic acid with an equal volume of 1·5 M sodium acetate. Then 50 ml of the dianisidine solution and 50 ml of the buffer are mixed just before use, and 0·4 ml hydrogen peroxide (100 vol.) is added. The gel is immersed in this solution for 5–10 minutes, during which the hemoglobin stains dark brown. The stain is then poured off and replaced by a 1:250 Teepol solution. During the next few hours, a green color appears and intensifies in the solution; eventually, the catalase zone takes on the green color. Under these conditions, Hb A and Hb A_2 are dark brown, and the green catalase band migrates well behind Hb A_2.

A much simpler method can be used if it is desired simply to locate the site of catalase activity in starch gel without regard to its pattern. Since the enzyme rapidly releases oxygen from hydrogen peroxide, exposure of the gel to 10 per cent H_2O_2 is followed by the appearance of many small bubbles in the starch within the region of the enzyme's location.

<h1 style="text-align:center">GALACTOSE-1-PHOSPHATE
URIDYL TRANSFERASE</h1>

ENZYME VARIABILITY IN GALACTOSEMIA

Galactose-1-phosphate uridyl transferase, or UDP glucose: α-D galactose-1-phosphate uridyl transferase, catalyzes the reaction:

$$\text{Gal-1-P} + \text{UDPG} \rightarrow \text{G-1-P} + \text{UDPGal}$$

Absence of the enzyme is associated with an accumulation of galactose and Gal-1-P in the tissues, resulting in the syndrome of galactosemia (Isselbacher, 1966). Individuals with this disease are homozygous for a gene, *gt*, which behaves essentially as an amorph and has a frequency of about 0·006 (Beutler *et al.*, 1966). However, Gt^+, the usual allele at this locus, does not have the expected frequency of over 0·99 because a second variant allele, Gt^D (D for Duarte variant), occurs in 10–13 per cent of the population, and thus has a frequency of about 0·06 (Beutler *et al.*, 1966). Homozygotes for this variant have about half of the normal enzyme activity and heterozygotes, three-quarters

(Beutler *et al.*, 1965); thus, there is no associated clinical disorder in either state.

Methods of detecting the variant genes by enzymatic assay or spot testing have been presented by Beutler *et al.* (1964, 1965), Beutler & Baluda (1966a,b) and Ng *et al.* (1967). In addition, Mathai & Beutler (1966) have shown that the Duarte variant has a slightly faster mobility than the normal enzyme on starch gel electrophoresis, due to a change in charge, rather than size of the molecule. It is possible that modifications of their method may permit a clear distinction of the various phenotypes by direct visualization rather than by assay. Although such a procedure is not currently feasible, their technique is briefly described.

METHODS

Electrophoresis of hemolysates is performed for 16 hours at 4°C and 4 volts/cm with 0·01 M phosphate buffer in the starch gel and 0·25 M phosphate buffer in the electrode compartment, both having a pH of 7·0. Visualization of enzyme activity is based on the conversion of Gal-1-P to G-1-P, which is further converted to G-6-P by phosphoglucomutase. Then, in the presence of G6PD, NADP and 6 PGD, the G-6-P is oxidized, and the NADP reduced to NADPH, which is made visible by its fluorescence in ultraviolet light.

The reaction mixture contains 1·42 mM UDPG, 5·65 mM Gal-1-P, 12·4 mM cysteine, 1·2 mM NADP, 8 mM $MgCl_2$, 91 mM tris-acetate buffer at pH 8·0, one unit of G6PD per ml, one unit of PGM per ml and 0·005 unit of 6PGD per ml.

Mathai & Beutler (1966) found that the slight increase in mobility of the Duarte variant was not altered by changes in pH between 6·5 and 8·0 nor by incorporation of urea into the gel. Furthermore, in the hemolysates of heterozygotes, it was very difficult to demonstrate the presence of the Duarte variant because of its decreased activity and overlapping migration.

'INDOPHENOL OXIDASE'

THE ACHROMATIC REGIONS IN
TETRAZOLIUM-STAINED GELS

This and the previous chapters have shown that the staining methods devised for demonstrating several enzymes, such as LDH, G6PD,

6PGD, PGM, etc., employ a combination of phenazine methosulfate (PMS) and a tetrazolium dye, such a MTT. The site of enzyme activity appears as a dark blue, insoluble formazan, resulting from the transfer of electrons from NADH or NADPH via PMS to the soluble MTT. In such gels, the background stains light blue as the result of non-enzymatic reduction of MTT on exposure to light. However, achromatic zones, appearing as white bands against the blue background, are usually observed, and can lead to some distortion of the stained enzyme pattern if the sites overlap.

Brewer (1967) observed that neither a substrate nor a coenzyme needs to be present for the achromatic regions to appear; only PMS and MTT are required. Besides hemolysates, many other tissue extracts have components with the property of preventing the development of the background blue color.

Brewer (1967) employed the pH 7·4 buffer system used for separating PGM (see p. 508), and stained the starch gels with an agar overlay containing only PMS and MTT in the proportions used for PGM staining. Two achromatic bands, the leading band the fainter of the two, were usually seen migrating ahead of the region occupied by the PGM g band. The indophenol oxidase and dichloroindophenol oxidase activity of these achromatic bands was demonstrated with appropriate staining procedures. The sites of oxidase activity did not coincide with other enzymes demonstrated by staining the gels for hexokinase, G6PD, 6PGD, acid phosphatase, AK, PGM, diaphorase and catalase. Non-erythrocytic tissue contained the bands observed in the hemolysates as well as additional, slower-moving zones.

A VARIANT PHENOTYPE

One individual of thousands tested by Brewer (1967) was found to have a variant pattern consisting of the two usual bands and two slower migrating bands. Several tissues were obtained during the patient's autopsy, and they contained the variant observed in his hemolysate. The additional bands migrating near the origin had the same pattern as that observed in the tissues of other individuals, so the two sets of isozymes appeared to be under independent genetic control.

In the family of the propositus, three other individuals were found to have the variant oxidase pattern, and the additional bands appeared to be inherited as a codominant character. X linkage seemed very unlikely, because both the usual gene and the variant gene products appeared in the hemolysates of males (as well as females).

METHODS

The method of demonstrating the achromatic bands after electrophoresis at pH 7·4 has been described above. Two additional methods used to characterize the isozymes as indophenol oxidase and as DCIP oxidase were also described by Brewer (1967). The first employed the 'Nadi reaction', in which the gels are incubated at 37°C in 10^{-3} M α-naphthol and 10^{-3} M N,N-dimethyl phenylenediamine in 0·05 M tris buffer, pH 8·0, for 30 minutes. In the second method, gels are incubated with a 1 per cent agar overlay containing $1·76 + 10^{-4}$ M NADH and in a 0·05 M tris buffer at pH 8·0. Both methods reveal the enzymatic sites as dark blue bands. The second stain probably also reveals the sites of NADH diaphorase activity, as demonstrated by West *et al.* (1967). However, this enzyme appears as an achromatic region under these circumstances, and its migration rate at pH 7·4 is probably considerably less anodal than that of the indophenol oxidase.

REFERENCES

AEBI H. & CANTZ M. (1966) Über die celluläre Verteilung der Katalase im Blut homozygoter und heterozygoter Defektträger (Akatalasie). *Humangenetik* **3**, 50.

AEBI H., JEUNET F., RICHTERICH R., SUTER H., BÜTLER R., FREI J. & MARTI H.R. (1963) Observations in two Swiss families with acatalasia. *Enzymol. biol. clin.* **2**, 1.

ALLEN J.M. (1961) Multiple forms of lactic dehydrogenase in tissues of the mouse: their specificity, cellular localization and response to altered physiological conditions. *Ann. N.Y. Acad. Sci.* **94**, 937.

APPELLA E. & MARKERT C.L. (1961) Dissociation of lactate dehydrogenase into subunits with guanidine hydrochloride. *Biochem. biophys. Res. Commun.* **6**, 171.

ARMSTRONG J.M., MYERS D.V., VERPOORTE J.A. & EDSALL J.T. (1966) Purification and properties of human erythrocyte carbonic anhydrases. *J. biol. Chem.* **241**, 5137.

BAUGHAN M.A., VALENTINE W.N., PAGLIA M.D., WAYS P.O., SIMON E.R. & DeMARSH Q.B. (1968) Hereditary hemolytic anemia associated with glucose-phosphate isomerase (GPI) deficiency—a new enzyme defect of human erythrocytes. *Blood* **32**, 236.

19

BAUMGARTEN A. (1963) Identification of catalase following electrophoresis on acrylamide gels. *Blood* **22**, 466.

BAUR E.W. (1963) Catalase abnormality in a Caucasian family in the United States. *Science* **140**, 816.

BEUTLER E. & BALUDA M.C. (1966a) Improved method for measuring galactose-1-phosphate uridyl transferase activity of erythrocytes. *Clin. chim. Acta* **13**, 369.

BEUTLER E. & BALUDA M.C. (1966b) A simple spot screening test for galactosemia. *J. Lab. clin. Med.* **68**, 137.

BEUTLER E., BALUDA M. & DONNELL G.N. (1964) A new method for the detection of galactosemia and its carrier state. *J. Lab. clin. Med.* **64**, 694.

BEUTLER E., BALUDA M.C., STURGEON P. & DAY R. (1965) A new genetic abnormality resulting in galactose-1-phosphate uridyl transferase deficiency. *Lancet* **i**, 353.

BEUTLER E., BALUDA M., STURGEON P. & DAY R.W. (1966) The genetics of galactose-1-phosphate uridyl transferase deficiency. *J. Lab. clin. Med.* **68**, 646.

BOYER S.H., FAINER D.C. & WATSON-WILLIAMS E.J. (1963) Lactate dehydrogenase variant from human blood: evidence for molecular subunits. *Science* **141**, 642.

BREWER G.J. (1967) Achromatic regions of the tetrazolium stained starch gels: inherited electrophoretic variation. *Am. J. hum. Genet.* **19**, 674.

BREWER G.J., EATON J.W., KNUTSEN C.S. & BECK C.C. (1967) A starch-gel electrophoretic method for the study of diaphorase isozymes and preliminary results with sheep and human erythrocytes. *Biochem. biophys. Res. Commun.* **29**, 198.

CAHN R.D., KAPLAN N.O., LEVINE L. & ZWILLING E. (1962) The nature and development of lactic dehydrogenase. *Science* **136**, 962.

CARSON P.E., BREWER G.J. & ICKES C. (1961) Decreased glutathione reductase with susceptibility to hemolysis. *J. Lab. clin. Med.* **58**, 804.

CHILSON O.P., KITTO G.B. & KAPLAN N.O. (1965) Factors affecting the reversible dissociation of dehydrogenases. *Proc. natn. Acad. Sci.* **53**, 1006.

CHRISTIE G.S. & JUDAH J.D. (1953) Intracellular distribution of enzymes. *Proc. R. Soc. Biol.* **141**, 420.

CORTNER J.A. & DAVIDSON R.G. (1967) A genetic variant of mitochondrial glutamic oxaloacetic acid transaminase. *Program Am. Soc. hum. Genet.* Toronto, p. 45.

DAVIDSON R.G. & CORTNER J.A. (1967a) Genetic variant of human erythrocyte malate dehydrogenase. *Nature* **215**, 761.

DAVIDSON R.G. & CORTNER J.A. (1967b) Mitochondrial malate dehydrogenase: a new genetic polymorphism in man. *Science* **157**, 1569.

DAVIDSON R.G., FILDES R.A., GLEN-BOTT A.M., HARRIS H. & ROBSON E.A. (1965) Genetical studies on a variant of human lactate dehydrogenase (subunit A). *Ann. hum. Genet.* **29**, 5.

DETTER J.C., WAYS P.O., GIBLETT E.R., BAUGHAN M.A., HOPKINSON D.A., POVEY S. & HARRIS H. (1968) Inherited variations in human phosphohexose isomerase. *Ann. hum. Genet.* (in press).

DEWEY M.M. & CONKLIN J.L. (1960) Starch gel electrophoresis of lactic dehydrogenase from rat kidney. *Proc. Soc. exp. Biol. Med.* **105**, 492.

HAUT A., TUDHOPE C.R., CARTWRIGHT G.E. & WINTROBE M.M. (1962) The nonhemoglobin erythrocytic proteins, studied by electrophoresis on starch gel. *J. clin. Invest.* **41**, 579.

HUENNEKENS F.M., CAFFREY R.W., BASFORD R.E. & GABRIO B.W. (1957) Erythrocyte metabolism. IV. Isolation and properties of methemoglobin reductase. *J. biol. Chem.* **227**, 261.

HUENNEKENS F.M., CAFFREY R.W. & GABRIO B.W. (1958) The electron transport sequence of methemoglobin reductase. *Ann. N.Y. Acad. Sci.* **75**, 167.

ISSELBACHER, K.J. (1966) Galactosemia, *in* STANBURY J.B., WYNGAARDEN J.B. & FREDRICKSON D.S. (eds.) *The Metabolic Basis of Inherited Disease*, p. 178. McGraw-Hill, New York.

JACOB H.S., INGBAR S.H. & JANDL J.H. (1965) Oxidative hemolysis and erythrocyte metabolism in hereditary acatalasia. *J. clin. Invest.* **44**, 1187.

JAFFÉ E.R. (1963) The reduction of methemoglobin in erythrocytes of a patient with congenital methemoglobinemia, subjects with erythrocyte glucose-6-phosphate dehydrogenase deficiency, and normal individuals. *Blood* **21**, 561.

JAFFÉ E.R. & HELLER P. (1964) Methemoglobinemia in man, *in* MOORE C.V. & BROWN E.B. (eds.) *Progress in Hematology*, vol. 4, p. 48. Grune & Stratton, New York.

JAFFÉ E.R., NEUMANN G., ROTHBERG H., WILSON T., WEBSTER R.M. & WOLFF J.A. (1966) Hereditary methemoglobinemia with and without mental retardation. A study of three families. *Am. J. Med.* **41**, 42.

KAPLAN J.C. & BEUTLER E. (1967) Electrophoresis of red cell NADH- and NADPH- diaphorases in normal subjects and patients with congenital methemoglobinemia. *Biochem. biophys. Res. Commun.* **29**, 605.

KAPLAN J.C. & BEUTLER E. (1968a) Electrophoretic study of glutathione reductase in human erythrocytes and leukocytes. *Nature* **217**, 256.

KAPLAN N.O. (1964) Lactate dehydrogenase: structure and function. *Brookhaven Symp. Biol.* **17**, 131.

KAPLAN N.O., CIOTTI M.M., HAMOLSKY M. & BIEBER R.E. (1960) Molecular heterogeneity and evolution of enzymes. *Science* **131**, 392.

KITTO G.B., WASSARMAN P.M. & KAPLAN N.O. (1966) Enzymatically active conformers of mitochondrial malate dehydrogenase. *Proc. natn. Acad. Sci.* **56**, 578.

KLEIHAUER E. & BETKE K. (1963) Elution procedure for the demonstration of methaemoglobin in red cells of human blood smears. *Nature* **199**, 1196.

KLEIHAUER E. & BRANDT G. (1964) Qualitative estimation of erythrocyte catalase by starch gel electrophoresis. *Nature* **204**, 478.

KRAUS A.P. & NEELY C.L. (1964) Human erythrocyte lactate dehydrogenase: four genetically determined variants. *Science* **145**, 595.

LAURENT G., GARÇON D., MARRIQ C., CHARREL M. & DERRIEN Y. (1965) Composition en amino acides et hydrolyse trypsique des anhydrases carboniques érythrocytaires humaines A (X_2) et B (X_1). *C. r. Acad. Sci.* **258**, 6557.

LAURENT G., MARRIQ C., GARÇON D., LUCCIONI C. & DERRIEN Y. (1967) Sur les anhydrases carboniques érythrocytaires humaines. VI. Enchainement N-terminal de l'enzyme B. *Bull. Soc. chim. Biol.* **49**, 1035.

LEWIS W.H.P. & HARRIS H. (1967) Human red cell peptidases. *Nature* **215**, 351.

LIE-INJO L.E. (1967) Red cell carbonic anhydrase Ic in Filipinos. *Am. J. hum. Genet.* **19**, 130.

LIE-INJO L.E., HOLLANDER L. & FUDENBERG H.H. (1967) Carbonic anhydrase and fetal hemoglobin in thyrotoxicosis. *Blood* **30**, 442.

LIE-INJO L.E. & TARAIL R. (1966) Carbonic anhydrase deficiency with persistence of foetal haemoglobin. *Nature* **211**, 47.

LÖHR G.W. & WALLER H.D. (1962) Eine neue enzymopenische hämolytische Anämie mit Glutathionreduktase-Mangel. *Med. Klin.* **57**, 1521.

LONG W.K. (1967) Glutathione reductase in red blood cells: variant associated with gout. *Science* **155**, 712.

MARKERT C.L. (1963) Lactate dehydrogenase isozymes: dissociation and re-combination of subunits. *Science* **140**, 1329.

MARKERT C.L. & APPELLA E. (1963) Immunochemical properties of lactate dehydrogenase isozymes. *Ann. N.Y. Acad. Sci.* **103**, 915.

MARKERT C.L. & MØLLER F. (1959) Multiple forms of enzymes: tissue, ontogenetic and species specific patterns. *Proc. natn. Acad. Sci.* **45**, 753.

MARKERT C.L. & URSPRUNG H. (1962) The ontogeny of isozyme patterns of lactate dehydrogenase in the mouse. *Devl. Biol.* **5**, 363.

MATHAI C.K. & BEUTLER E. (1966) Electrophoretic variation of galactose-1-phosphate uridyltransferase. *Science* **154**, 1179.

MATSUBARA S., SUTER H. & AEBI H. (1967) Fractionation of erythrocyte catalase from normal, hypocatalatic and acatalatic humans. *Humangenetik* **4**, 29.

NANCE W.E., CLAFLIN A. & SMITHIES O. (1963) Lactic dehydrogenase: genetic control in man. *Science* **142**, 1075.

NG W.G., BERGREN W.R. & DONNELL G.N. (1967) An improved procedure for the assay of hemolysate galactose-1-phosphate uridyl transferase activity by the use of ^{14}C-labelled galactose-1-phosphate. *Clin. chim. Acta.* **15**, 489.

NYMAN P.O. & LINDSKOG S. (1964) Amino acid composition of various forms of bovine and human erythrocyte carbonic anhydrase. *Biochim. biophys. Acta* **85**, 141.

NYMAN P.O., STRID L. & WESTERMARK G. (1966) Carboxyl-terminal amino acid sequence of human and bovine erythrocyte carbonic anhydrase. *Biochim. biophys. Acta* **122**, 554.

RICKLI E.E., GHAZANFAR S.A.S., GIBBONS B.N. & EDSALL J.T. (1964) Carbonic anhydrases from human erythrocytes. Preparation and properties of two enzymes. *J. biol. Chem.* **239**, 1065.

RIEDER D.F. & WEATHERALL D.J. (1964) A variation in the human electrophoretic pattern of erythrocyte carbonic anhydrase. *Nature* **203**, 1364.

SAGER R. (1964) Non chromosomal heredity. *New Engl. J. Med.* **271**, 352.

SALTHE S.N., CHILSON O.P. & KAPLAN N.O. (1965) Hybridization of lactic dehydrogenase *in vivo* and *in vitro*. *Nature* **207**, 723.

SCOTT E.M. (1960) The relation of diaphorase of human erythrocytes to inheritance of methemoglobinemia. *J. clin. Invest.* **39**, 1176.

SCOTT E.M. & GRIFFITH I.V. (1959) Enzymic defect of hereditary methemoglobinemia: diaphorase. *Biochim. biophys. Acta* **34**, 584.

SCOTT E.M. & McGRAW J.C. (1962) Purification and properties of diphosphopyridine nucleotide diaphorase of human erythrocytes. *J. biol. Chem.* **237**, 249.

SHAW C.R., SYNER F.N. & TASHIAN R.E. (1962) New genetically determined molecular form of erythrocyte esterase in man. *Science* **138**, 31.

SHOWS T.B. (1967) The amino acid substitution and some chemical properties of a variant human erythrocyte carbonic anhydrase: carbonic anhydrase $Id_{Michigan}$. *Biochem. Genet.* **1**, 171.

SHRAGO E. (1965) Cytoplasmic characteristics of human erythrocytic malic dehydrogenase. *Archs Biochem. Biophys.* **109**, 57.

STARKWEATHER W.H., COUSINEAU L., SCHOCH H.K. & ZARAFONETIS C.J. (1965) Alterations of erythrocyte lactate dehydrogenase in man. *Blood* **26**, 63.

TAKAHARA S. (1952) Progressive oral gangrene probably due to lack of catalase in the blood (acatalasemia), report of nine cases. *Lancet* **ii**, 1101.

TAKAHARA S., HAMILTON H.B., NEEL J.V., KOBARA T.Y., OGURA Y. & NISHIMURA E.T. (1960) Hypocatalasemia: a new genetic carrier state. *J. clin. Invest.* **39**, 610.

TAKAHARA S., MIHARA S., TSUGAWA K & DOI M. (1952) Acatalasemia. II., Contents of catalase in blood and tissues of men and animals. *Proc. Japan Acad.* **28**, 383.

TASHIAN R.E. (1961) Multiple forms of esterases from human erythrocytes. *Proc. Soc. exp. Biol. Med.* **108**, 364.

TASHIAN R.E. (1965) Genetic variation and evolution of the carboxylic esterases and carbonic anhydrases of primate erythrocytes. *Am. J. hum. Genet.* **17**, 257.

TASHIAN R.E., PLATO C.C. & SHOWS T.B. (1963) Inherited variant of erythrocyte carbonic anhydrase in Micronesians from Guam and Saipan. *Science* **140**, 53.

TASHIAN R.E., RIGGS S.K. & YU Y.S.L. (1966) Characterization of a mutant human erythrocyte carbonic anhydrase: carbonic anhydrase Ic_{Guam}. *Archs Biochem. Biophys.* **117**, 320.

TASHIAN R.E. & SHAW M.W. (1962) Inheritance of an erythrocyte acetylesterase variant in man. *Am. J. hum. Genet.* **14**, 295.

THORNE C.J.R., GROSSMAN L. & KAPLAN N.O. (1963) Starch gel electrophoresis of malate dehydrogenase. *Biochim. biophys. Acta* **73**, 193.

THORUP O.A., CARPENTER J.T. & HOWARD P. (1964) Human erythrocyte catalase: demonstration of heterogeneity and relationship to erythrocyte aging *in vivo*. *Br. J. Haemat.* **10**, 542.

THORUP O.A., STROLE W.B. & LEAVELL B.S. (1961) A method for the localization of catalase on starch gels. *J. Lab. clin. Med.* **58**, 122.

TUDHOPE G.R. (1965) Detection of catalase after electrophoresis of haemolysates on starch gel. *Nature* **205**, 404.

VERPOORTE K.A., MEHTA S. & EDSALL J.T. (1967) Esterase activities of human carbonic anhydrases B and C. *J. biol. Chem.* **242**, 4221.

VESELL E.S. (1965a) Genetic control of isozyme patterns in human tissues, *in* STEINBERG A.G. & BEARN A.G. (eds.) *Progress in Medical Genetics*, vol. 4. p. 128. Grune & Stratton, New York.

VESELL E.S. (1965b) Formation of human lactate dehydrogenase isozyme patterns *in vitro*. *Proc. natn. Acad. Sci.* **54**, 111.

VESELL E.S. (1965c) Polymorphism of human lactate dehydrogenase isozymes. *Science* **148**, 1103.

VESELL E.S. & BEARN A.G. (1961a) Significance of the heterogeneity of lactic dehydrogenase activity in human tissues. *Ann. N.Y. Acad. Sci.* **94**, 877.

VESELL E.S. & BEARN A.G. (1961b) Isozymes of lactic dehydrogenase in human tissues. *J. clin. Invest.* **40**, 586.

VESELL E.S. & BEARN A.G. (1962) Localization of a lactic dehydrogenase isozyme in nuclei of young cells of the erythrocyte series. *Proc. Soc. exp. Biol. Med.* **111**, 100.

VESELL E.S. & BRODY I.A. (1964) Biological applications of lactic dehydrogenase isozymes. *Ann. N.Y. Acad. Sci.* **121**, 544.

VESELL E.S. & PHILIP J. (1963) Isozymes of lactic dehydrogenase: sequential alterations during development. *Ann. N.Y. Acad. Sci.* **111**, 243.

WALLER H.D., LÖHR G.W., ZYSNO E., GEROK W., VOSS D. & STRAUSS G. (1965) Glutathionreduktasemangel mit hämatologischen und neurologischen Störungen (autosomal dominant vererbliche Bildung eines pathologischen Enzyms) *Klin. Wschr.* **43**, 8.

WEST C.A., GOMPERTS B.D., HUEHNS E.R., KESSEL I. & ASHBY J.R. (1967) Demonstration of an enzyme variant in a case of congenital methaemoglobinaemia. *Br. Med. J.* **4**, 212.

WHITNEY P.L., FÖLSCH G., NYMAN P.O. & MALSTRÖM B.G. (1967a) Inhibition of human erythrocyte carbonic anhydrase B by chloroacetyl sulfonamides with labeling of the active site. *J. biol. Chem.* **242**, 4206.

WHITNEY P.L., NYMAN P.O. & MALSTRÖM B.G. (1967b) Inhibition and chemical modifications of human erythrocyte carbonic anhydrase B. *J. biol. Chem.* **242**, 4212.

WILSON A.C., KAPLAN N.O., LEVINE L., PESCE A., REICHLIN M. & ALLISON W.S. (1964) Evolution of lactic dehydrogenases. *Fed. Proc.* **23**, 1258.

PART 3

MISCELLANY

CHAPTER 17

CONTRIBUTIONS OF BLOOD GENETIC MARKERS TO STUDIES OF HUMAN BIOLOGY

Mutation Mechanisms	557	Origin of Developmental Abnormalities	567
Point mutation and the genetic code	557	X chromosome anomalies	567
Small deletions from intragenic crossing-over	563	Cellular mosaicism	568
Unequal crossing-over with hybrid molecule formation	563	Blood group chimerism	568
Duplications	563	Generalized tissue mosaicism	568
Molecular Organization	565	Artificial mosaicism from surgical grafting or in cell cultures	569
Association of polypeptide subunits	565	Genetic Linkage	570
Association of oligosaccharide subunits and a peptide or lipid 'backbone'	566	Non-identity and Intra-family Relationships	571
X Chromosome Inactivation	566	Genetic Polymorphism and Selection	572
Origin of tumors	567	References	573

The study of genetics in microbial organisms has advanced much farther than the studies in mammalian species (see, for example, the splendid monographs of Stent, 1963 and Hayes, 1964). However, the extent to which certain of the mechanisms known to exist in micro-organisms can be applied to man is currently unknown. Investigations of the molecular structure, inheritance and geographic distribution of the human genetic variants in blood have made several contributions to the understanding of biological mechanisms. Some of these studies and their applications are briefly summarized in this chaper.

MUTATION MECHANISMS

POINT MUTATION AND THE GENETIC CODE

The sequence of amino acids in any peptide chain is dictated by the 64 purine and pyrimidine base triplets which comprise the code words (or codons) of the genetic code. Figure 17.1 shows the messenger RNA codons as presented by Crick (1966, 1967). The four RNA bases: uracil, cytosine, adenine and guanine, are complementary to those of DNA. Thus, uracil pairs with adenine, cytosine with guanine, adenine

2nd →	U	C	A	G	3rd
1st ↓					↓
U	Phe	Ser	Tyr	Cys	U
	Phe	Ser	Tyr	Cys	C
	Leu	Ser	ochre	?	A
	Leu	Ser	amber	Tryp	G
C	Leu	Pro	His	Arg	U
	Leu	Pro	His	Arg	C
	Leu	Pro	GluN	Arg	A
	Leu	Pro	GluN	Arg	G
A	Ileu	Thr	AspN	Ser	U
	Ileu	Thr	AspN	Ser	C
	Ileu	Thr	Lys	Arg	A
	Met	Thr	Lys	Arg	G
G	Val	Ala	Asp	Gly	U
	Val	Ala	Asp	Gly	C
	Val	Ala	Glu	Gly	A
	Val	Ala	Glu	Gly	G

FIGURE 17.1. The genetic code. The purine bases are adenine (A) and guanine (G); the pyrimidine bases are uracil (U) and cytosine (C). The first base of any messenger RNA triplet is indicated by the large letters at the left; the second base at the top, and the third base at the right of the figure. The amino acids signified by the abbreviations are listed in the glossary. The 'ochre' and 'amber' triplets probably signal polypeptide chain termination. (From Crick, 1967. Reprinted with permission from the *Proc. R. Soc.* B, **167**, 331).

with thymine and guanine with cytosine. Since each amino acid, with the exception of methionine and tryptophane, has two or more codons, the code is said to be 'degenerate', but the relative efficiency of alternative codons and the availability of their corresponding transfer RNA molecules are not yet known. For a very clear account of the mechan-

isms involved in gene action, the reader is referred to the book by
Watson (1965).

Point mutation consists of a substitution of one base for another
within a triplet. As shown in Fig. 17.1, substitution of the third member
of the triplet often produces no change in the corresponding amino
acid, but when the first or second base is substituted, a different amino
acid is very likely to be produced.

An example of point mutation is given in Fig. 17.2, which shows one

Hb$_\beta^A$ DNA segment	– – – – TGA GGT CTC CTC – – – –
mRNA segment	– – – – ACU CCA GAG GAG – – – –
HbA β peptide segment	– – – – Thr Pro Glu Glu – – – –
	4 5 6 7

Hb$_\beta^S$ DNA segment	– – – – TGA GGT C|A|C CTC – – – –
mRNA segment	– – – – ACU CCA G|U|G GAG – – – –
HbS β peptide segment	– – – – Thr Pro |Val| Glu – – – –
	4 5 6 7

FIGURE 17.2. An example of point mutation. Possible DNA and RNA
codons for amino acids 4 through 7 of the hemoglobin beta chain are shown
at the top. A change from thymine to adenine in the DNA coding for the
sixth amino acid alters the messenger so that valine, instead of glutamic
acid, occupies that position in the sequence. The result is sickle hemoglobin (Hb S).

of the possible base sequences in a short segment of DNA near the
beginning of the Hb_β^A gene. Also shown is its complementary segment
of messenger RNA and the corresponding amino-acid sequence in
positions 4 through 7 of the γ globin chain. In this example, a DNA
substitution of adenine for thymine results in the substitution of uridine for adenine in an RNA codon, so that valine, rather than glutamic
acid, occupies the sixth position of the chain. Comings (1966) suggested
that apparent point mutation with single amino-acid substitution
might in some instances be the result of unequal homologous crossing-
over. Since this mechanism requires the prior existence of gene

duplication by unequal, non-homologous crossing-over, it is likely to be a less common event.

In the example given in Fig. 17.2, the point mutation changed Hb A to Hb S, the crystalline structure of which was discussed on p. 372. There are several abnormal hemoglobins in which the substitution of one amino acid for another resulted in molecular instability and/or alterations in reversible oxygenation. However, in many other instances, the altered hemoglobin shows little or no difference in overall structure or function (see Chapter 10). Apparently analogous variations occur in other polymorphic proteins, such as the G6PD and PHI enzymes (see Chapters 12 and 16).

Since the amino-acid substitution in nearly all of the abnormal hemoglobins so far detected is known, it has been possible to check the validity of the genetic code derived from studies of micro-organisms with the human hemoglobin amino-acid substitutions, assuming a single-step mutation in each instance. Using Fig. 17.2 as an example, the RNA code word for glutamic acid (Glu) is either GAA or GAG, while that for valine (Val) is GUU, GUC, GUA or GUG. The first member of the triplet is the same in all instances, but the second letter is A for Glu and U for Val. Thus, a change from Glu to Val requires a change from A to U, involving the second member of the triplet: either GAA to GUA or GAG to GUG.

Figure 17.3 depicts the genetic code as given in Fig. 17.1 but with the names of the amino acids omitted. It shows the 36 different amino-acid changes in 82 abnormal hemoglobins (see Table 10.2 of Chapter 10). Crick (1967) pointed out that it is possible to deduce the base change responsible for the amino-acid substitution in all instances except when the change is within a single square of the code or, sometimes, when it involves one of the three amino acids (leucine, serine and arginine) which occupy more than one square. In Fig. 17.3, a change in the first position of a codon is indicated by a vertical arrow; if it is in the second position, the arrow is horizontal. Third base changes remain within one square. If there were a diagonal arrow, it would imply at least two base changes; however, none is seen.

Since most of the abnormal hemoglobins have been detected by altered electrophoretic mobility due to changes in electrical charge, there is a heavy concentration of arrows going in or out of the shaded areas of Fig. 17.3, which denote the charged amino acids: aspartic acid, glutamic acid, arginine, lysine and histidine. However, the more recent finding of neutral for neutral amino acid replacements is reflected by the arrows situated entirely in the unshaded squares.

Some of the substitutions occur in several different hemoglobins. For example, there are nine instances of Glu → Lys substitution and

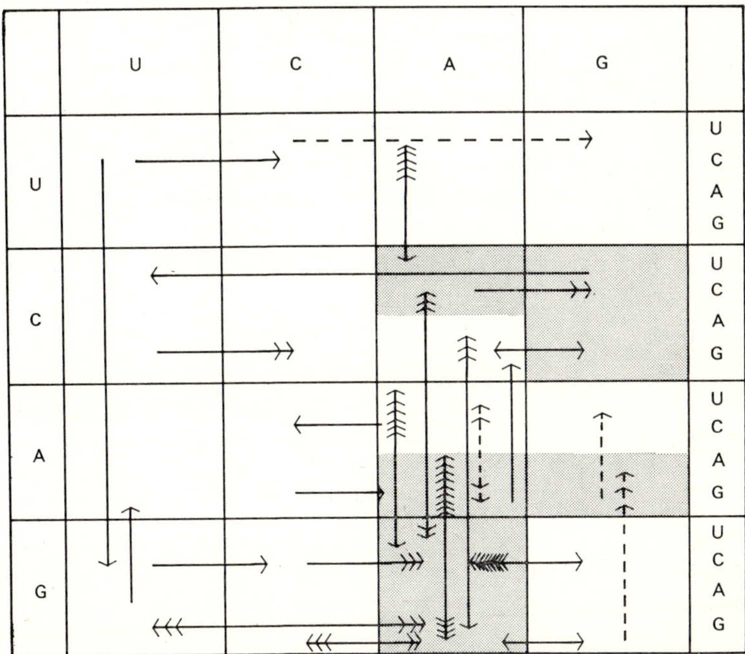

FIGURE 17.3. Diagram of single amino acid changes in the abnormal hemoglobins as they appear on the genetic code chart. The head of each arrow represents an individual hemoglobin variant. The interrupted arrows indicate that the base change is uncertain. Names of amino acids have been excluded for clarity. The shaded area represents the charged amino acids, identified in Fig. 17.1. (Adapted from Crick, 1967. Reprinted with permission from the *Proc. R. Soc.* B. **167**, 331).

four of Lys → Glu. In eight hemoglobins, glycine is substituted by aspartic acid, but the reverse (Asp → Gly) has been found only once. Transitional changes (U ⇌ C and A ⇌ G) have about the same frequency as transversions (U or C ⇌ A or G).

In one instance, it was possible to use the genetic code to correct the biochemical data (Beale & Lehmann, 1965). Hb I was thought to represent an alpha chain substitution of aspartic acid for lysine, which would have required two point mutations. However, reinvestigation of Hb I revealed that the actual change was lysine to glutamic acid, which requires only one mutational step.

Beale & Lehmann (1965) also provided evidence that the codon for a particular amino acid at a given position on the globin chain is not necessarily the same as that coding for the same amino acid at a different position on the chain. For example, in Hb Zürich, there is a change at position 63β from histidine, coded by CAC or CAU, to arginine, coded by CGU, CGC, CGA, CGG, AGA or AGG. A single base change requires that the arginine codon would be either CGU or CGC. On the other hand, there is a glutamine (CAA or CAG) to arginine substitution at 54α in the Hb Shimonoseki, in which the codon for arginine (assuming single base substitution) is either CGA or CGG.

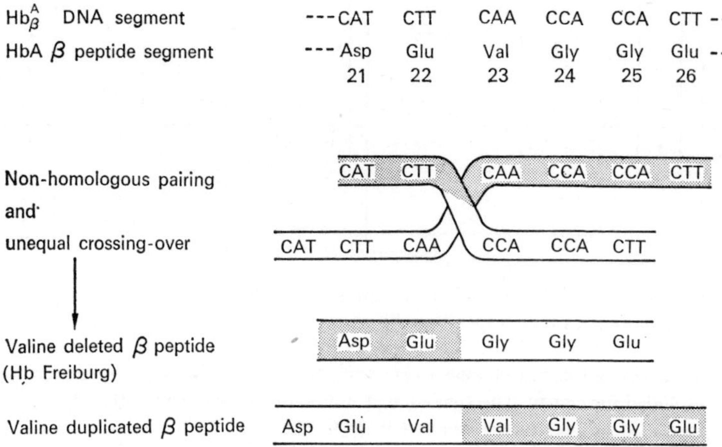

FIGURE 17.4. An example of partial gene deletion (and duplication) due to unequal crossing over. Possible DNA triplets coding for the amino acid sequence at positions 21 through 26 of the hemoglobin A beta chain are given at the top. Slight mis-pairing during meiosis followed by crossing-over results in deletion of valine at position 23 β. This deletion is known to exist in Hb Freiburg. A second beta chain with duplicated valine would theoretically be expected from this genetic event. (Suggested by drawings in Jones *et al.*, 1966.)

SMALL DELETIONS FROM INTRA-GENIC
CROSSING-OVER

Two of the abnormal hemoglobins discussed in Chapter 10 have missing amino acids: Hb Freiburg lacks one (Jones *et al.*, 1966), while Hb Gun Hill lacks five (Bradley *et al.*, 1967). Explanations for the latter deletion have not been proposed, although it could have occurred in a manner similar to that proposed for Hb Freiburg, outlined in Fig. 17.4. A possible sequence of bases in the DNA coding for amino acids 21 through 26 in the β globin chain is shown at the top of the figure. The middle of the figure shows a slight misalignment of the DNA segment with that of its homologous chromatid during meiosis, so that crossing-over is shifted by one triplet. Thus, in the products of this crossing-over (bottom of Fig. 17.4) one has a valine deletion, such as that found in Hb Freiburg, while the other has a valine duplication. Such a duplication of amino acids has not yet been reported in the abnormal hemoglobins.

UNEQUAL CROSSING-OVER WITH
HYBRID MOLECULE FORMATION

The so-called Lepore hemoglobins, discussed in Chapter 10, provide an example of globin chain hybridization, probably due to unequal crossing-over between the Hb_β and Hb_δ genes during meiosis (Baglioni 1962). As shown in Fig. 17.5, these two genes have the same length and are adjacent on one of the autosomes. In normal synapsis, they line up in pairs. Thus, if any crossing-over occurred, it would be between identical genes. However, displaced synapsis could line up a beta with a delta chain gene. These two genes differ at ten positions, so that crossing-over at one of several different homologous regions could result in one of several possible hybrid, non-alpha, chains. On each occasion, there would be a hybrid with 146 amino acids, as is characteristic of the Lepore hemoglobins. A second hybrid would consist of a duplicated chain, containing both parent chains and the reciprocal portions of the 'deletion' hybrid.

DUPLICATIONS

The kind of duplication hybrid indicated in Fig. 17.5 was suggested by Nance (1963) and by Smithies (1964). It has not yet been identified among any of the proteins. However, a partially duplicated hybrid was described in Chapter 2. The Hp^2 gene product consists of most of the Hp^{1S} and Hp^{1F} gene products joined end to end, with a small deletion of both at their junction (Smithies *et al.*, 1962). The partial duplication was probably the result of intragenic crossing-over be-

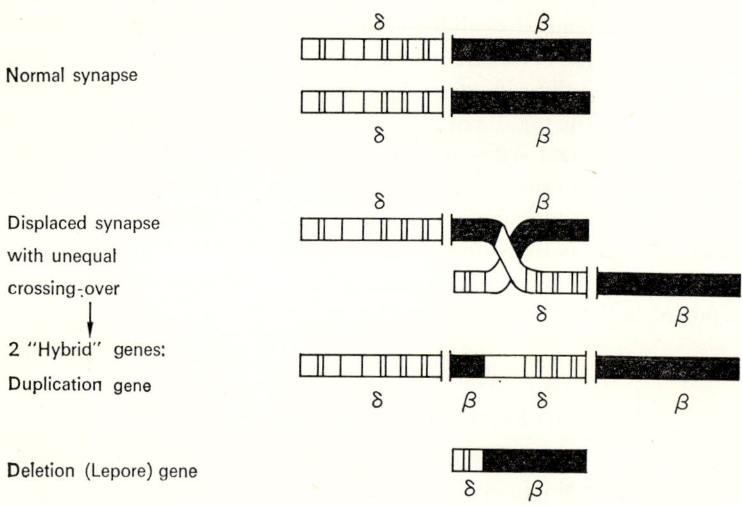

FIGURE 17.5. An example of gene hybridization. The genes for the δ and β chains of hemoglobin are closely linked, and they differ in base sequence at only ten sites (indicated by vertical lines in the δ chain gene). If, during synapsis, the δ gene lines up with the β gene of the homologous chromosome, crossing-over can occur at one of many homologous base sequences. Of the two kinds of hybrid genes expected from such a crossover, one, the so-called deletion gene, is represented by the non-alpha chain of the Lepore hemoglobins. A product of the theoretical duplicated hybrid has not yet been identified among the abnormal hemoglobins. (Suggested by drawings in Baglioni, 1962).

tween two widely separated regions of Hp^1 with similar base sequences (see Chapter 2).

Smithies *et al.* (1962) also showed that synapsis of two homologous genes of unequal size, such as Hp^2 and Hp^1, would be a very likely source of new genes arising from unequal crossing-over. Furthermore, displaced synapsis and crossing-over between two partially duplicated genes (i.e. in a Hp^2/Hp^2 homozygote) could result in partial triplication, as shown in Chapter 2. The evolutionary consequences of gene mutation and duplication were briefly discussed in Chapters 1, 2 and 10.

MOLECULAR ORGANIZATION

ASSOCIATION OF POLYPEPTIDE SUBUNITS

The fact that the four polypeptide chains of hemoglobin are held together only by non-covalent bonds provides the Hb molecule with the ease of dissociation necessary for its allosteric property. In some of the Hb variants, alteration of this property can be traced to a specific amino acid substitution (see Chapter 10).

Another example of a tetramer composed of two similar-sized subunits linked by non-covalent bonds is lactate dehydrogenase. However, in this case, the two subunits, A and B, when present in equal amounts, tend to polymerize at random, forming five molecular species instead of one (see Chapter 16). Mutation of one or the other of the genes coding for these subunits results in the formation of an additional set of isozymes. Some tissues synthesize mainly the A subunit and other tissues synthesize mainly the B subunit, indicating that the homotetramers (A_4 and B_4) probably have functional characteristics which are advantageous in specific tissues. When an individual is heterozygous at either the A or B structural gene locus, products of both homologous genes (A and A' or B and B') are found in these tissues, where they probably perform the same function.

A third kind of tetramer is characteristic of the immunoglobulin and haptoglobin molecules. Here, the two kinds of subunits differ considerably in size, and two of the like chains, as well as the two unlike chains, are linked by both covalent (disulfide) and non-covalent bonds. In the case of the immunoglobulins, there is considerable variability among the heavy and light chains produced by a given individual. However, each molecule consists of two identical heavy chains and two identical light chains. As indicated in Chapter 1, this is a consequence of autosomal inactivation involving the structural genes synthesizing immunoglobulin heavy and light chains. Since no such inactivation occurs at the haptoglobin structural locus, products of both *Hp¹* and *Hp²* genes are made within single cells, where they form the series of polymers characteristic of the Hp 2-1 phenotype.

ASSOCIATION OF OLIGOSACCHARIDE SUBUNITS AND A PEPTIDE OR LIPID 'BACKBONE'

As described at some length in Chapter 9, the soluble blood group substances, H, A, B, Le^a and Le^b, consist of a peptide 'backbone' and a large number of short carbohydrate chains. The antigenic specificity of these chains resides in the sugars of their non-reducing ends. On the red cells, similar or identical oligosaccharides attached to a lipid 'backbone' endow the molecule with its blood group specificity. Unfortunately, the function of these substances is not yet known, so it is impossible to speculate about the functional advantage of such a structure.

Information about the sequence of sugars in the individual oligosaccharide chains has provided considerable insight into the interactions of genes which control the synthesis of transglycosyllases involved in the construction of the chains (Watkins, 1967). However, the association of ABO blood group and secretor status with disease (Clarke, 1961; Muschel, 1966) presents a major challenge which has so far eluded explanation.

X CHROMOSOME INACTIVATION

The Lyon hypothesis, briefly considered in Chapters 9 and 12, states that the cells of females consist of two populations in which one or the other of the maternally-derived and paternally-derived X chromosomes is partially or completely inactive. The random inactivation, thought to occur at an early stage in embryogenesis, accounts for the so-called 'dosage compensation', i.e. the quantity of X-linked gene products made in cells of females with two X chromosomes is no greater than that made in cells of males with one X chromosome.

Studies on the X-linked blood group gene, *Xg*^{a}, have provided equivocal evidence about inactivation of its locus (see Chapter 9). However, the X-linked G6PD locus almost certainly undergoes inactivation, so that in any given cell, one or the other, but not both of the Gd genes is active in females (see Chapter 12).

ORIGIN OF TUMORS

The establishment of Gd inactivation made possible the study of the cellular origin of tumors. Thus, in a woman heterozygous for the Gd^A and Gd^B genes, tumors arising from a single cell contain G6PD of either type A or type B, while tumors arising from more than one cell may contain both enzymes (Linder & Gartler, 1965).

Fialkow *et al.* (1967) took advantage of this characteristic of G6PD to study the clonal origin of chronic myelocytic leukemia. Most patients with this disease have the so-called Philadelphia chromosome, which is a partial deletion of the long arm of (probably) the twenty-first chromosome. Fialkow *et al.* (1967) reasoned that if this abnormality is present in all the granulocytes of leukemic patients, and if both red cells and granulocytes normally arise from the same stem cell precursor, then all of the red cells and granulocytes in the blood of these patients might arise from a single cell clone. They studied three Negro female patients heterozygous at the Gd locus, in whom extracts of cultured skin indicated the presence of both A and B types of G6PD. In all three cases, electrophoresis of red cell and white cell extracts showed only the A enzyme. Column chromatography performed in one case confirmed the electrophoretic findings.

The authors concluded that their data could be best explained by malignant transformation of a single stem cell in which there was normal inactivation of that part of one X chromosome carrying the Gd locus. Thus, all of the daughter cells, differentiating into either red cells or granulocytes, produced a single species of G6PD molecules.

ORIGIN OF DEVELOPMENTAL
ABNORMALITIES

X CHROMOSOME ANOMALIES

The distribution of the Xg^a blood group gene is favorable to the use of its red cell marker to trace the parental origin of extra or missing X chromosomes and, in some instances, to assign parental non-disjunction to the first or second meiotic division. This subject is covered extensively in the most recent edition of Race & Sanger (1968).

CELLULAR MOSAICISM

BLOOD GROUP CHIMERISM

The anastomosis of blood vessels between non-identical twins *in utero* is associated with mutual grafting of red cell and white cell precursors. This rare event was first described in human subjects by Dunsford *et al.* (1953), and several other cases have subsequently been reported. Unlike the sheep and cattle chimeras, human subjects grafted by a twin of the opposite sex are not freemartins, and there is no evidence that any tissue besides blood is involved either directly or indirectly. The proportion of XX and XY cells in the blood thus has no relation to the sex characteristics (Uchida *et al.*, 1964).

Detection of blood group chimerism depends upon the demonstration of two red cell populations with different antigens, such as O and A, M and N, C and c, etc. Such cells appear as 'mixed fields' in which one cell population is agglutinated by a given antibody and the other is not agglutinated. The two populations can be separated by differential agglutination (Booth *et al.*, 1957) and individually characterized according to their blood group antigens and cell enzyme types.

GENERALIZED TISSUE MOSAICISM

In very rare instances, two egg nuclei are fertilized by two different sperm, and the result, instead of twins, is the formation of a single individual with two euploid but genetically different cell lines. The first human example of this phenomenon, described by Gartler *et al.* (1962) and Giblett *et al.* (1963) was a little girl with heterochromia simplex (one eye blue, the other hazel) and an enlarged clitoris. Cultured leukocytes were euploid, but about half had the XX karyotype, while the others were XY. At laparotomy, an apparently normal ovary was found on the left. The ovotestis found on the right was removed and shown to contain both ovarian follicles and seminiferous tubules. Skin from the left side of the abdomen was mainly XX; from the right it was mainly XY. In the blood, two red cell populations with different Rh and MNSs phenotypes were present in approximately equal proportions.

In the second case, described by Beattie *et al.* (1964) and Zuelzer *et al.* (1964), the 19-year-old male patient had patchy skin pigmentation and slight unilateral gynecomastia. Leukocyte cultures showed a predominance of XY over XX cells. The blood contained major

(90 per cent) and minor (10 per cent) red cell populations separable by their ABO and Kidd blood groups, and by the presence of Hb S in the minor population. The authors concluded that in their case, fertilization of a polar body nucleus by one of the two spermatozoa may have been responsible for the disparity in the proportion of cells of each phenotype.

In the case described by Myhre *et al.* (1965), there was a 85:15 per cent ratio of two red cell populations, but the peripheral leukocyte culture contained about 50:50 XX and XY cells.

The child found by Corey *et al.* (1967) was male by appearance, but an ovary was found on the right side at laparotomy. The ratio of XX to XY cultured leukocytes was 60:40, and the same approximate proportion of two red cell populations was found, one of which contained phosphoglucomutase of type 2, and the other, PGM 2-1 (or 1). The serum haptoglobin had the appearance of a mixture of Hp 2-1 and Hp 1-1 types, consistent with mosaicism of the cells of origin (probably in the liver). The hermaphrodite described by de Grouchy *et al.* (1964) also had an unusual Hp phenotype, resembling a mixture of Hp 2-2 and 2-1. Aside from the XX/XY karyotype, the blood provided no further clues to the mosaicism in that instance.

ARTIFICIAL MOSAICISM FROM SURGICAL GRAFTING
OR IN CELL CULTURES

In cases of aplastic anemia, the grafting of bone marrow from a healthy identical twin can be a life-saving procedure (Robins & Noyes, 1961; Thomas *et al.*, 1964; Pegg *et al.*, 1964; Mills *et al.*, 1964; Pillow *et al.*, 1966). However, since there are no genetic differences between the tissues of donor and recipient, the success of the graft can only be assessed on the basis of quantitative measurements.

Eventually, it may be possible to obtain permanent grafts from non-isogeneic marrow in such patients and in patients with marrow malignancies after treatment by irradiation or marrow-suppressing drugs. In that event, the red cell antigens will provide excellent markers to follow repopulation. However, when tissues other than marrow are grafted, the only genetic markers available are products of those tissues released into the peripheral blood. For example, if the haptoglobin, Gc or Tf phenotypes of a liver donor differs from that of the recipient, the production of these proteins by the grafted tissue can be followed (after the Hp, Gc and Tf protein in the blood transfused during surgery has been catabolized). Of course, if the grafted tissue itself is sampled by biopsy, its cells would also be expected to contain the isozyme patterns of the donor. For example, white cells produced from grafted white cell precursors would have the donor's enzyme phenotypes as well as his white cell antigens.

A somewhat related use of blood markers is the detection of either a mixture of isozymes from two different phenotypes or exchange of one phenotype for another in an apparently pure culture of cells. For example, Gartler (1968) found in a survey of human heteroploid cell lines that all of the 20 tested had the G6PD A isozyme and the series of isozymes characteristic of phosphoglucomutase (PGM) type 1. The probability that 20 random specimens would have these phenotypes was of the order of 6×10^{-5}. Thus, it seemed almost certain that the cultures were heavily contaminated with one cell line, in this case, HeLa.

GENETIC LINKAGE

Compared with the elaborate chromosome maps prepared for some micro-organisms and even for the mouse, knowledge of gene linkage in man is greatly deficient. Although a number of markers have been assigned to the human X chromosome (McKusick, 1962), most of these are very rare diseases which provide little opportunity for obtaining linkage data. However, as indicated in Chapters 9 and 12, some mapping information is now available from the use of the blood markers Xg^a and G6PD.

From the attempts to assign the loci of autosomal markers to specific chromosomes, the most convincing place the haptoglobin locus on chromosome 13 (see p. 88) and the Duffy blood group locus on chromosome one (see p. 312). Some negative evidence is also being accumulated from the finding of two allelic gene products (i.e. heterozygosity) of various loci in individuals with autosomal deletions. However, in many instances, precise identification of the abnormal chromosome is difficult, so that extensive data are required to rule out specific autosomal regions as the sites of particular loci.

Several examples of autosomal linkage have been detected by studies of blood group antigens (see Chapter 9). In addition, evidence for linkage of other genetic markers has been presented; for example, albumin with the Gc protein (Chapters 4 and 8), transferrin with E_1 pseudocholinesterase (Chapters 3 and 6), adenylate kinase with ABO (Chapter 15) and phosphoglucomutase with PTC tasting (Chapter 14). Further evidence is required to substantiate all but the first of these linkage pairs.

NON-IDENTITY AND
INTRA-FAMILY RELATIONSHIPS

The usefulness of the blood group antigens for medico-legal purposes, particularly to establish non-paternity, has been acknowledged for many years. Race & Sanger (1962) pointed out that when the usually available antibodies of seven different blood group systems are employed in blood tests, an Englishman has about a 60 per cent chance of being exonerated of a false charge of paternity brought by an Englishwoman. Extension of the calculations to include Gm, Hp, Gc, acid phosphatase and phosphoglucomutase phenotypes increases that chance to over 80 per cent, since exclusion probabilities of these systems are 0·17 (testing for Gm(1) and Gm(5)), 0·18, 0·15, 0·22 and 0·15, respectively.

The relative usefulness of 17 different blood genetic systems in distinguishing between any two randomly selected people in an average western European population is presented in Table 17.1. The data were taken from Race & Sanger (1962), Harris (1966) and portions of this book. In all of the systems listed, there are two or more alleles with a frequency greater than 0·01. The combined data indicate that less than one in 350,000 people would be expected to have the same combination of phenotypes in these 17 systems. Calculations for the Inv and β lipoprotein polymorphisms are not included, because the antibodies required for testing are not generally available.

Other sets of probabilities would be required for compiling a table applicable to members of different ethnic groups. Furthermore, additional markers, such as transferrin, albumin, ceruloplasmin and G6PD could be included for certain populations.

The availability of increasing numbers of useful genetic systems is especially valuable in laboratories called upon to determine the likelihood of monozygosity in twins. The most pressing need for this kind

TABLE 17.1. Seventeen blood genetic systems listed in order of their usefulness (i.e. MNSs is the most useful) for distinguishing between two random samples of blood from western Europeans. Parentheses denote the antigens tested in a given system. The Inv and lipoprotein systems are omitted because of limited availability of reagents for testing blood specimens. The figure at the bottom of the third column indicates that less than one in 350,000 people would be expected to have the same combinations of phenotypes in these 17 systems. (Modified from Race & Sanger, 1962 and Harris, 1966)

Genetic System	Probability that two randomly selected people have the same phenotype	Combined probability
MNSs	0·16	0·16
Rh (CC^{w}cDEe)	0·20	0·032
ABO (A$_1$A$_2$B)	0·33	0·011
Acid phosphatase	0·34	0·0037
Kidd (Jk^a Jk^b)	0·38	0·0014
Duffy (Fy^{a}Fy^b)	0·38	0·0005
Haptoglobin	0·39	$1·95 \times 10^{-4}$
Gm (1,5)	0·40	$7·8 \times 10^{-5}$
Gc	0·45	$3·5 \times 10^{-5}$
PGM	0·47	$1·6 \times 10^{-5}$
Lewis (Le^{a}Le^b)	0·57	$9·3 \times 10^{-6}$
P (P$_1$P$_2$)	0·67	$6·2 \times 10^{-6}$
Adenylate kinase	0·82	$5·1 \times 10^{-6}$
Pseudocholinesterase, E$_2$	0·82	$4·2 \times 10^{-6}$
Kell (Kk)	0·84	$3·5 \times 10^{-6}$
Lutheran (Lu^{a}Lu^b)	0·86	$3·0 \times 10^{-6}$
6PGD	0·92	$2·8 \times 10^{-6}$

of information arises when tissue or organ grafting is being considered. Methods of calculating the probability are described in Race & Sanger (1968).

GENETIC POLYMORPHISM
AND SELECTION

The remarkable diversity which sets each person apart from any other is not only obvious from casual inspection of facial characteristics, pigmentation, body contour, etc, but also from analysis of tissue components at a molecular level. On the basis of blood group and enzyme gene frequencies, Lewontin (1967) has estimated that about a third of structural gene loci may be polymorphic. The fact that the genes for these molecular variants are maintained at different frequencies within various racial groups suggests the existence of strong selective factors within the environment. There is little doubt that a dwindling balanced polymorphism exists in certain areas of the world between falciparum malaria and the gene for sickle hemoglobin as well as (probably) the X-linked enzyme, G6PD. Some selective effects on gene frequencies within the ABO and Rh systems were discussed in Chapter 9. However, the identity of selective factors influencing the other polymorphic systems remains obscure.

The physiological functions of some proteins with genetic variation are known, at least in part. For example, transferrin is vitally important as the iron-binding protein, and one might expect that selective effects on the frequency of its variants would involve the efficiency of the molecule in binding and/or transporting iron to the bone marrow. However, the Tf variants do not appear to differ in these properties from the common Tf type (see Chapter 3). It is perhaps more likely that the functions of Tf and of other proteins which are presumably influenced by selection are not those associated with their major roles in metabolism, but rather with some other unidentified effects, possibly related to interaction with other genes. Such interactions might affect the response of the total organism to variations in the environment and/or susceptibility to metabolic, neoplastic and infectious disease.

At the present time, we are faced with the unfortunate fact that much more is known about the nature of human variation than about the reasons for its existence. Nevertheless, it may be that further study of the polymorphic systems will be rewarded by some answers to questions which we must now consider to be completely unresolved.

REFERENCES

BAGLIONI C. (1962) The fusion of two peptide chains in hemoglobin Lepore and its interpretation as a genetic deletion. *Proc. natn. Acad. Sci.* **48**, 1880.

BEALE D. & LEHMANN H. (1965) Abnormal haemoglobins and the genetic code. *Nature* **207**, 259.

BEATTIE K.M., ZUELZER W.W., McGUIRE D.A. & COHEN F. (1964) Blood group chimerism as a clue to generalized tissue mosaicism. *Transfusion* **4**, 77.

BOOTH P.B., PLAUT G., JAMES J.D., IKIN E.W., MOORES P., SANGER R. & RACE R.R. (1957) Blood group chimaerism in a pair of twins. *Br. med. J.* **i**, 1456.

BRADLEY T.B., WOHL R.C. & RIEDER R.F. (1967) Hemoglobin Gun Hill: deletion of five amino acid residues and impaired heme-globin binding. *Science* **157**, 1581.

CLARKE C.A. (1961) Blood groups and disease, *in* STEINBERG A.G. & BEARN, A.G. (ed.) *Progress in Medical Genetics*, vol. 1, p. 81. Grune & Stratton, New York.

COMINGS D.E. (1966) Single amino acid substitutions as a result of unequal crossing-over. *Nature* **212**, 545.

COREY M.J., MILLER J.R., McLEAN J.R. & CHOWN B. (1967) A case of XX/XY mosaicism. *Am. J. hum. Genet.* **19**, 378.

CRICK F.H.C. (1966) The genetic code—yesterday, today, and tomorrow. *Cold Spring Harbor Symposia on Quantitative Biology*, **31**, 3.

CRICK F.H.C. (1967) The genetic code. *Proc. R. Soc. B.* **167**, 331.

DUNSFORD I., BOWLEY C.C., HUTCHISON A.M., THOMPSON J.S., SANGER R. & RACE R.R. (1953) A human blood group chimaera. *Br. med. J.* **ii**, 81.

FIALKOW P.J., GARTLER S.M. & YOSHIDA A. (1967) Clonal origin of chronic myelocytic leukemia in man. *Proc. natn. Acad. Sci.* **58**, 1468.

GARTLER S.M. (1968) Apparent HeLa cell contamination of human heteroploid cell lines. *Nature* **217**, 750.

GARTLER S.M., WAXMAN S.H. & GIBLETT E.R. (1962) An XX/XY human hermaphrodite resulting from double fertilization. *Proc. natn. Acad. Sci.* **48**, 332.

GIBLETT E.R., GARTLER S.M. & WAXMAN S.H. (1963) Blood group studies on the family of an XX/XY hermaphrodite with generalized tissue mosaicism. *Am. J. hum. Genet.* **15**, 62.

GROUCHY J. DE, MOULLEC J., SALMON C., JOSSO N., FREZAL J. & LAMY M. (1964) Hermaphrodism avec caryotype XX/XY étude génétique d'un cas. *Annls Génét.* **7**, G25.

HARRIS H. (1966) Enzyme polymorphism in man. *Proc. R. Soc. B.* **164**, 298.

HAYES W. (1964) *The Genetics of Bacteria and Their Viruses. Studies in Basic Genetics and Molecular Biology*. Blackwell, Oxford.

JONES R.T., BRIMHALL B., HUISMAN T.H.J., KLEIHAUER E. & BETKE K. (1966) Hemoglobin Freiburg: abnormal hemoglobin due to deletion of a single amino acid residue. *Science* **154**, 1024.

LEWONTIN R.C. (1967) An estimate of average heterozygosity in man. *Am. J. hum. Genet.* **19**, 681.

LINDER D. & GARTLER S.M. (1965) Glucose-6-phosphate dehydrogenase mosaicism: utilization as a cell marker in the study of leiomyomas. *Science* **150**, 67.

McKUSICK V.A. (1962) On the X chromosome of man. *Quart. rev. Biol.* **37**, 69.

MILLS S.D., KYLE R.A., HALLENBECK G.A., PEASE G.L. & CREE I.C. (1964) Bone-marrow transplant in an identical twin. *J. Am. med. Ass.* **188**, 1037.

MUSCHEL L.H. (1966) Blood groups, disease and selection. *Bact. Rev.* **30**, 427.

MYHRE B.A., MEYER T., OPITZ J.M., RACE R.R., SANGER R. & GREENWALT T.J. (1965) Two populations of erythrocytes associated with XX/XY mosaicism. *Transfusion* **5**, 501.

NANCE W.E. (1963) Genetic control of hemoglobin synthesis. *Science* **141**, 123.

PEGG D.E., FLEMING W.J.D. & COMPSTON N. (1964) Case of aplastic anaemia treated by isologous bone marrow infusion. *Postgrad. med. J.* **40**, 213.

PHILLIPS M.E. & THORBECKE G.J. (1966) Studies on the serum proteins of chimeras. I. Identification and study of the site of donor type serum proteins in adult rat-into-mouse chimeras. *Int. Archs Allergy appl. Immun.* **29**, 553.

PILLOW R.P., EPSTEIN R.B., BUCKNER C.D., GIBLETT E.R. & THOMAS E.D. (1966) Treatment of bone-marrow failure by isogeneic marrow infusion. *New Engl. J. Med.* **275**, 94.

RACE R.R. & SANGER R. *Blood Groups in Man*, 4th edn. 1962; 5th edn. 1968. Blackwell, Oxford.

ROBINS M.M. & NOYES W.D. (1961) Aplastic anemia treated with bone-marrow transfusion from an identical twin. *New Engl. J. Med.* **265**, 974.

SMITHIES O. (1964) Chromosomal rearrangements and protein structure. *Cold Spring Harbor Symposia on Quantitative Biology* **29**, 309.

SMITHIES O., CONNELL G.E. & DIXON G.H. (1962) Chromosomal rearrangements and evolution of haptoglobin genes. *Nature* **196**, 232.

STENT G.S. (1963) *Molecular Biology of Bacterial Viruses*. W.H. Freeman, San Francisco.

THOMAS E.D., PHILLIPS J.H. & FINCH C.A. (1964) Recovery from marrow failure following isogeneic marrow infusion. *J. Amer. med. Ass.* **188**, 1041.

UCHIDA I.A., WANG H.C. & RAY M. (1964) Dizygotic twins with XX/XY chimerism. *Nature* **204**, 191.

WATKINS W.M. (1967) The possible enzymic basis of the biosynthesis of blood-group substances. *Proc. 3rd Cong. int. Soc. hum. Genet.* Chicago, p. 171. Johns Hopkins Press, Baltimore.

WATSON J.D. (1965) *Molecular Biology of the Gene*. W.A. Benjamin, New York.

WOODRUFF M.F.A., BUCKTON K.A., FOX M. & JACOBS P.A. (1962) The recognition of human blood chimaeras. *Lancet* **i**, 192.

ZUELZER W.W., BEATTIE K.M. & REISMAN L.E. (1964) Generalized unbalanced mosaicism attributable to dispermy and probable fertilization of a polar body. *Am. J. hum. Genet.* **16**, 38.

GLOSSARY

◈

Allele (*allelomorph*). One of two or more alternative forms of a gene occupying the same locus on homologous chromosomes. The expressed characters of allelic genes are antithetical, because they are never (normally) inherited together from a single parent.

Allosteric effect in proteins. Interaction between a protein and a small molecule which changes the protein conformation so that its interaction with another molecule is altered.

Allotypes; allotypic specificity. Genetically determined molecular differences within a single mammalian species which are detected by their antigenic properties; thus 'alloantigens' are detected by 'alloantibodies.' The alternative terms 'isoantigens' and 'iso-antibodies' are also widely used, although the prefix 'iso' implies similarity rather than difference. For example, 'isogeneic' tissue refers to an individual's own tissue or that of an identical twin, and 'isotypic' specificities are shared by all members of a species.

Amino acids. The building blocks of protein. The abreviations in general use are:

Glycine	Gly	Alanine	Ala	Valine	Val
Leucine	Leu	Isoleucine	Ileu	Serine	Ser
Threonine	Thr	Tyrosine	Tyr	Phenylalanine	Phe
Tryptophan	Try	Aspartic acid	Asp	Glutamic acid	Glu
Lysine	Lys	Arginine	Arg	Histidine	His
Cysteine	Cys	Methionine	Met	Asparagine	Asn
Glutamine	Gln	Proline	Pro		(or AspN)
(or GluN)					

Each amino acid contains a terminal carboxyl ($COOH$) group and a subterminal carbon atom to which are attached a hydrogen atom, an amino (NH_2) group, and a side chain. The chemical nature of the side chain determines the different characteristics of the amino acids:

Non-polar: Gly, Ala, Val, Leu, Ileu, Met, Cys, Phe
Basic: Lys, Arg, His
Acidic: Asp, Glu

Alcoholic: Ser, Thr, Tyr.

Aromatic, non-alcoholic: Phe, Try

Amides: Asn, Gln

Imino: Pro

Amino-terminal (N-terminal) and Carboxy-terminal (C-terminal). The two ends of a polypeptide chain, the beginning of which has a free α-amino group, and the end of which has a free α-carboxyl group.

Amorph. A gene which has no detectable product and is thus (apparently) inactive.

Aneuploidy. Deviation from the normal (euploid) number of chromosomes within a cell.

Anisocytosis. Irregularity in size of red cells in the peripheral blood.

Anomers. Sugars existing as ring structures in which there is a difference in steric configuration at the first carbon (position C-1).

Antibody. An immunoglobulin molecule with specific receptor sites formed in response to an antigenic stimulus. The term is usually used collectively to refer to molecules with similar specificity within a serum specimen.

Antigen. Any substance which can stimulate neutralizing antibody production when introduced into a vertebrate.

Antigenic determinant. The individual antigenic site on an antigen molecule which combines with a specific antibody.

Assortment. The random distribution of genes within the haploid set of chromosomes in the gamete which accounts for the fact that non-allelic genes from a given parent are inherited at random. To be differentiated from *segregation.*

Autoimmunity. The formation of antibodies with which one or more of the individual's own antigens can react.

Autosome (human). One member of the 22 pairs of chromosomes which are neither X nor Y (sex) chromosomes.

Base-pairing. In a nucleic acid double helix, adenine always pairs with thymine or uracil, and guanine pairs with cytosine. This fundamental rule permits DNA and RNA to function in replication and in protein synthesis.

Bence-Jones protein. Immunoglobulin light chains (usually dimers) found in the urine of some patients with myeloma or macroglobulinemia.

Bohr effect. Changes in the oxygen affinity of hemoglobin accompanying changes in pH. For example, a decrease in pH shifts the oxygen dissociation curve to the right, so that more oxygen is given up by hemoglobin at a given oxygen tension (i.e. the affinity of hemoglobin for oxygen is decreased).

Carrier. An individual who is heterozygous for a normal gene and an abnormal gene which is not clinically expressed but may be detected by appropriate laboratory tests.

Centromere. A small mass of heterochromatin by which the chromosome is attached to the spindle and which holds together the two chromatids.

Chromatid. One of two chromosome strands visible during prophase and metaphase which are held together by the centromere.

Clone. Cells in cultured tissue or in certain tumors which are strictly homogeneous because they are the descendents of a single cell.

Codon or coding triplet. A triplet of adjacent bases in DNA and its complementary triplet in RNA which specifies which of the twenty amino acids occupies a given position in a polypeptide chain.

Coenzyme. An organic cofactor which acts as the acceptor or the donor of an atom or a specific grouping from or to the enzyme substrate.

Complement system. Components of a complex series of enzymatic reactions involved in several immunological events, such as hemolysis, phagocytosis, immune adherance and conglutination.

Concordance. The presence of a given characteristic in both members of a pair of twins. When the trait differs, there is said to be *discordance.*

Coupling or cis position. The presence on the same chromosome of two genes at different loci or two different genetic determinants within a 'genetic region'.

Crossing-over (crossover). The process by which genetic material is exchanged between homologous chromosomes. The physical evidences of crossing-over are the chiasmata, seen during the diplotene phase of cell division. Both intergenic and intragenic crossing-over is known to occur. In both instances, an equivalent segment of DNA is usually involved. Unequal crossing-over can result in both deleted and duplicated segments.

Degeneracy of the code. The term applied to the existence of two or more codons which code for the same amino acid.

Deletion. Loss from a chromosome of a segment which varies in size from a single nucleotide or codon triplet to one or more genes.

Differentiation. Process by which cells develop and maintain special properties of structure and function.

Disulfide bond. Covalently linked sulfide groups, –S—S–, important for intra- and inter-chain bonding in proteins, via the sulfhydryl (SH) groups of cysteine.

Dizygotic twins. Twins produced by two separate ova fertilized by different sperm.

DNA (Deoxyribonucleic acid). A nucleic acid in which the backbone sugar unit is deoxyribose, and the principle pyrimidines are thymine and cytosine. As in RNA, the principal purines are adenine and guanine. DNA usually occurs as a double helix, in which the paired bases are linked by hydrogen bonds, three for guanine-cytosine, and two for adenine-thymine.

Dominant trait. A trait is dominant if it is expressed in the heterozygote when the allele of its gene is not expressed. A trait is co-dominant when both genes at the same locus of homologous chromosomes are expressed.

Endoplasmic reticulum. Branching channels for intracellular transport which form a membranous network in the cell interior, running from the nuclear membrane to the outer surface.

Epistasis. The effect of one gene upon the expression of another, non-allelic, gene.

Gamete. The egg (in females) and sperm (in males) which contain a haploid number of chromosomes and together provide all of the genetic material inherited by an individual.

Gene (structural). A segment of chromosomal DNA which codes for the synthesis of a single polypeptide. The position of the gene is its *locus*.

Gene flow. Changes in gene frequencies within a population due to immigration or intermarriage with other racial groups.

Genetic determinant. That segment of DNA within a gene which determines the structure of a specific region on a synthesized molecule.

Genetic drift. Changes in gene frequency within a population which are not due to selection, gene flow or mutation, but reflect chance variations in expected distribution. The smaller the population, the greater the genetic drift.

Genetic system or region. A DNA sequence determining a series of phenotypes which are not randomly distributed among various populations and which are inherited in a manner consistent with a series of alleles. Particularly applicable to the Rh, Kell and MNSs blood group systems and the Gm serum system.

Genotype. This term may be applied to the entire set of genes, but it is more often applied to the paired set of alleles at a single locus.

Golgi body or complex. A membranous organelle which has the probable function of storing products secreted by the cell.

Haploid. The number of chromosomes in a gamete, which contains only half of the somatic cell diploid number because only one of each chromosome pair is present. In man, the haploid number is 23 (22 autosomes and one sex chromosome).

Hapten. A non-protein substance which reacts with a specific combining site on an antibody molecule but which by itself cannot stimulate the formation of antibodies.

HeLa cells. Cultured cells originating from a human cervical carcinoma, carried in tissue culture for a longer period than any other known human cell strain.

Hemizygous. A term applied to the genes on the X chromosome of males, who are hemizygous because they have only one X chromosome.

Heterogametic. The sex which produces unlike sex chromosomes, as opposed to homogametic, in which the sex chromosomes are of only one kind. In humans, men are heterogametic and women are homogametic.

Heterozygous. The presence of two different alleles at a given locus on paired (homologous) chromosomes.

Homotetramer. A molecule composed of four identical subunits.

Homozygous. The presence of two apparently identical alleles at a given locus on paired (homologous) chromosomes.

Hydrogen bond. A weak inter- or intra-molecular force (stronger than van der Waals force) between an electronegative atom such as O^- and a H^+ atom which is co-valently linked to a second electronegative atom.

Hydrophobic bond. A bond between non-polar groups in an aqueous medium which is due to the tendency of water molecules to exclude non-polar molecules.

Idiotypes; idiotypic specificity. Antigenic determinants which occur uniquely on the protein product of a single cell or clone of cells

and which differentiate one molecule from another within a given
subclass.

Immunological tolerance. Inability to respond to a specific antigenic
stimulus due to previous exposure, especially during embryonic
development, to that specific antigen.

Ineffective erythropoiesis. Premature destruction within the bone
marrow of developing red cells; characteristic of diseases in which
there is a defect in synthesis of heme, globin, or nucleic acid.

Isotypes; isotypic specificity. Antigenic determinants common to all
members of a species but characteristic of a molecular class or
subclass within a family of proteins such as the immunoglobulins.

Isozymes (isoenzymes). Multiple molecular forms of an enzyme in a
single tissue.

Karyotype. The chromosomes of an individual, usually displayed as a
microphotograph, with the chromosomes arranged, according to
standard nomenclature, into eight different groups:

A:	chromosomes 1–3		E:	chromosomes 16–18
B:	chromosomes 4–5		F:	chromosomes 19–20
C:	chromosomes 6–12		G:	chromosomes 21–22
D:	chromosomes 13–15		Sex:	chromosomes X and Y

Lectin. Plant extracts which agglutinate red cells in accordance with
their blood group specificity, such as H, A, A_1 and N.

Linkage. Genes are linked when their loci are on the same chromosome. The greater the distance between their loci, the more
frequently they are separated by recombination. Two genes on the
same chromosome are linked in *coupling* phase; if they are on
homologous chromosomes, they are in *repulsion* phase.

Mapping a chromosome. Determining the linear sequence of loci on a
chromosome by assessing the results of recombination between the
loci, which indicate their relative positions as well as the distances
separating them.

Marker (genetic). A genetically determined trait which is subject
to accurate classification and whose inheritance is usually unequivocal.

Meiosis. The process of cell division in which haploid gametes are
produced from diploid cells in the gonads. The first division is
reductive (diploid to haploid) while the second division results in
four haploid descendents of the original diploid cell.

20

Messenger RNA. RNA with a base sequence complementary to a DNA strand. Its molecular weight is proportional to the size of the polypeptide chain for which it acts as a template.

Michaelis constant (K_m). The concentration of an enzyme substrate (in moles per liter) at half the maximal velocity, V_{max}.

Microsomes. Ribosomal particles attached to the endoplasmic reticulum.

Mitochondrion. The 'powerhouse', a complex organelle which provides energy to the cell through ATP production. Its biogenesis is independent of the other cell components.

Mitosis. Cellular division resulting in the production of two daughter cells identical to the parent cell. The four stages of active mitosis are called prophase, metaphase, anaphase and telophase. A fifth, inactive stage is called interphase.

Monozygotic twins. Twins which are the identical products of a single ovum fertilized by a single sperm.

Mosaicism. The presence of two or more cell lines with different genetic composition involving part or all of an organism. According to some definitions, *chimerism* differs from mosaicism in that the cells are derived from different zygotes. However, it is difficult to apply strict distinctions, particularly when the origin of the anomaly is in doubt.

MTT. 3-(4,5-dimethyl-thiazolyl-2)-2,5-diphenyl tetrazolium bromide. A tetrazolium salt which acts as an electron acceptor from an oxidized substrate, forming an insoluble, dark blue formazan.

Myeloma. A malignant disease associated with the proliferation of monoclonal plasma cells producing IgG, IgA or IgD molecules with homogenous structure.

Neuraminidase (*Receptor-destroying enzyme, RDE*). An enzyme which breaks the glycoside linkage which binds the sialic acid, N-acetyl-neuramic acid (NANA) to an adjacent sugar. The cell receptor for certain viruses is destroyed by this enzyme.

Non-disjunction. Failure of paired chromosomes to disjoin during anaphase so that one daughter cell contains both, and the other contains neither.

Nucleic acid. A polymeric chain of nucleotides, in which a backbone of repeating sugar units is connected by phosphate bridges. Each sugar is attached to a purine or pyrimidine base.

Nucleotide. A sugar-phosphate with an attached purine or pyrimidine base. Examples are thymidylic, cytidilic and uridylic acids (formed from pyrimidines) and adenylic and guanylic acids (from purines). The nucleic acids, DNA and RNA, are composed of long sequences of these nucleotides.

Oligosaccharide. A short chain of 5–7 sugars held together by glycoside linkage.

Operator. A region on the chromosome which interacts with a specific repressor, thus controlling the function of an adjacent operon.

Operon. Genes adjacent on a chromosome which function coordinately under the control of an operator and a repressor.

Oxygen dissociation curve. The curve obtained by plotting the oxygen content measured at various partial pressures of oxygen.

Peptide bond. The covalent bond linking two amino acids, in which the α-amino group of one is bonded to the α-carboxyl group of the other, with the elimination of water.

Phenotype. The measurable characteristics of an organism which reflect the genotype in cooperation with the environment.

Poikilocytosis. The occurrence of irregularly shaped red cells in the peripheral blood, commonly associated with ineffective erythropoiesis.

Polymorphism. The occurrence in a population of two or more genetically determined alternative phenotypes with frequencies greater than could be accounted for by mutation or drift.

Polynucleotide. Characteristic structure of nucleic acid, which contains a sequence of nucleotides in which there is $3' \rightarrow 5'$ phosphate linkage between the sugars of adjacent nucleotides.

Polypeptide. A long chain of amino acids linked by peptide bonds.

Polyploidy. The presence of more than two sets of chromosomes within a cell. Thus triploidy implies a haploid number of three, tetraploidy a haploid number of four, and so forth.

Polysome (polyribosome). The site of protein synthesis, the polysome consists of a variable number of ribosomes which move along the template of messenger RNA during the construction of a polypeptide.

Propositus or proband. A family member whose disease, trait or phenotype signals the presence of a given inherited characteristic in a pedigree.

Prosthetic group. The non-polypeptide portion of a complex protein,

20*

such as the heme moiety of hemoglobin. The prosthetic group is usually involved in the active site of the protein.

Protein structural organization

Primary: peptide bonding of amino acids in linear sequence in a polypeptide chain.

Secondary: coiling of a polypeptide chain into α-helices of varying length, stabilized by internal hydrogen bonding.

Tertiary: bending and folding of the helical and non-helical portions into a shape stabilized by hydrogen, covalent, ionic, non-polar and van der Waals linkages.

Quaternary: the spatial arrangement of two or more polypeptide subunits of a protein molecule.

Purine bases. Derivatives of purine, containing a six-membered pyrimidine ring and a five-membered imidazole ring. DNA and RNA contain the same purine bases: adenine and guanine. Other biologically important purine bases are xanthine, hypoxanthine and uric acid.

Pyrimidine bases. Derivatives of pyrimidine formed by substitution of amino, hydroxyl and methyl groups for hydrogen atoms. The principle pyrimidine bases in RNA are uracil and cytosine; in DNA they are thymine and cytosine.

Reaginic antibodies. Skin-sensitizing antibodies found in allergic (atopic) individuals.

Recessive trait. A trait is recessive when it is expressed clinically in the homozygous, but not the heterozygous state. Clinically recessive traits are often 'biochemically codominant', because their gene products are demonstrable in the laboratory. The term recessive is often applied to genes rather than to traits.

Recombinant. An offspring who has two or more inherited traits not found together in either parent.

Recombination. The redistribution of genes on a chromosome due to crossing-over between their loci. The frequency of recombination is proportional to the distance between loci.

Regulator genes. Genes which control the rate at which the products of other genes are synthesized by making a substance which represses the operator.

Repulsion or trans-position. The presence of two genes on different chromosomes. True alleles are always in repulsion because they cannot co-exist on the same chromosome. Their respective characters are said to be antithetical.

Ribosomal RNA. RNA of two molecular weight classes (1·2 million and 0·5 million) which comprises the bulk of cellular RNA. Its function in the ribosome is unknown; it may dictate the structure of ribosomal protein.

Ribosome. Particles of protein and ribosomal RNA which compose the polysome. As the ribosomes move along the template of messenger RNA, they 'read' the message and construct the polypeptide in accordance with the arrival of activated amino acids transported by transfer RNA.

RNA (Ribonucleic acid). A nucleic acid in which the backbone sugar unit is ribose and the principal pyrimidines are uracil and cytosine. As in DNA, the principal purines are adenine and guanine. Three kinds of RNA are recognized: ribosomal, messenger and transfer (or soluble) RNA.

Sedimentation coefficient. Rate in seconds of sedimentation of a solute in cm per second per unit centrifugal field of force (dynes per gram).

Sedimentation constant. Sedimentation coefficient corrected to conditions in water at 20°C (usually) and expressed in Svedberg (S) units.

Segregation. The non-random distribution of allelic genes to different gametes due to the separation of homologous chromosomes during meiosis. Thus, alleles segregate and non-alleles assort.

Selection. The effects of selective forces within the environment such as temperature, food, elevation, predators and diseases which determine the relative fitness of a gene within a population.

Somatic mutation. Mutation occurring in a non-gametic cell.

Synapsis. The process of chromosome pairing which occurs in the prophase of the first meiotic division and which precedes the exchange of genetic material (crossing-over).

Transcription. The formation of messenger RNA by complementary base pairing with a DNA strand which contains the genetic information for the amino acid sequence of a polypeptide. The subsequent translation of the messenger occurs on the polysome.

Transfer RNA (tRNA) or soluble RNA (sRNA). Any of 20 or more structurally similar RNA species with a molecular weight of about 25,000. Each of these species can combine through covalent linkage with a specific amino acid and transfer it to a specific position on messenger RNA by forming a hydrogen bond with a nucleotide triplet of the latter molecule.

Tris. Tris (hydroxymethyl) aminomethane. An organic base frequently used in buffer systems. The term 'tris buffer' implies a tris-HCl buffer, unless otherwise specified.

Trisomy. The occurrence of three, rather than the normal two, homologous chromosomes within a cell. When a large proportion of cells are involved, the individual is clinically abnormal. For example, Down's syndrome is usually associated with trisomy of the twenty-first chromosome.

Van der Waals force. A weak attractive inter- or intra-molecular force due to attraction of induced dipoles.

Zygote. The fertilized egg or ovum, resulting from fusion of nuclei of male and female gametic cells (i.e. the sperm and the egg).

ADDENDA
September, 1968

CHAPTER 1 · THE IMMUNOGLOBULINS

Frangione and Milstein (*J. Molec. Biol.* **33**, 893, 1968) reported that in contrast to γG1 and γG4 (both of which have two interheavy chain disulfide bridges), the γG3 molecule has five such bridges, all apparently in or near the hinge region. The γ3 heavy chain contains two sequences which are very similar to the single sequences characteristic of γ1 and γ4, respectively, at the site of light chain disulfide bonding. However, only the γ4-like sequence binds the light chains to the heavy chains.

A pseudosymmetrical relationship of the light chain and Fd portions of the Fab fragment, proposed by Singer and Thorpe (*Proc. nat. Acad. Sci.* **60**, 1371, 1968), was derived from considering the placement of disulfide bonds and of affinity-labelled tyrosine residues in the active site of IgG antibody molecules.

Of 332 monoclonal IgA proteins studied by Vaerman, Heremans and Laurell (*Immunology* **14**, 425, 1968), 6·6 per cent belonged to the α chain subclass He, and the remainder to subclass Le. The possible origin of salivary 11S IgA from the 7S IgA of serum was discounted by Selner, Merrill and Claman (*Pediatrics* **40**, 452, 1967), who found no IgA in the saliva of newborn infants after exchange transfusion. Studies on IgA nasal secretions in adults, reported by Butler, Rossen and Waldmann (*J. clin. Invest.* **46**, 1883, 1968) indicated that there is active synthesis of the 11S polymer in tissues of the upper respiratory tract.

Further evidence for classifying γE globulin with the immunoglobulins was presented by Ishizaka, Ishizaka and Terry (*J. Immunol.* **99**, 849, 1967). Molecules of γE globulin were shown to have both K and L types of light chain, a unique heavy chain, and reaginic antibody activity. A WHO memorandum (Bull. WHO, vol. 38, no. 1, 1968) has officially recognized γE globulin (or IgE) as an immunoglobulin class.

The suspected non-allelic behaviour of the lambda chain Oz groups has now been confirmed by Ein (*Proc. nat. Acad. Sci.* **60**, 982, 1968). From combined studies, he reported that 75% of 107 Bence Jones lambda chains tested were Oz(−) and 25 per cent were Oz (+). Nevertheless, analysis of the Oz peptides from lambda chains isolated from whole serum showed the presence of both Oz(+) and Oz(−) peptides in all of 10 normal subjects. Assuming gene frequencies of 0·75 and 0·25 from the Bence Jones studies, the calculated probability of finding ten consecutive heterozygotes would be one in 20,000. Thus, non-allelism seems very likely. Ein noted that the presence of both peptides suggests either gene duplication or a translational ambiguity. The same mechanisms were proposed by Schroeder and his colleagues to explain the normal co-existence of two different hemoglobin gamma chains (see Addenda for Chapter 10). In the latter instance, point mutation following gene duplication appears more likely, on the basis of inheritance studies. The same mechanism may apply to the immunoglobulin lambda chains.

Vyas, Fudenberg, Pretty and Gold (*J. Immunol.* **100**, 274, 1968) described a new method for Gm and Inv typing using red cells coated with specific proteins by means of treatment with chromic chloride. The rapidity of this method was considered to provide an improvement over the tanned red cell method.

CHAPTER 2 · HAPTOGLOBIN

Nance, Empson, Bennett and Larson (*Science* **160**, 1230, 1968) presented preliminary evidence for linkage between the haptoglobin (alpha chain) gene locus and a catalase structural locus, from a study of four Brazilian families in which there were two electrophoretic variants of red cell catalase.

Brunori, Alfsen, Saggese, Antonini and Wyman (*J. biol. Chem.* **243**, 2950, 1968) studied the half-saturated and saturated complexes of haptoglobin with hemoglobin, and found that the changes in oxidation-reduction potential indicated a difference in the binding of haptoglobin by the hemoglobin alpha and beta chains. These findings were extended by Chiancone, Alfsen, Ioppolo, Vecchini, Agrò, Wyman and Antonini (*J. Molec. Biol.* **34**, 347, 1968), who confirmed that haptoglobin has a much greater affinity for alpha than for beta chains of hemoglobin. They also showed that unliganded single globin chains, unlike unliganded hemoglobin, are able to combine with haptoglobin. Either chain can be displaced by oxyhemoglobin.

An additional physiological role of haptoglobin was suggested by Snellman and Sylvén (*Nature* **216**, 1033, 1967). They found that the proteolytic activity of cathepsin B was appreciably inhibited by haptoglobin, and proposed that the rise in plasma haptoglobin associated with tissue injury may provide a protective mechanism against active proteolysis.

CHAPTER 3 · TRANSFERRIN

Greene and Feeney (*Biochemistry* **7**, 1366, 1968) reported their studies on transferrin from chicken, rabbit and human sources. Attempts to dissociate the molecule into subunits were completely unsuccessful, and the authors concluded that transferrin, in all three species, is a monomer.

Wang, Shuster, Epstein and Fudenberg (*Biochem. Genet.* **1**, 347, 1968) compared the degree of immunological cross-reactivity of human transferrin with the transferrins of 23 different primate species. The results indicated that the evolution of antigenic determinants on the transferrin molecule is in good agreement with the taxonomic classification.

CHAPTER 4 · Gc GROUPS

A new Gc variant, designated Gc-Bangkok, was found in a Thai family by Rucknagel, Shreffler and Halstead (*Am. J. hum. Genet.* **20**, 478, 1968). Its migration rate was between that of Gc-2 and Gc-Z. The authors stated that all of the known Gc proteins are equally spaced electrophoretically, suggesting that each may differ from its neighbors by a single amino acid substitution.

An expansion of their earlier reports on the genetic linkage between the Gc and albumin loci was presented by Weitkamp, Robson, Shreffler and Corney (*Am. J. hum. Genet.* **20**, 392, 1968).

CHAPTER 5 · BETA LIPOPROTEIN ALLOTYPES

Vierucci, Dettori, Morganti, Beolchini and Bütler (*Vox Sang.* **14,** 151, 1968) studied the disappearance of lipoprotein molecules bearing the donor's specific Ag determinants after exchange transfusion of newborn infants. By 12 hours, the donor's phenotype could no longer be determined. Thus, catabolism of injected Ag molecules and their replacement by molecules of the infant's type was very rapid.

Vierucci, Blumberg, Dettori, Borgatti and Levene (*J. Pediat.* **72,** 776, 1968) compared the production of antibodies to Ag and Gm antigens in multi-transfused children with thalassemia. They concluded that the likelihood of developing antibodies of both specificities was increased if transfusions were given within the first year of life. Also, they reported that some children with anti-Ag antibodies had fairly severe febrile reactions when transfused with blood having Ag antigens of corresponding specificity.

Hirschfeld (*Vox Sang.* **14,** 95, 1968) reported that the Ag(x) antigen may be a useful genetic marker in medico-legal investigations, since the theoretical paternity-exclusion rate is 8·14 per cent (in the Swedish population). Hirschfeld, Contu, Rittner and Geserich (*Vox Sang.* **14,** 124, 1968) confirmed the apparent allelism of Ag^x and Ag^y genes, and reported that if tests are performed for both of their respective antigens, the theoretical paternity exclusion rate is increased to 14·2 per cent. Segregation of the Ag(z) antigen in 42 families was studied by Vierucci, Blumberg, and Morganti (*Nature* **216,** 1231, 1967). They confirmed its usefulness as a genetic marker and found a similar gene frequency in two populations (northern Italy and northeastern United States): 0·140 and 0·129 respectively.

A study of Ag inheritance in 75 families by Hirschfeld and Rittner (*Vox Sang.* in press) showed that Ag(a_1), Ag(z) and Ag(t) are closely associated genetically with Ag(x) and Ag(y), since no recombinations were observed. To fit the data, they proposed some genetic models similar to those suggested for the Rh and Gm systems.

Schultz, Harvie and Shreffler (Abstr. No. 5, *Am. Soc. hum. Genet.* Meeting, 1968) isolated the lipoprotein fractions from human serum by ultacentrifugation and found that the fraction with Lp(a) activity, which migrates with a distinct peak in Lp(a+) sera, is not detectable in the serum of Lp(a−) individuals. They suggested that Lp phenotypic variability represents the presence or absence of this specific protein.

CHAPTER 6 · PSEUDOCHOLINESTERASE

Differential inhibition of serum cholinesterase by alkyl alcohols was described by Whittaker (*Acta genet.* **18**, 325 and 335, 1968), who used n-butyl alcohol to determine the 'alcohol number' and reported a new, rare variant found by this technique.

Nineteen cases of pseudocholinesterase deficiency were found in eleven Alaskan Eskimo families by Gutsche, Scott, and Wright (*Nature* **215**, 322, 1967). In four families, there was no detectable enzyme, while in five others, there was a small amount of enzyme activity. No evidence was obtained for the E_1^a or E_1^f allele in a large Eskimo population sample.

Neumann and Walter (*Nature* **219**, 950, 1968) reported the pseudocholinesterase phenotypes of Icelanders, Pakistanis, Germans and Greeks. Of the 128 Islandic serum samples, three had the UF phenotype, but none had I or IF types.

CHAPTER 7 · PLASMA ALKALINE PHOSPHATASE

Langman, Constantinopoulos and Bouchier (*Nature* **217**, 863, 1968) measured the alkaline phosphatase concentration in small intestinal mucosa strips obtained during surgery from patients of known ABO group and secretor status. They found the highest mean value among group O and B secretors, an intermediate mean in group A secretors, and the lowest mean among non-secretors, although there was considerable overlapping. One possible interpretation of these findings, if they are really significant, is that the blood group and secretor status influences the rate of enzyme production within the cell or its rate of removal from the cell. The question remains unresolved.

CHAPTER 8 · OTHER MARKERS IN PLASMA

Melartin (*Acta path. microbiol. scand.* suppl. **191**, 1967) reviewed her extensive studies on albumin variants in North American populations, giving further data on the Naskapi and Mexico variants, the genetic linkage of albumin and Gc, and the binding properties of variant albumin molecules.

Albumin Máku, a rapidly migrating variant, was found in a family of South American Yanomama Indians by Weitkamp and Chagnon (*Nature* **217**, 759, 1968); and Albumin Syracuse was reported in one of 1,000 specimens tested by Schneiderman, Berger and Krieg (*Nature* **218**, 1159, 1968). The latter variant may be the same as one of the previously reported variants.

Shokeir, Rucknagel, Shreffler, Na-Nakorn and Wasi (Abstr. No. 79, *Am. Soc. hum. Genet.* Meeting, 1968) found a common fast-moving ceruloplasmin variant, similar but not identical to Cp A, in sera obtained from residents of Thailand. The allele of this variant, called Cp Th, had an overall frequency of 0·157 in the 2,979 specimens tested.

Preliminary evidence suggesting a linkage relationship between the Xm serum protein system and the X-linked form of Hurler's syndrome (Hunter's syndrome) was presented by Berg, Danes and Bearn (*Am. J. hum. Genet.* **20**, 398, 1968). They also mentioned a previous paper by Berg and Bearn (*Trans. Ass. Amer. Phycns.* **79**, 165, 1966) in which some family studies suggested linkage between the loci for Xm and those for hemophilia A, hemophilia B and deutan color vision, but not for Xg^a. (For their recent summary, see *Ann. Rev. Genet.*, 1968.)

Fibrinogen, a molecule composed of three paired polypeptide chains α, β, and γ), has a molecular weight of 340,000 and, during clotting, undergoes polymerization to fibrin after the enzymatic cleavage of fibrinopeptides from the α and β chains. Inherited variation in fibrinogen structure has been reported in eight unrelated families, reviewed by Forman, Ratnoff and Boyer (*J. Lab. clin. Med.* **72**, 455, 1968). However, the amino acid substitution responsible for the protein abnormality has been determined in only one instance. Blombäch, Blombäch, Mammen and Prasad (*Nature* **218**, 134, 1968) found that in Fibrinogen Detroit a neutral amino acid, probably serine, replaces arginine at position 19 of the fibrinogen α chain. This position is located in the so-called 'disulfide knot', which is the site of interchain disulfide bonding within the fibrinogen molecule. The resultant conformational change probably interferes with polymerization, thus accounting for the clotting abnormality of Fibrinogen Detroit.

Although several inherited variations in the components of complement are known, the most promising as a genetic marker appears to be C'_3 (β_{1c} globulin). Using high voltage electrophoresis, Azen and Smithies (*Science*, **162**, 905, 1968) found five different starch gel patterns in the serum of 113 people chosen at random, and family studies indicated the existence of three alleles at the β_{1c} locus. Alper and Propp (*J. clin. Invest.* **47**, 2181, 1968), using agarose electrophoresis, defined eight different phenotypes and postulated as many as five alleles. Comparisons of selected specimens by the two methods are now required, although the results appear to be essentially in agreement.

CHAPTER 9 · BLOOD GROUP ANTIGENS

Kobata, Grollman and Ginsberg (*Arch. Biochem.* **124**, 609, 1968) detected an N-acetyl-D-galactosaminoyl transferase in the breast milk of women with blood groups A and AB, but not with B or O. Presumably, this enzyme is the *A* gene product.

Hakomori and Strycharz (*Biochemistry* **7**, 1279, 1968) isolated glycolipids from human red cells and found, contrary to expectation, Le^b activity in addition to A, B and H associated with the ceramide pentasaccharide 'skeleton' of the molecules. The relationship of red cell antigens to membrane structure was reviewed by Zahler (*Vox Sang.* **15**, 81, 1968).

Zaluski, Bar-Shang, Cuttner and Rosenfield (Abstr. No. DD-26, *12th Cong. int. Soc. Hemat.* 1968) studied a 26 year old woman with Rh$_{null}$ red cell phenotype. Her mild anemia, accentuated during pregnancy, was found to be due to an intrinsic red cell defect. No abnormalities of enzymes, fats or ion exchange mechanisms could be demonstrated. Survival of the abnormal cells was shortened (T 1/2 11 days) in the patient and in a normal subject, but in a splenectomized recipient, the T 1/2 was not decreased.

Prevention of Rh isosensitization by passive antibody injection continues to be a spectacular success, summarized by Clarke (*Lancet* **ii**, 1, 1968) and Pollack and his associates (*Transfusion* **8**, 151, 1968). Some excellent papers providing further insight into the mechanisms involved in immune suppression are those of Uhr and Möller (*Advanc. Immunol.* **8**, 81, 1968) and Siskind (*Transfusion* **8**, 127, 1968).

CHAPTER 10 · HEMOGLOBIN

Recent papers by Perutz and his colleagues have provided much more detailed information about the three-dimensional structure of oxy-hemoglobin (*J. Molec. Biol.* **33**, 283, 1968; *Nature* **219**, 131, 1968; *Nature* **219**, 902, 1968). Invariance of the amino acids surrounding the heme moiety was re-emphasized, and 60 interactions between atoms of heme and atoms of each globin chain were described. In examining the $\gamma^1\beta^1$ interchain contact, the authors found about 110 atomic interactions, involving 34 amino acid residues, opposed to 80 interactions of 19 residues in the $\alpha^1\beta^2$ contact. However, the latter contact contains more side chain bridges touching heme groups, is invariant at all positions but one, and undergoes greater movement than $\alpha^1\beta^1$ during deoxygenation. It was concluded that the hemoglobin tetramer, rather than the $\alpha\beta$ dimer, is actually the functional unit of hemoglobin.

Stereoscopic drawings of the hemoglobin atomic model were presented, as well as a detailed explanation for the structural effects of amino acid replacement in 35 abnormal hemoglobins. Several mechanisms were clarified. For example, the alkaline resistance of Hb Rainier was explained on the basis of hydrogen bonding between the imidazole ring of the substituted histidine at H 23 and the carbonyl group of leucine at H 19. The reader is urged to consult this paper by Perutz and Lehmann (*Nature* **219**, 902, 1968) as an indispensable source of current information on hemoglobin molecular structure.

Schroeder, Huisman, Shelton, Shelton, Kleihauer, Dozy and Robberson (*Proc. nat. Acad. Sci.* **60**, 537, 1968) presented evidence which strongly suggests the existence of two or more non-allelic genes coding for normal human γ globin chains. In all specimens of infant blood studied, they found a major and a minor HbF, differing in at least one amino acid: glycine in the major, and alanine in the minor component at position 136 γ. Gamma chains from abnormal fetal hemoglobins were found to have either glycine or alanine at 136 γ. Also, in a family with hereditary persistence of fetal hemoglobin, the HbF of the mother and her two daughters had only glycine at 136 γ; however, both kinds of chain were found in the HbF of another case studied. The authors felt that their data were best explained by the presence of two or more gamma chain non-allelic structural genes, but the precise number and their mode of interaction remains to be determined. Somewhat similar observations have also been made for the human lambda immunoglobulin chain (see Addenda for Chapter 1) and the hemoglobins of several animal species. Their mechanistic interpretation provides a vital problem for contemporary genetics.

The existence of embryonic hemoglobins in sheep, calf, and pig was demonstrated by Kleihauer and Stöffler (*Molec. gen. Genet.* **101**, 59, 1968). During ontogenesis, epsilon chains are formed first, followed by alpha chains, and finally, gamma chains. Presumably a similar sequence occurs in human embryonic development.

In another important paper, Gilman and Smithies (*Science* **160**, 885, 1968) presented excellent evidence for close linkage of the loci for a fetal and an adult mouse globin chain, analogous to the γ and β chains of human subjects. In man, linkage of the γ chain gene to the closely linked δ and β chain genes has been previously suggested, and this paper offers the first proof that β-γ linkage does exist in at least one mammalian species.

Several new hemoglobins have been reported (see below), including the unstable variants, Santa Ana, Sabine, Philly and Wien. In Hb Santa Ana, a proline for leucine substitution at beta chain position F4 removes the leucine-heme contacts, and the loss of heme opens a molecular crevice into which water may enter. Hb Sabine has the same amino acid substitution, but at position F7. In Hb Philly, the phenylalanine for tyrosine beta chain substitution at position C1 causes weakening of the bonds between the β^1 and α^1 chains, and all six cysteines are reactive. The beta chain of Hb Wein contains a replacement of tyrosine by aspartic acid at position H8, but there is no alteration in electrophoretic mobility. Presumably the creation of an internal negative charge by aspartic acid necessitates the displacement of a positively charged group created by a rise in pK of either histidine at NA 2 or valine at NA 1. The result is distortion of the A helix.

From a genetic standpoint, a particularly interesting variant is Hb Korle-Bu, described by Konotey-Ahulu, Gallo, Lehmann and Ringlehann (*J. med. Genet.* **5**, 107, 1968). This slow-moving hemoglobin has the molecular formula $\alpha_2^A \beta_2^{73}$ $^{Asp \rightarrow Asn}$. The beta chain of Hb C-Harlem also contains this amino acid substitution, in addition to the β^{6} $^{Glu \rightarrow Val}$ characteristic of HbS. It was postulated that the origin of the Hb C-Harlem gene may have been intragenic crossing-over in an individual heterozygous for Hb_β^S and $Hb_\beta^{Korle-Bu}$.

The new hemoglobins and their references include:

Hb Etobicoke	$\alpha_2^{84\ \text{Ser}\to\text{Arg}}\ \beta_2^A$	Submitted by D. Beale to *Atlas of Protein Structure* (ed. Dayhoff & Eck), 1968.
Hb Dakar	$\alpha_2^{112\ \text{His}\to\text{Gln}}\ \beta_2^A$	Rosa, Oudart, Pagnier, Belkhodja, Biogné & Labie (Abstr. T-6, p. 72, *12th Cong. int. Soc. Hemat.*, New York, 1968).
Hb Manitoba	$\alpha_2^{102\ \text{Ser}\to\text{Arg}}\ \beta_2^A$	Attributed to Crookston, Goldstein, Lehmann & Beale by Perutz & Lehmann (*Nature* **219**, 902, 1968).
Hb Chiapas	$\alpha_2^{114\ \text{Pro}\to\text{Arg}}\ \beta_2^A$	Jones, Brimhall & Lisker (*Biochim. biophys. Acta* **154**, 488, 1968).
Hb Sogst	$\alpha_2^A\ \beta_2^{14\ \text{Leu}\to\text{Arg}}$	Submitted by Monn, Gaffney & Lehmann to *Atlas of Protein Structure* (ed. Dayhoff & Eck), 1968.
Hb G-Saskatoon	$\alpha_2^A\ \beta_2^{22\ \text{Glu}\to\text{Lys}}$	Vella, Isaacs & Lehmann (*Canad. J. Biochem.* **45**, 351, 1967).
Hb G-Taiwan-Ami	$\alpha_2^A\ \beta_2^{25\ \text{Gly}\to\text{Arg}}$	Blackwell & Liu (*Biochim. biophys. Acta* **154**, 488, 1968).
Hb Philly	$\alpha_2^A\ \beta_2^{35\ \text{Tyr}\to\text{Phe}}$	Attributed to Rieder, Oski & Clegg by Perutz & Lehmann (*Nature* **219**, 902, 1968).
Hb Korle-Bu	$\alpha_2^A\ \beta_2^{73\ \text{Asp}\to\text{Asn}}$	Konotey *et al.* (see above)
Hb Santa Ana	$\alpha_2^A\ \beta_2^{88\ \text{Leu}\to\text{Pro}}$	Attributed to Green by Perutz & Lehmann (*Nature* **219**, 902, 1968).
Hb Sabine	$\alpha_2^A\ \beta_2^{91\ \text{Leu}\to\text{Pro}}$	Schneider, Veda, Alperin, Brimhall & Jones (Abstr. No. 27, *Amer. Soc. hum. Genet.* Meeting, Oct. 1968).

Hb Wien $\alpha_2^A \beta_2^{130\ \text{Tyr}\to\text{Asp}}$ Attributed to Pietschmann, Lorkin, Lehmann & Braunsteiner by Perutz & Lehmann (*Nature* **219**, 902, 1968).

The α chain variant referred to as Hb Sinai-Sealey in this book was independently described in a third family as Hb Hasharon by Halbrecht, Isaacs, Lehmann and Ben-Porat(*Israel J. med. Sci.* **3**, 827, 1967). The name, Hasharon, should probably take precedence, because of the prior publication of their paper.

Several preliminary reports mentioned in this book have now been expanded into more detailed papers. For example, further data are available on the structure and function of two abnormal hemoglobins: Hb Kempsey, discussed by Reed, Hampson, Gordon, Jones, Novy, Brimhall, Edwards & Koler (*Blood* **31**, 623, 1968) and Hb Kansas, by Bonaventura & Riggs (*J. biol. Chem.* **243**, 980, 1968). The mechanism underlying instability of the hemoglobins associated with Heinz body anemia was more fully elaborated by Jacob, Brain, Dacie, Carrell and Lehmann (*Nature* **218**, 1214, 1968). A detailed review of the HbM diseases in Japan by Shibata, Miyaji, Iuchi, Ohba and Yamamoto appeared in the *Bulletin of the Yamaguchi Medical School* **14**, 141, 1967. A comparison of δ chain and β^s chain synthesis was presented by Kazazian and Itano (*J. biol. Chem.* **243**, 2048, 1968). Results of electron microscopy of Hb S in red cells were described by Döbler and Bertles in *J. exp. Med.* **127**, 711, 1968.

The rate of appearance of these and many other papers reflects the very rapid expansion in the study of hemoglobin, and makes it impossible to present a truly up-to-date summary.

CHAPTER 11 · RED CELL ACID PHOSPHATASE

A high frequency of the P^b gene and absence of P^c were reported in Australian aborigines by Lai (*Nature* **217**, 1186, 1968). In 92 natives at Cairns, the P^b gene frequency was 0·956. A similar frequency of 0·984 was found in 153 Darwin aborigines.

Proceedings of a conference on multiple molecular forms of enzymes, held by the New York Academy of Sciences in December, 1966, has only recently been published in the *Ann. N.Y. Acad. Sci.* **151**,

1-689, 1968. It contains several papers especially concerned with the phenomenon of two or more isozyme products of single genes, a characteristic of acid phosphatase and several other enzymes discussed in this book. Included are the papers by Markert (p. 14), Epstein & Schechter (p. 85), Kirkman & Hanna (p. 133), Allen (p. 190), Harris, Hopkinson & Luffmann (p. 232), Kaplan (p. 382), Poulik (p. 476), and Cann & Goad (p. 638).

CHAPTER 12 · GLUCOSE-6-PHOSPHATE DEHYDROGENASE

Piomelli, Corash, Davenport, Miraglia and Amorosi (*J. clin. Invest.* **47**, 940, 1968) measured G6PD activity in red cells of different ages, obtained by differential centrifugation of blood from normal and G6PD deficient subjects. They found that the activity of normal G6PD had a half-life of about 62 days. Activity of the Gd^{A-} product had a T $1/2$ of about 13 days; while the T $1/2$ of the very unstable Gd^{B-} product could not be measured.

Several new variants of G6PD have been reported. Azevedo, Kirkman, Morrow and Motulsky (*Ann. hum. Genet.* **31**, 373, 1968) found G6PD deficiency in five of 101 unrelated Asiatic Indians. In three, the enzyme was indistinguishable from the Mediterranean (Gd B−) variant. However, the fourth individual had a new variant called Gd Kerala, with chemical characteristics similar to Gd Seattle, except for its slower electrophoretic migration. Another variant, Gd West Bengal, was found in the fifth subject. It had slow electrophoretic mobility, higher than normal K_m TPN, and lower than normal K_m G6P. The characteristics of two other variants (Ijebu-Ode and Ita-Bale), found in western Nigerians, were described by Luzzatto and Afolayan (*J. clin. Invest.* **47**, 1833, 1968). A slow-moving variant (Madrona) with kinetic characteristics similar to Gd B was the subject of a report by Hook, Stamatoyannopoulos, Yoshida and Motulsky (*J. Lab. clin. Med.* **72**, 404, 1968).

A method of staining G6PD in red cells using a tetrazolium-linked system was described by Fairbanks and Lampe (*Blood* **31**, 589, 1968). It was found to be particularly valuable in identifying females heterozygous for G6PD deficiency, and it further corroborated their red cell mosaicism.

CHAPTER 13 · 6-PHOSPHOGLUCONATE DEHYDROGENASE

Partially purified enzyme was prepared from red cells of individuals with 6PGD phenotypes I, II, and III by Shih, Justice and Hsia (*Biochem. Genet.* **1**, 359, 1968). The transitional temperature and K_m of the three molecular forms were found to differ in a manner consistent with their inheritance as the homozygous and heterozygous states of two allelic genes. In a study of 42 patients with chronic myelocytic leukemia, Fialkow, Rubin, Detter, Giblett and Zavala (Abstr. No. 28, *Amer. Soc. hum. Genet.* Meeting, 1968) found hemizygous expression of 6PGD in the blood cells, but not the fibroblasts, of a 6PGD heterozygote. Deletion was considered the most likely explanation, but the Philadelphia chromosome was only questionably involved.

CHAPTER 14 · PHOSPHOGLUCOMUTASE

The occurrence of a rare phenotype indicating the existence of a PGM_2^4 allele was reported by Monn (*Acta Genet.* **18**, 123, 1968); it presumably represents the allele, also called PGM_2^4, whose products were shown in Fig. 14.6 of this book. That figure has now been published by its authors, Hopkinson and Harris, in *Ann. hum. Genet.* **31**, 359, 1968. The major content of their paper concerns the PGM_3 locus and its products, which are best demonstrated in placental extracts.

Another allele at the PGM_2 locus, called PGM_2^{Pyg}, was found by Santachiara-Benerecetti and Modiano (personal communication from Doctor Modiano). Of 328 Babinga pygmies tested, 43 had a slow-moving variant different from the other variants which have been described at this locus.

Study of a large family with individuals heterozygous at each of the three PGM loci was reported by Parrington, Cruickshank, Hopkinson, Robson and Harris (*Ann. hum. Genet.* **32**, 27, 1968). The data showed that PGM_2 and PGM_3 loci are not closely linked, and supported previous evidence that the PGM_1 locus is not closely linked to the PGM_2 or PGM_3 loci. Thus, if the three loci arose during evolution by a process of duplication from a common ancestral locus, subsequent translocations must also have occurred to account for their separation onto (presumably) different chromosomes.

CHAPTER 16 · OTHER ENZYMES IN BLOOD CELLS

Lewis, Corney and Harris (*Ann. hum. Genet.* **32**, 35, 1968) described two rare peptidase variants: Pep A 5-1 and Pep A 6-1. On storage, the isozyme pattern of Pep A 5-1 changed in a characteristic manner and could be restored by addition of mercaptoethanol. The authors presented evidence that the *Pep A*^{5} allele codes for a peptide chain in which arginine is replaced by cysteine, while in the *Pep A*^{6} gene product, there may be a substitution of lysine by glutamic acid.

Another structural variant of phosphohexose isomerase (PHI) associated with severe hemolytic anemia was found by Paglia, Holland, Baughan & Valentine (*New Engl. J. Med.*, in press, 1968). The affected members in this family appear to be homozygous for a gene whose enzyme product has only a slight retardation in electrophoretic mobility. The heterozygote pattern is not distinguishable from normal, but the enzyme activity is reduced by about fifty per cent.

A new human polymorphism, involving the enzyme adenosine deaminase, was described by Spencer, Hopkinson and Harris (*Ann. hum. Genet.* **32**, 9, 1968). Electrophoresis of hemolysates revealed a common phenotypic pattern called ADA 1, consisting of three overlapping components. In about ten per cent of English people tested, a second pattern, called ADA 2-1, was encountered. A third pattern, ADA 2, was found in only one of 580 specimens. Family studies showed that *ADA*^{1} and *ADA*^{2} are allelic genes with a frequency of about 0·94 and 0·06, respectively. The frequency of *ADA*^{2} in 147 Negroes and 213 Asiatic Indians was 0·03 and 0·11, respectively.

CHAPTER 17 · APPLICATIONS OF GENETIC MARKERS

Using the genetic markers of red cells and placentae, Corney, Robson and Strong (*Ann. hum. Genet.* **32**, 89, 1968) studied the zygosity of 326 twin births and compared the results with the kind of placentation of each twin pair. They found that all dizygotic twins were dichorionic, but between 15 and 20 per cent of probably monozygotic twins were also dichorionic. A table was constructed from the enzyme gene frequencies for calculating the probability of dizygosity when the parents of twins are not available for testing.

Fialkow, Rubin, Detter, Giblett and Zavala (Abstr. No. 28, *Am. Soc. hum. Genet.* Meeting, 1968) studied the genetic markers of 42 patients with chronic myelocytic anemia. They confirmed the clonal origin of this disease by demonstrating hemizygous expression of 6PGD in the blood cells, but not the fibroblasts, of a 6PGD heterozygote. They also ruled out inactivation as a normal occurrence at twelve different autosomal loci.

A new method of calculating mutation rates in the human species has been proposed by H. Harris (*Am. J. hum. Genet.*, in press). In this method, the rare mutant genes belonging to the genetic marker systems are considered to have no selective properties. Thus, their frequencies in large population samples can be used as a measure of the mutation rate.

21

INDEX

A, B and H antigens
A_1 and A_2 269, 272, 279
A_m 291
A_x 291–2
activity on cultured cells 300–1
activity on tissue cells 281
altered reactions in disease 292
as transplantation antigens 281
biosynthesis of 288–90
carbohydrate specificity 270, 272, 285–90, 291
crossreactivity with bacterial antigens 270, 292
determinants of 285–9
expression in rare phenotypes 290–2
function in organogenesis 281
immaturity in infants 279
inheritance of 278–80, 291
in body fluids 280, 281
masking of Lewis antigens by 288
on red cell glycolipid molecules 281, 290, 291, 593
on secreted glycoprotein molecules 280, 284–90
A, B and H substances
as secreted glycoproteins 280, 284
biosynthetic pathway of 288–90
carbohydrate specificity of 270, 285–8
H as precursor to A and B 280, 289–90
in body fluids 280
in cyst fluid 280, 284
in plasma 281
on tissue cells 280, 281
secretion of 280–6, 288–91
Abnormal hemoglobins (*see also* the individual Hemoglobins)
amino acid substitution in 363–85, 595–7
and the genetic code 559–62
associated with methemoglobinemia 375–7
causing red-cell distortion 371–5
electrophoretic migration of 364–9

Abnormal hemoglobins—*cont.*
geographic distribution of 405
homotetramers 360–1, 396, 580
list of 364–9, 596–7
nomenclature of 370
polymer formation of 385
rate of synthesis 387
unstable hemoglobins 380–5, 595
with decreased oxygen affinity 380
with deleted amino acids 360, 383–5, 562–3
with increased oxygen affinity 377–380
ABO blood group system 278–301
antibodies (*see* Anti-A and anti-B)
antigens (*see* A and B antigens)
disease associations 292, 313–4
gene frequencies in 318
genetic interaction
with alkaline phosphatase system 228–30, 292–6
with Hh locus genes 288–90
with Ii system 297
with Lewis system 282, 283, 288–90
with P system 300–1
genetic linkage and 311, 515, 570
inheritance of alleles 278–80
nomenclature of 272
phenotypes of 272, 278
alterations in disease 292
rare phenotypes 290–2
selective effects on 313–6
soluble substances (*see* A, B and H substances)
Acatalasia 533, 544
Acetazolamide 530, 531
Acetylcholinesterase 192, 193, 197
Achromatic zones (*see* 'Indophenol oxidase')
Acid phosphatase in red cells 424–42
alleles of
geographic distribution 434–7, 597
inheritance 426–8

Acid phosphatase in red cells—*cont.*
  alleles of—*cont.*
    rare alleles 427, 428
    relative enzyme activity 428–31
  chromatography of 425, 433
  compared with other phosphatases
    424
  electrophoretic patterns 425–8
    effect of additives 427, 431
    effect of heat 434
    effect of storage 426
  enzyme activity 425, 428–32
    effect of GSSG 431
    effect of genotype 428–31
    in G6PD deficiency 431–2, 454
    in malarious regions 430
    in megaloblastic anemia 425, 431
    in reticulocytes 425, 432
    in thalassemia 432
    normal range 425
    unimodal distribution curve 430
    with different substrates 425,
      432–4
  function of 424
  genetic variation
    common phenotypes 426
    frequencies of alleles 434–7
    inheritance of variants 426–8
    laboratory detection of 435,
      438–9
    rare phenotypes 427–8
  geographic distribution of alleles
    434–7
  in glucose-6-phosphate dehydrogen-
    ase deficiency 431–2
  inhibition of activity 424–5, 433
  instability of 424
    protection by SH compounds 424
  isozymes of 426, 432–4
    effect of heat 434
    effect of storage 426
    reaction kinetics 433
    relative activity 432
    relative isoelectric points 433
    substrate specificity 432–3
  laboratory methods 435, 438–9
  linkage studies 426
  nomenclature 426
  phenotypes of 425–6
    enzyme activity of 428–31
    inheritance of 426–8
    rare types 427–8

Acid phosphatase in red cells—*cont.*
  phenotypes of—*cont.*
    variation with buffer system
      426–7
  physico-chemical properties of 424,
    432–4
Adenosine deaminase 600
Adenosine triphosphate (ATP)
  and hemoglobin oxygen dissociation
    357
  and shunt pathway activity 446
  generated during glycolysis 445,
    490
  production mediated by adenylate
    kinase 512, 518
Adenosine triphosphatase (ATPase)
  310
Adenylate kinase (myokinase) 512–9
  distribution in tissues 512
  electrophoretic patterns 512, 513–4
  function of 512
  genetic variation
    geographic distribution of genes
      515–7
    inheritance 514–5
    linkage studies 515
    possible linkage with ABO locus
      515, 570
  laboratory demonstration of 512,
    517–8
  molecular weight 512
  phenotypes of 513–4
  physico-chemical characteristics
    512
Adrenalectomy, interference with hap-
    toglobin production 65
Affinity labelling of antibodies 16, 587
Ag (lipoprotein) system 178–82
  Ag alleles 180–2
  Ag genetics 178–81, 590–1
    allelism of Ag^x and Ag^y 179, 180,
      182, 590
    complexity of 178–80, 590
  Ag phenotypes 179–80
    geographic distribution of 181,
      590
    laboratory determination of
      186–7
  anti-Ag, sources of 179, 185, 187,
    590
Agar diffusion test for pseudocholin-
    esterase variants 213–4

Albumin 240–7
  Albumin Mexico 242
  Albumin Naskapi 241, 246
  genetic linkage with Gc locus 161, 245–6
  genetic variants of 240–7, 592
    allelism of genes 246
    amino acid substitution in 241
    binding constants of 241
    classified by electrophoretic mobility 242–3
    geographic distribution of 241–2
    inheritance of 241
    laboratory detection of 246–7
  unusual variant of 243–5
    effect of mercaptoethanol 245
    electrophoretic studies 243–5
    inheritance 244
    linkage with Gc locus 161, 246
    tendency to dimerize 245
Alkaline phosphatases 226–40, 292–6
  antigenic classes of 226
  heterogeneity of 226
  in intestine 227–31, 292–6
    measurements of 232, 295
    subcellular localization of 231
    susceptibility to L-phenylalanine 227, 228, 235
  isozymes of alkaline phosphatase in placenta 226, 232
    genetic variation of 232
    physico-chemical characteristics of 233
  isozymes of alkaline phosphatase in plasma 226–40, 292–6
    age effects on 230, 231
    biological significance of phenotypes 231–2, 294
    effect of fat ingestion on 228–30, 293
    effect of neuraminidase on 227, 235
    effect of L-phenylalanine on 227, 228, 235
    electrophoretic patterns of 227, 234, 293
    factors influencing inheritance of 230–1, 294
    genetic mechanisms and 231–2, 294–6
    genetic types, ABO and secretor status 228–31, 292–6

Alkaline phosphatases—cont.
  isozymes of alkaline phosphatase in plasma—cont.
    heterogeneity of phenotypes 227
    inheritance of phenotypes 230–1
    laboratory demonstration of 233–5
    origin of A and B components 226–8, 232
    phenotypes of twins 230
    quantitative variation of 227–30, 293
    substrate affinity of 228
Allele exclusion
  in G6PD synthesis 460–3, 566
  in immunoglobulin synthesis 8–9, 27
Allosterism 353–5, 576
  absence of, in Hb homotetramers 360
  of hemoglobin molecule 353–5, 357
Allotypes and allotypy 7, 15, 16, 23, 177, 576
Alpha$_1$-acid glycoprotein (orosomucoid) 257–9
  electrophoretic patterns 258
    effect of neuraminidase 258
    in twins 258
  phenotypes 258
  physico-chemical properties 257
Alpha$_1$-antitrypsin (*see* Protease inhibitor)
Alpha chain of haptoglobin (*see under* Haptoglobin)
Alpha chain of hemoglobin
  amino acid sequence of 348, 350
  compared with beta chain 350
  compensatory formation of 396
  3-dimensional structure of 350, 353
  evolution of 401–3
  excess in beta-thalassemia 392, 393
  in hemolysed blood 360
  proximal and distal histidines of 351, 376–7
  site of haptoglobin binding 71, 588–9
  structural gene of 386
  structural variants of 364–5, 596
  sulfhydryl groups of 348
  synthesis in thalassemia 395–9
  synthesis, normal 388–90

Alpha helices in hemoglobin   348, 351
Alpha- or beta-naphthyl acetate
   as substrate for alkaline phosphatase
      234
   as substrate for pseudocholin-
      esterases
      E_1 locus products   196–200,
         213–4
      E_2 locus product   218–9
   as substrate for red-cell esterases
      533
Alpha thalassemia   395–9
   genetic mechanisms   397, 398, 399
   globin synthesis in   395, 397
   hematological changes   396
   hemoglobin Bart's (γ_4) in   396
   hemoglobin H (β_4) in   396
   hemoglobin H disease   381, 397–8
   heterogeneity of   396–7
   heterozygous states   396–8
   homozygous states   396–8
   red cell inclusions in   396, 397
   with beta thalassemia   398
Amino acid abbreviations   576
Amino acid deletions in hemoglobin
   genetic origin
      hemoglobin Freiburg   366, 371,
         383, 562–3
      hemoglobin Gun Hill   367, 371,
         383–4, 563
   non-genetic origin
      hemoglobin Koelliker   360, 365
Amino acid, N-terminal
   of α_1-acid glycoprotein   257
   of β-lipoprotein   177
   of globin chains   349, 350, 356
   of haptoglobin chains   76
   of immunoglobulin light chains   15
   of transferrin   128
Amino acid substitutions
   in α_1-acid glycoprotein   259
      in albumin   241
   in carbonic anhydrase   531–1
   in Gc protein   165
   in glucose-6-phosphate dehydrogen-
      ase   452–3
   in haptoglobin   75, 79
   in hemoglobin   363–85, 595–7
   in immunoglobulin   15, 21, 31
   in transferrin   140–2
Anhaptoglobinemia   (hypohapto-
      globinemia)   86–9

Anhaptoglobinemia   (hypohapto-
      globinemia)—cont.
   association with hemolytic states
      67–8, 86, 99
   association with sickle Hb trait   87
   definition of   86
   frequency in American Negroes   68,
      86
   frequency in children   65, 68, 86
   frequency in Nigerians   86
   genetic basis of   87
   Hp^0 allele and   88
   inheritance studies   85, 87–8
   relationship to Hp 2-1 (Mod)   88
Anomers   285, 577
Anti-A and anti-B
   as agglutinating antibodies   321
   as selective agents   314–7
   causing hemolytic disease of the
      newborn   314–5
   effect on response to Rh stimulus
      316
   in cervical secretions   316
   in serum   279
Anti-A_1
   failure to react with A_2 cells   279
   in absorbed B serum   279
   in *Dolichos biflorus* extract   279
   'pseudo' anti-A_1, 297
Anti-Ag   178–82, 185, 187, 590
Anti-Gc   164, 168
Anti-Gm   25, 34–8, 44
Anti-H   279, 300, 322
Anti-I and anti-i
   association with disease   296
   auto- and iso-antibodies   296, 298
   cause of red cell destruction   296
   light chain type of   14, 15
   sources of   296
Anti-Inv   38, 40, 44
Anti-Le^a and anti-Le^b
   failure of placental transfer   284
   occurrence during pregnancy   284
   reactions with A_1, A_2 and O cells
      282
   specificity of anti-Le^{bH} and -Le^{bL}
      284
   use in phenotyping   282, 284
Anti-Lp   182, 185–7
Anti-LW   304, 305
Anti-M and anti-N   301–3
Anti-pseudocholinesterase   201, 205

Anti-Rh
as cause of hemolytic disease of the newborn 304, 314–6
as cell 'coat' for Gm and Inv testing 23, 41
as suppressor of Rh antibody synthesis 315–6
formation suppressed in ABO incompatibility 315
Anti-S, anti-s and anti-U 302–3, 314
Anti-transferrin 130, 133, 140, 143
Antibodies (*see also* Immunoglobins, Anti-A, Anti-Gm, etc.)
absorption of 324
antigen-binding sites 17
complete and incomplete 41, 321, 322
definition 577
diversity of structure 4, 12–3, 22
labelled with radioisotopes 310
light chains of 14
metabolism of 9–11
neutralization of 25, 40, 323–4
of primitive vertebrates 20–1
production of specific 8–9
red cell agglutination by 321, 322
relative frequency in pregnancy 314–5
specificity related to IgG subclass 19, 20
structure of 4, 5, 11–22
suppression of synthesis 9, 315–6, 593
synthesis of 7–9
Antigens (*see* specific antigens and individual genetic systems)
Antiglobulin (Coombs) test 322–3
Arylesterase 195, 196
Atransferrinemia (hypotransferrinemia) 134–5
Auto-analyser for blood grouping 324
Autosomal allele exclusion (*see* Autosomal inactivation)
Autosomal inactivation 9, 28, 565, 601

Balanced polymorphism 404, 465, 573
Bence-Jones proteins (*see also* Light chains of immunoglobulins)
definition of 14, 577
idiotypy of 15

Bence-Jones proteins—*cont.*
Inv activity on 17, 40
Oz types of 39, 40, 588
source of 13
structure of 14–5
use in determining light chain structure 13–6, 77
Benzoylcholine
as substrate for pseudocholinesterase 193–4, 198–9, 216
in determining dibucaine number 202, 217
in determining fluoride number 218
Beta chain of haptoglobin (*see under* Haptoglobin)
Beta chain of hemoglobin
amino acid sequence of 348, 350, 356
compared with δ and γ chains 356
3-dimensional structure of 349–50, 353
evolution of 402–3
excess in α-thalassemia 396, 397
genetic linkage with β-thalassemia 399
genetic linkage with delta chain 399
genetic linkage with gamma chain 399, 595
homotetramer β_4 (Hb H) 360–1, 396–8
proximal and distal histidines of 351, 376–7
structural gene of 386
structural variants of 366–8, 595–7
sulfhydryl groups of 348, 361, 381, 385
synthesis in thalassemia 391–3
synthesis, normal 388–90
Beta, delta-thalassemia (Thalassemia F) (*see under* Thalassemia)
Beta lipoprotein
allotypes of 176–91
serum levels and Lp phenotypes 183
structure of 177–8
synthesis of 183
Beta lipoprotein allotypes (*see* Ag *and* Lp systems)
Beta-thalassemia
characteristics of 391–3
genetic mechanisms in 399–401
globin chain synthesis in 391–3
hematological changes in 391

Beta-thalassemia—*cont.*
  hemoglobin A$_2$ in   391
  hemoglobin F in 391–3
  heterogeneity of   391–2
  heterozygous states   391–2
  homozygous states   391–2
  ineffective erythropoiesis in   392
  red cell inclusions in   392
  red cell survival in   392
  selection and   404
Biological   applications   of   genetic
      markers   557–75, 600–1
Bisalbuminemia   (alloalbuminemia)
      240–7
Blood group antigens (*see also* specific
      antigens)
  as selective agents   314–7
  cross-reactivity with bacteria   270
  definition of   268, 270
  determinants of   268–71, 285–7, 289
  disease associations of   292, 296,
      313–7
  dosage effects and   276, 324
  inheritance of   276–8
  list of   269
  'public' and 'private'   268
  specificity on protein moiety   272,
      309–10
  specificity on sugar moiety   270,
      272, 285–90
Blood group chimerism   277, 568
Blood group substances (*see also* A, B
      and H substances *and* Lewis
      substances)
  as secreted glycoproteins   280, 284
  biosynthesis of   270, 288–90
  chemical structure of   284–90
  crossreactivity with bacterial anti-
      gens   270
  definition of   270
  enzymatic degradation of   288
Blood group systems (*see also* ABO,
      Rh, etc.)
  assignment of antigens to   268
  chromosomal location of loci   312
  gene frequencies   317–20
  genetic linkage   311–3, 515, 570
  list of systems   269
  nomenclature   272–6
  racial distribution of alleles   317–20
Blood   grouping   (typing)   methods
      320–4

Blue RR salt   533
Bohr effect
  definition of   577
  of haptoglobin-hemoglobin complex
      70
  of hemoglobin Chesapeake   378
  of hemoglobin homotetramers   360
'Bombay' phenotype   278, 282, 290–1,
      295
Bone marrow grafting   569
Brilliant cresyl blue (BCB)
  in screening for G6PD deficiency   466
  inclusion bodies   361, 396, 398
Buffer systems for electrophoresis
  acetate-EDTA, pH 4·95   246–7
  borate, pH 8·0   533
  borate, pH 8·5   102
  borate, pH 9·5   255
  formate, pH 5·0   439
  histidine - citrate,   discontinuous
      517–8
  lithium-borate-Tris,   discontinuous
      168, 234
  lithium-citrate-Tris,   discontinuous
      218
  phosphate, pH 6·2   438
    pH 7·0   438, 467, 523, 527, 547
    pH 7·4   217
  Tris-borate, discontinuous   102, 147
    acid modified   252
  Tris-citrate, pH 8·0   529
  Tris-citrate-borate, pH 9·4   545–6
  Tris-citrate-succinate, discontinuous
      438
  Tris-EDTA, pH 9·3   543
    pH 9·6   541
  Tris-EDTA-borate, pH 6·4   247
    pH 8·6   523
  Tris-maleate, pH 7·4   539
  Tris-maleate-EDTA, pH 7·4   508

Carbonic anhydrase
  amino acid substitution in   530–1
  deficiency of 531–2
    associated with elevated HbF   532
    associated with hyperthyroidism
      532
    in cord blood   531
  electrophoretic variants   530–1
    analysis of   530–1
    inheritance of   530

Carbonic anhydrase—*cont.*
  function　529
  in cord blood　531
  inhibitors of　530
  laboratory demonstration of　533
  molecular forms of　529–30
  molecular weight of　530
Carboxypeptidase A, effect on Hp-Hb
    binding　71
Catalase　543–5
  'acatalasia'　543–4
  electrophoretic variant of　545
  function　446, 543
  genetic linkage with haptoglobin?
    588
  inheritance　543, 544
  laboratory demonstration of　545–6
Cellular mosaicism　568–9, 582
  artificial　569–70
  blood group chimerism　277, 568
  generalized　86, 277, 568–9
Cellular origin of leukemia　567, 601
Cellular origin of tumors　462, 567
Ceruloplasmin　252–5
  as copper-binding protein　252
  biological half-life　253
  concentration in plasma　253
  genetic variants of 253–5, 592
    alleles　254–5
    electrophoretic patterns　253–5
    frequencies of alleles　254, 255,
     592
    laboratory demonstration of　255
  heterogeneity of　253
  oxidase activity of　252, 255
  physico-chemical properties　253
  subunit structure　253
Chimerism (*see* Cellular mosaicism)
Chromosomal assignment of loci (*see*
    also Genetic linkage)
  Duffy locus　312, 570
  G6PD locus　452, 459
  Haptoglobin locus　88–9, 570
  Kell locus　312
  Kidd locus　312
  Xg locus　312
  Xm locus　256
Clone (*see also* Monoclonal origin)
  clonal origin of Bence-Jones protein
    13
  clonal origin of myelocytic leukemia
    567, 601

Clone—*cont.*
  clonal origin of myeloma protein
    7, 13
  clonal origin of tumors　462, 567
  definition of　578
Codons (*see also* Genetic code)
  defined　578
  Hp and Ig light chain codons com-
    pared　77
  in globin chain synthesis　559–63
  of the genetic code　557–8
  sequence of Hp chain codons　81
Complement fixation, association with
  anti-H-like component　322
  anti-Lewis and anti-Kidd　323
  Fc fragment of immunoglobulins　11
  IgG subclass　19
  immunoglobulin class　5, 6
Complement, molecular variants of
  259, 593
Conalbumin　129
Congenital non-spherocytic anemia
  due to glucose-6-phosphate dehydro-
    genase deficiency　459–60
  due to glutathione deficiency　451
  due to glutathione reductase
    deficiency　540
  due to phosphohexose isomerase
    deficiency　528–9, 600
Controller genes (*see* Regulator genes)
Coombs test (*see* Antiglobulin test)
Cooley's anemia (*see* Beta thalassemia)
Crossing-over　578 (*see also* Intragenic
    crossing-over *and* Unequal
    crossing-over)
Cytochrome as hemoglobin precursor
  403

D_1 chromosome as possible site of Hp
  locus　88–9, 570
DPN (*see* Nicotinamide adenine dinucleotide)
Deletion (*see* Gene deletion)
'Deletion' genes
  at haptoglobin locus　75, 82
  at hemoglobin locus　563–4
Delta chain of hemoglobin (*see also*
  Hb A_2)
  amino acid sequence of　356
  compared with beta and gamma
    chains　356

Delta chain of hemoglobin—*cont.*
  evolution of 402
  genetic linkage with beta chain 399
  structural gene of 386
  structural variants of 369
  synthesis of 390, 394–5
Delta-thalassemia 395
Deoxyhemoglobin (reduced hemo-globin)
  and molecular respiration 353
  binding of 2,3-diphosphoglycerate 357
  exchange of oxygen by 354
  failure to bind haptoglobin 71, 360
  sickling tendency of reduced Hb S 372–3
Deoxyribonucleic acid (DNA)
  and the genetic code 557–8
  definition of 579
  in a genetic system 32
  in globin chain genes 398–400, 559–65
  in immunoglobulin chain genes 8, 22
  in mutation mechanisms 271, 559–65
o-dianisidine (Fast blue B) (*see under* Diazonium salts)
Diazonium salts for dye coupling
  Blue RR 533
  Fast blue B (*o*-dianisidine) 105, 218, 255, 540, 545–6
  Fast blue RR 218, 219, 234
  Fast blue VRT 234
  Fast red TR 214, 218
  Garnet GBC 234
  Variamine blue B 234
Dibucaine inhibition of pseudocholin-esterase 199, 200
Dibucaine number
  definition of 202
  laboratory determination of 216–8
  of pseudocholinesterase E_1 variants 203–4, 206–7
  of pseudocholinesterase E_2 variant 211
2,6 dichlorophenol indophenol (DCIP), 361, 541, 543
Diego blood group system 269, 276
Di-isopropyl-fluorophosphate (DFP)
  differential inhibition of pseudo-cholinesterase 199, 213

Di-isopropyl-fluorophosphate—*cont.*
  effect on carbonic anhydrase activity 530
  inhibition of esterases by 192, 199, 533
2,3-diphosphoglycerate 357
Disulfide bridges
  in albumin molecule 245
  in fibrinogen molecule 592
  in Gc molecule 165
  in haptoglobin molecules 76, 77
  in Hb Porto Alegre molecule 385
  in immunoglobulin molecules 4, 11–3, 15–8, 587
  in peptidase molecule 537
  in phosphohexose isomerase mole-cule 528
  in Rh antigen molecule 309
Dosage compensation 452, 566
Dosage effects 276, 324
Duffy blood group system
  antigens 269, 276
  association with Rh 312
  chromosome locus of 312
  Fy (a–b–) phenotype 278, 320
  inheritance of antigens 278
  linkage with cataract locus 311
Dupanol 214, 215
Duplication (*see* Gene duplication)

Electrophoresis in starch gel
  general methods 101–5
  methods to demonstrate polymorph-ism of
    acid phosphatase 438–9
    adenylate kinase 517–8
    albumin 246
    alkaline phosphatase 233–4
    carbonic anhydrase 533
    catalase 545–6
    ceruloplasmin 255
    esterases 535
    galactose-1-phosphate uridyl transferase 547
    Gc protein 168
    glucose-6-phosphate dehydrogen-ase 467–9
    glutathione reductase 541
    haptoglobin 101–6
    'indophenol oxidase' 549
    lactate dehydrogenase 523

Electrophoresis in starch gel—*cont.*
  methods to demonstrate polymorph-
      ism of—*cont.*
    malate dehydrogenase   527
    methemoglobin reductase   543
    peptidases   539–40
    phosphoglucomutase   508–9
    phosphogluconate dehydrogenase
      494
    phosphohexose isomerase   529
    protease inhibitor   252
    pseudocholinesterase   218–9
    transferrin   147–8
Embryonic   hemoglobins   358–60,
    386–9, 595
Enzymes (*see* individual enzymes by
    name)
Epistasis   282, 579
Epsilon chain of hemoglobin   358–60,
    386, 389, 595
Erythremia, due to abnormal hemo-
    globins   377–80
Eserine (physostigmine)   192,   196,
    199, 533
Esterases (red cell)   533–5
  electrophoretic patterns of   533–4
  genetic variants of   534–5
  isozyme classification   534
  laboratory demonstration of   535
Evolution
  of haptoglobins   79–84, 564–5
  of hemoglobins   401–4
  of immunoglobulins   20–2
  of proteins   21, 401–4
  of 'serine enzymes'   498

F-thalassemia (*see* βδ thalassemia)
Fab and Fab' fragments
  antigen-binding by   11
  chain contents of   12
  difference between   12–3, 18, 33
  fingerprints of   17
  Gm activity on   33
  Inv activity on   38
  molecular weight of   12
  variable portion of (*see* Fd fragment)
Fc fragment of IgG
  allotypic (Gm) sites on   29–31, 32–3
  functions of   11
  in 'heavy chain disease'   13
  isotypic sites on   19

Fc fragment of IgG—*cont.*
  location on IgG molecule   12, 18
  molecular weight of   12
  peptide mapping of   19, 34
  structure of   12, 13
  susceptibility to digestion   33
Fd fragment of IgG
  allotypic (Gm) sites on   29–31, 33,
      39
  heterogeneity of   17
  idiotypic sites on   17, 40
  in cross-reactivity with IgM   20
  isotypic sites on   19
  location on IgG molecule   12, 18
Fast blue B (*see o*-dianisidine)
Fast blue RR   218, 219, 234
Fast blue VRT   234
Fast red TR salt   214, 218
Fat ingestion, effect on plasma alkaline
      phosphatase   228–30, 293
Favism (*see under* Glucose-6-phosphate
      dehydrogenase, deficiency of)
Fetal hemoglobin (*see* Hemoglobin F)
Fibrinogen   259, 592
Fluoride inhibition of pseudocholin-
      esterase   199–200
Fluoride number
  definition of   203
  laboratory determination of   218
  of pseudocholinesterase E_1 variants
      203–4, 206–7
  of pseudocholinesterase E_2 variant
      211
2′-fucosyllactose secretion   286, 288

Galactose-1-phosphate   uridyl   trans-
      ferase   546–7
  absence in galactosemia   546
  Duarte variant   546, 547
  function of   546
  laboratory demonstration of   547
Galactosemia 546
Gamma chain of hemoglobin *see also*
      Hemoglobin F)
  amino acid sequence of   356, 357
  compared with β and δ chains   356
  evolution of   402
  genetic linkage with β chain   400,
      595
  homotetramer γ_4 (Bart's Hb)   360,
      389, 396

Gamma chain of hemoglobin—*cont.*
  in Hb F₁ 359
  non-allelic genes of 594
  structural locus of 369, 387, 594
  structural variants of 369, 387
  synthesis of, in hereditary persistence of HbF 394–5
  synthesis of, in thalassemia 39, 391, 393, 394
  synthesis of, normal 389–90, 391, 393
Gamma globulins (*see* Immunoglobulins)
Garnet GBC 234
Gc protein (group-specific component) 160–76
  alleles of 161–4
  as post-albumin 161
  concentration in serum 160, 162
  distribution in body fluids 160
  electrophoretic patterns of 160, 163, 164
  function of 160
  genetic linkage with albumin 161, 245, 571
  genetic types of
    chemical characteristics of 164, 165
    electrophoretic migration of 161–4
    geographic distribution of 165–7
    immunoelectrophoretic patterns of 160, 161
    inheritance of 161
    laboratory detection by electrophoresis 168–9
    laboratory detection by immuno-electrophoresis 167–8
    rare types 162–4, 589
  peptide fingerprints of 165
  physico-chemical characteristics of 160–1, 164–7
  'silent allele' of 164
  subunit structure of 164–5
  susceptibility to degradation 166, 167
  synthesis of 160
Gene deletion
  at hemoglobin β-δ locus 401
  definition of 578
  partial, at Hb β locus 383, 562–3

Gene duplication
  at blood group loci 277
  at Hb β-δ locus
    in Lepore Hb gene formation 363, 564–5
    in thalassemia gene formation 400
  at haptoglobin locus 79–84, 564–5
  at immunoglobulin loci 21, 22–3
  at PGM locus 599–600
  derivation of albumin and Gc genes by 246
  in evolution of hemoglobins 401–4
  in evolution of immunoglobulins 21, 588
  in evolution of proteins 21, 401–4
  in evolution of 'serine enzymes' 498
Gene frequencies (*see* Geographic distributions)
Gene triplication
  as source of *Hp¹* gene 84, 89, 90
  due to misplaced synapsis 83, 84
Genetic code 557–62 (*see also* Codons)
  abnormal hemoglobins and 559–62
  degeneracy of 558, 578
  table of codons 558
Genetic evolution (*see* Evolution)
Genetic linkage (*see also* Chromosomal assignment of loci *and* Linkage studies)
  autosomal linkage
    adenylate kinase and ABO 515, 570
    blood group linkages 311–2, 515, 570–1
    definition of linkage 581
    Gc and albumin linkage 161, 245–6, 571
    globin chain linkage 399, 595
    haptoblobin and catalase linkage 588
    non-linkage of PGM loci 502, 599–600
    PGM₁ and PTC taster linkage 501, 571
    Tf and E₁ linkage 139–40, 210, 571
  X chromosomal linkage
    Gd and colorblindness linkage 458, 463–4

Genetic linkage—*cont.*
 X chromosomal linkage—*cont.*
  Gd and hemophilia A linkage 464
   Xg linkages 312–3, 464, 570
   Xm linkage 592
Genetic markers
 definition 581
 uses 557–75, 600–1
Genetic polymorphism 572–3 (*see also* individual genetic markers)
Genetic system (*see also* individual systems)
 definition of 32, 579
 relative use in distinguishing individuals 571–2
Geographic distributions
 of acid phosphatase 434–7
 of adenylate kinase 515–7
 of Ag 181
 of albumin 241, 242
 of alpha₁-acid glycoprotein 258
 of blood group antigens 317–9
 of ceruloplasmin 254, 255
 of esterases 535
 of Gc 165–7
 of glucose-6-phosphate dehydrogenase 453–4, 456–7
 of Gm 28–32, 36–7
 of haptoglobin 93–100
 of Inv 40
 of Lp 184–5
 of peptidases 538–9
 of phosphoglucomutase 505–8
 of phosphogluconate dehydrogenase 490–3
 of phosphohexose isomerase 528
 of protease inhibitor 251
 of pseudocholinesterase 207–10, 212
 of transferrin 144–6
 of Xm 257
Globin chains (*see also* Hemoglobin *and* the individual chains of hemoglobin)
 as αβ dimer 353, 361, 388
 combination with heme 388
 electrophoresis of chains 380
 interchain contacts 353, 378, 379, 594–5
 synthesis of 388

Glucose-6-phosphate dehydrogenase 444–82
 A and A- enzymes compared 453
 A and B enzymes compared 452, 458
 A- and B- enzymes compared 458, 598
 B and B- enzymes compared 458
 cellular mosaicism in females 460–3
 detection by staining 461
  detection in cell culture 462–3
  detection in sickle trait 461
 chromatography of 448, 453, 458
 deficiency of G6PD
  clinical course 450
  comparison in Negroes and 'non-Negroes' 454, 455, 464
  diseases causing hemolysis in 452
  drugs causing hemolysis in 452, 454–5
  effect of infection 452
  effect on acid phosphatase 454
  favism in 444, 454–5
  frequency of 444, 459
  geographic distribution of 453–4
  glucose tolerance in 448
  glutathione stability in 447, 451, 452
  Heinz body formation in 451
  in infants 455
  in malarious regions 465
  in Negroes 450–3, 454–7
  in 'non-Negroes' 453–8
  in red cells and white cells 454
  in twins 462
  inheritance of 452, 459
  laboratory tests for 466–9
  methemoglobin reduction in 447, 451
  red cell destruction in 450, 451–2, 459
  with congenital anemia 459–60
  X-linkage of 452, 459, 463–4
 dosage compensation 452, 566
 electrophoretic patterns of G6PD
  in red *vs.* white cells 447–8, 453
  in tumor cells 567
 enzyme activity
  dependence on coenzyme 449
  effect of red cell age 448, 450, 453, 460
  in males and females 452
  in metabolic disorders 448

Glucose-6-phosphate dehydrogenase
—*cont.*
  enzyme activity—*cont.*
    in X chromosome disorders 461
    measurement of 466–7
    variation within phenotype 458, 464
  function in glycolysis 445–6
  genetic linkage of G6PD locus 458, 463–4
    non-linkage with Xg^a 464
    with color blindness 458, 463–4
    with hemophilia 464
  genetic variants of G6PD
    evidence for allelism 458, 463
    geographic distribution 453, 454, 456–7, 598
    inheritance 452
    list of 456–7, 598
    physico-chemical characteristics 456–7, 458
    separation by electrophoresis 450, 452, 459, 467–9
    with increased enzyme activity 456, 459
    X-linkage 452, 459, 463–4
  heterogeneity of gene expression 458, 464
  hybridization of A and B enzymes 462
  in white cells 447, 448, 453, 454
  inactivation by stromal NADPase 449
  indicator of cell culture contamination 570
  indicator of monoclonal origin 567
  indicator of X chromosome inactivation time 463
  instability of A- and B- enzymes 453, 458, 598
  laboratory tests for G6PD 466–9
    electrophoretic separation of variants 467–9
    measurement of enzyme activity 466–7
    'screening' for deficiency 466
  locus name 458
  Lyon hypothesis and 460–3, 566
  molecular structure of 444, 448–9
  molecular weight of 449
  natural selection and 444, 465, 573
  peptide fingerprints of 452–3

Glucose-6-phosphate dehydrogenase
—*cont.*
  'primaquine sensitivity' 444, 445, 450–1
  staining of, in red cells 598
  tissue distribution of 447, 448, 454
Glucose phosphate isomerase (*see* Phosphohexose isomerase)
Glutathione, oxidized (GSSG)
  effect on acid phosphatase 431
  increase in G6PD deficiency 431, 451
  influence on glycolytic rate 446, 447
  reduction by glutathione reductase 445, 446, 540–1
Glutathione peroxidase 446, 447, 544
Glutathione, reduced (GSH)
  binding to HbA 359
  binding to HbH 361
  in G6PD deficiency 451–2
    effect of acetylphenylhydrazine 451
    effect of primaquine 451, 452
  influence on glycolytic rate 446, 447
  oxidation to GSSG by peroxidase 446, 447
  role in maintaining reduced thiols 451
  stability in G6PD deficiency 451–2, 461
  in 6-PGD deficiency 489
Glutathione reductase
  function 445, 446, 540–1
  genetic variation 540–1
  laboratory demonstration 541
Glutamic oxaloacetic acid transaminase 526
Glutathione synthesis 444
Glycolysis in red cells 444–7
Gm allotypes 23–38
  assignment to IgG subclass 28–31
  comparison of Gm and Rh inheritance 27
  dependence on molecular conformation 32, 33, 34
  determination of 24–6, 31, 40–4
  effect of enzyme digestion on 33
  genetic linkage 27
  genetic theories of inheritance 31–3
  geographic distribution 36–7
  heterogeneity in single serum specimen 27, 28

Gm allotypes—*cont.*
  influence on subclass concentrations 33
  inheritance of 24, 26–33
  list of antigens 24
  laboratory methods 40–44
  location on γG chain 25, 28–31, 38
  nomenclature 24–5
  of myeloma proteins 28, 29
  racial distribution 28–32
  restoration by light chains 33–4
  unusual alleles 29, 36, 37
Gm(1) and Gm(–1) peptides 34
Group-specific component (*see* Gc protein)

H and *h* genes
  absence of *H* gene in 'Bombay' phenotype 289, 291
  function in H substance biosynthesis 288–91, 295
Haptoglobin 63–125
  alpha chains of 74–6, 85, 89
    homologies with Ig light chain 77
    hp α^{1F} compared with hp α^{1S} 74, 75
    hp α^1 compared with hp α^2 75, 76
    molecular weight of 75
    peptide fingerprints of 75, 91
    rare variants of 89
    structure of 75–6, 80
    variation in common types 74
  alleles of 72–4, 79–85, 88–92
  antigenic sites of 78–9
  beta chains of 74, 76, 79, 87, 91–3
    as binding site of hemoglobin 79
    molecular weight of 75
    variation in rare types 91–3
  binding site for hemoglobin 78, 79, 588–9
  biological role of 68–70, 589
    in heme catabolism 68, 69, 70
    in iron conservation 68, 69
    in preventing methemalbumin formation 69
  carbohydrate moiety of 72, 73–4, 76
  catabolism of 66–8
  combination with hemoglobin (*see also* HpHb complex)

Haptoglobin—*cont.*
  combination with hemoglobin—*cont.*
    dependence on Hb conformation 71
    Hb binding by Hp polymers 77–8
    molecular ratio of complexes 72, 77, 78
    nature of intermolecular bond 71
    sites of binding 79, 588–9
  concentration of, in serum 65–8 (*see also* Anhaptoglobinemia)
    as indicator of hemolysis 66, 67–8
    association with phenotype 78
    decrease in hemolytic states 66, 67–8
    effects of liver disease on 65
    in infants 65–6
    in pregnancy 87
    in twins 66
    increase in disease 65
    influence of age on 87
    influence of hormones on 65, 87
    normal range of 66
  depolymerization of 76
  determination of phenotypes 101–6
  electrophoretic patterns of 73, 81–3, 90
  genetics of 79–93, 564–5
  geographic distribution of genes 93–100
  hemoglobin binding of 70–2, 79
  inheritance of phenotypes 72, 84
  laboratory procedures 100–7
  linkage with catalase? 588
  locus on D_1 chromosome? 88–9, 570
  mechanism of Hp^2 formation 79–81, 564–5
  metabolism of 66–8
  molecular structure of 73–6, 565–6
  molecular weight of 76, 77
  phenotypes of (*see also* Haptoglobin, variants)
    Hp 1-1 72–9, 88
    Hp 2-2 72–9, 85, 86, 88
    Hp 2-1 72–9, 82–6, 89
    Hp 2-1 Bellevue 90–1
    Hp 2-1 D 90, 91
    Hp 2-1 (Haw) 86, 88
    Hp 2-1 Mb and 1-1 Mb 90, 91

Haptoglobin—*cont.*
  phenotypes of—*cont.*
    Hp 2-1 (Mod) 84–5, 87, 88
    Hp 2-1 (trans) 86
    Hp 1-B and 2-B 90
    Hp 1-P and 2-P 90, 92, 93
    Hp 1-H, 2-H, and 2-L 90, 92, 93
    Hp Ab 90, 93
    Hp Ca (Carlberg) 85–6, 87, 89
    Hp Johnson (1J and 2J) 84, 89, 90
  phenotypic patterns in cellular mosaicism 86, 569
  polymerization of 73, 76–7
  presence in body fluids 66
  quantitative variants of 84–9
  selective factors and 82, 83, 93
  'silent allele' of 82, 88, 89
  similarity to immunoglobulins 76, 77, 565
  structure of molecules 72–6, 565–6
    dissociation into subunits 74
    glycopeptides 73, 74
    polymeric forms 76
  subtypes of 75, 82–6, 89, 93
    determination of 100, 101
    molecular basis of subtyping 74, 75
  synthesis of 64–5
    effect of hemoglobin injection on 65
    effect of inflammation on 65
    site of 64
  variants of (*see also* Haptoglobin, phenotypes of)
    'qualitative' variants 89–93
    'quantitative' variants 84–9
Haptoglobin-hemoglobin complex 70–2 (*see also* Haptoglobin, combination with hemoglobin)
  'intermediate' complex 71, 72
  nature of bonding 71, 588–9
  physico-chemical properties 70–1
  rate of removal from blood 67
  saturated complex 72
  sites of removal from blood 68
Heavy chain disease 13
Heavy chains of immunoglobulins (*see also* Immunoglobulins, Fc fragment, Fd fragment *and* Myeloma proteins)
  amino acid sequence of 17, 34, 35

Heavy chains of immunoglobulins —*cont.*
  C-terminus of myeloma heavy chains 34–5
  classes of 4, 5, 20, 587
  common regions of 20
  cross reactivity of 20
  disulfide bridges of 11, 18, 587
  electrophoretic mobility of 17
  evolution of 21
  fragments of (*see* Fc *and* Fd fragments)
  Gm determinant on γ chain 25
  Gm specificity of γ chain subclasses 28–31
  'hinge region' of 18
  homologies with light chains 21
  in Fab and F(ab')$_2$ fragments 12
  in molecular formulae 5, 11
  inter-heavy chain bonds 11, 587
  invariant portion of 17
  isotypy of 18
  molecular weight of 5
  peptide mapping of 34
  sequences of classes compared 21–2
  subclasses of IgA heavy chain 20, 587
  subclasses of IgG heavy chain 18–20, 28–31, 34
  synthesis of 7–9, 20
  variable portion of (*see* Fd fragment)
Heinz inclusion bodies 380–5, 392, 451
HeLa cells 570, 580
Helix distortion by proline substitution 382, 385, 595
Heme
  as prosthetic group of catalase 543
  as prosthetic group of hemoglobin 349, 351
  atomic interaction with globin chains 354, 594–5
  catabolism of 70
  combination with globin 388, 389
  exchange between albumin and methemoglobin 69–70
  'heme-heme' interaction 353, 355, 378, 383, 402
  linkage to globin chain histidines 349, 351, 376, 377
  location in hemoglobin molecule 349, 351, 352, 371

Heme—*cont.*
  loss from abnormal hemoglobins 377, 381, 383
  staining of 106
  structure of 351
  synthesis of 399
    feedback control 399
    in thalassemia 399
Heme-α-methyl oxygenase 70, 82
Hemoglobin 347–423 (*see also* individual hemoglobins)
  abnormal (*see* Abnormal hemoglobins)
  allosterism of 353–5
  as an enzyme 353
  dissociation into subunits 353–5, 360, 378
  effect of hemolysis on 360
  exchange of subunits 355
  functional unit of 354, 594
  general features of 347–53
  homotetramers of 360–1, 396, 580
  hybrid molecules of 355, 379, 380, 387
  inheritance of 386–8
  interchain contacts of 353, 378, 379, 594–5
  invariant residues of 379, 382, 403, 404
  molecular dimensions of 352
  nomenclature of 370
  oxygenation of 353 (*see also* Oxygen dissociation curve)
  polypeptide chains of (*see* Alpha *and* Beta chains of hemoglobin)
  subunit structure of 347–53, 565, 594
  synthesis
    during fetal life 389–90
    in thalassemia 390–401
    of abnormal chains 387, 390
    of normal chains 388–90
  thalassemia (*see* α and β thalassemia)
Hemoglobin A
  globin chains of 347–53
  heterogeneity of 359–60
  structure of 347–53
  synthesis of 388–9, 390–401
Hemoglobin A$_{Ic}$ and A$_{Id}$ 359
Hemoglobin A$_2$ 355–6
  electrophoretic migration 356
  in normal blood 355, 390

Hemoglobin A$_2$—*cont.*
  in thalassemia 391–5, 396, 398
  molecular formula 355
  oxygen dissociation curve 356
Hemoglobin A$_3$ 359
Hemoglobin Bart's (γ4) 360
  failure to combine with haptoglobin 71, 360
  in alpha thalassemia 396
  in fetal blood 360, 389
  physiological properties of 360, 396
  structure of 360
Hemoglobin Beilinson (L-Ferrara) 364, 384
Hemoglobin Bibba 365, 371, 384–5
Hemoglobin C 366, 370, 371, 375, 404, 405
Hemoglobin C$_{Georgetown}$ and C$_{Harlem}$ 366, 371, 374, 595
Hemoglobin Chesapeake 365, 371, 377–8
Hemoglobin D$_{Punjab}$ 368, 374, 375, 405
Hemoglobin E 366, 371, 375, 399, 405
Hemoglobin F 356–8
  Hb F thalassemia (*see* Thalassemia)
  heterogeneity of 359, 594
  in acquired hematological diseases 358
  in hereditary persistence of fetal Hb 358, 394–5
  in thalassemia 358, 391–4, 396, 398
  in thyrotoxicosis 532
  intracellular distribution of 373, 392
  molecular formula of 357
  oxygen affinity 357
  physiological properties of 357
  protective effect of 373, 392
  resistance to denaturation 357
  structure of 357
  synthesis of 390
Hemoglobin F$_I$ 359
Hemoglobin Freiburg 366, 370–1, 383, 562–3
Hemoglobin Genova 366, 371, 381–2
Hemoglobin G-Philadelphia 365, 374
Hemoglobin Gower-1 and Gower-2 358, 360, 386, 387, 389
Hemoglobin Gun Hill 367, 371, 383–4, 563

Hemoglobin H (β_4) 360–1, 396–8
  failure to combine with haptoglobin 71, 360
  in acquired diseases 361
  in alpha thalassemia 361, 396–8
  inclusion bodies 361
  instability of 361
  physicochemical properties 71, 361
  physiological properties 360, 396
  removal by spleen 361
  structure of 360
  synthesis of 361
Hemoglobin H disease 361, 381, 397–9
Hemoglobin Hammersmith 367, 371, 382, 385
Hemoglobin I 364, 374–5, 380, 562
Hemoglobin J$_{Capetown}$ 365, 371, 378
Hemoglobin Kansas 368, 371
Hemoglobin Kempsey 368, 371, 379
Hemoglobin Koelliker 360, 365
Hemoglobin Köln 367, 370, 371, 381–2
Hemoglobin Korle-Bu 595, 596
Hemoglobin Lepore 362–3, 390, 393, 563
  clinical states 390, 393–4
  mechanism of gene formation 277, 363, 563–4
  structure of 362
Hemoglobins M 371, 375–7, 382
  amino acid substitutions in 365, 367, 376
  as cause of methemoglobinemia 376–7
  characteristics of 376–7
Hemoglobin New York 368, 370
Hemoglobin Philly 595, 596
Hemoglobin Portland-1 358–60
Hemoglobin Porto Alegre 245, 366, 385
Hemoglobin Pylos (see Hb Lepore)
Hemoglobin Rainier 368, 371, 594
Hemoglobin S
  effect of methemoglobin on sickling 461
  effect on cell shape 372–3
  in cellular mosaicism 569
  interaction with other hemoglobins 373–4
  molecular stacking of 373

Hemoglobin S—cont.
  point mutation in allele 559
  selection and 404–5, 573
Hemoglobin Sabine 595, 596
Hemoglobin Santa Ana 595, 596
Hemoglobin Seattle 367, 384
Hemoglobin Shimonoseki 365, 562
Hemoglobin Sinai (Sealy, Hasharon) 364, 371, 384, 597
Hemoglobin St. Mary's 384
Hemoglobin Sydney 367, 371, 382
Hemoglobin Tacoma 366, 371, 384
Hemoglobin Torino 364, 371, 385
Hemoglobin Ube-1 384
Hemoglobin Ube-2 374
Hemoglobin Wien 595, 597
Hemoglobin Yakima 368, 371, 378–9
Hemoglobin Ypsi 355, 379–80
Hemoglobin Zürich 367, 371, 380–2, 562
Hemoglobin-haptoglobin complex (see Haptoglobin-hemoglobin complex)
Hemolytic disease of the newborn, immune origin
  blood group incompatibility in 284, 304, 314–6
  due to IgG class antibodies 6
  mimicked by alpha thalassemia 396
  mimicked by G6PD deficiency 455, 458
  prevention of 315–6, 593
Hemopexin 69
Hereditary persistence of fetal hemoglobin
  accompanied by α or β thalassemia 394, 398
  analogy with i antigen persistence 297
  and regulator gene hypothesis 400
  as a form of thalassemia 390
  characteristics of 394–5
  with decreased carbonic anhydrase 532
Heterogametic sex 311, 580
Hexose monophosphate shunt pathway
  effect of endogenous agents on 452, 459
  effect of glutathione oxidation on 446, 447
  effect of redox catalysts on 447, 451, 459

Hexose monophosphate shunt pathway
    —*cont.*
  enzymatic steps of   445, 446, 483
  methemoglobin reduction via   446,
    447, 451
'High hemoglobin F' syndrome (*see*
    Hereditary persistence of fetal
    hemoglobin)
'Hinge' region of IgG   18, 77, 587
Histidines, proximal and distal
  as site of globin-heme binding   349,
    351, 388
  effect of nearby amino acid substitu-
    tion   377, 378, 381
  peptide bond alteration in Hb Gun
    Hill   383
  substituted in four M hemoglobins
    376–7
  substituted in Hb Zürich   380
Homoalleles   32
Homotetramers of hemoglobin
    360–1, 396, 580
Hp 2-1 (mod) phenotypes
  hetereogeneity of   84, 85
  inheritance of   84, 85
Hyperthyroidism
  glucose-6-P dehydrogenase levels in
    448
  red cell carbonic anhydrase in   532

Idiotypy   7, 15, 17, 580
IgA, IgD, IgE, IgG, IgM (*see* Immuno-
    globulins, classes of)
Ii system
  antibodies (*see* Anti-I and anti-i)
  antigens, I and i
    activity associated with age   296
    association with ABH antigens
      297
    heterogeneity of   296–7
    inheritance of   297
    in leukemia   298
    in marrow stress   297, 306
    on cultured tissue   300
    reciprocal relationship of   296–7
Immunodiffusion method   186–7
Immunoelectrophoresis for Gc typing
    167–8

Immunoglobulins (*see also* Antibodies)
  antigenic specificities
    allotypic   7, 17
    idiotypic   7, 15, 17
    isotypic   7, 13, 18, 19, 588
  carbohydrate moiety of   6
  catabolism of IgG   10–1
  classes of   4, 5
    comparison of   4–6, 21–2
    concentration in plasma   5, 7
    distribution in fluid compart-
      ments   5, 10
    functional differences of   5–6
    heavy chain differences of   5, 20
    metabolism of   5, 9–11, 587
    physicochemical   properties   of
      4–6
  diagram of IgG structure   12
  enzyme cleavage products of   11, 12,
    13, 18
  fetal synthesis of   7
  genetic basis of variability   22–3
  general properties of   4–6
  metabolism of   9–11
  polypeptide chain structure   12–22
    (*see also* Heavy chains *and*
    Light chains)
    structural interdependence of H
    and L chains   17
  structure of   4, 5, 11–22, 565–6, 587
  susceptibility to sulfhydryl reagents
    6
  subclasses of γA and γM chains   20,
    587
  subclasses of   γG chains   18–20,
    28–31
    antibody   specificity   differences
      19, 20
    C-terminal sequences of   34, 35
    concentration in plasma   18, 28
    functional differences of   19
    Gm specificity of   28–31
    nomenclature   18
    structural similarities of   18, 32,
      34
  synthesis   7–9
    feedback effect of IgG   9
  turnover   9–11
'Indophenol oxidase'   547–9
  achromatic zones of   547–8
  inherited variant of   548–9
  laboratory demonstration of   549

'Indophenol oxidase'—*cont.*
  non-identity with other enzymes 548
Ineffective erythropoiesis 67, 392, 581
Intra-genic crossing-over
  at globin chain loci 562–4, 595
  at haptoglobin α chain locus 79–84, 564–5
  at immunoglobulin gene loci 23
Inv allotypes of κ chains 38–40
  absence from γG2 and γG4 40
  assignment to kappa chains 17, 38
  associated amino acid substitution 39, 40
  dependence on chain interaction 40
  effect of heavy chain on 40
  geographic distribution of 40
  inheritance of 38
  laboratory determination of 40–4
Invariant residues of globin chains 379, 382, 403–4, 594
Iron absorption 133, 135
Iron stain 148
Iron turnover 132–4, 135
Isotypy 7, 13, 18, 19, 581
Isozymes (Isoenzymes)
  definition of 581
  electrophoretic pattern of
    acid phosphatase 426–8
    adenosine deaminase 600
    adenylate kinase 513–4
    alkaline phosphatase 227–8
    carbonic anhydrase 530–1
    catalase 545
    esterases (red cell) 533–5
    galactose-1-phosphate uridyl transferase 547
    glucose-6-phosphate dehydrogenase 450–3, 456–7
    glutathione reductase 540–1
    'indophenol oxidase' 548–9
    lactate dehydrogenase 521–2
    malate dehydrogenase 524–6
    methemoglobin reductases 542–3
    peptidases 537–8
    phosphoglucomutases 498–505
    6-phosphogluconate dehydrogenase 485–9
    phosphohexose isomerase 528–9
    pseudocholinesterases 196, 205, 210–1

Kappa chains (*see under* Light chains of immunoglobulins)
Kell blood group system
  antibodies 314, 315
  antigens 269, 271, 275
  chromosome locus of 312
  inheritance of antigens 271
  nomenclature 275, 318
Kidd blood group system 269, 276, 312, 314
Klinefelter's syndrome 313

Laboratory methods (*see under* individual genetic systems)
Lactate dehydrogenase 521–3
  electrophoretic pattern of 521, 522
  function of 521
  genetic variants of 522–3, 565
    gene frequencies 523
    linkage studies 523
  laboratory demonstration of 523
  subunit structure of 521, 565
Lambda chains (*see* Light chains of immunoglobulins)
Lectins
  definition of 581
  from *Dolichos biflorus* 279
  from *Lotus tetragonolobus* 279
  from *Ulex europeus* 279
  from *Vicia graminea* 301
  use in blood grouping 321
Lepore hemoglobin (*see* Hemoglobin Lepore)
Lepore thalassemia 391, 393–4
Leukemia
  ABO phenotypic changes in 292
  clonal origin of 567, 601
Lewis antigens: Le^a and Le^b
  biosynthesis of Lewis substances 288–90
  carbohydrate specificity 270, 284–90
  expression during infancy 282, 284
  expression in 'Bombay' phenotype 282, 291
  expression influenced by ABO type 282
  influence of sex hormones and pregnancy 284
  masking effect of A, B, and H 288

Lewis antigens: Leᵃ and Leᵇ—*cont.*
  secretion of Lewis substance 281–4, 288–91
Lewis substance (*see also* Lewis antigens)
  as glycoprotein molecule 281, 284–90
  biosynthesis of 288–90
  secretion of 282–4, 291
Lewis system 281–4 (*see also* Lewis antigens *and* Lewis substance)
  adsorption of antigens to red cells 270, 281, 291
  antibodies (*see* Anti-Leᵃ and anti-Leᵇ)
  antigens (*see* Lewis antigens)
  biosynthesis of Lewis substances 288–90
  developmental changes in phenotype 282, 297
  epistasis with ABO locus 282
  expression in 'Bombay' phenotype 282, 289, 291
  genes: *Le* and *le* 282–4, 286–91
  genetic interaction with ABO system 282, 283, 289
  genetic interaction with Hh system 282, 283, 289–91
  genetic interaction with secretor system 282, 283, 286, 289
  phenotypes of 228–30, 281–4
    association with alkaline phosphatase 228–30
    association with secretor types 282, 283
    changes during development 282
    changes following transfusion 281–2
Light chains of immunoglobulins (*see also* Bence Jones proteins *and* Immunoglobulins)
  amino acid sequence of 15–7
  compared with haptoglobin light chain 77
  diagram of 16
  disulfide bridges in 15, 16
  electrophoretic patterns of 14
  evolution of 21
  Gm activity restored by 33–4
  homologies of κ and λ chains 15, 16
  idiotypy of 15
  in molecular formulae 5, 11

Light chains of immunoglobulins —*cont.*
  in specific antibodies 14–5
  Inter- and intra-species comparison of 16
  linkage groups in 23
  molecular weight of 4
  ratio of κ to λ chains 13
  site of Inv determinant on κ chain 17, 38, 39
  site of Oz determinant on λ chain 17, 39
  synthesis of 8, 9
  types in cold agglutinins 14, 15
  variable portion of 15–7, 39
'Linkage groups' in immunoglobulins 23
Linkage studies (*see also* Genetic linkage)
  of acid phosphatase 426
  of adenylate kinase 515
  of alkaline phosphatase 231
  of blood groups 310–3
  of E_1 pseudocholinesterase 210
  of E_2 pseudocholinesterase 211
  of Gc 161, 246, 589, 592
  of glucose-6-phosphate dehydrogenase 463–4
  of Gm 27
  of haptoglobin, 88–9, 92, 588
  of hemoglobin, 386, 595
  of lactate dehydrogenase 523
  of Lp lipoprotein 182
  of phosphoglucomutase 501, 502, 599
  of transferrin 139, 210
  of Xm 592
Linked gene theory
  in blood group genetics 271–2, 273–5
  in immunoglobulin genetics 31–3
Lipoprotein molecular classes 176
Liver
  clearance of excess iron by 133
  haptoglobin levels in diseases of 65, 86
  site of synthesis of
    Gc protein 160
    haptoglobin 64
    pseudocholinesterase 193
    transferrin 131

Locus assignments (*see* Chromosomal assignment of loci)
Lp (lipoprotein) system 182–5
  Lp antigens 182–3, 184
  Lp genetics 182
  Lp phenotypes 182
    association with disease 183
    association with lipoprotein level 183
    geographic distribution 184–5
    heterogeneity of 183, 184
    laboratory determination of 186–7
  preparation of anti-Lp sera 186, 187
Lutheran blood group system 269, 276, 278, 310
LW antigen 304–9
  absence in rare phenotypes 304, 305
  relationship with Rh antigens 304, 308, 309
LW and *lw* genes
  inheritance of 305
  low frequency of *lw* 305
  rare phenotypes: LW negative 305–9
Lyon hypothesis (*see* X chromosome inactivation)

Macroglobulinemia of Waldenstrom 13
Malaria
  and acid phosphatase types 430
  and G6PD deficiency 465
  and sickle hemoglobin 404, 573
Malate dehydrogenase, (MDH) 524–7
  cytoplasmic MDH 524–5
    electrophoretic pattern of 524
    inherited variant of 525
    laboratory demonstration of 527
  mitochondrial MDH 525–7
    electrophoretic pattern 526
    inherited variant of 525, 526
    laboratory demonstration 527
'Master' gene 23
Medico-legal uses of genetic markers 571–2
Methemalbumin 68, 69, 106
Methemoglobin (ferrihemoglobin)
  and heme transfer 69

Methemoglobin—*cont.*
  binding to haptoglobin 70, 71
  failure of sickling by methemoglobin S 461
  formation by M hemoglobins 375–7
  formation by unstable hemoglobins 381–8
  reduction of, by NADH diaphorase 445–6, 542–3
  reduction of, by NADPH diaphorase 447, 542–3
  reduction test in G6PD deficiency 447, 451, 461, 466
  reduction test in PGD deficiency 489
  staining of, in red cells 461, 544
Methemoglobin reductase (NADH and NADPH diaphorases)
  deficiency as cause of methemoglobinemia 542
  function of 445–7, 542
  inherited variation of 542
  laboratory demonstration of 543
Methemoglobinemia
  due to abnormal hemoglobins 375–7
  due to NADH diaphorase deficiency 542
'Minus-minus' blood group phenotypes 278, 308, 320
Mitochondrial enzymes
  glutamic oxaloacetic acid transaminase 526
  malate dehydrogenase 525
Mixed tetramers of hemoglobin 355, 359, 279–80, 387
MNSs system
  antibodies (*see* Anti-M and anti-N)
  inheritance of antigens 271, 301
  M and N antigens 269, 301–3, 319, 320
    carbohydrate specificity 270, 302, 303
    chemical structure of 302
    complexity of 271, 319–20
    cross reactivity of 301, 303
    myxovirus receptor activity of 302
    nomenclature of 272, 320

MNSs system—*cont.*
S, s and U antigens 302–3
activity on Rh_{null} cells 303, 306, 308
S^u allele 302
U-negative cells; lack of MN crossreaction 303, 306, 308
Monoclonal origin (*see also* Clone)
as determined by G6PD phenotype 567
of Bence Jones proteins 13
of myelocytic leukemia cells 567, 601
of myeloma cells 7, 13
of various tumors 462, 567
Mosaicism (*see* Cellular mosaicism)
MTT tetrazolium salt
mode of action in electron transfer 468, 518, 582
use in staining isozymes of
adenylate kinase 518
glucose-6-phosphate dehydrogenase 468
glutathione reductase 541
'indophenol oxidase' 549
lactate dehydrogenase 523
malate dehydrogease 527
methemoglobin reductase 543
phosphoglucomutase 509
phosphogluconate dehydrogenase 494
phosphohexose isomerase 529
Multiple allele theory
in blood group genetics 271–2, 273–5
in immunoglobulin genetics 31–3
Mutation mechanisms 557–65
hybrid molecule formation (*see also* Gene duplication *and* Unequal crossing-over)
of blood group genes 277
of haptoglobin genes 79–84, 564–5
of hemoglobin genes 363, 400, 564–5
of immunoglobulin genes 21–3
partial deletion (*see also* Gene deletion)
of hemoglobin genes 383, 562–3
point mutation (*see also* Amino acid substitution)
in carbonic anhydrase gene 530–1

Mutation mechanisms—*cont.*
point mutation—*cont.*
in G6PD gene 452–3
in haptoglobin gene 75, 79
in hemoglobin genes 363–85, 559–62, 595–7
in immunoglobulin genes 15, 21, 31
in transferrin gene 140–2
Myeloma proteins (*see also* Immunoglobulins, Heavy *and* Light chains of immunoglobulins)
as monoclonal products 7, 13
C-terminal amino acid sequence of 34–5
correspondence of Bence Jones type with 14
Gm determinants on 25, 28–31
idiotypy of light chains 15
Inv activity of 40
peptide maps of components 34
Myoglobin 70, 348, 360

'Nadi reaction' 549
Neuraminidase, effect on
alkaline phosphatase 227
$alpha_1$-acid glycoprotein 167, 258
Gc protein 161, 167
haptoglobin 70, 167
M and N blood group antigens 301, 302
pseudocholinesterase 195, 196
transferrin 140, 143, 167
Nicotinamide adenine dinucleotide (NAD or DPN)
diaphorase, in methemoglobin reduction 446, 542
reduction of, during anaerobic glycolysis 445–6
use in LDH and MDH staining 523, 527
Nicotinamide adenine dinucleotide phosphate (NADP or TPN)
diaphorase, in methemoglobin reduction 447, 542
effect on PGD isozyme patterns 486–7, 488
in stains for enzyme activity (*see under* Staining methods)
reduction of, by shunt pathway 446–50, 466–7, 483

Non-covalent bonding of Ig chains
      11, 565
Non-paternity determination   571

Operon   399, 400, 583
Orosomucoid (*see* α_1-acid glycoprotein)
Oxygen affinity of hemoglobin (*see also*
      Oxygen dissociation curve)
   hemoglobin   with   decreased
      affinity   380
   hemoglobins with increased affinity
      360, 377–80, 383
Oxygen dissociation curve of hemo-
      globin
   associated molecular configuration
      353, 354
   definition of   583
   effect of 2,3 diphosphoglycerate   357
   hemoglobin H and Bart's   360, 396
   in haptoglobin-hemoglobin complex
      70
   in   structurally   abnormal   hemo-
      globins   377–80, 383
   normal sigmoid shape   353
Oz types of λ chains   17, 39–40, 588

P system
   antibody   specificity in cold hemo-
      globinuria   300
   antigens P, P_1 and P^k 269, 299–301
      biosynthesis of   299–300
      carbohydrate   specificity of   270,
         300
      in hydatid cyst fluid   300
      on cultured tissues   300
      relationship   with   ABH   and   I
         antigens   300–1
Pentose phosphate shunt pathway (*see*
      Hexose monophosphate shunt
      pathway)
Peptidases   535–40
   electrophoretic patterns of   537, 538
   genetic variation of   537–9
      A peptidase variants   537–9, 600
      B peptidase variants   537, 539
      inheritance of   538, 539
   laboratory   demonstration of   539–
      40
   racial distribution of variants   538,
      539
   substrates of   535, 536

Peptide fingerprints
   of alpha globin chain   392
   of carbonic anhydrase   529
   of Gc protein   165
   of glucose-6-phosphate dehydrogen-
      ase   452, 453
   of haptoglobin chains   75
   of immunoglobulin chains   18, 34
   of transferrin   128, 140, 141
Phagosome attachment of Fc fragment
      11
Phenazine methosulfate (PMS)
   description of   548
   use in electron transport (*see under*
      Staining   methods   of   G6PD,
      LDH, etc.)
L-Phenylalanine inhibition of intestinal
      alkaline phosphatase   227, 228,
      235
Philadelphia chromosome   567, 599
Phosphatase, acid (*see* Acid phospha-
      tase)
Phosphatase, alkaline (*see* Alkaline
      phosphatase)
Phosphoglucomutase   497–511
   active center of   498
   alleles of   499–503
   as marker in tissue culture   570
   Atkinson phenotypes of   500–3
   Babinga phenotypes   599
   chromatography of   498
   electrophoretic patterns (*see* isozyme
      patterns)
   function of   497
   genetic types
      common phenotypes   499–500
      comparison of   500, 503, 504
      inheritance of   499, 502
      of PGM_1 locus   499–501
      of PGM_2 locus   501–4
      of PGM_3 locus   504–5
      rare variants   500–2, 599
   genetic linkage with PTC tasting
      501, 571
   geographic   distribution   of   types
      505–8
   in placental tissue   499, 504
   indicator of cell culture contamina-
      tion   570
   isozyme patterns of   499–505
      comparison of   500, 503, 504
      in cellular mosaicism   569

Phosphoglucomutase—*cont.*
  isozyme patterns of—*cont.*
    of red cells *vs.* white cells 504, 505
    of reticulocytes 504
    PGM$_1$ locus isozymes 499–501
    PGM$_2$ locus isozymes 501–4
    PGM$_3$ locus isozymes 504–5
  laboratory determination of phenotypes 498, 508–9
  linkage studies 501, 502, 504, 599
  molecular weight of 498
  Mozambique type 500, 502, 503
  PGM$_1$ locus 499–501
  PGM$_2$ locus 501–4, 599
  PGM$_3$ locus 504–5, 599
  PTC and PGM$_1$ locus: possible linkage 501, 571
  phospho- and dephospho- forms 497, 498, 499
  Palmer phenotype 502, 503
  tissue distribution of 504
Phosphogluconate dehydrogenase 483–96
  autosomal locus of 485
  Canning variant (*see* phenotype III)
  common variant (*see* phenotype II)
  Dalston variant 484, 489
  deficiency of 489–90
    inheritance 489, 490
    with hemolytic anemia 490
  electrophoretic patterns of 486–8
    demonstration of 485, 494
    effect of NADP 486, 487, 488–9
    in red cells *vs.* white cells 486, 488
  enzyme activity measurement 491, 494
  Friendship variant (*see* phenotype VI)
  function in glycolysis 445, 446, 483
  genetic locus, autosomal in animals 485
  genetic variants 485–90
    effect of iodoacetate and urea on 487, 489
    electrophoretic patterns of 486–8
    enzyme activity of 484, 487, 489, 490
    geographic distribution of 490–3
    inheritance of 486, 487
    laboratory demonstration of 494
    list of 484
    'qualitative' variation 485–9

Phosphogluconate dehydrogenase—*cont.*
  genetic variants—*cont.*
    quantitative variation 489–90
    stability of 484, 487, 489
  geographic distribution of genes 490–3
  Hackney variant (*see* phenotype V)
  hemizygous expression in leukemia 599
  Ilford variant 484, 489
  molecular weight 483
  Newham variant 484, 489
  PGD alleles 486–93
  phenotypes I, II and III 484–7, 492–3
  phenotypes IV, V and VI 484, 487–9
  Richmond variant (*see* phenotype IV)
  subunit structure 486, 490
  Whitechapel variant 484, 489, 490
  X-linkage in *Drosophila* 485
Phosphohexose isomerase (glucose-phosphate isomerase) 527–9
  deficiency of 528–9, 600
    hemolytic anemia in 528, 600
  electrophoretic patterns of 528, 529
    effect of hemolysate age 528
    effect of mercaptoethanol 528
  function of 527
  genetic variants of 527
    geographic distribution 528
    inheritance 527–8
  laboratory demonstration of 529
  subunit structure of 528
Physostigmine (eserine) 192, 196, 199, 533
Placental enzymes
  alkaline phosphatase (*see under* Alkaline phosphatases)
  phosphoglucomutase (*see under* Phosphoglucomutase)
Plasma cells (plasmacytes) 6, 7, 8, 28
Pneumococcus polysaccharide, Type XIV 286, 287, 288
Point mutation
  mechanism of 557–61
  results of (*see* Amino acid substitutions)
Polycythemia (*see* Erythremia)
Polymer formation
  of albumin molecules 245

Polymer formation—*cont.*
  of haptoglobin molecules 73, 76–7
  of hemoglobin Porto Alegre 385
Polymorphism 572–3, 583 (*see also*
      individual genetic markers)
Polypeptide chains 583 (*see also*
      specific chains by name)
Polyribosomes (polysomes)
  definition of 583
  in hemoglobin chain synthesis
      388–9, 399, 400
  in immunoglobulin synthesis 9
Population distributions (*see* Geo-
      graphic distributions)
Post-albumin protein (*see* Gc protein)
Pre-albumin (*see* Protease inhibitor,
      alpha$_1$ anti-trypsin)
'Primaquine sensitivity' (*see under*
      Glucose-6-phosphate dehydro-
      genase)
Primary immune response, suppression
      of 315–6
Protease inhibitor (α_1 anti-trypsin)
      247–52
  alleles 249, 250
  deficiency of 248, 249
    association with pulmonary
        disease 248, 250
    inheritance 248
  functional role of 252
  genetic variants of 248–50
    electrophoretic patterns of 248,
        249, 250
    frequencies of Pi alleles 250, 251
    inheritance of 248
    laboratory detection of 252
  physicochemical properties of 247
  recognition as pre-albumin 248
  serum concentration of 247
    alterations in disease 248
  two-dimensional electrophoresis of
      249
Pseudocholinesterase 192–225
  anenzymia 194–5, 205
  anionic and esteratic sites of 197,
      199, 200
  antiserum 201, 205
  'atypical' variant 198, 200–4, 213,
      214
  biological half-life of 194
  chloride salts as inhibitors of 199,
      200

Pseudocholinesterase—*cont.*
  DFP as inhibitor 192, 199, 213
  dibucaine number of 203–4, 206–7
  differentiation from acetylcholin-
      esterase 193
  E_1 locus of pseudocholinesterase
      197–210
    geographic distribution of variants
        207–10, 591
    inheritance of E_1^a gene 203, 211
    inheritance of E_1^f gene 203
    inheritance of E_1^s gene 204–5
    inheritance of E_1^u gene 203, 206,
        211
    linkage with transferrin locus
        139, 210, 571
    nomenclature 204, 206–7
    'silent' allele at 204–5
  E_2 locus of pseudocholinesterase
      210–3
    C_5 component 210–2
    detection of C_5 210, 218–9
    effect of E_1 locus genes 211, 212
    electrophoretic appearance of C_5
        210
    enzyme activity of C_5 211
    gene frequencies 212
    inheritance of genes 211
    other possible E_2 locus variants
        212–3
  electrophoretic migration of com-
      ponents 196, 205, 210–3
    effect of neuraminidase on 196
    effect of pH on 210, 211
    in infant serum 196
    in two-dimensional electrophoresis
        196, 210–1
  evidence of structural variability
      196–7, 198–200
  fluoride inhibition of 199, 200, 203
  fluoride number of 203–4, 206–7
  fluoride-resistant variant of 203–4
  function of 194
  geographic distribution of variants
      207–10, 212, 591
  heterogeneity of 195–8
  inhibitors of 199–204
  laboratory tests for pseudocholin-
      esterase 213–9
    dibucaine number 216–8
    enzyme activity 217
    fluoride number 218

Pseudocholinesterase—*cont.*
  laboratory tests for pseudocholin-
    esterase—*cont.*
    'screening' tests 213–6
    tests for E_2 locus variants 218–9
  levels in plasma 193–8, 200–1, 213
    age differences 194
    disease association 194, 195
  metabolism of 193–4
  molecular weight of 197
  phenotypes of pseudocholinesterase
    characteristics of 204, 206, 207
    laboratory tests for differentiation
     204, 206, 207
    list of 204
  physico-chemical properties of
    195–7
  RO2-0683 as inhibitor of 199, 201,
    204
  separation of components 196
  'silent gene' 204–5
  structure of pseudocholinesterase
    195–7
  suxamethonium (succinyl dicholine)
    apnea due to sensitivity 195, 198,
     199, 205, 207, 212
    inactivation by pseudocholin-
     esterase 195
  synthesis of pseudocholinesterase
    193
  tissue distribution 193
  two-dimensional electrophoresis of
    196, 210–1
  types (*see* phenotypes)
  variability of E_1^u activity 206
Pulmonary emphysema, association
  with Pi deficiency 248, 250

Radioautography of Fe^{59} 148
Ragg and SNagg agglutinators 25, 37,
  40, 42, 44
Reaginic activity of IgE 4, 6, 587
Recombination
  definition of 584
  recombination fraction of
    ABO-adenylate kinase linkage
     515
    albumin-Gc linkage 246
    Gd and deutan linkage 463
    Gd and hemophilia A linkage 464
    Gd and protan linkage 463

Recombination—*cont.*
  recombination fraction of—*cont.*
    PGM_1 and PTC taster linkage
     501
    transferrin and E_1 linkage 140,
     210
    sex difference in frequency of 311
Red cell acid phosphatase (*see* Acid
    phosphatase in red cells)
Red cell destruction mechanisms
  in G6PD deficiency 451–2
  in Hb Köln disease 381
  in Hb S disease 371–3
  with increased level of Hb H 361
Reduced hemoglobin (*see* Deoxyhemo-
    globin)
Regulator (controller) genes
  definition of 584
  in haptoglobin synthesis 87
  in hemoglobin synthesis 399, 400
Rh antigens (*see also* Rh system)
  as protein-phospholipid complex
    272, 309–10
  chemical studies of 309–10
  complexity of 271–4, 304, 319, 320
  inheritance of 271, 277–8, 304–9
  nomenclature of 273–5, 303, 304,
    317, 320
  number of antigenic sites 271, 310
  resemblance to ATPase 310
Rh system 273–8, 303–10, 314–6
  antibodies (*see* Anti-Rh)
  antigens (*see* Rh antigens)
  association with Duffy system
    311–2
  cis and *trans* position effect 276–7
  genetic linkage 311
  genetic theories of inheritance
    272–5, 277–8, 306–9
  nomenclature 273–5, 303, 304, 317,
    320
  rare phenotypes 277, 278, 305–9
  relationship to LW 308, 309
Rh_{null} phenotypes 305–9 (*see also*
    Rh system)
  as red cell membrane defect 309,
    593
  description of 305–6, 308–9
  inheritance of 305–9
  reactions with other antibodies 306,
    308
  'silent' allele inheritance 308–9

Ribonucleic acid (RNA)
  and the genetic code 557–8
  haptoglobin and immunoglobulin codons compared 77
  in hemoglobin chain synthesis 388, 398–400, 559–65
  in immunoglobulin chain synthesis 8, 9, 22
  in mutation mechanisms 559–65
  sequence of haptoglobin codons 81
Ribosomes (*see* Polyribosomes)
Rivanol (2-ethoxy-6,9-diaminoacridine lactate) 127, 148, 160
Ro2-0683
  differentiation of E_1 types by 203–4, 206–7
  inhibitory effect on pseudocholinesterases 199, 200, 211
  use in screening test for E_1 types 214, 215

'Scrambler' gene 23
Secretor system
  ABH secretors and non-secretors 280–91, 314, 316
  association with alkaline phosphatase 228–30, 292–6
  association with Lewis phenotypes 282–4, 286, 289–90
  genes of 280
  linkage with Lutheran system 310
  Se as a regulator gene 280, 288, 289, 295
  secretion of 2′-fucosyllactose in milk 286, 288
Selection
  and blood group antigens 313–6
  and G6PD types 465, 573
  and Gm types 32
  and Hb types 404–5, 573
  and Hp types 82, 83, 93
  and Tf types 142, 146, 573
'Serine' enzymes 498
Sialic acid
  in alkaline phosphatase 227
  in alpha$_1$-acid glycoprotein 257
  in blood group active substance 285, 302
  in haptoglobin 72, 74
  in pseudocholinesterase 195
  in transferrin 140

Sickle-cell anemia 373
Sickle hemoglobin (*see* Hemoglobin S)
'Silent' alleles (*see also* 'Minus-minus' blood group phenotypes)
  at blood group loci 278, 280, 281, 308
  at haptoglobin locus 82, 88, 89
  at immunoglobulin loci 33, 39
  at pseudocholinesterase E_1 locus 204, 205
Splenectomy, effect on red cell inclusions 361, 380
Staining methods, for
  acid phosphatase 439
  adenylate kinase 518
  Ag and Lp 187
  albumin 247
  alkaline phosphatase 234–5
  carbonic anhydrase 533
  catalase 545–6
  ceruloplasmin 255
  esterases 535
  galactose-1-P uridyl transferase 547
  Gc protein 168, 169
  glucose-6-P dehydrogenase 468–9
  glutathione reductase 541
  haptoglobin 105–6
  heme 106
  'indophenol oxidase' 549
  lactate dehydrogenase 523
  malate dehydrogenase 527
  methemoglobin reductase 543
  peptidases 539–40
  phosphoglucomutase 509
  phosphogluconate dehydrogenase 494
  phosphohexose isomerase 529
  protease inhibitor 252
  pseudocholinesterase 218–9
  transferrin 148
Starch gel electrophoresis (*see* Electrophoresis in starch gel)
Structure-rate hypothesis 390
Subclasses of IgG molecules (*see under* Immunoglobulins)
Subunit structure of molecules
  of blood group antigens 280, 281, 566
  of Gc 165
  of glucose-6-phosphate dehydrogenase 449
  of haptoglobin 74–7, 565–6

628   INDEX

Subunit structure of molecules—*cont.*
  of hemoglobin 347–53, 565
  of immunoglobulin 11–7, 565
  of lactate dehydrogenase 521–2, 565
  of 6-phosphogluconate dehydrogenase 486, 490
  of phosphohexose isomerase 528
  of transferrin 128–9, 141, 589
Suppression of immune response 9, 315–6, 593
Sutter antigens (*see* Kell blood group system, antigens of)
Suxamethonium (succinyl dicholine) (*see under* Pseudocholinesterase)

T chain of IgA 5, 6
TPN (*see* Nicotinamide adenine dinucleotide phosphate)
Tetra-ethylpyrophosphate (TEPP) 192, 199
Thalassemia (*see also* Alpha and Beta thalassemia)
  acid phosphatase levels and 432
  analogy with enzyme deficiency 449
  beta-delta thalassemia
    as indicator of gene deletion 401
    characteristics of 391, 393, 398
  with beta thalassemia 393
  characteristics of 390–401
  definition of 390
  delta thalassemia 395
  geographic distribution 404–5
  hemoglobin synthesis in 390–401
  iron kinetics in 134
  Lepore thalassemia 391, 393–4
  presence of anti-Ag in 185–6
  theories of inheritance 399–401
Thyrotoxicosis (*see* Hyperthyroidism)
Tissue culture markers 570
Tissue grafting 569–70
Transferases (*see* Transglycosylases)
Transferrin 126–59
  alleles of 136, 139, 141
  amino acid substitutions in 140–2, 143
  as a carrier protein 126, 132, 133
  atransferrinemia 134, 135
  bacteriostatic effect of 134, 143
  binding of metals besides iron 130

Transferrin—*cont.*
  carbohydrate moiety of 127, 128, 140
  catabolism of 131–2
  concentration in plasma 131–2
    as indicator of iron-binding capacity 130
    effect of albumin injection 131
    in adults 131
     in disease 132
    in infants 131
    in pregnancy 131, 132
    measurement of 130
  physiological role of 133, 134
  demonstration of polymorphism 135
  determination of phenotypes 147–9
  effect of selection on 146, 573
  electrophoretic migration of 135–40
    effect of iron 143
    effect of neuraminidase 140, 143, 144
  function in iron metabolism 126, 132–4
  genetic linkage with E_1 locus 139–40, 210, 571
  geographic distribution of variants 144–6
  inheritance of variants 136, 139, 140
  iron binding by 128–30, 140, 142, 144
  laboratory methods 147–9
  metabolism of 131–2
  migration rate of 135, 140
  molecular structure of 128–9, 140–2, 589
  molecular weight of 127
  peptide fingerprints of 128, 140, 141
  physico-chemical properties of 126–30
  preparation and purification of 127
  role in infection 134
  role in iron metabolism 126, 132–4
  structure of 128–9, 140–2
  synthesis of 131, 132
  Tf C 136, 138–42, 144
  turnover of 131–2
  uptake by reticulocytes 133, 142
  variants of 135–46
    chemical structure of 140–2

Transferrin—*cont.*
  variants of—*cont.*
    diagram of electrophoretic patterns 138
    difficulties in identifying 143
    geographic distribution of 144–6
    inheritance of 136, 139
    list of 137
    nomenclature of 136, 137
    physiological behavior of 142, 143
    population frequencies of 144–6
    rates of electrophoretic migration 136, 137, 138
    selective effects on 146, 573
    Tf B_{0-1} 137–9, 144, 146
    Tf B_2 136, 138–41, 144, 146
    Tf D_1 136, 138–41
    Tf D_{Chi} 137–9, 141, 144, 146
Transglycosylases (sugar transferases)
  as blood group gene products 270–1
  effect on alkaline phosphatase 295, 296
  in blood group biosynthesis 270–1, 290
  suppression of activity 279
  with two recognition sites 277
Turner's syndrome 313
Turnover
  of haptoglobin molecules 66–8
  of immunoglobulin molecules 9–11
  of iron 132–4, 135
  of transferrin molecules 131–2
Twin zygosity 571–2, 600–1
Two-dimensional electrophoresis
  of albumin 244
  of E_1 locus enzymes 196
  of E_2 locus enzymes 210–1
  of protease inhibitor 249
  of transferrin 135

Unequal crossing-over
  as cause of amino acid substitution 559
  as cause of gene deletion 383, 562–3

Unequal crossing-over—*cont.*
  as cause of gene duplication 79–84, 400, 563–4
  between blood group genes 277
  between globin chain genes 363, 394, 400, 562–4
  between haptoglobin genes 79–84
  between immunoglobulin genes 21, 22–3
  in formation of albumin and Gc genes 246
  in formation of thalassemia genes 400
Unstable hemoglobins 380–5, 595
Uses of genetic markers 557–75, 600–1

Variamine blue B 234

Wilson's disease 253

X chromosome inactivation
  at G6PD locus 460–3, 566
  at Xg locus 313, 566
X-linked genetic markers in blood
  G6PD 452, 463–4
  Xg 312–4
  Xm 256, 592
Xg^a red cell antigen 276, 312–4
  inactivation of locus 313–4
  linkage to other X-linked loci 312–3, 464, 570
  locus on X chromosome 312
Xm(a) serum antigen 256–7, 592
  association with α_2 macroglobulin 256
  frequencies of alleles 257
  inheritance of 256
    as X chromosome marker 256, 592
    late manifestations in females 256
  linkage studies 592
  phenotype frequencies 257